T0300939

QUANTUM MEASUREMENT THEORY
AND ITS APPLICATIONS

Recent experimental advances in the control of quantum superconducting circuits, nano-mechanical resonators and photonic crystals have meant that quantum measurement theory is now an indispensable part of the modeling and design of experimental technologies.

This book, aimed at graduate students and researchers in physics, gives a thorough introduction to the basic theory of quantum measurement and many of its important modern applications. Measurement and control is explicitly treated in superconducting circuits and optical and optomechanical systems, and methods for deriving the Hamiltonians of superconducting circuits are introduced in detail. Further applications covered include feedback control, metrology, open systems and thermal environments, Maxwell's demon, and the quantum-to-classical transition.

KURT JACOBS is an Associate Professor of Physics at the University of Massachusetts at Boston. He is a leading researcher in quantum measurement theory and feedback control, and applications in nano-electromechanical systems. He is author of the textbook *Stochastic Processes for Physicists: Understanding Noisy Systems* (Cambridge University Press, 2010).

QUANTUM MEASUREMENT THEORY AND ITS APPLICATIONS

KURT JACOBS

University of Massachusetts at Boston

CAMBRIDGE
UNIVERSITY PRESS

CAMBRIDGE
UNIVERSITY PRESS

University Printing House, Cambridge CB2 8BS, United Kingdom

One Liberty Plaza, 20th Floor, New York, NY 10006, USA

477 Williamstown Road, Port Melbourne, VIC 3207, Australia

314-321, 3rd Floor, Plot 3, Splendor Forum, Jasola District Centre, New Delhi - 110025, India

79 Anson Road, #06-04/06, Singapore 079906

Cambridge University Press is part of the University of Cambridge.

It furthers the University's mission by disseminating knowledge in the pursuit of
education, learning and research at the highest international levels of excellence.

www.cambridge.org
Information on this title: www.cambridge.org/9781107025486

© Kurt Jacobs 2014

This publication is in copyright. Subject to statutory exception
and to the provisions of relevant collective licensing agreements,
no reproduction of any part may take place without the written
permission of Cambridge University Press.

First published 2014

A catalogue record for this publication is available from the British Library

Library of Congress Cataloging in Publication data
Jacobs, Kurt (Kurt Aaron), author.
Quantum measurement theory and its applications / Kurt Jacobs, University of Massachusetts at Boston.
pages   cm
Includes bibliographical references and index.
ISBN 978-1-107-02548-6 (hardback)
1. Quantum measure theory.   I. Title.
QC174.17.M4J33   2014
530.801–dc23   2014011297

ISBN 978-1-107-02548-6 Hardback

Cambridge University Press has no responsibility for the persistence or
accuracy of URLs for external or third-party internet websites referred to in
this publication, and does not guarantee that any content on such websites is,
or will remain, accurate or appropriate.

To my mother, Sandra Jacobs, for many things.
Not least for the Nelson Southern Link Decision, a great triumph unsung.

Contents

Preface

I would like to thank here a number of people to whom I am indebted in one way or another. To begin, there are five people from whose insights I especially benefited in my formative years in physics. In order of appearance: *Sze M. Tan,* for teaching me classical measurement theory, and introducing me to information theory and thermodynamics; *Howard M. Wiseman,* for teaching me about quantum measurement theory; *Salman Habib,* for teaching me about open systems and classical chaos; *Tanmoy Bhattacharya,* for enlightenment on a great variety of topics, and especially for the insight that measurement is driven by diffusion gradients; *Gerard Jungman,* for mathematical and physical insights, and for introducing me to many beautiful curiosities.

I am very grateful to a number of people who helped directly to make this book what it is: *Os Vy, Luciano Silvestri, Benjamin Cruikshank, Alexandre Zagoskin, Gelo Tabia, Justin Finn, Josh Combes, Tauno Palomaki, Andreas Nunnenkamp,* and *Sai Vinjanampathy* who read various chapters and provided valuable suggestions that improved the book. *Xiaoting Wang* who derived Eqs. (G.43)–(G.48). *Jason Ralph* who enlightened me on some superconductor facts that were strangely difficult to extract from the literature. Jason also helped me with the brief history of superconductivity and quantum superconducting circuits in Chapter 7. *Justin Guttermuth* who saved our asses when we had a house to move into, rooms to paint, a new baby, and I had this book to finish. My colleagues in the UMass Boston physics department for their support — especially *Maxim Olchanyi, Bala Sundaram, Vanja Dunjco,* and *Steve Arnason.* And last but not least, my wonderful wife *Jacqueline,* who helped me with the figures and the cover, and put up with the long hours this book required.

I apologize in advance for any errors and inadvertent omissions in this book. I would be most grateful to be notified of any errors that you may find. I will include corrections to all errors, as they are found, in an errata file on my website. I am very grateful to a number of readers who sent me corrections for my previous book, *Stochastic Processes for Physicists,* all of which have been made in the current printing. I was also able to acknowledge these readers in the current printing, which due to their efforts now appears to be largely error-free.

While I have endeavored to cite a fairly comprehensive and representative set of research papers on the topics I have covered in this text, it is likely that I have omitted some that deserve to be included. If you discover that your important paper on topic X has been missed, please send me the reference and I will be glad to correct this omission in any further edition.

Finally, it is a pleasure to acknowledge an ARO MURI grant, W911NF-11-1-0268, that was led by Daniel Lidar and administered by Harry Chang. This grant provided partial support for a number of research projects during the writing of this book, and from which it greatly benefited.

1

Quantum measurement theory

1.1 Introduction and overview

Much that is studied in quantum measurement theory, and thus in this book, is very different from that studied in its classical counterpart, Bayesian inference. While quantum measurement theory is in one sense a minimal extension of the latter to quantum states, quantum measurements cause dynamical changes that never need appear in a classical theory. This has a multitude of ramifications, many of which you will find in this book. One such consequence is a limit to the information that can be obtained about the state of a quantum system, and this results in relationships between information gained and dynamics induced that do not exist in classical systems. Because of the limits to information extraction, choosing the right kinds of measurements becomes important in a way that it never was for classical systems. Finding the best measurements for a given purpose, and working out how to realize these measurements, is often nontrivial.

The relationship between information and disturbance impacts the control of quantum systems using measurements and feedback, and we discuss this application in Chapter 5. The ability to extract information about quantum systems is also important in using quantum systems to make precise measurements of classical quantities such as time, acceleration, and the strengths of electric and magnetic fields. This subject is referred to as quantum metrology, and we discuss it in Chapter 6.

The dynamics generated by quantum measurements is as rich as the usual unitary evolution, and quite distinct from it. While the latter is linear, the former is both nonlinear and random (stochastic). The nonlinear dynamics due to quantum measurement, as we will see in Chapter 4, exhibits all the chaotic dynamics of nonlinear classical systems. This is made more interesting by the fact that all processes involving measurements, as we will see in Chapter 1, can be rewritten (almost) entirely as unitary and thus linear processes.

The role that measurements can play in thermodynamics is also an interesting one, especially because of the close connection between the entropy of statistical mechanics and the entropy of information theory. While this is not a special feature of quantum, as opposed to classical measurements, we spend some time on it in Chapter 4 both because of its fundamental nature, and because it is an application of measurement theory to the manipulation of heat and work.

1

There is another important connection between quantum measurement theory and thermodynamics: thermal baths that induce thermalization and damping also carry away a continuous stream of information from the system with which they interact. Because of this, thermal baths can mediate continuous measurements, and there is a useful overlap between the descriptions of the two. Further, the continuous measurements induced by a system's environment are sufficient to induce the quantum-to-classical transition, in which the trajectories of classical dynamics emerge as a result of quantum mechanics.

It is now possible to construct and manipulate individual quantum systems in the laboratory in an increasingly wide range of physical settings. Because of this, we complete our coverage of measurement theory by applying it to the measurement and control of a variety of concrete mesoscopic systems. We discuss nano-electromechanical systems, in which superconducting circuits are coupled to nano-mechanical resonators, and optomechanical systems in which superconducting circuit elements are replace by the modes of optical resonators. In Chapter 7 we introduce these systems and show how to determine their Hamiltonians and the interactions between them. In this chapter we also explain how the continuous quantum measurements introduced in Chapter 3 are realized in these systems, and rephrase them in the language usually used by experimentalists, that of amplifiers. In Chapter 8 we consider a number of examples in which the above systems are controlled, including realizations of the feedback control techniques described in Chapter 5.

Despite its importance, quantum measurement theory has had an uneasy relationship with physicists since the inception of quantum mechanics. For a long time the tendency was to ignore its predictions as far as measurements on individual systems were concerned, and consider only the predictions involved with averages of ensembles of systems. But as experimental technology increased, and experimentalists were increasingly able not only to prepare and measure individual quantum systems in well-defined states with few quanta but to make repeated measurements on the same system, physicists were increasingly forced to use the full predictions of measurement theory to explain observations.

The uneasiness associated with measurement theory stems from a philosophical issue, and this has disturbed physicists enough that many still seem uncomfortable with using the language of measurement in quantum mechanics, and assigning the knowledge gained to an observer. In fact, as will be explained in due course, all processes that involve quantum measurements can be described without them, with the possible exception of a single measurement relegated to the end of the process. In using this second description the language changes, but the two descriptions are completely equivalent. Thus the use of the language of measurement theory is entirely justified. What is more, analysis that makes explicit use of measurement theory is often vastly more convenient than its unitary equivalent, providing a strong practical reason for using it. This should not be taken to mean that the philosophical problem has gone away – far from it. In fact, while the focus of this book is applications, we feel that no education in quantum measurement theory is quite complete without an understanding of "the measurement problem," and because of this we include a discussion of it at the end of Chapter 4.

We do not cover all the applications of quantum measurement theory, most notably those to do with quantum information theory. The latter is now itself a large subfield of physics, and there are a number of textbooks devoted to it. While our goal is to focus on applications that are not usually covered in books on quantum information, given the fundamental connection between measurement and information the two subjects are intertwined. A number of concepts from information theory are therefore invaluable in broader applications of measurement, and so we discuss information theory in Chapter 2. We do not cover "foundational" questions in quantum theory, with the sole exception of the so-called "quantum measurement problem." Another topic within measurement theory that we do not discuss is that termed "weak values," not to be confused with weak measurements which we treat in detail.

Quantum measurement theory is a weird and wonderful subject. I hope I have whetted your appetite, and that you find some topics in this book as stimulating and fascinating as I have.

A guide to this book

Different readers will want to take different pathways through the text. Many experimentalists, for example, may wish to learn how to describe weak or continuous measurements in their experiments without wading through Chapters 1 and 3. Such readers can start with Appendix B, and possibly Appendix C, from which a number of parts of the book are accessible. Experimentalists working with mesoscopic systems can then proceed to Section 7.7.1 that describes measurements using the language of amplifiers, and Section 3.1.4 that shows how to calculate the spectrum of a measurement of a linear quantum system using an equivalent classical model. In fact, Section 7.7.1 can be read first if desired, as it does not require the theoretical language used for continuous measurements. Section 5.4.3 is also accessible from the above appendices, and deals with the feedback control of a single qubit. Experimentalists working with quantum and atom optics could instead proceed to Sections 3.3 and 3.4 that deal with optical cavities, photon counting, and optical homodyne detection, although these sections do also require Chapter 1 up to and including Section 1.3.3.

The first parts of Chapter 7 and much of Chapter 8, introducing mesoscopic circuits and optomechanical systems, do not require any previous material. Sections 7.7.3 and 8.2 are the exceptions, requiring either Appendix B or Chapters 1 and 3.

Students who want to learn the broad fundamentals of measurement theory should start with Chapters 1 and 2, and proceed to Chapter 3 if they are interested in the details of continuous measurements. Section 4.3 of Chapter 4 is also core reading on the relationship between continuous measurements and open systems.

Having studied Chapters 1, 2, and 3 all of the remainder of the text is accessible and topics can be selected on preference. We now list the latter sections that do not require Chapter 3, since these can be included in a course of study that does not involve continuous measurements: Sections 4.1.2 and 4.1.3 on Landauer's erasure principle and Maxwell's demon; parts of Section 5.3 on feedback control; all but Section 6.2.2 in Chapter 6 on metrology; Sections 7.1 through 7.6, and Sections 8.1 and 8.3 on controlling mesoscopic systems.

Some assumptions and terminology

This text is for graduate physics students, so we assume that the reader is familiar with quantum mechanics, the basics of probability theory, and various mathematical concepts such as Fourier transforms and δ-functions. Everything that the reader needs to know about probability theory and Fourier transforms can be found in Chapter 1 of reference [288] or Chapter 4 of reference [592] and Chapter 1 of reference [591]. We also recommend Jaynes' landmark book on probability theory and its role in reasoning and science [311].

In mathematics texts it is usual to denote a random variable as a capital letter, say X, and the variable denoting one of the values it can take as the corresponding lower case letter, x. This provides technical precision, since the concept of a random variable, and the concept of one of the values it can take, are distinct. However, physicists tend to use the same symbol to denote both things, because it causes no confusion in practice. This is the style I prefer, so I use it here.

We physicists often use the term "probability distribution" as synonymous with "probability density," whereas mathematicians use the former term to mean the anti-derivative of the latter. Defining "probability distribution" to mean "probability density" is useful, because standard English usage prevents us from using "probability density" for discrete distributions. For this reason we will use the term *probability distribution* to mean probability density.

To refer to a set of quantities, for example a set of probabilities p_j indexed by j, in which j takes integer values in some range, we will use the shorthand $\{p_j\}$. The locations of the definitions of all acronyms we use can be found in the index.

1.2 Classical measurement theory

Before learning quantum measurement theory, it is valuable to understand classical measurement theory, as the two are very closely connected. Classical measurement theory, also known as Bayesian statistical inference, tells us how our knowledge about the value of some quantity, x, changes when we obtain a piece of data relating to x. To understand how this works, we need to know first how to describe the knowledge we have regarding x. We will assume that x is some variable that can take any real value, but more generally it could be a vector of discrete or continuous variables. Since we need to make a measurement to determine x, we must be uncertain about its value. Our knowledge about x is therefore captured by a probability distribution, $P(x)$, for the values of x. This probability distribution tells us, based on the information currently available, the likelihood that x will have various values, and overall how certain, or uncertain, we are about x. This distribution is called our ***state-of-knowledge*** of x.

To determine how our state-of-knowledge changes when we obtain a piece of data y, we have to know how y is related to x. To be able to describe measurements, this relationship must be probabilistic: if y were deterministically related to x (meaning that y was a function of x), then we could determine x precisely once we knew y, simply by inverting the function. This is not what happens when we make measurements; measurements are never perfectly precise. After we have made a measurement of x, giving a result y, we are always left with some uncertainty about x.

Consider measuring the length of an object with a ruler. In this case the result of the measurement is equal to the true length plus a random error. Thus, given the true length of the object (the value of x) there is a probability distribution for the result y. The probability distribution for y is peaked at $y = x$. Because this probability distribution for y depends on (is conditional upon) the value of x, it is called a *conditional* probability distribution for y, and is written $P(y|x)$ (this is read as "P of y given x"). The conditional probability for the measurement result (the data) y, given the value of the quantity to be measured, x, completely defines the measurement. This conditional probability is determined by the physical procedure used to make the measurement, and is referred to as the **likelihood function** for the measurement. So if one wishes to obtain the likelihood function for a given measurement process, one must first work out how the value of the thing to be measured leads to the measurement result. If y is the measurement result, and x is the quantity to be measured, then the process by which y leads to x can often be written in the form $y = f(x) +$ a random variable, where f is a deterministic function. The likelihood function can then be determined directly from this relationship.

To determine how our state-of-knowledge regarding x, $P(x)$, changes when we obtain the value of y, we use the relationships between the joint probability for two random variables x and y, $P(x, y)$, and the conditional probabilities $P(y|x)$ and $P(x|y)$. These relationships are

$$P(x,y) = P(x|y)P(y) = P(y|x)P(x). \tag{1.1}$$

Here $P(y)$ is the probability distribution for y irrespective of the value of x (also called the *marginal* distribution for y), and is given by

$$P(y) = \int_{-\infty}^{\infty} P(x,y)\,dx. \tag{1.2}$$

Now $P(x)$ is our state-of-knowledge of x prior to making the measurement, and it is therefore the probability density for x irrespective of the value of y. It is therefore also the marginal probability for x, and thus given by

$$P(x) = \int_{-\infty}^{\infty} P(x,y)\,dy. \tag{1.3}$$

While the relationships in Eq. (1.1) are fairly intuitive, they are explained further in, for example, references [288] and [592].

Rearranging Eq. (1.1) we obtain the famous relationship known as *Bayes' theorem*, being

$$P(x|y) = \frac{P(y|x)P(x)}{P(y)}. \tag{1.4}$$

Upon examining Eq. (1.4) we will see that it tells us exactly how to change our state-of-knowledge when we obtain the measurement result y. First note that since $P(x|y)$ must be normalized (that is, its integral over x must be unity), the value of $P(y)$ on the bottom line is completely determined by this normalization. We can therefore write Bayes' theorem as

$$P(x|y) = \frac{P(y|x)P(x)}{\mathcal{N}}, \quad \text{where} \quad \mathcal{N} = \int_{-\infty}^{\infty} P(y|x)P(x)\,dx = P(y). \tag{1.5}$$

We see that on the right-hand side (RHS) we have our state-of-knowledge of x before we obtain the data y, and on the left-hand side (LHS) the probability for x *given* that value of y. The LHS is therefore our state-of-knowledge after obtaining the value of y. In Eq. (1.5) $P(x)$ is called the **prior** probability, and $P(x|y)$ the **posterior** probability. So Bayes' theorem tells us that to obtain our new state-of-knowledge once we have made our measurement: we simply multiply our current state-of-knowledge by the likelihood function $P(y|x)$, and normalize the result. Note that the prior is simply the marginal (overall) distribution of x. The relationship given by Eq. (1.5), Bayes' theorem, is the fundamental theorem of classical measurement theory.

1.2.1 Understanding Bayes' theorem

While Bayes' theorem (Eq. 1.5) is simple to derive, to obtain a direct understanding of it requires a bit more work. To this end consider a measurement of a discrete variable, x, in which x has only two values. We will call these values **0** and **1**. Our measurement will also have only two outcomes, which we will denote by $y = 0$ and $y = 1$. In our discussion of Bayes' theorem above, we assumed that x was continuous, so let us take a minute to re-orient our thinking to discrete variables. In the present case our state-of-knowledge (our prior), $P(x)$, has only two values, being $P(0)$ and $P(1)$ (and, of course, $P(0) + P(1) = 1$). The conditional probability, $P(y|x)$, also has only two values for each value of x. If we make the measurement and obtain the result $y = 1$ (for example), then Bayes' theorem tells us that our posterior is given by

$$P(x|1) = \frac{P(1|x)P(x)}{\sum_{x'} P(1|x')P(x')} = \frac{P(1|x)P(x)}{P(1|0)P(0) + P(1|1)P(1)}. \tag{1.6}$$

We now wish to obtain a better understanding of this expression.

To do so let us choose a specific likelihood function for the measurement. This likelihood function is given in Table 1.1, and contains the parameter α. If α is close to unity, then the two values of x give very different distributions for the measurement result y, and in this

Table 1.1. *The likelihood function for a simple "two-outcome" measurement. The body of the table gives the probability for y given x.*

$P(y\|x)$	$x = 0$	$x = 1$
$y = 0$	α	$1 - \alpha$
$y = 1$	$1 - \alpha$	α

case we would expect the measurement to tell us a lot about x. Conversely, if α is close to $1/2$, then the reverse is true. For the sake of concreteness let us choose $0.5 < \alpha < 1$. With this choice, if $x = \mathbf{0}$ then it is more likely that we will get the result $y = 0$, and if $x = \mathbf{1}$ it is more likely that we will get $y = 1$.

Now assume that we initially know nothing about x, so that our prior state of knowledge is $P(\mathbf{0}) = 0.5 = P(\mathbf{1})$. What happens when we make the measurement and get the result $y = 1$? Since our prior is flat, by which we mean that it does not change with x, it cancels on the top and bottom lines of Bayes' theorem, telling us that our posterior is simply the likelihood function, normalized if necessary:

$$P(x|y) = \frac{P(y|x)}{\sum_{x'} P(y|x')} = \frac{P(y|x)}{P(y|0) + P(y|1)}. \tag{1.7}$$

Our posterior is thus $P(\mathbf{0}|1) = 1 - \alpha$ and $P(\mathbf{1}|1) = \alpha$. So it is now more likely that $x = \mathbf{1}$ than $x = \mathbf{0}$. This is indeed intuitively reasonable. The likelihood function tells us that if x were to be $\mathbf{0}$ then it would be less likely that we would get the result 1, so it is reasonable that since we obtained the result 1, it is more likely that $x = \mathbf{1}$ than that $x = \mathbf{0}$.

The above discussion shows that if the prior is uniform, Bayes' theorem tells us that *the values of x that are more likely are those for which the result we have obtained is the more likely outcome.*

Now let us examine why we need to include the prior, in addition to the likelihood function, when calculating our posterior. This is clearest when the prior is strongly weighted toward one value of x. Consider a situation in which the prior is $P(\mathbf{0}) = 0.999$ and $P(\mathbf{1}) = 0.001$. This means that in the absence of any further data, on average x will only be equal to $\mathbf{1}$ one time in a thousand cases. We now consider a slightly different two-outcome measurement from the one above, as this will make the logic simple. The likelihood function for the new measurement is given in Table 1.2, and in words is as follows: if $x = \mathbf{1}$ then y is always equal to 1. If $x = \mathbf{0}$, then $y = 0$ with probability 0.999 and $y = 1$ only one time in a thousand. This means that, if the prior is flat, upon obtaining the result $y = 1$ the value of x would be equal to $\mathbf{1}$ approximately nine hundred and ninety-nine times out of a

Table 1.2. *The likelihood function for our second "two-outcome" measurement. The body of the table gives the probability for y given x.*

$P(y|x)$	$x = 0$	$x = 1$
$y = 0$	0.999	0
$y = 1$	0.001	1

thousand:

$$P(x = 1|y = 1) = \frac{P(1|1)}{P(1|0) + P(1|1)} = \frac{0.999}{1.001}. \tag{1.8}$$

So what is the case when the prior is highly weighted toward $x = 0$, as described above? Well, in this case x is equal to **1** only one time in a thousand. Now, if we get the result $y = 1$ there are two possibilities. Either $x = 1$, which happens one time in one thousand, or $x = 0$ and we got the result $y = 1$ anyway, which also happens approximately one time in a thousand. (The precise figure is the frequency of $x = 0$ multiplied by the frequency of the result $y = 1$ given $x = 0$, which is $0.999 \times 0.001 \approx 1/1001$.) Thus the result $y = 1$ happens approximately one time in 500, and half of these are due to $x = 0$, and half due to $x = 1$. So when we obtain the result $y = 1$, there is only a 50% chance that $x = 1$. This is, of course, exactly what Bayes' theorem tells us; by multiplying the likelihood function that weights $x = 1$ very highly, by the prior that weights $x = 0$ very highly, we obtain an approximately flat posterior.

The example we have just considered applies directly to a real and very important situation: testing for the HIV virus. Each test is pretty reliable, giving a false positive only about one time in a thousand. On that basis alone one might think that when a result comes back positive, there is little reason to perform a follow-up test to confirm it. But this is very wrong. Since very few patients have HIV, false positives come up just as frequently as real positives. Thus, whenever a positive test result comes back, it is essential to do a follow-up test to check that it is not a false positive. Bayesian inference, and thus measurement theory, is therefore crucial in real-world problems.

To complete our discussion of Bayes' theorem it is worth noting that our state-of-knowledge does not necessarily become more certain when we make a measurement. To take the example of the HIV test above, before we obtain the result of the test we are almost certain that the patient is HIV negative, since the vast majority of patients are. However, upon obtaining a positive test result, there is an approximately fifty-fifty chance that the patient is HIV positive. Thus, after obtaining the measurement result, we are less certain of the HIV status of the patient. Even in view of this, all classical measurements do have the property that, upon making the measurement, we become more certain *on average* of the value of the measured quantity (where the average is taken over all the possible measurement results). We will be able to make this statement precise once we have described how to quantify the concept of information in Chapter 2.

1.2.2 Multiple measurements and Gaussian distributions

Multiple measurements

Having made a measurement of x, what happens when we make a second measurement? We might expect that we simply repeat the process of multiplying our current state-of-knowledge, now given by $P(x|y)$, by the likelihood function for the new measurement. This is correct so long as the results of the two measurements are independent, and is simple to show as follows. Let us say that we make N measurements, and the results (the data) obtained from these measurements are y_i for $i = 1,\ldots,N$. Bayes' theorem tells us that

$$P(x|y_1,\ldots,y_N) = \frac{P(y_1,\ldots,y_N|x)P(x)}{\mathcal{N}}, \tag{1.9}$$

with $\mathcal{N} = \int_{-\infty}^{\infty} P(y_1,\ldots,y_N|x)P(x)\,dx$. The fact that all the measurement results are independent means that

$$P(y_1,\ldots,y_N|x) = P(y_1|x)P(y_2|x)\cdots P(y_N|x), \tag{1.10}$$

and with this Bayes' theorem becomes

$$P(x|y_1,\ldots,y_N) = \frac{P(y_1,\ldots,y_N|x)P(x)}{\mathcal{N}} = \frac{P(y_N|x)\cdots P(y_1|x)P(x)}{\mathcal{N}}$$
$$= \frac{P(y_N|x)}{\mathcal{N}_N}\cdots\frac{P(y_1|x)P(x)}{\mathcal{N}_1}, \tag{1.11}$$

where $\mathcal{N}_n = P(y_n)$. So we see that each time we make another independent measurement we update our state-of-knowledge by multiplying it by the likelihood function and normalizing the result.

Pooling independent knowledge

It turns out that classical measurement theory provides us with a simple way to pool the knowledge of two observers so long as their information has been obtained independently. If two observers, A and B, have the respective states-of-knowledge $P_1(x)$ and $P_2(x)$ about a quantity x, then we can write each of these as

$$P_A(x) = \frac{P(\mathbf{y}_A|x)P_{\text{prior}}(x)}{\mathcal{N}_A}, \tag{1.12}$$

$$P_B(x) = \frac{P(\mathbf{y}_B|x)P_{\text{prior}}(x)}{\mathcal{N}_B}, \tag{1.13}$$

where $\mathbf{y}_A$ and $\mathbf{y}_B$ are the vectors of data obtained by the respective observers. Since we intend the data the observers have obtained to represent *all* the information they each have about x, $P_{\text{prior}}(x)$ is the prior that describes having no initial knowledge about x. The problem of determining such a prior can be surprisingly difficult, and we discuss it further below. If we assume for now that we know what P_{prior} is, and we choose the measure for integration over x so that $P_{\text{prior}}(x) = 1$ (that is, we absorb P_{prior} into the measure), then an observer who has access to the data of both A and B has the state-of-knowledge

$$P(x) = \frac{P(\mathbf{y}_A|x)P(\mathbf{y}_B|x)}{\mathcal{N}'} = \frac{P_A(x)P_B(x)}{\mathcal{N}}. \tag{1.14}$$

So all we have to do to pool the knowledge of two (or more) observers is to multiply their states-of-knowledge together, and normalize the result.

The ubiquity of Gaussian measurements

Now consider applying classical measurement theory to the simple example discussed above, that of measuring the length of an object with a ruler. To describe this measurement we need to decide how the result that we get, y, depends upon the true length x. We can think of y as being equal to the true length plus a random "error." The error in a measurement is often described well by a Gaussian distribution. The reason for this is that the error is usually the result of random contributions from many different sources. When we add many independent random variables together, the central limit theorem tells us that the resulting random variable has an approximately Gaussian distribution. If the error in our measurement of x is a Gaussian with mean zero and variance V, then the probability distribution for y, given x, is a Gaussian centered at x:

$$P(y|x) = \frac{1}{\sqrt{2\pi V}} e^{-(y-x)^2/(2V)}. \tag{1.15}$$

This is the likelihood function for the measurement. If we have absolutely no knowledge of x before the measurement (never the case in reality, of course), then we can set $P(x) = 1$. Our knowledge of x after making the measurement is then simply the likelihood function, normalized so that it is a valid probability distribution over x for each value of y. In this case the normalization is already correct, and so

$$P(x|y) = \frac{1}{\sqrt{2\pi V}} e^{-(x-y)^2/(2V)}. \tag{1.16}$$

This tells us that the value of x is a Gaussian centered at y, with variance V, and thus that the most likely value of x is y, and the expectation value of x is also equal to y. It is customary to quote the error on a measured quantity as twice the standard deviation. Thus we write the value of x as $y \pm 2\sqrt{V}$.

The ubiquity of Gaussian states-of-knowledge

It is not only measurement errors that tend to be Gaussian, but also states-of-knowledge. The reason for this is that when one makes many measurements of a given quantity, the resulting state-of-knowledge is the result of multiplying together many likelihood functions. One can show that even if these likelihood functions are not themselves Gaussian, the result usually is. This is because there is a central limit theorem for multiplication that follows from the usual central limit theorem for summing random variables. To see how this works we first recall that if we have two independent random variables x and y, the probability distribution for their sum, $z = x + y$, is the *convolution* of the distributions for x and y:

$$P_z(z) = \int_{-\infty}^{\infty} P_x(u)P_y(z-u)\,du \equiv P_x(x) * P_y(y). \tag{1.17}$$

The central limit theorem states that so long as the variances of all the convolved distributions are finite, then as one convolves more distributions together, the closer the result is to a Gaussian.

To obtain a similar result for multiplying a number of probability distributions together, we take the Fourier transforms of these probability distributions. Multiplying two functions together is the same as convolving their Fourier transforms. That is, the Fourier transform of the product of two functions is the convolution of their Fourier transforms. So the central limit theorem tells us that if we convolve the Fourier transforms of a set of distributions together, then the result is close to a Gaussian. Since the inverse Fourier transform of this approximate Gaussian is the product of all the probability distributions, and since the Fourier transform of a Gaussian is also Gaussian, the product of the distributions is approximately Gaussian.

So when we make multiple measurements to learn about some quantity, or when our knowledge about a quantity is derived from many different sources, then our state-of-knowledge will be approximately Gaussian. Just as Gaussian errors are common in nature, Gaussian states-of-knowledge are also common.

Adding noise and making measurements can be regarded as opposite processes. The more noise (random variables) we add to a quantity, the more uncertain it becomes. The more measurements we make on a quantity, the more certain (on average) it becomes. It is interesting to note that the mathematical operations that describe each of these processes, convolution for adding noise, and multiplication for making measurements, are dual to each other from the point of view of the Fourier transform.

1.2.3 Prior states-of-knowledge and invariance

Bayes' theorem makes it clear how to update our state-of-knowledge whenever we obtain new data. So long as we know the overall distribution for the unknown quantity (the prior), then Bayes' theorem is strictly correct and open to no interpretation. However, for many situations it is not clear what the prior should be, since the observer usually knows something before making measurements, but it is not always possible to quantify this knowledge. A sensible solution to this problem is to assume that the observer has no initial knowledge, as this prevents any biasing of the final state-of-knowledge with unwarranted prejudice. The problem is that it is not always easy to determine the prior that quantifies "no knowledge." Here we will discuss a useful method for obtaining such priors (and give references for further methods), but before we do this it is worth noting that in many situations the choice of prior is, in fact, unimportant. This is the case when we can obtain enough data that the likelihood function is much sharper (has much less uncertainty) than any "reasonable" prior. In this case the prior has very little influence on the final expectation value of the measured quantity, and is therefore irrelevant. In this case we can ignore the prior completely and set the posterior equal to the (correctly normalized) likelihood function for the set of measurements.

The question of what it means to know nothing is actually quite subtle. In the case of measuring the position of an object in space, it seems quite obvious that the prior should be constant and extend from $-\infty$ to ∞. This prior is not normalizable, but this does not matter, since for any measurement that provides a reasonable amount of data, the likelihood function, and thus the posterior, will be normalizable. On the contrary, the question of what it means to know nothing about a positive quantity, such as the volume of water in a lake, is not at all obvious. As will become clear, a prior that is constant on the interval $(0, \infty)$ is not the correct choice.

The problem of what it means to know nothing about a positive quantity is solved by using the idea that the prior should be invariant under a transformation. The key is to realize that if one knows nothing about the value of a positive quantity, λ, then multiplying λ by any positive number should not change one's state-of-knowledge. That is, if one is completely ignorant of λ, then one must be ignorant of the overall *scale* of λ. There is only one state-of-knowledge that is invariant under a scaling transformation, and that is

$$P(\lambda) = a/\lambda, \qquad\qquad (1.18)$$

where a is any dimensionless constant. This prior is also not normalizable, but this is fine so long as the likelihood function is normalizable. The set of all scaling transformations, which is the set of all positive real numbers, forms what is called a *transitive group*. It is a group because it satisfies the properties of a group (e.g., the consecutive application of two scaling transformations applied consecutively is another scaling transformation), and the term "transitive" means that we can transform any value of our positive quantity to any other value by using one of these transformations.

The powerful technique of identifying transformations, in particular transformations that are transitive groups, over which our prior state-of-knowledge must be invariant, was developed by Jeffreys [312, 314, 313] and Jaynes [522, 311]. Returning to the problem of the prior for the location of an object, we see that the flat prior is the one that is invariant under all translations, being the transitive group for that space. Transitive groups for a given quantity must map all admissible values of the quantity to admissible values. That is why the translation transformation is not appropriate for quantities that can only be positive. An excellent example of the use of invariance for determining priors is given by Jaynes' solution to "Bertrand's problem," and can be found in [311].

Using the notion of invariance under a transitive group it is simple to determine priors for quantities that have such groups. The simplest case is a "six-sided" die. If the die is perfectly symmetric, then our prior should be invariant under all permutations of the six faces of the die. The result is a prior that assigns equal probability to all the faces. A less trivial example is a prior over the possible states of an N-dimensional quantum system. In this case the set of possibilities is an N-dimensional complex vector space, and the transitive group is that of all the unitary transformations on this vector space. The zero-knowledge prior is the one that is invariant under all unitaries.

Instead of specifying the prior in terms of a set of coordinates that parametrizes the space of possibilities, one usually specifies the measure over which to integrate in terms of these coordinates, and chooses this measure so that it is invariant under the required transformations. With this choice for the measure, the invariant prior is merely equal to unity. For unitary transformations the invariant measure is called the *Haar measure*, and is given in Appendix G. The definition of a measure, using a minimum of technical jargon, can be found in reference [288].

The method of transitive groups fails when the set of possibilities has no such group. The interval $[0, 1]$ is one such example, and applies when one is inferring a probability. More sophisticated methods have been devised to address these situations. We will not discuss these methods here, but the reader can find further details in [52, 479, 363].

Priors: a tricky example

For the sake of interest we pause to consider an example that shows the kinds of problem that can arise when choosing priors. An apparent paradox occurs in the "two-envelopes problem," in which we are given two envelopes, one of which contains twice as much money as the other. We are allowed to open only one of the envelopes, and then we get to choose which envelope we will take home to spend. Before we open one of the envelopes, clearly there is no reason to favor either envelope. Assume that we now open one of the envelopes and find that it contains x dollars. We now know that the second envelope (the one we haven't opened) contains either $x/2$ dollars or $2x$ dollars. We conclude that if we choose to keep the second envelope instead of the first, then we stand to lose $x/2$ dollars, or to gain x dollars. We thus stand to gain more than we will lose by choosing the second envelope, and so it appears that we should do so. There is clearly something wrong with this reasoning, because if we had instead chosen to open the second envelope, then precisely the same reasoning would now tell us to take home the first envelope. Because the reasoning always tells us to do the same thing, we can apply the reasoning without actually opening the envelope! Thus opening the envelope provides us with no information, and so it cannot tell us which envelope we should take home. We have a paradox.

What is wrong with the above analysis is the hidden assumption we have made about what it means to initially "know nothing" about the amount of money in the envelopes. That is, we have arbitrarily chosen a prior, and this causes problems with our reasoning. Now let us consider the problem with the prior made explicit. Let us call the envelopes A and B, and say that envelope A contains x dollars, and envelope B contains $2x$ dollars. To perform any reasoning about probabilities we must assign a probability distribution to x to start with, so let us call this distribution $P_x(x)$. Now we pick one envelope at random. Call the envelope we pick envelope 1, the amount of money we find in it y, and denote by z the unknown amount of money in envelope 2. Since the probability that envelope 1 is envelope A is independent of which envelope is which, Bayes' theorem tells us that the relative probabilities that envelope 2 has amounts $z = y/2$ and $z = 2y$ are determined by the prior alone. Thus

$$\text{Prob}(z = y/2) = \frac{P_x(y/2)}{P_x(y/2) + P_x(2y)}, \quad \text{Prob}(z = 2y) = \frac{P_x(2y)}{P_x(y/2) + P_x(2y)}. \quad (1.19)$$

If we chose to keep envelope 1 then we will have y dollars. The expectation value of the amount of money we will have if we choose to keep envelope 2 instead is

$$\langle z \rangle = \frac{(y/2)P_x(y/2) + 2yP_x(2y)}{P_x(y/2) + P_x(2y)}. \quad (1.20)$$

We will want to keep envelope 2 if $\langle z \rangle > y$.

Now that we have made the prior explicit, we can see that our initial reasoning about the problem corresponds to choosing the prior to be flat, from which it follows that the probability that $z = 2y$ is equal to the probability that $z = y/2$. But our discussion in the previous section makes us suspicious of this prior. If x were merely an unknown positive quantity it would have $P_x(x) \propto 1/x$. The present case is a little different, but we might expect a somewhat similar prior. Let us first consider a prior in which the expectation value of x is finite, as this allows us to see how opening envelope 1 provides us with information. If you put in an explicit function for $P_x(x)$ such that $\langle x \rangle$ is finite, you will find the following. If y is large enough then we are sufficiently sure that envelope 2 is the one with less money, and we keep envelope 1. If y is small enough then it is sufficiently likely that envelope 2 is the one with more money, and so we keep it instead. For every P_x with finite $\langle x \rangle$ there is a threshold value y_c such that when $y > y_c$ we keep envelope 1, and if not we keep envelope 2.

Now let us return to the problem of what it might mean to know nothing about x prior to opening an envelope. Let us denote the "zero-knowledge" distribution that we seek by $P_0(x)$. From our discussion in the previous section, $P_0(x)$ should have no scale, since a scale, such as a given value for $\langle x \rangle$, means that we know something about x. Because $P_0(x)$ must be scale-free, when we open an envelope it cannot provide a number y_c that determines which envelope we should keep. The only remaining option is that when we open an envelope we learn nothing about which envelope to keep. This in turn requires that no matter what value we find for y, it will always be true that $\langle z \rangle = y$, giving us no preference for either envelope. You can determine for yourself from Eq. (1.20) that the only prior distribution satisfying this requirement is $P_0(x) = 1/\sqrt{x}$. While this is not strictly scale-invariant, it is scale-free as desired. The lesson is that you cannot just pick any prior and assume that it captures what it means to know nothing.

1.3 Quantum measurement theory

1.3.1 The measurement postulate

The state of a quantum system, $|\psi\rangle$, is a vector in a complex vector space. If the set of vectors $\{|n\rangle\}, n = 0, \dots, N-1$ (where N may be ∞) is an orthonormal basis for this space, then we can always express $|\psi\rangle$ as

$$|\psi\rangle = \sum_n c_n |n\rangle \tag{1.21}$$

for some complex coefficients c_n, where $\sum_n |c_n|^2 = 1$. In analyzing quantum measurements we will often assume that the system is finite-dimensional. This simplifies the analysis while losing very little in the way of generality: since all systems ultimately have bounded energy, any real system can always be approximated arbitrarily well using a finite number of states.

The basis of quantum measurement theory is the following postulate: We can choose any basis, and look to see which one of these basis states the system is in. When we do so, we *will* find the system to be in one of these basis states, even though it may have been in *any* state $|\psi\rangle$ before the measurement. Which basis state we find is random. If the system is initially in the state $|\psi\rangle$ then the probability that we will find state $|n\rangle$ is given by $|c_n|^2$. A measurement like this, for which the result is one of a set of basis states, is called a *von Neumann* measurement. Before we use this basic measurement postulate to derive quantum measurement theory (which has, necessarily, a very similar structure to classical measurement theory), we need to know how to describe states-of-knowledge about quantum systems.

1.3.2 Quantum states-of-knowledge: density matrices

Quantum states already contain within them probabilities – once we express a quantum state in some basis, the coefficients for that basis determine the probabilities for finding the system in those basis states. However, these probabilities are not enough to describe all the possible states-of-knowledge that we might have about a quantum system. Even though a system may actually *be* in a given state $|\psi\rangle$, we may not know what this state is. In general then, our state-of-knowledge about a quantum system can be described by a probability density over all the possible states $|\psi\rangle$. We might refer to this probability density as describing our *classical* uncertainty about the system, and the coefficients c_n as describing the quantum uncertainty inherent in a given quantum state vector.

While a complete state-of-knowledge of a quantum system is a probability density over all the possible states $|\psi\rangle$, for most purposes one can use a more compact representation of this state-of-knowledge. This compact representation is called the *density matrix*. It was devised independently by von Neumann [642] and Landau [351] in 1927.

To obtain the density matrix formalism, we first recall that the expectation value of a physical observable, for a system in state $|\psi\rangle$, is given by

$$\langle X \rangle = \langle \psi |X| \psi \rangle = \mathrm{Tr}[X|\psi\rangle\langle\psi|], \tag{1.22}$$

where X is the operator corresponding to that observable. So while the expectation value of an observable is quadratic in the vector $|\psi\rangle$, it is linear in the operator (matrix) $|\psi\rangle\langle\psi|$. Expectation values are also linear in classical probabilities. If our state-of-knowledge is a probability distribution over the M states $\{|\phi_m\rangle\}$, where the probability of the system being in the state labeled by m is p_m, then the expectation value of X is

$$\langle X \rangle = \sum_m p_m \langle \phi_m |X| \phi_m \rangle = \sum_m p_m \mathrm{Tr}[X|\phi_m\rangle\langle\phi_m|] = \mathrm{Tr}[X\rho], \tag{1.23}$$

where

$$\rho \equiv \sum_m p_m |\phi_m\rangle\langle\phi_m|. \tag{1.24}$$

So we see that the matrix ρ is sufficient to calculate the expectation value of any operator. This is precisely because expectation values are a linear function of each of the matrices $|\phi_m\rangle\langle\phi_m|$, and thus all we have to do to include classical probabilities is to weight each of the $|\phi_m\rangle\langle\phi_m|$ by its classical probability.

We will see below that ρ is also sufficient to calculate the results of any measurement performed on the system. Note that it is only the results of measurements on quantum systems that determine events in the macroscopic world. This is because it is measurements that determine a set of mutually exclusive outcomes, and since the macroscopic world behaves classically, it is only sets of mutually exclusive possibilities that appear in it. (See also the discussions in Sections 1.5, 4.5, and 4.4.) We can conclude that questions about the future behavior of a quantum system are ultimately questions about the results of measurements. Thus ρ is sufficient to fully characterize the future behavior of a quantum system, and this is why it is a sufficient description of one's state-of-knowledge for many purposes.

In the absence of measurements the evolution of ρ is very simple. This evolution is given by evolving each of its component states. Since the evolution of a quantum state $|\psi\rangle$ is given by applying to it a unitary operator, $U(t)$, the evolution of ρ is

$$\rho(t) = \sum_m p_m |\phi_m(t)\rangle\langle\phi_m(t)| = \sum_m p_m U(t)|\phi_m(0)\rangle\langle\phi_m(0)|U^\dagger(t)$$

$$= U(t)\rho(0)U^\dagger(t). \tag{1.25}$$

The density matrix, ρ, is therefore a simple and compact way to describe our state-of-knowledge of a quantum system. The term *density matrix* comes from its role as a quantum equivalent of a probability density for classical systems, and also because its diagonal elements constitute a probability distribution. Specifically, if we express our states in the basis $\{|j\rangle\}$, so that

$$|\phi_m\rangle = \sum_j c_{jm}|j\rangle, \tag{1.26}$$

then the elements of ρ are

$$\rho_{jk} = \langle j|\rho|k\rangle = \sum_m p_m c_{jm} c_{km}^*. \tag{1.27}$$

So the jth diagonal element of ρ is

$$\rho_{jj} = \langle j|\rho|j\rangle = \sum_m p_m |c_{jm}|^2. \tag{1.28}$$

Since $|c_{jm}|^2$ is the (conditional) probability of finding the system in the state $|j\rangle$ given that it is initially in the state $|\phi_m\rangle$, ρ_{jj} is the total probability of finding the system in the state $|j\rangle$.

If the density matrix consists of only a single state, so that $\rho = |\psi\rangle\langle\psi|$, then it is described as being **pure**. If it consists of a sum over more than one state, then it is described as being **mixed**, and the system is said to be in a *statistical mixture*, or simply a *mixture*, of states. As an example, if the system is in the pure state $|\psi\rangle = \sum_j c_j|j\rangle$, then

$$\rho = |\psi\rangle\langle\psi| = \sum_{jk} c_j c_k^* |j\rangle\langle k| \tag{1.29}$$

and the system is said to be in a superposition of the basis states $|j\rangle$. If the system is in the state

$$\rho = \sum_j p_j |j\rangle\langle j|, \tag{1.30}$$

then it is said to be in a mixture of the basis states $|j\rangle$. In the latter case, we can think of the system as really being in one of the basis states, and the density matrix merely describes the fact that we do not know which one (although some may be more likely than others). On the other hand, if the system is in a superposition of two states, then we cannot describe the system as really being in one state or the other, and we discuss this further in the next section.

The density matrix is said to be *completely mixed* if it is equal to I/N, where N is the dimension of the system. If this is the case, then each of its eigenstates are equally likely. Further, *every* state in the Hilbert space (the complex vector space) is equally likely, since

$\langle \psi | (I/N) | \psi \rangle = 1/N$ for every state $|\psi\rangle$. In this case we have no information about the system.

It is not difficult to show that the density matrix is Hermitian ($\rho = \rho^\dagger$), and has unit trace ($\mathrm{Tr}[\rho] = 1$). Since the density matrix is Hermitian it has a complete set of eigenvectors (eigenstates), and can always be diagonalized (that is, written in the basis of its eigenstates so that it is diagonal). If the eigenvalues are $\lambda_j, j = 1, \ldots, N$, and the corresponding eigenvectors are $|\lambda_j\rangle$, the density matrix is

$$\rho = \sum_j \lambda_j |\lambda_j\rangle \langle \lambda_j|. \tag{1.31}$$

So a system is never in a superposition of the eigenstates of the density matrix – it is either in a mixture of the eigenstates, or in a single eigenstate, in which case the density matrix is pure. Note that the eigenvalues are the probabilities of finding the system in the corresponding eigenstates.

The difference between a superposition and a mixture

In case the physical distinction between a mixture of two or more quantum states, and a superposition of those states is not yet clear to you from your physics training to date, we pause here to explain it. A mixture of two states describes a situation in which a system really is in one of these two states, and we merely do not know which state this is. On the contrary, when a system is in a superposition of two states, it is *definitely not* in either of these states. The truth of this statement, while somewhat incredible, is easy to show. Consider the famous double-slit experiment, in which an electron is fired toward a metal sheet with two slits in it. The slits are very close to each other, and the electron is fired at the sheet so that it has a high probability of passing through the slits and thus reaching the other side. After the electron has passed through the slits, it is then detected on a screen some distance from the metal sheet. Consider what happens if one slit is blocked up. The electron passes through the open slit, and is detected on the screen. If the state of the electron after passing through the slit and reaching the screen is $|\psi_1\rangle$, then the probability density that the electron is detected on the screen at position x is $P_1(x) = |\langle \psi_1 | x \rangle|^2$, where $|x\rangle$ denotes the state in which the electron is on the screen at position x. Similarly, if the state of the electron after having passed through the other slit is $|\psi_2\rangle$, then the probability distribution for the electron to land on the screen at position x is $P_2(x) = |\langle \psi_2 | x \rangle|^2$.

Now we open both slits and fire the electron through them. If the electron *really* goes through one slit or the other, then we can immediately determine the probability distribution for the electron on the screen. Each electron that goes through the slits can be assigned as having gone through slit 1 or slit 2. Those going through slit 1 have probability distribution $P_1(x)$, and those going through slit 2 have distribution $P_2(x)$. If half of the electrons go through each slit (actually, some of the electrons will not pass through the slits, and instead scatter back from the metal sheet, but we ignore those) then the distribution of electrons on the screen will be

$$P(x) = \frac{1}{2}P_1(x) + \frac{1}{2}P_2(x).\tag{1.32}$$

This is the probability distribution we get if we assume that after going through the slits, the electron is in a *mixture* of states $|\psi_1\rangle$ and $|\psi_2\rangle$. In this case the state of the electron when it reaches the screen is $\rho = (1/2)(|\psi_1\rangle\langle\psi_1| + |\psi_2\rangle\langle\psi_2|)$, and the probability distribution for being in state $|x\rangle$ is

$$P_{\text{mix}}(x) = \text{Tr}[|x\rangle\langle x|\rho] = \frac{1}{2}P_1(x) + \frac{1}{2}P_2(x) = P(x).\tag{1.33}$$

In fact, after the electron passes through the slits, it is *not* in a mixture of states 1 and 2. To see this, note that the passage of the electron through the slits is described simply by evolving the wavefunction of the electron in the potential formed by the barrier (the metal sheet) with the slits in it. So after the electron passes through the slits, it is still described by a single wavefunction. The electron is therefore in a pure state after going through the slits. So what is this pure state? Well, it is too complex for us to determine the exact solution, but it turns out that far from the slits the electron's wavefunction is approximately given by the sum (superposition) of the states it would have been in if either of the slits had been blocked.[1] Thus the state of the electron is $|\psi\rangle = (|\psi_1\rangle + |\psi_2\rangle)/\sqrt{2}$. If we now calculate the probability density for the electron to be at position x on the screen, we immediately see the difference between this "superposition" state and the mixture above. The probability density for being in state $|x\rangle$ is

$$P_{\text{sup}}(x) = |\langle\psi|x\rangle|^2 = \frac{1}{2}P_1(x) + \frac{1}{2}P_2(x) + \text{Re}[\langle x|\psi_1\rangle\langle\psi_2|x\rangle]$$

$$\neq P(x).\tag{1.34}$$

Since $P_{\text{sup}}(x)$ is not equal to $P(x)$, and $P(x)$ is a necessary result of the electron *really* being in state $|\psi_1\rangle$ or $|\psi_2\rangle$, a superposition cannot correspond to the electron really being in either of these two states. For want of better terminology, one usually refers to a system that is in a superposition of two states as being in *both states at once*, since this seems more reasonable than saying that it is in neither.

A useful theorem: the most likely state

The following theorem answers the question, what is the most likely pure state of a system for a given density matrix ρ? Note that the probability of finding the system in state $|\psi\rangle$ is given by

$$P(|\psi\rangle) = \langle\psi|\rho|\psi\rangle = \text{Tr}[|\psi\rangle\langle\psi|\rho].\tag{1.35}$$

---

[1] Note that this is also what happens to water waves if they encounter a barrier with two small openings. While Schrödinger's wave-equation is not the same as that for water waves, all wave-equations have this feature in common.

Theorem 1 If the state of a quantum system is ρ, then the most likely pure state is the eigenstate of ρ with the largest eigenvalue. This eigenvalue is the probability that the system will be found in that state.

Proof We need to find the state $|\psi\rangle$ that maximizes $\langle\psi|\rho|\psi\rangle$. Let us denote the eigenbasis of ρ as $\{|\lambda_n\rangle\}$. Writing the density matrix in terms of its eigenbasis (Eq. 1.31), and writing $|\psi\rangle = \sum_n c_n|\lambda_n\rangle$, we have

$$\langle\psi|\rho|\psi\rangle = \sum_n |c_n|^2\lambda_n. \tag{1.36}$$

Since $\sum_n |c_n|^2 = 1$, the above expression is the average of the eigenvalues over the probability distribution given by $p_n = |c_n|^2$. Thus to maximize it we must place all of the probability on the largest eigenvalue. If we denote this eigenvalue by λ_j, then this means that $|c_n|^2 = \delta_{nj}$, and therefore $|\psi\rangle = |\lambda_j\rangle$. $\qquad\square$

A density matrix consists of a set of states, $\{|\psi_n\rangle\}$, in which each state has an associated probability, p_n. Such a set of states and probabilities is called an *ensemble*. In fact, there is no reason why every state in an ensemble must be a pure state – a set of states $\{\rho_n\}$ with associated probabilities $\{p_n\}$ is also an ensemble, and generates a density matrix $\rho = \sum_n p_n\rho_n$. If an ensemble contains only pure states, then it is called a *pure-state* ensemble. We will often write an ensemble for the set of states $\{\rho_n\}$, with associated probabilities $\{p_n\}$, as $\{\rho_n, p_n\}$.

Since every pure-state ensemble corresponds to some density matrix, a natural question to ask is, for a given density matrix, can one determine all the possible pure-state ensembles that correspond to it? It turns out that not only is the answer yes, but the collection of ensembles can be characterized very simply. We leave these facts to Chapter 2, however, because one of the characterizations uses the concept of *majorization*. This is a very simple concept which captures the intuitive notion of uncertainty, and it will be defined in the next chapter.

1.3.3 Quantum measurements

In Section 1.3.1 we introduced the measurement postulate, and defined the concept of a von Neumann measurement. Given a state-of-knowledge, ρ, we can describe a von Neumann measurement of the system in the following way. Given a basis $\{|n\rangle\}$, we define the set of projection operators

$$P_n \equiv |n\rangle\langle n|. \tag{1.37}$$

Now, if we make a von Neumann measurement in the basis $\{|n\rangle\}$, then after the measurement we will find the system to be in one of these basis states, say $|m\rangle$, with probability $p_m = \langle m|\rho|m\rangle$. To reduce an arbitrary initial density matrix, ρ, to the final density matrix

$|m\rangle\langle m|$, we must sandwich ρ between two copies of the projector P_m, and normalize the result. That is, the final state is given by

$$\tilde{\rho}_m = |m\rangle\langle m| = \frac{P_m \rho P_m}{\text{Tr}[P_m \rho P_m]}. \tag{1.38}$$

We include the tilde on top of ρ_m to indicate that it is a state that results from a measurement. This convention will help to make our expressions clearer later on. Note that the expression on the bottom line is actually the probability p_m:

$$p_m = \langle m|\rho|m\rangle = \text{Tr}[P_m \rho P_m] = \text{Tr}\left[(P_m)^2\rho\right] = \text{Tr}[P_m\rho], \tag{1.39}$$

where we have used the cyclic property of the trace: $\text{Tr}[ABC] = \text{Tr}[CAB] = \text{Tr}[BCA]$. The reason why we chose to write the expression for the final state using the projectors P_n is that this form will make it clear later on that von Neumann measurements are a special case of the more general measurements that will be derived below.

Note that a von Neumann measurement induces a nonlinear change in the density. The operation of applying the projector P_n is linear, since it is matrix multiplication, but the operation of rescaling the density matrix so that the final state is normalized is not linear. Recall that an operation $Q(\cdot)$ on a vector (a set of numbers) is linear if $Q(\alpha\mathbf{v} + \beta\mathbf{u}) = \alpha Q(\mathbf{v}) + \beta Q(\mathbf{u})$, for all real numbers α, β and vectors $\mathbf{v}$ and $\mathbf{u}$. In our case the density matrix is the "vector," and it is not hard to see that the above linearity condition is not satisfied by von Neumann measurements. This nonlinearity is a significant departure from the linear evolution of Schrödinger's equation. In classical measurement theory measurements also change the observer's state-of-knowledge in a nonlinear way, but they do not affect the dynamics of the system itself. In Section 1.4.3 we will see that quantum measurements do cause dynamical changes. Because of this they cause the dynamics of quantum systems to be nonlinear.

Von Neumann measurements are certainly not the only kinds of measurements one can make on a quantum system. To gain an understanding of quantum measurements we need to examine different kinds of measurements, and what kind of information they extract, and how they affect the systems being measured. We could start by considering a number of physical examples, and follow this by deriving a general formalism that describes them all. We have chosen instead to do it the other way around, as we feel that this is quicker. But instead of following this path, if you prefer, you can skip the derivation and go straight to Theorem 2 (p. 24), which presents the concise mathematical description of (almost) all measurements. From Theorem 2 you can then go to Section 1.4 which explains, via some examples, how this description relates to classical measurements and the classical likelihood function, and why quantum measurements induce a dynamical change in the system being measured.

Deriving general quantum measurements from the measurement postulate

Fortunately we can derive all possible measurements that can be made on a quantum system from von Neumann measurements. To do this we consider a quantum system we wish to measure, which we will call the *target*, and a second system we will call the *probe*. We will denote the dimension of the target system by N, and that of the probe by M. The probe is prepared in some state independently of the target, and then the two systems are allowed to interact. After the interaction we perform a von Neumann measurement on the probe. As a result of the interaction, the probe is correlated with the target, and so the measurement on the probe provides us with information about the target. This procedure gives us a much more general kind of measurement on the target.

We will denote the basis in which we measure the probe system as $\{|n\rangle\}, n = 0, \ldots, M-1$. Since the interaction between the systems may be any unitary operator that acts in the joint space of both systems, we can start the probe in the state $|0\rangle$ without loss of generality. The combined initial state of the target and probe is therefore

$$\rho_{\mathrm{comb}} = |0\rangle\langle 0| \otimes \rho, \tag{1.40}$$

where ρ is the initial state of the target. We will always write the state of the probe on the left-hand side of the tensor product, and that of the target on the right. To understand the following it is essential to be familiar with how a composite quantum system, consisting of two subsystems, is described by the tensor product of the spaces of these two subsystems. If you are not familiar with how to combine two quantum systems using the tensor product, denoted by '$\otimes$', the details are presented in Appendix A.

The form of the target/probe interaction operator

The interaction between the target and probe is described by some unitary operator U that acts in the space of both systems. The subsequent von Neumann measurement on the probe is described by a projection onto one of the probe states $|n\rangle$, followed by a normalization. In analyzing the measurement process, it will make things clearer if we first know some things about the structure of U.

Since U acts in the tensor-product space, it can be written as the matrix

$$U = \sum_{nn'kk'} u_{nk,n'k'} |n\rangle |s_k\rangle \langle n'| \langle s_{k'}|, \tag{1.41}$$

where $|s_k\rangle$ are a set of basis states for the target, and $u_{nk,n'k'}$ are the matrix elements of U. For each pair of probe states $|n\rangle$ and $|n'\rangle$, there is a sub-block of U that acts in the space of the target. We can alternatively write U in terms of these sub-blocks as

$$U = \sum_{nn'} |n\rangle\langle n'| \otimes A_{nn'}, \tag{1.42}$$

where the operators $A_{nn'}$ are given by

$$A_{nn'} = \sum_{kk'} u_{nk,n'k'} |s_k\rangle \langle s_{k'}|. \tag{1.43}$$

The relationship between the matrices $A_{nn'}$ and U is most clear when they are written as matrices, rather than in Dirac notation. This relationship is

$$U = \begin{pmatrix} A_{00} & A_{01} & \cdots & A_{0M} \\ A_{10} & A_{11} & \cdots & A_{1M} \\ \vdots & \vdots & \ddots & \vdots \\ A_{M0} & A_{M1} & \cdots & A_{MM} \end{pmatrix} \tag{1.44}$$

Recall that the system has dimension N and so each matrix $A_{nn'}$ is N-dimensional. In what follows we will use the notation A_n as shorthand for the matrices A_{n0}. These matrices constitute the first column of the sub-blocks of U. Let us now denote the $M \times M$ sub-blocks of the matrix $U^\dagger U$ by $B_{nn'}$. Because U is unitary, $U^\dagger U = I$. This means that each of the sub-blocks B_{nn} (those on the diagonal of $U^\dagger U$) must be equal to the identity. Further, the sub-block B_{00} is the product of the first column of the sub-blocks of U, multiplied by the first row of the sub-blocks of $U^\dagger$. This gives us immediately the relationship

$$I = B_{00} = \sum_n A_n^\dagger A_n. \tag{1.45}$$

The final important fact about the structure of U is that, apart from the restriction given by Eq. (1.45), we can choose the sub-blocks $A_n \equiv A_{0n}$ to be *any* set of operators. This is because Eq. (1.45) is alone sufficient to ensure that there exists a unitary U with the operators A_n as its first column of sub-blocks. To see why this is true, note that the operators A_n, being the first column of sub-blocks, constitute the first N columns of U, where N is the dimension of the target. To show that Eq. (1.45) ensures that U is unitary, we need merely show that Eq. (1.45) implies that these first N columns are mutually orthonormal. So long as this is the case, we can always choose the remaining columns of U so that all the columns of U form an orthonormal basis, making U unitary. This task is not especially difficult, and we leave it as an exercise.

Using a probe system to make a measurement

We now apply the unitary interaction U to the initial state of the two systems, and then project the probe system onto state $|n\rangle$. The final state of the combined system is then

$$\check{\sigma} = (|n\rangle\langle n| \otimes I)U(|0\rangle\langle 0| \otimes \rho)U^\dagger(|n\rangle\langle n| \otimes I). \tag{1.46}$$

We have placed a caron (ˇ) over the final state, σ, to indicate that it is not necessarily normalized, due to the fact that we have projected the system onto a subspace. Writing U

in terms of its sub-blocks, we immediately obtain a simple form for the final state:

$$\breve{\sigma} = |n\rangle\langle n| \otimes A_n \rho A_n^\dagger. \tag{1.47}$$

The final state of the *target* is therefore

$$\tilde{\rho}_n = \frac{A_n \rho A_n^\dagger}{\mathrm{Tr}[A_n^\dagger A_n \rho]}, \tag{1.48}$$

where we have normalized it by dividing by its trace.

Now that we have a simple form for the final state for each outcome n, we need to know the probability, p_n, that we get each of these outcomes. This is given by the probability of finding the probe in state $|n\rangle$ after the interaction U, and this is in turn the sum of all the diagonal elements of the density matrix that correspond to the probe being in state $|n\rangle$. The density matrix after the interaction is given by $\rho_U = U(|0\rangle\langle 0| \otimes \rho)U^\dagger$. The sum of its diagonal elements that correspond to the probe state $|n\rangle$ is therefore

$$\begin{aligned}
p_n &= \mathrm{Tr}[\,(|n\rangle\langle n| \otimes I)\rho_U(|n\rangle\langle n| \otimes I)\,] \\
&= \mathrm{Tr}[\,(|n\rangle\langle n| \otimes I)\,U(|0\rangle\langle 0| \otimes \rho)U^\dagger\,(|n\rangle\langle n| \otimes I)\,] \\
&= \mathrm{Tr}[\breve{\sigma}] = \mathrm{Tr}[|n\rangle\langle n| \otimes A_n \rho A_n^\dagger] = \mathrm{Tr}[|n\rangle\langle n|]\,\mathrm{Tr}[A_n \rho A_n^\dagger] \\
&= \mathrm{Tr}[A_n^\dagger A_n \rho]. \tag{1.49}
\end{aligned}$$

We now have a complete description of what happens to a quantum system under a general measurement process. Further, we know that *every* set of operators $\{A_n\}$ that satisfies Eq. (1.45) describes a measurement that can be realized by using an interaction with a probe system. For emphasis we state this result as a theorem.

Theorem 2 The fundamental theorem of quantum measurement: Every set of operators $\{A_n\}$, $n = 1,\ldots,N$, that satisfies $\sum_n A_n^\dagger A_n = I$, describes a possible measurement on a quantum system, where the measurement has n possible outcomes labeled by n. If ρ is the state of the system before the measurement, $\tilde{\rho}_n$ is the state of the system upon obtaining measurement result n, and p_n is the probability of obtaining result n, then

$$\tilde{\rho}_n = \frac{A_n \rho A_n^\dagger}{p_n}, \tag{1.50}$$

$$p_n = \mathrm{Tr}[A_n^\dagger A_n \rho]. \tag{1.51}$$

We refer to the operators $\{A_n\}$ as the *measurement operators* for the measurement.

In fact, the form given for quantum measurements in the above theorem, in which each outcome is associated with a single measurement operator, does not cover all possible measurements. But the additions required to describe every possible measurement are purely classical. We need merely include classical uncertainty about the result of the measurement, or about what measurement was performed. Because of this all the purely *quantum* features of any measurement are captured by a set of operators $\{A_n\}$, with $\sum_n A_n^\dagger A_n = I$. We show how to add additional classical uncertainty to quantum measurements in Section 1.6 below. Measurements that have no additional classical uncertainty are referred to as *efficient* measurements, and we explain the reason for this terminology in Section 1.6.

Theorem 2 provides a simple mathematical description of quantum measurements, but we don't yet know the physical meaning of the operators $\{A_n\}$. That is, we don't know how the form of a particular operator A_n is related to the physical properties of the measurement. We turn to this question in Section 1.4, but before we do, we discuss a few useful concepts and topics.

The partial inner product

We discuss in detail how to treat systems that consist of more than one subsystem in Appendix A, and we have already used some of this machinery in our analysis above. Nevertheless we feel that it is worth pausing to discuss one particular part of this machinery here. Let us denote the basis states of two subsystems A and B by $\{|n\rangle_A\}$ and $\{|n\rangle_B\}$, respectively. In our analysis above we used a projector that projected out a state of one subsystem only, namely $P_n \equiv |n\rangle_A\langle n|_A \otimes I_B$, where I_B is the identity operator for system B. If the joint state of the two systems is $|\psi\rangle = \sum_{mk} \alpha_{mk}|m\rangle_A|k\rangle_B$, then the action of this partial projector is to leave only the joint states that contain $|n\rangle_A$. That is,

$$P_n|\psi\rangle = |n\rangle_A\langle n|_A \otimes I_B \sum_{mk} \alpha_{mk}|m\rangle_A|k\rangle_B = \sum_k \alpha_{nk}|n\rangle_A|k\rangle_B. \tag{1.52}$$

It is useful to define a "partial inner product" that is closely related to the partial projector P. We define the inner product $\langle n|_A|\psi\rangle$ as

$$\langle n|_A|\psi\rangle = \langle n|_A \sum_{mk} \alpha_{mk}|m\rangle_A|k\rangle_B \equiv \sum_{mk} \alpha_{mk}\langle n|m\rangle_A|k\rangle_B, \tag{1.53}$$

which is a state of system B alone. The partial inner product takes the inner product with the states of one system, and leaves the states of the other alone. One way to define this inner product is to say that when we have an expression of the form $\langle a|_A|\psi\rangle$, where $\langle a|_A$ is a state of one subsystem, and $|\psi\rangle$ is a joint state of both systems, then $\langle a|_A$ is shorthand for the unnormalized vector $\sum_k \langle a|_A\langle k|_B$.

To make clearer the relationship between the projector P_n and the partial inner product, denote the state of system B that is produced by the partial inner product as $|\psi_n\rangle_B = \langle n|_A|\psi\rangle$. If we apply $P_n = |n\rangle_A\langle n|_A \otimes I_B$ to $|\psi\rangle$ then this contains within it the application of the partial inner product:

$$P_n|\psi\rangle = |n\rangle_A\langle n|_A \otimes I_B|\psi\rangle = |n\rangle_A \otimes |\psi_n\rangle_B. \tag{1.54}$$

With this definition of the partial inner product, if we have a joint operator X of the two subsystems, then $\langle n|_A X|n\rangle_A$ is a sub-block of the matrix X that operates only on system B. If you are not familiar with this fact then you can show it as an exercise.

We also note that the partial trace, described in detail in Appendix A, can be written in terms of the partial inner product. If the joint density matrix is ρ, the partial trace over system A is $\rho_b = \sum_n \langle n|_A \rho|n\rangle_A$.

Discarding a system

Let us say we have a quantum system S that is initially in a state ρ. It then interacts with a second system B over which we have no control, and to which we have no access. What is our state-of-knowledge of S following the interaction? One way to answer this question is to realize that another observer could measure B, just as we measured the probe in the scenario above. Since we do not know the outcome of this measurement, our state-of-knowledge is given by averaging over all the outcomes of the measurement on B. It is therefore

$$\tilde{\rho} = \sum_n p_n \tilde{\rho}_n = \sum_n A_n \rho A_n^\dagger = \sum_n \langle n|U(|0\rangle\langle 0| \otimes \rho)U^\dagger|n\rangle$$

$$\equiv \mathrm{Tr}_B[U(|0\rangle\langle 0| \otimes \rho)U^\dagger], \qquad (1.55)$$

where we have used the partial-inner-product notation, $|n\rangle$ denotes a basis state of the probe, $|0\rangle\langle 0|$ is the initial state of the probe, and Tr_B denotes the partial trace over system B, which is defined and discussed in detail in Appendix A.

So our state-of-knowledge after the interaction is given by the partial trace over the second system. In fact, the partial trace is completely independent of the measurement that the observer makes on the second system. Of course, this is crucial, because we cannot know what measurement the observer will make, so our state-of-knowledge must be independent of this measurement. The invariance of the partial trace under operations on the traced system is proved in Appendix A.

The operator-sum representation

Consider an evolution of a quantum system that takes every initial density matrix, ρ, to a single final density matrix $\mathcal{L}(\rho)$. This is a deterministic evolution in the sense that there is only one final density matrix for every initial density matrix. Such an evolution is often referred to as a "trace-preserving" quantum operation, because there is no need to normalize the final density matrix as is required when there are multiple measurement outcomes.

There are certain properties that a deterministic map from initial to final quantum states must have. It must be linear, and every final density matrix must be positive. We will call a map that produces only positive density matrices a positive map. In fact this positivity condition contains a subtlety. Since a quantum system might initially be in joint state with a second quantum system, the evolution of the first system, by itself, must be such that the evolution of the joint system is positive. To obtain the map for the first system we start with the map for the joint system, and take the trace over the second system. The subtlety is that not all positive linear maps for the first system are derivable from positive linear maps for the joint system. Only the subset of positive maps that can be derived in this way are valid evolutions. These maps are referred to as being *completely positive*. We note that the definition of a "quantum operation" is any completely positive linear map that may or may not preserve the trace of the density matrix.

Every realizable quantum evolution, $\mathcal{L}(\rho)$, can be written in the form

$$\mathcal{L}(\rho) = \sum_n A_n \rho A_n^\dagger, \tag{1.56}$$

where $\sum_n A_n^\dagger A_n = I$. This follows immediately from our analysis of a quantum measurement above. The evolution of every quantum system is ultimately the result of a unitary evolution on a larger system of which the system is a subsystem, and the evolution for the subsystem is given by tracing out the rest of the larger system. This trace can be written as the average over the results of a von Neumann measurement on the rest of the larger system, and this results in the form in Eq. (1.56), being the average over the results of a measurement on the system.

The form given in Eq. (1.56) is called an "operator-sum" representation for a quantum evolution. The result that any linear map is completely positive if and only if it can be written in the operator-sum representation was first proved by Choi [124]. Another useful fact that is not obvious is that the evolution of an N-dimensional quantum system can always be written using no more than N^2 operators $\{A_n\}$. This was first shown by Hellwig and Kraus [247, 345], and the measurement operators $\{A_n\}$ are often referred to as Kraus operators. Schumacher proved this result in an elegant and more direct way by defining a linear map using the partial inner product, and we give his proof in Appendix H.

Positive-operator valued measures (POVMs)

General measurements on quantum systems are often referred to in the literature as "positive-operator valued measures." While the usefulness of this term is unclear, and should probably be replaced simply with "quantum measurement," its origin is as follows. A *measure* is a map that associates a number with all subsets of a given set. A positive-operator valued measure is thus a map that associates a positive operator with every subset. For the purposes of quantum measurements, the set in question is the set of outcomes of the measurement. Let us label the outcomes by n. If we pick some subset of the outcomes, call it $\mathcal{M}$, then the probability that we get an outcome from this subset is given by

$$\mathrm{Prob}(n \in \mathcal{M}) = \sum_{n \in \mathcal{M}} p_n = \sum_{n \in \mathcal{M}} \mathrm{Tr}[A_n^\dagger A_n \rho] = \mathrm{Tr}\left[\left(\sum_{n \in \mathcal{M}} A_n^\dagger A_n\right)\rho\right]. \tag{1.57}$$

Since the operators $A_n^\dagger A_n$ are all positive, the operator in the parentheses is also positive. This is the positive operator associated with the set $\mathcal{M}$, and is thus the positive operator of the "positive-operator valued measure."

1.4 Understanding quantum measurements

1.4.1 Relationship to classical measurements

While we have shown that a measurement on a quantum system can be described by a set of operators $\{A_n\}$, we do not yet understand how the form of these operators relates to the nature of the measurement. We can gain such an understanding by examining various key forms that the A_n can take, and the measurements to which they correspond. It is most informative to begin by examining under what conditions quantum measurements reduce to classical measurements. This reveals the relationship between the measurement operators and the classical likelihood function.

Consider a state-of-knowledge that is diagonal in the basis $\{|n\rangle\}$, so that

$$\rho = \sum_n p_n |n\rangle\langle n|. \tag{1.58}$$

Note that in this case the system is not in a superposition of the basis states. It actually *is* in one of the basis states; we merely do not know which one. Recall that we refer to this situation by saying that the system is in a *statistical mixture* (or simply a *mixture*) of the states $\{|n\rangle\}$. Now, if we make a measurement on the system purely to obtain information about the basis state it is in, then we might expect this to be described perfectly well by a classical measurement. This is because the basis states are distinguishable, just like classical states, and our state-of-knowledge is a probability distribution over these states, just like a classical state of knowledge.

With the above thoughts in mind, consider a quantum measurement in which all the measurement operators A_j are diagonal in the same basis as the density matrix. Note that we can think of the diagonal elements of the measurement operators, and of ρ, simply as functions of the index n. That is, we can write

$$\rho = \sum_n P(n)|n\rangle\langle n|, \tag{1.59}$$

$$A_j = \sum_n A(j,n)|n\rangle\langle n|, \tag{1.60}$$

where $P(n)$ and $A(j,n)$ are the functions in question. The final (or posterior) state-of-knowledge that results from our measurement is

$$\tilde{\rho}_j = \frac{A_j \rho A_j^\dagger}{\mathrm{Tr}[A_j^\dagger A_j \rho]} = \frac{\sum_n A^2(j,n)P(n)|n\rangle\langle n|}{\mathcal{N}}. \tag{1.61}$$

This is precisely the same form as the update rule for classical measurement theory. All we have to do is write the posterior state $\tilde{\rho}_j$ as

$$\tilde{\rho}_j = \sum_n P(n|j)|n\rangle\langle n|, \tag{1.62}$$

and identify $A^2(j,n)$ as the likelihood function, $P(j|n)$, and we have

$$P(n|j) = \frac{P(j|n)P(n)}{\mathcal{N}}. \tag{1.63}$$

So, at least in this case, quantum measurements are merely classical measurements, and the diagonal of the measurement operator A_j corresponds to the square-root of the likelihood function evaluated for the jth outcome.

Two operators A and B are diagonal in the same basis if and only if they commute with one another. In view of the above results, we will refer to quantum measurements for which

$$[A_j, A_k] = 0, \quad \forall j, k \tag{1.64}$$

as *semiclassical measurements*, since they are completely equivalent to their classical counterparts when made on a system in a state that is in a mixture of the eigenstates of the measurement operators.

Incomplete measurements and Bayesian inference

As an example we consider a simple semiclassical measurement that provides only partial information about the final state. Such measurements have measurement operators that are not projectors onto a single state, and are referred to variously as "incomplete" measurements [47], "finite-strength" measurements [188], or "weak" measurements [548].

Consider a two-state system in the state

$$\rho = p_0|0\rangle\langle 0| + p_1|1\rangle\langle 1|, \tag{1.65}$$

and a measurement with two possible outcomes, described by the operators

$$A_0 = \sqrt{\kappa}|0\rangle\langle 0| + \sqrt{1-\kappa}|1\rangle\langle 1|, \tag{1.66}$$

$$A_1 = \sqrt{1-\kappa}|0\rangle\langle 0| + \sqrt{\kappa}|1\rangle\langle 1|. \tag{1.67}$$

If $\kappa = 1$ (or $\kappa = 0$) then both measurement operators are projectors, and after the measurement we are therefore certain about the state of the system. If $\kappa = 1/2$ then both operators are proportional to the identity. They therefore have no action on the system, and we learn nothing from the measurement. For $\kappa \in (1/2, 1)$ the measurement is an "incomplete" measurement. It changes the probabilities p_0 and p_1, but does not reduce either to zero.

As discussed in the previous section, since the measurement operators commute with the density matrix, their action on the diagonal elements of the density matrix is simply that of Bayesian inference. The measurement operators we have chosen above correspond precisely to the simple two-outcome measurement that we discussed in Section 1.2.1, with the identification $\alpha = \kappa$. So, for example, if $\kappa > 1/2$ and we get the outcome 0, then the diagonal elements of ρ are re-weighted so as to make the state $|0\rangle$ more likely than it was before. Finite-strength measurements thus provide some information about which state the system is in, but not complete information.

We will see in Section 1.4.4 that we can, in fact, understand the action of all quantum measurements, to a large extent, in terms of Bayesian inference.

1.4.2 Measurements of observables and resolving power

Measurements of observables are a subset of semiclassical measurements. This is because a measurement of an observable should not change the state of the system if it is already in an eigenstate of the observable, but merely provide information about the eigenvalue. This condition demands that all the measurement operators must commute with the observable. We speak of the measurement as obtaining information in the basis of the observable, and we refer to this basis as the *measurement basis*. But measuring a specific observable requires more than merely having a set of measurement operators that commute with it. To define a measurement of an observable we introduce the concept of *resolving power*. Let us say that a system is prepared in an equal ("fifty-fifty") mixture of two different states of the measurement basis. The resolving power of a measurement, with respect to these two states, is the amount by which, on average, the probabilities of the two states are made more different by the measurement. In particular, we define the resolving power as the absolute value of the difference between the probabilities of the two basis states following the measurement, averaged over the measurement results. As an exercise, if you calculate this quantity for a von Neumann measurement you will find that it is unity, and this is clearly the maximum possible value.

A measurement obtains information about an observable X if the probabilities of the measurement results depend on the eigenvalues of X. If these probabilities do not change when we change from one initial state to another, then the measurement has no resolving power for those two states. If two states have the same eigenvalue for X then a measurement of X should not distinguish between them, and the resolving power for these two states should be zero. Further, for pairs of states that have the same separation between their respective eigenvalues, the resolving power should be the same. The resolving power should also increase monotonically with the eigenvalue separation (it would be very strange if it were harder to distinguish two objects that were a meter apart than a millimeter apart!). This gives us a minimal definition of a measurement of an observable.

If we wish we can be even more specific about what constitutes a measurement of an observable by specifying how the resolving power between two states should scale in some way with the separation between their eigenvalues. For example, for continuous measurements, which we introduce in Chapter 3, it is natural to define a measurement of an observable as one in which the resolving power scales with the eigenvalue separation in the same way as it scales with the rate at which the measurement extracts information.

There are many semiclassical measurements that do not correspond to measurements of any observable. This is because the possible values of any physical observable lie on a single line, so that a measurement of an observable must have more resolving power for some pairs of states than for others.

An example of a measurement of an observable X is $\{A_\alpha\}$, where

$$A_\alpha = f(\alpha)e^{-k(X-\alpha)^2}, \tag{1.68}$$

and where $f(\alpha)$ is the scaling required for each operator A_α so that $\sum_\alpha A_\alpha = I$. The larger the parameter k, the higher the resolving power of the measurement.

1.4.3 A measurement of position

We now give an example of a measurement of position that illustrates the dynamical change caused by a measurement, something that we have not yet discussed. Quantum measurements induce dynamical changes because the measurement of one observable will usually change the distribution of other observables that do not commute with it. In classical systems different dynamical variables are independent of each other. Thus if we measure the position of a classical object, this does not change the object's momentum. The measurement does change the distribution of the position, but this is just a change in our state-of-knowledge about the position, not a change to the position itself. The reason that we know a classical measurement does not change a classical object's position is that if we average our state-of-knowledge over all the measurement results, we are left precisely with the position distribution that the object had prior to the measurement. We know that a change is dynamical (that is, a real change to an observable of the system) if the change remains when we average over the measurement results. This averaging eliminates any change that is due purely to a change in our state-of-knowledge.

Position is a continuous variable, rather than the discrete variables we have considered so far. So sums over the eigenstates of position will become integrals. Of course, the "eigenstates" of position are not real quantum states, as they are not normalizable. They are just a notational convenience introduced by Dirac. We must also remember that we cannot measure position exactly, since a state that is infinitely sharp in position has an infinite variance for the momentum, and thus infinite energy. But we can make measurements that reduce the position variance, even though we are always left with some uncertainty.

Recall that in measuring a classical variable like position, we often assume that the error in the measurement has a Gaussian distribution, and in Section 1.2.2 we showed that this is described by a Gaussian likelihood function. For a quantum measurement, since it will induce a dynamical change in the system, it is worth considering the interaction that mediates the measurement. Let us say that we measure the position of an object by bouncing a photon off it, and recording the time it takes for the photon to reach our detector. If the photon was in a superposition of waves of different frequency, so that its spatial profile was Gaussian, then upon detecting the time of arrival we would still be uncertain about the position. If we knew very little about the position to begin with, then we would be left with a Gaussian distribution for the position. This is one example of the quantum equivalent of a measurement with a Gaussian error. As another example we could measure the phase of a pulse of light reflected from the object. If the pulse was a superposition of coherent states of different frequencies (coherent states are defined in Appendix D), then the initial phase would be uncertain, and the distribution of this phase would determine that of the error. It is especially important to note that the states of light in both these examples are pure states. This means that the uncertainty in the measurement is purely due to quantum uncertainty, not to any additional classical uncertainty about the initial state of the probe or the measurement result. Any classical uncertainty would introduce extra errors, and this is the subject of *inefficient measurements* discussed in Section 1.6. Note also that the uncertainty inherent in the initial state of the light is enforced by quantum mechanics, just as is the uncertainty in the position of the object: we cannot prepare a pulse of light with a perfectly defined time or phase without infinite energy.

We saw above in Section 1.4.1 that a quantum measurement of an observable x, with result denoted by y, is equivalent to a classical measurement with a likelihood function $P(y|x)$, when it has a set of measurement operators $\{A_y\}$ given by

$$A_y = \int_{-\infty}^{\infty} \sqrt{P(y|x')}|x'\rangle\langle x'|\,dx'. \tag{1.69}$$

Here the states $|x'\rangle$ are the eigenstates of the operator x with eigenvalue x'. In our case x is position, and for a Gaussian measurement we therefore have

$$A_y = \int_{-\infty}^{\infty} \frac{e^{-(x'-y)^2/(4V)}}{(2\pi V)^{1/4}}|x'\rangle\langle x'|\,dx' = \frac{e^{-(x-y)^2/(4V)}}{(2\pi V)^{1/4}}. \tag{1.70}$$

This is the quantum equivalent of the Gaussian classical measurement discussed in Section 1.2.2.

Now let us make a measurement of the position of an object in the Gaussian state

$$|\psi\rangle = \int_{-\infty}^{\infty} \frac{e^{-(x'-x_0)^2/(4V_0)}}{(2\pi V_0)^{1/4}}|x'\rangle\,dx'. \tag{1.71}$$

If the object is macroscopic, then its position coordinate is the center-of-mass degree of freedom of all the particles that make up the object. Upon making the measurement, and obtaining result y, the new state is

$$|\tilde{\psi}\rangle = \frac{1}{\mathcal{N}}A_y|\psi\rangle = \int_{-\infty}^{\infty} \frac{e^{-(x'-x_1)^2/(2V_1)}}{(2\pi V_1)^{1/4}}|x'\rangle\,dx', \tag{1.72}$$

where the new mean, x_1, and new variance V_1 are given by

$$x_1 = x \left(\frac{V}{V_0 + V} \right) + y \left(\frac{V_0}{V_0 + V} \right), \tag{1.73}$$

$$V_1 = \frac{V_0 V}{V_0 + V}. \tag{1.74}$$

Curiously, for a Gaussian measurement of a Gaussian state the change in the variance is independent of the result y. As we expect, the less the error in the measurement, the larger the reduction in the initial variance of x. We also see that the change in the mean position resulting from the measurement is weighted both by the initial uncertainty and the error in the measurement.

As far as the position of the object is concerned, the above quantum measurement exactly parallels a classical measurement of position. But we see a difference as soon as we examine the momentum distribution before and after the measurement. To obtain the wavefunction of the system in momentum space, which we will denote by $\phi(p)$, we take the Fourier transform of the position-space wavefunction, $\psi(x)$:

$$\phi(p) = \int_{-\infty}^{\infty} \psi(x) e^{ixp/\hbar} dx. \tag{1.75}$$

The result of this relationship between position and momentum is that the initial state $|\psi\rangle$ has a mean momentum of zero, and a momentum variance of $V_p = \hbar^2/(4V_x)$. The spread of the momentum is inversely proportional to the spread of the position, which is Heisenberg's uncertainty relation. Since the position variance is reduced by the measurement, the momentum variance is *increased*. The measurement therefore causes a *dynamical* change to the object, since the increase in the momentum variance is the same for every outcome y, and thus will not disappear when the state of the system is averaged over all possible measurement results (all values of y). Note that the measurement does not change the mean momentum. In this sense the change is completely random: if the mean momentum was changed then we could speak of the change as having a deterministic component. Note also that since the average kinetic energy of the object is proportional to the square of the momentum, the measurement increases the kinetic energy, on average. This means that the interaction with the probe, the pulse of light in our examples above, necessarily imparts energy to the object, on average.

The dynamical change caused by a measurement of position is an example of a very general effect. Except for specially chosen initial states, a measurement of any observable will induce a dynamical change in all observables that do not commute with it. This change is often referred to as the *back-action* of the measurement.

1.4.4 The polar decomposition: bare measurements and feedback

The polar decomposition theorem states that any operator A may be written as the product of a positive operator, $\mathcal{P}$, and a unitary operator U [266, 61]. (The unitary operator may be on the right or the left of the positive operator.) Positive operators have a complete set of eigenvalues, and all their eigenvalues are greater than or equal to zero (the density matrix is thus a positive operator). As a result $\mathcal{P} = \mathcal{P}^\dagger$. When a unitary operator, U, is applied to the density matrix ρ via $U\rho U^\dagger$, it gives a possible time evolution for the system. In doing so it does not change the eigenvalues of the density matrix, but only the eigenvectors. This means that it does not change the amount of our uncertainty about a system, just *where* in the state space it is most likely to be. The eigenvalues of a unitary matrix are, in general, complex but have unit norm, and $U^\dagger U = UU^\dagger = I$. As we will see, the polar decomposition theorem provides tremendous insight into the structure of quantum measurements.

To apply the polar decomposition theorem to a quantum measurement described by a set of operators $\{A_n\}$, we write each of these operators as $A_n = U_n \mathcal{P}_n$. With this replacement the probability of obtaining the nth result becomes

$$p_n = \text{Tr}[(\mathcal{P}_n)^2 \rho], \tag{1.76}$$

and the final state can be written as

$$\tilde{\rho}_n = U_n \left(\frac{\mathcal{P}_n \rho \mathcal{P}_n}{p_n} \right) U_n^\dagger. \tag{1.77}$$

The requirement upon the operators $A_n = U_n \mathcal{P}_n$ for them to represent a measurement becomes

$$\sum_n A_n^\dagger A_n = \sum_n (\mathcal{P}_n)^2 = I. \tag{1.78}$$

This representation of a quantum measurement has a number of important ramifications. First note that since the unitary operators U_n do not change the eigenvalues of the density matrix, they play no role in describing the information extraction by the measurement. The information-gathering properties of the measurement are completely captured by the positive operators $\mathcal{P}_n$.

The unitary operators U_n are not constrained in any way. Note further that the form of Eq. (1.77) is exactly the same form we would obtain if we made a measurement described by the positive operators $\{\mathcal{P}_n\}$, and applied forces to the system to generate the evolution U_n upon getting the nth result. Applying forces to a system conditional upon the result of a measurement is often referred to as *feedback* and is the basis of the subject of feedback control.

The polar decomposition shows us that to understand the process of information gathering in any quantum measurement, we need only examine the action of positive operators on quantum states. We will see shortly that this allows us to understand much of the action of quantum measurements in terms of Bayesian inference. Before we examine positive operators in more detail, it is worth making a closer connection between the structures of quantum and classical measurement theory.

The parallel structure of classical and quantum measurements

As introduced in Section 1.2, classical measurement theory involves multiplying our initial probability distribution by the likelihood function (and normalizing) to obtain our final state-of-knowledge. The reason that we do not include any other action on the system is because classical measurements need not disturb the system at all. A classical system *really is* in one of its possible states n, and the only thing that changes during a measurement is our state-of-knowledge regarding the system. Of course, we could if we wanted include the effects of forces applied to the system dependent upon the result of the measurement.

In general the evolution of a classical system is given by a symplectic operator acting on a state in the system's phase space. But for our purposes here we can think of our classical system as having a discrete set of states; a six-sided die, for example, where the state is the number that is face-up. We can represent the states of this system (the die) as the elementary basis vectors of a six-dimensional vector space. With this representation, the difference between a quantum and classical system is that the former can be in any superposition of the basis states, whereas the latter can only be in one of the basis states. With this representation for a classical system we can use the formalism of quantum measurement theory: our state-of-knowledge is given by a density matrix that is diagonal in the basis defining the classical states, and all our measurement operators (the likelihood function) are also diagonal in this basis. With this representation, the only actions we can perform on our classical system are permutations of the classical states. Permutations of the basis vectors are given by applying permutation matrices. These are unitary matrices whose elements are either 0 or 1, and each row and column contains exactly one 1.

Defining $\{T_n\}$ to be a set of permutation matrices, then a classical measurement followed by a deterministic action on the system is given by

$$\tilde{\rho}_n = T_n \left(\frac{\mathcal{P}_n \rho \mathcal{P}_n}{p_n} \right) T_n^\dagger, \qquad (1.79)$$

where $[\mathcal{P}_n, \mathcal{P}_m] = 0$, $\forall n, m$. This exactly parallels a quantum measurement. This suggests that for quantum measurements one can view the positive operators as taking the place of the likelihood functions of classical measurement theory. The unitary part of the operators that describe quantum measurements plays the role of the permutation matrices (or symplectic evolutions) in the classical theory (and for this reason are not usually included in the classical theory). When we consider the subject of information in the next chapter, we will see that there are further reasons to view the positive operators as describing the quantum analogue of classical measurements.

The action of bare measurements

Since it is the positive operators that characterize the information extraction in a quantum measurement process, we will refer to quantum measurements that are described purely by a set of positive operators, $\{\mathcal{P}_n\}$, as **bare** measurements. This name is intended to indicate that there is a sense in which they capture the notion of information extraction without additional actions.[2]

Now let us examine the action of bare measurements (positive operators) on a quantum system. First recall that while a semiclassical measurement changes our state-of-knowledge of a quantum system, it need not change the underlying state vector. To see this, consider a semiclassical von Nuemann measurement. This is a von Neumann measurement that projects the system onto the basis in which the density matrix is diagonal. If the density matrix is diagonal in the basis $\{|n\rangle\}$, then the system could really be in one of the basis states. If this is the case, the action of the von Neumann measurement is merely to select the basis state that the system is already in. In this case it does not change the state of the system, merely our information about it.

Contrary to the behavior of classical measurements, however, von Neumann measurements, and thus bare measurements in general, *can* change the state of a system. If we prepare a system in a pure state that is *not* one of the basis states onto which a von Neumann measurement projects, then the measurement necessarily changes the state of the system. More generally, a positive operator will change the state of a system whenever the operator does not commute with the density matrix. Nevertheless, the action of a positive operator on a quantum system can be understood in quite simple terms. To do this we go to the basis in which the positive operator, $\mathcal{P}$, is diagonal. Let us call this the $\mathcal{P}$ basis. In general the density operator will not be diagonal in this basis. Recall that even so, the diagonal elements of ρ do still give the probability distribution over the $\mathcal{P}$ basis states. Now, if we examine the action of $\mathcal{P}$ on the diagonal elements of ρ in the $\mathcal{P}$ basis, we find that this is exactly the same as it would be if ρ were diagonal in this basis. That is, as far as the measurement is concerned, the action of $\mathcal{P}$ on ρ is precisely to apply Bayes' theorem to the distribution over the $\mathcal{P}$ basis. (In doing so the action of $\mathcal{P}$ also changes the off-diagonal elements of ρ, re-weighting them in a manner which is consistent with the way it re-weights the diagonal elements.)

We see from the above discussion that we can think of each of the positive operators, $\mathcal{P}_n$, of a bare measurement as extracting information about the basis in which they are diagonal (their eigenbases). The difference between a bare quantum measurement and a classical measurement is that the operators of a bare measurement may each extract information about a *different* basis.

Summary

The polar decomposition theorem shows that the measurement operators describing a given measurement can be factored into the product of a positive operator and a unitary operator.

---

[2] We note that "bare" measurements have also been referred to in the literature as "minimally disturbing" measurements [671, 35]. While these two terms are similar in spirit, we prefer the former for two reasons. The first is that whether or not bare measurements are minimally disturbing does depend on one's definition of disturbance. (For various notions of disturbance see Section 2.3.2.) The second is its brevity.

It is the positive operators that describe the information extraction. Since the unitary operators describe the application of forces to a system, and since in general they will depend upon the outcome of the measurement, they can be interpreted as a feedback process. Consequently we can think of the polar decomposition theorem as providing an interpretation of all measurements, and thus all quantum evolutions, as a combination of information extraction and feedback.

Since the actions of positive operators are identical to classical measurements when the density operator is diagonal in their eigenbasis, one can regard them as describing information extraction without additional actions. That is, the effect on the quantum state due to a positive measurement operator can be regarded, if one wishes, as only that which is a direct result of information extraction, and nothing more.

1.5 Describing measurements within unitary evolution

Measurement does introduce an extra feature into quantum mechanics, over and above the unitary (and thus reversible) evolution given by Schrödinger's equation. But this addition is more subtle than is at first apparent. In fact, all processes that involve sequences of measurements can be rewritten as unitary processes with the exception of a single measurement performed at the end. Further, a measurement may be thought of as merely selecting which of a set of mutually orthogonal subspaces the observer finds themselves in (or if we want to eliminate explicit reference to an observer, which subspace turns out to be reality). Let us examine this second claim first. Any measurement can be made by using a probe system, and making a von Neumann measurement on the probe. A von Neumann measurement is a projection onto one of the basis states of the probe. But each basis state of the probe defines a subspace for the joint system consisting of the probe and the measured system, and all the spaces defined by these basis states are mutually orthogonal. Further, the probability that we will find the joint system in a given subspace is simply the trace of the joint density matrix after it has been projected onto this subspace. The nonlinear evolution that comes from a measurement is therefore purely a result of selecting a subspace, setting the probabilities for all states in the rest of the Hilbert space to zero, and scaling the probabilities of those within the chosen subspace so that they sum to unity.

Now consider our claim that all processes involving measurements can be rewritten as unitary evolution. If the process is merely a sequence of repeated measurements, then each measurement involves bringing up a new probe system, and then selecting a subspace defined by a basis state of this probe. It is not difficult to show that the repeated subspace selections can be moved to the end of the process, and this does not change the statistics of the final outcome, or the final state for each outcome. Since the subspace selection is the only non-unitary element, the entire process is unitary, with a single measurement at the end. If we now include unitary evolution between each measurement, we can still move the subspace selection to the end, and the claim is proved. While this analysis shows that measurement processes can always be written as unitary evolution, with the sole addition of the selection of a subspace at the end, it does not provide us with a great deal of insight into how to do this for a given process. To provide further insight we now consider an example of a feedback process.

If you think about it for a moment, the only reason for making a measurement is to do something conditional upon the answer. Measuring a system, and then acting on it to change its state based on the result of the measurement, is often referred to as feedback. Consider using feedback to prepare a system in the pure state $|\psi\rangle$. To do so we make a measurement $\{A_n\}$, and upon getting result n apply the unitary operation U_n. For outcome n the final state is $\tilde{\rho}_n = U_n A_n \rho A_n^\dagger U_n^\dagger / p_n$, where ρ is the initial state and $p_n = \text{Tr}[A_n^\dagger A_n \rho]$. Thus the probability that the system is in the desired state is $Q_n = \langle\psi|\tilde{\rho}_n|\psi\rangle$. To obtain the overall probability that the feedback procedure will put the system in the desired state we must average over all the possible outcomes. Thus the overall probability of success is

$$P = \sum_n p_n \langle\psi|\tilde{\rho}_n|\psi\rangle = \sum_n \langle\psi|U_n A_n \rho A_n^\dagger U_n^\dagger|\psi\rangle. \tag{1.80}$$

Now instead let us make the measurement by using an interaction U with a probe system, as described in Section 1.3.3. In this case the joint state of the two systems following the interaction is $w = U\rho\otimes\sigma U^\dagger$. The unitary U is chosen so that if we project the probe onto the state $|n\rangle$ then the joint state becomes $\tilde{w}_n = A_n\rho A_n^\dagger \otimes |n\rangle\langle n|$. Note that each state $|n\rangle$ of the probe defines a subspace of the joint system, and in the subspaces corresponding to $|n\rangle$ the state of the system is that given by the measurement result n.

Instead of making a projection measurement on the probe we could apply to the joint system the unitary operation

$$V = \sum_n U_n \otimes |n\rangle\langle n|. \tag{1.81}$$

This applies the operation U_n to the system in the subspace corresponding to $|n\rangle$. The actions of U and V together therefore apply the measurement and the unitary operation that depends on the measurement result, even though no measurement has been made. If we now calculate the probability that the system is in the state $|\psi\rangle$, we have

$$P = \langle\psi|\text{Tr}_{\text{probe}}[VU\rho\otimes\sigma U^\dagger V^\dagger]|\psi\rangle = \langle\psi|\sum_n U_n A_n \rho A_n^\dagger U_n^\dagger|\psi\rangle, \tag{1.82}$$

and this is precisely the expression above obtained for the measurement with the "feedback" operations $\{U_n\}$.

Note that in the above example of a feedback process, we did not even need to invoke a measurement at the end of the process, because the observer is only interested in the final *unconditional* result, and not the results of any measurements that may be involved in the process. Any action that can be taken as a result of a measurement can be implemented in the above way, by employing a Hamiltonian that is different in the subspaces defined by a set of basis states of the probe.

One way to prepare a system in a desired state is to make a measurement in which one of the outcomes prepares this state. If the wrong result is obtained, then we throw away the system and measure another identical system. We proceed in this way until we get the desired measurement result. This procedure is referred to as *post-selection*. Since post-selection is merely an example of an action conditional on a measurement result, it can also be rewritten as unitary evolution. We explore post-selection in Exercise 25.

1.6 Inefficient measurements

We stated in Section 1.3.3 that there are measurements not covered by the form given in Theorem 2, in which each outcome corresponds to a single measurement operator A_n. Given the von Neumann measurement postulate, the measurement of a system via a probe is the most general measurement procedure, with the exception that the observer may be uncertain about the result of the measurement, or what measurement was performed. It is only this additional classical uncertainty that is missing from the measurements of Theorem 2. This uncertainty can be included in the description of a measurement using a probe either by choosing the initial state of the probe to be mixed, or by having the probe interact with a third system prepared in a mixed state before the von Neumann measurement is made.

Measurements that contain additional classical uncertainty are referred to as being *inefficient*. This terminology expresses the fact that uncertainty about the result of the measurement reduces the information that the observer obtains, but does not reduce the back-action from the measurement. The measurement is inefficient in that there is more back-action on the system than is required by quantum mechanics for the information rendered.

We can describe inefficient measurements by labeling the measurement operators with two indices: $\{A_{jk}\}$. We give the observer access to the value of j, but not the value of k. The observer's final state-of-knowledge is then given by averaging over the index k, and is

$$\tilde{\rho} = \sum_k P(k|j)\tilde{\rho}_{jk} = \sum_k \frac{P(j,k)}{P(j)}\tilde{\rho}_{jk} = \frac{1}{P(j)}\sum_k A_{jk}\rho A_{jk}^\dagger, \qquad (1.83)$$

where

$$\tilde{\rho}_{jk} = \frac{A_{jk}\rho A_{jk}^\dagger}{P(j,k)}, \qquad (1.84)$$

and $P(j,k) = \mathrm{Tr}[A_{jk}^\dagger A_{jk}\rho]$.

Of course, the most general situation is when we have just a single index for the measurement result, n, and the observer has some general classical state-of-knowledge regarding this result. We can describe this situation by saying that the observer makes a classical measurement of n, and gets result m, where this classical measurement is described by the likelihood function $P(m|n)$. The formulation above, while appearing to be simpler, actually covers this most general scenario. To see this, note that in this general case the observers final state-of-knowledge is given by

$$\tilde{\rho} = \sum_n P(n|m)\tilde{\rho}_n = \sum_n \frac{P(m|n)P(n)}{P(m)}\tilde{\rho}_n$$

$$= \frac{1}{P(m)} \sum_n P(m|n)B_n\rho B_n^\dagger, \tag{1.85}$$

where the measurement operators are $\{B_n\}$. Now we observe that we can obtain this result using the two-index formulation above by setting $m = j$, $n = k$, and

$$A_{mn} = \sqrt{P(m|n)}B_n. \tag{1.86}$$

If the observer lacks information about which of a set of measurements were made, the state resulting from each measurement result once again involves averaging over more than one measurement operator. Determining the final state in this case is Exercise 26.

1.7 Measurements on ensembles of states

It is sometimes useful to know not only the density matrix for a system, but also the full set of possible pure states that the system might be in, along with their respective probabilities. This may be important if we know that someone has prepared the system in one of a specific set of states, and we wish to make a measurement to determine which one. A set of states $\{|\psi_j\rangle\}$, along with their corresponding probabilities, $\{P(j)\}$, is referred to as an *ensemble*, and we will often write it as $\{|\psi_j\rangle, P(j)\}$. Of course, there is no reason why an ensemble need contain only pure states, and so in general we will write our ensembles as $\{\rho_j, P(j)\}$. So far we have examined how the action of a measurement changes ρ, but it is also useful to know how it changes an ensemble.

If the initial ensemble is $\{\rho_j, P(j)\}$, then the density matrix is

$$\rho = \sum_j P(j)\rho_j. \tag{1.87}$$

If we now make a measurement described by the measurement operators $\{A_n\}$, and obtain the outcome n, then the final state is

$$\tilde{\rho}_n = \frac{A_n \sum_j P(j)\rho_j A_n^\dagger}{\mathcal{N}}$$

$$= \frac{1}{\mathcal{N}} \sum_j P(j)\text{Tr}[A_n^\dagger A_n \rho_j] \left(\frac{A_n \rho_j A_n^\dagger}{\text{Tr}[A_n^\dagger A_n \rho_j]} \right). \tag{1.88}$$

Now let us examine the various elements in the above expression. The quantum state in the final parentheses is correctly normalized. It is the state the system would be in following the measurement if the system had initially been in the state $|\psi_n\rangle$. We shall therefore denote it by $\tilde{\rho}_{j,n}$, meaning the state of the system given the values of both n and j.

The quantity $\text{Tr}[A_n^\dagger A_n \rho_j]$ is the probability that we obtain result n if the system is initially in the jth state, so we can write this as the likelihood function $P(n|j)$. With these definitions the final state is now

$$\tilde{\rho}_n = \sum_j \left[\frac{P(n|j)P(j)}{\mathcal{N}} \right] \tilde{\rho}_{j,n}. \tag{1.89}$$

We can also ask, what does the measurement tell us about the state that the system was in *before* the measurement? In classical measurement theory these two questions are the same, since classical measurements do not change the state of the system. But for quantum systems the two questions are distinct. Because we know the initial ensemble, we know that the system was initially in one of the states $\{|\psi_n\rangle\}$. To determine what the measurement tells us about the initial preparation, all we have to do is to apply Bayesian inference to our classical state of knowledge regarding which of the possible states was prepared, being $P(n)$. After the measurement our state-of-knowledge regarding which state was prepared is thus

$$P(j|n) = \frac{P(n|j)P(j)}{\mathcal{N}}, \tag{1.90}$$

and this is also the expression appearing in Eq. (1.89). So we can also write the state of the system after the measurement as

$$\tilde{\rho}_n = \sum_j P(j|n)\tilde{\rho}_{j,n}, \tag{1.91}$$

which, if you think about it, makes perfect sense.

Summary

An important message of the above discussion is that when we make a measurement on a quantum system, we can consider two distinct kinds of information that the measurement extracts. The first is information about the *final* state of the system – this concerns how certain we are of the state of the system following the measurement. This is the kind of information that is useful when we are trying to control a system. The second kind of information is information about the *initial* state of the system. As we will see in the next chapter (specifically, Section 2.3), this is useful in a scenario in which the quantum system is being used to *communicate* information from one person to another. If Alice wishes to send a message to Bob, then she prepares the system in one of a set of states, where each state represents a possible message, and sends the system to Bob. Bob then makes a measurement on the system to determine the initial state, and hence the message. In the next chapter we will learn how both kinds of information can be quantified.

History and further reading

Bayes' theorem, the cornerstone of classical measurement theory, was introduced by Bayes in 1763 [39]. Further details regarding Bayesian inference and methods for finding priors are given in Jaynes' book on probability theory [311]. Many examples of the application of Bayesian inference can be found in [81]. The polar decomposition theorem can be found in *Matrix Analysis* by Horn and Johnson [266], and *Matrix Analysis* by Rajendra Bhatia [61]. Our modern understanding of the structure of quantum measurement theory was pioneered by Ludwig [379] and Hellwig and Krauss [247, 345], among others.

The problem of finding the optimal measurement to distinguish between the states of a given ensemble appears to have no analytic solution. A number of properties of the optimal measurement are known, however, including approximately optimal measurements. Conditions that the optimal measurement must satisfy were first obtained by Holevo [257] and Yuen *et al.* [704]. Further powerful results on optimal measurements are given by Tyson in [617, 618]. The former provides a numerical method to find optimal measurements, and the latter includes an overview of the approximately optimal measurements that are presently known.

Exercises

1. There are two boxes, one of which contains two red balls and the other contains one red ball and one blue ball. You put your hand into one of the boxes (without looking at the contents) and withdraw a red ball. What is the probability that the ball that remains in the box is blue?

2. There is a variable x that has one of two values, being 0 or 1. An observer knows that there is a 20% probability that the value of the variable is 0 (that is, the observer's state-of-knowledge is that $x = 0$ with probability 20%, and this state-of-knowledge is objective in the sense that the observer knows the physical process that generates the value of x). The observer then makes a measurement of the variable that has two possible outcomes, A and B. Upon obtaining outcome B, the observer then knows that the probability that $x = 0$ is 50%. What is the likelihood function for the measurement? Has the observer's uncertainty regarding x increased or decreased as a result of making the measurement?

3. (i) Show that the probability distribution $P(x) = a/x$, defined on the interval $x \in [0, \infty)$, is invariant under a scaling transformation. You do this by determining the probability distribution for $y = sx$, where s is a positive real number, and showing that this distribution is $Q(y) = a/y$. (ii) Show that all probability densities of the form $P(x) = 1/x^\alpha$, where α is any real number, are not normalizable. (iii) Show that all probability distributions of the form $P(x) = 1/x^\alpha$ change only by a constant factor under a scaling transformation.

4. Use Bayes' theorem to derive Eq. (1.19). Then use Eq. (1.19) to show that the prior $P_x(x) = 1/\sqrt{x}$ is the only prior consistent with the notion that we no nothing about the

money in the envelopes except that one contains twice as much as the other. Finally, consider a new problem in which there are three identical envelopes which contain, respectively, x dollars, $2x$ dollars, and $3x$ dollars. In this case is there any prior that is consistent with the notion that we know nothing about the value of x before we open any of the envelopes?

5. Show that all the eigenvalues of the density matrix are non-negative. Hint: consider the probability that the system is found in an arbitrary state $|\psi\rangle$, which is $\langle\psi|\rho|\psi\rangle$, and show that this is never negative.

6. Show that $\text{Tr}[\rho^2]$ depends only on the eigenvalues of ρ. Hint: use the cyclic property of the trace ($\text{Tr}[ABC] = \text{Tr}[CAB] = \text{Tr}[BCA]$), and the fact that the diagonal form of ρ is given by $\rho_d = U\rho U^\dagger$ for some unitary U.

7. Show that the sum of the square moduli of all the elements of a Hermitian matrix is independent of the basis in which the matrix is written. (Hint: write $\text{Tr}[A^\dagger A]$ in terms of the matrix elements of A.) Show that this sum is therefore equal to the sum of the squares of the eigenvalues of the Hermitian matrix.

8. Consider a state-of-knowledge (density matrix) in which every pure state is equally likely. Show that in this case the density matrix is proportional to the identity.

9. The polar decomposition theorem states that every operator A can be written as UP, or QV, where P and Q are positive, and U and V are unitary. Show that $P = \sqrt{A^\dagger A}$ and $Q = \sqrt{AA^\dagger}$.

10. Show that if a matrix A has a complete set of orthogonal eigenvectors (that is, A is diagonalizable) then $A^\dagger A = AA^\dagger$. (Thus, operators that do not satisfy this condition cannot be diagonalized.) In fact, the converse is also true – every matrix A for which $A^\dagger A = AA^\dagger$ has a complete set of orthogonal eigenvectors [266, 447]. If $A^\dagger A = AA^\dagger$ then A is referred to as being *normal*.

11. Diagonalizing a matrix is a nice method of solving a time-independent linear differential equation. If one has the equation $d\mathbf{x}/dt = A\mathbf{x}$, then one can obtain the solution by finding the eigenvalues and eigenvectors of A [288]. (Note that Schrödinger's equation with a time-independent Hamiltonian, $d|\psi\rangle/dt = -i(H/\hbar)|\psi\rangle$ is exactly of this form.) The following example shows that if the matrix A is not a normal matrix ($A^\dagger A = AA^\dagger$) and thus does not have a complete set of eigenvectors, it may be possible to transform it into a normal matrix merely by scaling one or more the dynamical variables (the elements of $\mathbf{x}$). Consider the equation of motion for the density matrix of a single qubit, with the Hamiltonian

$$H = \begin{pmatrix} 0 & i\mu \\ -i\mu & 0 \end{pmatrix}. \tag{1.92}$$

(i) Show that if we write the density matrix as

$$\rho = \begin{pmatrix} P & c \\ c & Q \end{pmatrix}, \tag{1.93}$$

then the equation of motion for the vector $\mathbf{x}^{\mathrm{T}} = (P, Q, c)$ is $d\mathbf{x}/dt = A\mathbf{x}$ where

$$A = \mu \begin{pmatrix} 0 & 0 & 2 \\ 0 & 0 & -2 \\ -1 & 1 & 0 \end{pmatrix}. \tag{1.94}$$

(ii) Show that A is not normal. (iii) Guess the right scaling of the variable c so that the matrix A becomes normal. (iv) Find the eigenvalues and eigenvectors of the new version of A.

12. Consider a measurement that has two possible outcomes. Show that in this case the positive operators associated with each of the measurement operators commute with one another.

13. Consider the unitary operator U defined in Eq. (1.44). Show that Eq. (1.45) implies that the first M columns of U are orthonormal vectors.

14. Consider a two-state system with basis states $|0\rangle$ and $|1\rangle$. The system is prepared in the state $|\psi_1\rangle = |1\rangle$, or the state $|\psi_2\rangle = (|0\rangle + |1\rangle)/\sqrt{2}$, with equal probability. We now make a two-outcome measurement on the system to determine which of the two states it is in. We choose one of the measurement operators to be the projector $P_0 = |0\rangle\langle 0|$. This ensures that, if this outcome occurs, we can be *certain* that the initial state was $|\psi_2\rangle$ (why?). Determine the probability that, following the measurement, we will be uncertain of the initial preparation.

15. Show that performing any two measurements one after the other is equivalent to a single measurement, and determine the measurement operators for the single measurement in terms of the first two. This result clearly generalizes for a sequence of many measurements.

16. Consider an efficient measurement performed on a system in a pure state. Show that by using feedback (that is, by applying to the system a unitary operator that depends upon the measurement result), one can prepare the system in the same final state for every outcome of the measurement, and this can be done for *any* efficient measurement. Is the same true if the initial state is mixed? If not, why not?

17. Show that if a system is prepared in one of two states that are not orthogonal to one another, there is no measurement that can guarantee to determine with certainty which state the system was prepared in. (You can use the construction involving a probe system, and the fact that unitary evolution does not change the inner product between two states.)

18. Consider a measurement on a qubit (a quantum system with two states) made by using a probe system, and making a von Neumann measurement on the probe. Determine the unitary operator that must be applied to the system and probe to realize the measurement $\{A_0, A_1\}$, where

$$A_0 = \sqrt{1/2 + k}|0\rangle\langle 0| + \sqrt{1/2 - k}|1\rangle\langle 1|, \tag{1.95}$$

$$A_1 = \sqrt{1/2 - k}|0\rangle\langle 0| + \sqrt{1/2 + k}|1\rangle\langle 1|. \tag{1.96}$$

19. Consider a system prepared in one of two non-orthogonal states, where the probability of each is $1/2$. In this case we can describe the system entirely using a two-dimensional state space (why?). Denoting our basis-states as $|0\rangle$ and $|1\rangle$, we can write the two states as $|\pm\rangle = \alpha|0\rangle \pm e^{i\phi}\sqrt{1-\alpha}|1\rangle$. To determine which measurement will most accurately determine which of these two states the system is in, we must quantify what we mean by "most accurate." In fact, different goals will motivate different definitions, and thus different optimal measurements. One such optimal measurement, first determined by Levitin [367], is the following. It has two outcomes, and the corresponding measurement operators, $A_\pm$, are, respectively, the projectors onto the two states $|\pm\rangle_m = (|0\rangle \pm e^{i\phi}|1\rangle)/\sqrt{2}$. Let us say that if the outcome is $+$, then the observer guesses that the state was $|+\rangle$, and vice versa. (i) Draw all the above states on the Bloch sphere. (ii) Calculate the probability that the observer's guess is incorrect (that the observer fails to correctly distinguish the states). (Note: this measurement maximizes the "mutual information," to be discussed in Chapter 2, and does not maximize the probability of a successful guess. Can you think of how the measurement could be altered to increase the probability that the observer's guess is correct?)

20. A system is prepared in one of the two orthogonal states $|0\rangle$ and $|1\rangle$, each with probability $1/2$. A measurement is made on the system which has two outcomes given by the operators $A_\pm$. These operators project the system onto the states $|\pm\rangle = \alpha|0\rangle \pm \sqrt{1-\alpha}|1\rangle$, respectively. (i) Calculate the change in the observer's state-of-knowledge when he gets each of the two outcomes. (ii) Which value of α provides the observer with no information about the initial state?

21. Find an example of an inefficient measurement that leaves the system in a completely mixed state, regardless of the initial state.

22. Consider a two-state system in state $|0\rangle$ undergoing evolution given by the Hamiltonian $H = \hbar\mu\sigma_y$. The state of the system as a function of time is then $|\psi(t)\rangle = \cos(\mu t)|0\rangle + \sin(\mu t)|1\rangle$. An observer now makes a measurement that projects the system either onto state $|0\rangle$ or state $|1\rangle$. Consider that the observer makes a measurement after a short time interval $t = \Delta t \ll 1/\mu$. (i) What is the probability that the observer will project the system into state $|1\rangle$? (Hint: expand the sine function as a power series in Δt, and drop all terms higher than second-order in Δt. This will be useful for the next part.) (ii) Now consider what happens if the observer makes repeated measurements, interrupting the evolution at intervals Δt apart. What is the probability that the observer will find the system to be in state $|1\rangle$ after N measurements? (iii) Finally, consider an interval T, in which the observer makes N measurements, each separated by the interval $\Delta T = T/N$. Calculate the probability that the system is found to be in state $|1\rangle$ for every measurement. Now determine this probability in the limit in which $\Delta t \to 0$ ($N \to \infty$). The result is that the system is never found to be in state $|1\rangle$: the evolution of the system is frozen by the measurement. This is the *quantum Zeno effect*.

23. *The "anti"-quantum-Zeno effect* (it is best to do Exercise 22 first): Consider a two-state system that starts in state $|0\rangle$. After a time Δt an observer makes a two-outcome measurement on the system, whose measurement operators project the system onto the

states $|+\rangle = \cos(\Delta t)|0\rangle + \sin(\Delta t)|1\rangle$, and $|-\rangle = \sin(\Delta t)|0\rangle - \cos(\Delta t)|1\rangle$. Calculate the probability that the system is found to be in the state $|-\rangle$ after the measurement. This result tells you that, to first order in Δt, the system will *always* be found in state $|+\rangle$ after the measurement. The measurement thus "drags" the system from state $|0\rangle$ to state $|+\rangle$. This is called the anti-quantum-Zeno effect, but is really just the Zeno effect used in a different way (see Exercise 22 above).

24. Weak, or *incomplete*, measurements are those that do not project an initially mixed state only onto pure states – that is, they are measurements whose measurement operators are not all rank 1, and thus do not provide the observer with full information about the state of the system following the measurement. Consider a two-state system that is initially in the state

$$\rho = p|0\rangle\langle 0| + (1-p)|1\rangle\langle 1|. \tag{1.97}$$

Now consider the two-outcome measurement with projection operators given by

$$A_+ = \sqrt{k}|+\rangle\langle+| + \sqrt{1-k}|-\rangle\langle-|, \tag{1.98}$$

$$A_- = \sqrt{1-k}|+\rangle\langle+| + \sqrt{k}|-\rangle\langle-|, \tag{1.99}$$

where

$$|+\rangle = \sqrt{\alpha}|0\rangle + \sqrt{1-\alpha}|1\rangle, \tag{1.100}$$

$$|-\rangle = \sqrt{1-\alpha}|0\rangle + \sqrt{\alpha}|1\rangle. \tag{1.101}$$

The purity of a quantum state ρ is defined as $\mathrm{Tr}[\rho^2]$. (This measure of how pure the quantum state is – loosely, how much information we have about the state of the system – will be discussed in Chapter 2.) (i) Write a computer program in your favorite language (e.g., Matlab) to calculate the increase in the purity of the state due to the measurement, averaged over the two measurement outcomes. For fixed values of p and k, what value of α gives the maximum average increase in the purity? (ii) Draw the initial state ρ on the Bloch sphere, along with the states $|+\rangle$ and $|-\rangle$ for the measurement that gives the maximum increase in purity.

25. Any process that involves measurements can be accomplished by replacing the measurements with unitary interactions with additional systems, and applying a single measurement only at the end of the process. This single measurement serves merely to delineate a set of mutually orthogonal subspaces which correspond to a set of mutually exclusive outcomes. That is, we can think of an observer as "being in" one of the mutually orthogonal subspaces. We gave an example of replacing measurements with unitary operations in Section 1.5. Now consider the following process. An observer wishes to prepare a two-state system (a qubit) in state $|0\rangle$. To do this she makes a projective measurement onto the basis $\{|0\rangle, |1\rangle\}$. If she obtains $|0\rangle$, then she is done. If not she takes another qubit and repeats the procedure. If the second measurement gives

the right result, she uses a unitary operation to swaps the state of the second qubit into the first qubit, thereby preparing the first qubit in state $|0\rangle$. Assume that (1) the initial state of both qubits is completely mixed, (2) the measurement prepares the qubit in the wrong state with probability p (that is, if the measurement result is 0, the qubit is prepared in the state $|1\rangle$ with probability p). (i) What is the probability that after the two measurements the observer does not have a qubit in state $|0\rangle$? (ii) Now consider a unitary process that is equivalent to the above post-selection process. To make the errors the same, after a probe system has been correlated with one of the qubits, you can apply an operation that flips the state of the qubit with probability p. What is the final state of the first qubit at the end of the unitary "post-selection" procedure?

26. Consider a situation in which one of M measurements is made on a system. Each measurement has the possible results $1, \ldots, N$. For the mth measurement the operator corresponding to result n is denoted by A_{mn}. The observer knows that the measurement result k has been obtained, but doesn't have full information about what measurement was made. The observer knows that the person who made the measurement chose randomly which measurement to make, and chose the mth measurement with probability q_m. If the observer knows that the initial state of the system was ρ, determine (i) the observer's state-of-knowledge regarding which measurement was made, now that she knows that result k was obtained (hint: use Bayesian inference), and (ii) the observer's state-of-knowledge of the system.

2

Useful concepts from information theory

2.1 Quantifying information

2.1.1 The entropy

In 1948 Claude Shannon realized that there was a way to quantify the intuitive notion that some messages contain more information than others. He approached this problem by saying that a message provides information when it reveals which of a set of possibilities has occurred. This concept certainly makes sense given our intuitive notion of uncertainty: information should reduce our uncertainty about a set of possibilities. He then asked, how long must a message be to convey one of M possibilities?

To answer this question we must first specify the alphabet we are using. Specifically, we need to fix the number of letters, or symbols, in our alphabet. It is simplest to take the smallest workable alphabet, one that contains only two symbols, and we will call these 0 and 1. Imagine now that there is a new movie showing, and you are trying to decide whether to go see it. You call a friend who has seen it, Alice, and ask her whether she liked it. To answer your question yes or no she need only send you one symbol, so long as you have agreed beforehand that 1 will represent yes and 0 no. Note that two people must always agree beforehand on the meaning of the symbols they use to communicate – the only reason you can read this text is because we have a prior agreement about the meaning of each word.

Now think about the above scenario a little further. What happens if we want to ask Alice her opinion of N movies? She can reply with a message that is a sequence of N symbols. (We only ask her about movies we know she has seen.) If Alice likes, on average, half the movies she watches then each symbol is equally likely to be 1 or 0, and so every possible sequence that she could send you is equally likely. But if she likes most of the movies she watches, then most of the symbols in the sequence she sends you will be 1s. Let us say that the probability Alice likes each movie is p. If $p = 0.99$ then on average only one in every hundred symbols will be equal to 0, making it mind-bogglingly unlikely that all the symbols in the message will be 0. We now realize that if certain sequences are sufficiently unlikely, then Alice will never have to send them. This means that the total number of different sequences that she needs to be able to send, to inform you of her movie

48

opinions, will be less than 2^N. This would in turn mean that she could represent each of these sequences using *less* than N 0s and 1s.

We now ask about what happens in the limit in which $N \to \infty$. The key fact that we need is that when we have a long sequence of symbols, where each is chosen randomly to be 1 with probability p, and 0 with probability $1 - p$, the fraction of symbols in the sequence that are 1 will be very close to p. This is called Borel's law of large numbers. In fact, as $N \to \infty$, the probability that the number of 1s in the sequence is not equal to pN drops to zero. This means that if we want to send a message that conveys which of the possible sequences has occurred, we only need to consider the sequences that have exactly pN 1s. These sequences are referred to as the *typical* sequences, and the rest are called *atypical*. As N gets larger, the probability that a randomly generated sequence is atypical gets smaller.

We can determine the number of typical sequences Alice would need to send using some well-known combinatorics. Since every typical sequence has pN 1s, the number of these sequences is the number of ways we can place pN 1s in N boxes. This is

$$\mathcal{N} = \frac{N!}{(pN)!([1-p]N)!} \approx 2^{N[-p\log_2 p - (1-p)\log_2(1-p)]}, \tag{2.1}$$

where $\log_2$ denotes the logarithm to the base 2. To obtain the right-hand side of the above equation we have used Stirling's approximation, which is

$$\ln(N!) \approx N(\ln N - 1) \quad \text{for} \quad N \gg 1. \tag{2.2}$$

To send a message that conveys which typical sequence Alice has chosen we will once again use an alphabet with two symbols, but this time we will refer to the symbols as *bits*, short for *binary digits*. If we send a message containing M bits then we can send 2^M possible messages. Note that since each typical sequence is equally likely, the probability that each bit will be equal to 1 is 0.5. This is, in fact, what defines a bit. The number of bits required to convey which typical sequence Alice has chosen is

$$M = \log_2(\mathcal{N}) = N\left[-p\ln_2 p - (1-p)\log_2(1-p)\right]. \tag{2.3}$$

We define the amount of information in the message as the number of bits it contains, M. Because M is proportional to the number of symbols in Alice's sequence, N, we can define the amount of information contained in each of her symbols as the total amount of information divided by N, which is

$$\frac{M}{N} = -p\log_2 p - (1-p)\log_2(1-p). \tag{2.4}$$

The amount of information per symbol depends on the probability distribution. It is maximal when $p = 0.5$, in which case $M = N$, and in this case we speak of each of the symbols as containing one whole bit of information. This notion of the information per symbol also makes intuitive sense from the point of view of the change in our uncertainty caused by receiving one symbol: if p is not equal to 0.5 then we are less uncertain before we receive the answer, and so the answer causes less of a change to our state-of-knowledge.

The difference between the symbols in Alice's original sequence, and the bits in the message we constructed, is that the probability distribution for the latter is exactly 50/50. The reason for this is that each of Alice's typical sequences has the same probability, since they all have the same number of 1s. The crucial point is that if the symbols in the original sequence have $p \neq 0.5$, then they contain, on average, less than 1 bit of information. We can refer to the symbols in our message as bits, precisely and only because they have a 50/50 distribution. If they did not, then they would not generate 2^N typical messages.

In the above analysis we considered how to communicate as efficiently as possible the values of N samples of a symbol (a random variable) that had two states, 0 and 1. To make the analysis more general, consider communicating the values of N samples of a random variable with K states. Thus we consider a sequence of identical and independent random variables, each with states labeled 1 through K, and each of which has the probability p_k of being in the kth state. Borel's law of large numbers now tells us that in a sequence of N samples of the variable, exactly $p_k N$ of the samples will have the state k, for every k, when $N \to \infty$. We can once again use combinatorics to calculate the total number of sequences for which this is true. Applying Stirling's approximation to the appropriate combinatorial formula reveals that the number of message bits needed to convey which sequence has occurred is

$$M = N \left[-\sum_k p_k \log_2 p_k \right]. \tag{2.5}$$

Up until now we have been using base-2 logarithms because we have been asking how many bits are necessary to send a message. If we had an alphabet with 26 symbols, then the appropriate logarithm would be base-26. In fact we prefer to use the natural logarithm to calculate the length of a message, and the resulting unit of information is called the *nat* (a fictitious alphabet with e symbols).

Switching to natural logarithms, we define the *Shannon entropy* of the probability distribution $\{p_k\}$ as

$$H[\{p_k\}] \equiv -\sum_k p_k \ln p_k. \tag{2.6}$$

We will usually refer to this quantity simply as the *entropy*. It quantifies the amount of information contained in a random variable having the probability distribution $\{p_k\}$, in the sense that $M = NH[\{p_k\}]$ nats are required to convey the values of N samples of the variable.

The result given in Eq. (2.5) is the heart of "Shannon's noiseless coding theorem." The term "coding" refers to the act of associating a sequence of symbols (also called a code word) with each possible message we might need to send. (The term noiseless means that the messages are sent to the receiver without any errors being introduced *en route*.) To further relate this result to real communication, consider sending English text across the internet. Each English letter (symbol) that is sent comes from an effectively random distribution that we can determine by examining lots of typical text. While each letter in English prose is not actually independent of the next (given the letter "t," for example, the letter "h" is more likely to follow than "s," even though the latter appears more frequently overall), it is not that bad an approximation to assume that they are independent. In this case Shannon's theorem tells us that if we send N symbols we must send on average H bits per symbol, where H is the entropy of the distribution (relative frequency) of English letters. In a real transmission protocol N can be very large (e.g., we may be sending the manuscript of a large book). However, since N is never infinite, we sometimes have to send atypical messages. Real coding schemes handle this by having a code word for every possible message (not merely the typical messages), but they use short code words for typical messages, and long code words for atypical messages. This allows one to get close to the smallest number of bits per letter (the ultimate compression limit) set by the Shannon entropy.

As a simple example of a real coding scheme, consider our initial scenario concerning the N movies. If Alice likes almost all the movies, then our typical messages will consist of long strings of 1s, separated by the occasional 0. In this case a good coding scheme is to use a relatively long sequence for 0 (say 0000), and for each string of 1s send a binary sequence giving the *number* of 1s in the string. (To actually implement the coding and decoding we must remember that the binary numbers used for sending the lengths of the strings of 1s are not allowed to contain the sequence that codes for 0.)

Typical elements and typical subsets

The concept of a "typical" element of a set is useful not only in information theory but also in statistical mechanics, the subject of Chapter 4. It is therefore worth pausing to define it more broadly. In the example above we had a set of objects, where the objects were the possible sequences that Alice might wish to send. There was also a probability associated with each object, and a natural way in which size of the set could be increased (by increasing the length, N, of Alice's sequences).

The notion of a typical element applies to any situation in which (i) we have a set in which each element has a probability, and (ii) there is an integer N with which the size of the set increases. In this case an element is typical if (i) it shares a certain property with other elements in the set, and (ii) the combined probability of all the elements that share that property tends to unity as $N \to \infty$. In this case elements that share the given property are collectively referred to as a typical subset.

Entropy and uncertainty

While the entropy was derived by considering the resources required for communication, it also captures the intuitive notion of uncertainty. Consider first a random variable with just two possibilities, 0 and 1, where the probability for 0 is p. If p is close to zero or close to unity, then we can be pretty confident about the value the variable will take, and our uncertainty is low. Using the same reasoning, our uncertainty should be maximum when $p = 1/2$. The entropy has this property – it is maximal when $p = 1/2$, and minimal (specifically, equal to zero) when $p = 0$ or 1. Similarly, if we have a random variable with N outcomes, the entropy is maximal, and equal to $\ln N$, when all the outcomes are equally likely.

If we have two independent random variables X and Y, then the entropy of their joint probability distribution is the sum of their individual entropies. This property, which is simple to show (we leave it as an exercise) also captures a reasonable notion of uncertainty – if we are uncertain about two separate things, then our total uncertainty should be the sum of each.

A less obvious, but also reasonable property of uncertainty, which we will refer to as the "coarse-graining" property, is the following. Consider an object X with three possible states, $m = 1, 2$, and 3, with probabilities $p_1 = 1/2$, $p_2 = 1/3$, and $p_3 = 1/6$. The total entropy of X is $H(X) = H[\{1/2, 1/3, 1/6\}]$. Now consider two ways we could provide someone (Alice) with the information about the state of X. We could tell Alice the state in one go, reducing her uncertainty completely. Alternatively, we could give her the information in two steps. We first tell Alice whether m is less than 2, or greater than or equal to 2. The entropy of this piece of information is $H[\{1/2, 1/2\}]$. If the second possibility occurs, which happens half the time, then Alice is still left with two possibilities, whose entropy is given by $H[\{2/3, 1/3\}]$. We can then remove this uncertainty by telling her whether $m = 2$ or 3. It turns out that the entropy of X is equal to the entropy of the first step, plus the entropy of the second step, weighted by the probability that we are left with this uncertainty after the first step. More generally, if we have a set of states and we "coarse-grain" it by lumping the states into a number of mutually exclusive subsets, then the total entropy of the set is given by the weighted average of the entropies of all the subsets.

It turns out that the following three properties are enough to specify the entropy as the unique measure of uncertainty for a probability distribution $\{p_m\}$, $m = 1, \ldots, M$ (the proof is given in [556]):

1. The entropy is continuous in each of the p_m.
2. If all the M possibilities are equally likely, then the entropy increases with M.
3. The entropy has the coarse-graining property.

To summarize, the amount of communication resources required to inform us about something is also a measure of our initial uncertainty about that thing. In information theory, the entropy of the thing we are trying to send information about is called the "entropy of the source" of information. We can therefore think of the entropy as measuring both the uncertainty about something, and the amount of information "contained" by that thing. In information theory the notion of information is a precise one, and is motivated by a task involving communication; the task is sending a message picked from a distribution of possible messages, and the "information" is precisely the number of symbols required to send the message in the limit of a long message. It is usually simpler to prove relationships between information-theoretic quantities in the long-message limit, and information theory is often concerned with this limit. It is important to remember that in this text we use the term "information" in a general and imprecise sense to mean "reduction in uncertainty," as well as in the precise information-theoretic sense.

2.1.2 The mutual information

Now we have a quantitative measure of information, we can quantify the information obtained by a classical measurement. If we denote the unknown quantity as x, and the measurement result as y, then as we saw in Section 1.2, the measurement is described by the conditional probability $P(y|x)$. We will define x as having a discrete set of values labeled by n, with $n = 1,\ldots,N$, and the measurement as having a discrete set of outcomes labeled by j, with $j = 1,\ldots,M$. If our state-of-knowledge regarding x prior to the measurement is $P(n)$, then upon obtaining the result j, our state-of-knowledge becomes

$$P(n|j) = \left[\frac{P(j|n)}{P(j)} \right] P(n). \tag{2.7}$$

Our uncertainty before the measurement is the entropy of $P(n)$, $H[x] = H[\{P(n)\}]$, and our uncertainty after the measurement is $H[\{P(n|j)\}]$. The reduction in our uncertainty, given outcome j, is therefore $\Delta H_j = H[\{P(n)\}] - H[\{P(n|j)\}]$. A reasonable measure of the amount of information the measurement provides about x is the reduction in the entropy of our state-of-knowledge about x, *averaged* over all the possible measurement results j. This is

$$\langle \Delta H \rangle = H[\{P(n)\}] - \sum_j P(j) H[\{P(n|j)\}]. \tag{2.8}$$

Like the entropy, this new quantity is important in information theory, and we explain why below. Before we do, however, it is worth examining the expression for $\langle \Delta H \rangle$ a little further. First note that the quantity that is subtracted in Eq. (2.8) is the entropy of x after the measurement, averaged over the measurement results. This is called, reasonably enough, the *conditional entropy* of x given y, and written as $H[x|y]$:

$$H[x|y] \equiv \sum_j P(j) H[\{P(n|j)\}]. \tag{2.9}$$

Next, the joint entropy of x and y, being the total uncertainty when we know neither of them, is the entropy of their joint probability density, $P(n,j)$. We write this as $H[x,y] \equiv H[\{P(n,j)\}]$. It turns out that by rearranging the expression for $\langle \Delta H \rangle = H[x] - H[x|y]$, we can alternatively write it as

$$\langle \Delta H \rangle = H[x] + H[y] - H[x,y]. \tag{2.10}$$

One way to interpret this expression is the following. The random variables x and y both contain uncertainty. However, if they are correlated, then some of this uncertainty is shared between them. So when we add the entropies $H[x]$ and $H[y]$, we are counting the entropy that they have in common *twice*. This in turn means that the total uncertainty of x and y together, $H[x,y]$, is in general *less* than $H[x] + H[y]$. The difference $(H[x] + H[y]) - H[x,y]$ is therefore a measure of how correlated y is with x (how much y knows about x), and it turns out that this difference corresponds precisely to what we learn about x when we know y. Since the expression in Eq. (2.10) is a measure of the uncertainty that is common to x and y, it is called the *mutual information* between x and y, and written as $H[x{:}y]$. Thus

$$\langle \Delta H \rangle = H[x{:}y] = H[x] + H[y] - H[x,y] = H[x] - H[x|y]. \qquad (2.11)$$

The expression for $\langle \Delta H \rangle$ in Eq. (2.10) is also useful because it shows us that $\langle \Delta H \rangle$ is symmetric in x and y. That is, the amount of information we obtain about x upon learning the value of y, is exactly the same as the amount of information we would obtain about y if we learned the value of x.

The mutual information and communication channels

In the context of measurements, we think of two correlated variables x and y as describing, respectively, the unknown quantity and the measurement result. In this context the conditional probability $P(y|x)$ characterizes the measurement. To instead relate the mutual information to communication, one considers x to be the input to a communication channel, and y to be the output. Thus the sender encodes a message into some alphabet, and the values of x are the symbols of this alphabet. The output of the channel, however, may not be a faithful copy of the input. In general the channel may be noisy, so that the output of the channel, just like the result of a measurement, is related probabilistically to the input. In the context of information theory, the conditional probability $P(y|x)$ characterizes a *channel*.

We can regard the average reduction in entropy as being the *information* provided by the measurement because of the following result. The mutual information between the input and output of a channel is precisely the amount of information, per symbol, that can be reliably sent down the channel. Specifically, consider that a sender has a message of N symbols, and for which the entropy per symbol is H. If she encodes the message appropriately, and sends it down a channel with mutual information $H[x{:}y]$, then in the limit of large N, the sender will be able to decode the message without error so long as $H \leq H[x{:}y]$. This result is called Shannon's noisy coding theorem, and along with the noiseless coding theorem makes up the fundamental theorems of information theory. Note that while the conditional probability $p(x|y)$ depends only on the channel, the mutual information depends on the probability distribution of the input to the channel, x. So to obtain the maximal transmission rate, the sender must encode her message so that the input symbols have the distribution that maximizes the mutual information for the given channel. There is no analytic form for this maximum in general, and it must often be computed numerically. The maximum of the mutual information for a given channel is called the *capacity* of the channel. Proofs of the noisy coding theorem can be found in most textbooks on information theory.

2.2 Quantifying uncertainty about a quantum system

2.2.1 The von Neumann entropy

Up until now we have usually written a probability distribution over a finite set of states, n, as the set of probabilities $\{p_n\}$. It is often useful to use instead a vector notation for a discrete probability distribution. An important probability distribution associated with a quantum system is the set of eigenvalues of the density matrix. We will denote the vector whose elements are the eigenvalues of a density matrix ρ as $\boldsymbol{\lambda}(\rho)$. We will write the elements of the vector $\boldsymbol{\lambda}$ as λ_n, so that $\boldsymbol{\lambda} = (\lambda_1, \ldots, \lambda_N)$. From now on vectors will be synonymous with probability distributions.

A useful measure of our uncertainty about the state of a quantum system is provided by the Shannon entropy of the eigenvalues of the density matrix, being

$$H[\boldsymbol{\lambda}(\rho)] = -\sum_n \lambda_n \ln \lambda_n. \tag{2.12}$$

This quantity is called the *von Neumann entropy* of the state ρ. One of the main reasons that the von Neumann entropy is useful is because there is a sense in which it represents the *minimum* uncertainty that we have about the future behavior of a quantum system. Before we explain why, we note that this entropy can be written in a very compact form. To write this form we must be familiar with the notion of an arbitrary function of a matrix. Consider a matrix ρ that is diagonalized by the unitary operator U. That is,

$$\rho = UDU^\dagger = U \begin{pmatrix} \lambda_1 & 0 & \cdots & 0 \\ 0 & \lambda_2 & & \vdots \\ \vdots & & \ddots & 0 \\ 0 & \cdots & 0 & \lambda_N \end{pmatrix} U^\dagger. \tag{2.13}$$

Consider also a function $f(x)$ of a number x that has the Taylor series $f(x) = \sum_n c_n x^n$. Then two natural and completely equivalent definitions of the function $f(\rho)$ are [288]

$$f(\rho) \equiv U \begin{pmatrix} f(\lambda_1) & 0 & \cdots & 0 \\ 0 & f(\lambda_2) & & \vdots \\ \vdots & & \ddots & 0 \\ 0 & \cdots & 0 & f(\lambda_N) \end{pmatrix} U^\dagger \equiv \sum_n c_n \rho^n \tag{2.14}$$

With this definition, the von Neumann entropy of ρ is given by

$$S(\rho) \equiv -\mathrm{Tr}[\rho \ln \rho] = -\mathrm{Tr}[U\rho U^\dagger U \ln \rho U^\dagger] = -\mathrm{Tr}[D \ln D]$$
$$= -\sum_n \lambda_n \ln \lambda_n = H[\lambda(\rho)]. \tag{2.15}$$

There are two senses in which $S(\rho)$ represents the minimum uncertainty associated with the state-of-knowledge ρ. The first is the following. If we make a measurement that provides us with complete information about the system – that is, a measurement in which each measurement operator projects the system onto a pure state – then the von Neumann entropy is the minimum possible entropy of the measurement outcomes. Specifically, if we make any complete measurement that has M outcomes labeled by $m = 1, \ldots, m$, then their respective probabilities, p_m, will always satisfy

$$H[\{p_m\}] = -\sum_m p_m \ln p_m \geq S(\rho). \tag{2.16}$$

Since the measurement is complete,

$$p_m = \alpha_m \langle \psi_m | \rho | \psi_m \rangle \quad \text{and} \quad \sum_m \alpha_m |\psi_m\rangle \langle \psi_m| = I, \tag{2.17}$$

for some set of states $|\psi_m\rangle$ and positive numbers α_m. The value of the lower bound is achieved by a von Neumann measurement in the eigenbasis of the density matrix. This lower bound is certainly not obvious, and is a special case of a deep result regarding measurements that we will describe later in Section 2.3.1. Since ultimately questions about the future behavior of a quantum system are questions about the results of future measurements that will be performed on it, the von Neumann entropy measures our maximum predictability about future behavior.

The second sense in which $S(\rho)$ represents the minimum uncertainty inherent in ρ regards the pure-state ensembles that generate ρ: if $\rho = \sum_n p_n |\psi_n\rangle \langle \psi_n|$, then the ensemble probabilities p_n always satisfy

$$H[\{p_n\}] = -\sum_n p_n \ln p_n \geq S(\rho). \tag{2.18}$$

This result is also not obvious, and will be discussed in Section 2.2.3.

There is a further reason why the von Neumann entropy is an important quantity for quantum states: because it is an entropy, it is intimately connected with various information-theoretic questions one can ask about measurements on quantum systems. We will consider some of these questions in Section 2.3.1. The von Neumann entropy also has a number of very useful properties. Some of these are

1. For a system of dimension N: $\ln N \geq S(\rho) \geq 0.$
2. For $p \in [0,1]$: $S(p\rho_1 + (1-p)\rho_2) \geq pS(\rho_1) + (1-p)S(\rho_2).$

3. For $\rho = \sum_n p_n \rho_n$: $\qquad\qquad\qquad\qquad S(\rho) \leq \sum_n p_n S(\rho_n) + H[\{p_n\}]$.
4. If ρ_A is the state of system A, ρ_B is the state of system B, and ρ is the joint state of the combined systems, then
   (i) $S(\rho) \leq S(\rho_A) + S(\rho_B)$ with equality if and only if $\rho = \rho_A \otimes \rho_B$.
   (ii) $|S(\rho_A) - S(\rho_B)| \leq S(\rho)$.
5. Consider three systems A, B and C, where we denote the states of each by ρ_X, with $X = A, B$, or C, the joint states of any two systems X and Y by ρ_{XY}, and the joint state of all three by ρ_{ABC}. The von Neumann entropy satisfies

$$S(\rho_{ABC}) + S(\rho_B) \leq S(\rho_{AB}) + S(\rho_{BC}).$$

The first property gives the range of possible values of the von Neumann entropy: it is zero if and only if the state is pure, and is equal to $\ln N$ if and only if the state is completely mixed. Property two is called *concavity*, property 4 (i) is *subadditivity*, properties 4 (i) and (ii) together are called the Araki–Lieb inequality [15, 653] and property 5 is *strong subadditivity*. The last of these is extremely powerful. Many important results in quantum information theory follow from it, and we will encounter one of these in Section 2.3.1. The proofs of all the above properties, except for the last, are quite short, and all can be found in [447]. The simplest proof of strong subadditivity to date is due to Ruskai, and can be found in [523].

Two complementary definitions of information

As described above, the entropy (Shannon for classical and von Neumann for quantum) measures the information content of a set of possibilities. In the case of the von Neumann entropy this is a set of possible orthogonal states. However, since entropy also characterizes uncertainty about a system, in a different context it can be appropriate to define entropy as an observer's *lack of information*. If a sender sends a message with Shannon entropy S, then we can speak of the message as containing S nats of information. But now think of this situation from the point of view of the receiver. The entropy of the receiver's state-of-knowledge regarding the message is S nats only until she reads it. Before she reads the message she *lacks* S nats of information, and when she reads the message she gains S nats. Her information increases as the entropy of her state-of-knowledge decreases. Thus, if we are interested in the information content of a message then the entropy is the appropriate measure (for quantum systems this question is a little more subtle, and the appropriate measure is the "accessible" information; see Section 2.3.1). If instead one is concerned with the information that an observer possesses about an N-dimensional system, then $I = \ln(N) - S$ is the appropriate measure. If the observer's state-of-knowledge is completely mixed, so that $S = \ln(N)$, then she has zero knowledge about the system.

2.2.2 Majorization and density matrices

While the Shannon entropy is a measure of uncertainty, it is also more than that – it is the *specific* measure of uncertainty that corresponds to the resources required to send a message that will eliminate the uncertainty. There is, in fact, a more basic way to capture the notion of uncertainty alone, without the extra requirement that it provide a measure of information. This is the concept of *majorization* [404, 61], and it turns out to be very useful in quantum measurement and information theory.

Consider two vectors $\mathbf{p} = (p_1, \ldots, p_N)$ and $\mathbf{q} = (q_1, \ldots, q_N)$ that represent probability distributions. To define majorization we always label the elements of our probability vectors so as to put these elements in decreasing order. This means that $p_1 \geq p_2 \geq \cdots p_N$. With this ordering, the vector $\mathbf{p}$ is said to *majorize* $\mathbf{q}$ if (and only if)

$$\sum_{n=1}^{k} p_n \geq \sum_{n=1}^{k} q_n, \ \text{for } k = 1, \ldots, N. \tag{2.19}$$

In words, this means that the largest element of $\mathbf{p}$ is greater than the largest element of $\mathbf{q}$, the sum of the largest two elements of $\mathbf{p}$ is greater than the sum of the largest two elements of $\mathbf{q}$, and so on. If you think about this, it means that the probability distribution given by the elements of $\mathbf{p}$ is more narrowly peaked than $\mathbf{q}$. This in turn means that the $\mathbf{p}$ probability distribution has *less uncertainty* than that of $\mathbf{q}$. The relation "$\mathbf{p}$ majorizes $\mathbf{q}$" is written as

$$\mathbf{p} \succ \mathbf{q} \quad \text{or} \quad \mathbf{q} \prec \mathbf{p}. \tag{2.20}$$

Majorization also has a connection to the intuitive notion of what it means to randomize, or mix. Consider a probability vector $\mathbf{p}$, and some set of permutation matrices $\{T_j\}$, with $j = 1, \ldots, M$. (Multiplying a vector by a permutation matrix permutes the elements of the vector, but does nothing else.) If we choose to perform at random one of the permutations T_j on the elements of $\mathbf{p}$, and choose permutation T_j with probability λ_n, then our resulting state of knowledge is given by the probability vector $\mathbf{q} = \sum_n \lambda_n T_n \mathbf{p}$. It turns out that for any set of permutations

$$\mathbf{q} = \sum_n \lambda_n T_n \mathbf{p} \prec \mathbf{p}. \tag{2.21}$$

A proof of this result is given in Chapter 2 of [61].

It turns out that it is very useful to extend the concept of majorization so as to compare probability vectors that have different lengths. To do this, one merely adds new elements, all of which are zero, to the shorter of the two vectors so that both vectors are the same length. We then apply the usual definition of majorization to these two vectors of equal length. For example, to determine the majorization relation between the vectors $\mathbf{p} = (2/3, 1/3)$ and $\mathbf{q} = (1/2, 1/4, 1/4)$, we first add a zero to the end of $\mathbf{p}$, giving the vector

$\tilde{\mathbf{p}} = (2/3, 1/3, 0)$. Now applying the majorization criteria to $\tilde{\mathbf{p}}$ and $\mathbf{q}$ gives us

$$\mathbf{p} = \begin{pmatrix} 2/3 \\ 1/3 \end{pmatrix} \succ \begin{pmatrix} 1/2 \\ 1/4 \\ 1/4 \end{pmatrix} = \mathbf{q}. \tag{2.22}$$

The concept of majorization imposes a *partial* order on probability vectors. The order is partial because there are many pairs of vectors for which neither vector majorizes the other. This property reflects the fact that under this basic notion of uncertainty, some vectors cannot be considered to be strictly more or less uncertain than others.

One of the reasons that majorization is an important concept is that it does have a direct connection to the entropy. This is a result of a theorem connecting a certain class of concave functions to majorization. Note that the entropy of a vector $\mathbf{p}$, $H[\mathbf{p}]$, is a function of N inputs (being the N elements of $\mathbf{p}$), and it is concave in all these inputs. This means that, for all vectors $\mathbf{p}$ and $\mathbf{q}$,

$$H[\alpha \mathbf{p} + (1-\alpha)\mathbf{q}] \geq \alpha H[\mathbf{p}] + (1-\alpha)H[\mathbf{q}], \quad \alpha \in [0,1]. \tag{2.23}$$

The entropy is also *symmetric* in its inputs. This means that its value does not change when its inputs are permuted in any way. Functions that are concave in all their inputs, as well as being symmetric, are called *Schur-concave* functions.

The strong relationship between majorization and Schur-concave functions is that

$$\mathbf{p} \succ \mathbf{q} \implies f[\mathbf{p}] \leq f[\mathbf{q}] \quad \text{for } f \text{ Schur-concave}. \tag{2.24}$$

If we apply this to the entropy H, it shows us that if $\mathbf{p}$ majorizes $\mathbf{q}$, then $\mathbf{p}$ has less entropy than $\mathbf{q}$, and is thus less uncertain than $\mathbf{q}$. The relation given by Eq. (2.24) is a direct result of the inequality in Eq. (2.21).

We can apply the concept of majorization to density matrices via their eigenvalues. If the vector of eigenvalues of ρ is $\boldsymbol{\lambda}(\rho)$, and that of σ is $\boldsymbol{\mu}(\sigma)$, then we write

$$\rho \succ \sigma \quad \text{meaning that} \quad \boldsymbol{\lambda}(\rho) \succ \boldsymbol{\mu}(\sigma). \tag{2.25}$$

The reason that majorization is especially useful in quantum mechanics is a relation, called *Horn's lemma*, that connects majorization to unitary transformations. To present this relation we need to define something called a *stochastic matrix*. A stochastic matrix is a matrix whose elements are all greater than or equal to zero, and for which the elements in each column sum to unity. If we define $\mathbf{q} = S\mathbf{p}$, where $\mathbf{p}$ is a probability vector and S is a stochastic matrix, then $\mathbf{q}$ is another probability vector. The fact that the sum of each of the columns of S is unity means that $\mathbf{q}$ is formed in the following way: each of the elements of $\mathbf{p}$ is broken up into pieces, and these pieces are distributed among the elements of $\mathbf{q}$. While this transformation preserves the total probability (the elements of $\mathbf{q}$ sum to unity), it does not imply any particular majorization relationship between $\mathbf{q}$ and $\mathbf{p}$. For example, the definition of S is flexible enough to allow all the elements of **p** to be dumped into a single element of **q**, so **q** can be less mixed than **p**. The term *stochastic matrix* comes from the fact that the action of these matrices describes probabilistic transitions, or "jumps," between states in a stochastic process. We are not concerned with this here, but further details can be found in [288, 197].

It turns out that it is simple to define a subset of stochastic matrices for which **q** is never less mixed than **p**. If the *rows* of S also sum to unity, then each element of **q** is some weighted average over the elements of **p**, and this is enough to guarantee that $\mathbf{p} \succ \mathbf{q}$. In this case S is called a *doubly stochastic* matrix. If D is a doubly stochastic matrix then

$$\mathbf{q} \prec \mathbf{p} \quad \text{if and only if} \quad \mathbf{q} = D\mathbf{p}. \tag{2.26}$$

The proof of this can found in, e.g., [12, 266, 61]. The condition that the rows and columns of a matrix, D, sum to unity is equivalent to the following two conditions: (1) As mentioned above, D is "probability preserving", so that if **p** is correctly normalized, then so is $D\mathbf{p}$, and (2) that D is "unital", meaning that it leaves the uniform distribution unchanged. Note that Eq. (2.26) implies that if any elements of a probability vector are replaced by a weighted average of its elements, its entropy does not decrease.

Now consider a unitary transformation U, whose elements are u_{ij}. If we define a matrix D whose elements are $d_{ij} \equiv |u_{ij}|^2$, then D is doubly-stochastic. When the elements of a doubly stochastic matrix can be written as the square moduli of those of a unitary matrix, then D is called *unitary-stochastic*, or *unistochastic*. Not all doubly stochastic matrices are unitary-stochastic. Nevertheless, it turns out that if $\mathbf{q} \prec \mathbf{p}$, then one can always find a unitary-stochastic matrix, D_u, for which $\mathbf{q} = D_u\mathbf{p}$. In addition, since D_u is doubly stochastic, the relation $\mathbf{q} = D_u\mathbf{p}$ also implies that $\mathbf{q} \prec \mathbf{p}$. This is Horn's lemma:

Lemma 1 (Horn [265]) If u_{ij} are the elements of a unitary matrix, and the elements of the matrix D_u are given by $(D_u)_{ij} = |u_{ij}|^2$, then

$$\mathbf{q} \prec \mathbf{p} \quad \text{if and only if} \quad \mathbf{q} = D_u\mathbf{p}. \tag{2.27}$$

A simple proof of Horn's lemma is given by Nielsen in [445].

The following useful fact about density matrices is an immediate consequence of this lemma.

Theorem 3 The diagonal elements of a density matrix ρ in any basis are majorized by its vector of eïgenvalues, $\boldsymbol{\lambda}(\rho)$. This means that the probability distribution for finding the system in any set of basis vectors is always broader (more uncertain) than that for the eigenbasis of the density matrix.

Proof Denote the vector of the diagonal elements of a matrix A as **diag**$[A]$, the density operator when written in its eigenbasis as the matrix ρ, and the vector of eigenvalues of ρ

as $\lambda = (\lambda_1, \ldots, \lambda_N)$. Since ρ is diagonal we have $\lambda = \mathbf{diag}[\rho]$. Now the diagonal elements of the density matrix written in any other basis are $\mathbf{d} = \mathbf{diag}[U\rho U^\dagger]$. Denoting the elements of U as u_{ij}, and those of $\mathbf{d}$ as d_i, this becomes

$$d_i = \sum_j |u_{ij}|^2 \lambda_j, \tag{2.28}$$

so that Horn's lemma gives us immediately $\mathbf{d} \prec \lambda$.

Concave functions and "impurity"

While in an information-theoretic context the von Neumann entropy is the most appropriate measure of the uncertainty of the state of a quantum system, it is not the most analytically tractable. The concept of majorization shows us that any Schur-concave function of a probability vector provides a measure of uncertainty. Similarly, every Schur-*convex* function provides a measure of certainty. (A Schur-convex function is a symmetric function that is convex in all its arguments, rather than concave.)

A particularly simple Schur-convex function of the eigenvalues of a density matrix ρ, and one that is often used as a measure of certainty, is called the *purity*, defined by

$$\mathcal{P}(\rho) \equiv \mathrm{Tr}[\rho^2]. \tag{2.29}$$

This is useful because it has a much simpler algebraic form than the von Neumann entropy.

One can obtain a positive measure of uncertainty directly from the purity by subtracting it from unity. This quantity, sometimes called the "linear entropy," is thus

$$L(\rho) \equiv 1 - \mathrm{Tr}[\rho^2]. \tag{2.30}$$

We will call $L(\rho)$ the *impurity*. For an N-dimensional system, the impurity ranges from 0 (for a pure state) to $(N-1)/N$ (for a completely mixed state).

2.2.3 Ensembles corresponding to a density matrix

It turns out that there is a very simple way to characterize all the possible pure-state ensembles that correspond to a given density matrix, ρ. This result is known as the *classification theorem for ensembles*.

Theorem 4 (Jaynes [310]/Hughston, Josza, and Wootters [276]) Consider an N-dimensional density matrix ρ, that has K non-zero eigenvalues $\{\lambda_j\}$, with the associated eigenvectors $\{|j\rangle\}$. Consider also an ensemble of M pure states $\{|\psi_m\rangle\}$, with associated probabilities $\{p_m\}$. It is true that

$$\rho = \sum_m p_m |\psi_m\rangle \langle \psi_m| \tag{2.31}$$

if and only if

$$\sqrt{p_m}|\psi_m\rangle = \sum_{j=1}^{K} u_{mj}\sqrt{\lambda_j}|j\rangle, \qquad m = 1,\ldots,M, \qquad (2.32)$$

where the u_{mj} are the elements of an M dimensional unitary matrix. We also note that the ensemble must contain at least K linearly independent states. Thus M can be less than N only if $K < N$. There is no upper bound on M.

Proof Let ρ be an N-dimensional density matrix. Let us assume that the ensemble $\{|\phi_n\rangle, p_n\}$, with $n = 1,\ldots,M$, satisfies $\rho = \sum_n p_n|\phi_n\rangle\langle\phi_n|$. We first show that each of the states $|\phi_n\rangle$ is a linear combination of the eigenvectors of ρ with non-zero eigenvalues. (The space spanned by these eigenvectors is called the *support* of ρ.) Consider a state $|\psi\rangle$ that is outside the support of ρ. This means that $\langle\psi|\rho|\psi\rangle = 0$. Now writing ρ in terms of the ensemble $\{|\phi_n\rangle, p_n\}$ we have

$$0 = \langle\psi|\rho|\psi\rangle = \sum_n |\langle\phi_n|\psi\rangle|^2. \qquad (2.33)$$

This means that every state, $|\psi\rangle$, that is outside the support of ρ is also outside the space spanned by the states $|\phi_n\rangle$, and so every state $|\phi_n\rangle$ must be a linear combination of the eigenstates with non-zero eigenvalues. Further, an identical argument shows us that if $|\psi\rangle$ is any state in the space of the eigenvectors, then it must also be in the space spanned by the ensemble. Hence the ensemble must have at least K members ($M \geq K$). We can now write $\sqrt{p_n}|\phi_n\rangle = \sum_j c_{nj}\sqrt{\lambda_j}|j\rangle$, and this means that

$$\sum_j \lambda_j|j\rangle\langle j| = \rho = \sum_n p_n|\phi_n\rangle\langle\phi_n| = \sum_{jj'}\left(\sum_n c_{nj}c_{nj'}^*\sqrt{\lambda_j\lambda_{j'}}\right)|j\rangle\langle j'|. \qquad (2.34)$$

Since the states $\{|j\rangle\}$ are all orthonormal, Eq. (2.34) can only be true if $\sum_n c_{nj}c_{nj'}^* = \delta_{jj'}$. If we take j to be the column index of the matrix c_{nj}, then this relation means that all the columns of the matrix c_{nj} are orthonormal. So if $M = K$ the matrix c_{nj} is unitary. If $M > K$ then the relation $\sum_n c_{nj}c_{nj'}^* = \delta_{jj'}$ means that we can always extend the $M \times K$ matrix c_{nj} to a unitary $M \times M$ matrix, simply by adding more orthonormal columns.

We have now shown the "only if" part of the theorem. To complete the theorem, we have to show that if an ensemble is of the form given by Eq. (2.32), then it generates the density matrix $\rho = \sum_j \lambda_j|j\rangle\langle j|$. This is quite straightforward, and we leave it as an exercise. $\square$

The classification theorem for ensembles gives us another very simple way to specify all the possible probability distributions $\{p_m\}$ for pure-state ensembles that generate a given density matrix:

Theorem 5 (Nielsen [445]) Define the probability vector $\mathbf{p} = (p_1,\ldots,p_M)$, and the vector of eigenvalues of ρ as $\lambda(\rho)$. The relation

$$\rho = \sum_m p_m |\psi_m\rangle\langle\psi_m| \tag{2.35}$$

is true if and only if $\lambda(\rho) \succ \mathbf{p}$.

Proof The proof follows from the classification theorem and Horn's lemma. To show that the "only if" part is true, we just have to realize that we can obtain p_m by multiplying $\sqrt{p_m}|\psi_m\rangle$ by its Hermitian conjugate. Using Eq. (2.32) this gives

$$p_m = \left(\langle\psi_m|\sqrt{p_m}\right)\left(\sqrt{p_m}|\psi_m\rangle\right) = \sum_{j=1}^K\sum_{k=1}^K u^*_{mk}u_{mj}\sqrt{\lambda_k\lambda_j}\langle k|j\rangle = \sum_{j=1}^K |u_{mj}|^2\lambda_j. \tag{2.36}$$

Now if $M > K$, we can always extend the upper limit of the sum over j in the last line from K to M simply by defining, for $j = K+1,\ldots,M$, the "extra eigenvalues" $\lambda_j = 0$. We can then apply Horn's lemma, which tells us that $\lambda(\rho) \succ \mathbf{p}$.

We now have the "only if" part. It remains to show that there exists an ensemble for *every* $\mathbf{p}$ that is majorized by $\lambda(\rho)$. Horn's lemma tells us that if $\lambda(\rho) \succ \mathbf{p}$ then there exists a unitary transformation u_{mj} such that Eq. (2.36) is true. We next define the set of ensemble states $\{|\psi_m\rangle\}$ using Eq. (2.32), which we can do since the u_{mj}, λ_j and $|j\rangle$ are all specified. We just have to check that the $|\psi_m\rangle$ are all correctly normalized. We have

$$p_m\langle\psi_m|\psi_m\rangle = \sum_{j=1}^K\sum_{k=1}^K u^*_{mk}u_{mj}\sqrt{\lambda_k\lambda_j}\langle k|j\rangle = \sum_{j=1}^K |u_{mj}|^2\lambda_j = p_m,$$

where the last line follows because the u_{mj} satisfy Eq. (2.36). That completes the proof. $\square$

Theorem 5 tells us, for example, that for any density matrix we can always find an ensemble in which all the states have the same probability. This theorem also provides the justification for a claim we made in Section 2.2.1: the entropy of the set of probabilities (the probability distribution) for an ensemble that generates a density matrix ρ is always greater than or equal to the von Neumann entropy of ρ.

2.3 Quantum measurements and information

The primary function of measurements is to extract information (reduce uncertainty). Now that we have quantitative measures of classical information, and also of our uncertainty about a quantum system, we can ask questions about the amount of information extracted by a quantum measurement. We will also see how the properties of quantum measurements differ in this regard from those of classical measurements.

2.3.1 Information-theoretic properties

Information about the state of a quantum system

As we discussed in Section 1.7, there are two kinds of information that a measurement can extract from a quantum system. The first concerns how much we know about what state the system is in after we have made the measurement. We can now use the von Neumann entropy to quantify this information: we will define it as the average reduction in the von Neumann entropy of our state-of-knowledge, and denote it by I_{sys}. Thus

$$I_{\text{sys}} \equiv \langle \Delta S \rangle = S(\rho) - \sum_n p_n S(\tilde{\rho}_n), \qquad (2.37)$$

where, as usual, ρ is the initial state of the system and $\tilde{\rho}_n$ are the possible final states. The quantity I_{sys} is sometimes called Groenewold's information, as he was the first to consider it [231]. We will refer to it here as the *entropy reduction*. It is useful in applications such as feedback control, because it tells us how much a measurement reduces an observer's uncertainty of the *state* of the system (as opposed to uncertainty about an ensemble in which the system is prepared). It is also useful in thermodynamics as we will see in Chapter 4. Note that if the measurement is *semiclassical*, meaning that the initial density matrix commutes with all the measurement operators, then I_{sys} is precisely the mutual information between the measurement result and the ensemble made up of the eigenstates of the density matrix.

It is worth remembering in what follows that if one discards the results of a measurement, then the state following the measurement is $\tilde{\rho} = \sum_n p_n \tilde{\rho}_n$. For semiclassical measurements one always has $\tilde{\rho} = \rho$, since the system itself is not affected by the measurement, and throwing away the result eliminates any information that the measurement obtained. However, there are many quantum measurements for which

$$\sum_n p_n \tilde{\rho}_n \neq \rho. \qquad (2.38)$$

This fact actually embodies a specific notion of the *disturbance* caused by a quantum measurement, and we will discuss this in Section 2.3.2. We will also consider below two results regarding the relationship between $\sum_n p_n \tilde{\rho}_n$ and the initial state, ρ.

If a measurement is efficient (that is, we have full information about the result of the measurement) then intuition would suggest that, on average, we should never know less about the system after the measurement than we did beforehand. That is, I_{sys} should never be negative. This is in fact true, as shown in the following theorem.

Theorem 6 (Ozawa [458]) Efficient measurements, on average, always increase the observer's information about the system. That is

$$\sum_n p_n S(\tilde{\rho}_n) \leq S(\rho), \qquad (2.39)$$

where $\tilde{\rho}_n = A_n \rho A_n^\dagger / p_n$, $p_n = \text{Tr}[A_n^\dagger A_n \rho]$, and $\sum_n A_n^\dagger A_n = I$.

Proof The following elegant proof is due to Fuchs [188]. First note that for any operator A, the operator $A^\dagger A$ has the same eigenvalues as $AA^\dagger$. This follows immediately from the polar decomposition theorem: writing $A = PU$ we have $AA^\dagger = P^2$, as well as $A^\dagger A = UP^2 U^\dagger = UAA^\dagger U^\dagger$, and a unitary transformation does not change eigenvalues. We now note that

$$\rho = \sqrt{\rho} I \sqrt{\rho} = \sum_n \sqrt{\rho} A_n^\dagger A_n \sqrt{\rho} = \sum_n p_n \sigma_n, \tag{2.40}$$

where σ_n are density matrices given by

$$\sigma_n = \frac{\sqrt{\rho} A_n^\dagger A_n \sqrt{\rho}}{\mathrm{Tr}[A_n^\dagger A_n \rho]} = \frac{\sqrt{\rho} A_n^\dagger A_n \sqrt{\rho}}{p_n}. \tag{2.41}$$

From the concavity of the von Neumann entropy we now have

$$S(\rho) \geq \sum_n p_n S(\sigma_n). \tag{2.42}$$

But σ_n has the same eigenvalues as $\tilde{\rho}_n$: defining $X_n = \sqrt{\rho} A_n$ we have $\sigma_n = X_n X_n^\dagger$ and $\tilde{\rho}_n = X^\dagger X$. Since the von Neumann entropy depends only on the eigenvalues, $S(\sigma_n) = S(\tilde{\rho}_n)$, and the result follows. $\qquad\square$

The fact that the average entropy never increases for an efficient measurement is a result of a stronger fact regarding majorization. This stronger result implies that *any* concave function of the eigenvalues of the density matrix will never increase, on average, under an efficient measurement. The majorization relation in question is

$$\lambda(\rho) \prec \sum_n p_n \lambda(\tilde{\rho}_n). \tag{2.43}$$

This result can be obtained by modifying the proof of Theorem 6 only a little, and we leave it as an exercise.

For inefficient measurements the entropy reduction $\langle \Delta S \rangle$ can be negative. This is because quantum measurements can disturb (that is, change the state of) a system. In general a measurement will disturb the system in a different way for each measurement result. If we lack full information about which result occurred, then we lack information about the induced disturbance, and this reduces our overall knowledge of the system following the measurement. This is best illustrated by a simple, and extreme, example. Consider a two-state system, with the basis states $|0\rangle$ and $|1\rangle$. We prepare the system in the state

$$|\psi\rangle = \frac{1}{\sqrt{2}} (|0\rangle + |1\rangle). \tag{2.44}$$

Since the initial state is pure, we have complete knowledge prior to the measurement, and the initial von Neumann entropy is 0. We now make a measurement that projects the system into one of the states $|0\rangle$ or $|1\rangle$. The probabilities for the two outcomes are $p_0 = p_1 = 1/2$.

But the catch is that we throw away all our information about the measurement result. After the measurement, because we do not know which result occurred, the system could be in state $|0\rangle$, or state $|1\rangle$, each with probability $1/2$. Our final density matrix is thus

$$\tilde{\rho} = \frac{1}{2}(|0\rangle\langle 0| + |1\rangle\langle 1|). \tag{2.45}$$

This is the completely mixed state, so we have no information regarding the system after the measurement. The disturbance induced by the measurement, combined with our lack of knowledge regarding the result, was enough to completely mix the state.

An information-theoretic meaning for I_{sys}

As mentioned above, the quantity I_{sys} is a quantum version of the classical mutual information. But in our discussion so far, when we refer to I_{sys} as quantifying the information gained by an observer, we are using the term information in the qualitative sense, to mean a reduction in the observer's uncertainty. The classical mutual information also characterizes the reduction in uncertainty, but has in addition a precise information-theoretic meaning: it quantifies the ability of the correlation between two variables to transmit information (when using many copies of the variables). In fact, I_{sys} has at least two information-theoretic meanings, in that it quantifies the amount of information in two specific tasks involving an asymptotic limit (many copies). These meanings would take more time to describe than we wish to devote to them here, however. The interested reader can find further details in [58, 667, 48, 260, 261].

Information about an encoded message

The second kind of information is relevant when a system has been prepared in a specific ensemble of states. In this case, the measurement provides information about *which* of the states in the ensemble was actually prepared. As mentioned in Section 1.7, this is useful in the context of a communication channel. To use a quantum system as a channel, an observer (the "sender") chooses a set of states $|\psi_j\rangle$ as an alphabet, and selects one of these states from a probability distribution $P(j)$. Another observer (the "recipient") then makes a measurement on the system to determine the message. From the point of view of the recipient the system is in the state

$$\rho = \sum_j P(j)|\psi_j\rangle\langle\psi_j|. \tag{2.46}$$

The result, m, of a quantum measurement on the system is connected to the value of j (the message) via a likelihood function $P(m|j)$. The likelihood function is determined by the quantum measurement and the set of states used in the encoding, as described in Section 1.7. But once the likelihood function is determined, it describes a purely classical measurement on the classical variable j. Thus the information extracted about the message, which we will call I_{ens} (short for ensemble), is simply the mutual information:

$$I_{\text{ens}} = H[P(j)] - \sum_m P(m)H[P(j|m)] = H[J:M].\tag{2.47}$$

On the right-hand side J denotes the random variable representing the message, and M the random variable representing the measurement result. When the encoding states are an orthogonal set, and the measurement operators commute with the density matrix, ρ, then the measurement reduces to a semiclassical measurement on the set of states labeled by j. It is simple to show that the result of this is

$$I_{\text{sys}} = \langle \Delta S \rangle = H[J:M] = I_{\text{ens}}, \quad \text{(semiclassical measurement)}.\tag{2.48}$$

For a fixed density matrix, determining the measurement that maximizes I_{sys} is a nontrivial task (see, e.g., [554]), and the same is true for I_{ens}. For a given ensemble, the maximum possible value of I_{ens} has been given a special name – the *accessible information* [190, 548].

Bounds on information extraction

There are a number of useful bounds on I_{ens} and I_{sys}. The first is a bound on I_{sys} in terms of the entropy of the measurement results.

Theorem 7 (Lanford and Robinson [353]) If a measurement is made on a system in state ρ, for which the final states are $\{\tilde{\rho}_n\}$, and the corresponding probabilities are $\{p_n\}$, then

$$I_{\text{sys}} = S(\rho) - \sum_n p_n S(\tilde{\rho}_n) \leq H[\{p_n\}],\tag{2.49}$$

where the equality is achieved only for semiclassical measurements.

The proof of this result is Exercise 11 at the end of this chapter. In Chapter 4 we will see that this bound is intimately connected with the second law of thermodynamics. The bound shows that quantum measurements are often less efficient, from an information-theoretic point of view, than classical measurements. The amount of information extracted about the quantum system is generally less than the randomness (the information) of the measurement results. If we want to communicate the result of the measurement, we must use $H[\{p_n\}]$ nats in our message, but the amount this measurement tells us about the system cannot be more than $H[\{p_n\}]$ nats, and is often less. The above bound is also the origin of our claim, in Section 2.2.1, that the von Neumann entropy was the minimum possible entropy of the outcomes of a complete measurement on the system. To see this all we have to do is note that for a complete measurement all the final states are pure, so that $S(\rho_n) = 0$ for all n. In this case Eq. (2.49) becomes

$$H[\{p_n\}] \geq S(\rho).\tag{2.50}$$

Another important bound extends the above theorem to the situation in which the probe system is mixed to begin with. The probe interacts with the system via a joint unitary U, and the final states of the system are given by projecting the probe into one of a set of basis states. If we average the state of the probe over these projections, then its von Neumann entropy is simply the entropy of the measurement results, $H[\{p_n\}]$. We will denote this entropy by $\langle S_{\text{fin}}^{\text{probe}} \rangle$. The more general theorem is the following:

Theorem 8 (Entropy-exchange between system and probe) The entropy of the measurement results, $H[\{p_n\}]$, which is also the von Neumann entropy of the probe after averaging over the measurement results, $\langle S_{\text{fin}}^{\text{probe}} \rangle$, is greater than the initial von Neumann entropy of the probe by at least I_{sys}. (Recall that I_{sys} is the average reduction in the entropy of the system due to the measurement.) As an equation this statement is

$$\langle \Delta S^{\text{probe}} \rangle \equiv \langle S_{\text{fin}}^{\text{probe}} \rangle - S_{\text{start}}^{\text{probe}}$$
$$= H[\{p_n\}] - S(\sigma^{\text{probe}})$$
$$\geq S(\rho^{\text{sys}}) - \sum_n p_n S(\tilde{\rho}_n^{\text{sys}}) = I_{\text{sys}}. \tag{2.51}$$

Here, to avoid confusion, we have written the states of the system with the superscript "sys."

Proof The initial state of the system and probe is $\sigma^{\text{probe}} \otimes \rho$. First this joint state is acted upon by a unitary operator, U, to give the state $U[\sigma^{\text{probe}} \otimes \rho]U^\dagger$, and then a projection measurement is performed on the probe. This measurement projects the system onto one of a complete set of basis states for the probe. We now need a non-trivial result by Ando, Theorem 11 below, that states that the entropy of the averaged state of a system following a bare measurement is never less than the initial entropy. Since the unitary U does not change the entropy, Ando's theorem tells us that

$$S(w) \geq S(U[\sigma^{\text{probe}} \otimes \rho]U^\dagger) = S(\sigma^{\text{probe}}) + S(\rho), \tag{2.52}$$

where w is the final joint state of the two systems, averaged over the measurement results. It has the form

$$w = \sum_n p_n \tilde{\rho}_n \otimes |n\rangle\langle n|, \tag{2.53}$$

where the set of states $\{|n\rangle\}$ are a basis for the probe. We can now diagonalize each of the states ρ_n without affecting the total von Neumann entropy of the joint state. Note that each of these density matrices will have a different eigenbasis, but this is irrelevant since the state w is block-diagonal, where each of the density matrices ρ_n is one of the blocks. With w now in a diagonal form, the entropy of w is that of a classical joint probability distribution, where the two subsystems are discrete classical variables. The variable for the probe is the label n, and the marginal distribution for the probe is the set of probabilities

p_n. Let us denote the classical variable for the system by j, and the probability distribution for j given by state ρ_n by $p_{j,n}$. For a joint classical distribution, the relation

$$S(w) = H(w) = H(\{p_n\}) + \sum_n p_n H(\{p_{j,n}\}) = H(\{p_n\}) + \sum_n p_n S(\rho_n)$$

is just an algebraic rearrangement. The last equality comes from the fact that diagonalizing the density matrices ρ_n does not change their entropy. Combining the above relation with Eq. (2.52) gives us the result. □

The next set of bounds are all special cases of the Schumacher–Westmoreland–Wootters (SWW) bound, a bound on the mutual information (I_{ens}) in terms of the initial ensemble and the final states that result from the measurement. This bound looks rather complex at first, and is most easily understood by introducing first one of the special cases: Holevo's bound.

Theorem 9 (Holevo [258]) If a system is prepared in the ensemble $\{\rho_j, p_j\}$, then

$$I_{\text{ens}} \leq \chi = S(\rho) - \sum_j p_j S(\rho_j), \qquad (2.54)$$

where $\rho = \sum_j p_j \rho_j$ is the initial state. The equality is achieved if and only if $[\rho, \rho_j] = 0, \forall j$. The quantity χ is called the *Holevo χ-quantity* (pronounce "kai quantity") for the ensemble.

Since the accessible information is defined as the maximum of I_{ens} over all possible measurements, Holevo's bound is a bound on the accessible information.

Recall that an incomplete (and possibly inefficient) measurement will in general leave a system in a mixed state. Further, as discussed in Section 1.7, it will also leave the system in a new ensemble, and this ensemble is different for each measurement result. We are now ready to state the bound on the mutual information that we alluded to above.

Theorem 10 (Schumacher, Westmoreland, and Wootters [548]) A system is prepared in the ensemble $\{\rho_j, p_j\}$, whose Holevo χ-quantity we denote by χ. A measurement is then performed having N outcomes labeled by n, in which each outcome has probability p_n. If, on the nth outcome, the measurement leaves the system in an ensemble whose Holevo χ-quantity is χ_n, then the mutual information is bounded by

$$I_{\text{ens}} \leq \chi - \sum_n p_n \chi_n. \qquad (2.55)$$

The equality is achieved if and only if $[\rho, \rho_n] = 0, \forall n$, and the measurement is semiclassical. This result is also true for inefficient measurements. In this case the measurement will in general have operators $\{A_{nk}\}$, such that the observer knows n but not k. The measurement outcome labeled by n is the result of summing over the outcomes for the index k.

The proof of this theorem uses the strong subadditivity of the von Neumann entropy, and we give the details in Appendix H. The SWW bound can be understood easily, however, by realizing that the maximal amount of information that can be extracted by a subsequent measurement, after we have obtained the result n, is given by χ_n (Holevo's bound). So the average amount of information we can extract using a second measurement is $\sum_n p_n \chi_n$. Since the total amount of accessible information in the ensemble is χ, the first measurement must be able to extract, on average, no more than $\chi - \sum_n p_n \chi_n$.

In the measurement considered in Theorem 10 above, the states in the initial ensemble are labeled by j, and the measurement results by n. By using the fact that the conditional probabilities for j and n satisfy $P(j|n)P(n) = P(n|j)P(j)$, it is possible to rewrite the SWW bound as

$$I_{\text{ens}} \leq \langle \Delta S(\rho) \rangle - \sum_j p_j \langle \Delta S(\rho_j) \rangle. \tag{2.56}$$

Here $\langle \Delta S(\rho) \rangle$ is, as usual, the average entropy reduction due to the measurement. The quantity $\langle \Delta S(\rho_j) \rangle$ is the entropy reduction that *would have* resulted from the measurement if the initial state had been the ensemble member ρ_j. For efficient measurements we know that $\langle \Delta S(\rho) \rangle$ is positive. In this case we are allowed to drop the last term, and the result is Hall's bound:

$$I_{\text{ens}} \leq \langle \Delta S(\rho) \rangle = I_{\text{sys}}, \quad \text{(efficient measurements)}, \tag{2.57}$$

which says that for efficient measurements the mutual information is never greater than the average entropy reduction. For inefficient measurements, as we saw above, I_{sys} can be negative. Thus the last term in Eq. (2.56) is essential, because I_{ens} is always positive.

Information gain and initial uncertainty

In practice one tends to find that the more one knows about a system, the *less* information a measurement will extract. We will certainly see this behavior in many examples in later chapters. This is only a tendency, however, as counter-examples can certainly be constructed. Because of this it is not immediately clear what fundamental property of measurements this tendency reflects. It is, in fact, the bound in Eq. (2.56) that provides the answer. Since I_{ens} is always non-negative, Eq. (2.56) implies that

$$I_{\text{sys}}[\rho] = \langle \Delta S(\rho) \rangle \geq \sum_j p_j \langle \Delta S(\rho_j) \rangle = \sum_j p_j I_{\text{sys}}[\rho_j]. \tag{2.58}$$

But $\rho = \sum_j p_j \rho_j$, and thus Eq. (2.58) is nothing more than the statement that the average reduction in entropy is *concave* in the initial state. Choosing just two states in the ensemble, Eq. (2.58) becomes

$$I_{\text{sys}}[p\rho_1 + (1-p)\rho_2] \geq p I_{\text{sys}}[\rho_1] + (1-p)I_{\text{sys}}[\rho_2]. \tag{2.59}$$

This tells us that if we take the average of two density matrices, a measurement made on the resulting state will reduce the entropy *more* than the reduction that it would give, on average, for the two individual density matrices. Of course, taking the average of two density matrices also results in a state that is more uncertain.

It is the concavity of I_{sys} that underlies the tendency of information gain to increase with initial uncertainty. A result of this concavity property is that there are special measurements for which I_{sys} *strictly* increases with the initial uncertainty. These are the measurements that are completely symmetric over the Hilbert space of the system. (That is, measurements whose set of measurement operators is invariant under all unitary transformations.) We will not give the details of the proof here, but it is simple to outline: the $I_{sys}[\rho]$ for symmetric measurements is completely symmetric in the eigenvalues of ρ. Since $I_{sys}[\rho]$ is also concave in ρ, this implies immediately that it is Schur-concave. This means that if $\rho \prec \sigma$, then $I_{sys}[\rho] \geq I_{sys}[\sigma]$.

Information, sequential measurements, and feedback

The information extracted by two classical measurements, made in sequence, is always less than or equal to the sum of the information that each would extract if it were the sole measurement. This is quite simple to show. Let us denote the unknown quantity by X, the result of the first measurement as the random variable Y, and the result of the second measurement as Z. The information extracted when both measurements are made in sequence is the mutual information between X and the joint set of outcomes labeled by the values of Y and Z. Another way to say this is that it is the mutual information between X and the random variable whose probability distribution is the joint probability distribution for Y and Z. We will write this as $I(X : Y, Z)$. Simply by rearranging the expression for $I(X : Y, Z)$, one can show that

$$I(X : Y, Z) = I(X : Y) + I(X : Z) - I(Y : Z). \tag{2.60}$$

The first two terms on the right-hand side are, respectively, the information that would be obtained about X if one made only the first measurement, and the corresponding information if one made only the second measurement. The final term is a measure of the correlations between Y and Z. Since the mutual information is never negative we have

$$I(X : Y, Z) \leq I(X : Y) + I(X : Z). \tag{2.61}$$

The above inequality ceases to be true if we manipulate the state of the classical system between the measurements, in a manner that depends on the result of the first measurement (that is, use feedback). Briefly, the reason for this is as follows. Performing a "feedback" process allows us to interchange the states of the classical system so that those states that we are most uncertain about following the first measurement can be most easily distinguished by the second. Because we can do this for *every* outcome of the first measurement, it allows us to optimize the effect of the second measurement in a way that we could not do if we made the second measurement alone. Explicit examples of this will be described in Chapter 5, when we discuss feedback control and adaptive measurements.

Like a classical feedback process, the entropy reduction for general quantum measurements does not obey the inequality in Eq. (2.61). However, it is most likely that bare quantum measurements do. In fact, we can use Eq. (2.58) above to show that this is true when the initial state is proportional to the identity. We leave this as an exercise. The question of whether bare measurements satisfy this classical inequality for *all* initial states is still open. If the answer is affirmative, then it would provide further motivation for regarding bare measurements as the quantum equivalent of classical measurements.

2.3.2 *Quantifying disturbance*

Quantum measurements often cause a disturbance to the measured system. By this we do not mean that they merely change the observer's state-of-knowledge. All measurements, classical and quantum, necessarily change the observer's state-of-knowledge as a result of the information they provide. What we mean by disturbance is that they cause a random change to some property of the system that increases the observer's uncertainty in some way. Below we will discuss four distinct ways in which quantum measurements can be said to disturb a system. All but the first are related to one of the two kinds of information, I_{sys} and I_{ens}, that we discussed in the previous section. Further, each disturbance is in some way complementary to its related notion of information, in that one can identify situations in which there is a trade-off between the amount of information extracted, and the disturbance caused.

Disturbance to conjugate observables

The notion of disturbance that is taught in undergraduate physics classes is that induced in the momentum by a measurement of position, and vice versa. This is not a disturbance that increases the von Neumann entropy of our state-of-knowledge, since in this case both the initial and final states can be pure. It is instead a disturbance that increases the Shannon entropy of the probability density for a *physical observable*.

One can show that measurements that extract information about an observable X will tend to reduce our information about any observable that does not commute with X. This is a result of an uncertainty relation, obtained by Robertson in the early days of quantum mechanics [520], that connects the variances of any two observables that do not commute. The derivation of this relation is quite short, and uses the Cauchy–Schwartz inequality. Let us assume that the commutator of two observables, X and Y, is $[X, Y] = Z$, for some operator Z. For convenience we first define the operators $A = X - \langle X \rangle$ and $B = Y - \langle Y \rangle$. We then have

$$(\Delta X)^2 (\Delta Y)^2 = \langle A^2 \rangle \langle B^2 \rangle = ((\langle \psi | A)(A | \psi \rangle))((\langle \psi | B)(B | \psi \rangle))$$

$$\geq |\langle \psi | AB | \psi \rangle|^2$$

$$= \frac{1}{4i}(\langle\psi|[A,B]|\psi\rangle)^2 + \frac{1}{4}(\langle\psi|AB+BA|\psi\rangle)^2$$

$$\geq \frac{1}{4i}(\langle\psi|[X,Y]|\psi\rangle)^2 = \frac{1}{4i}\langle Z\rangle^2. \tag{2.62}$$

Here the second line is given by the Cauchy–Schwartz inequality, which states that for two vectors $|a\rangle$ and $|b\rangle$, $\langle a|a\rangle\langle b|b\rangle \geq |\langle a|b\rangle|^2$. The third line is simply the fact that for any complex number z, $|z|^2 = (\text{Re}[z])^2 + (\text{Im}[z])^2$. The last line follows because the second term on the fourth line is non-negative, and $[A,B] = [X,Y]$. For position and momentum this uncertainty relation is the famous Heisenberg uncertainty relation $\Delta x \Delta p \geq \hbar/2$.

The above uncertainty relation tells us that measurements that increase our knowledge of X enough that the final state has $\Delta X \leq \sqrt{\langle Z\rangle^2/(4i)}/\zeta$ for some constant ζ, must also leave us with an uncertainty for Y which is greater than ζ. This constitutes, loosely, a trade-off between information extracted about X and disturbance caused to Y, since the smaller ΔX, the larger the lower bound on ΔY. However, it should be noted that the uncertainty relation really tells us about the relation between the momentum and position uncertainty at any single instant, and does not directly quantify the relationship between the amount of information obtained and the *change* caused to the system. It is possible to quantify this relationship by precisely defining a notion of disturbance to a variable, and the interested reader can find such an analysis in [459, 460]

Classically, an increase in an observer's uncertainty is caused by a random change to the state of a system, about which the observer has limited information (such random changes are usually referred to as *noise*). Note that while the disturbance caused to Y is a result of the change induced in the state of the system by the quantum measurement, it is not induced because this change is random: if the initial state is pure, then on obtaining the measurement result we know exactly what change the measurement has made to the state of the system. As we will see in Chapter 5, we can often interpret the increase of our uncertainty of Y as (classical) noise that has affected Y, but this is not always the case. A detailed discussion of this point is given in [685].

One can also obtain uncertainty relations that involve the Shannon entropies of two noncommuting observables, rather than merely the standard deviations. Consider two non-commuting observables, X and Y for a system of dimension N, and denote their respective eigenstates by $\{|x\rangle\}$ and $\{|y\rangle\}$, with $x,y = 1,\ldots,N$. The probability distributions for these two observables are given by

$$P_x = \langle x|\rho|x\rangle \quad \text{and} \quad P_y = \langle y|\rho|y\rangle, \tag{2.63}$$

where ρ is the state of the system. The sum of the entropies of the probability distributions for X and Y obeys the uncertainty relation

$$H[\{P_x\}] + H[\{P_y\}] \geq \max_{x,y} |\langle x|y\rangle|, \tag{2.64}$$

which was proved by Maassen and Uffink [382], building on the work of Deutsch [140] and Kraus [346]. The proof of this is much more involved than that for the elementary uncertainty relation in Eq. (2.62). The Maassen–Uffink relation is also different in kind from that in Eq. (2.62) because the lower bound does not depend on the state of the system. The lower bound in the elementary uncertainty relation depends on the state whenever $[X, Y]$ is not merely a number.

It is also possible to obtain entropic uncertainty relations for observables for infinite-dimensional systems, and observables with a continuous spectrum. The Shannon entropy of a continuous random variable x, whose domain is $(-\infty, \infty)$, is defined as

$$H[P(x)] \equiv \int_{-\infty}^{\infty} P(x) \ln P(x) \, dx. \tag{2.65}$$

However, in this case the entropy can be negative. As a result, the entropy quantifies the relative difference in uncertainty between two distributions, rather than absolute uncertainty.

To state entropic uncertainty relations between observables with a continuum of eigenvalues, we must scale the observables so that their product is dimensionless. If we scale the position x and momentum p so that $[x, p] = i$, then the uncertainty relation is

$$H[P(x)] + H[P(p)] \geq \ln(e\pi). \tag{2.66}$$

This relation was proved by Beckner [40] and by Bialynicki-Birula and Mycielski [65], building on the work of Hirschman [255]. Entropic uncertainty relations for angle and angular momentum have also been obtained and can be found in [467, 66, 64].

Noise added to a system: the "entropy of disturbance"

Semiclassical measurements, in providing information, reduce, on average, the entropy of the observer's state-of-knowledge. They do not, however, affect the state of the system in the sense of applying forces to it. Because of this, if the observer is ignorant of the measurement outcome, then the measurement cannot change his or her state-of-knowledge. This means that

$$\rho = \sum_n p_n \tilde{\rho}_n, \tag{2.67}$$

for all semiclassical measurements. Here, as usual, ρ is the initial state, and the $\tilde{\rho}_n$ are the final states that result from the measurement. This relation is simple to show using the fact that all the measurement operators commute with ρ for a semiclassical measurement (see Section 1.4.1).

If one applies forces to a system, dependent on the measurement result, then the final state, averaged over all measurement results, can have less entropy than the initial state. This is the heart of the process of feedback control – one learns the state of a system by making a measurement, and then applies forces that take the system to the desired state. This means that different forces are applied to the system for each possible measurement outcome. The result is a deterministic mapping from a probability distribution over a range of states to a single state. This takes the system from a state with high entropy to one with zero entropy (in fact, this entropy does not vanish from the universe, but is transferred to the control system, something that will be discussed in Chapter 4 when we examine Maxwell's demon.)

We recall from Section 1.4.4 that the definition of a quantum measurement includes both information extraction and the application of forces. If we consider only measurements whose operators contain no unitary part (bare measurements), then the measurement operators are all positive. A fundamental fact about bare quantum measurements is revealed by Ando's theorem:

Theorem 11 (Ando [12]) Consider a bare measurement described by the set of positive measurement operators $\{\mathcal{P}_n\}$. If the initial state of the system is ρ, then it is always true that

$$S\left(\sum_n p_n \tilde{\rho}_n\right) \geq S(\rho), \tag{2.68}$$

where $\tilde{\rho}_n = \mathcal{P}_n \rho \mathcal{P}_n / \mathrm{Tr}[\mathcal{P}_n^2 \rho]$ is the final state for the nth measurement result, and p_n is its probability.

Proof This proof is remarkably simple, but does require that we set up some notation first. To do so we note that the map from ρ to $\tilde{\rho} = \sum_n p_n \tilde{\rho}_n$ is linear, and it takes the identity to the identity: if $\rho = I$ then $\sum_n p_n \tilde{\rho}_n = \sum_n \mathcal{P}_n I \mathcal{P}_n = \sum_n \mathcal{P}_n^2 = I$. Let us denote this map by Θ, so that $\tilde{\rho} = \Theta(\rho)$. Since the entropy depends only on the eigenvalues of the density matrix, what we are really interested in is the map that takes $\boldsymbol{\lambda}(\rho)$ to $\boldsymbol{\mu}(\tilde{\rho})$. Let us denote by Diag[**v**] the diagonal matrix that has the vector **v** as its diagonal. Conversely, let us write the vector that is the diagonal of a matrix M by **diag**(M). If U is the unitary that diagonalizes ρ, and V is the unitary that diagonalizes $\tilde{\rho}$, then diag[$\boldsymbol{\lambda}$] = $U\rho U^\dagger$ and diag[$\boldsymbol{\mu}$] = $V\tilde{\rho}V^\dagger$. With this notation, we can now write the map, Φ, that takes $\boldsymbol{\lambda}$ to $\boldsymbol{\mu}$ as

$$\boldsymbol{\mu} = \Phi(\boldsymbol{\lambda}) \equiv \mathbf{diag}[V\Theta(U^\dagger \mathrm{Diag}[\boldsymbol{\lambda}]U)V^\dagger]. \tag{2.69}$$

It is simple to verify from the above expression that Φ is linear, and thus $\boldsymbol{\mu} = D\boldsymbol{\lambda}$ for some matrix D. Each column of D is given by $\boldsymbol{\mu}$ when $\boldsymbol{\lambda}$ is chosen to be one of the elementary basis vectors. From the properties of the map Θ, it is not difficult to show that all the elements of D are positive, and that D is probability-preserving and unital (see Section 2.2.2 for the definitions of these terms). Thus D is doubly stochastic, $\boldsymbol{\lambda} \succ \boldsymbol{\mu}$, and the result follows. $\qquad\square$

Theorem 11 tells us that if we discard the measurement results, a bare quantum measurement cannot decrease the entropy of a system, just like a classical measurement. This provides further motivation for regarding bare measurements as the quantum equivalent of classical measurements. It is also simple to verify that, quite unlike classical measurements, bare quantum measurements can *increase* the uncertainty (von Neumann entropy) of a system. This reflects the fact that quantum measurements can cause disturbance; if they did not disturb the system, then averaging over the measurement results must leave the initial state unchanged. The disturbance caused by a quantum measurement can therefore be quantified by

$$S_D \equiv S\left(\sum_n p_n \tilde{\rho}_n\right) - S(\rho), \tag{2.70}$$

which we will refer to as the *entropy of disturbance.*

Once we allow our measurement operators to contain a unitary part, then, like classical measurements with feedback, quantum measurements can deterministically reduce the entropy of a system. The notion of disturbance captured by the "entropy of disturbance" is relevant for feedback control. One of the primary purposes of feedback is to reduce the entropy of a system. The entropy of disturbance measures the amount by which a quantum measurement acts like noise driving a system, and thus interferes with an attempt to control it. We will examine feedback control in Chapter 5.

Disturbance to encoded information

The third notion of disturbance is that of a change caused to the states of an ensemble in which the system has been prepared. To illustrate this, consider a von Neumann measurement that projects a three-dimensional quantum system onto the basis $\{|0\rangle, |1\rangle, |2\rangle\}$. If we prepare the system in the ensemble that consists of these same three states, then the measurement does not disturb the states at all; if the system is initially in ensemble state $|n\rangle$, then it will still be in this state after the measurement. The only change is to our state-of-knoweldge – before the measurement we do not know which of the ensemble states the system has been prepared in, and after the measurement we do.

What happens if we prepare the system instead in one of the three states

$$|a\rangle = \frac{|0\rangle + |2\rangle}{\sqrt{2}}, \quad |b\rangle = \frac{|1\rangle + |2\rangle}{\sqrt{2}}, \quad |c\rangle = |2\rangle? \tag{2.71}$$

This time, if the initial state is $|a\rangle$ or $|b\rangle$, the measurement is guaranteed to change it. In fact, because the three states are not all mutually orthogonal, there is *no* complete measurement that will not change at least one of the states in the ensemble, for at least one outcome. In this case we cannot extract the accessible information without causing a disturbance at least some of the time.

To quantify the notion of disturbance to an ensemble, we must have a measure of the *distance* between two quantum states; otherwise we cannot say how *far* a measurement has moved the initial state of a system. While we will describe such distance measures next, we will not explore further the present notion of disturbance. The interested reader can find more on this kind of disturbance in the articles by Barnum [35] and Fuchs [186].

There is another interesting trade-off relation regarding ensembles, but this time relating to the amount learned about encoded information by making measurements of different observables. This relation was derived by Hall [238]. If we encode information in a quantum system using the ensemble of states $\{p_i, \rho_i\}$, then we can choose this ensemble so that a measurement of an observable X gives us the maximum information, or choose a different ensemble so that a different observable, Y, provides the maximum information. Hall's trade-off relation states that however we encode the information, the amount we would learn by making a measurement of X, ΔI_X (quantified by the mutual information), and similarly for Y, ΔI_Y, satisfies

$$\Delta I_X + \Delta I_Y \leq 2\ln\left(N \max_{x,y} |\langle x|y\rangle|\right), \tag{2.72}$$

where all the symbols are as defined above Eq. (2.64). Here a measurement of X means a von Neumann measurement that projects onto the eigenbasis of X, and similarly for a measurement of Y.

Hall also derived what might be referred to as the "reverse" of the above bound, in which one makes a single general measurement, and asks how much information this can provide about each of two noncommuting observables, X and Y. This new trade-off relation has precisely the same form as that above in Eq. (2.72), but in this case ΔI_X and ΔI_Y denote instead the mutual information obtained about X and Y by making the single measurement. In Exercise 23 you can show that this second relation can be derived from that in Eq. (2.72), elucidating the close connection between them.

Disturbance to quantum information

We can consider a quantum state *itself* as information. We can encode a sequence of quantum states of an N-dimensional system into a sequence of two-dimensional systems, and this idea has led to quantum versions of Shannon's coding theorems [549, 142]. In this context, a two-dimensional quantum system is called a *qubit*, a term coined by Schumacher [549].

Let us say that Alice has a means of sending an N-dimensional quantum system to Bob. This means is called a *quantum channel*, since it can be used by Alice to send quantum states to Bob. If Alice has a quantum state with N^M dimensions, she can write this state into M N-dimensional systems, and use the channel M times to send the entire state to Bob. We now ask what happens if the channel is noisy – that is, if it changes the state in some unknown way en route to Bob? Such noise can always be described by having the system carrying the state interact with another quantum system, where the latter is subsequently lost. Because of this we can always model a noisy channel by saying that a measurement $\mathcal{M}$, described by the measurement operators $\{A_n\}$, acts on the system en route, the results of which are lost.

The key question regarding a noisy channel is, how much quantum information (what dimension of quantum state) can be sent so that it is essentially error-free, by encoding the state across M uses of the channel, in the limit $M \to \infty$. For classical channels this question is answered by the mutual information. For quantum channels there is an equivalent quantity called the *coherent information* [36, 551, 142]. The coherent information is quantified in terms of entropy, and is dependent on the input state, ρ, and the measurement, $\mathcal{M}$, and is usually denoted by $I(\rho,\mathcal{M})$ Without going into further detail, it is the entropy of the input state that plays the role of the entropy of the probability distribution of the input states for a classical channel. For the input state ρ, the coherent information for a perfect channel is $S(\rho)$.

The coherent information provides a way to characterize the disturbance caused by a measurement to quantum information: the disturbance, δ, is how much the measurement $\mathcal{M}$ reduces coherent information that can be transmitted by the state ρ. That is, $\delta(\rho,\mathcal{M}) \equiv S(\rho) - I(\rho,\mathcal{M})$. Buscemi, Hayashi, and Horodecki have shown that for efficient measurements one has the fundamental information disturbance relation [94]

$$\delta(\rho,\mathcal{M}) \equiv S(\rho) - I(\rho,\mathcal{M}) = \Delta I_{\text{sys}}. \tag{2.73}$$

This states that an efficient measurement that extracts information ΔI_{sys} (in the sense defined by ΔI_{sys}), causes disturbance $\delta(\rho,\mathcal{M})$. In fact, the relation obtained by Buscemi *et al.* is more general than Eq. (2.73), and also covers inefficient measurements, in which case it involves a generalization of ΔI_{sys} that is non-negative for all measurements. This is a beautiful information-disturbance relation, but is probably not yet the end of the story.

2.4 Distinguishing quantum states

Consider the following scenario: Alice prepares a quantum system in one of two predetermined states and does not tell Bob which one. Bob then makes a measurement on the system to determine which of the two states the system is in. The question of how well Bob can achieve this goal is motivated by two quite different situations. The first is when Alice is trying to communicate a message to Bob, and in this case she would want to choose the states so that Bob could distinguish between them as reliably as possible.

The second situation is when Alice is trying to control the state of a system. In this case there is a specific "target" state that Alice would ideally like the system to be in. In the presence of noise her control algorithm will not be able to achieve the target state perfectly, but will be able to get close to it. The questions are, how close, and what do we mean by close anyway? To answer the second, we mean that the effect of the system on the rest of the world is similar to the effect it has when it is in the target state. While the exact nature of this effect will depend on the situation, we can obtain a good general-purpose measure of the similarity of two states by considering the similarity of the results of measurements performed on them. This is directly related to the question of how well an observer, Bob, can distinguish those states, and this is the original question we posed above.

To distinguish the two states ρ and σ an observer must make a measurement on the system, and decide upon one of the two states based on the result of the measurement. This is, of course, a special case of the more general situation we discussed in Section 1.7, in which an observer extracts information about which of a set of N states have been prepared. It would be natural, therefore, to measure the distinguishability of two states as the mutual information between the measurement result and the variable denoting which of the states had been prepared, maximized over all measurements. Unfortunately this measure of distinguishability does not provide a closed-form expression in terms of ρ and σ.

The trace distance

Fortunately there is a measure of the "distance" between two quantum states that has a simple form, *and* which is directly related to a specific notion of how surely the two states can be distinguished. Given two states described by the density matrices ρ and σ, this measure is given by

$$D(\rho,\sigma) \equiv \frac{1}{2}\text{Tr}[|\rho - \sigma|], \qquad (2.74)$$

and is called the *trace distance*. The factor of a half included in the definition of the trace distance is so that its value lies between zero and unity. The absolute value of a matrix A, denoted $|A|$, is defined as $\sqrt{A^2}$, or equivalently as the matrix obtained from A by making all its eigenvalues equal to their absolute values.

There are two sensible notions of the "difference" between two quantum states to which the trace distance corresponds. The first, which is most relevant to problems in quantum control, is as follows. Let us say we make a measurement that has M outcomes, described by the measurement operators $\{A_m\}$. If the system is in state ρ, then each outcome has probability $p_m = \text{Tr}[A_m^\dagger A_m \rho]$, otherwise these probabilities are $q_m = \text{Tr}[A_m^\dagger A_m \sigma]$. It is the measurement result that constitutes the realized impact of the system on the rest of the world. Thus a reasonable measure of how similar the two states are is the absolute value of the difference between p_m and q_m, summed over all m. It turns out that the trace distance is half the maximum possible value of this sum, where the maximum is taken over all possible measurements. The proof of this is actually quite short:

Theorem 12 (Helstrom [248]/Fuchs [186]) Given two density matrices, ρ and σ, and a measurement $\mathcal{M}$, described by the operators $\{A_m\}$, then

$$D(\rho,\sigma) \equiv \frac{1}{2}Tr[|\rho - \sigma|] = \frac{1}{2}\max_{\mathcal{M}} \sum_m |p_m - q_m|, \qquad (2.75)$$

where $p_m = \text{Tr}[A_m^\dagger A_m \rho]$ and $q_m = \text{Tr}[A_m^\dagger A_m \sigma]$. The maximum is taken over all possible measurements $\mathcal{M}$.

Proof We show first that the operator $X = \rho - \sigma$ can always be written as the difference of two positive operators that have support on orthogonal subspaces. (The support of an operator is the space spanned by those of its eigenvectors with non-zero eigenvalues.) To show this we note that since X is Hermitian all its eigenvalues are real. If the eigenvectors and eigenvalues of X are, respectively $\{|x_n\rangle\}$ and $\{x_n\}$, then $X = x_n|x_n\rangle\langle x_n|$. If we denote the positive eigenvalues of X as y_n, and the negative eigenvalues as z_n, we can define the positive operators $P \equiv y_n |y_n\rangle \langle y_n|$ and $Q \equiv -z_n |z_n\rangle \langle z_n|$. Then P and Q have support on orthogonal spaces ($PQ = 0$), and $X = P - Q$. This also means that $|\rho - \sigma| = P + Q$.

We now show that the trace distance is an upper bound on the sum in Eq. (2.75). We have

$$\sum_m |p_m - q_m| = \sum_m |\text{Tr}[A_m^\dagger A_m (\rho - \sigma)]|$$

$$= \sum_m |\text{Tr}[A_m^\dagger A_m P] - \text{Tr}[A_m^\dagger A_m Q]|$$

$$\leq \sum_m \text{Tr}[A_m^\dagger A_m P] + \text{Tr}[A_m^\dagger A_m Q]$$

$$= \text{Tr}\left[\left(\sum_m A_m^\dagger A_m \right) |\rho - \sigma| \right]$$

$$= \text{Tr}[|\rho - \sigma|] = 2D(\rho, \sigma). \tag{2.76}$$

All we have to do now is to find a measurement for which equality is reached in Eq. (2.75). We leave this as an exercise.   □

The second sensible measure of distinguishability to which the trace distance corresponds is the minimum probability that Bob will make an error when trying to determine whether the state is ρ or σ. To decide which state Alice has prepared, Bob makes a measurement, and uses Bayesian inference, as described in Section 1.7, to determine the probability that the system was initially prepared in each of the states. This depends upon the prior, which in this case is given by the probabilities with which Alice chose to prepare each of the states. For our purposes here, Alice chose each state with equal probability. Once Bob has determined the probabilities for each of the two states, his best guess is simply the one that is most likely. By averaging over all the measurement outcomes, we can calculate the total probability that Bob's best guess is wrong, and this is the "probability of error." Some measurements will give a smaller probability of error than others. Assuming that Alice prepares each of the two states with equal probability the expression for the error probability (which we leave as an exercise) is

$$P_{\text{err}} = \frac{1}{2} \sum_m \min(p_m, q_m), \tag{2.77}$$

where $\min(p_m, q_m)$ denotes the smaller of the two values p_m and q_m. It turns out that we can rewrite P_{err} as

$$P_{\text{err}} = \frac{1}{2} - \frac{1}{4} \sum_m |p_m - q_m|. \tag{2.78}$$

But we now know from Theorem 12 that the minimum value of this, minimized over all measurements, is

$$\min_{\mathcal{M}} P_{\text{err}} = \frac{1}{2}[1 - D(\rho, \sigma)]. \tag{2.79}$$

Note that when the probability of error is $1/2$, Bob has no information about which of the two states the system is in.

To be a proper measure of distance (a metric) between two points x and y in some space, a measure $d(x, y)$ must satisfy three properties. It must be equal to zero if and only if $x = y$, it must be symmetric in its arguments (that is, $d(x, y) = d(y, x)$), and it must satisfy the triangle inequality, $d(x, y) \leq d(x, a) + d(a, y)$, where a is any third point in the space. It is clear that $D(\rho, \sigma)$ satisfies the first two. A proof of the third property can be found in [506].

The trace distance can be expressed in a particularly simple way for two-state systems. A density matrix for a two-state system can be represented by a three-dimensional vector, **a**, via the relation

$$\rho = \frac{1}{2}[I + \mathbf{a} \cdot \boldsymbol{\sigma}] = \frac{1}{2}[I + a_x \sigma_x + a_y \sigma_y + a_z \sigma_z]. \tag{2.80}$$

The vector **a** is called the *Bloch vector*, and $\boldsymbol{\sigma}$ is the vector of Pauli matrices. For pure states the Bloch vector has unit length (and thus lies on the surface of a unit sphere), and for mixed states the Bloch vector lies within the sphere. The Pauli matrices, Bloch vector, and Bloch sphere are described in more detail in Appendix D.

The trace distance between two two-state density matrices, ρ and σ, whose Bloch vectors are, respectively, **a** and **b**, is

$$D(\rho, \sigma) = \frac{|\mathbf{a} - \mathbf{b}|}{2}. \tag{2.81}$$

That is, half the geometrical distance between the two Bloch vectors.

The fidelity

There is another measure of the similarity between two quantum states, called the *fidelity*. This is often used in quantum information theory in place of the trace distance because it has a simpler algebraic form. Further, a connection can be made between the trace distance and the fidelity. The fidelity between two density matrices is defined as

$$F(\rho, \sigma) \equiv \text{Tr}\left[\sqrt{\sigma^{1/2} \rho \sigma^{1/2}}\right]. \tag{2.82}$$

While it is not obvious, the fidelity is, in fact, symmetric in its two arguments. Its value lies between zero and unity, being zero if and only if ρ and σ have orthogonal supports, and unity if and only if $\rho = \sigma$. If σ is the pure state $|\psi\rangle$, then the fidelity reduces to the very simple form

$$F(\rho, |\psi\rangle) = \sqrt{\langle \psi | \rho | \psi \rangle}. \tag{2.83}$$

The connection between the fidelity and the trace distance is that, for every state ρ and σ,

$$1 - F(\rho,\sigma) \le D(\rho,\sigma) \le \sqrt{1 - [F(\rho,\sigma)]^2}. \tag{2.84}$$

When one of the states is pure, the lower inequality may also be replaced with

$$1 - [F(\rho,|\psi\rangle)]^2 \le D(\rho,|\psi\rangle). \tag{2.85}$$

The proofs of the above properties of the fidelity may be found in [447], and the references at the end of this chapter.

2.5 Fidelity of quantum operations

For the purposes of controlling a quantum system, one may want to characterize how close the state of the system is to some "target" state, as discussed above. Further, one may wish to control not only the state of a system, but the entire mapping from a set of initial states to a set of final states. That is, one may wish to ensure that a system undergoes a specified unitary transformation. To determine how close the actual evolution is to the target unitary transformation, one needs a measure of the similarity of evolutions.

A transformation that specifies how all the initial states of a system will map to the final states is called a *quantum operation*. We can distinguish two kinds of quantum operations. Those that are probabilistic (those that depend upon the outcome of a measurement) and those that are deterministic (those for which a given input state always gives the same output state).

In the most general case, for a deterministic quantum operation, our system may interact with another "auxiliary" system, which is subsequently discarded. As we saw in Chapter 1, the effect of discarding the auxiliary system is the same as that of averaging over the results of a von Neumann measurement made on it. This, in turn, is the same as averaging over the results of some generalized measurement made on the primary system. Thus, if our system starts in the state ρ, the most general deterministic quantum operation may be written as

$$\varepsilon(\rho) = \sum_m A_m \rho A_m^\dagger, \tag{2.86}$$

where $\sum_m A_m^\dagger A_m = I$. (For a probabilistic quantum operation, the most general case is that of an inefficient measurement, as described in Section 1.6).

The standard measure of the similarity of two quantum operations is the fidelity between their output states, averaged over all pure input states. The average is, of course, taken with respect to the Haar measure (the definition of this measure is given in Appendix G). One is usually interested in engineering evolutions that are unitary operations. These are the most useful, because any other kind of operation involves discarding an auxiliary system, resulting in a loss of information. If we assume that our target evolution is given by the unitary operator U, and the actual evolution is given by the map ε, then the average fidelity is

$$\overline{F}(\varepsilon, U) = \int \langle \psi | U^\dagger \varepsilon(\rho) U | \psi \rangle \, d|\psi\rangle, \qquad (2.87)$$

where $\rho = |\psi\rangle\langle\psi|$, and $\int d|\psi\rangle$ denotes integration with respect to the Haar measure.

Most often one needs to evaluate the fidelity of an operation that is simulated numerically, or implemented experimentally. In the above form the fidelity would be prohibitively difficult to calculate, since it involves obtaining a continuum of output states. Due to the fact that the space of operations is finite-dimensional, it is possible to rewrite the fidelity as a sum over a discrete set of initial states, and this makes numerical evaluation feasible.

To present the alternative expression for the fidelity, we need first to define a number of operators. Let us denote a basis for an N-dimensional system as $|n\rangle$, $n = 0, \dots, N-1$. First we define an operator X that shifts each of the basis states "up by one":

$$X|n\rangle = |n+1\rangle, \quad n = 0, \dots, N-1, \qquad (2.88)$$

where it is understood that $n = N$ is identified with $n = 0$. We also define the operator Z by

$$Z|n\rangle = e^{2\pi i n/N}|n\rangle, \quad n = 0, \dots, N-1. \qquad (2.89)$$

Using these operators we define a set of N^2 unitary operators by

$$V_{jk} = X^j Z^k, \quad j, k = 0, 1, \dots N-1. \qquad (2.90)$$

Using this set of operators, the expression for the average fidelity is

$$\overline{F}(\varepsilon, U) = \frac{\sum_{j=1}^N \sum_{k=1}^N \text{Tr}[UV_{jk}^\dagger U^\dagger \varepsilon(V_{jk})]}{N^2(N+1)}. \qquad (2.91)$$

To evaluate this expression, one simulates the actual evolution of a system (the map ε) for the set of N^2 initial density matrices given by V_{jk}. Of course, since these matrices are unitary, they are not real initial states. Nevertheless one can still use them as initial states when performing the simulation, and this provides the "output states" $\varepsilon(V_{jk})$. One then substitutes these output states into Eq. (2.91).

History and further reading

The concept of entropy first arose in thermodynamics and then in statistical mechanics, and the information-theoretic quantity that Shannon discovered turned out to have the same form. The fact that Shannon's entropy is identical to the entropy of statistical mechanics is not coincidental, but fundamental. The essential connection between information theory and statistical mechanics will be discussed in Chapter 4. Further details regarding information theory can be found in, for example, *Elements of Information Theory* by Cover and Thomas [133].

The powerful and highly non-trivial subadditivity property of the von Neumann entropy was first conjectured by Lanford and Robinson [353] in 1968, after Robinson and Ruelle [521] noted the importance for statistical mechanics of the subadditivity of the classical entropy. It was finally proven in 1973 by Ruskai and Lieb [371, 372], using Lieb's theorem which was obtained in the same year [370]. Lieb's original proof of his eponymous theorem was rather involved, and it was some time before simple proofs of strong subadditivity were obtained. The first were given by Narnhofer and Thiring in 1985 [440], and Petz in 1986 [480] (a straightforward presentation is given in Nielsen and Petz [448]). The simplest proof to date was obtained by Ruskai in 2006 [523].

The classification theorem for ensembles was obtained independently by Jaynes [310], and Hughston, Josza, and Wootters [276]. The majorization result regarding the ensemble probabilities was obtained by Nielsen [445].

The information gain regarding the state of a quantum system, being the average reduction in the von Neumann entropy, was first considered by Groenewold [231]. The intuitive (but nontrivial) result that efficient measurements never increase this information was first obtained by Ozawa in 1986 [458]. The more general result regarding majorization, and thus all concave functions, was obtained by Nielsen in 2000 [446] (published 2001). The very simple proof that we use here was obtained by Fuchs, and appears in Fuchs and Jacobs [188]). In his beautiful paper in 2001 Nielsen also re-derives in a simple and unified way a number of majorization inequalities, one of which gives the bound whose derivation we include as Exercise 11 (it was originally obtained by Lanford and Robinson [353]). The bound on the entropy exchange between a system and probe was proved in [301].

The famous Holevo bound on the accessible information was obtained by Holevo in 1973 [258]. Hall's bound is proved in a slightly different form in [239], and was written in the form we presented here in [283]. The stronger bound on the mutual information (Theorem 10) was obtained for incomplete measurements by Schumacher, Westmoreland and Wootters (SWW) [548] in 1996. The extension to inefficient measurements is quite straightforward and appears in [286]. Barchielli and Lupieri also derived the general form of the SWW theorem using more sophisticated methods [30].

The coherent information, the quantity that characterizes the ability of a quantum channel to transmit quantum information, was introduced by Shumacher and Nielsen in [551]. It was finally proved that it quantified the rate at which a quantum channel could transmit quantum information by Shor [561] and Devetak [142], independently in 2004.

The minimum error probability was first obtained by Helstrom, and the connection with the trace distance was made by Fuchs. Proofs can be found in [248, 186]. The fidelity was devised by Jozsa in 1995 [320], and the inequalities connecting the trace distance with the fidelity were obtained by Fuchs and van der Graaf [189]. Properties of the trace distance and fidelity may be found in [185, 189, 619, 447].

The formula for the fidelity of a quantum operation described in Section 2.5 was obtained by Nielsen [449].

Exercises

1. Show that for two independent random variables X and Y, the entropy of their joint probability distribution is the sum of their individual entropies.

2. Derive Eq. (2.10) by rearranging the expression for the mutual information, $H[x:y] = H[x] - H[x|y]$.

3. Consider all the possible sequences of N bits. (Recall that each bit has a 50/50 chance of being 0 or 1.) There are a total of 2^N sequences. Not all sequences are typical, for example the sequence containing only 0s is atypical. In the limit as $N \to \infty$, the total number of typical sequences is (effectively) 2^N. Explain.

4. Let $f(x)$ be a function with Taylor series expansion $f(x) = \sum_{n=0}^{\infty} c_n x^n$. If A is a matrix, and $D = UAU^\dagger$, where D is diagonal and U is unitary, show that

$$\sum_n c_n A^n = U \begin{pmatrix} f(\lambda_1) & 0 & \cdots & 0 \\ 0 & f(\lambda_2) & & \vdots \\ \vdots & & \ddots & 0 \\ 0 & \cdots & 0 & f(\lambda_N) \end{pmatrix} U^\dagger, \qquad (2.92)$$

where the numbers λ_i are the diagonal elements of D. Hint: use the fact that $U^\dagger U = UU^\dagger = I$.

5. Consider encoding a message in a single qubit by using the states $|\psi_0\rangle = \cos\theta|0\rangle + \sin\theta|1\rangle$ and $|\psi_0\rangle = \cos\theta|0\rangle - \sin\theta|1\rangle$. (i) What von Neumann measurement provides the maximal mutual information between the measurement result and the encoded information, assuming that each coding state is chosen with probability $1/2$. To answer this question it may help to make an intuitive guess by picturing the states on the Bloch sphere (the Bloch sphere is described in Appendix D.) (ii) What von Neumann measurement gives the minimum disturbance, as defined by the "entropy of disturbance" (Eq. (2.70))? (iii) This time we again ask about the entropy of disturbance, but before we calculate this disturbance we are allowed to perform a unitary operation on the system based upon the measurement result. Since a von Neumann measurement has two outcomes, this means that we can apply a unitary U_1 when we get outcome 1, and another unitary U_2 when we get outcome 2. Recall that all unitary operations for a single qubit are merely rotations on the Bloch sphere. What von Nueman measurement will give the minimum entropy of disturbance in this case?

6. Consider a measurement on a qubit that has the measurement operators

$$\Omega_\pm = \sqrt{\kappa}|\pm\rangle_\theta\langle\pm|_\theta + \sqrt{1-\kappa}|\mp\rangle_\theta\langle\mp|_\theta. \qquad (2.93)$$

with $|+\rangle_\theta = \cos\theta|0\rangle + \sin\theta|1\rangle$ and $|-\rangle_\theta = \sin\theta|0\rangle - \cos\theta|1\rangle$. We now make this measurement on a system with density matrix $\rho = p|0\rangle\langle0| + (1-p)|1\rangle\langle1|$. (i) What is ΔI_{sys} as a function of θ and p? (ii) For what value of θ is ΔI_{sys} greatest? (iii) Assuming that information has been encoded in the system using the states $|0\rangle$ and $|1\rangle$, what is ΔI_{ens} as a function of θ and p? (iv) For what value of θ is ΔI_{ens} greatest?

7. Consider two systems A and B whose joint density matrix is ρ. We can define a quantum version of the mutual information between A and B by $I_{\text{Q}}(A,B) \equiv S(\rho_A) + S(\rho_B) - S(\rho)$. The Araki–Lieb inequality, $|S(\rho_A) - S(\rho_B)| \le S(\rho) \le S(\rho_A) + S(\rho_B)$ shows that this "quantum mutual information" is non-negative. (We will not concern ourselves here with what $I_{\text{Q}}(A,B)$ might mean.) Now consider the process of reducing the entropy of one system by "transferring" this entropy to another system. Systems A and B start in the state $\rho_0 = \rho_A \otimes \rho_B$. If we apply a unitary transformation to the joint system that reduces the entropy of A by the (positive) amount ΔS_A, show that the (positive) increase in the entropy of B is $\Delta S_B = \Delta S_A + I$, where I is the quantum mutual information for the final joint state of the two systems. What can we say about ΔS_B when we decrease the entropy of A by applying a joint operation that is not unitary, and that increases the entropy of the joint density matrix? The relation between the entropy changes in A and B may remind you of the entropy exchange referred to in Theorem 8. Elucidate the difference between the present notion of entropy exchange and that in Theorem 8.

8. Consider a quantum system with two subsystems A and B. The *Schmidt decomposition* says that if the system is in a pure state, then this state can always be written in the form

$$|\psi\rangle = \sum_{n=1}^{N} c_n |a_n\rangle |b_n\rangle, \tag{2.94}$$

where N is the dimension of the smaller of the two systems, the set of states $\{|a_n\rangle\}$ is a set of orthonormal states for system A, and $\{|b_n\rangle\}$ is a set of orthonormal states for system B. The form above is called the Schmidt decomposition of the state $|\psi\rangle$, and the bases $\{|a_n\rangle\}$ and $\{|b_n\rangle\}$ are called the *Schmidt bases*. This decomposition is very useful in obtaining results regarding entropy and information in interactions between two systems.

(i) Show that the Schmidt decomposition reveals that (1) the entropy of the density matrix of A (with B traced out) is equal to the entropy of B when A is traced out; (2) the maximum entropy of either system is $\ln N$, where N is the dimension of the smaller system.

(ii) Now consider two qubits in a joint pure state. Use the Schmidt decomposition to show that in parametrizing all the possible joint states, one can define two real parameters that remain unchanged by a unitary operation that acts on either of the two systems. Such a unitary operation is referred to as being *local*. Next consider a state with specified values of these locally invariant parameters. If we want to encode information in one of the systems, and retrieve this information by making a measurement on the other system, what are the best bases to use to encode and to measure? Are they the Schmidt bases?

9. Show that $\langle \Delta S \rangle$ reduces to the mutual information for a classical measurement.

10. Let **p** be a probability vector, and $\mathbf{q} = S\mathbf{p}$. (i) Show that if S is a stochastic matrix then **q** is a probability vector. (ii) Show that if S is doubly stochastic, then each of the elements of **q** is a weighted average of the elements of **p**.

11. We are going to show that if $\rho = \sum_i p_i \rho_i$, then

$$S(\rho) \leq H[\{p_i\}] + \sum_i p_i S(\rho_i). \tag{2.95}$$

(i) First consider the case when all the ρ_i are pure states. Show that for this case, Theorem 4 implies that

$$S(\rho) \leq H[\{p_i\}].$$

(For pure states this result is the same as Eq. (2.95), because then $S(\rho_i) = 0, \forall i$.)

(ii) Now consider the case in which the ρ_i are mixed. First we write all of the ρ_i in terms of their eigenbases, so that for each ρ_i we have $\rho_i = \sum_j \lambda_{ij} |v_{ij}\rangle \langle v_{ij}|$. This means that the state is given by

$$\rho = \sum_i \sum_j p_i \lambda_{ij} |v_{ij}\rangle \langle v_{ij}|.$$

Now use the result in (i), and the fact that for each i, $\sum_j \lambda_{ij} = 1$, to show that Eq. (2.95) is true.

12. Show that a conditional unitary operation can always be used, following a measurement, to return a system to its initial state, from the point of view of an observer who does not know the measurement outcome. That is, there always exist a set of unitary operators $\{U_n\}$ such that

$$\rho = \sum_n p_n U_n \tilde{\rho}_n U_n^\dagger, \tag{2.96}$$

where $\tilde{\rho}_n = A_n \rho A_n^\dagger / p_n$, $p_n = \mathrm{Tr}[A_n^\dagger A_n \rho]$, and $\sum_n A_n^\dagger A_n = I$. Hint: use results from the analysis in the proof of Theorem 6.

13. Use the inequality proved in Exercise 2, above, and the trick used to prove Theorem 6, to prove Theorem 7.

14. Use the classification theorem for ensembles (Theorem 4) to show that any two pure-state ensembles that generate the same density matrix are related by a unitary transformation.

15. Use the concavity relation for I_{sys} (Eq. (2.58)) to show that for bare measurements on an initially completely mixed state, I_{sys} satisfies the classical relation for sequential measurements (Eq. 2.61).

16. Complete the proof of Theorem 12.

17. In Section 2.4 we considered the probability of error when trying to distinguish between two probability distributions from a single sample. (i) Show that this error probability is given by Eq. (2.77). (ii) Show that the expression in Eq. (2.77) is equal to that given in Eq. (2.78).
18. Complete the proof of Theorem 4.
19. Prove the inequality in Eq. (2.43) by using the method in the proof of Theorem 6.
20. Rewrite Eq. (2.55) in the form given by Eq. (2.56).
21. Show that the fidelity is symmetric in its two arguments. Hint: the simplest way to do this is to realize that if one defines $A = \sigma^{1/2}\rho^{1/2}$, then the polar decomposition theorem can be used to show that the eigenvalues of $\sigma^{1/2}\rho\sigma^{1/2}$ are the same as those of $\rho^{1/2}\sigma\rho^{1/2}$. An alternative method that can be used to show that the eigenvalues of these two operators are the same, is to show that if $|\psi\rangle$ is an eigenvector of $\sigma^{1/2}\rho\sigma^{1/2}$, then $\rho^{1/2}\sigma^{1/2}|\psi\rangle$ is an eigenvector of $\rho^{1/2}\sigma\rho^{1/2}$ with the same eigenvalue.
22. Imagine that there is a system in state ρ, to which we can apply any unitary transformation. Our goal is to obtain a final state $U\rho U^\dagger$ that has the maximum possible fidelity with a "target" state σ. That is, we wish to find a unitary U that maximizes $\mathrm{Tr}\left[\sqrt{\sigma^{1/2}U\rho U^\dagger \sigma^{1/2}}\right]$. To determine the desired U, first note that

$$\max_V |\mathrm{Tr}[AV]| = \max_V |\mathrm{Tr}[\sqrt{A^\dagger A}V'V]| = \max_V |\sum_j \sigma_j(A)e^{i\theta_j}| = \mathrm{Tr}[\sqrt{A^\dagger A}],$$

where we have used the polar decomposition theorem for A ($A = \sqrt{A^\dagger A}V'$), and the $\sigma_j(A)$ are the eigenvalues of $\sqrt{A^\dagger A}$. Note that V and V' are unitary. Secondly, there is a result proved by von Neumann that says

$$\max_V |\mathrm{Tr}[srV]| \le \sum_j \lambda_j(s)\lambda_j(r), \qquad (2.97)$$

where r and s are density matrices, and V is unitary. The $\lambda_j(s)$ and $\lambda_j(r)$ are, respectively, the eigenvalues of s and r, ordered from largest to smallest.

Use the above results to determine the unitary, U, that maximizes the above fidelity. Hint: the explicit construction of U will involve the eigenvectors of ρ and σ.

23. In Hall's information trade-off relation given in Eq. (2.72), information is encoded in a system using an ensemble $\{p_j, \rho_j\}$. We then consider making one of two von Neumann measurements on the system. The first projects the system onto the eigenstates of an observable X, and the second onto the eigenstates of Y. If we label the outcomes for X by $\{x_j\}$, and those for Y by $\{y_k\}$, then ΔI_X (in Eq. (2.72)) is the mutual information between $\{p_i\}$ and $\{x_j\}$ and ΔI_Y is that between $\{p_i\}$ and $\{y_k\}$. Now consider a situation in which there is a single initial state, and we make a measurement $\{A_n\}$ on the system. There is an initial probability distribution for the eigenvalues of X given by $P_X(j) = \{\langle x_j|\rho|x_j\rangle\}$, where $\{|x_j\rangle\}$ are the eigenstates of X, and similarly $P_Y(k)$ for Y. Let us denote the mutual information between the measurement results $\{n\}$ and the $\{x_j\}$ by

$\Delta \tilde{I}_X$ and that between $\{n\}$ and $\{y_k\}$ by $\Delta \tilde{I}_X$. Your task is to show that Eq. (2.72) implies that

$$\Delta \tilde{I}_X + \Delta \tilde{I}_Y \leq 2 \ln \left(N \max_{x,y} |\langle x|y \rangle| \right). \tag{2.98}$$

To do this, one has to realize that there is an ensemble that makes $\Delta \tilde{I}_X = \Delta I_X$ and $\Delta \tilde{I}_Y = \Delta I_Y$. To show this, use Fuchs's trick in Theorem 6 to write the states that result from the measurement $\{A_n\}$ as an ensemble that gives the initial state ρ.

24. *Horn's lemma and Ky Fan's maximum principle*: Recall that Horn's lemma implies that the diagonal elements of a diagonal N-dimensional matrix, A, majorize the diagonal elements of the matrix $B = UAU^\dagger$ for every possible unitary U. This means that the sum of the largest $k \leq N$ eigenvalues of B are greater than or equal to the diagonal elements of B, for every unitary U. We can use this result to prove the following relation, which is known as Ky Fan's maximum principle:

$$\sum_{j=1}^{k} \lambda_j(A) = \max_P \text{Tr}[AP], \tag{2.99}$$

where $\lambda_j(A)$ is the jth largest eigenvalue of A, and the maximum is taken over all rank-k projectors P. Show that this relation is true by (i) showing that there is a rank-k projector Q so that $\sum_{j=1}^{k} \lambda_j(A) = \text{Tr}[AQ]$, and (ii) use the results of Horn's lemma described above, and the fact that all rank-k projectors P can be written as $P = UQU^\dagger$, to show that for all rank-k projectors P it is true that $\text{Tr}[AP] \leq \sum_{j=1}^{k} \lambda_j(A)$.

25. Use Ky Fan's maximum principle, Eq. (2.99), to show that for any two N-dimensional Hermitian matrices A and B, $\lambda(A + B) \prec \lambda(A) + \lambda(B)$, where $\lambda(C)$ denotes the vector of eigenvalues of the matrix C.

3

Continuous measurement

In Chapter 1 we introduced the most general form that a quantum measurement can take. Even so, something is still missing from this description. Dynamical processes are never discontinuous. But so far in our treatment of quantum measurements, the measurement acts instantaneously: one moment we may know nothing about the state of a system, and the next moment – pizzam! – we know everything. While this may be an adequate description for some purposes, we must expect that there are many physical situations in which it is not. In this chapter we show how to describe measurement processes that take an appreciable time. Such measurements are referred to as *continuous* measurements. This will allow us to examine how the state of a system changes *while* information is being extracted. We will first introduce classical continuous measurements, for which the description is a little simpler. With this theory in hand, the extension to quantum continuous measurements will be very natural and straightforward.

3.1 Continuous measurements with Gaussian noise

3.1.1 Classical continuous measurements

If a measurement takes time to extract information, then it follows that as the duration of the measurement tends to zero, the amount of information obtained also tends to zero. This latter statement is our *definition* of a continuous measurement. To describe a classical continuous measurement, we first return to a simple instantaneous measurement introduced in Chapter 1. Consider a measurement of a quantity x, where x takes a continuum of values. Recall that because of the central limit theorem, most measurements of such a quantity can be expected to have a Gaussian error. This means that the measurement result, y, is equal to the true value, x_{true}, plus a Gaussian random variable. We will call this variable ξ, and assume that the measurement result is *unbiased*, meaning that the mean of ξ is zero. If the variance of ξ is Δ, then the conditional probability distribution for the measurement result (the likelihood function) is

$$P(y|x) = \frac{1}{\sqrt{2\pi\Delta}}e^{-(y-x)^2/(2\Delta)}, \tag{3.1}$$

and we can write the measurement result as

$$y = x_{\text{true}} + \xi,\tag{3.2}$$

where the probability distribution for ξ is

$$P(\xi) = \frac{1}{\sqrt{2\pi\Delta}} e^{-\xi^2/(2\Delta)}.\tag{3.3}$$

To describe a continuous measurement, we divide the duration of the measurement, T, into small intervals of duration Δt. In each interval we make the simple measurement above, and assume that the error ξ for the measurement in each interval is independent of the error in all the other intervals. Finally, we then take the limit in which the duration of each interval tends to zero, while the error in each interval tends to infinity, so that the overall error for the total time T remains fixed.

In order to take the continuum limit, in which the number of measurements in time T goes to infinity, while the overall error remains fixed, we need to know how this overall error is related to the errors for the measurements. Let there be N intervals in the time T, so that $\Delta t = T/N$, and denote the nth interval by Δt_n, where $n = 0,\ldots,N-1$, the nth measurement result by y_n, the nth error by ξ_n, and the variance of ξ_n by $\Delta_n = \Delta$. To determine our state-of-knowledge of x after the N independent measurements, we calculate the posterior probability density for x. This is straightforward, and we leave it as Exercise 1. This posterior is Gaussian, and is peaked at

$$x_{\text{mean}} = \frac{1}{N}\sum_n y_n,\tag{3.4}$$

with a variance of

$$\Delta_{\text{tot}} = \frac{1}{N^2}\sum_n \Delta_n = \frac{\Delta}{N}.\tag{3.5}$$

This result for the variance follows from the fact that the variance of the sum of N independent random variables (in this case the ξ_n), is merely the sum of their respective variances. To make this clear, since our estimate of x is x_{mean}, we have

$$x_{\text{mean}} = \frac{1}{N}\sum_n y_n = \frac{1}{N}\sum_n (x_{\text{true}} + \xi_n) = x + \frac{1}{N}\sum_n \xi_n = x_{\text{true}} + \xi_{\text{tot}}.\tag{3.6}$$

So the error is given by ξ_{tot}, which is the sum of all the ξ_n (divided by N), and Eq. (3.5) follows.

Of course, the above results merely tell us something that you are already familiar with: that after making N independent measurements, the best estimate of x is the average of the N measurement results, and the standard deviation of this estimate, $\sqrt{\Delta_{\text{tot}}}$, is the standard deviation of the error in a single measurement divided by $\sqrt{N}$.

We will call the final standard deviation of our estimate, $\sqrt{\Delta_{\text{tot}}}$, the *error* of our measurement. Note that this is a different use of the word "error" from that used above, where we referred to the random variable added to the measurement result as the "error," rather than its standard deviation. We will use both meanings of the word "error" below.[1]

Now Eq. (3.5) shows us what we need to do to keep the error in our measurement the same as we increase the number of intervals N: we must make the variance of the error of each measurement, Δ, scale as N, or equivalently as $1/\Delta t$. Note that it is not the random variable added to each measurement that must increase linearly with N, but the *variance* of these random variables. This is because it is the variances of the errors of the N measurements that add together to give the total variance, whereas the standard deviations do not add together in this way.

We can now define the measurements in each time-interval Δt so that the continuum limit (the limit as $N \rightarrow \infty$) is well defined. Since Eq. (3.4) tells us that the final estimate of x is related in a simple way to the sum of the measurement results, it makes sense to define the final measurement result simply as this sum. With this definition, each of the measurement results y_n can be regarded as the increments of the final result, y, in each time interval Δt. We therefore denote the y_n now as Δy_n, and define the final result as

$$y \equiv \sum_n \Delta y_n. \tag{3.7}$$

We now define a Gaussian random variable with mean zero and variance Δt, which we will call ΔW. We redefine each measurement result to be

$$\Delta y_n = x_{\text{true}} \Delta t + \beta \Delta W_n, \tag{3.8}$$

where each random variable ΔW_n has the same probability distribution as ΔW. This gives us an error for the measurement of x that has the right dependence on Δt: upon obtaining Δy_n, the best estimate of x is

$$\frac{\Delta y_n}{\Delta t} = x_{\text{true}} + \beta \frac{\Delta W_n}{\Delta t}, \tag{3.9}$$

and the variance of the error, $\beta \Delta W_n / \Delta t$, is $1/\Delta t$.

We can now take the continuum limit, in which $N \rightarrow \infty$, and so Δt becomes infinitesimal, and the sums become integrals. The measurement result after time T is then

$$y \equiv \lim_{\Delta t \to 0} \sum_n \Delta y_n = \lim_{\Delta t \to 0} \left[\sum_n x_{\text{true}} \Delta t + \beta \sum_n \Delta W_n \right] = \int_0^T x \, dt + \beta \int_0^T dW, \tag{3.10}$$

---

[1] When reporting the accuracy of a measurement result, for example when presenting results in a research article, one uses the notation $x \pm 2\sqrt{\Delta_{\text{tot}}}$, or $x(2\sqrt{\Delta_{\text{tot}}})$, where x is the estimated value, and $\sqrt{\Delta_{\text{tot}}}$ is the standard deviation of this estimate. In this case one refers to $2\sqrt{\Delta_{\text{tot}}}$ as the *error* in the measurement.

The integral over the infinitesimal random variables dW (of which each has variance dt) is simply a random variable, and is usually denoted by $W(T)$. Thus

$$W(T) = \int_0^T dW \equiv \lim_{N \to \infty} \sum_n \Delta W_n. \tag{3.11}$$

The integral $\int_0^T dW$ is defined purely by being the limit of the sum of the ΔW_n. This integral is called a *stochastic integral*, and being the sum of Gaussian variables, is also a Gaussian random variable. Even though the stochastic integral looks like a strange object, we can calculate the mean and variance of $W(T)$, and since it is Gaussian, this fully determines its probability distribution.

The mean of $W(T)$ is clearly zero, since the means of each of the ΔW_n are zero. We see from Eq. (3.10) that the variance of the error in our final estimate of x is equal to the variance of $\beta W(T)$. The variance of $W(T)$ is the sum of all the variances of the ΔW_n, in the limit when $N \to \infty$. But now recall that we specifically chose the variance of each measurement so that the variance of the error after making all the measurements was independent of the number of measurements. That is, we chose the variances of the ΔW_n so that the sum of these variances is always equal to $N\Delta t = N(T/N) = T$. Thus the variance of $W(T)$ is

$$V[W(T)] = T, \tag{3.12}$$

and so the variance of our final estimate of x is $V[\beta W(T)] = \beta^2 T$.

Even though we have determined the probability distribution of $W(T)$, the stochastic integral is actually a rather subtle entity. In particular, differential equations that involve the random (stochastic) increment dW require a new rule that deviates from the usual rules of calculus. This new rule is the subject of stochastic calculus, also called Ito calculus. We give a quick introduction to Ito calculus in Appendix C, enough to show you how to manipulate differential equations that contain dW, referred to as *stochastic* differential equations. A full introduction to Ito calculus can be found in [288], for example. To have a thorough understanding of continuous quantum measurements, you must understand the content of chapters 3 and 4 of reference [288]. In a nutshell, Ito calculus boils down to the surprising rule that $(dW)^2 = dt$. Noise consisting of independent random increments proportional to dW at each time-step is called *Wiener noise*. Stochastic differential equations that are linear, and in which the noise increment does not multiply any of the dynamical variables, can be solved using a simplified method. These equations can be treated as standard linear differential equations driven by a function with a flat spectrum. This function, effectively the derivative of $W(t)$, is so-called *white noise*, and we describe this method of analysis below in Section 3.1.4.

Now we have a description of a continuous measurement, in that we have a measurement that produces an outcome in each infinitesimal time-step dt, being given by

$$dy = x_{\text{true}} dt + \beta dW, \tag{3.13}$$

where the probability density for dW (called the *Wiener increment*) is

$$P(dW) = \frac{1}{\sqrt{2\pi\,dt}} e^{-(dW)^2/(2dt)}.$$ (3.14)

We will use the term "measurement record," or "measurement signal," for the stream of measurement results dy.

To complete our analysis of a continuous measurement, we need to determine how our state-of-knowledge, the probability distribution for x, evolves with time as we obtain the continuous stream of measurement results. To do this we use Bayes' theorem, naturally, and Ito calculus. To obtain a differential equation that describes how our state-of-knowledge of x, $P(x)$, changes with time, all we need to do is to determine the change in $P(x)$ caused by the measurement in a single time-step dt. If our state-of-knowledge before the measurement is $P(x)$, then Bayes' theorem tells us that after the measurement it becomes

$$P(x|dy) = \frac{1}{\mathcal{N}} P(dy|x) P(x),$$ (3.15)

where $\mathcal{N}$ is merely chosen to normalize $P(x|dy)$. From Eq. (3.13) we see that the conditional probability density for dy given x is Gaussian, with mean equal to $x\,dt$ and variance $\beta^2\,dt$. So we have

$$P(x|dy) = \frac{1}{\mathcal{N}} \exp\left\{ -\frac{(dy - x\,dt)^2}{2\beta^2\,dt} \right\} P(x)$$

$$= \frac{1}{\mathcal{N}} \exp\left\{ -\frac{([x_{\text{true}} - x]\,dt + \beta\,dW)^2}{2\beta^2\,dt} \right\} P(x).$$ (3.16)

We now expand the exponential to first order in dt. This is appropriate because dt is infinitesimal, but we must remember that $(dW)^2 = dt$. Doing this, and dropping constant factors, we obtain

$$P(x|dy) = \frac{1}{\mathcal{N}} \left[1 + \beta^{-1}(x - x_{\text{true}})\,dW \right] P(x),$$ (3.17)

where $\mathcal{N}$ is the normalization. Calculating $\mathcal{N}$ gives

$$\mathcal{N} = 1 + \beta^{-1}(\langle x \rangle - x_{\text{true}})\,dW.$$ (3.18)

Substituting this expression for $\mathcal{N}$ into Eq. (3.17), and using the binomial expansion

$$(1 + x)^p = 1 + px + \frac{p(p-1)}{2} x^2 + \cdots$$ (3.19)

with $p = -1$, we obtain

$$P(x|dy) = \left[1 + \beta^{-2}(x - \langle x \rangle)[\beta\,dW + (x_{\text{true}} - \langle x \rangle)\,dt] \right] P(x).$$ (3.20)

If $P(x)$ is the distribution for x at time t, then $P(x|dy)$ is the distribution at time $t + dt$. A differential equation for $P(x)$ tells us how $P(x)$ changes in each time interval dt. Thus, to derive this differential equation we obtain the change in $P(x)$ in the time interval dt, which we call $dP(x)$, by writing the left-hand side of the above equation as

$$P(x|dy) = P(x, t + dt) = P(x) + dP(x). \tag{3.21}$$

The stochastic differential equation for $P(x)$ is then

$$\begin{aligned} dP(x) &= \beta^{-2}(x - \langle x \rangle)[\beta\, dW + (x_{\text{true}} - \langle x \rangle)\, dt]\, P(x) \\ &= \beta^{-2}(x - \langle x \rangle)(dy - \langle x \rangle\, dt) P(x). \end{aligned} \tag{3.22}$$

In the last line we have written the differential equation in terms of dy. It is essential that the equation can be written in this form, since it is only dy that the observer has access to. If you are not familiar with writing differential equations in this fashion – that is, using differentials instead of derivatives – then refer to chapter 2 (and/or chapter 3) in reference [288].

It is worth noting that there is an alternative way to write the measurement result dy. To do this we note first that the expectation value of dy is

$$\langle dy \rangle = \langle x_{\text{true}} \rangle\, dt + \beta \langle dW \rangle = \left[\int_{-\infty}^{\infty} x P(x)\, dx \right] dt = \langle x \rangle\, dt. \tag{3.23}$$

Secondly, dy has a Gaussian distribution, because (in the infinitesimal limit) the variance of dW is infinitely larger than the variance of $x\, dt$, and thus dominates the distribution. We can therefore obtain the correct distribution for dy by setting

$$dy = \langle x \rangle\, dt + \beta\, dV, \tag{3.24}$$

where dV has the same distribution as dW (dV is Wiener noise, just like dW). The only difference between dW and dV is that in any given realization of the measurement, since x_{true} will not usually be equal to $\langle x \rangle$, the sampled values of dW will be different to those of dV: in each time-step

$$dV = dW + (x_{\text{true}} - \langle x \rangle)\, dt/\beta. \tag{3.25}$$

Note: while it is clear that $\langle x_{\text{true}} \rangle = \langle x \rangle$ at $t = 0$, it is not quite so clear that this remains true as data is obtained. We will not give a detailed analysis of this fact here, but by playing with conditional probabilities you should be able to convince yourself that it is true. The crucial point is that the two definitions of dy must generate the same set of possible realizations for the measurement signal with the same probability distribution.

Equation (3.22) is called the *Kushner–Stratonovich equation*, and along with Eqs. (3.13) and (3.24), constitutes the theory of classical continuous measurement. This theory is used extensively in classical control theory; the Kalman filter, for example, is the special case of the Kushner–Stratonovich (K-S) equation for linear systems [414]. It is relatively straight-forward to extend the derivation of the K-S equation to the general case in which one is measuring a vector of variables, where this vector is the state of a dynamical system possi-bly driven by noise, and in which the measurement noise (the measurement error) in each measurement may be correlated across the various variables. This general version of the K-S equation is given in Appendix G.

3.1.2 Gaussian quantum continuous measurements

To describe continuous quantum measurements, we do the same thing as we did above for classical measurements. The reason we qualify the resulting quantum measurements as "Gaussian" is that there are natural continuous quantum measurements in which the random component of the measurement result is not Gaussian, and we will discuss these in Section 3.3.

As in our classical analysis above, we make a measurement in each time-step Δt. We use the same likelihood function (recall from Chapter 1 that the likelihood function defines the set of measurement operators), but this time the measured quantity is a quantum observ-able, X. This means that we choose a quantum measurement for which the measurement result, dy, is related to X by

$$dy = \langle X \rangle \, dt + \frac{dW}{\sqrt{8k}}. \tag{3.26}$$

Here we have defined $k = 1/(8\beta^2)$ for future convenience. We will usually refer to the noise on the measurement result, the term $dW/\sqrt{8k}$ in the above equation, as the *measurement noise*, or alternatively as the *shot noise* or *imprecision noise*.

Recall that, in terms of k the likelihood function for the classical measurement in each time-step was

$$P(dy|x) = \sqrt{\frac{4k}{\pi \, dt}} \exp\left\{ -4k \frac{(dy - x \, dt)^2}{dt} \right\}, \tag{3.27}$$

where dy is the measurement result, and x is the measured variable. The set of quantum measurement operators corresponding to this likelihood function (see Section 1.4.1) are

$$
\begin{aligned}
A(dy) &= \int_{-\infty}^{\infty} \sqrt{P(dy|x)} \, |x\rangle \langle x| \, dx \\
&= \left(\frac{4k}{\pi \, dt} \right)^{1/4} \int_{-\infty}^{\infty} \exp\left\{ -2k \frac{(dy - x \, dt)^2}{dt} \right\} |x\rangle \langle x| \, dx \\
&= \left(\frac{4k}{\pi \, dt} \right)^{1/4} \exp\left\{ -2k \frac{(dy - X \, dt)^2}{dt} \right\},
\end{aligned}
\tag{3.28}
$$

where $|x\rangle$ is an eigenstate of X, defined by $X|x\rangle = x|x\rangle$. Each operator $A(dy)$ is a Gaussian-weighted sum of projectors onto the eigenstates of X.

First let us verify that the measurement result dy is indeed as given by Eq. (3.26). If the initial state $|\psi\rangle = \int \psi(x)|x\rangle\,dx$ then

$$P(dy) = \text{Tr}\left[A(dy)^\dagger A(dy)|\psi\rangle\langle\psi|\right] = \text{Tr}\left[A^2(dy)|\psi\rangle\langle\psi|\right]. \qquad (3.29)$$

Simplifying the notation by defining $\alpha \equiv dy$, the expectation value of dy is therefore

$$\langle dy\rangle = \int_{-\infty}^{\infty} \alpha\, P(\alpha)\,d\alpha = \int_{-\infty}^{\infty} \alpha\,\text{Tr}\left[A^2(\alpha)|\psi\rangle\langle\psi|\right]d\alpha$$

$$= \sqrt{\frac{4k}{\pi\,dt}} \int_{-\infty}^{\infty}\left[\int_{-\infty}^{\infty} \alpha\,e^{-4k(\alpha-x\,dt)^2/dt}\,d\alpha\right]|\psi(x)|^2\,dx$$

$$= dt\int_{-\infty}^{\infty} x|\psi(x)|^2\,dx = \langle X\rangle\,dt \qquad (3.30)$$

as required. To obtain $P(dy)$ we now write

$$P(dy) = \text{Tr}\left[A^2(dy)|\psi\rangle\langle\psi|\right] = \sqrt{\frac{4k}{\pi\,dt}} \int_{-\infty}^{\infty} |\psi(x)|^2 e^{-4k(dy-x\,dt)^2/dt}\,dx. \qquad (3.31)$$

Since dt is infinitesimal, the Gaussian in the integral is much broader than $\psi(x)$. This means we can approximate $|\psi(x)|^2$ by a delta function. This delta function must be centered at the expectation value of X so that $\langle dy\rangle = \langle X\rangle\,dt$ as calculated above. We therefore have

$$P(dy) \approx \sqrt{\frac{4k}{\pi\,dt}} \int_{-\infty}^{\infty} \delta(x - \langle X\rangle)\,e^{-4k(dy-x\,dt)^2/dt}\,dx$$

$$= \sqrt{\frac{4k}{\pi\,dt}}\,e^{-4k(dy-\langle X\rangle\,dt)^2/dt}, \qquad (3.32)$$

and this is precisely the distribution for dy when dy is given by Eq. (3.26).

We can now easily derive the stochastic differential equation that describes how the quantum state $|\psi\rangle$ changes in each time-step dt. We apply the measurement operator $A(dy)$ to $|\psi\rangle$, and expand the exponential to first order in dt. For ease of calculation we drop all parts of $A(dy)$ that change only the scaling of $|\psi\rangle$, since we can normalize $|\psi\rangle$ afterwards. (For example, we drop the factor $\exp[-2k(dy)^2/dt]$ that appears when we multiply out the square in the exponential. This affects only the scaling of $|\psi\rangle$ because it contains only numbers, not operators such as X.) Doing this, and using Ito's rule to expand the exponential to first order in dt, we obtain

$$|\hat\psi(t+dt)\rangle \propto A(dy)|\psi(t)\rangle$$

$$\propto \{1 - [kX^2 - 4kX\langle X\rangle]\,dt + \sqrt{2k}X\,dW\}|\psi(t)\rangle. \qquad (3.33)$$

Here the "hat" over ψ indicates that the state is not normalized. To determine the evolution equation for $|\psi\rangle$, we must normalize $|\hat{\psi}(t+dt)\rangle$. We do this by first calculating the square norm

$$\langle\hat{\psi}(t+dt)|\hat{\psi}(t+dt)\rangle = 1 + 8k\langle X\rangle X\,dt + \sqrt{8k}X\,dW, \qquad (3.34)$$

and then taking the square root of this (to first order in dt) using the binomial expansion

$$(1+x)^p = 1 + px + \frac{p(p-1)}{2}x^2 + \cdots. \qquad (3.35)$$

Dividing $|\psi(t+dt)\rangle$ by its norm, and expanding to first order in dt gives the stochastic differential equation for $|\psi\rangle$, which is

$$d|\psi\rangle = \{-k(X - \langle X\rangle)^2\,dt + \sqrt{2k}(X - \langle X\rangle)\,dW\}|\psi(t)\rangle. \qquad (3.36)$$

This is called a *stochastic Schrödinger equation* (SSE). Note that while the Schrödinger equation is linear, the stochastic Schrödinger equation is not, due to the term containing $\langle X\rangle$; once one adds measurements to the dynamics of a quantum system it becomes nonlinear.

Of course, in general the observer's state-of-knowledge will be a density matrix, not a pure state. From the SSE we can immediately obtain the equation of motion for the density matrix, which is called a *stochastic master equation* (SME). To obtain the SME we write the evolution of ρ in terms of the evolution of $|\psi\rangle$:

$$\rho(t+dt) = \rho + d\rho = (|\psi\rangle + d|\psi\rangle)(\langle\psi| + d\langle\psi|)$$
$$= \rho + (d|\psi\rangle)\langle\psi| + |\psi\rangle(d\langle\psi|) + (d|\psi\rangle)(d\langle\psi|). \qquad (3.37)$$

Thus

$$d\rho = (d|\psi\rangle)\langle\psi| + |\psi\rangle(d\langle\psi|) + (d|\psi\rangle)(d\langle\psi|)$$
$$= -k[X,[X,\rho]]\,dt + \sqrt{2k}(X\rho + \rho X - 2\langle X\rangle\rho)\,dW. \qquad (3.38)$$

If the quantum system is evolving under the Hamiltonian H while it is being measured, then the full evolution is given simply by adding the Hamiltonian evolution in the time-step dt to the evolution induced by the measurement. The two contributions to the evolution add together because, since the Hamiltonian contribution is first-order in dt, any complications that could arise from the fact that H does not commute with X are *higher* than first order, and thus make no contribution to the solution of the differential equation.

The full evolution of a quantum system with Hamiltonian H, subjected to a continuous measurement of the observable X, is given by the stochastic master equation

$$d\rho = -\frac{i}{\hbar}[H,\rho]\,dt - k[X,[X,\rho]]\,dt + \sqrt{2k}(X\rho + \rho X - 2\langle X\rangle\rho)\,dW. \qquad (3.39)$$

The constant k is called the measurement strength, and it scales the rate at which the measurement extracts information. This equation, together with Eq. (3.26), is one of the primary equations of continuous quantum measurement theory. It is the quantum equivalent of the classical Kushner–Stratonovich equation.

Note that the SME has many possible final states. The final state depends on the value that dW takes at each time-step. A given sequence of sample values of dW for a given time period is called a *realization* of the noise. The sequence of states that results (that is, the evolution of $\rho(t)$ over that time period) is called a *quantum trajectory*. Each possible trajectory, and thus each final state, has a different probability, being the probability of the corresponding noise realization. The solution to the SME at some final time T is thus a probability distribution over a continuum of possible final states. We will see an example of such a solution in Section 3.2.

The observer's initial state of knowledge

In our discussion so far, we have assumed implicitly that the initial state-of-knowledge of the observer, $P(x)$ in the classical case, and ρ in the quantum case, is equal to the true probability distribution (or density matrix) for the state of the system. So what happens if this is not the case? Consider first a classical measurement of a random variable x. For example, the observer might think that x has a Gaussian distribution with a certain mean and variance, but perhaps someone is actually choosing the value of x to be always either 0 or 1. It is clear what happens in this case: in any given realization of the measurement, x always has some fixed value regardless of how this fixed value is chosen. So the measurement will cause the observer's state-of-knowledge to become increasingly peaked about the true value regardless of how these values are really chosen. This is simply a reflection of that fact that as more data is obtained, the observer's prior becomes increasingly irrelevant, and the observer's state-of-knowledge is determined by the data. The one exception is when the observer's initial state-of-knowledge is zero for the true value of the variable. That is, if the observer's state-of-knowledge states that the value of a variable can *never* be in some range, then the K-S equation can never produce a state-of-knowledge peaked about those values. So in choosing a prior state-of-knowledge, it is important that this prior be non-zero for all values that the variable could take.

For quantum measurements the equivalent reasoning must also hold, or else quantum measurement theory would be fundamentally flawed. That is, so long as the observer's initial state ρ gives a non-zero probability for the system to be in any pure state, then the measurement will (on average) continually evolve ρ toward the true state of the system. Of course, in quantum mechanics, the measurement changes the true state of the system, so the observer's state-of-knowledge evolves in parallel with the true state. The statement that the prior becomes increasingly irrelevant as data is obtained is therefore equivalent to saying that these two states converge together. Further, any two initial states-of-knowledge, so long as they are consistent with the true state, will converge together as data is obtained.

That is,

$$\lim_{t\to\infty} \rho_1(t) = \lim_{t\to\infty} \rho_2(t) \quad \text{given } y(t). \tag{3.40}$$

Obtaining a rigorous proof of this result for a stochastic master equation is mathematically non-trivial, even though it is intuitively clear that such a result must hold. Few rigorous results in this direction exist as yet, although a proof by van Handel goes some way to confirming Eq. (3.40) [626]. Similar results for classical continuous measurements are given in [455].

One can also go some way to understanding why quantum measurements satisfy Eq. (3.40) from the following simple observation. Note that the result of a continuous measurement performed for some time t is merely a quantum measurement, and is therefore captured by some set of measurement operators A_n. If a continuous measurement converts an initially mixed state into a final state that is nearly pure, then the corresponding measurement operators A_n must be close to projectors onto pure states. Trivially, a projector onto a pure state projects *every* initial state onto the *same* pure state (so long as these initial states have a non-zero overlap with the projector). While this makes it fairly obvious that SMEs that produce pure states as $t \to \infty$ will satisfy equation Eq. (3.40), this is not the end of the story. Many SMEs involve sources of noise that compete with the measurement process, so that the observer's state-of-knowledge does not become perfectly pure over time. Still, in these cases, the consistency of quantum measurement theory requires that all initial states must converge for a given measurement record as $t \to \infty$.

If we believe that quantum measurement theory is consistent as it stands, then it follows that in making quantum measurements, and thus in performing continuous parameter estimation (Section 3.8) or quantum feedback control (Chapter 5), the observer does not need to know the "true" density matrix for the system. She merely needs to make sure that all possible initial states are given some non-zero probability. However, the observer should try to make her initial state a reasonably accurate reflection of reality. The reason is that the closer the observer's prior is to the true density matrix, the faster on average will her state collapse to the true state. This is because in this case the observer will assign high initial probabilities to those states that actually do appear more frequently. Another way to say this is that on fewer occasions will the measurement signal have to radically change the observer's state-of-knowledge, a process which takes time.

The efficiency of a measurement

Real measuring devices are not as perfect as we have been imagining so far. As we discussed in Chapter 1, a measurement involves a probe system interacting with the system to be measured and carrying away some information about it. The probe can then be measured to obtain this information. Once the system has interacted with the probe, it has been affected by the measurement procedure. Even if the probe is left unmeasured, the state of the system is now given by tracing out the probe, and because the system is correlated with the probe, this new state is not the same as the original state.

Of course, the amount of information that is obtained about the system *does* depend on whether or not the probe is measured. If the measuring device does not manage to extract all the information from the probe, then the measurement is described as *inefficient*. One can think of this terminology in the following way. Extracting information is desirable, but the effect on the system (the measurement "back-action") may not be. Inefficient measurements are inefficient in the sense that they do not get all the information that is possible for the amount of back-action that they induce.

For continuous measurements, inefficiency can be described in the following way. We first divide all the time intervals Δt into two equal parts. We now make a measurement with strength k_1 in the first half of every time interval, and make the same measurement, but with the measurement strength k_2, in the second half of every time interval. The resulting stochastic equation for the system is

$$dp = -k_1[X[X,\rho]]\,dt + \sqrt{2k_1}(X\rho + \rho X - 2\langle X\rangle\rho)\,dW_1$$
$$-k_2[X[X,\rho]]\,dt + \sqrt{2k_2}(X\rho + \rho X - 2\langle X\rangle\rho)\,dW_2, \qquad (3.41)$$

and the measurement signals for the two measurements are

$$dy_1 = \text{Tr}[X\rho]\,dt + \frac{dW_1}{\sqrt{8k_1}} \qquad (3.42)$$

$$dy_2 = \text{Tr}[X\rho]\,dt + \frac{dW_2}{\sqrt{8k_2}}. \qquad (3.43)$$

Since it is the variances of the noise increments that are proportional to $\Delta t/2$, and thus sum together to give the variance of the increment in each full time interval Δt, the two interlaced measurements are completely equivalent to a single measurement of X with measurement strength

$$k = k_1 + k_2 \qquad (3.44)$$

and noise increment

$$dW = \sqrt{\frac{k_1}{k}}\,dW_1 + \sqrt{\frac{k_2}{k}}\,dW_2. \qquad (3.45)$$

To describe an inefficient measurement we now give the observer access only to the results of the measurement with strength k_1, and not to the results of the other measurement. The observer, who we will call Alice, must then average over the results of the other measurement, and so her SME becomes

$$d\tilde\rho = -(k_1 + k_2)[X[X,\tilde\rho]]\,dt + \sqrt{2k_1}(X\tilde\rho + \tilde\rho X - 2\langle X\rangle\tilde\rho)\,dV, \qquad (3.46)$$

and her measurement signal is

$$dy = \text{Tr}[X\tilde\rho]\,dt + \frac{dV}{\sqrt{8k_1}}, \qquad (3.47)$$

where dV is, as usual, some Wiener noise increment (but we will see shortly that this increment is not equal to dW_1). We have used the tilde to distinguish Alice's state-of-knowledge from that of an observer with access to both measurements. We will call the latter observer Charlie.

Let us now explain why we have replaced dW_1 by dV in Eq. (3.46). Since Alice's state-of-knowledge is different to Charlie's, her measurement record involves a different average, namely $\mathrm{Tr}[X\tilde{\rho}]$, from Charlie's $\mathrm{Tr}[X\rho]$. Of course, Alice's and Charlie's measurement results must have the same probability distribution, or our description would be inconsistent. They have the same probability distribution because Alice's state-of-knowledge, $\tilde{\rho}$, is averaged over all the possible measurement results that Charlie could get for dy_2, and hence takes into account all the possible probability distributions for Charlie's dy_1 on the next time-step. Alice's distribution for dy_1 is a coarse-grained version of Charlie's. Further, on any given realization of the measurement, Alice's results dy will be equal to Charlie's dy_1. But since Alice's mean is different from Charlie's, when we equate dy with dy_1, on any given run the realization of dV is not the same as dW_1:

$$dV = \sqrt{8k_1}(\mathrm{Tr}[X\rho] - \mathrm{Tr}[X\tilde{\rho}])\,dt + dW_1. \tag{3.48}$$

The SME for an inefficient measurement is usually written in the form

$$d\rho = -k[X[X,\rho]]\,dt + \sqrt{2\eta k}(X\rho + \rho X - 2\langle X\rangle\rho)\,dW, \tag{3.49}$$

where η is called the *measurement efficiency*, and lies between 0 and 1. Comparing this form with that given in Eq. (3.46), we see that

$$k = k_1 + k_2 \quad \text{and} \quad \eta = \frac{k_1}{k_1 + k_2}. \tag{3.50}$$

Simultaneous measurements and multiple observers

Now that we know how to treat inefficient detection, treating simultaneous measurements by multiple observers is straightforward. We can describe two simultaneous measurements of different observables X and Y by interlacing continuous measurements of each in the same way that we did for inefficient detection. The reason that this is reasonable is because, even though X and Y may not commute, the commutator of $X\,dt$ and $Y\,dt$ is second-order in dt, and thus vanishes in the continuum limit. Further, the noise increments from the measurements of X and Y are independent Wiener noises, and so their product also vanishes in the continuum limit. Because of these things, it does not matter in which order we make our measurements of X and Y in any interval Δt. If the order does not matter, then there cannot be any difference between a "simultaneous" measurement of X and Y in any interval Δt, and one that is interleaved. If you are still concerned with the correctness of this approach, the matter is considered in detail in [553].

The SME for simultaneous measurements of X and Y is thus

$$dp = -k_1[X[X,\rho]]dt + \sqrt{2k_1}(X\rho + \rho X - 2\langle X\rangle\rho)dW_1$$
$$- k_2[Y[Y,\rho]]dt + \sqrt{2k_2}(Y\rho + \rho Y - 2\langle Y\rangle\rho)dW_2, \tag{3.51}$$

and the measurement signals for the two measurements are

$$dy_1 = \text{Tr}[X\rho]dt + \frac{dW_1}{\sqrt{8k_1}} \tag{3.52}$$

$$dy_2 = \text{Tr}[Y\rho]dt + \frac{dW_2}{\sqrt{8k_2}}. \tag{3.53}$$

The state $\rho(t)$ is the state-of-knowledge of an observer Charlie who has access to all the measurement results. We can now consider two observers: Alice, who has access only to the measurement signal dy_1, and an observer Bob, who has access only to dy_2. Using the same reasoning as in our treatment of inefficient measurements above, the SMEs for Alice and Bob are respectively

$$d\rho_1 = -k_1[X[X,\rho_1]]dt + \sqrt{2k_1}(X\rho_1 + \rho_1 X - 2\langle X\rangle\rho_1)dV_1, \tag{3.54}$$
$$d\rho_2 = -k_2[Y[Y,\rho_2]]dt + \sqrt{2k_2}(Y\rho_2 + \rho_2 Y - 2\langle Y\rangle\rho_2)dV_2, \tag{3.55}$$

with the measurement signals

$$dy_1 = \text{Tr}[X\rho_1]dt + \frac{dV_1}{\sqrt{8k_1}}, \tag{3.56}$$

$$dy_2 = \text{Tr}[Y\rho_2]dt + \frac{dV_2}{\sqrt{8k_2}}. \tag{3.57}$$

Once again following the arguments in the previous section, equating Alice's and Bob's measurement signals with Charlie's, the relationship between the noise increments on any given realization are

$$dV_1 = \sqrt{8k_1}(\text{Tr}[X\rho] - \text{Tr}[X\rho_1])dt + dW_1, \tag{3.58}$$
$$dV_2 = \sqrt{8k_2}(\text{Tr}[Y\rho] - \text{Tr}[Y\rho_2])dt + dW_2. \tag{3.59}$$

Continuous measurements and physical measuring devices

We have now derived the equations that give the evolution of a quantum system under a natural kind of continuous measurement. What we have not yet done is to see how this evolution is obtained from an analysis of a real measuring device interacting with a quantum system. In the classical case we do not feel any particular need to do so – we realize that the measurement error in each time-step is due to random forces, and it seems unlikely that elucidating these forces would provide further theoretical insight. In the quantum mechanical case it is not nearly so clear how to analyze a measuring device that continually monitors a system, or exactly where the measurement errors will come from. This is at least partly because some of these errors will come from the fundamental and irreducible uncertainties inherent in all quantum states, and the way this plays out through the interaction of the measuring device with the system.

There is a further subtlety with analyzing measurement devices. Note that when we realize a general measurement on a quantum system, achieved using a probe system, we assume that a von Neumann measurement is made on the probe. However, the actual "collapse" (projection) of the wavefunction given by the postulated von Neumann measurement does not actually correspond to the application of any physical forces. This statement will be clarified in Section 4.5, but for now all we need to know is that this lack of a direct connection between the action of a von Neumann measurement, and the forces on a quantum system (captured by the Hamiltonian), does not cause us any problems. There is still a clear prescription as to when one is allowed to apply the von Neumann projection postulate. Once the system has interacted with the probe, and we can be confident that this interaction will not be undone (to be explained further in Section 4.3), then we can apply the von Neumann measurement operators. The reason for this is that, once the interaction is complete, the state of the system is given by tracing over the probe, which, as shown in Section 1.3.3, puts it in a mixture of the various possible outcomes of the von Neumann measurement. As a result we are safe to pick one of these states as the one that has "actually occurred"; all future predictions about the behavior of the system, and those systems with which it interacts in the future, will be consistent with it being in one of the measurement outcomes.

We do not wish to break the flow here to analyze a physical measuring apparatus, but we will give the details of experimental implementations of continuous measurements in Sections 3.3, 3.4, and 7.7.3. At the end of this chapter you will also find a list of references that treat specific realizations of continuous measurements.

3.1.3 When the SME is the classical Kalman–Bucy filter

When the dynamical variables of a quantum system are canonical coordinates (a set of positions and momenta), and the equations of motion for these coordinates are linear, then the quantum dynamics is the same as that of a linear classical system. If one continuously measures a linear combination of the coordinates and momenta of a linear quantum system, and the initial state is Gaussian in the coordinates, then the dynamics is still reproduced by an equivalent measured classical system. In this case the classical system must be driven by the right amount of noise, so as to preserve Heisenberg's uncertainty relation. We will discuss this equivalence in more detail in Section 5.5.2. Here we can quite quickly make the connection between continuously measured linear quantum and classical systems, state-estimation, and *filters*.

If the initial state of a linear quantum or classical system is Gaussian in the coordinates, then the state remains Gaussian throughout the evolution. Because of this fact the stochastic master equation (SME) reduces to equations of motion for the means and variances of the

Gaussian state. As an example, if the system is a single harmonic oscillator, and the position is continuously measured, then the SME is

$$d\langle x \rangle = \frac{\langle p \rangle}{m} dt + \sqrt{8k}\, V_x\, dW, \tag{3.60}$$

$$d\langle p \rangle = -m\omega^2 \langle x \rangle\, dt + \sqrt{8k}\, C_{xp}\, dW, \tag{3.61}$$

and

$$\dot{V}_x = (2/m)C_{xp} - 8kV_x^2, \tag{3.62}$$

$$\dot{V}_p = 2\hbar^2 k - 8kC_{xp}^2 - 2m\omega^2 C_{xp}, \tag{3.63}$$

$$\dot{C}_{xp} = (1/m)V_p - 8kV_x C_{xp} - m\omega^2 V_x \tag{3.64}$$

where ω is the oscillator frequency, m is its mass, k is the measurement strength, V_x and V_p are the variances of the position and momentum, and $C_{xp} = \langle XP + PX \rangle/2 - \langle X \rangle\langle P \rangle$ is the symmetric covariance. The key features of these equations of motion are (i) that those for the variances are not coupled to the means, are completely deterministic, and reach a steady-state, and (ii) once the variances have reached a steady state, the equations of motion for the estimates of the means are just the (linear) equation of motion for the system, with fixed driving noise.

The above equations of motion are also the *classical* equations of motion for the estimates of the means and variances for a noise-driven oscillator whose dynamics is given by

$$dx = \frac{p}{m}\, dt, \tag{3.65}$$

$$dp = -m\omega^2 x\, dt + \sqrt{2k\hbar}\, dW_c, \tag{3.66}$$

and whose position is continuously measured with strength k. The Wiener noise driving the oscillator, dW_c, is uncorrelated with the measurement noise, dW. In the context of classical continuous measurement, the equations of motion for the estimates are referred to as the *Kalman–Bucy filter* [324, 414].

The state-estimation equations of motion, Eqs. (3.60)–(3.64), are referred to as a filter for the following reason. A filter is defined as a device that takes an input signal and gives an output whose power spectrum is that of the input multiplied by some function of the frequency (for readers unfamiliar with the power spectrum we give an introduction below). The spectrum of the output is $S_{out}(\omega) = G(\omega)S_{in}(\omega)$, where $S_{in}(\omega)$ is the spectrum of the input, and $G(\omega)$ is called the *transfer* function of the filter. A linear dynamical system can be used as a filter. To do so, one drives the system with the input signal, and then takes the output signal to be one of the dynamical variables of the system. The relationship between the input and output is then given exactly by the filter relationship, for some transfer function $G(\omega)$. Since the measurement record is what drives the equations of the motion for the estimates $\langle x \rangle$ and $\langle p \rangle$, the estimates are a *filtered* version of the measurement record. To see this directly, we write the estimator equations in terms of the measurement record. Since $dW = \sqrt{8k}(dy - \langle x \rangle\, dt)$, the estimator equations become

$$d\begin{pmatrix} \langle x \rangle \\ \langle p \rangle \end{pmatrix} = \begin{pmatrix} -8k & 1/m \\ -m\omega^2 & 0 \end{pmatrix} \begin{pmatrix} \langle x \rangle \\ \langle p \rangle \end{pmatrix} dt - 8k \begin{pmatrix} V_x^{\mathrm{ss}} \\ C_{xp}^{\mathrm{ss}} \end{pmatrix} dy. \qquad (3.67)$$

where V_x^{ss} and C_{xp}^{ss} are the steady-state values for the respective variances. This form makes it clear that the estimator equations are simply linear differential equations driven by the measurement record, y. We can write the estimates $\langle x \rangle$ and $\langle p \rangle$ explicitly in terms of the measurement record. If we write Eq. (3.67) more compactly as $\dot{\mathbf{x}} = A\mathbf{x} + \mathbf{v}y(t)$, then the solution is

$$\mathbf{x}(t) = e^{At}\mathbf{x}(0) + \int_0^t e^{A(t-t')}\mathbf{v}y(t')\, dt'. \qquad (3.68)$$

This solution shows that the means, $\mathbf{x}(t)$, are given by convolving the measurement record with a (vector) function of time, and this is the time-domain representation of the action of a filter [25].

3.1.4 The power spectrum of the measurement record

If we calculate the *Fourier transform* of a time-varying function, or *signal*, then we obtain a representation of the function in terms of the frequencies that it contains. Fourier transforms can also be used to determine how the time-averaged variance of a noisy signal is made up from the variance of the different frequency components. The variance of a signal as a function of frequency is called the *power spectral density* of the signal, or simply the *power spectrum*. The reason for this name is that if the mean of a signal is zero, then the variance is the average of the square of the signal, and for electrical signals this square is the power in the signal.

For a linear classical system, defined as one in which the equation of motion for the variables is linear, one can calculate the power spectrum of the measurement record analytically. This is also true for a linear measurement on a linear quantum system. Recall that a linear quantum system is a system whose dynamical variables are canonical coordinates, and for which the Heisenberg equations of motion of these coordinates are linear. A linear measurement is a measurement of a linear combination of the canonical coordinates. There is more than one way to calculate the power spectrum of a continuous measurement of a linear quantum system. As mentioned in the section above, if a Hermitian operator is being measured there is an equivalent classical model of the dynamics, and this can be used to determine the measurement record. One can alternatively use the Heisenberg-picture, or *input–output* formulation of the measured system, to calculate the power spectrum, using essentially the same algebraic steps. This latter method is valid for measurements involving non-Hermitian operators such as homodyne detection, and we describe this method in

Section 3.6. Here we define the Fourier transform and the power spectrum of a noisy signal, and show how to calculate the power spectrum for a measurement of a linear quantum system using the equivalent classical model.

The power spectrum

Throughout this text we will use the following definition for the Fourier transform. For a function $f(t)$, the Fourier transform is

$$F(\omega) = \frac{1}{\sqrt{2\pi}} \int_{-\infty}^{\infty} f(t)e^{-i\omega t}\,dt. \tag{3.69}$$

The inverse transform is then given by

$$f(t) = \frac{1}{\sqrt{2\pi}} \int_{-\infty}^{\infty} F(\omega)e^{i\omega t}\,d\omega. \tag{3.70}$$

The form of the inverse transform shows that $F(\omega)$ gives the amplitude and phase with which the sine wave with frequency ω contributes to the signal $f(t)$. Note that the Fourier transform of $f(t) = 1$ is the Dirac delta-function, $\delta(\omega)$, satisfying $\int_{-\infty}^{\infty} \delta(\omega)d\omega = 1$.

Now consider what happens if $f(t)$ is noisy, so that it contains a random component. In this case we can think of $f(t)$ as a *random function*, meaning that it represents a set of possible functions, where there is a probability distribution over these functions. In this case the Fourier transform of $f(t)$ is also a random function, where each of its possible values, which are functions, is the Fourier transform of one of the possible functions for $f(t)$. Expressions like $\langle f(t) \rangle$ and $\langle F(\omega) \rangle$ therefore make sense.

The average power in a random signal, $f(t)$, is defined as

$$P = \lim_{T \to \infty} \frac{1}{T} \int_{T/2}^{T/2} \left\langle f^2(t) \right\rangle dt, \tag{3.71}$$

where the average is taken over the set of possible functions for $f(t)$. This definition comes to us from electronics, in which the power is proportional to the square of the voltage. But the definition is useful more generally because it is the mean square (the second "moment") of $f(t)$ averaged over time. If the average value of the function is zero, the average power is the variance of the function averaged over time.

The *power spectrum* of a real random signal $f(t)$ is a function, $S(\omega)$, that tells us how the total power in a signal – the mean-square of $f(t)$ – is distributed over the different frequencies that make up the signal. Usually this would mean that the power in a frequency band $[\omega_1, \omega_2]$ is the integral of S over this band, but this is not quite the case. Recall that for a real signal, $f(t)$, the Fourier transform $F(\omega)$ has both negative and positive frequencies, and $F(\omega) = F^*(-\omega)$. This means that for a real signal, the positive frequencies must also come with negative frequencies in the same amount, in the Fourier representation. The power spectrum is similarly defined to have the positive and negative frequencies, and the power is similarly equally divided between the positive and negative frequencies. Thus

$S(\omega)$ is symmetric about $\omega = 0$. In addition, the spectrum is defined so that it is scaled by 2π (this scaling gives it a simple relationship to the transfer function of a system – see Eq. (3.82) below). So the contribution to the total power that is contained in the band $[\omega_1, \omega_2]$ is in fact given by

$$P_{[\omega_1,\omega_2]} = \frac{1}{2\pi} \int_{\omega_1}^{\omega_2} [S(\omega) + S(-\omega)] \, d\omega = \frac{1}{\pi} \int_{\omega_1}^{\omega_2} S(\omega) \, d\omega. \tag{3.72}$$

If the mean of a signal $f(t)$ is independent of time, and the two-time correlation function, defined as

$$g(\tau) \equiv \langle f(t)f(t+\tau) \rangle, \tag{3.73}$$

depends only on the time difference, τ, then the signal is called *wide-sense stationary*. It turns out that if $f(t)$ is wide-sense stationary, then the power spectrum is related to f and its Fourier transform, $F(\omega)$, in two very useful ways. The first, called the Wiener–Kinchin theorem, is

$$S(\omega) = \int_{-\infty}^{\infty} g(\tau)e^{-i\omega\tau} \, d\tau, \quad g(\tau) = \langle f(t)f(t+\tau) \rangle. \tag{3.74}$$

This equation states that the power spectrum is the Fourier transform of the autocorrelation function. Because the factor of $\sqrt{1/(2\pi)}$ that we use in our definition of the Fourier transform is missing from the expression for $S(\omega)$, the inverse relationship is

$$g(\tau) = \frac{1}{2\pi} \int_{-\infty}^{\infty} S(\omega)e^{i\omega\tau} \, d\omega. \tag{3.75}$$

The steady-state mean-square of f, being the total power in f, is therefore

$$\langle [f(t)]^2 \rangle = g(0) = \frac{1}{2\pi} \int_{-\infty}^{\infty} S(\omega) \, d\omega = \frac{1}{\pi} \int_{0}^{\infty} S(\omega) \, d\omega. \tag{3.76}$$

The second useful relationship satisfied by the power spectrum is as follows. For any wide-sense stationary signal $f(t)$,

$$\langle F(\omega)F(\omega') \rangle = G(\omega, \omega')\delta(\omega + \omega') = G(\omega, -\omega)\delta(\omega + \omega') = S(\omega)\delta(\omega + \omega'), \tag{3.77}$$

for some function $G(\omega, \omega')$, where $F(\omega)$ is the Fourier transform of $f(t)$. We will not derive the above relationships here, but the reader can find more information about Fourier transforms, autocorrelations, and power spectra in [591, 272, 416].

The nature of the measurement record: white noise

Recall that we defined the measurement record $r(t)$ by its increment, $dr = \langle x \rangle \, dt + dW/\sqrt{8k}$. In fact, we used infinitesimal increments rather than derivatives to describe continuous measurement because of the rather pathological nature of the Wiener process. Since the size of the Wiener increment is proportional to $\sqrt{dt}$, if we try to calculate its derivative we find that $\langle dW/dt \rangle \sim 1/\sqrt{dt}$, and this tends to infinity as $dt \to 0$. The Wiener process has no derivative, reflecting the fact that it changes randomly, and is thus jagged, on even the smallest timescale.

So how do we handle the fact that the measurement record, $r(t)$, is given by

$$r(t) = \langle x(t) \rangle + \frac{1}{\sqrt{8k}} \xi(t), \tag{3.78}$$

where $\xi(t) \equiv dW/dt$ is the derivative of the Wiener process? The Wiener process is actually an idealization, as no real process changes on infinitely short timescales. It is the fact that we have defined each increment of the Wiener process to be completely independent of the previous increment that makes the Wiener process so jagged. If we demand that it be smooth below some timescale Δt, then the increments for shorter times could not be independent of each other.

Let us discretize time into increments of length Δt, and denote the Wiener increment in the nth interval by ΔW_n; then the correlation function of $\xi(t)$ is given by

$$\langle \xi(n\,\Delta t)\xi(m\,\Delta t) \rangle = \left\langle \frac{\Delta W_n}{\Delta t} \frac{\Delta W_m}{\Delta t} \right\rangle = \frac{\langle \Delta W_n \Delta W_m \rangle}{(\Delta t)^2} = \frac{\delta_{nm}}{\Delta t}. \tag{3.79}$$

So as $\Delta t \to 0$ this correlation function becomes the Dirac δ-function:

$$\langle \xi(t)\xi(t+\tau) \rangle = \delta(\tau). \tag{3.80}$$

Since the spectrum is the Fourier transform of the autocorrelation function, the spectrum of $\xi(t)$ is $S(\omega) = 1$. Because of this flat spectrum $\xi(t)$ is called *white noise*. Such a spectrum further implies that (i) $\xi(t)$ contains infinitely high frequencies, and (ii) it has infinite power.

Since the measurement noise, $\xi(t)$, is uncorrelated with the expectation value of the measured observable, the autocorrelation function of the measurement record is

$$r(t) = \langle x_{\mathrm{av}}(t+\tau)x_{\mathrm{av}}(t) \rangle + \frac{\delta(\tau)}{8k} \quad \text{with} \quad x_{\mathrm{av}}(t) \equiv \langle x(t) \rangle = \mathrm{Tr}[x\rho(t)]. \tag{3.81}$$

Because of the Fourier transform relationship between the autocorrelation function and the power spectrum, the narrower the autocorrelation function, the broader the spectrum. Thus $\xi(t)$ corresponds to the limiting case in which the correlation function is infinitely narrow, and the spectrum infinitely broad. So why is it that an impossible signal, something with infinite power, is a useful approximation to real noise? The answer has to do with the phenomenon of resonance. Recall that if a harmonic oscillator is driven at a frequency much higher than its own frequency, the driving force hardly moves the oscillator at all. It is the same with all dynamical systems. Every system has some range of frequencies over which it will respond, and signals with frequencies outside this range have little effect. Since infinitely high driving frequencies have no effect on a dynamical system, driving a system with white noise does not produce an unphysical effect. The resulting dynamics of the system is the same as if it were driven with real noise with a broad spectrum.

For linear dynamical systems there is a very simple relationship between the power spectrum of the driving signal and that of any one of the system variables. If the power spectrum of the driving signal is $S(\omega)$, and that of a system variable x is $X_S(\omega)$, then

$$X_S(\omega) = |G_x(\omega)|^2 S(\omega), \tag{3.82}$$

where $G_x(\omega)$ is called the *transfer function* for the system variable x. The transfer function completely characterizes the response of the variable x to the driving signal. The above relation shows explicitly that if a system is driven by white noise, the power spectrum of any system variable is the square modulus of its transfer function.

The spectrum of a position measurement for a damped oscillator

Determining the power spectrum of a linear classical system driven by an external signal or signals is straightforward. All one has to do is to transform the linear equations of motion for the system using the Fourier transform, solve the resulting algebraic equation, and then the power spectrum is obtained from the correlation function(s) of the external signal(s). We illustrate the procedure here with a simple but useful example.

In Section 3.1.3 above we presented the classical model that reproduces the dynamics of a quantum harmonic oscillator under a continuous position measurement. This model can be extended to include damping and heating of the oscillator. Damping is due to friction, and friction is caused by the interaction of the oscillator with its environment. In the case of a mechanical oscillator this damping is likely to be due to the air in the room, and could also be due to loss of energy in a spring due to dissipation into the spring's internal degrees of freedom. Environments contain many degrees of freedom, and as such obey the laws of thermodynamics. Thus a source of damping is also a source of heating, if the environment is hot enough. We usually think of oscillators being damped by their environments, rather than being excited by them, but that is because macroscopic oscillators are relatively very heavy, and thus hardly excited by the thermal energy in the air at room temperature. But a very tiny oscillator will experience both damping and heating from the environment.

The dynamics of a weakly damped (and heated) quantum oscillator, under continuous position measurement, is reproduced by the following classical equation:

$$\frac{d}{dt}\begin{pmatrix} \tilde{x} \\ \tilde{p} \end{pmatrix} = \begin{pmatrix} -\gamma/2 & \omega_0 \\ -\omega_0 & -\gamma/2 \end{pmatrix}\begin{pmatrix} \tilde{x} \\ \tilde{p} \end{pmatrix} + \begin{pmatrix} \sqrt{\gamma\tilde{n}}\,\xi_x(t) \\ \sqrt{\gamma\tilde{n}}\,\xi_p(t) + \sqrt{8\tilde{k}}\,\zeta(t) \end{pmatrix}. \tag{3.83}$$

Here we have scaled x and p so that they are dimensionless. The real position, momentum, and measurement strength are given respectively by $x = \sqrt{\hbar/(2m\omega_0)}\,\tilde{x}$, $p = \sqrt{\hbar m\omega_0/2}\,\tilde{p}$, and $k = (2m\omega_0/\hbar)\tilde{k}$. Here m, ω_0, and γ are respectively the mass, frequency, and damping rate of the oscillator, and k is the measurement strength. This model is only a good approximation for the damping and heating so long as the damping rate, γ, is much smaller than the frequency of the oscillator, ω. The dynamics of a weakly damped oscillator is derived in Section 4.3.1. The parameter $\tilde{n} = 2n_T + 1$, where $n_T = 1/(e^{\hbar\omega/k_B T} - 1)$ is the average energy of the resonator, in units of $\hbar\omega_0$, when it is at thermal equilibrium at temperature T (see Chapter 4). The noise sources $\xi_x(t)$, $\xi_p(t)$, and $\zeta(t)$ are all white-noise sources with spectra $S(\omega) = 1$ and mutually uncorrelated.

The classical model above is obtained by (i) using the classical model for a linear quantum system under a continuous measurement of a Hermitian operator given in Section 5.5.2; (ii) writing down the classical equivalent of the quantum description of a weakly damped and heated oscillator by inspection of the quantum equations of motion given in Section 3.6; (iii) combining the equations of motion for the two models.

The measurement record is given by

$$\tilde{r}(t) = \langle \tilde{x}(t) \rangle + \frac{1}{\sqrt{8\tilde{k}}} \xi(t), \quad \langle \xi(t)\xi(t+\tau) \rangle = \delta(\tau) \tag{3.84}$$

and since the noise driving the oscillator is uncorrelated with the measurement noise, $\xi(t)/\sqrt{8\tilde{k}}$, the spectrum of the record is simply the spectrum of $x(t)$ plus that of the measurement noise:

$$\tilde{S}_r(\omega) = \tilde{S}_x(\omega) + \frac{1}{8\tilde{k}}. \tag{3.85}$$

To determine the spectrum of $\tilde{x}(t)$, we first take the Fourier transform of both sides of Eq. (3.83). Using the fact that the Fourier transform of dx/dt is $i\omega X(\omega)$, where $X(\omega)$ is the Fourier transform of $\tilde{x}(t)$, and taking all terms containing the system variables to the left-hand side, we get

$$\begin{pmatrix} i\omega + \gamma/2 & -\omega_0 \\ \omega_0 & i\omega + \gamma/2 \end{pmatrix} \begin{pmatrix} X(\omega) \\ P(\omega) \end{pmatrix} = \begin{pmatrix} \sqrt{\gamma\tilde{n}}\, E_x(\omega) \\ \sqrt{\gamma\tilde{n}}\, E_p(\omega) + \sqrt{8\tilde{k}}\, Z(\omega) \end{pmatrix}. \tag{3.86}$$

Here $X(\omega)$, $P(\omega)$, $E_x(\omega)$, $E_p(\omega)$, and $Z(\omega)$ are the respective Fourier transforms of x, p, ξ_x, ξ_p, and ζ. We can now obtain the explicit expression for $X(\omega)$ in terms of the noise sources merely by inverting the matrix in the above equation. The result is

$$X(\omega) = \frac{\sqrt{\gamma\tilde{n}}(\gamma/2 + i\omega) E_x(\omega) + \omega_0[\sqrt{\gamma\tilde{n}}\, E_p(\omega) + \sqrt{8\tilde{k}}\, Z(\omega)]}{(\gamma/2 + i\omega)^2 + \omega_0^2}. \tag{3.87}$$

We now substitute this expression into the frequency-space correlation function for x, namely $\langle X(\omega)X(\omega') \rangle$, giving

$$\langle X(\omega)X(\omega') \rangle = f_1(\omega,\omega')\langle E_x(\omega)E_x(\omega') \rangle + f_2(\omega,\omega')\langle E_p(\omega)E_p(\omega') \rangle$$

$$+ f_3(\omega,\omega')\langle Z(\omega)Z(\omega') \rangle \tag{3.88}$$

for some functions $f_i(\omega,\omega')$. Here we have used the fact that since the noise sources are mutually uncorrelated, the cross terms such as $\langle E_x(\omega)Z(\omega') \rangle$ vanish. From the relationship given by Eq. (3.77), and the fact that the spectrum of each noise source is $S(\omega) = 1$, we have

$$\langle E_x(\omega)E_x(\omega')\rangle = \langle E_p(\omega)E_p(\omega')\rangle = \langle Z(\omega)Z(\omega')\rangle = \delta(\omega+\omega'). \tag{3.89}$$

Substituting these into Eq. (3.88) we have

$$\langle X(\omega)X(\omega')\rangle = \left[f_1(\omega,\omega')+f_2(\omega,\omega')+f_3(\omega,\omega')\right]\delta(\omega+\omega'). \tag{3.90}$$

Now we use Eq. (3.77) once more, and the fact that $\delta(\omega + \omega')$ is non-zero only when $\omega' = -\omega$, to obtain the spectrum of x:

$$\begin{aligned}
\tilde{S}_x(\omega) &= f_1(\omega,-\omega)+f_2(\omega,-\omega)+f_3(\omega,-\omega) \\
&= \frac{\gamma(2n_T+1)(\gamma^2/4+\omega_0^2+\omega^2)+8\tilde{k}\omega_0^2}{(\gamma^2/4+\omega_0^2-\omega^2)^2+(\omega\gamma)^2}.
\end{aligned} \tag{3.91}$$

The model is only valid for weak damping, so we may approximate the above expression using $\gamma \ll \omega_0$. Although it is not obvious at first sight, in this case the spectrum drops to zero rapidly when $|\Delta\omega| = |\omega - \omega_0| \geq \gamma$. Substituting $\Delta\omega$ into the spectrum, and using the approximation $|\Delta\omega| \ll \omega_0$, we obtain the Lorentzian

$$\tilde{S}_x(\omega) = \frac{\gamma(2n_T+1)+4\tilde{k}}{2(|\omega|-\omega_0)^2+\gamma^2/2}. \tag{3.92}$$

Note that there are two Lorentzian peaks, one at ω_0 and the other at $-\omega_0$.

We can now check that the spectrum of x is consistent with the variance of the position of the resonator when it is in a thermal state. To do so we first determine the steady-state mean of x. The equations of motion for the mean values of x and p can be obtained by taking averages on both sides of Eq. (3.86). Since all the noise sources have zero mean, this gives

$$\frac{d}{dt}\begin{pmatrix} \langle\tilde{x}\rangle \\ \langle\tilde{p}\rangle \end{pmatrix} = \begin{pmatrix} -\gamma/2 & \omega_0 \\ -\omega_0 & -\gamma/2 \end{pmatrix}\begin{pmatrix} \langle\tilde{x}\rangle \\ \langle\tilde{p}\rangle \end{pmatrix}. \tag{3.93}$$

The steady-state solutions for the means are therefore $\langle\tilde{x}\rangle = \langle\tilde{p}\rangle = 0$. Since the steady-state mean of x is zero, the steady-state variance of x is equal to the mean-square. Using Eq. (3.76), the mean-square is

$$\langle[\tilde{x}(t)]^2\rangle = \frac{1}{\pi}\int_0^\infty \tilde{S}_x(\omega)d\omega = 1+2n_T+\frac{4\tilde{k}}{\gamma}. \tag{3.94}$$

If we set $\tilde{k} = 0$ then this agrees with the thermal variance of $\tilde{x}$ for a quantum resonator at temperature T, and if we also set $T = 0$ it agrees with the variance of $\tilde{x}$ in the ground state. Note that the steady-state mean-square is also the total power in the fluctuations of $\tilde{x}$.

From Eq. (3.85), the power spectrum of the record, now in units of the real position x, is

$$S_r(\omega) = \frac{1}{8k} + \frac{\gamma V_x(2n_T + 1) + 4V_x^2 k}{2(|\omega| - \omega_0)^2 + \gamma^2/2},$$ (3.95)

Here $V_x = \sqrt{\hbar/(2m\omega_0)}$ is the variance of the position for the oscillator's ground state.

3.2 Solving for the evolution: the linear form of the SME

The SSE and SME describing continuous measurement are nonlinear, and this makes it difficult to obtain analytic solutions for the evolution of continuously observed systems. Remarkably, the SSE and SME can be rewritten in a form in which they are linear, and this makes it possible to obtain analytic solutions in a number of simple cases.

To see how it is possible to recast the SME as a linear differential equation, consider a sequence of identical measurements on a quantum system. If each of the measurements is described by the set of measurement operators $\{A_n\}$, then after one measurement the state of the system is

$$\tilde{\rho}_n = \frac{A_n \rho A_n^\dagger}{\text{Tr}\left[A_n \rho A_n^\dagger\right]},$$ (3.96)

and after two measurements is

$$\tilde{\rho}_{nm} = \frac{A_m \tilde{\rho}_n A_m^\dagger}{\text{Tr}\left[A_m \tilde{\rho}_n A_m^\dagger\right]} = \frac{A_m A_n \rho A_n^\dagger A_m^\dagger}{\text{Tr}\left[A_m A_n \rho A_n^\dagger A_m^\dagger\right]} = \frac{A_m A_n \rho A_n^\dagger A_m^\dagger}{P_{nm}},$$ (3.97)

where P_{nm} is the probability of getting both results n and m, and thus the probability of getting the final state ρ_{nm}. The above expressions for ρ_{nm} tell us two things about the effect of a sequence of an arbitrary number of measurements. First, since the top line of the final equation is a *linear* function of ρ, it tells us that, up to a normalization, the final state is a linear function of the initial state. Second, the trace of this linear function of the initial state (the normalization) is the probability of obtaining the given final state.

The above result shows that it must be possible to obtain a linear differential equation for the evolution of ρ under a continuous measurement, so long as ρ is not normalized, since this evolution is merely a sequence of measurements. It also tells us that the information about the probability of getting the final state is contained in the norm.

Consider now the unnormalized differential equation for ρ, Eq. (3.33), written in terms of the measurement result dy:

$$|\hat{\psi}(t + dt)\rangle = \{1 - kX^2 dt + 4kX dy\}|\hat{\psi}(t)\rangle.$$ (3.98)

We see that the nonlinearity in this equation is due solely to the fact that dy contains $\langle X \rangle$. So if we replaced dy by dW, the equation would be linear. Since both dW and dy have the entire real line as their range, we would still get all the correct final states if we used dW in place of dy. However, we would not get the various final states with the correct *probabilities*. This is because the distribution for dy is peaked at $\langle X \rangle dt$, and that for dW is peaked at zero. Nevertheless, the information regarding the probability of getting each final state should be contained in the norm.

To derive a linear differential equation for $|\psi\rangle$, we start with the normalized SSE, Eq. (3.36), written in terms of dy:

$$|\psi(t+dt)\rangle = \{1 - k(X - \langle X \rangle)^2 \, dt + 4k(X - \langle X \rangle)(dy - \langle X \rangle \, dt)\}|\psi(t)\rangle. \tag{3.99}$$

We now replace dy by $dW/\sqrt{8k}$, thus making the equation linear. We then multiply this equation by the square root of the probability for getting dy, so as to obtain the unnormalized equation in which the norm records this probability (we note that the norm given by Eq. (3.33) does not have this property). However, to obtain a valid stochastic equation, when we multiply by the square root of the probability for dy, we must also divide by the square root of the probability distribution for dW. This is because the ratio of $P(dy)$ to $P(dW)$ is well-defined to first order in dt, but neither of these is by itself. With this additional normalization, the true probability for a final state $\hat{\rho}_{\text{fin}}$ generated by the SSE will be the *product* of the trace of $\hat{\rho}_{\text{fin}}$ with the probability that the equation generated $\hat{\rho}_{\text{fin}}$ (that is, the probability for the noise realization $\int dW$ that gives $\hat{\rho}_{\text{fin}}$).

To first order in dt the square root of the ratio of the probabilities for dy and dW is

$$\sqrt{\frac{P(dy)}{P(dW)}} = 1 + \sqrt{2k}\langle X \rangle \, dW - k\langle X \rangle^2 \, dt. \tag{3.100}$$

Replacing dy by $dW/\sqrt{8k}$ in Eq. (3.99), multiplying by $\sqrt{P(dy)/P(dW)}$, and including the evolution of a Hamiltonian H, we obtain the linear SSE:

$$|\hat{\psi}(t+dt)\rangle = \{1 - (i/\hbar)H \, dt - kX^2 \, dt + \sqrt{2k}X \, dW\}|\hat{\psi}(t)\rangle. \tag{3.101}$$

From this we can immediately obtain the linear SME, which is

$$d\hat{\rho} = -\frac{i}{\hbar}[H, \hat{\rho}] \, dt - k[X[X, \hat{\rho}]] \, dt + \sqrt{2k}(X\hat{\rho} + \hat{\rho}X) \, dW. \tag{3.102}$$

The solution to the linear SME for the case $[H, X] = 0$

There are a number of special cases in which the linear SME can be solved exactly. The simplest of these is that of "pure measurement," in which the Hamiltonian commutes with the measured observable, and thus plays no significant role. We now solve the linear SME for this case. At the end of the chapter we give references to the small number of more complex situations for which analytic solutions are known.

The SME is very easily solved by first writing it as multiplication by an exponential. The SME is

$$\hat{\rho}(t+dt) = e^{[-iH/\hbar - 2kX^2] \, dt + \sqrt{2k}X \, dW} \hat{\rho}(t) e^{[iH/\hbar - 2kX^2] \, dt + \sqrt{2k}X \, dW}, \tag{3.103}$$

and it is simple to show that this is equivalent to Eq. (3.102) by expanding the exponentials to first order in dt, as always using the Ito relation $(dW)^2 = dt$.

The evolution for a finite time t is easily obtained now by repeatedly multiplying the initial state on both sides by the exponentials. We then combine all the exponentials on each side in a single exponential, which is simple only because $[H, X] = 0$. This gives the solution to Eq. (3.102), which is

$$\hat{\rho}(t, W) = e^{[-iH/\hbar - 2kX^2]t + \sqrt{2k}XW(t)} \hat{\rho}(0) e^{[iH/\hbar - 2kX^2]t + \sqrt{2k}XW(t)}, \tag{3.104}$$

where W is the stochastic integral introduced in Section 3.1.1:

$$W(t) = \int_0^t dW(t'). \tag{3.105}$$

The different possible final states are parametrized by W, by which we mean that there is a different final state for each value of W. This also means that the final state depends on the noise realization *only* via W.

As discussed in Section 3.1.1, the probability density for W at time t is a zero mean Gaussian with variance t:

$$\hat{P}(W, t) = \frac{1}{\sqrt{2\pi t}} e^{-W^2/(2t)}. \tag{3.106}$$

Because we are using the linear SME, $\hat{P}$ is *not* the true probability for obtaining the final state $\hat{\rho}(t, W)$; it is merely the probability with which $\hat{\rho}(t, W)$ is generated by picking a noise realization.

To get the true probability for obtaining the state $\hat{\rho}(t, W)$, we must multiply the density $\hat{P}(W, t)$ by the trace of $\hat{\rho}(t, W)$. Thus, the actual probability for getting a final state $\rho(t, W)$ is

$$P(W, t) = \frac{1}{\sqrt{2\pi t}} e^{-W^2/(2t)} \text{Tr} \left[e^{[-4kX^2]t + \sqrt{8k}XW} \rho(0) \right], \tag{3.107}$$

and the normalized final state is

$$\rho(W, t) = \frac{e^{[-iH/\hbar - 2kX^2]t + \sqrt{2k}XW(t)} \hat{\rho}(0) e^{[iH/\hbar - 2kX^2]t + \sqrt{2k}XW(t)}}{\text{Tr} \left[e^{[-4kX^2]t + \sqrt{8k}XW} \rho(0) \right]}. \tag{3.108}$$

Measurement of a spin

In the solution above, the measured observable X could be anything. Let us now consider the case when X has a small number of discrete, equally spaced eigenvalues. A good example of this is

$$X = J_z, \tag{3.109}$$

where J_z is the z-component of the angular momentum, for some fixed value of total angular momentum, j. In this case X has $2j + 1$ eigenvectors $|m\rangle$, with eigenvalues $m = -j, -j+1, \ldots, j$. The solution is very simple if we have no initial information about the spin at the start of the measurement. This is captured by setting the initial state to be proportional to the identity, so that

$$\rho(0) = \frac{I}{2j+1}. \tag{3.110}$$

With this initial state, the density matrix remains diagonal in the J_z eigenbasis throughout the measurement. The diagonal elements of $\rho(t, W)$ are

$$\langle m|\rho(t)|m\rangle = \frac{e^{-4kt(m-z)^2}}{\mathcal{N}}, \tag{3.111}$$

where $\mathcal{N}$ is the normalization and we have defined $z \equiv W/(\sqrt{8kt})$. The true probability distribution for z is

$$P(z,t) = \frac{1}{2j+1} \sum_{n=-j}^{j} \sqrt{\frac{4kt}{\pi}} e^{-4kt(z-n)^2}. \tag{3.112}$$

From this we see that after a sufficiently long time, the distribution for z is sharply peaked about the $2j+1$ eigenvalues of J_z. This density is plotted in Fig. 3.1 for three values of t. At long times, z tends to one of these eigenvalues. Further, we see from the solution for $\rho(t, W)$ that when z is close to a given eigenvalue m, the state of the system is sharply peaked about the corresponding eigenstate $|m\rangle$. So after a sufficiently long time, the system is projected into one of the eigenstates of J_z, and the continuous measurement is equivalent to a von Neumann measurement of J_z.

The random variable z has a physical meaning. Since we replaced the measurement record dy by $dW/\sqrt{8k}$ to obtain the linear equation, when we transform from the raw

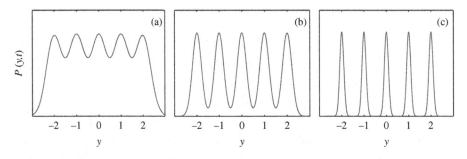

Figure 3.1 Here we show the probability distribution for the result of a continuous measurement of the z-component of a $j = 2$ angular momentum. This distribution is shown for three different measurement times: (a) $t = 1/k$; (b) $t = 3/k$; (c) $t = 15/k$, where k is the measurement strength.

probability density $\hat{P}$ to the true density P, this transforms the driving noise process dW back into $\sqrt{8k}\,dy = \sqrt{8k}\langle X(t)\rangle\,dt + dW$, being a scaled version of the measurement record. Thus, $z(t)$, as we have defined it, is actually the output record up until time t, divided by t. That is,

$$z(t) = \frac{y(t)}{t} = \frac{1}{t}\int_0^t dy = \frac{1}{t}\int_0^t \langle J_z(t)\rangle\,dt + \frac{1}{\sqrt{8k}\,t}\int_0^t dW. \qquad (3.113)$$

So z is really the result of the measurement. When making the measurement the observer integrates up the measurement record, and then divides the result by the final time. This result is z, and the closer z is to one of the eigenvalues, and the longer the time of the measurement, the more certain the observer is that the system has been collapsed onto the eigenstate with that eigenvalue. Note that as the measurement progresses the second, explicitly stochastic, term in the expression for z converges to zero, while the first term must evolve to the measured eigenvalue. (We will see *how* the dynamics of the SME ensures that $\langle J_z(t)\rangle$ evolves to one of the eigenvalues in Section 3.2.1.)

To summarize the key message of the above discussion, when we are using a linear SSE, expressions involving dW in the solution to the linear SSE are in reality expressions involving the *measurement record*. Thus, up to a scaling factor,

$$dW \to dy \quad \text{and so} \quad \int_0^t f(t')\,dW(t') \to \int_0^t f(t')\,dy(t'). \qquad (3.114)$$

The Zakai equation

Naturally the fact that we can derive a linear SME means that we can also derive a linear version of the classical Kushner–Stratonovich equation. This is called the Zakai equation [49]. The Zakai equation in which the classical variables are not undergoing any dynamics is a special case of the linear SME when all operators commute with the density matrix.

3.2.1 The dynamics of measurement: diffusion gradients

It is worth examining how the measurement result evolves for a given realization, and in particular, what kind of dynamics ensures that the state becomes increasingly "trapped" near (or "collapses" to) one of the eigenstates of the measured observable. To understand the dynamics of the measurement process, it is simplest to consider a measurement of a system with only two states – for example a spin-1/2 system. This is just a special case of the spin measurement we solved above, where J_z is the Pauli matrix σ_z. Since we are interested in understanding the dynamics of the measurement process itself, we set the Hamiltonian to zero. If our initial state is completely mixed, and there is no Hamiltonian dynamics, then the density matrix remains diagonal in the σ_z eigenbasis. In this case, for the spin-1/2 system, $\langle \sigma_z(t)\rangle$ determines completely the purity of the observer's state-of-knowledge, and thus tells us to what extent the measurement has collapsed the initial mixed state into a pure state. We will therefore examine the evolution of $\langle \sigma_z(t) \rangle$. The purity of the density matrix is related to $\langle \sigma_z(t) \rangle$ via

$$\text{Tr}[\rho^2] = \frac{1}{2} \left(1 - \langle \sigma_z \rangle^2 \right). \tag{3.115}$$

So when the state is completely mixed $\langle \sigma_z \rangle$ is zero, and as the measurement purifies the state $\langle \sigma_z \rangle$ tends to ± 1.

Setting $H = 0$ (or equivalently $H = \hbar \omega \sigma_z$), and $X = \sigma_z$ in the SME (Eq. 3.26) we can obtain the equation of motion for $\langle \sigma_z \rangle$, which is

$$d\langle \sigma_z \rangle = \sqrt{8k} \left(1 - \langle \sigma_z \rangle^2 \right) dW. \tag{3.116}$$

The startling thing about this equation is that there is *no* deterministic part – there is no "drift" term that will drive $\langle \sigma_z \rangle$ away from the origin, and toward ± 1. There is only random noise that kicks $\langle \sigma_z \rangle$ equally in both directions.

So how is it that the equation of motion will push $\langle \sigma_z \rangle$ to ± 1, and especially *pin it down* to one extreme value or the other as $t \to \infty$? It turns out that this is the effect of a *diffusion gradient*. Diffusion is the random motion induced by the noise term dW. We see from Eq. (3.116) that the magnitude (or "rate") of the diffusion depends on the value of $\langle \sigma_z \rangle$. The diffusion is greatest when $\langle \sigma_z \rangle = 0$ (so that the state is completely mixed, and the observer knows the least), and drops to zero as $\langle \sigma_z \rangle$ tends to ± 1 (when the state has collapsed to one of the eigenstates of σ_z, and the observer knows everything about σ_z). There is therefore a *gradient* in the rate of diffusion, with the diffusion decreasing monotonically away from $\langle \sigma_z \rangle = 0$. It is this gradient that pushes $\langle \sigma_z \rangle$ away from zero, and pins it to (one of) the extreme values as $t \to \infty$.

That a diffusion gradient produces an effectively deterministic motion in the direction in which the diffusion is *decreasing* can be seen by using the Fokker–Planck equation. If you are not familiar with this equation, it describes the evolution of the probability density of a variable driven by Gaussian noise. Details regarding the Fokker–Planck equation are given in chapter 6 of [288]. If the motion of a variable x is given by

$$dx = v\, dt + \sqrt{D(x)}\, dW \tag{3.117}$$

where v is a deterministic velocity, then the Fokker–Planck equation for the probability density of x is

$$\frac{\partial P}{\partial t} = -v \frac{\partial P}{\partial x} + \frac{1}{2} \frac{\partial^2}{\partial x^2} [D(x)P]. \tag{3.118}$$

From the Fokker–Planck equation one can calculate the *probability current*, $J(x)$, which is given by

$$\frac{\partial J}{\partial x} = -\frac{\partial P}{\partial t}. \tag{3.119}$$

Calculating J from Eq. (3.118) gives

$$J(x) = \left(v - \frac{1}{2}\frac{\partial D}{\partial x}\right)P - \frac{D(x)}{2}\frac{\partial P}{\partial x}. \tag{3.120}$$

From this we see that the gradient of the diffusion rate, $\partial D/\partial x$, generates a term in the probability current equivalent to a negative deterministic velocity.

In fact, a variable will only be pinned down to the points at which the diffusion is zero if the space of the variable is *bounded*. If the space is bounded, then the steady-state probability distribution for the variable is inversely proportional to the diffusion rate $D(x)$. Thus if there are points at which the diffusion vanishes, the variable will be precisely pinned down to one of these points as $t \to \infty$. The reason that this is only true if the space is bounded, is because to pin the system down to the points at which the diffusion vanishes, it must be able to explore *all* the available state space.

3.2.2 Quantum jumps

Quantum jumps are a phenomenon that emerge in continuous measurements of observables with discrete eigenvalues. The measurement process purifies the system so that the state is very close to a single eigenstate of the observable. If the system is subjected to another noise process that randomly kicks the measured observable, then this competes with the measurement process. That is, while the measurement is trying to reduce the uncertainty of the value of the observable, the noise is continually increasing this uncertainty. If the measurement dominates, then the combined effect of the two processes is as follows: the system spends the majority of its time close to one of the eigenstates of the observable, and this is punctuated every now and then by "jumps" in which the system rapidly switches to an adjacent eigenstate. During the switch the uncertainty in the value of the observable necessarily increases momentarily when it is "between" eigenstates. The jump phenomenon is due to the fact that while the measurement dominates, and thus manages to keep the uncertainty of the observable small, occasionally the noise has a large enough fluctuation to increase the uncertainty sufficiently that the measurement "collapses" the system to a different eigenstate.

We now give an explicit example of this phenomenon. Consider a measurement of the energy of a quantum harmonic oscillator. The Hamiltonian of the system is

$$H = \hbar\omega\left(a^\dagger a + \frac{1}{2}\right), \tag{3.121}$$

and the energy eigenvalues are

$$E_n = \hbar\omega(n + 1/2), \quad \text{for } n = 0, 1, 2, \ldots. \tag{3.122}$$

The eigenstates corresponding to the energies E_n, which we will denote by $|n\rangle$, are called the phonon, or Fock, states of the oscillator. If the oscillator is in state $|n\rangle$, it is referred to as having n phonons. If we make a continuous measurement of the phonon number, then this will collapse the oscillator to one of the energy eigenstates.

To observe quantum jumps we also need a source of noise that can change the phonon number. A simple noise source is a rapidly fluctuating force applied to the resonator. This is described by adding the term

$$H_{\text{noise}} = \hbar\sqrt{g}\xi(t)x \tag{3.123}$$

to the Hamiltonian. Here $x = (a + a^\dagger)/\sqrt{2}$ is the dimensionless position of the resonator, g is a rate constant, and $\xi(t)$ is white noise, having the autocorrelation function

$$\langle \xi(t)\xi(t')\rangle = \delta(t' - t). \tag{3.124}$$

The evolution of a pure quantum state due to this noise is

$$\frac{d}{dt}|\psi\rangle = -i\sqrt{g}\xi(t)x|\psi\rangle. \tag{3.125}$$

Since this white noise is intended to represent the broadband limit of a real noise process, and since the noise multiplies the state of the system in the evolution equation, it must be treated as Stratonovich noise, rather than Ito noise. (If you are not familiar with the distinction between Ito and Stratonovich noise, and how to convert between them, full details can be found in, e.g., chapter 5 of reference [288]. We also discuss Stratonovich noise further in Sections H.3 and F.6). Transforming Eq. (3.125) to an Ito equation gives

$$d|\psi\rangle = -\left[\frac{g}{2}x^2\,dt + i\sqrt{g}x\,dW\right]|\psi\rangle. \tag{3.126}$$

The equation of motion for the density matrix is then

$$\begin{aligned}
d\rho &= -\frac{g}{2}\left(x^2\rho - \rho x^2 + 2x\rho x\right)dt + i\sqrt{g}\,(x\rho - \rho x)\,dW \\
&= -\frac{g}{2}[x,[x,\rho]]\,dt + i\sqrt{g}\,(x\rho - \rho x)\,dW.
\end{aligned} \tag{3.127}$$

But we must remember that the observer who is making the measurement does not know what the fluctuations of the random force are, so she does not have access to the noise realization for dW in the above equation for ρ. The state-of-knowledge of the observer is thus given by averaging over dW at each time-step. Since the mean of dW is zero, the equation of motion for the observer's state-of-knowledge becomes

$$d\rho = -\frac{g}{2}[x,[x,\rho]]\,dt. \tag{3.128}$$

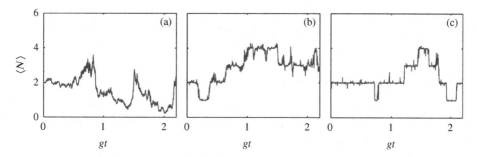

Figure 3.2 Here we show the evolution of the expectation value of the phonon number of a harmonic oscillator subject to a continuous measurement of phonon number, and driven by a white noise force. The strength of the force is $g = 0.1f$, and for the three plots the measurement strength is (a): $k = g/2$; (b) $k = 20g$; (c) $k = 100g$.

The evolution of the density matrix when a harmonic oscillator is subject both to a continuous measurement of $N = a^\dagger a$, as well as a rapidly fluctuating force, is therefore

$$d\rho = -i\omega[N, \rho]\,dt - \frac{g}{2}[x, [x, \rho]]\,dt$$

$$-k[N, [N, \rho]]\,dt + \sqrt{2k}(N\rho + \rho N - 2\langle N\rangle\rho)\,dW. \tag{3.129}$$

Simulating this SME with $\omega = 2\pi f$, $g = 0.1f$, and $k = 100g$ (where f is an arbitrary frequency) reveals the quantum jumps as advertised above. We display the results of this simulation in Fig. 3.2, where we plot the expectation value of N, $\langle N(t)\rangle = \mathrm{Tr}[N\rho(t)]$. The three plots show how the quantum jumps emerge as k is increased. We will address the question of how one might actually measure the phonon number of an oscillator in practice in Chapter 7.

To avoid confusion, it is important to note that the term "quantum jumps" is also used to refer to another closely related, but distinct quantum phenomenon, that of the sudden change in the state of a quantum system upon a discrete detection event, such as the detection of a photon by a photo-detector. We will describe the kinds of continuous measurements that involve a series of discrete events in Section 3.3.

Experimental realizations of the kind of quantum jumps discussed in this section may be found in [436, 535, 53, 37, 473, 213, 703], and theoretical analyses of further examples in [383, 215, 423, 296, 303, 196]. We note that in [215] the authors refer to the quantum jumps instead as "the quantum Zeno effect," and use the term "quantum jumps" in the second sense referred to above. In [383] and [303] the jumps are due to an energy measurement on an oscillator, although this measurement is performed indirectly. By this we mean that the oscillator interacts with a second system, and it is this second system to which the continuous measurement is applied. The information regarding the energy of the oscillator flows to the second system via the interaction, and is continuously extracted by the measurement. Constructing measurements using probe systems in this way can be very useful, and we will explore it further in Chapter 7.

3.2.3 Distinguishing quantum from classical

One may wish to make continuous measurements on a dynamical system to verify that it is behaving quantum mechanically, rather than classically, and there is a subtlety in this situation that we feel is worth discussing.

As an example, let us consider making a continuous measurement of phonon number (energy) on a harmonic oscillator. This certainly has the ability to distinguish between quantum and classical, since the energy levels of the quantum oscillator are quantized, whereas those of a classical oscillator are not.

Now consider the process of making the measurement and determining the motion of the oscillator. In the quantum case one obtains the stream of output results, $dy = \langle N(t) \rangle \, dt + dW/\sqrt{8k}$. To determine the evolution of the system the observer processes these results by using them to integrate the SME. Specifically, the observer uses dy at each time-step to determine dW via the relation

$$dW = \sqrt{8k}(dy - \langle N(t) \rangle \, dt) = \sqrt{8k}(dy - \text{Tr}[N\rho(t)] \, dt). \tag{3.130}$$

She then uses dW to solve the SME and thus obtain her full state-of-knowledge, $\rho(t)$, at the next time-step. From $\rho(t)$ she can calculate the expectation value of any observable (including $\langle N(t) \rangle$, of course, which she needs to determine dW for the next time-step). As we have seen above, when the measurement strength is much larger than the noise, the expectation value of the energy exhibits jumps between specific discrete values.

The procedure is essentially identical if the observer is measuring a classical oscillator. The only difference is that dW, obtained in the same way from the measurement record, is now used to solve the Kushner–Stratonovich equation instead of the SME. In this case the expectation value of the energy is not restricted in the values it can take – it will settle down to essentially any value on the real line, and the effect of the noise will be to cause this value to drift slowly in a random way.

While the observed evolution of the quantum and classical oscillators is quite different in the two cases described above, neither case represents the problem of distinguishing quantum from classical. When one is attempting to determine, via a measurement, whether a system is quantum or classical, the observer will have to select a single method, *either* quantum *or* classical, for processing the measurement results, because she does not know whether the system is quantum or classical. The problem of distinguishing quantum from classical dynamics therefore concerns the difference in the behavior of either the SME, or the K-S equation, when they are given a stream of measurement results from a quantum system as opposed to the equivalent classical system.

Distinguishing quantum from classical dynamics is similar to that of a measurement of a variable that has two values. We make a continuous measurement of the energy of the oscillator, and we need to use Bayesian inference to work out, based on the continuous stream of measurement results, which is more likely – that the system is undergoing the dynamics of a classical oscillator, or that of a quantum oscillator subjected to a measurement of energy. When the possibilities we wish to learn about are not merely values, but some more complex aspect of a system, the inference problem is often referred to as *hypothesis testing*. In this language we have two hypotheses about the system, one that it is classical and one that it is quantum, and we want to determine which is true from the results of our measurement. We will not discuss hypothesis testing from continuous quantum measurement further here, but this problem has been analyzed by Tsang [613], and he gives an extensive discussion of hypothesis testing via quantum measurements in [611].

3.2.4 Continuous measurements on ensembles of systems

In some situations it is natural to make a single continuous measurement on a large system that consists of a very large number of identical systems (an ensemble of systems), all of which are prepared in the same state. The classic example of this is a cloud of atoms in a vacuum chamber, or an array of atoms trapped in an optical lattice. In this situation, since all the atoms are prepared in the same state, one should be able to effectively make a measurement to determine the initial state without disturbing it. The reason for this is that one could, for example, make a projective measurement on one of the systems, and then throw it away. This measurement has provided information about the initial state, but we still have many systems left that are undisturbed. In this way we can obtain essentially full information about the state, by measuring a large number of the systems, and still have some systems that remain unaffected. Once we have determined the state, we can prepare more atoms in that state and replenish the ensemble. Thus, when we have many copies of the same quantum state, the state becomes essentially classical, because we can measure it without disturbing it. We now examine how this plays out when we make a continuous measurement on an ensemble.

For concreteness let us make a measurement on an ensemble of N two-level (spin-1/2) systems (the two levels could, for example, be two magnetic sublevels of the internal states of an atom). The z-component of the total spin for all the N systems, which we will call J_z, is simply the sum of the z-components for each system. Thus

$$J_z = \sum_{n=1}^{N} \sigma_z^{(n)}, \tag{3.131}$$

where $\sigma_z^{(n)}$ is the spin-z operator for the nth system.

Consider first what happens if we make a measurement of strength k/N separately on each spin-1/2 system. The measurement record for the nth spin is then

$$dy^{(n)} = \langle \sigma_z^{(n)} \rangle + \sqrt{\frac{N}{8k}} \, dW^{(n)}. \tag{3.132}$$

Since all the spins start in the same state, all the expectation values $\langle \sigma_z^{(n)} \rangle$ are equal, and we will call this value $\langle \sigma_z \rangle$. If we add all the measurement results together, and take the average, then we obtain the measurement result

$$dy = \frac{1}{N} \sum_{n=1}^{N} \left[\langle \sigma_z^{(n)} \rangle + \sqrt{\frac{N}{8k}} \, dW^{(n)} \right] = \langle \sigma_z \rangle + \sum_{n=1}^{N} \frac{dW^{(n)}}{\sqrt{8Nk}} = \langle \sigma_z \rangle + \frac{dW}{\sqrt{8k}}. \qquad (3.133)$$

So we obtain a measurement of σ_z of strength k, by making a measurement on each spin of only strength k/N. The back-action on each system is thus reduced by a factor of N in relation to the amount of information obtained about the initial state. Because there is a little bit of back-action in the first time-step dt, and because this back-action is different for each system, it will not be true that all the $\langle \sigma_z^{(n)} \rangle$ are equal for subsequent time-steps. However, if N is very large, then the back-action will be negligible, and we can assume that the evolution of each system is only that due to the Hamiltonian. In this case we can extract as much information about the state of the system as we want, without interfering with the evolution.

In practice one does not make a separate measurement on each system, but makes a single continuous measurement of the total spin J_z. If you write out the SME for N simultaneous measurements, one for each of the $\sigma_z^{(n)}$, and then write out the SME for a single measurement of J_z, you will see that the latter has a bunch of additional cross terms. The separate measurements do not entangle the subsystems, but a measurement of J_z will in general project the systems onto a subspace of the total system in which they are correlated and entangled. In the extreme case of a von Neumann measurement, the final value of J_z will be sharp, but there are many different ways that each of the spins can be aligned so as to give a single value for J_z. The state of the total system will be some superposition of these joint states of all the systems, with the result that the systems will usually be correlated.

A single measurement of J_z *does* act much like N simultaneous measurements, as far as the size of the back-action on each system is concerned. That is, if we make a measurement of J_z with strength k/N^2, and divide the measurement record by N, we get

$$dy = \frac{1}{N} \left[\langle J_z \rangle + \frac{N}{\sqrt{8k}} \, dW \right] = \langle \sigma_z \rangle + \frac{dW}{\sqrt{8k}}. \qquad (3.134)$$

We thus obtain a measurement of σ_z with strength k. The back-action of the measurement on each of the spins is the same order as a measurement of strength k/N (it is k/N as apposed to k/N^2 because of the N cross terms from the J_z measurement). So we can make a measurement of the total spin to extract information about the state of each spin-1/2 system, without any appreciable disturbance to the states of these systems, so long as N is large enough.

3.3 Measurements that count events: detecting photons

Counting events

So far we have considered continuous measurements in which the noise is Gaussian in any small time interval, and these are essentially the only kind of continuous measurements in which the stream of measurement results is a continuous function of time (for a further discussion of this point, see Section 3.5). There is another kind of continuous measurement in which the measurement results are a series of discrete events, where the randomness lies in the times at which these events occur. One could perform a classical measurement of this kind by standing beside a train track and observing the times at which trains pass. In quantum systems these kinds of measurements arise from the discrete nature of quantum mechanics, from which, of course, this theory gets its name. A classic example of these measurements is that of photon counting. In this case a beam of light hits a detector, and since light contains quanta called photons, the detector registers the arrival of a photon (a discrete event) at a sequence of random times. Noise of this kind is called a *point process*, or a *jump process*.

Consider a point process in which the random events have a constant probability per unit time of occurring, and call this probability rate λ. This means that in the time interval dt, the probability of an event is

$$\text{Prob(event)} = \lambda \, dt, \tag{3.135}$$

and thus the probability of no event in the same time interval is

$$\text{Prob(no event)} = 1 - \lambda \, dt. \tag{3.136}$$

It is useful to define a random variable $N(t)$ that counts the number of events up until time t. This means that if an event occurs in an interval dt the value of N jumps up by 1, and if no event occurs N remains the same. Thus the increment in N can take two values, $dN = 1$ or $dN = 0$, and the probability distribution for dN is $P(dN = 1) = \lambda \, dt$, and $P(dN = 0) = 1 - \lambda \, dt$. We can also write this distribution more compactly as a function of dN if we wish:

$$P(dN) = (1 - dN) + (2 \, dN - 1)\lambda \, dt. \tag{3.137}$$

If λ is constant (independent of time), the stochastic process $N(t)$ is called the *Poisson process*. One can derive the probability distribution for the number of events that occur in a time interval t. This is called the *Poisson distribution*, and is given by

$$P(N,t) = \frac{e^{-\lambda t}(\lambda t)^N}{N}. \tag{3.138}$$

The mean of the Poisson distribution is $\langle N(t) \rangle = \lambda t$, and the variance is equal to the mean. If you are not familiar with jump processes, they are described in detail in reference [288].

The equivalent of the Ito rule for jump processes is $(dN)^2 = dN$, but this is not nearly as useful for calculational purposes as the Ito rule, for reasons discussed in [288].

Quantum counting measurements

The simplest way to introduce counting measurements in quantum mechanics is to determine first what set of measurement operators is required to generate such measurements. In each time-step dt, there are two outcomes, and thus two measurement operators. Let us call them A_0 and A_1. One outcome must have probability λdt, so choosing this to be outcome 1, we have

$$p_1 = \text{Tr}\left[A_1 \rho A_1^\dagger\right] = \lambda dt. \tag{3.139}$$

This means that A_1 must be proportional to $\sqrt{dt}$. This seems rather strange, but it is reminiscent of the fact that the Wiener noise of Gaussian measurements has standard deviation equal to $\sqrt{dt}$. So let us define $A_1 = \Omega\sqrt{dt}$, where Ω is a constant operator. Since we must have $A_1^\dagger A_1 + A_0^\dagger A_0 = I$, we now have no option but to set

$$A_0^\dagger A_0 = I - \Omega^\dagger \Omega\, dt. \tag{3.140}$$

Now we need to factor the RHS, to determine A_0. While at first it looks difficult, it is actually easy because what we have on the RHS is a power series in dt. What is more, we can ignore all terms that are higher than first-order in dt because that is the continuum limit. So we replace A_1 with the power series $(I + X_1 dt)$, and solve for X_1:

$$(I + X_1^\dagger dt)(I + X_1 dt) = 1 + (X_1^\dagger + X_1)dt + \mathcal{O}(dt^2) = I - \Omega^\dagger \Omega\, dt. \tag{3.141}$$

So the result is

$$X_1^\dagger + X_1 = -\Omega^\dagger \Omega. \tag{3.142}$$

Since $\Omega^\dagger \Omega$ is Hermitian, we lose nothing by making X_1 Hermitian. Thus $2X_1 = -\Omega^\dagger \Omega$, and we have finally

$$A_1 = \Omega\sqrt{dt}, \tag{3.143}$$

$$A_0 = I - \frac{\Omega^\dagger \Omega}{2}\, dt. \tag{3.144}$$

This is the form that counting measurements have to take. The probabilities for the respective outcomes are

$$p_1 = \text{Tr}\left[\Omega^\dagger \Omega \rho(t)\right] dt, \tag{3.145}$$

$$p_0 = I - p_1. \tag{3.146}$$

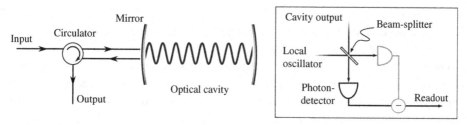

Figure 3.3 On the left is shown an optical cavity with input and output via the left-hand mirror. In the box are shown two ways to implement homodyne detection. In simple homodyne detection, the configuration discussed in Section 3.4, the gray photon-detector (or photon-counter) is absent. In balanced homodyne detection both photo-detectors are used, the measurement signal (the readout) is the difference between their respective outputs, and the beam-splitter is 50% reflecting.

So the instantaneous rate of counts (also called *jumps*), at time t is

$$\lambda(t) = \text{Tr}\left[\Omega^{\dagger}\Omega\rho(t)\right]. \tag{3.147}$$

A quick primer on optical cavities

The measurement scenario we wish to consider involves an optical cavity. Since optical cavities will appear in a number of chapters we take a moment now to introduce them. An optical cavity in its simplest form consists of two mirrors that face each other, so that a light beam can bounce backward and forward between them. The bouncing light forms standing waves between the mirrors, and is said to be trapped in the cavity. In practice both mirrors are curved slightly. If the mirrors were perfectly flat, any minute error in the angle would cause the beams to quickly walk out the sides of the cavity. With curved mirrors the beam forms a stable configuration, usually a Gaussian beam whose waist is thinner in the center of the cavity than at the mirrors. A common depiction of an optical cavity is shown in Fig. 3.3.

The standing waves inside the cavity can only exist at frequencies $\omega_n = c/(2L)$, where L is the distance between the mirrors, or "cavity length," and c is the speed of light. The standing waves at each of these discrete frequencies is called a cavity mode, and each has the same quantum dynamics as a harmonic oscillator: a mode can contain an integer number of photons, and the energy of each photon is $\hbar\omega$, where ω is the angular frequency of the light. Because of this the Hamiltonian for an optical mode is just that for a harmonic oscillator:

$$H = \hbar\omega a^{\dagger}a, \tag{3.148}$$

where we have dropped the vacuum energy of $\hbar\omega/2$, as it is not important for our purposes. The definitions of various states and operators for cavity modes are given in Appendix D.

Since no mirror is completely reflecting, the light trapped in the cavity will leak out of both mirrors. We often assume that one mirror has much less transmission than the other, and that light leaks out of only one of the mirrors. The rate at which the energy in a mode leaks out through a mirror is given by γE, where E is the energy in the mode. We refer to γ as the decay rate of the mirror, or *output channel*. The partial transmittance of the mirrors also provides a way to inject light into the cavity modes. When we shine a laser on one of the mirrors, some of the light will be transmitted into the cavity, and stored in a mode, so long as the frequency of the laser is not too far from that of one of the modes. Injecting light into a cavity mode is referred to as "driving the cavity."

Since the input beam comes through the same mirror as the output beam, in order to detect the output light we need some way to split the output beam off from the input. This requires a *non-reciprocal* device – something through which light takes a different path when its direction is reversed. It is possible to achieve this using a Faraday rotator, in which a magnetic field rotates the polarization of a light beam, along with a polarization beam-splitter. The non-reciprocal behavior is made possible because magnetic forces do not have time-reversal invariance. The non-reciprocal device depicted in Fig. 3.3 is called a circulator. This name indicates that light that enters via one of its three "ports" will exit via the port immediately to the left. Thus we can speak of light "going around" the circulator in only one direction.

Detecting photons leaking out of an optical cavity

We now consider a continuous quantum measurement that involves a jump process. This is the classic example of photon counting, useful in quantum optics, and a similar analysis applies to electron counting in mesoscopic circuits. To count photons we use a photo-detector, or photon-detector, to measure the light that leaks out of an optical cavity. Quantum mechanically the loss of energy from the cavity may be thought of as the emission of photons at random intervals. The rate at which photons will be detected is the same as the rate at which they escape from the cavity, assuming the detector is perfectly efficient. By reasoning what this rate should be, and matching it to the form that all counting measurements must take (Eqs. 3.143 and 3.144), we can determine the operators that describe a photon-counting measurement.

There are two arguments that give us the rate at which photons leak out of the cavity. The first is that since the energy in the mode is proportional to the photon number, the rate at which photons leak out should be the same as that at which the energy leaks out (divided by the energy of a single photon). This gives the average rate of photo-detections as $\gamma \langle a^\dagger a \rangle$. The second argument is more quantum mechanical in nature: if a single photon has a given probability per unit time of escaping, then if there are N photons in the cavity, the average rate of escape will be proportional to N. This implies that the average rate of counting will be $\gamma \langle a^\dagger a \rangle$, since $\langle a^\dagger a \rangle$ is the average number of photons in the cavity.

Matching the average rate of photon emissions to the expression for the average count rate in Eq. (3.147), we have

$$\lambda(t) = \text{Tr}\left[\Omega^\dagger \Omega \rho(t)\right] = \gamma \,\text{Tr}\left[a^\dagger a \rho(t)\right].$$

(3.149)

This means that $\Omega = \sqrt{\gamma} U a$, where U is an arbitrary unitary. Let us see what happens if we make the simplest choice, $U = I$. The operator that corresponds to a single detection event is then

$$A_1 = \sqrt{\gamma\,dt}\,a. \qquad (3.150)$$

This means that when a photon is detected the, effect on the state of the cavity is

$$\rho \rightarrow = \frac{A_1 \rho A_1^\dagger}{\mathrm{Tr}\left[A_1^\dagger A_1 \rho\right]} = a\rho a^\dagger. \qquad (3.151)$$

This make sense, because the action of the annihilation operator a is to reduce the number of photons in the cavity by one! As we expect, each photo-detection corresponds to one photon being removed from the cavity.

We will see in Chapter 4 that a model in which the damping of the cavity mode is derived from an interaction with the electromagnetic modes outside the cavity (effectively a thermal bath) results in the same operator U that we chose above (indeed, any other choice of U would be strange). We therefore conclude that photon counting (detecting photons emitted by an optical cavity) is described by the measurement operators

$$A_1 = \sqrt{\gamma\,dt}\,a, \qquad (3.152)$$

$$A_0 = I - \frac{\gamma}{2}a^\dagger a\,dt. \qquad (3.153)$$

Stochastic Schrödinger equation for photon counting

We can write the evolution of the state of the optical cavity, under photo-detection, as a stochastic equation using the Poisson stochastic increment dN. We merely have to construct the equation so that when $dN = 0$ it reduces to the action of operator A_0, and likewise when $dN = 1$ it reduces to that of A_1. If the state of the cavity is pure, then the action of the measurement operators is

$$\text{result 0:} \quad |\psi\rangle \rightarrow |\psi'\rangle = \frac{\left[1 - (\gamma/2)a^\dagger a\,dt\right]|\psi\rangle}{\sqrt{1 - \gamma\langle a^\dagger a\rangle\,dt}},$$

$$= \left[1 - (\gamma/2)(a^\dagger a - \langle a^\dagger a\rangle)dt\right]|\psi\rangle, \qquad (3.154)$$

$$\text{result 1:} \quad |\psi\rangle \rightarrow |\psi'\rangle = \frac{a|\psi\rangle}{\sqrt{\langle a^\dagger a\rangle}}, \qquad (3.155)$$

where $\langle a^\dagger a\rangle = \langle\psi|a^\dagger a|\psi\rangle$.

To write the SSE we determine $d|\psi\rangle$ for each of the two cases by setting $|\psi\rangle + d|\psi\rangle = |\psi'\rangle$ and solving for $d|\psi\rangle$. The result is

$$d|\psi\rangle = -\left(\frac{i}{\hbar}H + \frac{\gamma}{2}(a^\dagger a - \langle a^\dagger a\rangle)\right)dt|\psi\rangle + \left(\frac{a}{\sqrt{\langle a^\dagger a\rangle}} - 1\right)dN|\psi\rangle, \qquad (3.156)$$

where we have included the Hamiltonian evolution, and it is understood that the rate of the Poisson process at time t is given by $\lambda = \gamma \langle a^\dagger a \rangle$ at time t. Note that we do not have to include the Hamiltonian evolution for the case when $dN = 1$, because the jumps are instantaneous, and so no other evolution happens during that time. For the same reason we do not have to turn off the evolution for $dN = 0$ during that for $dN = 1$. Another way to state this is to apply the rule that $dN\,dt = 0$. Once you are familiar with Ito calculus this rule will not seem unusual, and it is not difficult to convince yourself that it is true in the continuum limit.

If we average the density matrix over the two values of the Poisson increment in a time-step dt, then we recover from the above SSE the master equation for a damped (lossy) optical cavity. We leave this derivation as Exercise 14. The resulting master equation is

$$d\rho = -\frac{i}{\hbar}[H, \rho]\,dt - \frac{\gamma}{2}(a^\dagger a\rho + \rho a^\dagger a - 2a\rho a^\dagger). \tag{3.157}$$

Here the Hamiltonian, $H = \hbar\omega(a^\dagger a + 1/2)$, gives the evolution of the light in the cavity, where ω is the angular frequency of the light.

Linear stochastic equations for Poisson noise

Like Ito stochastic equations, Poisson stochastic equations can sometimes be solved by first writing them in a linear form. To do this we use the same procedure as for Ito equations: we change the normalization, and we change the probability distribution for the noise to one that is independent of the state. In the case of Poisson noise it is the rate of the jumps, λ, that is dependent on the state, so we now simply make this rate a constant. Instead of normalizing the state after each time-step, we divide the state by the square root of the probability for whichever value of dN was chosen in that time-step. This means that the true probability that a given final state occurs is the product of the square norm of the state (the trace of the density matrix) with the probability density for the realization of the Poisson process.

Since we chose the Poisson process to have a constant rate λ, the probability for $dN = 0$ is $P(0) = 1 - \lambda\,dt$, and for $dN = 1$ is $P(1) = \lambda\,dt$. Using this new normalization, the two possible transformations of the state in each time-step are now

$$\text{jump:} \quad |\psi\rangle \to \frac{\sqrt{\gamma}a|\psi\rangle}{\sqrt{\lambda}} \qquad \text{[replaces Eq. (3.155)]} \tag{3.158}$$

$$\text{no jump:} \quad |\psi\rangle \to \frac{\left(1 - \frac{\gamma}{2}a^\dagger a\,dt\right)|\psi\rangle}{\sqrt{1 - \lambda\,dt}} \qquad \text{[replaces Eq. (3.154)]} \tag{3.159}$$

$$= \left[1 - \frac{1}{2}\left(\gamma a^\dagger a - \lambda\right)dt\right]|\psi\rangle.$$

Putting these two equations together, and setting $\lambda = \gamma$, we have the linear stochastic Schrödinger equation

$$d|\psi\rangle = \left[\frac{\gamma}{2}(1 - a^\dagger a) - \frac{i}{\hbar}H\right]dt|\psi\rangle + (a - 1)dN|\psi\rangle. \tag{3.160}$$

As an example we now solve this equation for the simplest case, in which the Hamiltonian of the cavity is just the free Hamiltonian given by Eq. (3.148). The linear SSE is then

$$d|\psi\rangle = \left[\frac{\gamma}{2} - \left(\frac{\gamma}{2} + i\omega\right)a^\dagger a\right]dt|\psi\rangle + (a - 1)dN|\psi\rangle. \tag{3.161}$$

To solve equations like this we first apply the relation

$$1 + A\,dN = e^{\ln(1+A)\,dN} = (A + 1)^{dN}, \tag{3.162}$$

which follows from the fact that $(dN)^2 = dN$, and also $dN\,dt = 0$. This gives us

$$|\psi(t + dt)\rangle = |\psi(t)\rangle + d|\psi(t)\rangle$$
$$= \exp\left\{\left[\frac{\gamma}{2} - \left(\frac{\gamma}{2} + i\omega\right)a^\dagger a\right]dt\right\}a^{dN}|\psi(t)\rangle. \tag{3.163}$$

We can now obtain the solution at time t by repeatedly applying this time-stepping equation to an initial state $|\psi(0)\rangle$. To put the result in a closed form, we must repeatedly swap the order of the operator a with an exponential of the form $\exp(\varepsilon a^\dagger a)$. Methods for obtaining reordering relations are discussed in Appendix G. To obtain the relation we need here we first determine the following transformation:

$$\exp(\varepsilon a^\dagger a)a\exp(-\varepsilon a^\dagger a) = a\exp(\varepsilon), \tag{3.164}$$

which can be done by solving a simple linear differential equation (see Appendix G). Rearranging the above equation gives us the order-swapping relation

$$\exp(\varepsilon a^\dagger a)a = a\exp(\varepsilon a^\dagger a)\exp(\varepsilon). \tag{3.165}$$

Using the above relation repeatedly, we obtain the closed-form solution to Eq. (3.161) as

$$|\psi(t)\rangle = \exp\left\{\left[\frac{\gamma}{2} - \left(\frac{\gamma}{2} + i\omega\right)a^\dagger a\right]t\right\}\exp\left\{-\left(i\omega + \frac{\gamma}{2}\right)Y(t)\right\}a^{N(t)}|\psi(0)\rangle. \tag{3.166}$$

This equation contains two random variables. The first is $N(t)$, whose distribution is given in Eq. (3.138). The second is

$$Y(t) = \int_0^t t'\,dN(t') = \sum_{n=0}^{N(t)} t_n, \tag{3.167}$$

where the t_n are the times at which the jumps occur. We can calculate the probability distribution for $Y(t)$ as follows. We first note that the probability distributions for each of the N photon-detection times, t_n, *given* that N detections have occurred, are uniform over the interval $[0, t]$. Thus

$$P(t_n) = \begin{cases} 1/t & 0 < t' < t \\ 0 & \text{otherwise} \end{cases}. \tag{3.168}$$

We also note that since the detection times are all independent, the joint probability density for N detections, $P(\{t_n\}, N, t)$ is simply the product $\Pi_{n=1}^{N} P(t_n)$. The probability density for Y, being the sum of the detection times, given that N photons have been detected, $P(Y|N, t)$, is the convolution of the probability distributions for each of them. The joint probability density for N and Y is then

$$P(N, Y, t) = P(N, t)P(Y|N, t). \tag{3.169}$$

Of course, to obtain the true probability for the final state parametrized by N and Y, we must also multiply by the square norm of $|\psi(t)\rangle$.

We can now simplify considerably the expression for the final state, because the random variable $Y(t)$ affects only the norm of the state, along with an overall phase factor that can be ignored. Thus after we have normalized the state, it will not depend on $Y(t)$ at all! We can therefore eliminate Y from the expression for $|\psi(t)\rangle$. The new final state is

$$|\psi(N, t)\rangle = e^{-\gamma Y(t)/2}|\psi(t)\rangle$$
$$= \exp\left\{[-(i\omega + \gamma/2)a^{\dagger}a + \gamma/2]t\right\}a^{N}|\psi(0)\rangle. \tag{3.170}$$

The true probability distribution for this final state is then given by integrating the true probability distribution over Y, to give the marginal for N. However, in doing this it is simpler to treat the t_n as the basic random variables, of which Y is merely a function. This means that we integrate over all the t_n rather than over Y. This procedure is easier because we have a simpler expression for the joint probability density for the t_n than we do for the function Y. Denoting the square norm of $|\psi(N, t)\rangle$ by $\mathcal{N}(N)$, we integrate out the detection times as follows

$$P(|\psi\rangle, t) = \int \left[\langle\psi(t)|\psi(t)\rangle \tilde{P}(N, \{t_n\}, t) \right] dt_1 \cdots dt_n$$
$$= \mathcal{N}(N)P(N, t)\int P(\{t_n\}|N, t)e^{-\gamma Y} dt_1 \cdots dt_n$$
$$= \mathcal{N}(N)\left[\frac{e^{-\gamma t}(1 - e^{\gamma t})^{N}}{N}\right]. \tag{3.171}$$

Since the action of both $a^{\dagger}a$ and a on coherent states is simple, it is easy to use the general solution, given by Eqs. (3.170) and (3.171), to determine the evolution if the initial state is written as a superposition or mixture of coherent states.

3.4 Homodyning: from counting to Gaussian noise

It turns out that a photo-detector can be used to perform a Gaussian continuous measurement on an optical mode. This is the optical measuring technique known as *homodyne detection*. It is possible for a jump process to become a Gaussian process when the effect on the system of each jump (each detection event) is very small. If the number of jumps N in a given time interval Δt is large, then the Poisson distribution approximates a Gaussian distribution, and in the limit in which $N \to \infty$, the Poisson is Gaussian. As we will see, a measurement can operate in this limit so long as the effect of each jump on the system also tends to zero in such a way that the total effect on the system in each interval Δt tends to zero as $\Delta t \to 0$.

Homodyne detection works as follows. Instead of directly measuring the output light from the cavity with a photo-detector, we first interfere this light with a beam of coherent light of the same frequency on a beam-splitter. This coherent beam is called the "local oscillator." A beam splitter takes two input modes, described by annihilation operators a and b, and transforms them to two output modes $\tilde{a}$ and $\tilde{b}$ via the relations

$$\tilde{a} = \sqrt{\eta}a + i\sqrt{1-\eta}b, \tag{3.172}$$

$$\tilde{b} = \sqrt{\eta}b + i\sqrt{1-\eta}a. \tag{3.173}$$

So if we detect output mode $\tilde{a}$, the operator that gives the effect of a photo-detection (the jump operator) is $\tilde{a}$, but as far as the cavity mode and local oscillator are concerned, this is now equal to $\sqrt{\eta}a + i\sqrt{1-\eta}b$. If we further choose the state of the b mode to be a coherent state with amplitude α, then we can replace the operator b with the number α, since the coherent state is an eigenstate of b. The jump operator $\tilde{a}$ is now equal to $\sqrt{\eta}a + i\sqrt{1-\eta}\alpha$. Finally, we make the beam-splitter highly transmitting, so that η is close to unity, and make α large so that the product $\beta \equiv \sqrt{1-\eta}\alpha$ is appreciable. We will also take β to be real. The jump operator can now be written as

$$\sqrt{\gamma}\tilde{a} = \sqrt{\gamma}(a + i\beta). \tag{3.174}$$

The average rate of detections is given by

$$\gamma\langle\tilde{a}^{\dagger}\tilde{a}\rangle = \gamma\langle a^{\dagger}a\rangle - i\gamma\beta\langle a - a^{\dagger}\rangle + \gamma|\beta|^2. \tag{3.175}$$

We assume that the amplitude of the local oscillator is much larger than that of the cavity mode, so that $\langle a^{\dagger}a\rangle$ is insignificant compared to β^2 and can be ignored. Apart from the offset $\gamma\beta^2$, which can be subtracted, the average rate of photo-detections is proportional to the observable

$$Y = \frac{-i(a - a^{\dagger})}{2}, \tag{3.176}$$

which is the "phase quadrature" of the light in the cavity. We therefore expect that homodyne detection will provide a continuous measurement of Y. (If one choses β to be imaginary, this will give a measurement of the amplitude quadrature, $X = (a + a^\dagger)/2$, instead. More generally, choosing $\beta = |\beta|e^{i\theta}$ will realize a measurement of an arbitrary quadrature $X_\theta = [ae^{-i(\theta+\pi/2)} + a^\dagger e^{i(\theta+\pi/2)}]/2$.)

We also need to know the evolution of the system during a time interval dt in which there is no detection (no jump). The simplest way to do this is to use the fact that since the photo-detector detects the light *after* it leaves the cavity, the mere presence or absence of the detector cannot change the evolution of the cavity. So if we don't know what the detector records, the evolution of the cavity mode must be the same as if there was no detector. Thus in the absence of the detector, the evolution of the cavity in the time dt is given by averaging the density matrix over the two possible detection outcomes. This gives the master equation

$$d\rho = -\frac{i}{\hbar}[H, \rho]\,dt - \frac{\gamma}{2}(a^\dagger a\rho + \rho a^\dagger a - 2a\rho a^\dagger). \qquad (3.177)$$

Further, since the beam-splitter and local oscillator are also added *after* the output light has left the cavity, then the average evolution must still be equal to Eq. (3.177) when performing homodyne detection. The evolution in the time interval dt when there is no jump must therefore be whatever is required so that the average evolution is still given by Eq. (3.177), when the jump operator is $\tilde{a} = a + i\beta$. To satisfy this requirement the evolution "between jumps" must be

$$d|\psi\rangle = \left[\frac{-i}{\hbar}H - \frac{\gamma}{2\mathcal{N}}(a^\dagger a - 2i\beta a + \beta^2)\right]dt|\psi\rangle, \qquad (3.178)$$

where $\mathcal{N}$ is the normalization.

It is important to note that the local oscillator is oscillating at the cavity frequency ω, so we should really have written β as $\beta e^{-i\omega t}$. However, if we move into the interaction picture with respect to $H_0 = \hbar\omega a^\dagger a$, then the oscillation of the mode operators a and $a^\dagger$ cancels the oscillation of the local oscillator. So by leaving out the time-dependence of β we have in fact been working in the interaction picture. We now make this explicit by replacing H in the above equation with the Hamiltonian of the cavity mode in the interaction picture, denoted by H_I. If the cavity mode is only evolving under H_0, then $H_I = 0$.

We now consider the evolution of the cavity mode during a time interval Δt that is short compared to the decay time $1/\gamma$, and in which the number of photo-detections is $M \gg 1$. The evolution of the system during this time is then given by M applications of the jump operator at M random times, and applications of the intervening evolution, Eq. (3.178), in between these times. Because $\gamma\Delta t \ll 1$ (and because we will take the continuum limit in which $\Delta t \to 0$), we can move all the jump operators to the right through the intervening evolution operators because they are all proportional to Δt, and thus commute to first-order in Δt. Doing this, and dropping terms that only affect the normalization of the state, the evolution in time Δt is

$$|\psi(t+\Delta t)\rangle = \exp\left\{-\left[\frac{i}{\hbar}H_{\mathrm{I}} + \frac{\gamma}{2}a^{\dagger}a - i\gamma\beta a\right]\Delta t\right\}\left(1 + i\frac{a}{\beta}\right)^{M}|\psi(t)\rangle. \qquad (3.179)$$

The average number of jumps in the interval Δt is approximately $\langle M\rangle = \gamma\beta^{2}\Delta t$. So to make M large we need to choose $\beta \gg 1/\sqrt{\gamma\Delta t}$. At the same time, to ensure that the change in the state is small in the interval Δt, we choose $\beta \ll 1/(\gamma\Delta t)$, and $\gamma \ll 1/\Delta t$.

Since $\langle M\rangle$ is large, the Poisson distribution for M can be approximated by a Gaussian with mean $\langle M\rangle = (\gamma\beta^{2} + 2\gamma\beta\langle Y\rangle)\Delta t$. It turns out the variance of this Gaussian is given by $V = \gamma\beta^{2}\Delta t$ (the precise reason for this is a little involved, and may be found in [679]). We can therefore write M as

$$M = \gamma\beta[\beta + 2\langle Y(t)\rangle]\Delta t + \sqrt{\gamma}\beta\Delta W, \qquad (3.180)$$

where ΔW is as usual an increment of the Wiener process. Substituting this expression for M into Eq. (3.179), and replacing Δt with dt, we get

$$d|\psi\rangle = \left[-\left(\frac{i}{\hbar}H_{\mathrm{I}} + \frac{\gamma}{2}a^{\dagger}a - i\gamma a\langle Y\rangle\right)dt + i\sqrt{\gamma}\,a\,dW\right]|\psi\rangle. \qquad (3.181)$$

Since we have ignored the normalization in our derivation, we must now normalize $|\psi(t+dt)\rangle = |\psi\rangle + d|\psi\rangle$ to first-order in dt, and the result is the SSE for the continuous measurement realized by homodyne detection:

$$d|\psi\rangle = \left\{-\left(\frac{i}{\hbar}H_{\mathrm{I}} + \frac{\gamma}{2}\left[a^{\dagger}a + i2\,a\overline{Y} + \overline{Y}^{2}\right]\right)dt + \sqrt{\gamma}\left(ia + \overline{Y}\right)dW\right\}|\psi\rangle, \qquad (3.182)$$

where we have defined $\overline{Y} = \langle Y\rangle$ for compactness. The measurement signal, dy, is the number of photo-detections in each time interval dt, which is given by M. Scaling this by $1/(2\gamma\beta)$, and removing the constant offset, we have

$$dy = \overline{Y}\,dt + \frac{dW}{\sqrt{4\gamma}}. \qquad (3.183)$$

So homodyne detection provides a continuous measurement of Y with measurement strength $k = \gamma/2$; but this is not the same kind of continuous measurement that we derived in Section 3.1.2. The reason for this is that homodyne detection does not only measure Y, but at the same time extracts energy from the cavity as photons leak out. The measurement we have derived here is a generalization of the measurement formalism we derived previously that includes dissipation.

The SSE for homodyne detection still provides a valid evolution for a quantum state if we replace ia with an arbitrary operator A. Further replacing γ with $2k$, and writing the resulting stochastic equation as an SME, we obtain

$$d\rho = -\frac{i}{\hbar}[H,\rho]\,dt - k(A^\dagger A\rho + \rho A^\dagger A - 2A\rho A^\dagger)\,dt$$
$$+\sqrt{2k}(A\rho + \rho A^\dagger - \langle A+A^\dagger\rangle\rho)\,dW,\tag{3.184}$$

and this has the corresponding measurement record

$$dr = \frac{\langle A+A^\dagger\rangle\,dt}{2} + \frac{dW}{\sqrt{8k}}.\tag{3.185}$$

This general SME reduces precisely to that for the measurement of an observable X, Eq. (3.102), if we set $A = X$, since $X^\dagger = X$.

The linear SSE that is equivalent to the above nonlinear SME is

$$d|\hat{\psi}\rangle = \left[\left(-\frac{i}{\hbar}H - kA^\dagger A\right)dt + \sqrt{2k}A\,dW\right]|\hat{\psi}\rangle.\tag{3.186}$$

From this one can immediately obtain the equivalent linear SME.

Same master equation, different unraveling

In the previous section we found that the master equation describing the average evolution of a damped optical cavity, namely,

$$d\rho = -\frac{\gamma}{2}(a^\dagger a\rho + \rho a^\dagger a - 2a\rho a^\dagger).\tag{3.187}$$

has not one, but *two* equivalent stochastic master equations. One of these, Eq. (3.182), contains the Wiener noise process, and the other, Eq. (3.156), contains the Poisson process. We have also seen that these different SMEs correspond to different ways of measuring the field coupled to the system. An SME with unit efficiency is often referred to as an *unraveling* of the master equation, a term coined by Carmichael [100].

Note that if we change the interaction between the system and the field, or more generally the system's environment, the master equation changes. But if we choose a different way to measure the field, it does not. Nevertheless, it is the system/field interaction that determines what information is extracted about the system: if the system interacts with the field via the Hermitian operator X, then the resulting continuous measurement extracts information about the value of X. Therefore the various unravelings that correspond to a given master equation all involve measurement of the same observable, but give different ways to extract it. This makes statements like "making a different measurement" somewhat ambiguous; it could mean changing the observable being measured, or merely changing the unraveling. In this book, since we will usually be concerned with application using Gaussian continuous measurement, when we refer to changing the measurement we mean changing the observable being measured. Similarly when we refer to a measurement of an observable X, we will mean a Gaussian continuous measurement given by the unraveling in Eq. (3.49) unless stated otherwise. Nevertheless, even if one is restricted to Gaussian measurements, there are many distinct ways that a master equation can be "unravelled." All the Gaussian unravelings for a given master equation have been characterized by Wiseman and Diosi [676] and Chia and Wiseman [118].

3.5 Continuous measurements with more exotic noise?

So far we have considered continuous measurements that involve Gaussian or Poisson noise. Both of these arise naturally in physical measurements and are applicable to many situations. So are there any continuous measurements that require other kinds of noise processes? The answer appears to be no, in a qualified sense explained below, at least for measurements in which the measurement noise is not correlated across different time-steps. Note that the Gaussian and Poissonian measurements that we have been considering both have this property – the noise on the measurement results, and hence the increments of the noise, dW and dN, are independent of each other at different times. The reasons we do not consider measurements in which the noise is correlated across different times are (1) that such measurements would be tremendously more difficult to analyze, and (2) measurements with independent noise increments are not only much simpler, but very useful, as they are applicable to many real situations.

If we restrict ourselves to noise processes in which all the infinitesimal increments are independent, and have identical distributions, then there is a lot one can say about them. This class of noise processes (the definition of which includes a third, somewhat technical condition [288]) are called the *Lévy processes*. Lévy has shown that such noise processes come in only three types: Gaussian (Wiener) noise, Poissonian noise (that is, noise with a finite rate of jumps), and a third kind, which has an infinite number of infinitesimal jumps in any finite time interval. While the possibility of measurements driven by the third kind of noise has not yet been fully investigated, this has been explored for a special class of these exotic noise processes called the *stable Lévy processes*. It has been shown that a continuous measurement cannot have the increments of a stable process as its noise process, with Gaussian noise as the one exception [292]. It therefore seems likely that measurements cannot be realized with any of the exotic noise processes.

The qualification to the above statements is that it is possible for continuous measurements to have noise that involves the distributions of stable Lévy processes, if these processes are combined in a specific way with Gaussian or Poissonian noise. In the case of Poissonian noise, this involves applying, each time a jump occurs, a single-shot measurement whose noise is distributed like the increment of a stable process. Further details regarding Lévy processes can be found in [288], and further details regarding their role in quantum measurement theory can be found in [292].

3.6 The Heisenberg picture: inputs, outputs, and spectra

Here we introduce the "input–output," or "quantum noise" formulation of open quantum systems. An *open system* is a system that is subject to damping or thermalization because it is coupled to a large, and therefore thermal, environment. In fact, continuous measurements are always mediated by such an environment, and if we discard the measurement results the resulting master equation describes an open system. The input–output formalism makes it easy to derive the equations of motion for the means and variances of the coordinates of an open system, and to calculate the power spectrum of the measurement record for a measured system. We showed how to calculate this spectrum using a slightly simpler method in Section 3.1.4 above, but this method was only applicable to measurements of Hermitian operators. In later chapters we will find that the input–output formalism allows us to inject the output from one system into the input of another (see Chapter 5), and provides a way to derive the Markovian master equation for open systems (see Appendix F).

We will explain how to describe damping and thermalization of quantum systems in Chapter 4, but it is worth discussing this subject a little now. Consider as an example of an irreversible process the damping of a single mode of an optical cavity, in which light leaks out via one of the mirrors. The reason the light leaks out can be thought of as a result of coupling between the light inside the cavity and the continuum of modes of the electromagnetic field outside. These outside modes constitute a system with many degrees of freedom, and therefore form a thermal environment. At room temperature there are no photons in the optical modes outside the cavity, and so the environment is effectively at zero temperature. That is why we do not need to worry about light leaking into the cavity from outside. If we place a photon-counter outside the cavity, then we can detect the photons as they leak out. The photon counting is a continuous measurement and gives us a measurement record. But if we take the photon-counter away, the situation is merely that of an optical cavity undergoing a damping process due to its coupling to a large thermal system. Before we can make a continuous measurement on the cavity mode, we need first to couple this mode to a large system. The photo-detector then measures this large system – in this case the electromagnetic modes outside the cavity – to realize the continuous measurement. Continuous measurement and coupling to a thermal system are therefore intimately related. The input–output formalism provides a way to explicitly determine the output that flows from the system into the bath, and to connect a measurement of the bath with the resulting continuous measurement of the system.

Quantum noise and the input–output formalism

We now present the equations that give the Heisenberg picture, or input–output description of an open quantum system, and explain their meaning. In this section we present the Heisenberg equations for a thermal environment at zero temperature. This means that the gaps between the energy levels of the system are large enough that the modes of the field to which it couples are all in the their ground state at the ambient temperature. It is this case that corresponds to the stochastic master equation for continuous measurement. We describe the modifications required to treat systems at non-zero temperature below. It is also important to note that the input–output formalism is only strictly valid when the damping rate or rates of the system are much smaller than the frequencies corresponding to the gaps between the system's energy levels. The quantum noise equations that can be used for a strongly damped harmonic oscillator are given in Section 4.3.5. We do not discuss the derivation of the input–output formulation here, but we give a full derivation in Appendix H. Treatments that give even more in-depth analysis can be found in [199] and [201].

As we have seen above, the SME for a measured system subjected to a Gaussian continuous measurement of the operator L is

$$d\rho = -\frac{i}{\hbar}[H,\rho]dt - \left[L^\dagger L\rho + \rho L^\dagger L - 2L\rho L^\dagger\right]dt$$
$$+ \sqrt{2}\left[L\rho + \rho L^\dagger - \langle L+L^\dagger\rangle\rho\right]dW, \qquad (3.188)$$

where the dW is an increment of Wiener noise. The measurement record, $r(t)$, is given by

$$dr = \sqrt{2}\langle L+L^\dagger\rangle\,dt + dW. \qquad (3.189)$$

In the Heisenberg picture the evolution of a quantum system is described by an equation of motion for the operators instead of the states. The Heisenberg equation corresponding to the above master equation is

$$dY = -\frac{i}{\hbar}[Y,H]dt - \left[YL^\dagger L + L^\dagger LY - 2L^\dagger YL\right]dt$$
$$+ \sqrt{2}\left(\left[Y,L^\dagger\right]db_{\text{in}} - db_{\text{in}}^\dagger[Y,L]\right). \qquad (3.190)$$

This equation needs some explaining. The operator Y is an arbitrary operator of the open quantum system, and db_{in} is an *operator version* of the Wiener noise increment dW. The fact that db_{in} is an operator is the origin of the term "quantum noise." The commutation relation for db_{in} is very similar to that for the annihilation operator, a, but captures the fact that db_{in} drives the system with noise. The commutation relation for the noise operator is

$$[db_{\text{in}}, db_{\text{in}}^\dagger] = dt, \qquad (3.191)$$

and the products of the noise operators are

$$db_{\text{in}}db_{\text{in}}^\dagger = dt, \quad db_{\text{in}}^\dagger db_{\text{in}} = 0. \qquad (3.192)$$

The commutator and products of the noise operators are the quantum noise equivalent of Ito's rule, $(dW)^2 = dt$. Equation (3.190) is called a quantum stochastic differential equation (QSDE) or a "quantum Langevin equation".[2]

In deriving the quantum Langevin equations, the environment is modeled as a continuum of modes, and thus as a quantum field. Because of this it is simplest to consider a concrete

---

[2] The QSDE is often referred to as a "quantum Langevin equation," after the person who first wrote down a classical differential equation driven by noise [354]. A good approximation to the pronunciation of Langevin is the following: *longe-e-varn*, where *longe* sounds like "plunge" with the u replaced with an o, *e* is the vowel sound in "get," and *varn* rhymes with "barn."

example, consisting of an optical cavity in which light leaks out through one of the mirrors. The electric field outside the cavity can be split into two parts, one containing waves that are traveling inward (toward the cavity), and the other the waves traveling outward (away from the cavity). The noise operator $db_{in}^{\dagger}(t)$ is the slice of the inward-traveling field that interacts with the cavity at time t. This noise operator therefore describes the *input* to the cavity. Similarly there is an output operator that describes the state of the field traveling away from the cavity, at a fixed location, as a function of time. Since the field is traveling away at speed c, the location at which the output field is evaluated is only important up to in a shift in time. If this location is the point at which the field leaves the cavity, then the output operator is

$$db_{out} = db_{in} - \sqrt{2}L(t)\,dt. \tag{3.193}$$

The operator L is the operator though which the system couples to the environment, and so we will call it the coupling operator.

Of course, the operator db_{in} is an *increment* that gives the effect on the system, rather than the input field itself. The input and output fields are actually

$$b_{in}(t) = \frac{db_{in}}{dt} \quad \text{and} \quad b_{out}(t) = \frac{db_{out}}{dt}, \tag{3.194}$$

and in terms of the fields the input–output relation is

$$b_{out}(t) = b_{in}(t) - \sqrt{2}L(t). \tag{3.195}$$

The commutation relations for the input fields are

$$[b_{in}(t+\tau), b_{in}^{\dagger}(t)] = \delta(\tau), \tag{3.196}$$

$$[b_{in}(t+\tau), b_{in}(t)] = [b_{in}^{\dagger}(t+\tau), b_{in}^{\dagger}(t)] = 0, \tag{3.197}$$

and thus the field operators commute with themselves at different times. While it is certainly not obvious from the expression for the output field given above, it has the same commutation relations as the input field. In fact this has to be true because, like the input field, it is a freely propagating field. So we have also

$$[b_{out}(t+\tau), b_{out}^{\dagger}(t)] = \delta(\tau), \tag{3.198}$$

$$[b_{out}(t+\tau), b_{out}(t)] = [b_{out}^{\dagger}(t+\tau), b_{out}^{\dagger}(t)] = 0. \tag{3.199}$$

It is also useful to note that at time t the system has not yet been influenced by the input field $b_{in}(t')$ for $t' > t$. Thus the system operators at time t commute with the input field at time $t' > t$. Similarly the output field at time t contains no information about the system state at time $t' > t$, and so the output field at time t commutes with the system operators at time $t' > t$:

$$[Y(t), b_{\text{in}}(t+\tau)] = [Y(t), b_{\text{in}}^\dagger(t+\tau)] = 0 \quad \text{for } \tau > 0, \qquad (3.200)$$

$$[Y(t+\tau), b_{\text{out}}(t)] = [Y(t+\tau), b_{\text{out}}^\dagger(t)] = 0 \quad \text{for } \tau > 0. \qquad (3.201)$$

In deriving the quantum Langevin equations above, the inward-traveling field modes are all taken to be in the vacuum state. This means that the expectation value of any product of operators in which the rightmost operator is b_{in}, or the leftmost operator is $b_{\text{in}}^\dagger$, is zero:

$$\langle \mathcal{O}(t') b_{\text{in}}(t) \rangle = \left\langle b_{\text{in}}^\dagger(t) \mathcal{O}(t') \right\rangle = 0 \qquad (3.202)$$

for any operator $\mathcal{O}$ and time t'. Thus $\langle b_{\text{in}}^\dagger(t) b_{\text{in}}(t+\tau) \rangle = \langle b_{\text{in}}^\dagger(t) b_{\text{in}}^\dagger(t+\tau) \rangle = 0$. Combining this with the commutator in Eq. (3.196) gives

$$\langle b_{\text{in}}(t) b_{\text{in}}^\dagger(t+\tau) \rangle = \delta(\tau). \qquad (3.203)$$

The input fields are δ correlated for the same reason that the derivative of Wiener noise is δ correlated (Eq. 3.80).

The "input–output" formulation of open systems was devised by Collett and Gardiner [128, 199, 200]. In comparing their equations with ours, it should be noted that we use a different sign convention. We describe the various sign conventions for input–output theory in Section F.5.

Multiple simultaneous measurements

If the master equation describes multiple measurements, so that it has the form

$$d\rho = -\frac{i}{\hbar}[H, \rho]\,dt - \sum_n \left[L_n^\dagger L_n \rho + \rho L_n^\dagger L_n - 2 L_n \rho L_n^\dagger \right] dt$$
$$+ \sqrt{2} \sum_n \left[L_n \rho + \rho L_n^\dagger - \langle L_n + L_n^\dagger \rangle \rho \right] dW_n, \qquad (3.204)$$

where the noise increments dW_n are uncorrelated, then the equivalent quantum Langevin equation is

$$dY = -\frac{i}{\hbar}[Y, H]\,dt - \sum_n \left[Y L_n^\dagger L_n + L_n^\dagger L_n Y - 2 L_n^\dagger Y L_n \right] dt$$
$$+ \sqrt{2} \sum_n \left(\left[Y, L_n^\dagger \right] db_{\text{in}}^{(n)} - db_{\text{in}}^{(n)\dagger} [Y, L_n] \right). \qquad (3.205)$$

Here the quantum noise operators $db_{\text{in}}^{(n)}$ are mutually uncorrelated, so that

$$[db_{\text{in}}^{(n)}, db_{\text{in}}^{(m)\dagger}] = db_{\text{in}}^{(n)} db_{\text{in}}^{(m)\dagger} = \delta_{mn}\,dt. \qquad (3.206)$$

The input–output relations for the fields are

$$b_{\text{out}}^{(n)}(t) = b_{\text{in}}^{(n)}(t) - \sqrt{2}L_n. \tag{3.207}$$

The input–output formalism for $T > 0$

It is not until Chapter 4 that we will discuss in detail the treatment of systems coupled to thermal baths at non-zero temperature. But we present the input–output equations for such systems here for ease of reference. These equations are derived in Appendix F. The meanings of all the symbols are given above.

The quantum Langevin equations corresponding to the master equation for a system at temperature T, given in Eq. (4.79), are

$$dY = \frac{1}{i\hbar}[Y,H]\,dt + \sum_n \sqrt{\gamma}\left([Y,L_n^\dagger]db_{\text{in}}^{(n)} - db_{\text{in}}^{(n)\dagger}[Y,L_n]\right)$$
$$+ \frac{\gamma}{2}\sum_n [n_T(\omega_n)+1]\left(2L_n^\dagger YL_n - L_n^\dagger L_n Y - YL_n^\dagger L_n\right)dt$$
$$+ \frac{\gamma}{2}\sum_n n_T(\omega_n)\left(2L_n YL_n^\dagger - L_n L_n^\dagger Y - YL_n L_n^\dagger\right)dt, \tag{3.208}$$

where Y is an arbitrary operator for the system. Note that the operators L_n in this equation for dY differ from those in Eq. (3.205) above by a factor of $\sqrt{\gamma/2}$. In the language of Section 4.3.1 they are the single-transition operators obtained from the elements of the system operator through which the system couples to the bath. The operator L_n is the downward-jump operator for all transitions with frequency ω_n, and

$$n_T(\omega_n) = \left(\frac{1}{\exp(\beta\hbar\omega_n) - 1}\right), \tag{3.209}$$

where $\beta = 1/kT$ is the Boltzmann factor. The Ito rules for the field-operator noise increment corresponding to L_n are

$$db_{\text{in}}^{(n)}db_{\text{in}}^{(n)\dagger} = [n_T(\omega_n)+1]\,dt, \quad \text{and} \quad db_{\text{in}}^{(n)\dagger}db_{\text{in}}^{(n)} = n_T(\omega_n)\,dt, \tag{3.210}$$

and $db_{\text{in}}^{(n)\dagger}db_{\text{in}}^{(n)\dagger} = db_{\text{in}}^{(n)}db_{\text{in}}^{(n)} = 0$. The correlation functions of the input fields are

$$\langle b_{\text{in}}^{(n)}(t)b_{\text{in}}^{(n)\dagger}(t+\tau)\rangle = [n_T(\omega_n)+1]\delta(\tau), \tag{3.211}$$
$$\langle b_{\text{in}}^{(n)\dagger}(t)b_{\text{in}}^{(n)}(t+\tau)\rangle = n_T(\omega_n)\delta(\tau). \tag{3.212}$$

The commutators of the field operators, and the input–output relations for the fields, are those given above in Eqs. (3.206) and (3.207).

An important example of the above quantum Langevin equations are those for a harmonic oscillator. If the oscillator feels a linear force from a thermal bath (that is, is coupled to a thermal bath through its position operator $x \propto a+a^\dagger$), then since the oscillator has only a single frequency, ω, there is only one transition operator $L_1 = a$. The quantum Langevin equations reduce to

$$\dot{a} = -i\omega a - \frac{\gamma}{2}a - \sqrt{\gamma}b_{\text{in}}(t) \tag{3.213}$$

with

$$\langle b_{\text{in}}(t)b_{\text{in}}^{\dagger}(t+\tau)\rangle = \langle b_{\text{in}}^{\dagger}(t)b_{\text{in}}(t+\tau)\rangle + \delta(\tau) = [n_T(\omega) + 1]\delta(\tau). \tag{3.214}$$

$b_{\text{out}} = b_{\text{in}} - \sqrt{\gamma}a$, and δ is the oscillator's damping rate.

Connecting the input–output formulation to continuous measurement

It is important to understand that the quantum Langevin equation is not precisely equivalent to the stochastic master equation, Eq. (3.204). This is because it does not give the evolution of the system conditioned on the observer's measurement results. It describes the evolution of both the system and the output field *without* a measurement having been made on the field. This is evident because the output is an operator, $db_{\text{out}}(t)$, that describes the state of the quantum field at a particular place and time, and not a real number corresponding to a measurement record.

The input–output formulation is more general than the stochastic master equation given in Eq. (3.204) because the latter is specific to Gaussian measurements. Since the input–output formalism does not include a measurement, it can be used to describe a variety of ways that the output field can be measured. For example, the light output from a cavity can be measured with a photon-counter, and this gives a stochastic master equation and measurement record containing the Poisson process. The output field could be measured using homodyne detection instead, resulting in a measurement record with Gaussian noise.

The operator db_{out} gives specifically that part of the field that will be measured by a detector in the time interval dt. To use the input–output method to calculate the spectrum of the measurement record for a given continuous measurement, we need to determine the field operator that must be measured to obtain that continuous measurement. For a continuous measurement of a Hermitian observable, x, the measurement record contains $\langle x \rangle$, the Lindblad operator is $L = \sqrt{k}x$, and k is the measurement strength. Since it is the field operator b_{out} that contains L, an observer will measure x if he or she measures the Hermitian field operator

$$R(t) = \frac{b_{\text{out}} + b_{\text{out}}^{\dagger}}{\sqrt{8k}} = x + \frac{b_{\text{in}} + b_{\text{in}}^{\dagger}}{\sqrt{8k}}. \tag{3.215}$$

We now examine the autocorrelation function of $R(t)$ to verify that it corresponds to the autocorrelation function of the measurement record for a measurement of x. This autocorrelation function is

$$\langle R(t+\tau)R(t)\rangle = \frac{1}{8k}\left\langle \left[b_{\text{out}}(t+\tau) + b_{\text{out}}^{\dagger}(t+\tau)\right]\left[b_{\text{out}}(t) + b_{\text{out}}^{\dagger}(t)\right]\right\rangle. \tag{3.216}$$

There are some tricks involved in manipulating the above expression to put it into a form that we can easily interpret. Simply substituting in $b_{\text{out}} = b_{\text{in}} - \sqrt{2k}x$ and expanding will give cross terms between x and the input fields that we cannot easily evaluate. We need to first reorder the output operators so that when we perform the substitution these terms disappear.

Expanding the correlation function we have four terms:

$$8k\langle R(t+\tau)R(t)\rangle = \langle b_{\text{out}}(t+\tau)b_{\text{out}}(t)\rangle + \left\langle b_{\text{out}}^\dagger(t+\tau)b_{\text{out}}(t)\right\rangle$$

$$+ \left\langle b_{\text{out}}(t+\tau)b_{\text{out}}^\dagger(t)\right\rangle + \left\langle b_{\text{out}}^\dagger(t+\tau)b_{\text{out}}^\dagger(t)\right\rangle. \qquad (3.217)$$

When we substitute in $b_{\text{out}} = b_{\text{in}} - \sqrt{2k}x$ the cross terms coming from the second term have b_{out} on the right and $b_{\text{out}}^\dagger$ on the left, and so will vanish. We can ensure that this happens with the third term by using the commutator in Eq. (3.198) to swap b_{out} and $b_{\text{out}}^\dagger$. This gives

$$\left\langle b_{\text{out}}(t+\tau)b_{\text{out}}^\dagger(t)\right\rangle = \left\langle b_{\text{out}}^\dagger(t)b_{\text{out}}(t+\tau)\right\rangle + \delta(\tau) = 2k\langle x(t)x(t+\tau)\rangle + \delta(\tau).$$

We cannot do the same thing with the first and fourth terms. Substituting $b_{\text{out}} = b_{\text{in}} - \sqrt{2k}x$ into the first term, and dropping the terms that vanish, we have

$$\langle b_{\text{out}}(t+\tau)b_{\text{out}}(t)\rangle = \sqrt{2k}\langle b_{\text{in}}(t+\tau)x(t)\rangle. \qquad (3.218)$$

But x at time t commutes with the input field at time $t+\tau$ so long as $\tau > 0$, as per Eq. (3.200). One can show that x does not contain white noise because its motion is given by b_{in} filtered through the response function of the system. Since the input field already gives us a δ correlation at $\tau = 0$, any contribution from $\langle b_{\text{in}}(t)x(t)\rangle$ is irrelevant. Thus

$$\langle b_{\text{in}}(t+\tau)x(t)\rangle = \langle x(t)b_{\text{in}}(t+\tau)\rangle = 0. \qquad (3.219)$$

If we substitute for the output fields in the fourth term in Eq. (3.217), then the cross term that remains contains $x(t+\tau)$ and $b_{\text{in}}^\dagger(t)$, and these do not commute for $\tau > 0$. However, we can fix this by swapping the output field operators in the fourth term *before* we substitute in the input fields. Using the relevant commutator in Eq. (3.199) we have

$$\left\langle b_{\text{out}}^\dagger(t+\tau)b_{\text{out}}^\dagger(t)\right\rangle = \left\langle b_{\text{out}}^\dagger(t)b_{\text{out}}^\dagger(t+\tau)\right\rangle \propto \left\langle x(t)b_{\text{in}}^\dagger(t+\tau)\right\rangle = \left\langle b_{\text{in}}^\dagger(t+\tau)x(t)\right\rangle,$$

and so this term vanishes too. The final expression for the correlation function is

$$\langle R(t+\tau)R(t)\rangle = \frac{\langle x(t+\tau)x(t)\rangle + \langle x(t)x(t+\tau)\rangle}{2} + \frac{\delta(\tau)}{8k}. \qquad (3.220)$$

We see that this correlation function has a white-noise term proportional to $1/(8k)$, just as does the continuous measurement record. We can also show that the symmetrized auto-correlation function for the operator x, being the first two terms on the right-hand side of the above equation, gives the autocorrelation function for the expectation value of $\langle x(t) \rangle$. Thus Eq. (3.220) is equal to the spectrum of the measurement record, given in Eq. (3.81). Showing this equality takes a bit of work, so we have relegated it to Section F.8.

To conclude, we can calculate the autocorrelation function and power spectrum of a measurement of any observable x by using the input–output formalism to calculate the symmetrized autocorrelation function of the operator x, which is straightforward for linear systems. In fact it is simplest to bypass the autocorrelation function and calculate the power spectrum directly by transforming the Langevin equations to the frequency domain. We explain how to do this below in Section 3.7.2.

3.7 Heisenberg-picture techniques for linear systems

Linear systems are defined as those whose Hamiltonians are polynomial functions containing only creation and annihilation operators, and for which each term in the polynomial contains no more than two of these operators. Linear systems are so called because the equations of motion for the annihilation operators, and thus for any linear combinations of these operators, are linear. When the system is open, it remains linear so long as the Lindblad operators are linear combinations of annihilation and creation operators.

Linear systems may also be thought of as those in which the dynamical variables are canonical coordinates, the Heisenberg equations of motion for these coordinates are linear, and any measurements are of linear combinations of the coordinates. The input–output formalism introduced in Section 3.6 provides very useful methods for linear systems.

3.7.1 Equations of motion for Gaussian states

For linear systems, initial states that have a Gaussian form maintain this form under the evolution. Further, open (measured) linear systems transform initially non-Gaussian states to Gaussian states over time. Since Gaussian states are completely characterized by the means and variances (and covariances) of the canonical coordinates, if a system is linear one can obtain closed equations for these means and variances. If the initial state is Gaussian, then these equations of motion are all that is required to determine the evolution of the system, greatly simplifying the analysis. Further, if one averages over the measurement results, then the quantum Langevin equations provide a very simple method to obtain these equations. To obtain the equations of motion for the means and variances of the evolution conditioned on the measurement results, we can use the method of Section 5.5.2.

To obtain the equations of motion of the unconditioned (average) means and variances, we first form a vector of the operators of the canonical coordinates. If we were dealing with a single particle, then this vector would be $\mathbf{v} = (x, p)^{\mathrm{T}}$. Alternatively we could form this vector from the annihilation and creation operators. If we were dealing with two modes described respectively by a and b, then this vector would be $\mathbf{v} = (a, a^{\dagger}, b, b^{\dagger})^{\mathrm{T}}$. The order of the elements is unimportant. Next we write down the quantum Langevin equations for the elements of $\mathbf{v}$. For a linear system these will always be in the form

$$\dot{\mathbf{v}} = A\mathbf{v} + \boldsymbol{\xi}(t). \tag{3.221}$$

Here A is a matrix of numbers, which could be time-dependent, and $\boldsymbol{\xi}$ is a vector of quantum noise sources.

Because the coordinates are noncommuting operators, the covariances must be symmetrized. For example, the covariance of x and p, which is classically defined by $C_{xp} = \langle xp \rangle - \langle x \rangle \langle p \rangle$, corresponds to the quantum covariance $C_{xp} = \langle xp + px \rangle - \langle x \rangle \langle p \rangle$. The reason for this is the following: if the quantum state is such that the Wigner function (to be discussed in Section 4.4) is non-negative, implying that the joint probability density of x and p is well defined, then the symmetrized covariance gives the covariance of this joint probability density (and thus the classical covariance) [252].

The unsymmetrized covariance matrix of $\mathbf{v}$, whose diagonal elements are the variances and off-diagonal elements the unsymmetrized covariances, is given by

$$C = \langle \mathbf{v}\mathbf{v}^{\mathrm{T}} \rangle - \mathbf{m}\mathbf{m}^{\mathrm{T}}, \tag{3.222}$$

where we have denoted the means by $\mathbf{m} = \langle \mathbf{v} \rangle$. The symmetrized covariance matrix is

$$C_{\mathrm{sym}} = \frac{1}{2}(C + C^{\mathrm{T}}). \tag{3.223}$$

The final quantity we need is the (unsymmetrized) covariance matrix for the noise sources. Since these noise sources have zero mean, the covariance matrix is just the matrix of second moments. The matrix of autocorrelation functions of the noise sources is given by

$$\langle \boldsymbol{\xi}(t)[\boldsymbol{\xi}(t+\tau)]^{\mathrm{T}} \rangle = G\delta(\tau), \tag{3.224}$$

for some matrix G whose elements are numbers. We call G the covariance matrix of the noise sources.

With the above definitions, the equations of motion for the means and unsymmetrized covariance matrix are not difficult to derive from the equations of motion for $\mathbf{v}$ (although one must remember to use Ito calculus). We have

$$\dot{\mathbf{m}} = A\mathbf{m}, \tag{3.225}$$

$$\dot{C} = AC + CA^{\mathrm{T}} + G. \tag{3.226}$$

3.7.2 Calculating the power spectrum of the measurement record

The input–output formalism provides possibly the simplest way to calculate the power spectrum of the measurement record for a linear quantum system. This is done using essentially the same procedure as that for a classical linear system shown in Section 3.1.4, except we must use the symmetrized correlation functions. You apply the Fourier transform to the Heisenberg equations of motion, which results in algebraic equations for the Fourier transforms of the dynamical variables. These Fourier-transformed variables are often referred to as the representation of the variables in "frequency space." You then solve these equations for the dynamical variables in frequency space, and this allows you to write the output field operators in terms of the input noise sources. Finally, one calculates the measurement spectrum, being the spectra of the output field operators, which is written entirely in terms of the spectra of the input noise sources. We now give these steps explicitly for a general linear system, and apply them to an example.

The quantum Langevin equations for any linear quantum system may be written as

$$\dot{\mathbf{v}} = M\mathbf{v} + J\mathbf{b}_{\text{in}}(t) + J^*\mathbf{b}_{\text{in}}^\dagger(t). \tag{3.227}$$

Here $\mathbf{v}$ is a vector of the Hermitian operators for the dynamical variables, M and J are constant matrices, and $\mathbf{b}_{\text{in}}(t)$ is a vector of quantum noise annihilation operators. We apply the Fourier transform to both sides of the above equation, and this gives

$$i\omega\mathbf{V}(\omega) = M\mathbf{V}(\omega) + J\mathbf{B}_{\text{in}}(\omega) + J^*\mathbf{B}_{\text{in}}^\dagger(\omega), \tag{3.228}$$

where $\mathbf{V}(\omega)$ is the Fourier transform of $\mathbf{v}(t)$, and $\mathbf{B}_{\text{in}}(\omega)$ is the Fourier transform of $\mathbf{B}_{\text{in}}(t)$. The solution of this set of algebraic equations requires calculating the inverse of the matrix M. This solution is

$$\mathbf{V}(\omega) = (i\omega I - M)^{-1}[J\mathbf{B}_{\text{in}}(\omega) + J^*\mathbf{B}_{\text{in}}^\dagger(\omega)], \tag{3.229}$$

where I is the identity matrix, and the negative exponent denotes the matrix inverse. The output operators in frequency space are

$$\mathbf{B}_{\text{out}}(\omega) = \mathbf{L}(\omega) + \mathbf{B}_{\text{in}} = C\mathbf{V}(\omega) + \mathbf{B}_{\text{in}}. \tag{3.230}$$

for some matrix C.

A linear Gaussian continuous measurement is obtained by measuring one of the elements of the output field vector $\mathbf{X}_{\text{out}} = \mathbf{B}_{\text{out}} + \mathbf{B}_{\text{out}}^\dagger$. Using the above equations, the operator $\mathbf{X}_{\text{out}}$ is

$$\begin{aligned}\mathbf{X}_{\text{out}}(\omega) &= 2\text{Re}[C]\mathbf{V}(\omega) + \mathbf{B}_{\text{in}}(\omega) + \mathbf{B}_{\text{in}}^\dagger(\omega) \\ &= [A(\omega)J + 1]\mathbf{B}_{\text{in}}(\omega) + [A(\omega)J^* + 1]\mathbf{B}_{\text{in}}^\dagger(\omega),\end{aligned} \tag{3.231}$$

where we have defined

$$A(\omega) = 2\text{Re}[C](i\omega I - M)^{-1}. \tag{3.232}$$

The output spectra corresponding to each of the elements of $\mathbf{X}_{\text{out}}$ (the measured output fields) are obtained from the diagonal elements of the symmetrized correlation matrix

$$C_{xx}(\omega, \omega') \equiv \frac{\langle \mathbf{X}_{\text{out}}(\omega)[\mathbf{X}_{\text{out}}(\omega')]^\dagger\rangle + \langle \mathbf{X}_{\text{out}}(\omega')[\mathbf{X}_{\text{out}}(\omega)]^\dagger\rangle}{2}, \tag{3.233}$$

where the superscript "t" denotes the matrix transpose. We can now calculate $C_{xx}(\omega, \omega')$ using Eq. (3.231) along with the correlation matrices for the input fields. These correlation matrices are

$$\langle \mathbf{B}_{\text{in}}^{\dagger}(\omega)[\mathbf{B}_{\text{in}}(\omega')]^{\text{t}} \rangle = N\delta(\omega + \omega'), \qquad \langle \mathbf{B}_{\text{out}}(\omega)[\mathbf{B}_{\text{out}}(\omega')]^{\text{t}} \rangle = 0, \qquad (3.234)$$

$$\langle \mathbf{B}_{\text{in}}(\omega)[\mathbf{B}_{\text{in}}^{\dagger}(\omega')]^{\text{t}} \rangle = (N+1)\delta(\omega + \omega'), \quad \langle \mathbf{B}_{\text{out}}^{\dagger}(\omega)[\mathbf{B}_{\text{out}}^{\dagger}(\omega')]^{\text{t}} \rangle = 0, \qquad (3.235)$$

for some diagonal real matrix N. We obtain $C_{xx}(\omega + \omega') = K_{xx}\delta(\omega + \omega')$, where

$$2K_{xx} = [A(\omega)J^* + 1]N[A(\omega')J + 1]^{\text{t}} + [A(\omega')J^* + 1]N[A(\omega)J + 1]^{\text{t}}$$

$$+ [A(\omega)J + 1](N+1)[A(\omega')J^* + 1]^{\text{t}} + [A(\omega')J + 1](N+1)[A(\omega)J^* + 1]^{\text{t}}.$$

We now recall from Eq. (3.77) that the spectrum is given by whatever function multiplies the δ-function on the right-hand side. Noting that we can set $\omega' = -\omega$, and that $A(-\omega) = A^*(\omega)$, the Hermitian matrix whose diagonal elements are the measurement spectra is

$$S(\omega) = \text{Re}\left[(A^*J + 1)N(AJ^* + 1)^{\text{t}}\right] + \text{Re}\left[(AJ + 1)N(A^*J^* + 1)^{\text{t}}\right],$$

$$(A^*J + 1)(AJ^* + 1)^{\text{t}}/2 + (AJ + 1)(A^*J^* + 1)^{\text{t}}/2, \qquad (3.236)$$

where we have written A as short hand for $A(\omega)$.

The spectrum of a position measurement for a damped oscillator

As an example we recalculate the measurement spectrum for the system we considered in Section 3.1.4, a damped quantum oscillator under a continuous position measurement, this time using the input–output formalism. The stochastic master equation for a damped oscillator in contact with a thermal bath at temperature T, under a position measurement is

$$d\rho = -i[\omega_0 a^{\dagger} a, \rho]\,dt - \tilde{k}[\tilde{x}, [\tilde{x}, \rho]]\,dt + \sqrt{2\tilde{k}}(\tilde{x}\rho + \rho\tilde{x} - 2\langle\tilde{x}\rangle\rho)\,dW \qquad (3.237)$$

$$- \frac{\gamma}{2}\left[(n_T + 1)(a^{\dagger} a\rho + \rho a^{\dagger} a - 2a\rho a^{\dagger}) - n_T(aa^{\dagger}\rho + \rho aa^{\dagger} - 2a^{\dagger}\rho a)\right]dt.$$

We will discuss the origin of this master equation later in Section 4.3.1. Here a is the annihilation operator for the oscillator, and ω_0 and γ are respectively its frequency and damping rate. The parameter $\tilde{k}$ is the dimensionless measurement strength and $\tilde{x}$ is a scaled version of the oscillator's position operator, defined as $\tilde{x} = a + a^{\dagger}$. The real position of the oscillator is $x = \sqrt{\hbar/(2m\omega_0)}\tilde{x}$, where m is the mass of the mirror, and the real measurement strength is $k = (2m\omega_0/\hbar)\tilde{k}$. The parameter $n_T = 1/(e^{\hbar\omega/k_{\text{B}}T} - 1)$ is the average number of phonons in the resonator at temperature T. The terms in the above equation proportional to γ give a good description of damping of the oscillator so long as $\gamma \ll \omega_0$ (weak damping).

We can now use the rules in Section 3.6 above. We need one input operator to describe the damping and heating due to the thermal bath, as per Eqs. (3.208), (3.211), and (3.212), and a second to describe the continuous measurement of position, as per Eqs. (3.204) and (3.190). The quantum Langevin equation is

$$\frac{dA}{dt} = -i\omega_0[A, a^\dagger a] - \tilde{k}[\tilde{x}, [\tilde{x}, A]] + \sqrt{2\tilde{k}}\left(c_{\text{in}}[\tilde{x}, A] + [A, \tilde{x}]c_{\text{in}}^\dagger\right)$$

$$- \frac{\gamma}{2}(n_T + 1)\left(Aa^\dagger a + a^\dagger aA - 2a^\dagger Aa\right)$$

$$- \frac{\gamma}{2}n_T\left(Aaa^\dagger + aa^\dagger A - 2aAa^\dagger\right) + \sqrt{\gamma}\left(b_{\text{in}}[a^\dagger, A] + [A, a]b_{\text{in}}^\dagger\right). \tag{3.238}$$

Replacing A in the above equation of motion with the oscillator's position and momentum operators we obtain the Langevin equations for the oscillator:

$$\dot{\tilde{x}} = \omega_0\tilde{p} - (\gamma/2)\tilde{x} + \sqrt{\gamma}\,\tilde{x}_{\text{in}}, \tag{3.239}$$

$$\dot{\tilde{p}} = -\omega_0\tilde{x} - (\gamma/2)\tilde{p} - \sqrt{\gamma}\,\tilde{p}_{\text{in}} - \sqrt{8\tilde{k}}\,y_{\text{in}}, \tag{3.240}$$

where $\tilde{p} = -i(a - a^\dagger)$ and we have defined the Hermitian noise sources

$$\tilde{x}_{\text{in}} = b_{\text{in}} + b_{\text{in}}^\dagger, \quad \tilde{p}_{\text{in}} = -i(b_{\text{in}} - b_{\text{in}}^\dagger), \quad y_{\text{in}} = -i(c_{\text{in}} - c_{\text{in}}^\dagger). \tag{3.241}$$

We can calculate the correlation functions of these noise sources from those of b_{in}, given in Eqs. (3.211) and (3.212), and c_{in}, given by Eq. (3.203). These correlation functions are

$$\langle\tilde{x}_{\text{in}}(t)\tilde{x}_{\text{in}}(t+\tau)\rangle = \langle\tilde{p}_{\text{in}}(t)\tilde{p}_{\text{in}}(t+\tau)\rangle = (2n_T + 1)\delta(\tau), \tag{3.242}$$

$$\langle\tilde{x}_{\text{in}}(t)\tilde{p}_{\text{in}}(t+\tau)\rangle + \langle\tilde{p}_{\text{in}}(t)\tilde{x}_{\text{in}}(t+\tau)\rangle = 0, \tag{3.243}$$

$$\langle y_{\text{in}}(t)y_{\text{in}}(t+\tau)\rangle = \delta(\tau). \tag{3.244}$$

Taking the Fourier transform of the Langevin equations for $\tilde{x}$ and $\tilde{p}$, and rearranging, we have

$$\begin{pmatrix} i\omega + \gamma/2 & -\omega_0 \\ \omega_0 & i\omega + \gamma/2 \end{pmatrix}\begin{pmatrix} X(\omega) \\ P(\omega) \end{pmatrix} = \begin{pmatrix} \sqrt{\gamma}\,X_{\text{in}}(\omega) \\ -\sqrt{\gamma}\,P_{\text{in}}(\omega) - \sqrt{8\tilde{k}}\,Y_{\text{in}}(\omega) \end{pmatrix}. \tag{3.245}$$

Here the capital letters denote the Fourier transforms of their lower-case counterparts, and we have used the fact that if $f(t) \leftrightarrow F(\omega)$ is a Fourier transform pair, then $df/dt \leftrightarrow i\omega F(\omega)$ is also. We can now solve these equations by calculating the inverse of the appropriate matrix, and obtain $X(\omega)$ in terms of the input noise sources. This gives

$$X(\omega) = \frac{\sqrt{\gamma}(\gamma/2 + i\omega)X(\omega) - \omega_0[\sqrt{\gamma}\,P(\omega) + \sqrt{8\tilde{k}}\,Y(\omega)]}{(\gamma/2 + i\omega)^2 + \omega_0^2}. \tag{3.246}$$

From Eq. (3.215) the measured field operator is

$$R(t) = \frac{c_{\text{out}} + c_{\text{out}}^\dagger}{\sqrt{8\tilde{k}}} = \langle\tilde{x}\rangle + \frac{c_{\text{in}} + c_{\text{in}}^\dagger}{\sqrt{8\tilde{k}}}, \tag{3.247}$$

and Eq. (3.77) tells us that $\langle R(\omega)R(\omega')\rangle_{\text{sym}} = S_r(\omega)\delta(\omega+\omega')$ where $S_r(\omega)$ is the spectrum of $R(t)$. If a signal f has a time-domain correlation function of the form $g(\tau) = \alpha\delta(\tau)$, then its Fourier transform $F(\omega)$ has the correlation function $\langle F(\omega)F(\omega')\rangle = \alpha\delta(\omega+\omega')$. This gives us the frequency-space correlation functions for all the noise sources. We can now calculate $\langle R(\omega)R(\omega')\rangle_{\text{sym}}$ using Eqs. (3.247) and (3.246), and finally extract the measurement spectrum. Writing this spectrum in units of the real position, and making the approximation $\gamma \ll \omega_0$, we obtain

$$S_r(\omega) = \frac{1}{8k} + \frac{\gamma V_x(2n_T+1)+4V_x^2 k}{2(|\omega|-\omega_0)^2+\gamma^2/2},$$

(3.248)

where $V_x = \sqrt{\hbar/(2m\omega_0)}$ is the variance of the position for the oscillator's ground state.

3.8 Parameter estimation: the hybrid master equation

We have seen above how an observer tracks the evolution of a system when making a continuous measurement. But the state of a system is not the only thing that may be unknown, and that we may therefore be interested in determining. We might wish to determine one or more parameters that define the Hamiltonian of the system, or we might want to know one or more parameters that specify the initial state. Determining such parameters is referred to as *parameter estimation*. Since a continuous measurement provides information about the evolution of the system, it provides information about both the Hamiltonian and the initial state. To perform parameter estimation using a continuous measurement we must derive a set of equations that determine the evolution of the observer's state-of-knowledge of the system, as well as the unknown parameters. Now that we are familiar with the SME and the K-S equation, it is not difficult to do this. We will call the resulting set of equations the hybrid master equation (HME), since it involves the simultaneous estimation of a quantum state and a set of classical parameters, and is therefore in a sense a hybrid between the SME and the K-S equation.

We will first consider the problem of estimating parameters in the Hamiltonian. Estimating parameters that define the initial state involves similar reasoning, and we will examine this via an example in the next section. To simplify the notation, we will assume that we are estimating a single real number – extending the analysis to a vector of parameters is quite straightforward. Let us call the unknown parameter λ, and write the Hamiltonian as $H(\lambda)$. The observer's initial state-of-knowledge is now given by an initial density matrix for the system, ρ, as well as an initial probability distribution for λ, $P(\lambda)$.

The evolution of the system, and thus the SME, now depends upon λ. But we do not know λ, so our state-of-knowledge regarding the system as it evolves is now given by averaging the density matrices that result from different values of λ. Specifically, let us say that we are measuring the observable X. Then the measurement signal is

$dy = \langle X \rangle \, dt + dW/\sqrt{8k}$, and for a given value of λ the evolution is

$$d\rho_\lambda = -\frac{i}{\hbar}[H(\lambda), \rho_\lambda] \, dt - k[X, [X, \rho_\lambda]] \, dt$$

$$+ 4k(X\rho_\lambda + \rho_\lambda X - 2\langle X \rangle_\lambda \rho_\lambda)(dy - \langle X \rangle_\lambda \, dt). \tag{3.249}$$

Here we have written the SME directly in terms of the measurement result, and we have put a subscript on the density matrix to indicate that it is the state-of-knowledge the observer would have if they knew the value of λ. For the same reason we have put a subscript on the expectation value of X, since $\langle X \rangle_\lambda = \text{Tr}[X\rho_\lambda]$.

The observer's state-of-knowledge is now given by averaging the matrices ρ_λ over the probability distribution for λ. So

$$\rho(t+dt) = \int_{-\infty}^{\infty} P(\lambda)\rho_\lambda(t+dt)d\lambda \; = \; \int_{-\infty}^{\infty}[P(\lambda)\rho_\lambda]\,d\lambda + \int_{-\infty}^{\infty}[P(\lambda)d\rho_\lambda]\,d\lambda$$

$$= \rho + \int_{-\infty}^{\infty}[P(\lambda)d\rho_\lambda]\,d\lambda, \tag{3.250}$$

and thus

$$d\rho = -\frac{i}{\hbar}\left[\int_{-\infty}^{\infty} P(\lambda)H(\lambda)d\lambda, \rho\right] dt - k[X, [X, \rho]] \, dt$$

$$+ 4k(X\rho + \rho X - 2\langle X \rangle \rho)(dr - \langle X \rangle \, dt) \tag{3.251}$$

This SME allows us to evolve our state-of-knowledge of the system, but only so long as we know how $P(\lambda)$ is also evolving with time. To determine this all we have to do is use Bayesian inference to calculate how the measurement result dy changes our knowledge about λ. Bayes' theorem tells us that the effect on $P(\lambda)$ of the measurement result dy is

$$P(\lambda|dy) = \frac{P(dy|\lambda)P(\lambda)}{\mathcal{N}}, \tag{3.252}$$

in which $P(dy|\lambda)$ is the probability density for dy given λ (the likelihood function for the measurement of λ), and $\mathcal{N}$ is a normalization constant. So we need to work out the probability density for dy for each value of λ. Now, let us say that the true value of λ is μ. In this case, even though from the observer's point of view the measurement result is given by $dy = \text{Tr}[\rho] + dW/\sqrt{8k}$, the measurement result is actually given by

$$dy = dy_\mu = \langle X \rangle_\mu \, dt + dV/\sqrt{8k}, \quad \text{where} \quad \langle X \rangle_\mu \equiv \text{Tr}[X\rho_\mu], \tag{3.253}$$

where dV is an increment of Wiener noise, and the evolution of ρ_μ is given by the SME in Eq. (3.249) with $\lambda = \mu$. Since dV is Gaussian distributed with mean 0 and variance dt, $P(dr|\lambda = \mu)$ is therefore

$$P(dy|\lambda = \mu) = \frac{\exp\left\{-4k(dy - \langle X \rangle_\mu \, dt)^2/dt\right\}}{\sqrt{2\pi \, dt}}. \tag{3.254}$$

The updated probability density for λ is thus

$$P(\lambda, t + dt) = P(\lambda | dy) \propto \exp\left\{-4k(dy - \langle X \rangle_\lambda dt)^2/dt\right\} P(\lambda, t), \qquad (3.255)$$

where dy is the measurement result obtained by the observer. We can now calculate the differential equation for $P(\lambda)$ by expanding the exponential to first order in dt. The result is

$$P(\lambda | dy) \propto \exp\left\{-4k\left(\langle X \rangle dt - \langle X \rangle_\lambda dt + dW/\sqrt{8k}\right)^2/dt\right\} P(\lambda)$$

$$\propto \exp\left\{-4k(\langle X \rangle_\lambda - \langle X \rangle)^2 dt + \sqrt{8k}(\langle X \rangle_\lambda - \langle X \rangle)dW\right\} P(\lambda)$$

$$\propto \left[1 + \sqrt{8k}(\langle X \rangle_\lambda - \langle X \rangle)dW\right] P(\lambda). \qquad (3.256)$$

The stochastic differential equation that updates the density $P(\lambda)$ is therefore

$$dP(\lambda) = \sqrt{8k}(\langle X \rangle_\lambda - \langle X \rangle)dW P(\lambda)$$

$$= 8k\mathrm{Tr}[X(\rho_\lambda - \rho)](dy - \langle X \rangle dt)P(\lambda). \qquad (3.257)$$

So we found above that the equation that evolves the observer's state-of-knowledge of the system, ρ, requires knowing how $P(\lambda)$ evolves, and now we have found that the equation of motion for $P(\lambda)$ requires knowing $\langle X \rangle_\lambda$. This means that we also need to simultaneously evolve each of the ρ_λ for every value of λ, because $\langle X \rangle_\lambda = \mathrm{Tr}[X\rho_\lambda]$. The hybrid master equation therefore constitutes the following set of coupled stochastic equations:

$$dy = \mathrm{Tr}[X\rho]dt + \frac{dW}{\sqrt{8k}},$$

$$d\rho = -\frac{i}{\hbar}\left[\left[\int_{-\infty}^{\infty} P(\lambda)H(\lambda)d\lambda, \rho\right]\right] dt - k[X,[X,\rho]]dt$$

$$+ 4k(X\rho + \rho X - 2\mathrm{Tr}[X\rho]\rho)(dy - \mathrm{Tr}[X\rho]dt),$$

$$d\rho_\lambda = -\frac{i}{\hbar}[H(\lambda), \rho_\lambda]dt - k[X,[X,\rho_\lambda]]dt$$

$$+ 4k(X\rho_\lambda + \rho_\lambda X - 2\mathrm{Tr}[X\rho_\lambda]\rho_\lambda)(dy - \mathrm{Tr}[X\rho_\lambda]dt),$$

$$dP(\lambda) = 8k\mathrm{Tr}[X(\rho_\lambda - \rho)](dy - \mathrm{Tr}[X\rho]dt)P(\lambda). \qquad (3.258)$$

Simulating the HME requires a great deal of numerical overhead, because it involves simulating the SME for many different values of λ. It is therefore useful to find approximate methods for parameter estimation that will reduce this overhead [500].

3.8.1 An example: distinguishing two quantum states

We now consider the situation in which an observer knows that a quantum system has been prepared in one of two, possibly non-orthogonal, quantum states, and makes a continuous measurement on the system to determine which one. This is an example of parameter estimation, where the parameter determines the initial state. In this case the parameter is discrete, as it has just two values corresponding to the two initial states. We will call these values 0 and 1, and the initial states to which they correspond $|\psi_0\rangle$, and $|\psi_1\rangle$.

Each of the states $|\psi_i\rangle$ is prepared with probability $P(i)$, so the initial state-of-knowledge of the observer is

$$\rho(0) = P(0)|\psi_0\rangle\langle\psi_0| + P(1)|\psi_1\rangle\langle\psi_1|, \tag{3.259}$$

and the evolution of ρ is of course given by the SME

$$d\rho = -\frac{i}{\hbar}[H,\rho]\,dt - k[X,[X,\rho]]\,dt$$
$$+ 4k(X\rho + \rho X - 2\text{Tr}[X\rho]\rho)(dy - \text{Tr}[X\rho]\,dt), \tag{3.260}$$

where H is the Hamiltonian of the system.

The observer's state-of-knowledge regarding which of the states was prepared is the probability distribution given by $\{P(0), P(1)\}$. To determine how this evolves with time we need to use Bayes' theorem to determine how the measurement result dy in each time-step changes these probabilities. This change is given by

$$P(i|dy) = \frac{P(dy|i)P(i)}{\mathcal{N}}, \tag{3.261}$$

where $P(dy|i)$ is the likelihood function for the measurement of i, and $\mathcal{N}$ is the normalization constant. The likelihood function is the probability distribution for the measurement result, dy, given that the initial state was really $|\psi_i\rangle$. If the initial state was $|\psi_i\rangle$, then the state of the system at time t would be given by $\rho_i(t)$, where $\rho_i(t)$ is the solution to the SME with $\rho_i = |\psi_i\rangle\langle\psi_i|$ as the initial state. The measurement result at time t would then be equal to

$$dy = dy_i = \text{Tr}[\rho_i(t)]\,dt + \frac{dV}{\sqrt{8k}}, \tag{3.262}$$

where dV is Wiener noise (even though as far as the observer is concerned, $dy = \text{Tr}[\rho(t)]\,dt + dW/\sqrt{8k}$). Eq. (3.262) tells us that the distribution for dy given i is a Gaussian with mean $\text{Tr}[\rho_i(t)]\,dt$ and variance $dt/(8k)$. That is,

$$P(dy|i) \propto \exp\left\{-4k(dy - \text{Tr}[X\rho_i]\,dt)^2/dt\right\}$$
$$\propto \left[1 + \sqrt{8k}\text{Tr}[X(\rho_i - \rho)]\,dW\right], \tag{3.263}$$

where in the second line we have expanded the exponential to first-order in dt. We can now obtain the differential equation for $P(i)$ by substituting the expression for $P(dy|i)$ into Eq.

(3.261). The result is

$$dP(i) = 8k(\text{Tr}[X(\rho_i - \rho)])(dy - \text{Tr}[X\rho]\,dt)P(i). \tag{3.264}$$

Since this equation depends on the ρ_i, to solve it we must also solve for the evolution of these density matrices. The full set of equations that the observer must solve to determine his or her knowledge of the initial state, is therefore

$$dy = \text{Tr}[X\rho]\,dt + \frac{dW}{\sqrt{8k}},$$

$$d\rho = -\frac{i}{\hbar}[H,\rho]\,dt - k[X,[X,\rho]]\,dt$$
$$+ 4k(X\rho + \rho X - 2\text{Tr}[X\rho]\rho)(dy - \text{Tr}[X\rho]\,dt),$$

$$d\rho_i = -\frac{i}{\hbar}[H,\rho_i]\,dt - k[X,[X,\rho_i]]\,dt$$
$$+ 4k(X\rho_i + \rho_i X - 2\text{Tr}[X\rho_i]\rho_i)(dy - \text{Tr}[X\rho_i]\,dt),$$

$$dP(i) = 8k(\text{Tr}[X(\rho_i - \rho)])(dy - \text{Tr}[X\rho]\,dt)P(i), \tag{3.265}$$

where the initial states are

$$\rho(0) = P(0)|\psi_0\rangle\langle\psi_0| + P(1)|\psi_1\rangle\langle\psi_1|, \tag{3.266}$$

$$\rho_i(0) = |\psi_i\rangle\langle\psi_i|. \tag{3.267}$$

Comparing this hybrid master equation to that given by Eqs. (3.258), we see that the only difference is that in the present case the Hamiltonian is the same for the evolution of all the density matrices, and it is instead the initial states that are different.

History and further reading

Gisin in 1984 [212], and Diosi in 1986 [151] were the first people to derive Gaussian stochastic evolution equations for measurements on pure quantum systems. The concept of the SME as the quantum equivalent of the K-S equation was first introduced by Belavkin in 1987 [41], and it was later obtained independently by Wiseman and Milburn in 1993 [679, 680], building directly upon the quantum trajectory work of Carmichael [100]. The stochastic evolution equation describing photon-counting was first developed in 1981 by Srinivas and Davies [573]. In 1992 Hegerfeldt and Wilser [245] and Dalibard, Castin, and Mølmer [136] showed that this approach could be used as an efficient method to simulate open quantum systems. This "wavefunction Monte Carlo" method, discussed in Section 4.3.4, is widely used today, and a number of extensions have been developed [194, 482, 277, 289]. Treatments of multiple observers can be found in [29] and [163].

The linear formulation of stochastic Schrödinger equations was first obtained by Goetsch and Graham using a Heisenberg-picture method [216], and formulated using quantum measurement theory by Wiseman [673]. Solutions to various linear SSEs and SMEs have been obtained by a number of authors, and can be found in [202, 216, 670, 673, 290, 282, 123].

Since the stream of measurement results in a continuous measurement is random, there is the possibility that a state can become partially projected, and then the projection undone by subsequent measurement results. This phenomenon is elucidated by Korotkov and Jordan in [344, 319].

The input–output (quantum Langevin) formalism for treating continuously measured quantum systems was devised by Collet and Gardiner [128, 199]. The resulting quantum stochastic differential equations were also developed independently, in a measure-theoretic way, by Hudson and Parthasarathy [275]. Goetsch and Graham subsequently developed a linear version of input–output theory from which they derived the linear version of quantum trajectories mentioned above [216]. We will discuss the input–ouput formalism further, and especially its uses in describing systems that are connected together via their inputs and outputs, in Chapter 5.

The experimental realization of continuous measurements on single quantum systems was initially only possible using optics, due to the lack of sufficiently low-noise amplifiers in other systems. Examples of these optical measurements may be found in [53, 436, 535, 37, 567, 19, 213], while [473] is somewhat of an exception. Quite recently such measurements have been realized in solid-state systems, and these experiments are reported in [703, 622, 444, 637].

Theory and recent experiments on continuous measurements on ensembles of systems can be found in [566, 564, 116, 539, 588, 589, 540, 366]. Various derivations of the hybrid master equation can be found in [632, 195, 500, 612]. We followed closely the derivation in [500].

Exercises

1. Consider a time interval T in which we make N independent measurements of a real quantity x. Each measurement gives the result $y = x + \varepsilon$, where ε is a zero mean Gaussian random variable with variance Δ. Calculate the likelihood function for this sequence of measurements and thus determine the posterior probability distribution for x, given that the prior for x is a Gaussian with mean x_{true} and variance V.
2. Expand the exponential in Eq. (3.16) to derive Eq. (3.17).
3. Normalize Eq. (3.17) to derive Eq. (3.20).
4. Calculate the norm of $|\hat{\psi}(t + dt)\rangle$ in Eq. (3.33), and use it to derive Eq. (3.36).
5. Show that the SME in Eq. (3.38) reduces to the K-S equation (Eq. 3.22) when X commutes with ρ.
6. Derive Eq. (3.97), and show that the probability P_{nm} is given as claimed.
7. Derive Eq. (3.100), and and use it to derive Eq. (3.101).
8. Use the linear SME to derive the Zakai equation.

9. Consider a continuous measurement of σ_z on a spin-1/2 system. Use the stochastic master equation to derive the equation of motion for $\text{Tr}[\rho^2]$, which is a measure of how pure the state is. To do this you need to (i) use the relation $d(\rho^2) = (\rho + d\rho)^2 - \rho^2$ to derive an expression for $d(\rho^2)$ in terms of $d\rho$, and (ii) use the Ito-calculus relation $(dW)^2 = dt$.

10. Consider a continuous measurement of σ_z on a spin-1/2 system. Use the (nonlinear) SME to obtain the equation of motion for $\langle\sigma_z\rangle$ for an arbitrary initial state. This can be done by using the Bloch vector representation for ρ. This representation is

$$\rho = \frac{1}{2}(I + \langle\sigma_x\rangle\sigma_x + \langle\sigma_y\rangle\sigma_y + \langle\sigma_z\rangle\sigma_z) = \frac{1}{2}(I + \mathbf{a}\cdot\boldsymbol{\sigma}), \quad (3.268)$$

where I is the two-by-two identity matrix, $\mathbf{a} \equiv \langle\boldsymbol{\sigma}\rangle$ is called the Bloch vector, and $\boldsymbol{\sigma} \equiv (\sigma_x, \sigma_y, \sigma_z)$. The equations of motion for the elements of the Bloch vector can be derived by substituting the above expression for ρ into the SME. More information on the Bloch vector is given in Appendix D.

11. Consider a continuous measurement of σ_z on a spin-1/2 system. (i) Use the linear SSE to obtain $\rho(t)$, given that the initial state is $\rho(0) = I/2$, where I is the identity matrix. (ii) Derive an expression for the linear entropy of $\rho(t)$, averaged over all noise realizations. This expression is an integral that cannot be solved analytically. (iii) Consider the long-time limit, in which $t \gg k$, and use the integral to derive an approximate analytic expression for the average linear entropy in this case.

12. Consider a continuous measurement of σ_z on a spin-1/2 system. For what quantum states ρ does the noise term (the part of the SME that is proportional to dW) vanish?

13. Verify that the operators A_0 and A_1 in Eqs. (3.143) and (3.144) satisfy $\sum_n A_n^\dagger A_n = 1$ to first-order in dt.

14. Show that the evolution of a cavity mode under photo-detection, when the observer does not have access to the measurement results, is given by Eq. (3.157).

15. Consider the evolution of a single cavity mode under photo-detection, in which the evolution is given by Eq. (3.170). What happens during each jump when the initial state is the "Schrödinger cat" state $|\psi\rangle = (|\alpha\rangle + |-\alpha\rangle)/\mathcal{N}$? The normalization is $\mathcal{N} = \sqrt{2(1 + \exp\{-2|\alpha|^2\})}$ and the state $|\alpha\rangle$ is a coherent state defined by $a|\alpha\rangle = \alpha|\alpha\rangle$.

16. Consider the amplitude of the local oscillator in homodyne detection. This determines the value of β in Eq. (3.174). (i) Show that if the complex amplitude of the local oscillator is $\alpha = Ae^{i(\theta-\pi/2)}$, then homodyne detection measures the quadrature $X_\theta = ae^{-i\theta} + a^\dagger e^{i\theta}$. (ii) Using Eq. (3.186), write down the linear SSE for this homodyne detection.

17. Derive Eq. (3.181) from Eq. (3.179). To do this use the fact that $\beta \gg 1$, and expand to first order in $1/\beta$.

18. The linear SSE for homodyne detection of the arbitrary quadrature $X_\theta = ae^{-i\theta} + a^\dagger e^{i\theta}$, is

$$d|\hat\psi\rangle = \left[-(\gamma/2)a^\dagger a\,dt + i\sqrt{\gamma}ae^{-i\theta(t)}\,dW\right]|\hat\psi\rangle, \quad (3.269)$$

where we have allowed θ to vary with time. The simplest way to solve this equation is to choose the initial state of the system to be a coherent state, since all the operators that appear in the equation act on coherent states in a simple way. In particular you need $a|\alpha\rangle = \alpha|\alpha\rangle$ and

$$\exp(-\gamma a^\dagger a t/2)|\alpha\rangle = \exp\left(\tfrac{1}{2}|\alpha|^2[1-e^{-\gamma t}]\right)|\alpha e^{-\gamma t/2}\rangle.$$

The linear SSE can now be reduced to an equation without operators, and so is simpler to solve. Since we can tell from the expression above that the coherent state will evolve as $|\alpha e^{-\gamma t/2}\rangle$, one way to the solve the linear SSE is to write the solution as $|\psi(t)\rangle = f(t)|\alpha e^{-\gamma t/2}\rangle$, determine the equation of motion for $f(t)$ and solve that. (i) Show that the solution to Eq. (3.269) with the initial coherent state $|\alpha\rangle$ is

$$|\psi(t)\rangle = \exp\left(\tfrac{1}{2}|\alpha|^2[1-e^{-\gamma t}]+\tfrac{1}{2}\alpha^2 S^*(t)+\alpha R^*(t)\right)|\alpha e^{-\gamma t/2}\rangle$$

with

$$S(t) = -\int_0^t e^{2i\theta(t')}e^{-\gamma t'}\,dt' \qquad R(t) = \int_0^t e^{i\theta(t')}dW(t'). \qquad (3.270)$$

If you need guidance, this problem is solved in [673]. One can now use the above solution to obtain the solution for any initial state: since we can write any state as a linear superposition of coherent states, the evolution operator will be the one that generates $|\psi(t)\rangle$ above when acting on a coherent state. (ii) Use Eq. (18) to show that the solution to Eq. (3.269) for any initial state is

$$|\psi(t)\rangle = \exp\left[\tfrac{1}{2}\gamma a^\dagger a t+\tfrac{1}{2}a^2 S^*(t)+aR^*(t)\right]|\psi(0)\rangle.$$

19. (i) Solve the linear SSE for photo-detection, Eq. (3.160), when the Hamiltonian includes a Kerr nonlinearity, so that $H = \hbar\omega a^\dagger a+\mu(a^\dagger a)^2$. You will need the operator relation

$$\exp[\varepsilon(a^\dagger a)^2]f(a)\exp[-\varepsilon(a^\dagger a)^2] = f(a\exp[\varepsilon a^\dagger a]),$$

which is valid for any function f. (ii) What is the state that results from applying the operator $\exp\{-i(\pi/2)(a^\dagger a)^2\}$ to a coherent state $|\alpha\rangle$? (iii) Using the answer to (ii), what is the state of the cavity with a Kerr nonlinearity under photo-detection when the initial state is a coherent state, and it evolves for a time $t = \pi/(2\mu)$? How does the final state depend on the number of photo-counts in that time?

20. One can use the linear SSE to obtain a fully analytic solution to the SME for inefficient homodyne detection. To do this we use the fact that inefficient detection is the same as making two measurements, where the observer has no access to the results of the second measurement. We therefore first solve the linear SSE for two simultaneous homodyne detection measurements. This SSE, which describes the case in which the

observer has access to both measurements, is

$$d|\hat{\psi}\rangle = \left[-(i\omega + v + \mu)a^\dagger a\,dt + i\sqrt{2v}\,a\,dW + i\sqrt{2\mu}\,a\,dV\right]|\hat{\psi}\rangle,$$

where dW and dV are independent sources of Wiener noise, so that $(dW)^2 = (dV)^2 = dt$, and $dW\,dV = 0$. Once you have solved this SSE, you then calculate the density matrix at time t for the observer with access only to the first measurement, but averaging over the measurement signal of the second measurement. If you need further guidance, this problem is solved in reference [123].

21. In Section 3.4 we analyzed simple homodyne detection which uses a single photo-detector. One can also realize the same continuous measurement by using *balanced homodyne detection*, depicted in Fig. 3.3. In this case both outputs of the beam-splitter are measured with photo-detectors, and the difference is taken between the output signals from the two detectors. The beam-splitter is also chosen to reflect 50% of the light ($\eta = 1/\sqrt{2}$). (i) In analogy with analysis in Section 3.4, show that the average of the output signal from balanced homodyne detection is a linear combination of a and $a^\dagger$ as required. (ii) Derive the stochastic Schrödinger equation for balanced homodyne detection. This is more complex than that for simple homodyne detection, because there are now two photo-detectors and thus two jump operators. If you need further guidance, the SSE for balanced homodyning is derived in [679].

22. The master equation for a continually measured system, in which the measurement results are ignored, can be derived from the quantum Langevin equation. Do this by using the fact that the equation of motion for the expectation value of an arbitrary operator Y can be calculated in two ways. The first is by taking expectation values on both sides of the quantum Langevin equation for Y. The second is by using the master equation and the fact that $\langle\dot{Y}\rangle = \text{Tr}[Y\dot{\rho}]$.

23. Consider an optical cavity mode damped at rate γ to a zero-temperature bath (in this case the bath is the electromagnetic field outside the cavity). There is one Lindblad operator $L = \sqrt{\gamma}a$, and the cavity is driven by a laser, so that its Hamiltonian is $H = \hbar\omega a^\dagger a + \hbar E(ae^{i\omega t} + a^\dagger e^{-i\omega t})$ where E is real. We assume that $E \ll \omega$, so that we can still use the thermal Langevin equation for an (undriven) harmonic oscillator, Eq. (3.190). (i) transform the Hamiltonian and the Lindblad operator L into the interaction picture with respect to the Hamiltonian $H_0 = \hbar\omega a^\dagger a$. (ii) Obtain the quantum Langevin equation for a using Eq. (3.190). (iii) By taking the expectation value of both sides of the quantum Langevin equation for a, determine the equation of motion for $\langle a\rangle$. (iv) Calculate the steady-state value of $\langle a\rangle$. (v) Use the quantum Langevin equations for a and $a^\dagger$ to derive the quantum Langevin equation for $a^\dagger a$. This is done by using the fact that $d(a^\dagger a) = a^\dagger da + (da^\dagger)a + (da^\dagger)(da)$. Check your answer by writing down the Langevin equation for $a^\dagger a$ directly using Eq. (3.190). (vi) Calculate the steady-state value of $\langle a^\dagger a\rangle$.

24. Consider an optical cavity mode damped at rate γ to a bath at non-zero temperature. (i) Obtain the quantum Langevin equations for a and $a^\dagger a$ using Eq. (3.208). Note

that in this case there is only one Lindblad operator, being $L_1 = \sqrt{\gamma}a$, where a is the annihilation operator for the mode. You should work in the interaction picture in which the Hamiltonian of the oscillator, $\hbar\omega a^\dagger a$ is removed. (ii) Obtain the equations of motion for $\langle a \rangle$ and $\langle a^\dagger a \rangle$ by taking the expectation value of both sides of the Langevin equations obtained in (i). (iii) Calculate the steady-state values of $\langle a \rangle$ and $\langle a^\dagger a \rangle$. (iv) Explain why one of the steady-state values is zero while the other is not.

25. Calculate the spectrum for a homodyne measurement of an optical cavity, where the cavity is driven by a coherent light beam with complex amplitude α. This driving is described by adding to the Hamiltonian of the optical mode the term $\alpha^* a e^{i\omega t} + \alpha a^\dagger e^{-i\omega t}$, where ω is the mode frequency.

26. Derive the hybrid SME for photo-detection measurements on a single cavity mode.

4

Statistical mechanics, open systems, and measurement

We now turn to the subject of thermodynamics, and to describing the process of thermalization of quantum systems. A quantum system that interacts with another system large enough to obey the usual rules of thermodynamics, is referred to as an *open system*. The thermal system with which it interacts is often referred to as a *thermal bath*, or simply as *the environment*. We will use these terms interchangeably. Modeling the behavior of open quantum systems has many applications. One of these is to model the noise that they are subjected to by their surroundings, noise that we will wish to control using measurement and feedback loops in Chapter 5. There are also two important connections between quantum measurements, information theory, and thermodynamics. The first of these is that the thermodynamic entropy of a macroscopic system can be identified with the information-theoretic entropy that characterizes the lack of knowledge of an observer about the microstates of the system. In turn, if an observer is allowed to make microscopic measurements on the system, then this modifies the thermodynamic rules regarding the conversion of heat to work. Such is the subject of *Maxwell's demon*. Curiously it took more than a century to understand how Maxwell's demon, essentially a microscopic measurement and feedback process, fits within the second law of thermodynamics. In another twist, it turns out that the process of erasing information requires an increase in the thermodynamic entropy of the environment, and this increase is directly related to the amount of information erased.

The second important connection between thermodynamics and measurement theory is that the interaction of a system with a large thermal bath continuously carries information away from the system. By this we mean that in interacting with the system, the environment obtains information about it (becomes correlated with it), and this correlation is continuously spread through the degrees of freedom of the bath, never to be undone by returning to the system. Because of this there is an intimate connection between continuous measurements and thermal baths: continuous measurements on a system are realized by measuring a bath with which the system interacts, and the evolution of a damped system can be reproduced by continuously measuring the system and forgetting the measurement results. Because of this the stochastic Schrödinger equations that describe continuous measurements, introduced in Chapter 3, provide an efficient numerical method for modeling the irreversible dynamics of open quantum systems. This technique is often referred to as the "wavefunction Monte Carlo" method.

Since this chapter deals with open systems, thermal behavior, and the relationship between thermodynamic and information-theoretic entropy, we begin with an introduction to thermodynamics, the fundamental postulates of statistical mechanics, and how they lead to irreversible behavior and the Boltzmann distribution. Since we must deal with both quantum mechanics and thermodynamics, we also go beyond the postulates of statistical mechanics to discuss how irreversible behavior and thermalization emerge from the quantum mechanics of large, many-body systems. This latter material is a result of new insights that have been obtained only recently. While much is now known regarding the quantum mechanical origins of thermodynamics, at the time of writing this area is still developing.

Continuous measurements, and thus thermal environments, induce the transition from quantum to classical dynamics, and we discuss this phenomenon in Section 4.4. At the end of this chapter, since we will have both measurement theory and irreversible behavior under our belts, we finish with a discussion of the so-called "quantum measurement problem."

4.1 Statistical mechanics

4.1.1 Thermodynamic entropy and the Boltzmann distribution

Elementary definition of the thermodynamic entropy

First let us remind ourselves of the standard concepts. The macroscopic state, or "macrostate," of a system is defined by those global physical properties that one can measure with simple macroscopic apparatus.[1] Examples of these properties are the total energy, pressure, and volume of a gas. We then consider the number of mutually orthogonal quantum states that the system could be in, given its macrostate. This is the dimension of the Hilbert space of the possible states of the system, given that the macroscopic variables have been fixed. Each of these possible states is referred to as a "microstate" of the system. Note that for large systems, consisting of many particles, the microstate cannot be determined by an observer; the observer has access only to the macrostate. For example, for a many-particle system the quantum states are given by all the possible positions and momenta (all the possible joint wavefunctions) of all the particles that make up the system. Each of the microstates is also referred to as a *configuration* of the particles in the system. For a system consisting of many subsystems, each macrostate has many microstates. The Hilbert space of a system is the space whose basis consists of all the possible microstates, and we will also refer to this space as the "state space" of the system.

The entropy of a system associated with a given macrostate is defined as the logarithm of the total number of microstates that give that macrostate, multiplied by an additional

---

[1] Curious factoid #267: In Latin "apparatus" is a 4th-declension noun, in which the plural is spelt the same, but pronounced with a long *u* (appara-*tuce*). The end result of this is that in English "apparatus" can also be used as its own plural without change. The officially correct alternative is *apparatuses*.

constant, called *Boltzmann's constant*, so as to give it units of energy divided by temperature. (The reason for this extra constant will be explained below.) So the thermodynamic entropy is

$$S_{th} = k \ln(\Omega), \tag{4.1}$$

where k is Boltzmann's constant, and Ω is the maximum number of mutually orthogonal quantum states that the system could be in, given the macrostate. (In other chapters we often use k to denote other quantities, but in this chapter k will always refer to Boltzmann's constant.) Note that if the system is closed, it will always be in a *single* microstate – the entropy does not refer to whether or not the system is in a single microstate, but merely how many states it has the *option* of being in, given its macrostate. This fact is crucial to understanding the theory.

If we have two systems, the first having accessible states Ω_1, and the second Ω_2, then the total number of accessible states of both systems combined is simply $\Omega = \Omega_1 \Omega_2$. Since the number of accessible states is the product of those of the two systems, the total entropy of the two systems is the sum of the entropies of each: entropy adds when we group independent systems together.

The fundamental postulate of statistical mechanics

The fundamental postulate of statistical mechanics, from which the rules of equilibrium thermodynamics follow, is that "all the accessible microstates are equally likely." Now what exactly does this statement mean? It could be interpreted to mean that if one starts a system in a given microstate, its evolution will take it on a path whereby it passes through all the microstates, and spends no more time in one region of state space than another. This would mean that, if there were subsets of the available states that were "typical" (that is, subsets that contain the majority of states – see Chapter 2), the system would spend almost all its time in states in these subsets.

The postulate could alternatively be taken to mean that the properties of the system are those given by assuming that it is in an equal mixture of all the available microstates. (For convenience we will refer to this state as the *canonical mixed state*.) We will see below that both the above interpretations are valid, and further, that they can themselves be derived from certain more fundamental properties of many-body quantum systems.

Thermal states verses states-of-knowledge

It has been common practice to replace the statement that "the properties of a system are those given by assuming that the system is in the canonical mixed state" by the statement that "the system *is* in the canonical mixed state." The problem with this latter statement is that mixed states describe states-of-knowledge, and this can lead to many kinds of confusion. At the simplest level, if one assigns to a large macroscopic system the mixed state given in which each accessible microstate is equally likely, then a problem arises for closed many-body quantum systems. If one starts such a system in a pure state, then it will thermalize even though its evolution is purely unitary. What happens is that it will go to a microstate for which the corresponding macrostate has the largest number of accessible microstates. But because the state of the system is still pure, the system is manifestly *not* in a mixed state, even though the thermodynamic entropy is still correctly given by the log of the number of accessible states. Further, if one asserts that the state of a many-body system is the canonical mixed state, then the thermodynamic entropy is equal to the von Neumann entropy of this state. This gives the erroneous idea that the von Nueman entropy of any state that one may assign to a system is equal to the thermodynamic entropy. But this is obviously not the case, because as we have seen above we can assign a pure state to a closed many-body system that is in thermal equilibrium. It is important not to confuse states-of-knowledge with objective properties, but it is also important to realize, as we will discuss below, that there are situations in which the thermodynamic entropy is given by the von Neumann entropy of the state-of-knowledge of an observer.

For the purposes of defining the thermodynamic entropy for large (many-body) quantum systems, it is safest to define the thermodynamic entropy as the log of the total number of accessible microstates, given the values of the macrosopic observables, irrespective of the density matrix. This definition would have to be modified, however, if an observer were to make measurements that revealed information about the microstate, since this information could be exploited to extract work. We will see how to treat such a situation when the measured system is mesoscopic, but we will not consider measurements on large (macroscopic) systems further here.

Refined definition of thermodynamic entropy

To make thermodynamic entropy a really useful concept, we need to extend the above basic definition to situations in which the microstates of a system are not all equally likely. It may seem that we are now contradicting the fundamental postulate, since we are implying that the microstates of some systems will *not* all be equally likely, so let us pause to explain. The fundamental postulate actually states that all the accessible microstates of a *closed* system are equally likely (something we omitted to mention above). If we have two systems that are interacting, but otherwise isolated, the microstates of the combined system will be equally likely, but this may not be true for those of the subsystems.

We want to define the thermodynamic entropy for a system that has different probabilities of being in different states, and thus whose density matrix has an arbitrary set of eigenvalues. It turns out that there is a way to do this while preserving the key property of the entropy, being that it counts the accessible microstates of an appropriate system. To see how this works, consider a system with two orthogonal states, $|0\rangle$ and $|1\rangle$. We want to define the entropy when the state of the system is an arbitrary mixture of the two, $\rho = P|0\rangle\langle0| + (1 - P)|1\rangle\langle1|$. Now consider that we have N of these systems, all with the same density matrix ρ, and N is large. We can now quantify the total number of microstates that the combined system, consisting of all the N systems, can be in. Recall from Chapter 2 that if we throw a coin N times, there are a number of *typical* sequences, one of which is very likely to occur, and the rest are atypical sequences that will almost never occur. When we have N two-state systems in a mixture, the state of the combined system is simply a mixture of states, where each state is given by one of the sequences of coin flips. In this case the coin has a probability of P for choosing state $|0\rangle$, and $(1-P)$ for choosing state $|1\rangle$. Each of the typical sequences of coin flips now corresponds to one of the possible states of the combined system, and all of them are *equally likely*. As $N \to \infty$, the probability of the non-typical states (the "atypical states") goes to zero. The typical states are therefore the only accessible states, and each is equally likely. Thus the entropy of the N systems, under our basic definition of entropy, is simply k times the logarithm of the number of typical states. From Chapter 2 this is

$$S_{\text{tot}} = -kN \sum_{i=0,1} P_i \ln(P_i). \tag{4.2}$$

Keeping the property that entropy adds when we add systems, this means that the entropy for each system is

$$S = \frac{S_{\text{tot}}}{N} = -k \sum_{i=0,1} P_i \ln(P_i). \tag{4.3}$$

This definition of the thermodynamic entropy, as we will explain shortly, is useful when dealing with operations on simple mesoscopic systems that are connected to thermal baths.

A note on counting states: Recall that when "counting" the number of accessible states for a system, we use the dimension of the space formed by all the accessible states. It is important to realize that this actually follows from assuming that every state in the accessible space, not merely every state of a chosen basis, is equally likely. If we make every state equally likely, then the probability density over the states is given by the Haar measure, the measure that is invariant under all unitary transformations. This measure captures the notion that all states are equally likely. We now note that when considering the relative probability that the state of a system will fall in one subspace or another, under the Haar measure this is just the ratio of the dimensions of the two spaces. Thus the probabilistic "weight" of a given space is just its dimension. If we form a density matrix from all the states using the Haar measure, then the density matrix is $\rho = I/N$, where N is the dimension. So this means that each of the basis states has probability $1/N$, which gives us our expression for the entropy. So using the Haar measure actually corresponds to "counting" the number of basis states, and assigning each a probability of one over this number. While any basis is equivalent for this purpose, we will see when we consider energy exchange between systems, that this exchange will pick out the energy basis as being special.

More than one definition of entropy?

In different situations we will use what appear to be different definitions of entropy. For example, in our definition of thermodynamic entropy above, we defined it as the (log of) the number of states a system could be in, given that its energy is fixed. What is more, we defined the entropy to be this quantity regardless of whether the system is in a pure or mixed state from the point of view of an observer. Earlier in this book we defined the entropy of a system as the von Neumann entropy of the density matrix. The fact that these definitions are apparently contradictory does not make them inconsistent, however. Different definitions are useful in describing different situations, in which different questions are being asked, and in which different sets of constraints are active. We do not have the space here to introduce a single precise definition, and to show that this definition breaks up correctly into the various definitions appropriate for all scenarios. In fact it is not clear that such an elucidation yet exists. But we do need to explain the relationships among the different notions of entropy that we will use.

First, the von Neumann entropy and the thermodynamic entropy of a many-body system are not equivalent. A large quantum system obeys the rules of thermodynamics, including irreversible evolution, even though its evolution is entirely unitary (we discuss this in detail in Section 4.2 below). The von Neumann entropy is by definition invariant under unitary transformations, and so is not the thermodynamic entropy. But this does not make the von Neumann and the thermodynamic entropy incompatible. It is clear that the two entropies can be unified by assigning sets of states that an observer can distinguish, and defining the thermodynamic entropy as the von Neumann entropy of the state-of-knowledge of this observer, given the pure state that the system is in. However, as emphasized above, it is important not to confuse the state-of-knowledge of this fictitious observer with that of any given real observer.

Second, we can use the von Neumann entropy as the thermodynamic entropy when we consider operations performed on simple microscopic or mesoscopic systems. The reason for this is that the unitary evolution of such systems is not a source of irreversibility, and thus does not change the thermodynamic entropy. All irreversibility is modeled by coupling the system to one or more large baths for which the temperature is well defined, and which have a well-defined effect on the system. This effect is a non-unitary transformation that takes the system from its present state to the Boltzmann state given by the temperature of the bath.

Third, for mesoscopic systems, when we are considering the work that can be extracted, we can consistently use a definition of the thermodynamic entropy that is dependent on the state-of-knowledge of an observer. That is, we can define the thermodynamic entropy as the von Neumann entropy of the density matrix of an observer even when this observer makes measurements on the system. The reason for this is that it is possible to show explicitly that for any initial state, ρ, of a mesoscopic system, the maximum average amount of work that can be extracted by an observer, given that we can couple the system to a thermal bath at temperature T, is given by $F \equiv \mathrm{Tr}[H\rho] - S(\rho)T$, where H is the initial system Hamiltonian. That is, the maximum average extractable work is given by the usual expression for the free energy, but with the thermodynamic entropy replaced by the von Neumann entropy of the density matrix of the observer [242, 590, 168].

The relationship between entropy and irreversibility

We explain now how the fundamental postulate of statistical mechanics connects the entropy with physical processes that are *irreversible*. The definition of a *physically irreversible* process is one that occurs spontaneously and is never observed to happen in reverse. A thermodynamic example of an irreversible process is the cooling of a cup of hot coffee when it is placed on a table – the coffee cools spontaneously, but the reverse process, in which some coffee, initially at room temperature, heats itself up by sucking energy from the air around it, never happens. Another example of physical irreversibility is when a vase that has been dropped smashes into a thousand pieces. A thousand pieces of a smashed vase are never observed to collect themselves together to form a whole vase. While the first example is thermodynamic, and the second is not, the cause of their irreversibility is the same.

Before we connect entropy with irreversible processes, it is worth pausing to discuss the relationship between physical irreversibility and *logical irreversibility*. A "process" is defined as something that takes one of a set of possible inputs and produces a distinct output for each distinct input. A logically reversible process is one in which it is possible to determine what the input was by looking at the output. That is, each input has a *different* output. Conversely, a logically irreversible process is one in which two or more inputs have the same output, so that it is not always possible to determine the input when one knows the output.

Thermodynamically reversible processes need not be logically reversible, but physically irreversible thermodynamic processes are logically irreversible, if one identifies the input and output states as the macrostates of the system. This is because these processes take many initial states to the same final "equilibrium" state. However, if one considers the microstates of the gas, then the evolution is always logically reversible.

As far as we have found to date, all fundamental physical processes, as determined by the standard model (we ignore general relativity and the processes involving black holes) are logically reversible. Consider, for example, the unitary evolution of quantum mechanics. Quantum evolution is given by applying a unitary operator U to the initial state, and unitary operators have an inverse given by $U^{\dagger}$. This tells us immediately that one can determine the initial state from the final state, since the initial state is given by applying the inverse operator $U^{\dagger}$ to the final state. All physical processes (all evolution) in quantum mechanics, and in fact in the standard model, are therefore logically reversible, at least with regard to the microscopic state. One of the key problems of quantum thermodynamics is therefore to understand how the irreversible behavior of the macrostates arises from the fundamentally reversible evolution of the microstates.

With the above prelude out of the way, we now turn to the relationship between entropy and irreversibility. The fundamental postulate says that all microstates are equally likely. First, we note that this simple postulate is enough to determine the equilibrium state of a macroscopic (many-body) system. The reason for this is that given the physical constraints on a system, there exists a single macrostate that has vastly more microstates than all the others combined: the total number of microstates for all the other possible macrostates combined is only a very tiny fraction of the total number of states. The fundamental postulate says that the evolution of the system takes it through all the available microstates, and that it spends an equal amount of time in each. If this is true, then the system must spend almost all its time in the macrostate that has almost all the microstates.

In fact, the probability that the evolving system will happen to find one of the microstates corresponding to a different macrostate is so small that it will never occur.

A simple example of the fact that one macrostate possesses almost all the microstates is provided by a gas of particles in a room. Consider first the macrostate in which the particles are evenly spread throughout the room, and second the macrostate in which the particles are confined to 1 cubic meter of the room. In this case the macrostate is defined by the density profile of the gas in the room. It is combinatorics alone that ensure that there are many more ways the particles can be distributed throughout the whole room than in any single cubic meter. The entropy (number of microstates) of any density profile that is significantly uneven is very much less than that of the even distribution.

The existence of irreversible processes now follows immediately from the fact that a system will spend almost all its time in a single macrostate, that being the macrostate with the most entropy (the most microstates). This means that if we start the system off in a different macrostate, it will soon find itself in the maximum entropy macrostate, and then it will stay there. This process is irreversible, because if we start the system in the maximum entropy macrostate it will not return to a lower entropy state. Note that the origin of irreversibility is not a time-asymmetry in the evolution of quantum mechanics, but the asymmetry between the macrostates, in that some have more microstates than others. Thus, if we start a system in a state that is out of equilibrium and run the evolution backward in time, the system will equilibrate in the same way that it equilibrates forward in time. Instead of viewing an out-of-equilibrium state as a special state, one can instead view it as a special *time* in evolution of the system. The fact that a system will always evolve to the macrostate with the highest entropy immediately gives us the connection between irreversibility and entropy: the entropy always increases for irreversible processes. Entropy will never decrease, and this determines completely which processes are irreversible and which are not: those that are reversible must not change the entropy, or they cannot be reversed.

The set of microstates that correspond to the maximum entropy state is a *typical set* in the space of accessible states, because it contains the vast majority of these states (typical sets are defined in Section 2.1.1). The members of a typical set are called typical elements, and since in this case they are pure quantum states we will call them *typical states*. Thus, the microstates that correspond to the maximum entropy state, which is the thermodynamic equilibrium state, are typical states.

We have now seen how the fundamental postulate, combined with the combinatorics of many-body systems, implies that evolution will be irreversible, and that entropy will never decrease. What we have not done yet is to go beyond the fundamental postulate to obtain a more detailed understanding of how this postulate arises from the quantum physics of many-body systems. This will be the subject of Section 4.2.

Thermal interaction and temperature

Now that we have our key concepts defined, we are able to determine the condition for thermal equilibrium. The energy of a many-particle system is the sum of the energies of all the particles. If we remove the energy of the center of mass of all the particles (that is, the energy associated with any *collective* motion that particles may have), then the remaining energy is called *heat*. Heat is the energy of the random motion of the particles. When two objects are brought into contact, the particles of each object collide with those of the other, mediating an exchange of energy between the objects. This is referred to as *thermal contact*. We are all aware that after some time the net flow of heat between the two objects will cease, and when this has happened the objects are said to be in thermal equilibrium.

If the postulate of statistical mechanics is to explain this phenomenon, then the state of the two objects when they have reached thermal equilibrium must have a much greater entropy than all the other states. It certainly makes sense that as energy is added to an object, there are more ways that this energy can be distributed among the particles, and so the number of available microstates, and therefore the entropy, will increase. If we assume that entropy increases with energy, then as energy flows from one object to the other, the entropy of the first falls, and that of the second rises. With the above assumptions, we can easily derive the condition under which an exchange of energy will not increase the total entropy. We do this by writing the total entropy as a function of the energies of both objects. If the energies of the two objects are E_1 and E_2, and their respective entropies S_1 and S_2, then the total entropy is

$$S = S_1(E_1) + S_2(E_2) = S_1(E_1) + S_2(E - E_1), \tag{4.4}$$

where E denotes the total energy of the two objects. If E_1 increases it means that energy has flowed from object 2, so that $E_2 = E - E_1$ must decrease. Now we merely differentiate S with respect to E_1, and note that this must be zero at the maximum. The resulting condition for the maximum is

$$\frac{dS_1}{dE_1} = \frac{dS_2}{dE_2}. \tag{4.5}$$

At the thermal equilibrium the rate of change of entropy with energy of the two objects must be the same. It is therefore this rate of change that determines whether heat flows and in what direction, and this rate of change must therefore be the *temperature* of an object. Specifically, temperature is defined as the inverse of the above rate, so that

$$T \equiv \frac{dE}{dS}. \tag{4.6}$$

It is worth noting that in this definition when the derivative dE/dS is evaluated, the volume must be held constant. This is because if work is done on the system, the energy will change, and this has nothing to do with the entropy, and thus the temperature. Strictly speaking, then, the temperature is $T \equiv (\partial E / \partial S)_V$, where the subscript V denotes constant volume.

To summarize, the fundamental postulate gives us a relation between temperature, energy, and entropy. This is why Boltzmann's constant is introduced in the definition of entropy. This constant scales the entropy to give it units of energy over temperature, so that it can appear by itself in the above definition of temperature.

The Boltzmann distribution and the energy basis

Arguably the central result in thermodynamics is the Boltzmann distribution, as it is from this that all the properties of systems in thermal equilibrium can be obtained. The Boltzmann distribution is very simple to derive from the fundamental postulate. To do so we consider placing a system in contact with a very large system, called a thermal bath, which has temperature T. The reason we make the bath very large is so that any heat that the system gives to it, or takes from it, has a negligible effect on its total energy, and thus does not change its temperature. While the system may exchange energy with the bath, the bath's temperature remains fixed. Let us denote the quantum microstates of the system by $|j\rangle$, and call their respective energies E_j (it does not matter in the following analysis if some of the states have the same energy). We now apply the fundamental postulate to the system and bath combined, which we will call the *universe*. If the total energy of the universe is E_{Tot}, then we assume that every microstate with that total energy is equally likely. With this assumption, we now ask, what are the probabilities for finding the system in each of its states $|j\rangle$? Since all the microstates of the universe are equally likely, the probability of finding the system in state $|j\rangle$ will be proportional to the number of states of the bath that have energy $E_{\text{Tot}} - E_j$: the more states of the bath that have this energy, the more likely we will find the system in state $|j\rangle$. More precisely, if we denote the number of states of the bath with energy $E_{\text{Tot}} - E_j$ by n_j, then the probability that the system is in state $|j\rangle$ is $P_j = n_j / \sum_i n_i$. So what is the number of states of the bath with a given energy? It turns out that we have already made a strong assumption about the numbers n_j. We did this when we assumed that the temperature of the bath was constant. If the temperature is constant, then dS/dE is constant, and so the entropy is a linear function of the energy: $S = S_0 + (dS/dE)E$, where S_0 and dS/dE are constants. But the entropy of the bath when it has energy E is the logarithm of the number of states it has with energy E. So if the number of states is the exponential of the entropy, and the entropy is linear in the energy, the number of states of the bath with energy E must *increase exponentially with E*. In particular, by using $S(E_{\text{Tot}} - E_j) = -k \ln(n_j)$, we obtain $n_j \propto \exp[(E_{\text{Tot}} - E_j)/(kT)]$.

From the analysis above we see that the assumption that temperature is a macroscopic quantity that varies smoothly with energy can be true only if the number of states per unit energy of large systems grows exponentially with their energy. This is indeed true. It is a direct consequence of the fact that large systems are collections of many identical systems, and stems from the resulting combinatorics.

Returning to the probability that our system will be in state $|j\rangle$, since $n_j \propto \exp[(E_{\text{Tot}} - E_j)/(kT)]$, we have

$$P_j = \frac{n_j}{\sum_i n_i} = \frac{\exp[(E_{\text{Tot}} - E_j)/(kT)]}{\sum_i \exp[(E_{\text{Tot}} - E_i)/(kT)]} = \frac{e^{-E_j/(kT)}}{\sum_i e^{-E_i/(kT)}}. \tag{4.7}$$

This probability distribution is the Boltzmann distribution.

To summarize, closed systems (systems with a fixed energy) have an equal probability of being in any of their microstates. In contrast, systems whose energy may vary, by virtue of being in contact with a bath at a fixed temperature T, have different probabilities of being in their various microstates, and these probabilities are given by the Boltzmann distribution. The equal probability distribution is referred to as the *microcanonical ensemble*, and the Boltzmann distribution is referred to as the *canonical ensemble*. Note that a large closed system composed of many subsystems will act as a bath for every subsystem; thus the microcanonical ensemble for the total system will lead to the Boltzmann distribution for the subsystems (recall that to derive the Boltzmann distribution we assumed that the system and bath were together a closed system, and that was in the microcanonical ensemble).

We have determined the probabilities (at equilibrium) for the energy eigenstates of a system interacting with a thermal bath. These are the diagonal elements of the density matrix of the system. To determine the full density matrix we now need the off-diagonal elements. To begin, we realize that the energy of the system is perfectly correlated with the energy of the bath. For each of the energy states of the system, the bath has a different energy, and thus is in a different subspace. Because these subspaces are orthogonal, tracing over the bath destroys the coherences between system states with different energy. So if the energy states of the system are non-degenerate, all the off-diagonal elements will be zero in the energy basis. What of the coherences between degenerate states of the system? Since all energy states of the universe within each degenerate subspace have equal probability, it follows that this is also true for the system. The density matrix is therefore proportional to the identity matrix on each subspace, and so once again all off-diagonal elements are zero.

We see from the above analysis that the energy basis is special, in that the density matrix is always diagonal in this basis when a system is at thermal equilibrium. We refer to the diagonal Boltzmann density matrix as the Boltzmann state, or the canonical thermal state. This is the state of a system when it is in thermal equilibrium at temperature T. We can write the Boltzmann state as

$$\rho_{\text{Boltz}} = \frac{e^{-H_{\text{sys}}/(kT)}}{\text{Tr}\left[e^{-H_{\text{sys}}/(kT)}\right]} = \frac{e^{-\beta H_{\text{sys}}}}{\text{Tr}\left[e^{-\beta H_{\text{sys}}}\right]}, \tag{4.8}$$

where H_{sys} is the Hamiltonian of the system, and we have defined $\beta \equiv 1/(kT)$.

An alternative derivation of the Boltzmann distribution

We now present another derivation of the Boltzmann state, because it allows us to show why this state has the maximum von Neumann entropy for a given average energy, and to show in more detail why the density of states of a many-body system increases exponentially with energy. This time we assume that the universe is a collection of N identical and distinguishable systems, where each system has energy eigenstates $\{|j\rangle\}$ with corresponding energy levels E_j. The system for which we wish to derive the Boltzmann distribution is merely one of these identical systems. Since all the systems are identical, and interact with each other in an identical way, we can assume that at equilibrium the reduced density matrices for each of the systems are all identical. Now let us assume that each system has probability P_j to be in state $|j\rangle$. Since the number of systems, N, is very large, we can use the concept of typical sequences to count the total number of states that the universe can access, given the probability distribution $\{P_j\}$ (typical sequences are explained in Section 2.1.1). In the limit of large N, it is almost certain that the number of systems that are in the state $|j\rangle$ is NP_j. This means that we can count the number of possible states of the universe by counting how many ways we can assign an eigenstate to each system so that there are a total of NP_j systems in each state $|j\rangle$. This is done in Section 2.1.1, where for large N we found that we can write the total number of states as $\exp[H(\{P_j\})]$, where $H(\{P_j\})$ is the entropy of the probability distribution $\{P_j\}$.

The fundamental postulate tells us that the state of the universe will be one of the microstates for which the total number of accessible states, $\Omega = \exp[H(\{P_j\})]$, is largest, and thus will be a state for which the entropy $H(\{P_j\})$ is largest. (If we assume that all the available microstates for a given distribution $\{P_j\}$ are equally likely, then the probability distribution for each system is $\{P_j\}$.) In finding the distribution that maximizes $H(\{P_j\})$, we must remember that the total energy of the universe is fixed. We must therefore maximize $H(\{P_j\})$ under the constraint that $E_{\text{unv}} = \sum_j NP_jE_j$ is constant. This maximization can be readily performed using the technique of Lagrange multipliers, and the result is that the distribution with the maximum entropy is $P_j = e^{-\beta E_j} / \sum_j e^{-\beta E_j}$ for some constant β that is determined by the total energy and the set of energy levels $\{E_j\}$. (As we saw above, β is the scaled inverse temperature.)

In our simple analysis all the accessible states have exactly the same energy, E_{unv}. However, the weak interactions between all the systems actually split the accessible states up into a narrow band of energies. If we assume that the width of the resulting band is independent of the number of accessible states, then the density of states with respect to energy is proportional to $\Omega(E_{\text{unv}})$. So what is Ω as a function of E_{unv}? Well, $E_{\text{unv}} = N\langle E \rangle$, where $\langle E \rangle$ is the average energy of each system, and

$$\langle E \rangle = \sum_j E_j P_j = \frac{\sum_j E_j e^{-\beta E_j}}{\sum_j e^{-\beta E_j}}. \tag{4.9}$$

Since $\Omega = \exp[H(\{P_j\})]$, we can determine Ω as a function of $\langle E \rangle$ if we can solve Eq. (4.9) for β. Unfortunately it is not possible to invert the relationship between $\langle E \rangle$ and β for an arbitrary set of energy levels. But we can do this if our systems are sufficiently simple. For example, if the systems are harmonic oscillators, so that $E_j = \hbar\omega(j+1/2)$, with $j = 0, 1, \ldots, \infty$, then we can calculate $\Omega(E_{\text{unv}})$ and see that Ω increases exponentially with E_{unv}. We leave this as Exercise 3.

4.1.2 Entropy and information: Landauer's erasure principle

Landauer's erasure principle says that when a single bit of information is erased, the entropy of the environment must increase by $k\ln(2)$. In fact, Landauer's erasure principle is not a separate principle but, like the Boltzmann distribution, can be derived from the fundamental postulate, and the reversibility of physical laws. It got the name "principle" because when it was first introduced by Landauer its connection to statistical mechanics was not well understood.

So first of all, what does it mean to erase information? For concreteness, let us consider a single qubit. When a qubit contains one bit of classical information, it must have the option of being in one of two orthogonal states. (From the point of view of someone who doesn't yet have access to the information, this means the qubit is in an equal mixture of its two states. If it is not in an equal mixture, then it contains less that $k \ln(2)$ bits of information – see Chapter 2.) To erase the information in the qubit means that we prepare the qubit in a single state. This erases the information because someone who now looks at the qubit can no longer tell what information was previously stored in it. Erasing information therefore means we take a qubit whose initial state is $|0\rangle$ or $|1\rangle$, and apply a process which leaves the qubit in state $|0\rangle$, for *both* initial states. The act of erasing information therefore means that the qubit must undergo a *logically irreversible* evolution, also referred to as a "many-to-one" mapping. It is important to note that at the start of the erasing process the qubit cannot be entangled with any other system, since for it to contain one whole bit of (classical) information an observer must have prepared it definitely in state $|1\rangle$ or state $|0\rangle$. Similarly, at the end of the erasing process the qubit cannot be entangled with any other system.

Landauer's erasure in a nutshell

We now present the key reasons for Landauer's principle in a nutshell, and then give a longer and more detailed derivation. Landauer's erasure principle follows because the erasing operation is logically irreversible, while all fundamental physical evolution is logically reversible. If the qubit starts in one of two orthogonal states, then after any physical evolution, there *must* be two orthogonal final states that correspond respectively to the two initial states. So what erasing *really* means, is that the two initial states of the qubit are mapped to two final states which the observer does not have access to. This in turn means that the two final states must be microstates of a macroscopic system, because these are by definition the only states that a macroscopic observer cannot access. It follows that for every initial microstate of the macroscopic system (from now on we will call this the *bath*), there must be *two* final microstates. This means that during the erasure process the macrostate of the bath must go from one with Ω microstates, to one with 2Ω microstates. Since the entropy is the log of the number of microstates, this means that the final entropy of the bath is

$$S_{\text{final}} = k \ln(2\Omega) = k \ln \Omega + k \ln 2 = S_{\text{initial}} + k \ln 2. \qquad (4.10)$$

So the entropy of the bath increases by $k \ln 2$. If the system that stores the information is larger than a single qubit, and has $\ln N$ information stored in its N orthogonal states, then the entropy of the environment increases by $k \ln N$.

Often Landauer's principle is stated by saying that when one bit of information is erased, heat equal to $kT \ln 2$ is dissipated in the environment. This is the case if the macroscopic system to which the qubit is coupled is a thermal bath at the fixed temperature T. In this case, the rate of change of energy of the bath with the entropy (at constant volume, so that the energy change is all heat and no work) is $T = (\partial E/\partial S)_V$. So when the entropy of the bath increases by $\Delta S = k \ln 2$, heat must have flowed to the bath to increase its energy by

$$\text{heat} = \Delta E = \left(\frac{\partial E}{\partial S}\right)_V \Delta S = kT \ln 2. \tag{4.11}$$

This is the origin of the alternative statement of the erasure principle.

Derivation of Landauer's principle

The qubit begins in one of the two states $|0\rangle$ or $|1\rangle$. The bath to which the system is coupled begins in any one of its N accessible microstates, which we will label $|n\rangle$. Since from the point of view of the observer the bath is in an equal mixture of all the states $|n\rangle$, for compactness we may as well write the state of the bath as this mixture, so $\rho_{\text{bath}} = \sum_n |n\rangle\langle n|/N$. The initial state of the system and bath combined (the universe), is therefore

$$\rho_0 = |0\rangle\langle 0| \otimes \left(\frac{1}{N}\sum_n |n\rangle\langle n|\right) \quad \text{or} \quad \rho_1 = |1\rangle\langle 1| \otimes \left(\frac{1}{N}\sum_n |n\rangle\langle n|\right). \tag{4.12}$$

Note that with the state of the bath defined as a mixture, the initial entropy of the bath is given by k times the von Neumann entropy of ρ_{bath}: $-k\text{Tr}[\rho_{\text{bath}} \ln \rho_{\text{bath}}] = k \ln N$.

The two final states of the universe after the erasing process are given by

$$\rho_i^{\text{fin}} = U \rho_i U^{\dagger} = |0\rangle\langle 0| \otimes \rho_{\text{bath}}^{(i)}, \tag{4.13}$$

where i can be 0 or 1. The fact that the evolution is unitary means that initial states that are orthogonal must result in final states that are also orthogonal. This is the statement of logical reversibility. Since the unitary must map each joint system–bath state $|0\rangle|n\rangle$ to an orthogonal state, the final states are of the form

$$\rho_i^{\text{fin}} = |0\rangle\langle 0| \otimes \left(\frac{1}{N}\sum_n |n,i\rangle\langle n,i|\right), \tag{4.14}$$

where $U|0\rangle|n\rangle = |0\rangle|n,0\rangle$, and $U|1\rangle|n\rangle = |0\rangle|n,1\rangle$. Since the states ρ_0 and ρ_1 are orthogonal, ρ_0^{fin} and ρ_1^{fin} are orthogonal, which requires that the states $|n,0\rangle$ are orthogonal to all the states $|n,1\rangle$:

$$\langle n,0|m,1\rangle = 0, \quad \text{for all } n, m. \tag{4.15}$$

We have now imposed the condition of logical reversibility, and it remains to impose the erasure condition. This is that the observer cannot tell which of the two states the qubit was originally in. This means that the bath microstates $|n,0\rangle$ and $|n,1\rangle$ must correspond to the *same* macroscopic state. The final macroscopic state must therefore contain a *minimum* of $2N$ microstates. Further, since the initial macrostate contains only N states, the final state of the bath must be a different macrostate, and thus all of the final bath microstates $|n,i\rangle$ must be orthogonal to the initial bath microstates:

$$\langle n|m,i\rangle = 0, \quad \text{for all } i,n,m. \tag{4.16}$$

The entropy of the final bath macrostate is therefore at least $k\ln(2N)$, and so the change in the entropy of the bath macrostate is

$$\Delta S \geq k\ln 2. \tag{4.17}$$

A note

We have performed our analysis of Landauer's erasure principle by considering each of the two initial states that must be brought to a single state. In doing so we avoided the need to say that the initial state of the qubit is the completely mixed state $\rho = 0.5|0\rangle\langle 0| + 0.5|1\rangle\langle 1|$ [723, 360, 358, 558, 481, 483]. Allowing the initial state of the qubit to be mixed simplifies the derivation, but is actually a kind of shorthand. We feel that this mixed-state approach, used by previous authors, can be confusing because to an observer who knows the information stored in the qubit, the initial state is not mixed at all. This can give the impression that Landauer's erasure is subjective, which is not the case. But we wish to point out that using our method one has to work a little harder to derive the entropy increase of the bath when the system does not contain one full bit of information. (That is, if the system is more likely to be one state than the other.) In this case, the entropy increase of the bath is equal to k times the amount of information that the qubit stores. To show this using our method, one employs the usual information-theoretic procedure: one considers erasing the information in N qubits, each of which has probability p to be in the state $|1\rangle$. Now one need only erase the *typical* sequences of the N qubits. It is then clear that the mixed-state approach, in which one takes the initial state of the system to be $\rho = (1-p)|0\rangle\langle 0| + p|1\rangle\langle 1|$ is a shorthand for this more precise analysis.

Another note: asymmetric storage states

It is possible to devise an "erasure" process in which the entropy of the bath increases by less than $k\ln 2$, by choosing one or more of the states that encodes the information, or the final state (the state after erasure) to be mixed [33, 402, 529]. For example, we could store a single bit of information in a system that has the three orthogonal states $|0\rangle, |1\rangle$, and $|2\rangle$, by encoding the bit using the two states $\rho_0 = |0\rangle\langle 0| + |2\rangle\langle 2|$, and $\rho_1 = |1\rangle\langle 1|$. Now the erasure operation involves mapping the two initial states to the single final state $\rho_0 = |0\rangle\langle 0| + |2\rangle\langle 2|$. Because the final state is mixed, and one of the initial states is not, the system effectively absorbs some of the entropy increase from the erasure. We will not examine these scenarios in detail here. Nevertheless, in all cases, if one examines a complete write–erase cycle, in which information is written to the system and then erased, one still finds that the total amount of entropy increase of the bath is $k\ln 2$ over the whole cycle. The use of mixed storage states is not what Landauer had in mind; whether one should view Landauer's principle as remaining valid for this situation seems to be a matter of opinion [33, 402, 529]. The author favors retaining Landauer's principle for all cases, since it emphasizes the essential point that erasing information is a thermodynamically non-trivial process. This arises from the fact that the basic physical laws are logically reversible, whereas erasure is logically irreversible.

4.1.3 *Thermodynamics with measurements: Maxwell's demon*

The conversion of heat to work, and the Helmholtz free energy

Recall that if a system is out of equilibrium, it spontaneously "relaxes" to equilibrium, and this involves an increase in the entropy. In fact, this process involves an increase not only in the entropy of the system but also in the overall entropy of the universe. This can be contrasted with a process in which the state of the system changes, but in doing so it always remains arbitrarily close to equilibrium. Such processes are referred to as quasi-equilibrium processes. In these processes, while the entropy of the system may change, it is always matched by an equal and opposite change of the system's environment (the environment is all other systems and/or thermal baths with which the system interacts). This means that quasi-equilibrium processes are also reversible processes, since the overall entropy of the universe remains the same, and there is therefore nothing to prevent the process being performed in reverse. An important part of the subject of thermodynamics is how energy in the form of heat can be converted to work, since work is of practical use. It turns out that in any process that takes a system from an equilibrium state to another equilibrium state, the work that is extracted is maximized by a quasi-equilibrium process – all irreversible processes that take the system between the same two states extract less work. The analysis that leads to the above conclusions can be found in introductory texts on thermodynamics.

It is important to note that the macroscopic variables that describe the state of a system, such as the pressure and temperature, are not necessarily well defined if the system is not in equilibrium. For example, when a gas is released from a container to fill a room that is initially a vacuum, as the gas is expanding its pressure varies across the room, so there is no single "pressure" that can be assigned to it. Similarly, a system only has a well defined temperature if the rate of change of entropy with energy is well defined. If we know that as the entropy changes, the system remains in a Boltzmann distribution, then if we change the average energy of the system (for large systems the average energy is the macroscopic energy) the change in the entropy is fixed, and thus the temperature is well defined. But if, when the average energy changes, the distribution over the energy states changes in some arbitrary way, then the resulting change in entropy could be essentially anything, and the temperature is therefore not defined. So it is only for quasi-equilibrium processes, in which the system remains in a Boltzmann distribution, that all the state variables are well defined during the process.

We now consider a system in contact with a thermal bath with a temperature T, and the amount of work that can be extracted by processes that take the system from one state to another. The system is restricted to exchanging heat only with the thermal bath, and no other system. While we also restrict the initial and final temperatures to be equal to T, the temperature of the system during the process may change or may be undefined. The total change in the energy of the system during any process is

$$\Delta E = Q - W, \tag{4.18}$$

where Q is the heat added to the system, and W is the work done *by* the system (we will define work done by the system as positive). Since the heat added is subtracted from the bath, which is always at temperature T, the change in entropy of the bath is $\Delta S_{\text{bath}} = -Q/T$. Since the entropy of the universe cannot decrease, we must have $\Delta S \geq -\Delta S_{\text{bath}}$, where ΔS is the change in the entropy of the system ($\Delta S = -\Delta S_{\text{bath}}$ when the process is reversible, so that the change in the entropy of the universe is zero). Substituting these into Eq. (4.18) we have

$$\Delta E = -T\Delta S_{\text{bath}} - W \leq T\Delta S_{\text{sys}} - W. \tag{4.19}$$

So the work done by the system during any process is

$$W \leq T\Delta S_{\text{sys}} - \Delta E. \tag{4.20}$$

We now define the function $F = E - TS$, called the *Helmholtz free energy*. Since the temperature of the initial and final states is the same, the right-hand side of Eq. (4.20) is the negative of the change in the free energy, $\Delta F = \Delta E - T\Delta S$. We therefore have

$$W \leq -\Delta F, \tag{4.21}$$

so that the maximum work that can be extracted from (done by) the system is the amount by which the free energy decreases. This is the reason for the name "free energy": this quantity gives the maximum amount of energy in the system that is free to be turned into useful work, when the system is in contact with a heat bath at a single temperature, and the initial and final states also have this temperature.

A note on heat and work in quantum mechanics

Work is a change in energy due to the application of forces. If we are treating the forces classically, then applying forces means changing the Hamiltonian. For example, if we consider a gas in a box, doing work on the gas means moving the walls of the box. The walls of the box determine the potential that the particles of the gas are subject to, and thus moving the walls means changing this potential. Conversely, a change in heat is a change in the probability distribution over the energy levels (e.g., changing the temperature changes the Boltzmann distribution). Now, if H is the Hamiltonian of a system, and ρ is its state, then the change in the average energy may be written as

$$dE = d(\text{Tr}[H\rho]) = \text{Tr}[d(H\rho)] = \text{Tr}[dH\rho] + \text{Tr}[Hd\rho]. \tag{4.22}$$

This is the traditional way of breaking down a change into work and heat for a quantum system. This breakdown is neat, but it is important to realize that it is inadequate for many purposes. For example, work can be done on a system by applying control fields that change the populations of the energy levels, not just the energy levels themselves. We will not explore this subject further here, save to present an example in which work is done by a bare measurement. Further information about quantum thermodynamics can be found in the references at the end of this chapter.

Maxwell's demon: work extraction with measurements

The laws of thermodynamics stem from the assumptions of statistical mechanics, whose basic tenet is that macroscopic observers do not have access to the microstates. If an observer could determine the microstate of a system, then she could turn heat directly into work, without the limits imposed by thermodynamics. The first person to write about this was Maxwell [359]. He imagined a tiny "demon" with the ability to determine the speed of individual molecules of a gas. The demon operates a tiny trapdoor in a wall separating two compartments containing the same gas. Each time molecules approach the door the demon lets the fast molecules past if they are going from compartment A into compartment B, and lets the slow molecules past if they are going the other way. In this manner the demon places all the fast molecules in one container, and the slow molecules in the other. This reduces the temperature of compartment A and increases the temperature of compartment B. The overall entropy of the gas decreases, and work can now be extracted from the two compartments because of the temperature difference. Both these things are in direct contradiction of the second law of thermodynamics. This famous scenario, referred to as "Maxwell's demon," raises the following question: is the second law only valid because macroscopic observers do not have fine-enough measuring devices, as appears to be implied by Maxwell's demon, or is it fundamental, and there is something missing in Maxwell's thought experiment?

It turns out that the answer is the latter; there is indeed something missing in the analysis of Maxwell's demon. What is missing is quite simple, but it took 111 years for it to be understood [46]. The missing ingredient is the fact that in order to make the measurements and then perform the feedback (opening and closing the trapdoor), the demon must store the result of the measurement in its memory (the result of the measurement is whether the incoming molecule is "fast" or "slow"). This is because the feedback is conditional on the measurement result, and the demon must store this result to be able to perform a conditional operation. The result is that the demon must have a large memory store, and this gets filled up with the random measurement results. This preserves the second law. One way to see this is that if the demon erases its memory, so as to reset itself to its initial state, by Landauer's erasure principle, the entropy put back into the environment turns out to be at least as great as the reduction in the entropy of the gas. Thus in a closed cycle the second law is preserved.

It is worth thinking about the above scenario a little further. What if the demon does not erase its memory? The memory of the demon is then in a typical state, corresponding to the typical state of the fast/slow gas molecules. (If the memory is stored in the demon's microstates, then by our definition of entropy, the demon's memory has at a minimum the entropy that the demon has removed from the gas.) Even if the demon's memory uses macrostates (such as a computer memory), the final random state of the memory is hardly useful. So perhaps we should, in fact, broaden our definition of entropy. In this case what the demon has actually done is to map a typical (micro-) state of the molecules of a gas to a typical (macro-) state of its memory. In order for that memory to be useful for storing information, we must be able to place it in a desired state – that is, we need to effectively reduce its entropy to zero. The implication is that regardless of whether typical (random) states are indistinguishable to an observer (microstates) or destinguishable (macrostates), they are not as useful as atypical states. Wherever there is entropy (a typical state drawn from a large set) this entropy can be moved around, but not destroyed, because all physical processes are microscopically reversible.

Quantum mechanical analysis of Maxwell's demon

We now analyze the action of a quantum mechanical Maxwell's demon, and show that the increase in entropy when the demon erases its memory is precisely that required to preserve the second law. We will also determine exactly how much extra work the demon can extract from a system by making measurements on it.

We begin with a system in equilibrium at temperature T. The state of the system is the Boltzmann state

$$\rho = \frac{e^{-H/(kT)}}{\text{Tr}\left[e^{-H/(kT)}\right]}, \tag{4.23}$$

where H is the Hamiltonian of the system. We will denote the energy eigenstates of the system as $|E_n\rangle$, where $H|E_n\rangle = E_n|E_n\rangle$.

The demon first makes a measurement on the system, and then performs an action on the system that depends on the result of the measurement. To describe properly the process by which the demon erases its memory, we will need to describe the demon quantum mechanically. Let us denote the set of basis states for the demon by $\{|j\rangle\}$. The demon can perform a different action on the system for each of its basis states if the demon and the system have an interaction Hamiltonian of the form

$$H_{\text{int}} = \sum_j |j\rangle\langle j| \otimes H_j. \tag{4.24}$$

In this case, if the demon is in state $|j\rangle$, then the system will undergo the evolution given by H_j. To perform an action that depends on the state of the *system*, the demon must correlate the states of the system with its own set of basis states, $|j\rangle$. Recall that this is precisely what a probe system must do to make a measurement on the system. Recall also that a general measurement is made by correlating the system with the probe, and then projecting the probe into one of its basis states. If the measurement is described by the set of operators $\{A_j\}$, then this means that for basis state $|j\rangle$ of the probe, the system must be in the state

$A_j \rho A_j^\dagger$. That is, the joint state of the system and probe after they have been correlated, but before the measurement on the probe, must be

$$\rho_{\text{joint}} = \sum_j |j\rangle\langle j| \otimes \left(A_j \rho A_j^\dagger \right) + \sum_{j,l} |j\rangle\langle l| \otimes \sigma_{jl}, \qquad (4.25)$$

for some matrices σ_{jl}.

Now the interaction Hamiltonian, H_{int}, is switched on, and the result is

$$\rho_{\text{joint}} = \sum_j |j\rangle\langle j| \otimes \left(U_j A_j \rho A_j^\dagger U_j^\dagger \right) + \text{stuff}$$

$$= \sum_j p_j |j\rangle\langle j| \otimes U_j \rho_j U_j^\dagger + \text{stuff}, \qquad (4.26)$$

where $U_j = \exp(-iH_j t/\hbar)$. This is how a quantum system performs feedback control on another quantum system, applying the operations U_j, that are (effectively) conditional upon the result of a measurement described by operators, A_j. In the last line of the above equation, p_j is the probability for getting the result j, if one were to make that measurement, and ρ_j is the density matrix for that result.

Now, neither the average energies nor the entropies of the states ρ_j are the same as the average energy of ρ, and neither are their entropies. Let us denote their respective average energies as E_j, and their entropies as S_j. (Note that here we will define the entropy of a quantum state as k times the von Neumann entropy, so it corresponds to the thermodynamic entropy. Since this is merely multiplication by a constant, all the properties of the von Neumann entropy still apply.) Recall from Theorem 6 in Chapter 2 that the average of the entropies S_j is less than or equal to the initial entropy, S:

$$\sum_j p_j S_j \leq S. \qquad (4.27)$$

To determine how much work the demon can extract using the above feedback process, we get the demon to use its conditional operations to transform the states ρ_j so that they are Boltzmann states, and therefore have a well-defined temperature. To do this the demon first performs the following two operations on each of the ρ_j: (i) it transforms the eigenbasis of ρ_j to the energy eigenbasis of the system, so that ρ_j is diagonal in this energy basis; (ii) it reorders the eigenstates of ρ_j, so that the eigenvalues for the energy eigenstates decrease monotonically with energy. The density matrix ρ_j can now be written as

$$\rho_j = \sum_n \lambda_{jn} |E_n\rangle\langle E_n|. \qquad (4.28)$$

The probabilities of the energy eigenstates are given by λ_{jn}, and these decrease with increasing energy. The only thing the demon has to do now, to turn ρ_j into a Boltzmann state, is to adjust the spacings between the energy levels of the system (that is, change the values of E_n) so that

$$\lambda_{jn} = \frac{\exp\{-E_n/(kT)\}}{Z}, \tag{4.29}$$

where $Z = \sum_m \exp\{-E_m/(kT)\}$. It may not be obvious that it is possible for the demon to adjust the E_n by performing a unitary operation. But parameters of a system Hamiltonian, which we usually regard as being classical variables, are ultimately determined by the states of other quantum systems: it is only as part of an *effective* theory that such parameters are classical. So the demon can ultimately adjust the Hamiltonian, and thus the energy levels of the system, by applying a unitary transformation to additional systems. We ignore these additional systems in our analysis, because they do not play a role in the entropy accounting that we need to do. Note that the act of turning off the interaction that correlates the demon with the system, as well as turning on H_{int}, also requires an additional quantum system. This system also has no effect on the entropy.

Note that the above operations on each state ρ_j leave the entropy S_j unchanged, but not the energy E_j. As a final step the demon now adds a constant energy shift to each of the Hamiltonians for the states ρ_j, so as to return the average energy of each of these states back to the original energy E. (Note that this does not change the energy gaps between the eigenstates, and thus preserves the Boltzmann distribution for each ρ_j.) By doing this the demon removes from the system any energy that was added by the measurement.

After the above operations, the joint state of the demon and system is now

$$\rho_{joint} = \sum_j p_j |j\rangle \langle j| \otimes \rho_j + \text{stuff}, \tag{4.30}$$

where the states ρ_j have entropy S_j, energy E, and temperature T.

The demon is now in a position to extract work from the system, by allowing the system to expand isothermally, so that the entropy of each state ρ_j returns to its original value, S (and thus each ρ_j returns to the initial state ρ). In doing so, for each state ρ_j the final state of the environment will have a different entropy, and so all these final states are macroscopically distinct. The final state of the demon will therefore be a *mixture* of the states $|j\rangle \langle j|$. Because of this, one way for the demon to perform the feedback is first to correlate itself with a classical apparatus, which then isothermally expands the system. If we denote the states of the classical apparatus as $|\bar{j}\rangle$, then when the demon correlates itself with the apparatus, the state of the apparatus + demon + system is

$$\rho_{joint} = \sum_j p_j |\bar{j}\rangle \langle \bar{j}| \otimes |j\rangle \langle j| \otimes \rho_j. \tag{4.31}$$

The "stuff" has disappeared, because the macroscopically distinguishable states of the apparatus cannot be in a superposition, and so the joint state becomes a mixture. Now the apparatus performs the feedback on the system, transforming each state ρ_j to ρ, and at the same time changing the bath state, the joint state becomes

$$\rho_{\text{joint}} = \left(\sum_j p_j |\bar{j}\rangle \langle \bar{j}| \otimes |j\rangle \langle j| \otimes \rho_j^{\text{bath}} \right) \otimes \rho. \qquad (4.32)$$

Here the states ρ_j^{bath} are the final states of the bath. Note that during the isothermal expansion, for each j the entropy of the system changes by $\Delta S_j = S - S_j$, and this entropy goes to the bath, so that the entropy of the bath changes by $-\Delta S_j$.

Since the probability that the system was in state ρ_j is p_j, the total work extracted by the demon's feedback operation, averaged over the possible "measurement outcomes," j, is

$$W_{\text{demon}} = \sum_j p_j (S - S_j) T = \langle \Delta S \rangle T, \qquad (4.33)$$

where $\langle \Delta S \rangle$ is the average entropy reduction of the system that is achieved by the demon's "measurement."

At this point the demon appears to have violated the second law, because it has extracted work from the system (turned heat into work) *without* increasing the thermodynamic entropy of the universe. We can see this because all the processes that were used are reversible, except for the correlation of the demon with the apparatus. This process is irreversible because it destroys the superposition over the states of the demon and system, turning it into a mixture. But the process does not increase the thermodynamic entropy, because the observer *knows* the state of the apparatus.

The apparatus and demon are in one of the states $|\bar{j}\rangle \langle \bar{j}| \otimes |j\rangle \langle j|$, for some value of j. Thus to reset the demon to its original state we must perform an irreversible mapping from these states to a single state: we must erase the information stored by the demon. This information is the Shannon entropy of the probability distribution $\{p_j\}$, denoted by $H(\{p_j\})$. From Landauer's erasure principle (see above), we know that this erasure process increases the thermodynamic entropy of the environment by $\Delta S_{\text{erase}} = k H(\{p_j\})$.

Once the demon (and apparatus) have been reset to their initial states, we have extracted the amount of work, $W_{\text{demon}} = \langle \Delta S \rangle T$, and the entropy of the universe has increased by ΔS_{erase}. Theorem 7 in Chapter 2, due to Lanford and Robinson, now guarantees that the second law is obeyed. This theorem says that in a quantum measurement, the Shannon entropy of the outcomes is greater than or equal to the average reduction in the von Neumann entropy of the system: $H(\{p_j\}) \geq S - \sum_j p_j S_j = \langle \Delta S \rangle$. Therefore the increase in the entropy of the universe is always greater than or equal to that required to extract the work W_{demon} under normal thermodynamics (that is, without measurement), and so the second law is still valid.

We note that it is only classical measurements for which $H(\{p_j\}) = \langle \Delta S \rangle$ [446], and thus for which the increase in the entropy of the universe is no more than it needs to be. Thus the demon must make classical measurements on the system if it is to be thermodynamically efficient.

Efficient bare measurements do non-negative work on a system in equilibrium

Consider making a bare measurement on a system initially in equilibrium. If the initial state is ρ, and we denote the set of measurement operators by $\{P_n\}$, then the final state is one of $\rho_n = P_n \rho P_n / p_n$, where $p_n = \text{Tr}[P_n^2 \rho]$ is the probability of obtaining outcome n. The average energy of the system after the measurement is

$$\langle E \rangle_{\text{fin}} = \sum_n \text{Tr}[H p_n \rho_n] = \text{Tr}\left[H \sum_n p_n \rho_n\right] = \text{Tr}[H\tilde{\rho}], \qquad (4.34)$$

where $\tilde{\rho}$ is the final state of the system averaged over the outcomes. Ando's theorem (Theorem 11, Chapter 2) now tells us that $S(\tilde{\rho}) \geq S(\rho)$. But the equilibrium state, ρ, has the maximum entropy for the given average energy, $\rangle E \langle_{\text{init}} = \text{Tr}[H\rho]$. Therefore, $S(\tilde{\rho}) \geq S(\rho)$ implies that

$$\Delta\langle E \rangle \equiv \langle E \rangle_{\text{fin}} - \langle E \rangle_{\text{init}} \geq 0. \qquad (4.35)$$

The work that can be extracted from the system prior to the measurement is the free energy, $F_{\text{init}} = \langle E \rangle_{\text{init}} - TS(\rho)$, and the average amount of work that can be extracted after the measurement is $\langle F \rangle = \langle E \rangle_{\text{fin}} - T\langle S \rangle_{\text{fin}}$, where $\langle S \rangle_{\text{fin}} = \sum_n p_n S(\rho_n)$. The increase in the extractable work is therefore

$$\Delta F \equiv \langle E \rangle_{\text{fin}} = \Delta\langle E \rangle - T\Delta\langle S \rangle, \qquad (4.36)$$

with $\Delta\langle S \rangle \equiv \langle S \rangle_{\text{fin}} - S(\rho)$. This means that any increase in the average energy, $\langle E \rangle$, over and above $T\Delta\langle S \rangle$, is an increase in the amount of work that can be extracted from the system. From Ozawa's theorem (Theorem 6, Chapter 2) we know that the average entropy after the measurement is less than or equal to the initial entropy, so $\Delta\langle S \rangle \leq 0$. This means that $\Delta F \geq \Delta\langle E \rangle$, and so all the energy given to the system by the measurement is extractable as work. Because of this we can consistently interpret the energy added by the measurement as work done on the system.

It is not surprising that measurements add or subtract energy from a system. The measurement involves a unitary interaction with a probe system, and this unitary, as well as the probe system, is the source of (or sink for) this energy. The fact that efficient measurements do not add heat to the system makes sense when considering the interaction with the probe: efficient measurements are performed using a probe in a pure state, and the probe is therefore not a source of entropy.

4.2 Thermalization I: the origin of irreversibility

4.2.1 A new insight: the Boltzmann distribution from typicality

In the section above we derived the Boltzmann distribution in the traditional manner, using the fundamental postulate. However, there is a much stronger way to derive it. It turns out that we do not need to average over all accessible states of the universe (assuming each is equally likely) to obtain the Boltzmann distribution for the system. It turns out that the result follows from *typicality*. Recall that typicality means that almost all the elements within some set share a given property. It turns out that almost all *pure* states of the universe that have a sufficiently narrow energy spread (that is, for which the universe has a fixed energy) have the property that the system is very close to the Boltzmann state. Since the Boltzmann state is a result of the "canonical ensemble" of traditional statistical mechanics, this fact is called *canonical typicality*, and at the time of writing was shown quite recently (in 2006) by Goldstein *et al.* [218] and Popescu, Short and Winter [487].

Canonical typicality is an immediate result of the following fact. Take a universe, and place it in a mixed state that is an equal mixture of all the states from a given large subspace. Calculate the resulting state of the system, by tracing out the bath, and call it ρ. The fact is this: it turns out that almost all *pure* states in that subspace, when one traces out the bath, give a state for the system that is very close to ρ. We will also refer to this more general fact as canonical typicality. When the partial trace of a universe state, $|\psi\rangle$, gives the state ρ for the system, we will refer to $|\psi\rangle$ as "placing" the system in a state ρ.

It is in fact simple to show, in a non-rigorous way, that canonical typicality is basically just the law of large numbers. For simplicity we chose the subspace to be the whole universe, but it is easy to extend the analysis to any subspace. If the universe is completely mixed, then the state of the system, ρ, will also be completely mixed. We wish to determine whether *typical* states of the universe are such that they place the system in – or very close to – the completely mixed state, $\rho = I/N$.

To begin, we note that almost all states of the universe will place the system close to the state I/N, if and only if a universe state picked at random places the system close to I/N, with very high probability. Selecting a state "at random" means picking a state from the distribution given by the Haar measure. However, since typicality means that *almost all* states have the given property, it should not matter precisely what distribution we use to pick our states. So we will choose a distribution that is mathematically simple to work with, and it is for this reason that our "proof" is not rigorous.

Let us denote the energy eigenstates of the system as $|n\rangle$, where $n = 1,\ldots,N$, and those of the bath as $|j\rangle$, where $j = 1,\ldots,N_{\text{bath}}$. A basis of the states for the universe is then $|n,j\rangle \equiv |n\rangle|j\rangle$. We now choose a random state of the universe, $|\psi\rangle$, by setting

$$|\psi\rangle = \frac{1}{\sqrt{\mathcal{N}}} \sum_{n,j} A_{nj} \exp(i\theta_{nj})|n,j\rangle \qquad (4.37)$$

where the A_{nj} are real numbers chosen from a zero-mean Gaussian distribution, and the θ_{nj} are random phases. The constant $\mathcal{N}$ is chosen so that the state is normalized. The density matrix corresponding to this state is

$$\rho = \frac{1}{\mathcal{N}} \sum_{n,j} \sum_{n',j'} A_{nj} A_{n'j'} \exp(i[\theta_{nj} - \theta_{n'j'}])|n,j\rangle\langle n',j'|. \qquad (4.38)$$

Let us denote the diagonal matrix elements of ρ^{sys} by $\rho_n^{\text{sys}} = \langle n|\rho^{\text{sys}}|n\rangle$, and the off-diagonal elements by $\rho_{nm}^{\text{sys}} = \langle n|\rho^{\text{sys}}|m\rangle$. Taking the partial trace over the bath, the diagonal elements of the system density matrix are

$$\rho_n^{\text{sys}} = \frac{1}{\mathcal{N}} \sum_j A_{nj}^2 \equiv \frac{A_n^2}{\mathcal{N}}, \tag{4.39}$$

and the off-diagonal elements are

$$\rho_{n,n'}^{\text{sys}} = \frac{1}{\mathcal{N}} \sum_j A_{nj} A_{n'j} \exp\left(i[\theta_{nj} - \theta_{n'j}]\right). \tag{4.40}$$

Now we need to evaluate these elements to determine if $\rho^{\text{sys}} \approx I/N$. To obtain the most probable value of the diagonal elements, and how likely they are to deviate from this value, we need the means and variances of $\mathcal{N}$ and the sum $\sum_j A_{nj}^2$. Since all the amplitudes A_{nj} are independent random variables, these means and variances are easily determined. The means are

$$\langle \mathcal{N} \rangle = \sum_{n,j} \langle A_{nj}^2 \rangle = \sum_{n,j} 1 = N_{\text{unv}}, \tag{4.41}$$

$$\langle A_n \rangle = \sum_j \langle A_j^2 \rangle = \sum_j 1 = \frac{N_{\text{unv}}}{N}, \tag{4.42}$$

and the variances are

$$V[\mathcal{N}] = \sum_{n,j} V\left[A_{nj}^2\right] = \sum_{n,j} \langle A_{nj}^4 \rangle = \sum_{n,j} 2 = 2N_{\text{unv}}, \tag{4.43}$$

$$V[A_n] = \sum_j V\left[A_{nj}^2\right] = \sum_j \langle A_{nj}^4 \rangle = \sum_j 2 = \frac{2N_{\text{unv}}}{N}. \tag{4.44}$$

The above expressions tell us that $\mathcal{N}$ is equal to N_{unv} plus a random fluctuation whose size (standard deviation) is on the order of $\sqrt{N_{\text{unv}}}$. Similarly, each of the A_n is equal to N_{unv}/N plus a random fluctuation with standard deviation $\sim \sqrt{N_{\text{unv}}}/N < \sqrt{N_{\text{unv}}}$. Thus the diagonal matrix elements are

$$\rho_n^{\text{sys}} = \frac{A_n}{\mathcal{N}} = \frac{N_{\text{unv}}/N + \mathcal{O}\left(\sqrt{N_{\text{unv}}}\right)}{N_{\text{unv}} + \mathcal{O}\left(\sqrt{N_{\text{unv}}}\right)} = \frac{1/N + \mathcal{O}\left(\sqrt{1/N_{\text{unv}}}\right)}{1 + \mathcal{O}\left(\sqrt{1/N_{\text{unv}}}\right)}$$

$$= \frac{1}{N} + \mathcal{O}\left(\sqrt{\frac{1}{N_{\text{unv}}}}\right) \tag{4.45}$$

We can conclude that when $N_{\text{unv}} \gg N$, or equivalently $N_{\text{bath}} \gg 1$, the diagonal elements of the system density matrix are very close to $1/N$, with very high probability.

Now we need to show that the off-diagonal elements tend to zero as $N_{\text{bath}} \to \infty$. We will skip the details of this analysis, but the off-diagonal elements converge to a single number for just the same reason that the diagonal elements do: it is the result of adding together many independent random variables. The off-diagonal elements are given by sums of complex numbers with random amplitudes and random phases. Because of the random phases, these complex numbers are symmetrically distributed around zero. As we add more and more of them together, with increasing probability the answer is closer and closer to zero. Once again the deviation from zero is on the order of $\sqrt{1/N_{\text{unv}}}$.

We see from the above analysis that almost all pure states, randomly picked from the Hilbert space of the universe, will place the system very close to the completely mixed state. And we have seen that this follows purely from the law of large numbers – when one adds together increasingly many independent random variables, the result fluctuates less and less about its mean. Therefore *typical* states of the universe have the property that they place any small subsystem in the state it would be in if the universe were in an equal mixture of all its eigenstates.

It is not difficult to extend the above analysis to any *subspace* of the universe. In this case, a *typical state of this subspace* places the system in the state it would be in if the universe were in an equal mixture of all the states within the subspace.

The above results show us immediately that the state of the universe does not need to be a mixture for the resulting state of the system to be the Boltzmann state. Recall that we showed in the previous section that if the universe is in an equal mixture of all its states within a small energy band, then the system will be in the Boltzmann state. We now see that if the universe is in almost any *pure state* within this same subspace (assuming the subspace is very large) the system will also be in the Boltzmann state. It follows from this that so long as the energy levels of the universe are very densely packed in energy, almost all states of the universe with a narrow energy uncertainty will place the system in the Boltzmann state. We leave it as an exercise to show this explicitly, by using our random-state analysis above, and the fact that the density of energy states of the bath increases exponentially with energy.

4.2.2 Hamiltonian typicality

The canonical typicality of pure states allows us to apply to quantum systems the traditional argument for irreversible behavior in classical systems: if most states give the thermodynamic steady state (for small subsystems), then a large system that starts in an atypical state will quickly evolve to a typical state, and then spend almost all its time in typical states. Hence the system will relax to equilibrium and stay there. But one can go much further than this simple traditional argument.

We now show that if the eigenstates of a Hamiltonian are all typical, then if the system does equilibrate, the steady state of the subsystems will be the Boltzmann state. We will refer to a Hamiltonian with typical eigenstates as a typical Hamiltonian. In fact, the conjecture that every energy eigenstate places a system in equilibrium was introduced prior to the notion of canonical typicality. This conjecture is called the eigenstate thermalization hypothesis (ETH); it was first suggested by Deutsch [141] and Srednicki [572] and has origins in the work of Shnirelman [559].[2]

To begin, let us denote a typical Hamiltonian of a large system as H_{typ}, and its (also typical) eigenstates as $|E_n\rangle$. We now write the initial state of the large system, $|\psi_0\rangle$, in terms of the energy basis: $|\psi_0\rangle = \sum_n c_n |E_n\rangle$. The density matrix for the universe, as a function of time, can now be written as

$$\rho_{\text{univ}}(t) = \sum_n \sum_m c_n c_m^* |E_n\rangle \langle E_m| e^{-i(E_n - E_m)t/\hbar}. \tag{4.46}$$

If we denote the trace over everything but a given small subsystem as $\mathbf{Tr}_u[\cdots]$, and define $\rho_n \equiv |E_n\rangle\langle E_n|$, the evolution of the subsystem is

$$\rho(t) = \sum_n \sum_m c_n c_m^* \mathbf{Tr}_u\big[|E_n\rangle\langle E_m|\big] e^{-i(E_n - E_m)t/\hbar}$$

$$= \sum_n |c_n|^2 \mathbf{Tr}_u[\rho_n] + \sum_{n \neq m} c_n c_m^* \mathbf{Tr}_u\big[|E_n\rangle\langle E_m|\big] e^{-i(E_n - E_m)t/\hbar}, \tag{4.47}$$

where in the second line we have separated the time-independent part of $\rho(t)$ (the diagonal elements) from the time-dependent part (the off-diagonal elements).

Now here is the key observation: The time-dependent part of $\rho(t)$ is composed entirely of a sum of oscillating terms, and each oscillating term extends equally positive and negative (is symmetric about zero). If $\rho(t)$ *does* reach a steady state, then the time-dependent part must lose its time-dependence. But the only way this can happen is that the oscillating terms cancel each other out and the whole sum vanishes (the symmetry of the oscillations precludes a non-zero average). So *if* the subsystem reaches a steady state, then this steady state *must* be given by the time-independent part of $\rho(t)$. It is certainly plausible that the time-dependent part of $\rho(t)$ cancels itself out after a short time: the universe has a large number of densely spaced eigenvalues, and so the time-dependent part contains a near-continuum of frequencies. As the oscillations of these many frequencies go out of phase we can expect them to cancel. We can also expect them to take an exceedingly long time to come back into phase, so for large enough systems, the relaxation to a steady state should be effectively irreversible.

The typicality of the eigenstates has transformed the problem of showing that the dynamics of the universe explores the state space in uniform manner, to how sine waves with different phases sum together. We can well imagine that there is typicality again at work here: for a random choice of the phase factors the sum of complex exponentials will be very small, and only for a very small subset of atypical states will the subsystem be out of equilibrium. To the author's knowledge, rigorous results regarding the properties of the phase factors as relates to equilibration have not been obtained at the time of writing, although a remarkable theorem of Kac [323, chap. 3, sec. 5] is certainly interesting in this regard [322].

---

[2] Berry's conjecture [572, 246, 57] and the work in [177] are also closely connected to the ETH.

We can conclude from the above analysis that the steady state of a small subsystem is given by

$$\rho_{ss} = \sum_n p_n \, \mathbf{Tr}_u \big[|E_n\rangle \langle E_n| \big], \tag{4.48}$$

where $p_n = |c_n|^2$ is the probability that the universe is initially in state $|E_n\rangle$. Since each of the eigenvectors $|E_n\rangle$ is typical, $\mathbf{Tr}_u \big[|E_n\rangle \langle E_n| \big]$ is the Boltzmann state, and so it follows immediately that ρ_{ss} is also the Boltzmann state.

We have shown that the equilibrium state of typical Hamiltonians will be thermal, if they equilibrate. Showing that typical Hamiltonians will equilibrate, which requires that their eigenvalues are sufficiently non-degenerate, is rather involved, and for this we refer the reader to [509, 375, 220, 219, 513]. At the time of writing, the question of how long a subsystem will take to equilibrate is not yet fully solved [562].

While it has been established that systems with typical Hamiltonians, and sufficiently non-degenerate eigenvalues, will thermalize, this only explains thermalization in many-body systems if many-body systems do have typical Hamiltonians and thus obey the ETH. Eigenstate thermalization was confirmed for the first time numerically for many-body systems by Rigol, Dunjko, and Olshanii [516] and further evidence for the typicality of many-body Hamiltonians has been obtained in [532, 514, 533, 443, 161, 60]. Evidence for the ETH in single-particle systems with chaotic classical counterparts can be found in [34, 643, 138, 710].

Random matrix theory and many-body systems

In the previous section we termed Hamiltonians with typical eigenvectors, *typical Hamiltonians*. While this was not intended to mean that these Hamiltonians were *themselves* typical, meaning that almost all Hamiltonians shared their properties, there is, in fact, a sense in which they do. The statement that almost all Hamiltonians have some property is only meaningful with respect to some probability distribution over the set of all Hamiltonians. We can define such a distribution by choosing all the independent matrix elements of a Hamiltonian to be independent, zero mean, Gaussian random numbers. The resulting matrices are called *random matrices*, and indeed the eigenvectors of typical random matrices are typical. We can therefor create typical Hamiltonians by picking them at random in the above way.

It turns out that many properties of large random matrices can be, and have been, determined [419, 420]. An example of this is the distribution of the spacing between their adjacent eigenvalues. This distribution is given by [234]

$$P(s) = \frac{\pi s}{2} e^{-\pi s^2/4}, \tag{4.49}$$

where s denotes the spacing between adjacent levels. This is referred to as the Wigner–Dyson distribution. Energy-level spacings have been calculated numerically for a number of many-body systems, confirming that they follow the above distribution. This provides further evidence that the Hamiltonians of many-body systems are indeed typical.

Exceptions: integrable systems and Anderson localization

Not all large many-body quantum systems thermalize. If a system has a few conserved observables in addition to the energy, then the system will still thermalize but to a generalized Boltzmann distribution [515]. There are many-body systems for which analytic solutions can be obtained, usually by using the Bethe ansatz [38, 326, 598, 13], and these are referred to as *integrable*, by analogy with integrable classical systems [587]. In these systems there are as many conserved observables as the number of subsystems, and the eigenstates do not overlap with large numbers of the elementary basis states (that is, the states that are the tensor products of the eigenstates of the subsystems). Because of this the eigenstates of the universe are not similar to random states picked from large spaces of the elementary basis states, and the generalized Boltzmann state is very different from a thermal state [515]. A second, and very different, situation in which many-body systems do not thermalize is that of *Anderson localization* [11]. In this case the interactions between the subsystems are such as to produce thermalization, but the Hamiltonians of the subsystems themselves are changed in a random way from subsystem to subsystem. Remarkably the fact that the subsystems are no longer interchangeable results in eigenstates that are localized, in that they involve only small groups of adjacent subsystems. Once again, since the eigenstates do not cut across large subspaces, the subsystems do not thermalize [644, 464, 217].

4.3 Thermalization II: useful models

We have discussed above how a small system, when coupled to an environment (a large system), will undergo an effectively irreversible evolution to an equilibrium state, in which it will remain. The subject of this section is how one can simulate this irreversible behavior *without* explicitly simulating the evolution of the bath. Simulating the dynamics of the bath in its entirety is computationally prohibitive (although powerful numerical techniques for exact simulation of the subsystem by a reduced simulation of the bath do now exist, and we discuss these in Section 4.3.5). It turns out that when the rate at which the thermalization/damping occurs is much slower than the evolution of the system (referred to as "weak damping"), a simple equation describing the irreversible dynamics can be derived for the system's density matrix *alone*. In particular, the evolution equation for the density matrix depends only on the density matrix at the present time, and as such is referred to as being *Markovian*. While it is usual for the future evolution of a physical system to depend only on the current state, the process of eliminating the environment from the dynamics will, in general, force the resulting equation for the system to be non-Markovian. This is because the environment might "remember" the state of the system at earlier times, and feed this back to the system at later times. If the environment only carries information away from the system, then the master equation can be Markovian.

The term "Markovian" comes from the theory of stochastic processes. Because the master equation for a density matrix is the quantum equivalent of an equation of motion for a classical probability density, it describes a stochastic process. A Markovian stochastic process for a variable $x(t)$ is one in which the future statistics of $x(t)$ are determined entirely by the present value of x.

When systems are weakly damped we can derive simple Markovian master equations for the system that describe the process of damping and thermalization. While these master equations represent a tremendous saving in numerical resources, if the "small" system is large enough, simulating a density matrix still requires significant resources, since it contains N^2 independent real numbers for an N-dimensional system. The "wavefunction Monte Carlo" or "quantum trajectory" technique, based on continuous measurement theory, provides efficient, easily parallelizable methods for simulating master equations, and we describe these in Section 4.3.4.

4.3.1 Weak damping: the Redfield master equation

It is important to realize that damping and thermalization have the same cause. In undergraduate mechanics one treats the damping of an oscillator by introducing a friction force that is proportional to the velocity, and the connection to thermodynamics is never made. However, friction is an irreversible process, and has the same origin as the irreversible process of thermalization. The friction that a mechanical system feels is due to its interaction with a system with many degrees of freedom (many particles), and this "bath" causes the damping in which the mechanical motion is turned into heat (thermal energy of the many particles of the bath). If the bath has a given temperature, then the result of the damping is that the mechanical system ends up in the Boltzmann distribution – that is, the oscillator is thermalized to the temperature of the bath. If we can derive equations of motion for a system in contact with a thermal bath, we obtain a model for damping in the same breath.

Prelude: damping of a classical oscillator

The purpose of this prelude is to explain how the equations for a damped oscillator become symmetric in position and momentum when the damping is slow compared to the oscillation frequency. This is the regime of weak damping. If you want to get straight into modeling damping in quantum systems, you can skip this section.

It turns out that modeling damping and thermalization of a classical oscillator, or in fact a single particle under the influence of *any* fixed potential, is very simple. All one needs to do is to add a friction force (a force proportional to the particle speed), and a white-noise force. If the potential that the particle feels is $V(x)$, then the equations of motion are

$$dx = (p/m)\,dt, \tag{4.50a}$$

$$dp = -V'(x)\,dt - \gamma p\,dt + \sqrt{\gamma mkT}\,dW. \tag{4.50b}$$

Here γ is the damping rate, k is Boltzmann's constant, and T is the temperature of the bath. The friction force is γp, and the change to the momentum in a time-step dt, due to the fluctuating (white noise) force, is given by $\sqrt{\gamma mkT}\,dW$, where dW is an increment of

the Wiener process [288]. If we set the potential to zero, then these equations describe
Brownian motion.

These equations correctly describe thermalization, because the steady-state probability
distribution for the particle is precisely the Boltzmann distribution. This can be shown by
rewriting the stochastic equations as a Fokker–Planck equation, and the interested reader
can find further details in [288, 197]. If the temperature is zero, then the fluctuating force
is zero and we are left only with the damping. For a macroscopic oscillator, and for room
temperature, the fluctuating force is very small compared to the potential. In this case it
can also be dropped, and this is why it is ignored in mechanics classes.

If the potential in the above equations is that of an oscillator with angular frequency
ω, then $V(x) = \omega^2 m x^2/2$. If we scale the position and momentum, so as to use the
dimensionless variables

$$\tilde{x} = x/\Delta x, \quad \text{and} \quad \tilde{p} = p/\Delta p, \tag{4.51}$$

where $\Delta x \equiv \sqrt{\hbar/(2m\omega)}$ and $\Delta p \equiv \sqrt{\hbar m\omega/2}$, then the energy of the oscillator is

$$E = \frac{p^2}{2m} + \frac{1}{2} m\omega^2 x^2 = \hbar\omega \left(\frac{\tilde{x}^2}{4} + \frac{\tilde{p}^2}{4} \right), \tag{4.52}$$

and the equations of motion are

$$d\tilde{x} = \omega \tilde{p}\, dt, \tag{4.53a}$$

$$d\tilde{p} = -\omega \tilde{x}\, dt - \gamma \tilde{p}\, dt + \sqrt{4\gamma n_T}\, dW, \tag{4.53b}$$

in which $n_T \equiv kT/(2\hbar\omega)$ is the average energy of the classical oscillator when it is at
temperature T, in units of $\hbar\omega$.

Now consider what happens if the damping is slow compared to ω, so that the oscillator has a high quality factor, or Q, defined by $Q \equiv \omega/\gamma$. In this case, in the time it
takes the damping to reduce the amplitude by an appreciable amount, the oscillation has
swapped the values of $\tilde{x}$ and $\tilde{p}$ many times. Any damping that is experienced by $\tilde{p}$ is therefore effectively shared by $\tilde{x}$, and it is the same for the noise. Because of this, we can
approximate the equations of motion for the damped oscillator by equations in which both
$\tilde{x}$ and $\tilde{p}$ are damped at the same rate, and both are driven by noise. This model is a good
approximation for large Q ($\gamma \ll \omega$) so long as the two noise sources are uncorrelated and
have identical strengths. The resulting symmetric equations of motion for the harmonic
oscillator are

$$d\tilde{x} = \omega \tilde{p}\, dt - \frac{\gamma}{2} \tilde{x}\, dt + \sqrt{2\gamma n_T}\, dW_1, \tag{4.54a}$$

$$d\tilde{p} = -\omega \tilde{x}\, dt - \frac{\gamma}{2} \tilde{p}\, dt + \sqrt{2\gamma n_T}\, dW_2, \tag{4.54b}$$

when $\gamma \ll \omega$. Here dW_1 and dW_2 are mutually independent Wiener increments.

While it is possible to derive Eqs. (4.54) from Eqs. (4.53) by assuming $\gamma \ll \omega$, the simplest way to show that these two sets of equations are equivalent in this regime is to calculate the power spectra of $\tilde{x}$ and $\tilde{p}$ for both sets of equations, and show that these spectra agree to first order in the small parameter γ/ω. You can do this in Exercise 9.

We can also write the symmetric equations for $\tilde{x}$ and $\tilde{p}$ in a more compact form by defining the complex variable

$$\alpha \equiv \frac{\tilde{x}}{2} + i\frac{\tilde{p}}{2}, \tag{4.55}$$

so that $\alpha^* \alpha$ is the energy of the oscillator, in units of $\hbar\omega$. The equations of motion for the damped oscillator become

$$d\alpha \equiv -i\omega\alpha\, dt - \frac{\gamma}{2}\alpha\, dt + \sqrt{\gamma}\, dZ, \tag{4.56}$$

where $dZ = \sqrt{n_T/2}(dW_1 + i\, dW_2)$ is a complex noise source with the correlation function

$$\langle Z^*(t)Z(t+\tau)\rangle = n_T\delta(\tau). \tag{4.57}$$

The reason we introduce the symmetric form of the damped oscillator equations here is that, when we derive equations that model the damping of *quantum* systems, it is this symmetric form that we will obtain for a quantum oscillator. Obtaining good models of damping and thermalization for quantum systems is much more difficult than for classical systems. The reason for this is that if we just "make up" a model, as we did above for the classical oscillator by adding damping and noise, we will usually break the rules of quantum mechanics; the resulting equations of motion will not preserve the commutation relations of the quantum operators. To obtain a consistent model, we must explicitly couple the quantum system to a large quantum bath, containing many particles or just many states, and then derive from the dynamics of both the system and bath an approximate equation of motion for the system alone. It turns out that this procedure provides a simple master equation only when the damping is slow compared to the system dynamics (weak damping, in which case we obtain symmetric damping equations), or when the system is linear (a free particle or harmonic oscillator). In the next section we consider the case of weak damping, which often applies, for example, to optical modes inside cavities, superconducting oscillators in mesoscopic circuits, and nano-mechanical resonators. These systems are the subject of Chapters 7 and 8.

The thermal master equation

To obtain equations that describe damping and thermalization for quantum systems, one must start by coupling the system to a much larger system, which we will refer to as the bath (short for "thermal bath"). We denote the Hamiltonian for the system by H_{sys}, that for the bath by H_{bath}, and that for the universe (system + bath) as H_{unv}; then a fairly general model of damping is given by

$$H_{\text{unv}} = H_{\text{sys}} + gX \otimes Y + H_{\text{bath}}. \tag{4.58}$$

Here the system couples to the bath via the system operator X, and the bath couples via the operator Y. The overall strength of the interaction is determined by g. The goal is now to derive an equation of motion for the system *alone*, by "tracing out" the bath degrees of freedom in some approximate way. The resulting equation of motion for the system, which now includes the irreversible evolution induced by the bath, is called a master equation. It turns out that the bath degrees of freedom can be eliminated, to a good approximation, if the bath has many energy levels, these levels are much more closely spaced than the energy levels of the system, and the elements of the interaction operator, $H_{int} = gX \otimes Y$, are small compared to the separation of the system energy levels. Note that these are essentially the requirements for a system to thermalize, as discussed above. Recall that a system will thermalize only if the interaction is weak enough so as to change the energy levels of the system only slightly. This also corresponds to weak damping, so that weak damping and thermalization of the system go together, and both are easily described by deriving a master equation. The difficulty arrises when one wants to describe strong damping, and we will examine this is Section 4.3.5.

There is more than one way to derive the master equation describing thermalization of a weakly damped system. This is due partly to the fact that there is more than one way to model the bath. Redfield was the first to derive this master equation for an arbitrary system [504, 505], and the method he used is still the standard approach. This involves modeling the bath as a large number of harmonic oscillators, which is essentially a quantum field. While this model is directly applicable to a system interacting with the continuum of modes of the electromagnetic field, it is not so satisfying from a fundamental point of view, because one must assume that the harmonic oscillators are already at thermal equilibrium. Instead one can use a bath that is defined by more fundamental properties, such as an exponential density of states.

Another problem with the oscillator-bath model is that the standard method of deriving the master equation from this model involves approximations whose content is at best unclear and at worst misleading. While this method can be improved using the projection operator technique (see, e.g., [519]), a better approach to the oscillator-bath model, in the author's opinion, is the input–output derivation, in which Heisenberg equations of motion for the operators of both the system and the bath are derived first. This method is also pedagogically important because it makes a clear physical connection between the damping induced by the bath (in this case the electromagnetic field) and the continuous measurement induced in the system by a measurement of the field (the light that is emitted by the system). For this reason we present the input–output theory for open quantum systems in considerable detail in Appendix F, and show how the thermal master equation is derived from it.

Here we choose to take a conceptually more direct approach to deriving the thermal master equation. This procedure is not a derivation, but a collection of facts that imply, in hindsight, what Markovian master equations should look like. The purpose is to provide insights and conceptual tools for thinking about open systems. We will draw upon Fermi's golden rule, simple elements of equilibrium statistical mechanics, and the continuous measurement that describes counting events from Section 3.3.

The thermal master equation using physical arguments

Our first key ingredient is conservation of energy. When the bath induces an energy change in the system, if the interaction is weak then approximate energy conservation implies that the bath changes energy by the same amount. This means that a change in energy of the system creates a correlation (and entanglement) between the system and bath. The second ingredient is that of irreversibility. We assume that once a correlation has been created between the system and bath, it will never be undone, due to the complexity of the bath evolution. The third ingredient is measurement. If the bath changes energy, and in doing so records for all future time the energy change of the system, then we can measure this change in the bath without changing the average evolution of the system. That is, tracing out the bath is equivalent to measuring all the energy changes, and averaging over the measurement outcomes. From the above arguments we can expect there to be a description of the evolution of the system that involves a sequence of energy jumps.

The fourth ingredient is Fermi's golden rule. If the system starts in an energy eigenstate, then when it exchanges energy with the system it goes from one energy eigenstate to another. But it also goes from one bath state to another. Since the states of the bath form a near-continuum in energy, the situation is precisely that covered by Fermi's golden rule. This rule states that the transition between the two states can be described by a *transition rate*, and not the oscillation that is typical of unitary evolution. Transition rates appear in the dynamics of probability distributions when the underlying dynamics is that of random jumps between the states (see for example standard treatments of random jump processes [288, 197]). This agrees with the conclusion we reached above.

Recall that the steady state of our master equation must be the thermal steady state for the system, which is the Boltzmann state. Fortunately the condition that the transition rates must satisfy to obtain the correct steady state is a simple one. If $|j\rangle$ and $|k\rangle$ are energy eigenstates of the system, where the energy of $|j\rangle$ is higher than that of $|k\rangle$ by ΔE, then the transition rate from $|j\rangle$ to $|k\rangle$ (the downward transition) must be R times greater than the rate from $|j\rangle$ to $|k\rangle$ (the upward transition), where $R = \exp(\beta \Delta E)$. This ensures the correct ratio between the steady-state populations of all the states, in which higher energy states have exponentially lower populations.

We know from our discussion in Section 4.1.1 that the density of states of the bath increases exponentially with energy. Since Fermi's golden rule says that the transition rate is proportional to the density of states at the destination state, it provides us with *exactly* the right relationship between the upward and downward rates. Because a downward transition for the system must be accompanied by an upward transition for the bath (and vice versa), each downward transition will be larger than its upward counterpart by the ratio of the densities of bath states at the energies of the two states. This ratio is precisely $\exp(\beta \Delta E)$.

If the element of the coupling operator X that couples levels $|j\rangle$ and $|k\rangle$ is X_{jk}, we assume that the coupling operator Y has effectively random elements, whose mean-square value, Y_{rms}, does not change with the energy of the bath, then Fermi's golden rule says that the transition rates from $|j\rangle \to |k\rangle$, and $|k\rangle \to |j\rangle$ are given by

$$\gamma_{kj} = \Gamma \langle d_{j\to k}\rangle |X_{kj}|^2 \qquad\qquad j \to k, \tag{4.59}$$

$$\gamma_{jk} = \Gamma \langle d_{j\to k}\rangle |X_{kj}|^2 e^{-\beta(E_j - E_k)} \qquad k \to j, \tag{4.60}$$

where $\Gamma = 2\pi Y_{rms}$, and $\langle d_{j\to k}\rangle$ is the average density of the bath states at the destination.

We will assume from now on that the energy spread of the bath is large enough that $\langle d_{j\to k}\rangle = \langle d\rangle$ is independent of the initial state of the system, and of the transition, so that we can absorb it into the overall rate constant Γ. We can therefore write the transition rates as

$$\gamma_{kj} = \gamma_{jk}e^{\beta(E_j - E_k)} = \gamma |X_{kj}|^2, \qquad \gamma \equiv \langle d\rangle\Gamma. \tag{4.61}$$

Assuming also that the times at which the jumps occur for different transitions are uncorrelated (this is natural because different transitions couple to different bath states), the resulting rate-equations for the populations of the energy eigenstates (the diagonal elements of the density matrix in the energy basis) are

$$\dot{\rho}_{jj} = \gamma \left[\sum_{E_k < E_j} e^{-\beta(E_j - E_k)} |X_{jk}|^2 \rho_{kk} + \sum_{E_k > E_j} |X_{jk}|^2 \rho_{kk} \right]$$

$$- \gamma \left[\sum_{E_k < E_j} |X_{kj}|^2 + \sum_{E_k > E_j} e^{-\beta(E_j - E_k)} |X_{kj}|^2 \right] \rho_{jj}, \tag{4.62}$$

where ρ_{jj} is the probability of being in state $|j\rangle$.

To derive a master equation for the whole density matrix, we need to examine what happens when a jump occurs. Recall that we can measure the energy of the bath to determine if a jump has occurred for any of the transitions. As soon as we see a jump for the transition from $|j\rangle$ to $|k\rangle$, we know that the state of the system was $|j\rangle$, and is now $|k\rangle$. This means the system is projected onto state $|j\rangle$, and then immediately flipped to state $|k\rangle$. This transformation of the density matrix is given by

$$\tilde{\rho} = J_{kj}\rho J_{kj}^\dagger, \quad \text{where} \quad J_{kj} = |k\rangle\langle j|. \tag{4.63}$$

We now note that this operator is applied each time the corresponding jump occurs, and the jump occurs with a certain probability per-unit-time. This probability is the transition rate multiplied by the probability of being in state $\langle j|$, ρ_{jj}. But this random measurement process is described *precisely* by the continuous measurement introduced in Section 3.3 for counting events. To obtain the right rate of jumps the measurement operators for the transition $|j\rangle \to |k\rangle$ are given by Eqs. (3.143) and (3.144) with

$$\Omega = \sqrt{\gamma} X_{kj} J_{kj}. \tag{4.64}$$

If we define

$$L_{kj} = |k\rangle\langle k|X|j\rangle\langle j| = X_{kj}J_{kj}, \quad j > k, \tag{4.65}$$

then the evolution of the system due to the measurement, averaged over the measurement results (the jumps), is

$$\dot{\rho} = -\gamma \left[L_{kj}^{\dagger}L_{kj}\rho + \rho L_{kj}^{\dagger}L_{kj} - 2L_{kj}\rho L_{kj}^{\dagger} \right]. \tag{4.66}$$

Assuming that all the jumps are independent, the total master equation for the system is the result of adding together the evolutions coming from each transition. Including also the Hamiltonian evolution of the system, the thermal master equation is

$$\dot{\rho} = -\frac{i}{\hbar}[H_{\text{sys}}, \rho] - \gamma \sum_{j>k} \left[L_{kj}^{\dagger}L_{kj}\rho + \rho L_{kj}^{\dagger}L_{kj} - 2L_{kj}\rho L_{kj}^{\dagger} \right]$$

$$- \gamma \sum_{j>k} e^{-\beta(E_j - E_k)} \left[L_{kj}L_{kj}^{\dagger}\rho + \rho L_{kj}L_{kj}^{\dagger} - 2L_{kj}^{\dagger}\rho L_{kj} \right]. \tag{4.67}$$

This is the thermal master equation when none of the transitions are degenerate, meaning that no two transitions share the same energy gap. We will show below why it needs to be modified if this is not the case.

The transition rates and the density of states: additional details

Recall that in the above derivation of the thermal master equation, we used the fact that the transition rates, due to Fermi's golden rule, are proportional to the density of states at the destination. It is worth discussing this in a little more detail, as there is a subtlety here: there may be many destination energies for a single transition. Let us define $\Delta E_{jk} \equiv E_j - E_k$. If the initial states of the bath has energy concentrated around E_0^{bath}, and the system is initially in a state with energy E_k^{sys}, then for the transition $j \rightarrow k$ the energy of the bath at the destination is $E_1^{\text{bath}} = E_0^{\text{bath}} + \Delta E_{jk}$. If, alternatively, the initial system state is E_j^{sys}, then to make the transition $j \rightarrow k$ the system must first jump up to state $|j\rangle$, giving the bath the energy $-\Delta E_{jk}$. In this case, when the system makes the transition $j \rightarrow k$, the energy of the bath at the destination is simply E_0^{bath}. This means that the transition rates can be different for different initial states of the system, although in both cases the ratio of the rates is the same, and gives the same thermal steady state. Now we ask what happens if the initial state of the system is a superposition of $|j\rangle$ and $|k\rangle$. The superposition will decohere, with the result that there will be two separate populations, one of which started in state $|j\rangle$, and the other in state $|k\rangle$. These two populations will jump between the two states at different rates, but since both have the same ratio of upward and downward rates, the steady state is unaffected. Finally, we ask what happens when the bath is in an initial state with an energy spread that is much larger than ΔE_{jk}. In this case, the total population breaks up into subpopulations, each of which starts with a different bath energy. Once again each of these populations jumps between states $|j\rangle$ and $|k\rangle$ with a different rate while giving the same steady state. But now, because the initial energy spread of the bath is large compared to the energy gap ΔE_{jk}, the total transition rate is independent of the initial state of the system. The total transition rate for the transition $j \to k$ is the average of all the transition rates for the different initial bath energies. It is therefore given by averaging the density of states over all the destination energies, and this is the meaning of $\langle d_{j \to k}\rangle$ in Eqs. (4.59) and (4.60).

A bath of harmonic oscillators: the Markovian Redfield master equation

In the previous derivation of the thermal master equation we assumed that the bath had an exponential density of states, and we gave the coupling operator, Y, a uniform structure. But there is a great deal of freedom in choosing Y while preserving the thermal steady state. For example, the size of the elements of Y could change with the energy gap, giving transitions with larger gaps faster rates.

It is common to model a thermal bath as a collection of harmonic oscillators. This is an accurate description of an atom or optical cavity interacting with the electromagnetic field, and is often used by default when a model of damping is required. To obtain a thermal master equation from a bath of non-interacting harmonic oscillators, we have to assume that all the oscillators are in thermal equilibrium. The master equation that results is equivalent to a specific choice of coupling operator Y, and in particular a specific choice of how the transition rates vary with the energy gap. We now derive this master equation.

The Redfield bath consists of a very large number of harmonic oscillators, none of which interacts with the others. Each oscillator has a different frequency, ω, and these frequencies are chosen so as to form a near-continuum on the real line. It is assumed that each oscillator is in a thermal equilibrium state at temperature T. Denoting the annihilation operator for the oscillator with frequency ω by a_ω, we choose the interaction between the system and each oscillator to be $H(\omega) = X(a_\omega + a_\omega^\dagger)$, where X is an operator of the system. Fermi's golden rule states that the transition rate from a state $|j\rangle$ to a state $|k\rangle$ in a near-continuum is proportional to the square modulus of the matrix element that couples the two states together, and to the density of states per unit energy in the near-continuum [86].

The thermal equilibrium (Boltzmann) state for the oscillator with frequency ω is

$$\rho(\omega) = \sum_n p_n(\omega)|n,\omega\rangle\langle n,\omega|, \quad \text{with} \quad p_n(\omega) = \frac{e^{-\beta n\hbar\omega}}{1 - e^{-\beta\hbar\omega}}. \tag{4.68}$$

Here the states $|n,\omega\rangle$ are the "number states," being the energy eigenstates of the oscillator with frequency ω. The energy of $|n,\omega\rangle$ is $\hbar\omega(n + 1/2)$, where n is a non-negative integer. This means that if the system is in state $|j\rangle$, the oscillator with frequency ω could be in any one of its number states. The overall transition rate from $|j\rangle$ to $|k\rangle$ will therefore be the average rate, where the average is taken over all the possible initial values of n.

Let us denote the energy of $|j\rangle$ as E_j and that of $|k\rangle$ as E_k. The energy lost (gained) by the system in the transition $|j\rangle \to |k\rangle$ will be (approximately) gained (lost) by the bath.

Thus for the system to make this transition, the bath must make a transition between two states with the energy difference $\Delta E_{jk} = E_j - E_k$. It is the oscillator with frequency $\omega = \Delta E_{jk}/\hbar \equiv \omega_{jk}$, and oscillators with neighboring frequencies, that have the right transitions. We now assume that $E_j > E_k$ so that $|j\rangle \to |k\rangle$ is a downward transition and $|k\rangle \to |j\rangle$ is an upward transition. The transitions that contribute to the total transition rate for these two transitions are then

$$|j\rangle|n,\omega\rangle \to |k\rangle|n+1,\omega\rangle, \quad \text{when } E_j > E_k, \tag{4.69}$$

$$|k\rangle|n,\omega\rangle \to |j\rangle|n-1,\omega\rangle, \quad \text{when } E_j > E_k, \tag{4.70}$$

The matrix elements that couple these transitions are, respectively,

$$M_{\text{down}}(\omega_{jk}) = \langle k|\langle n+1,\omega_{jk}| H(\omega_{jk}) |j\rangle|n,\omega_{jk}\rangle = X_{kj}\sqrt{n+1}, \tag{4.71}$$

$$M_{\text{up}}(\omega_{jk}) = \langle j|\langle n-1,\omega_{jk}| H(\omega_{jk}) |k\rangle|n,\omega_{jk}\rangle = X_{jk}\sqrt{n}. \tag{4.72}$$

If we choose the density of oscillators per unit frequency so that the density of states per unity energy is constant, then the above matrix elements give the full dependence of the transition rates on ω_{jk}, and thus on the energy gap ΔE_{jk}.

To obtain the rates for the downward and upward transitions we therefore average the square moduli of the above matrix elements over the probability distribution for n. This gives

$$\langle M_{\text{up}}(\omega_{jk})\rangle = |X_{jk}|^2 \sum_{n=0}^{\infty} n p_n(\omega_{jk}) = |X_{jk}|^2 n_T(\omega_{jk}), \tag{4.73}$$

where

$$n_T(\omega_{jk}) = \left(\frac{1}{\exp(\beta\hbar\omega_{jk}) - 1}\right). \tag{4.74}$$

Similarly we have $\langle M_{\text{down}}(\omega_{kj})\rangle = |X_{kj}|^2[n_T(\omega_{jk}) + 1]$. Multiplying these averages by a (frequency-independent) rate constant γ, the transition rates are

$$\gamma_{kj} = \gamma|X_{kj}|^2[n_T(\omega_{jk}) + 1], \quad (\text{downward, } j \to k), \tag{4.75}$$

$$\gamma_{jk} = \gamma|X_{jk}|^2 n_T(\omega_{jk}), \quad (\text{upward, } k \to j). \tag{4.76}$$

Note that $n_T(\omega_{jk})$ is the average number of excitations in a harmonic oscillator with frequency $\omega = \omega_{jk}/\hbar$ when it is at equilibrium at temperature T.

The crucial relationship between the upward and downward transition rates for each transition is satisfied, since $\gamma_{kj} = \gamma_{jk}\exp(\beta\hbar\omega_{jk})$. Note that this is not due to a fundamental property of the density of states of the bath, however, but because we assumed that all the bath oscillators are in thermal equilibrium.

Now that we have our transition rates, we use the procedure we used above to obtain Eq. (4.67) from the transition rates given in Eq. (4.61). The result is

$$
\dot{\rho} = -\frac{i}{\hbar}[H_{\text{sys}}, \rho] - \gamma \sum_{j>k} [n_T(\omega_{jk}) + 1] \left[L_{kj}^\dagger L_{kj} \rho + \rho L_{kj}^\dagger L_{kj} - 2L_{kj} \rho L_{kj}^\dagger \right]
$$

$$
- \gamma \sum_{j>k} n_T(\omega_{jk}) \left[L_{kj} L_{kj}^\dagger \rho + \rho L_{kj} L_{kj}^\dagger - 2L_{kj}^\dagger \rho L_{kj} \right], \tag{4.77}
$$

which is the Redfield master equation when no two transitions are degenerate (that is, share the same energy gap). There is some ambiguity as to the meaning of the term "Redfield master equation," since Redfield included both Markovian and non-Markovian versions in his 1957 paper. Here we will always mean the Markovian version of the Redfield equation, given in Eq. (3.15) of [504], and in Eq. (3.143) of [89].[3]

The Redfield equation including degenerate energy gaps

Recall that when a jump occurs in the system from one energy level to another, a jump occurs in the bath that approximately balances the energy lost or gained by the system. Energy jumps therefore occur in the system because pairs of energy levels of the system are coupled to pairs of energy levels of the bath. The jumps are merely an effective description, and are actually due to underlying unitary evolution between the system and bath.

Now consider what happens if there are two (or more) pairs of system energy levels that have the same energy gap, Δ. In this case, because the energy gaps for these transitions are the same, they couple to the *same* transitions of the bath. As a result, they experience the same coherent dynamics. If the probability amplitude for the bath to have gained energy Δ increases, then the probability amplitude for *all* the transitions to have made the downward jump increases by the same amount. Thus the bath state is correlated with all the transitions simultaneously. When a jump occurs in the bath with energy Δ, which can be thought of as due to a measurement that projects the bath onto the higher energy state, then all the transitions are simultaneously projected onto their respective lower energy states. Because of this we can no longer have a separate transition operator for each transition. We must instead have a *single* operator for all the transitions with the same gap, since this describes a single jumping process for all these transitions. For each energy gap the operator is simply the sum of all the operators L_{kj} for the transitions with *that gap*. These new operators are now labeled by the energy, Δ, of the gap to which they correspond. Labeling all the distinct energy gaps by ω_n, the definition of L_n is

$$
L_n = \sum_{E_j - E_k = \hbar\omega_n > 0} L_{kj}. \tag{4.78}
$$

---

[3] In established terminology, what we refer to as the Markovian Redfield equation involves making the weak-coupling, Born, Markov, and secular (rotating-wave) approximations.

The operator L_n is therefore upper-triangular (it describes downward transitions), but may have many non-zero matrix elements. The master equation is now

$$\dot{\rho} = -\frac{i}{\hbar}[H_{\text{sys}}, \rho] - \gamma \sum_n [n_T(\omega_n) + 1] \left[L_n^\dagger L_n \rho + \rho L_n^\dagger L_n - 2L_n \rho L_n^\dagger \right]$$

$$- \gamma \sum_n n_T(\omega_n) \left[L_n L_n^\dagger \rho + \rho L_n L_n^\dagger - 2L_n^\dagger \rho L_n \right]. \tag{4.79}$$

We can see immediately the difference between this master equation and that given by Eq. (4.77), by substituting into the former $L_n = L_{kj} + L_{ml}$. We see that the difference is the appearance of extra "cross-terms" in Eq. (4.79) that contain products of L_{kj} and L_{ml}. These affect only the equations of motion for the off-diagonal elements of ρ. They are due to "coherence" of the damping between the two transitions $j \leftrightarrow k$ and $l \leftrightarrow m$.

When are two transitions degenerate?

In classifying transitions as being degenerate or non-degenerate, we have begged a question: how close do the frequencies of two transitions, ω_1 and ω_2, have to be for them to be degenerate, in the sense that we must use the master equation in Eq. (4.79)? Since the only other timescales in the problem are the transition rates, the answer is that the two levels can be considered non-degenerate when $|\omega_1 - \omega_2| \gg \max(\gamma_1, \gamma_2)$, where γ_1 and γ_2 are the downward transition rates for the respective transitions (the downward rates characterize the interaction with the bath, since the upward rates are zero for $T = 0$). The two transitions are definitely degenerate when $|\omega_1 - \omega_2| \ll \min(\gamma_1, \gamma_2)$. When $|\omega_1 - \omega_2| \sim \gamma_1, \gamma_2$, so that neither condition is satisfied, then the master equations we have presented above break down. As far as Redfield's derivation of the master equation is concerned, it means that we cannot make the secular (rotating-wave) approximation, thus leading to a more complex master equation [504, 519].

The damping rate

Note that in the above master equations there is a single free parameter, γ. This is fixed by the overall strength of the coupling between the system and the bath, and is usually determined experimentally. For example, if we are describing the damping of an optical mode from a cavity, then γ is fixed by the measured rate at which energy is lost from the cavity, or determined from measurements of the reflectivity of the mirrors that form the cavity.

The quantum optical master equation

One of the most useful master equations is the Redfield master equation for a weakly damped harmonic oscillator. This is obtained by starting with a harmonic oscillator coupled to the bath via its position, so that the full Hamiltonian is

$$H_{\text{unv}} = \hbar \omega a^\dagger a + g\tilde{x} \otimes Y + H_{\text{bath}}. \tag{4.80}$$

Here ω is the frequency of the harmonic oscillator, a its annihilation operator, and

$$\tilde{x} = a + a^\dagger \qquad (4.81)$$

is the oscillator's dimensionless position. This coupling makes sense because we expect the bath to subject the oscillator to a fluctuating force, and a force is given by a term in the Hamiltonian proportional to the position.

By examining the non-zero matrix elements of $\tilde{x}$, we see that the transitions between Fock states that are induced by the bath are only those which involve adding or subtracting a single phonon. Thus *every* transition has the same energy gap, being $\hbar\omega$. It is therefore the second form of the master equation that we require, Eq. (4.79), and there is only a single L_n operator, $L_1 = a$. The resulting master equation is

$$\dot{\rho} = -i\omega[a^\dagger a, \rho] - \frac{\gamma}{2}[n_T(\omega) + 1]\left[a^\dagger a\rho + \rho a^\dagger a - 2a\rho a^\dagger\right]$$
$$- \frac{\gamma}{2}n_T(\omega)\left[aa^\dagger\rho + \rho aa^\dagger - 2a^\dagger\rho a\right]. \qquad (4.82)$$

Here γ is the damping rate of the resonator, defined as the rate at which the energy damps. The temperature of the bath, T, determines the value of, n_T, which is the average number of photons in the resonator when it is at that temperature. The expression for $n_T(\cdot)$ is given in Eq. (4.74). This master equation is often referred to as the "quantum optical" master equation because it describes the damping of an optical mode inside a cavity.

Since the oscillator in the above master equation experiences a fluctuating force, we would expect the momentum to be damped, in analogy to the model of classical Brownian motion given by Eqs. (4.53). However, the equation actually gives us damping in both x and p, just like the symmetrical equations of classical Brownian motion weak damping (Eqs. (4.54)) . This is easily seen by using the fact that $\langle\tilde{x}\rangle = \mathrm{Tr}[x\rho]$ to calculate the equations of motion for the mean position and momentum. For $T = 0$, so that one has pure damping, we obtain

$$\frac{d}{dt}\langle\tilde{x}\rangle = -\left(\frac{\gamma}{2}\right)\langle\tilde{x}\rangle, \qquad \frac{d}{dt}\langle\tilde{p}\rangle = -\left(\frac{\gamma}{2}\right)\langle\tilde{p}\rangle, \qquad (4.83)$$

where $\tilde{p} \equiv -i(a - a^\dagger)$. Thus the Redfield equation (and the quantum optical master equation) contain the quantum equivalent of the classical weak-damping approximation.

The "Lindblad form" and "output channels"

In 1976 Lindblad showed that all equations of motion for a density matrix that are time-independent and Markovian must have the form [374, 89]

$$\dot{\rho} = -\frac{i}{\hbar}[H, \rho] - \sum_n \left[L_n^\dagger L_n\rho + \rho L_n^\dagger L_n - 2L_n\rho L_n^\dagger\right] \qquad (4.84)$$

if they are to preserve the complete positivity of the density matrix. Here the operators L_n are completely arbitrary. Clearly the Redfield master equation has this form.

Each distinct operator L_n in the expression above for the Lindblad form is often referred to as a Lindblad operator, and as corresponding to a single "output channel." The latter refers to the fact that each term in the sum over n can be thought of as describing a continuous flow of information being carried away from the system. We discuss this further in Sections 3.6 and 4.3.3.

4.3.2 Redfield equation for time-dependent or interacting systems

When using the Redfield equation we are restricted to modeling systems whose damping rates are small compared to the gaps between their energy levels (weak damping). Even though the damping of a system may be weak, we may wish to describe its behavior when it is strongly coupled to another weakly damped system. We might also want to describe its evolution when the joint Hamiltonian of the two systems is changing with time. Since the Boltzmann distribution depends on the energy levels of a system, and thus on its Hamiltonian, the Lindblad operators L_n in the Redfield master equation also depend on the Hamiltonian. If the system Hamiltonian changes with time, then these Lindblad operators must also change with time.

First consider a single system whose Hamiltonian is time-dependent. In this case it is useful to distinguish two situations. Let us say that the Hamiltonian of a system is given by $H(t) = H_0 + H_1(t)$, where H_0 is constant. If the energy gaps of $H_1(t)$ remain very small compared to H_0 throughout the evolution, then H_1 does not significantly change the energy levels or energy eigenvectors of the system. In this case we can determine the Lindblad operators of the Redfield equation using H_0, and use the resulting Redfield equation throughout the evolution. In this case we can speak of the Hamiltonian as being *weakly time-dependent*.

If the Hamiltonian is strongly time-dependent, so that the energy eigenvalues or eigenvectors change with time, then we must take this time-dependence into account in determining the Redfield equation. Fortunately, if we assume that the bath is Markovian, so that its effect on the system at time t is not influenced by the state of the system at early times (the bath has no memory), then there is a simple prescription for generating the Redfield equation. Since the bath has no memory, the Lindblad operators in the Redfield equation are determined solely by the Hamiltonian of the system at time t. So the only modification to the Redfield master equation is that the Lindblad operators L_n are time-dependent.

Now consider what happens if we have two systems that are interacting. As an illustrative example, we will assume that the joint Hamiltonian of the two systems is given by

$$H(t) = H_0^A + H_I(t) + H_0^B. \tag{4.85}$$

Here H_0^A is the constant Hamiltonian of system A, H_0^B the constant Hamiltonian of system B, and $H_I(t)$ is a time-dependent coupling between the two. Once again it is useful to divide this scenario into two regimes. If $H_I(t)$ is only a small perturbation on the Hamiltonians

of the two systems, then we refer to the systems as being *weakly coupled*, and we can use
the Redfield equation for each system determined separately for each using their respective
Hamiltonians H_0^A and H_0^B.

If the systems A and B are strongly coupled, so that the energy levels or eigenstates
depend significantly on $H_I(t)$, then the systems can no longer be considered independent
from the point of view of the Boltzmann distribution and therefore the Redfield equation.
We must derive the Lindblad operators by using the joint Hamiltonian of the two systems,
$H(t)$, and the joint coupling operator by which the two systems are coupled to the bath.
If system A is coupled to the bath via X_A, and system B via X_B, then the joint coupling
operator is simply $X = X_A + X_B$.

The Lindblad operators are given by

$$L_n(t) = \sum_{E_j(t)-E_k(t)=\hbar\omega_n(t)>0} |k(t)\rangle\langle k(t)|X|j(t)\rangle\langle j(t)|, \quad j > k, \tag{4.86}$$

for each of the distinct transition frequencies $\omega_n(t)$, where $E_j(t)$ and $|j(t)\rangle$ are respectively
the eigenvalues and eigenvectors of $H(t)$. The Redfield equation is then

$$\dot{\rho} = -\frac{i}{\hbar}[H(t),\rho] - \gamma \sum_n [n_T(\omega_n(t)) + 1]\left[\left\{L_n^\dagger(t)L_n(t),\rho\right\} - 2L_n(t)\rho L_n^\dagger(t)\right]$$

$$- \gamma \sum_n n_T(\omega_n(t))\left[\left\{L_n(t)L_n^\dagger(t),\rho\right\} - 2L_n^\dagger(t)\rho L_n(t)\right], \tag{4.87}$$

in which ρ is the density matrix of the joint system, and the curly braces denote the anti-
commutator: $\{A,B\} \equiv AB + BA$.

Note that this time-dependent Redfield equation assumes that all the transition frequencies (energy gaps) ω_n that are distinct initially, remain distinct throughout the
evolution. Information on the accuracy of Markovian master equations for interacting and
time-dependent systems can be found in [519].

4.3.3 Baths and continuous measurements

There is a strong connection between continuous measurements and baths that induce weak
damping. This connection is that the dynamics of both are described by master equations in
the Lindblad form. The reason for this, as discussed in Section 3.6, is that continuous measurements are mediated by thermal baths that are effectively at zero temperature. But there
is a subtlety here that we also need to make clear. If we derive the Heisenberg equations
of motion for a continuous measurement via a thermal bath at zero temperature, under the
weak-coupling and rotating-wave approximations, the resulting continuous measurement
will generate damping of the system to the zero-temperature Boltzmann state. This is true
of homodyne detection and photon counting, but not true of continuous measurements of
Hermitian observables, such as position. In fact we will see below that the measurement of
position acts like a thermal bath at infinite temperature.

To derive the spectrum of a continuous measurement of position in Section 3.6 we simply used the Heisenberg equations for coupling to a bath at zero temperature, with the Lindblad operator L replaced with the Hermitian operator for position. This correctly models a continuous measurement of position, because the resulting Heisenberg equations are formally equivalent to the stochastic master equation that describes this measurement. But this can be confusing because these equations do not lead to thermalization at zero temperature, something that we would expect, because the bath is at zero temperature. The solution to this apparent contradiction is that while the Heisenberg equations correctly describe a continuous measurement when we replace L with a Hermitian observable, the resulting equations are not, in fact, derived directly from an interaction with a zero-temperature bath, as are the Heisenberg equations that give the Redfield master equation. To derive a measurement of a Hermitian operator, we connect the system to be measured to an intermediate system, where the intermediate system is coupled to a thermal bath. We do this in Section 7.7.3 where we show how to make a continuous measurement of position by coupling the object to an optical or superconducting cavity, and making a homodyne measurement on the signal output from the cavity.

Now let us examine the relationship between the dynamics induced by a continuous position measurement and that of thermalization. Recall from Section 3.1.2 that a Gaussian continuous measurement of an observable X induces the evolution

$$\dot{\rho} = -\frac{i}{\hbar}[H, \rho] - \mu[X, [X, \rho]] \tag{4.88}$$

if we average over (ignore) the measurement results. (This master equation is obtained from Eq. (3.39) by averaging over the noise term.) We can obtain Eq. (4.88) from the general Lindblad form by choosing just one Lindblad operator, L_1, and setting it equal to a Hermitian operator X.

Now let us see what happens if we measure the (scaled) position of a harmonic oscillator, so that $X = (a + a^\dagger)/\sqrt{2}$, and

$$H = \hbar\omega a^\dagger a. \tag{4.89}$$

If we now move into the interaction picture, defined by the Hamiltonian H, then the evolution of the operators a and $a^\dagger$ is

$$a(t) = ae^{-i\omega t}, \quad a^\dagger(t) = a^\dagger e^{i\omega t}. \tag{4.90}$$

If we substitute these time-dependent operators into the master equation given by Eq. (4.88), with $X = (a + a^\dagger)/\sqrt{2}$, then there are some terms that remain time-independent, and others that oscillate at $\pm 2\omega$. If $\mu \ll \omega$ then on the timescale of the measurement evolution, the oscillating terms will average close to zero. Dropping these terms, which is called the *rotating-wave approximation* (RWA), we obtain the master equation

$$\dot{\rho} = -i\omega[a^\dagger a, \rho] - \frac{\mu}{2}\left(a^\dagger a\rho + \rho a^\dagger a - 2a\rho a^\dagger + aa^\dagger \rho + \rho aa^\dagger - 2a^\dagger \rho a\right). \tag{4.91}$$

This is very similar to the quantum optical master equation, Eq. (4.82)! If we set $n_T \gg 1$ in the quantum optical master equation, and then make the approximation $n_T + 1 \approx n_T$ (the "high-temperature limit"), then we have exactly the equation we have just derived, with $\mu = \gamma n_T$.

So a continuous measurement of the position of a resonator when the measurement strength is much smaller than the oscillator frequency is just like coupling the resonator to a high-temperature thermal bath. In fact, the RWA that we used above is the same approximation that turns weak momentum damping into symmetric damping. The master equation for a position-measurement, Eq. (4.88) with $X = x$, describes an oscillator driven by a white-noise fluctuating force (this fluctuating force is the "quantum back-action," or "Heisenberg noise" from the position measurement). It therefore describes Brownian motion, but without the damping term. When we make the RWA we symmetrize the noise, and since there is no damping term, we end up with the quantum optical master equation, but without the damping. This is called the "high-temperature/weak-damping limit" of the quantum optical master equation.

We have seen that a weak continuous measurement of position affects an oscillator in a similar way to a high-temperature thermal bath. There is also a direct connection between a zero-temperature bath and photon counting. If we detect the photons that leak out of an optical cavity, then the evolution of the mode in the cavity is given by a stochastic master equation, Eq. (3.156). If we average over all the possible times at which we could have detected the photons, then we obtain the master equation that gives the evolution of our state-of-knowledge of the cavity mode when we have no detector. That is, we obtain the master equation for the damping of an optical cavity mode, and this is given in Eq. (3.177). This is exactly the quantum optical master equation with $n_T = 0$.

For weak damping, an interaction with a large environment looks to the system like a continuous measurement. The environment is therefore continuously carrying information away from the system. It is as if, in each small time interval dt, a new subsystem of the environment interacted with the system, thus correlating itself with it, and then was whisked away, never to return. This effect of the bath has a lot to do with the fact that the master equation is Markovian. For the master equation to be Markovian, the effect that the system has on the bath at time t must not change the effect that the bath has on the system at time $t + \tau$. If it did, then the master equation at time $t + \tau$ would depend on the state of the system at time t, as this is the only way it could record the effect on the bath at that time, and thus have it determine the motion at time t. So the fact that one obtains a Markovian master equation for weak damping means that the effect that the system has on the bath is "carried away," and does not act back on the system. This is consistent with the measurement picture, in which subsystems of the bath continually carry information away from the system.

The fact that a weak interaction with a large environment acts just like a continuous measurement is very useful in at least two ways. The first is that continuous measurement provides a theoretical tool to understand the action of environments on quantum systems. In particular, it is useful in analyzing how an environment will change the dynamics of a quantum system with sufficient "size" into that of the corresponding classical system. This explains how it is that the nonlinear, chaotic dynamics of the classical world emerges from the underlying quantum dynamics, a process referred to as the quantum-to-classical transition. This transition will be discussed in detail in Section 4.4. The second important use of continuous measurements for describing environments is as a numerical tool, and we turn to this in the next section.

4.3.4 Wavefunction "Monte Carlo" simulation methods

Master equations allow us to describe damping and thermalization of quantum systems without having to explicitly consider the environmental degrees of freedom. This is a tremendous saving in computational resources. Nevertheless, to numerically simulate a master equation means evolving a density matrix, rather than a state vector (wavefunction). If the state vector in question has dimension N, then the density matrix has dimension N^2. The latter therefore takes much longer to simulate, as well as requiring much more memory to store.

We now recall from Chapter 3 that a continuous quantum measurement is described by a stochastic Schrödinger equation, and when one averages over all the noise realizations the result is a master equation in the Lindblad form. This means that we can simulate a master equation by first simulating many trajectories of a stochastic Schrödinger equation for the state vector, and then averaging over them. To obtain a sufficiently accurate result, we will require on the order of 1000 or more individual noise realizations, which may require considerable numerical overhead. But even if this numerical overhead is the same as that required to simulate the master equation directly, there is still a significant gain from a computational point of view. Because each noise realization is independent, it is very simple to simulate many at the same time using a parallel computing cluster.

The technique of simulating a master equation by simulating many realizations of a stochastic Schrödinger equation (SSE) is referred to as using a "wavefunction Monte Carlo" method. This name derives from the name given to similar classical methods. To simulate a Lindblad-form master equation, one can use a stochastic Schrödinger equation driven either by Gaussian noise or Poissonian noise (jumps).

Wave-function Monte Carlo with Gaussian noise

We can simulate the general Lindblad master equation, Eq. (4.84), by using the stochastic Schrödinger equation

$$d|\psi\rangle = -\frac{i}{\hbar}H|\psi\rangle\, dt - \sum_n \left[L_n^\dagger L_n + L_n\langle L_n + L_n^\dagger\rangle + \frac{1}{4}\langle L_n + L_n^\dagger\rangle^2 \right] |\psi\rangle\, dt$$

$$+\sqrt{2}\left[L_n + \frac{1}{2}\langle L_n + L_n^\dagger\rangle \right] |\psi\rangle\, dW_n, \tag{4.92}$$

where

$$\langle L_n + L_n^\dagger \rangle = \langle \psi | (L_n + L_n^\dagger) | \psi \rangle, \tag{4.93}$$

and all the Wiener increments dW_n are mutually independent. The density matrix at time t is then given by

$$\rho(t) = \sum_j |\psi_j(t)\rangle \langle \psi_j(t)|, \tag{4.94}$$

where j indexes the noise realization. (A single noise realization is a sequence of values chosen for the random increments to simulate the differential equation up to time t.)

While the above method works, it is in fact simpler to use an un-normalized form of the SSE whose stochastic part is linear in the state vector. This is because it is simple to normalize the state vector after each time-step, and because having a linear stochastic part makes it simple to add a "Milstein" term to the equation [288, 338]. This extra term ensures that the numerical solution will be accurate to first-order in dt. The un-normalized form of the SSE is

$$d|\psi\rangle = -\frac{i}{\hbar}H|\psi\rangle \, dt + \sum_n \left[2\langle L_n + L_n^\dagger \rangle - L_n^\dagger \right] L_n |\psi\rangle \, dt + \sqrt{2}L_n|\psi\rangle \, dW_n. \tag{4.95}$$

With the addition of the Milstein term, this SSE becomes

$$d|\psi\rangle = -\frac{i}{\hbar}H|\psi\rangle \, dt + \sum_n \left[2\langle L_n + L_n^\dagger \rangle - L_n^\dagger \right] L_n |\psi\rangle \, dt + \sqrt{2}L_n|\psi\rangle \, dW_n$$

$$+ \frac{1}{2}(dW_n^2 - dt)L_n^2|\psi\rangle. \tag{4.96}$$

Wave-function Monte Carlo with Poisson noise

We can simulate the general Lindblad master equation, Eq. (4.84), by using the stochastic Schrödinger equation

$$d|\psi\rangle = -\sum_n \left(\frac{i}{\hbar}H + \frac{\gamma}{2}(L_n^\dagger L_n - \langle L_n^\dagger L_n \rangle) \right) dt|\psi\rangle + \left(\frac{L_n}{\sqrt{\lambda_n(t)}} - 1 \right) dN|\psi\rangle, \tag{4.97}$$

where $\lambda_n(t) = \langle L_n^\dagger L_n \rangle$. Here the increments dN_n are mutually independent increments of Poisson noise. The probabilities for $dN_n = 0$ and $dN_n = 1$ in the time-step dt are

$$P(dN_n = 1) = \lambda_n(t) \, dt, \tag{4.98a}$$

$$P(dN_n = 0) = 1 - \lambda_n(t) \, dt. \tag{4.98b}$$

The density matrix at time t is then given by

$$\rho(t) = \sum_j |\psi_j(t)\rangle \langle \psi_j(t)|, \tag{4.99}$$

where j indexes the noise realization. (A single noise realization is a sequence of values chosen for the random increments to simulate the differential equation up to time t.)

One could simulate the above stochastic equation by using the following procedure for each n, in each time-step: (i) calculate $P(dN_n = 1)$; (ii) choose a random number, r_n, uniformly distributed in the interval $[0, 1]$, and set $dN = 1$ if $r \le P(dN_n = 1)$. However, there is a more efficient way to perform the simulation. To simplify the presentation, we assume for now that there is just one Lindblad operator, and thus a single noise stream dN. Instead of considering the probability that there is a jump in each time-interval dt, we consider how long it will be until the next jump occurs. The time until the next jump is a random variable, and if we could determine its probability density then we could randomly pick this time. In that case we could integrate the deterministic evolution for $dN = 0$ up until this time, and then apply the jump.

It turns out that we can determine the probability density for the time to the next jump. To see this, consider the probability that there is no jump up until time t, and call this $P_0(t)$. The probability that there is no jump up until time $t + dt$ is the product of P_0 with the probability that there is no jump in the time-interval $[t, t+dt]$. Thus

$$P_0(t + dt) = P_0(t)P(dN = 0) = P_0(t)(1 - \lambda(t)dt). \tag{4.100}$$

Rearranging this equation gives us a differential equation for P_0, namely $dP_0/dt = -\lambda(t)P_0$. Solving this for P_0 we obtain

$$P_0(t) = \exp\left(-\int_0^t \lambda(t')dt'\right). \tag{4.101}$$

The $\lambda(t)$ that appears in this equation is the one that results from having no jump up until time t, and is therefore given by $\langle\psi(t)|L^\dagger L|\psi(t)\rangle$, where $|\psi(t)\rangle$ is the solution to the stochastic Schrödinger equation with dN set to zero:

$$\frac{d|\psi\rangle}{dt} = -\left(\frac{i}{\hbar}H + \frac{\gamma}{2}(L^\dagger L - \langle L^\dagger L\rangle)\right)|\psi\rangle. \tag{4.102}$$

We now note that probability that the next jump occurs in the time interval $[0, t]$ is $P_1(t) = 1 - P_0(t)$, and is also the integral of the probability density for the time until the next jump. That is, if we denote the probability density for the time until the next jump by $P(t)$, then

$$P_1(t) = \int_0^t P(t')dt'. \tag{4.103}$$

Now recall that we want to pick a random sample from $P(t)$. The simplest way to sample an arbitrary probability density $P(x)$, for $x \in [a, b]$, is the following procedure: (i) determine $D(x) = \int_a^x P(x')dx'$; (ii) pick a random number, r, uniformly distributed on the unit interval; (iii) pick the value of x for which $D(x) = r$.

The above results give us the following procedure for simulating Eq. (4.97):

1. Pick a random number, r, uniformly distributed on the interval $[0, 1]$.
2. Numerically solve the differential equation Eq. (4.102) for $|\psi(t)\rangle$, and use this solution to calculate $P_0(t)$ using Eq. (4.101).
3. Stop evolving $|\psi(t)\rangle$ when $P_0(t) = r$, apply the operator L to $|\psi(t)\rangle$, and normalize the result.
4. Repeat the sequence of steps 1 through 4.

If we have more than one Lindblad operator, and thus more than one kind of jump, then we need only modify the algorithm a little. If we have N kinds of jump, then we pick N random numbers $\{r_n\}$. We evolve $|\psi(t)\rangle$ with $dN_n = 0$ for all n, until one of our $P_0^{(n)}(t)$ hits its corresponding r_n. Let us say that $P_0^{(m)}$ reaches the number r_m. We then apply the jump operator L_m, choose a new value for r_m, and reset the probability $P_0^{(m)}$ to unity. But we leave all the other $P_0^{(n)}(t)$ as they were, and continue evolving $|\psi(t)\rangle$ with $dN = 0$ for all n, continuing also to evolve all the $P_0^{(n)}(t)$. So at each time when one of the $P_0^{(n)}(t)$ hits its value r_n, we apply the jump L_n, choose a new value for r_n, and reset the probability $P_0^{(n)}(t)$. The other probabilities are not affected immediately, and we continue to evolve all the $P_0^{(n)}(t)$ with $dN = 0$. The fact that a jump has occurred changes the state, and thus affects the future evolution of the $P_0^{(n)}(t)$ via this change. In this way we obtain the correct evolution for all the $P_0^{(n)}(t)$, and thus the correct statistics for the jump times.

Simulating measured systems that are also damped

Consider now the situation in which a system is continuously measured, but either this measurement is inefficient (see Section 3.1.2) or the system is simultaneously damped by an environment. In both these cases, even though the system is continuously measured, the state does not remain pure as time goes by, and must be described by a stochastic master equation for the density matrix. One would like to play the same Monte Carlo trick for simulating deterministic Lindblad master equations discussed above, so that we can evolve pure-states instead of a density matrix. This can be done, but the process is rather more involved. To determine this method, we draw upon our knowledge of the effect of a measurement on an ensemble of quantum states, which we examined in Section 1.7.

Recall that an ensemble of states describes a situation in which the observer knows that the system could be in any one of a number of states, $|\psi_j\rangle$, $j = 1, 2, \ldots, J$, each with some probability. If we denote the probability that the system is in state $|\psi_j\rangle$ by p_j, then the observer's density matrix is given by

$$\rho = \sum_j p_j |\psi_j\rangle\langle\psi_j|. \tag{4.104}$$

Now consider what happens when the observer makes a measurement on the system. The ensemble changes in two ways: the states $|\psi_j\rangle$ are changed, and the probabilities for the states, p_k, are also changed. This provides a method to simulate the evolution of a density matrix by evolving an ensemble of pure states, but unlike the previous Monte Carlo method, it requires evolving the probability for each state as well.

The fact that the probability for each state in the ensemble changes over time presents us with a new problem. Many of these probabilities will become very small as time goes by, and this effectively reduces the size of the ensemble. The smaller the ensemble, the less accurate will be our final answer for the density matrix. The solution to this problem is actually quite simple and was first used for simulations of classical continuous measurements. This solution is to periodically discard the states whose probabilities have become very small. We then replace these states in the following way. For each of the states that we have dropped, we pick one of the states that has a large probability and add a copy of it to the ensemble. We then halve the probability of the original state and give the half that was removed to the new state. This keeps the density matrix the same, but increases the size of the ensemble. So long as the total amount of probability that is periodically discarded is sufficiently small, the method will be accurate.

By determining the way in which a continuous measurement changes the probabilities, which can be done by applying the analysis in Section 1.7 to a continuous measurement, we arrive at the algorithm for Monte Carlo simulation of an SME.

Monte Carlo method for Stochastic Master Equations

The SME given by

$$dp = -\frac{i}{\hbar}H|\psi\rangle dt - \gamma(L^\dagger L\rho + \rho L^\dagger L - 2L\rho L^\dagger)dt \qquad (4.105)$$

$$-\mu(A^\dagger A\rho + \rho A^\dagger A - 2A\rho A^\dagger)dt + \sqrt{2\mu}(A\rho + \rho A^\dagger - \langle A + A^\dagger\rangle \rho)dW$$

describes a measurement of A and decoherence due to L. To simulate this SME, one performs the following algorithm:

Step I: Create a set of N pure states, $|\psi\rangle_n$ and N probabilities p_n, so that the desired initial value of ρ is approximately

$$\rho(0) = \sum_{n=1}^{N} p_n |\psi_n\rangle\langle\psi_n|, \quad \sum_{n=1}^{N} p_n = 1. \qquad (4.106)$$

From the point of view of the accuracy in obtaining ρ from averaging the states in the ensemble, the effective size of the ensemble will not usually be N. Instead it is determined by the set of probabilities, p_n. We can characterize the size of the ensemble, for example, as the exponential of the von Neumann entropy of the set $\{p_n\}$:

$$N_{\text{eff}} = \exp\left[-\sum_{n=1}^{N} p_n \ln p_n\right] \leq N. \qquad (4.107)$$

This effective size is maximized (equal to N) if and only if all the p_n are equal to $1/N$. It is therefore desirable to choose the initial ensemble so that all the p_n are equal, and Theorem 5 in Section 2.2.3 shows that this can be done for every choice of ρ.

Step II: Evolve each pure state by repeating the following seven steps. (Note that the $N+1$ Wiener increments used below, $dW_j, j = 0, 1, \ldots, N$, are mutually independent.)

1. Replace each of the states $|\psi_n\rangle$, $n = 1, \ldots, N$, by $|\psi_n\rangle + d|\psi_n\rangle$, where

$$d|\psi_n\rangle = -\gamma \left[L^\dagger - 2\langle L+L^\dagger\rangle_n \right] L|\psi_n\rangle \, dt + \sqrt{2\gamma} L|\psi_n\rangle \, dW_n, \qquad (4.108)$$

and $\langle L+L^\dagger\rangle_n \equiv \langle\psi_n|(L+L^\dagger)|\psi_n\rangle$.
2. Normalize each of the $|\psi_n\rangle$.
3. Replace each of the states $|\psi_n\rangle$ by $|\psi_n\rangle + d|\psi_n\rangle$, where

$$d|\psi_n\rangle = -\mu \left[A^\dagger - 2\langle\langle A+A^\dagger\rangle\rangle \right] A|\psi_n\rangle \, dt + \sqrt{2\mu} A|\psi_n\rangle \, dW_0, \qquad (4.109)$$

and

$$\langle\langle A+A^\dagger\rangle\rangle \equiv \sum_{n=1}^{N} p_n \langle\psi_n|(A+A^\dagger)|\psi_n\rangle. \qquad (4.110)$$

4. Multiply each of the probabilities p_n by $\langle\psi_n|\psi_n\rangle$.
5. Normalize the p_n. That is, divide each p_n by $\sum_{n=1}^{N} p_n$.
6. Normalize each of the $|\psi_n\rangle$.
7. Every few iterations, perform the following operation (which might be referred to as "splitting," "breeding," or "regenerating" the ensemble): For each pure state whose probability p_n is less than a fixed threshold $p_{\text{thresh}} \ll 1$, we pick the state $|\psi_m\rangle$ from the ensemble that has the largest probability. p_m. We then set $|\psi_n\rangle$ equal to $|\psi_m\rangle$, thus erasing the original $|\psi_n\rangle$ from the ensemble. We set both p_n and p_m equal to $p_m/2$. Thus we have "split" the highest probability state into two members of the ensemble, and p_n and p_m are (most likely) no longer the largest probabilities. After we have done this for each $p_n < p_{\text{thresh}}$, we then normalize all the p_n as per [v] above.

Step III: The density matrix at time t is (approximately)

$$\rho(t) = \sum_{n=1}^{N} p_n(t)|\psi_n(t)\rangle\langle\psi_n(t)|. \qquad (4.111)$$

When using a Monte Carlo method to simulate a deterministic (Lindblad) master equation, the size of ensemble is fixed; this is the only parameter in the method that determines the final accuracy. In this case one merely increases N until the desired accuracy is reached. When simulating a stochastic master equation, we have *two* parameters that affect the error. The first is the minimum effective ensemble size during the evolution, $\min(N_{\text{eff}})$, and the second comes from the regeneration step. In each regeneration we eliminate some states. If we denote the sum of the probabilities for these "lost" states as p_{lost}, then the maximum value of p_{lost} during the simulation will determine the error from the regeneration step. So to ensure numerical accuracy we require that

$$\min(N_{eff}) \gg 1,$$

$$\max(p_{lost}) \ll 1. \tag{4.112}$$

The values of these two quantities are determined jointly by N and p_{thresh}. For a given SME, and a choice of N, there will be some optimal value of p_{thresh} that ensures that $\min(N_{eff})$ is large while keeping $\max(P_{drop})$ small. For a given simulation it is not difficult to check whether N and P_{thresh} give sufficient accuracy. One merely runs the simulation a second time with the same realization for the measurement noise dW_0, and a different set of realizations for the noises that model the decoherence. The difference between the two simulations provides an estimate of the error.

4.3.5 Strong damping: master equations and beyond

Recall that weak damping is defined as $\gamma \ll \omega$ where γ is the rate of damping (the rate of the irreversible evolution induced by the bath), and ω is the frequency of the dynamics of the system. Another way to say this is that the minimum energy gap between any two energy eigenstates of the system is $\Delta E \geq \hbar\omega$. When one has weak damping the irreversible dynamics of the system can be described simply, and is given by a time-independent Markovian master equation in the Lindblad form. The general form of this equation is called the Redfield equation, of which the quantum optical master equation is a particular example. Weak damping is also the regime in which the system will thermalize to the Boltzmann distribution.

Describing strong damping is much more difficult. The problem is that memory effects in the bath become important, resulting in a non-Markovian master equation. A number of non-Markovian master equations have been derived by making a perturbative approximation, in which the interaction between the system and bath is treated only to second order [438, 725, 726, 115]. However the accuracy of these master equations has been called into question [180], and recently more sophisticated techniques have been developed to allow essentially exact numerical simulations of non-Markovian open systems. In this section we give a brief overview of these numerical techniques. We begin by discussing the only exactly solvable model of a system coupled to a bath, the Caldeira–Leggett model.

A harmonic oscillator coupled to a bath of harmonic oscillators

This model of a system and bath was first introduced by Caldeira and Leggett [97, 98], and finally fully solved by Hu, Paz, and Zhang in 1992 [273]. It is referred to as the "quantum Brownian motion" master equation, because it is the first model that explicitly derives the quantum version of Brownian motion (momentum damping of a single particle) from a microscopic model. This is achieved by coupling the particle (usually taken to be trapped in a harmonic potential) to a "bath" consisting of a large number of harmonic oscillators, all with different frequencies. Specifically, the model is given by

$$H = \hbar \omega a^\dagger a - x \sum_{n=1}^{\infty} g_n x_n + \sum_{n=1}^{\infty} \hbar \omega_n b_n^\dagger b_n, \tag{4.113}$$

where ω is the frequency of the damped oscillator, $x = (a + a^\dagger)/\sqrt{2}$ is the oscillator's (scaled) position, and $x_n = (b_n + b_n^\dagger)/\sqrt{2}$ is the scaled position of the nth bath oscillator. The frequency of the nth bath oscillator is ω_n, and g_n is the coupling strength between the damped oscillator and the nth bath oscillator. For reference, the physical position and momentum of the oscillator are respectively, $X = \sqrt{\hbar/(\omega m)}x$ and $P = \sqrt{\hbar \omega m}p$, where m is the oscillator's mass, and $p = -i(a - a^\dagger)/\sqrt{2}$ is the scaled momentum. Ultimately the harmonic oscillators of the bath are taken to be a continuum, so that the discrete index ω_n is replaced by a density of oscillators per unit frequency.

Because the dynamics of this model is entirely linear, an exact master equation can be derived, and is given by

$$\dot{\rho} = -i\Gamma(t)[x, \{p, \rho\}] - \xi(t)[x, [x, \rho]] + \zeta(t)[x, [p, \rho]], \tag{4.114}$$

where $\{\cdot, \cdot\}$ denotes the anti-commutator. The first term gives damping of the momentum, and the second gives diffusion of the momentum, and thus describes a fluctuating force. The last term is an additional term, not found in classical Brownian motion. It comes from the memory of the bath, and is only significant for a short period after the initial time. It is due to transient effects in which correlations build up between the system and bath, since they are initially uncorrelated.

At first glance the above Brownian motion master equation looks like it might be Markovian, since only the present time, t, appears explicitly. However, the memory of earlier times is hidden in the definitions of the time-dependent coefficients $\Gamma(t)$, $\xi(t)$, and $\zeta(t)$. Each of these is an integral, from the initial time up until t, of functions of time that are themselves the solutions of two non-Markovian differential equations. The Brownian motion master equation can certainly be solved numerically, but one must concurrently solve the two non-Markovian differential equations.

One can obtain a Markovian master equation by setting $\zeta(t) = 0$, and making the damping and noise terms constant. This gives an approximate description of quantum Brownian motion if we set

$$\Gamma = \frac{\gamma}{4}, \quad \xi = \frac{\gamma}{4}(2n_T + 1), \tag{4.115}$$

where

$$n_T = \frac{1}{\exp\left[\hbar \omega/(kT)\right] - 1}, \tag{4.116}$$

is the average number of phonons at temperature T. For high temperatures ξ is approximately

$$\xi = \gamma \left(\frac{kT}{2\hbar \omega}\right). \tag{4.117}$$

This approximate Markovian master equation is a simple analogue of the classical Brownian motion Langevin equations in Section 4.3.1 and does provide a reasonable model of a damped quantum oscillator for many purposes. Nevertheless it has some problems. Since it is not in the Lindblad form, it can break the positivity of the density matrix, although this only happens at very short times [214, 203]. For the same reason, it does not have the usual equivalent SSE and thus lacks the usual Monte Carlo method. (However, a non-Markovian SSE has been derived for the full non-Markovian Brownian motion master equation [581].) If one uses the Markovian form of the Brownian motion master equation to calculate noise spectra, it can generate unphysical terms [295, 211]. Finally, it only applies, of course, to damping of harmonic oscillators, and not to more general quantum systems.

The Heisenberg-picture equations for a strongly damped oscillator

While the approximate master equation for a strongly damped oscillator is not appropriate for calculating measurement spectra, it turns out that there is a quantum Langevin equation that is equivalent to the full quantum Brownian motion master equation, and can be used to obtain the power spectra of observables and measurement records, including when the resonator is weakly coupled to other systems, using the methods of Section 3.7. As we show in Section G.10, the power spectra can also be used to obtain the steady-state variances of the coordinates, and thus the steady-state energy of the resonator.

The quantum Langevin equations for a strongly damped harmonic oscillator are [211]

$$\dot{x} = p/m, \tag{4.118}$$

$$\dot{p} = -m\omega_0^2 x - \gamma p + f_{\text{in}}(t), \tag{4.119}$$

where ω_0 is the frequency, m the mass, and γ the damping rate of the oscillator. A detailed derivation of these Langevin equations can be found in [211, 201, 89]. The autocorrelation function of the noise, f_{in}, is not the simple δ-function of the weak-coupling input–output equations of Section 3.6. In fact it is rather complicated. Nevertheless, if we want to calculate power spectra using the methods in Section 3.7 then we need only the frequency-space correlation function, which is simple. Denoting the Fourier transform of $f_{\text{in}}(t)$ by $F_{\text{in}}(\omega)$ this correlation function is

$$\langle F_{\text{in}}(\omega) F_{\text{in}}(\omega') \rangle = 2\pi \gamma m \hbar \omega \left[1 + \coth\left(\frac{\hbar\omega}{2kT} \right) \right] \delta(\omega + \omega'). \tag{4.120}$$

An example of using these Langevin equations to determine the spectrum of a position measurement of a damped oscillator via an optical cavity is given in [211].

Exact numerical methods for non-Markovian systems

The treatment of non-Markovian open quantum systems is presently an active area of research. There are to date a number of methods that have been devised for treating these systems that go beyond perturbative master equations and are able to simulate the non-Markovian evolution essentially exactly. The first is a *non-perturbative* master equation derived by Ishizaki and Tanimura [278, 279, 150]. Their master equation is in fact an infinite hierarchy of master equations, which can be truncated to obtain a numerical solution. They have demonstrated that the method is practical by showing that the number of equations required to obtain an accurate solution is not infeasibly large. To determine how many equations need to be included in the truncation for a given problem, one increases this number until the solution stabilizes.

A second method starts with the Feynman–Vernon approach to deriving a master equation. We have not discussed this approach, but it involves writing the exact evolution for the system as a path integral with an influence functional determined by the environment. The method, devised by Makri and Makarov [393, 394, 395, 601], and called the quasiadiabatic propagator path integral method (QUAPI), provides a way of numerically evaluating the path integral. This has been applied, for example, to the decoherence of biomolecular excitons [602]

There are two methods that exploit renormalization-group (RG) techniques for simulating many-body systems. The first is a time-dependent extension of the numerical renormalization group (NRG) method, due to Wilson [662]. Here the bath, consisting of a continuum of harmonic oscillators, is discretized on a logarithmic scale and mapped to a semi-infinite chain of oscillators [93]. In this chain each system represents an exponentially decreasing energy scale, and the dynamics of the chain can be simulated numerically [10].

The second method is due to Plenio and collaborators, and exploits the time-dependent density-matrix RG (*t*-DMRG) methods. The trick that enables these methods is that local many-body Hamiltonians (in which each system interacts only with its few nearest neighbors) for lattices of systems usually satisfy an "area-law" for entanglement. This means that the entanglement between a contiguous block of the constituent systems, and the rest of the many-body system, scales as the size of the *boundary* of the block. If the lattice of spins is three-dimensional, and thus the block is three-dimensional, then the boundary of the block is an area. This is where the term *area-law* comes from. If the many-body system is one dimensional – for example, a spin-chain – then the entanglement remains fixed as the length of the block is increased. It turns out that this bound on the entanglement of local many-body systems means that their ground states, and in certain cases their time evolution, do not explore the whole state space. In fact, they explore only an exponentially small fraction of the state space, allowing these systems to be simulated using a representation whose size increases only polynomially in the number of spin systems [167]. A number of methods have been developed to achieve this, both for one-dimensional lattices (chains) [635, 636, 135, 727, 474] and lattices of two or more dimensions [435, 633, 634, 113, 406]. These are referred to collectively as DMRG methods [544].

The method developed by Plenio's group [120, 493] first uses orthogonal polynomials to exactly map the continuum of oscillators to a semi-infinite chain of oscillators. Then a one-dimensional t-DMRG method can be used to simulate the evolution of the chain. Depending on the frequency dependence of the coupling to the bath (the spectrum of the bath), sometimes the mapping can be performed analytically.

It is also possible to model the bath as a system with an exponentially increasing density of energy levels and random (typical) eigenstates. With a change of basis this translates to a bath that is diagonal in the elementary basis, with the interaction operator for the bath a random matrix. This model correctly describes thermalization, and for weak damping reproduces the dynamics of the thermal master equation. This may, in fact, be a better model of a many-body bath than a collection of oscillators, since we know already from Section 4.2.2 that random Hamiltonians correctly reproduce various features of many-body systems. It turns out that depending on the temperature required for the bath, accurate results can be obtained with baths containing 10^3–10^4 states. If the system has only a small number of states, then the joint Hamiltonian of the system and bath can be diagonalized numerically, and the dynamics solved exactly for arbitrarily long times [298]. The limitation of this method is that it becomes inefficient at low temperatures, due to the exponential density of energy states.

4.4 The quantum-to-classical transition

Quantum mechanics and classical chaos

Macroscopic objects, such as cars, baseballs, and pendulums, all obey Newton's laws. It is not until one looks at microscopic objects, such as individual atoms and electrons, that it becomes apparent that the laws governing motion are really those of quantum mechanics. But macroscopic objects are *composed* of microscopic particles: the motion of a macroscopic object is that of the center-of-mass coordinate of all of its constituent particles. The classical dynamics of a macroscopic object must therefore emerge from quantum mechanics, and this emergence is the subject of the "quantum-to-classical transition" (QCT). The transition is especially intriguing because classical dynamics is nonlinear, and thus chaotic, whereas quantum evolution is not.

We will find that large quantum environments, or thermal baths, to which all systems are coupled, not only induce damping and thermalization, but also transform quantum evolution to classical dynamics. And it turns out that continuous measurement theory is the key to understanding how this happens; recall from Chapter 4 that weak environmental interactions act precisely as continuous measurements, in which the information from the measurement is discarded. Continuous measurement theory provides a simple way to show how a weak interaction with a thermal environment transforms quantum dynamics to classical dynamics, in the absence of any measurement. It also shows why, and under what circumstances, an observer sees the smooth trajectories of classical motion.

The understanding of the QCT in terms of measurement theory is relatively new, and it is worth reviewing briefly the traditional explanation for the emergence of Newton's laws, which predated the theories of either chaos or continuous quantum measurement. The first ingredient in this traditional explanation is Ehrenfest's theorem. It states that if the uncertainty of a quantum system is sufficiently small in both position and momentum, then the equations of motion for the mean position and mean momentum of the quantum system are the same as the classical equations of motion [86]. To see this, consider the Heisenberg equations of motion for a single particle, whose Hamiltonian is $H = p^2/(2m) + V(x)$. These equations are

$$\dot{x} = \frac{p}{m} \quad \text{and} \quad \dot{p} = \frac{-i}{\hbar}[p, V(x)] = -\frac{dV}{dx} = F(x). \tag{4.121}$$

Here we have used the fact that $[p, f(x,p)] = -i\hbar \, \partial f/\partial x$ for any function f. This relation can be proved by using the fact that the commutator shares key properties with the derivative, such as the product rule. Without these properties the commutator could not mirror the Poisson bracket.[4]

The equations of motion for the mean values of x and p are obtained by taking expectation values on both sides of the equations of motion for x and p. This gives

$$\langle \dot{x} \rangle = \frac{\langle p \rangle}{m} \quad \text{and} \quad \langle \dot{p} \rangle = \langle F(x) \rangle. \tag{4.122}$$

Now, if the probability distribution for x, $P(x)$, is narrow compared to the scale on which the force $F(x)$ varies, then we can expand $F(x)$ as a power series about the mean, $\langle x \rangle$:

$$F(x) = F(\langle x \rangle) + F'(\langle x \rangle)(x - \langle x \rangle) + \frac{1}{2}F''(\langle x \rangle)(x - \langle x \rangle)^2 + \cdots \tag{4.123}$$

Substituting this into the equations of motion for the momentum, we have

$$\langle \dot{p} \rangle = F(\langle x \rangle) + \frac{1}{2}F''(\langle x \rangle)V_x + \cdots \tag{4.124}$$

where V_x is the variance of the position for the system. If this variance is small enough, so that

$$V_x \ll \frac{2F(\langle x \rangle)}{F''(\langle x \rangle)}, \tag{4.125}$$

then the equations of motion for the means are approximately

$$\langle \dot{x} \rangle = \frac{\langle p \rangle}{m} \quad \text{and} \quad \langle \dot{p} \rangle = F(\langle x \rangle) \tag{4.126}$$

and these are the classical equations of motion.

The second step in the traditional explanation is that if the variances of x and p are very small (compared to the scale upon which the force $F(x)$ varies), the rate at which they will increase over time is also very small. For macroscopic objects, if these variances are on the order of the minimums given by Heisenberg's uncertainty principle, then they are *exceedingly* small compared to the scale of the force. It appears from this that the variances will stay small, for times larger than the age of the universe, so that macroscopic objects will always effectively obey classical mechanics.

---

[4] In fact, if $[A,B] = c$, where c is a number and A and B are any operators, then $[A, f(A,B)] = c \, \partial f/\partial B$ for any function f.

While the above arguments contain valuable insight, they are unsatisfactory for a number of reasons. The first problem is a flaw in the argument that the uncertainties in the dynamical variables remain small. While their initial rates of increase are indeed small, this is irrelevant if the increase is governed by an equation that gives exponential growth. In fact, one can show that in systems whose classical dynamics exhibits chaos, the uncertainties will grow exponentially. One can see this as follows. If the initial state of the system is a Gaussian state (having a Gaussian probability density in both x and p), and these variances are small in the sense of Ehrenfest's theorem, then the dynamics of the means, as well as the variances of x and p, obey the same equations as they would if the system were classical, and the state was merely a classical probability distribution for x and p. This is not that difficult to show, and the details can be found in [63, 235].

Now consider a classical system, and two nearby initial points in phase space, (x_1, p_1) and (x_2, p_2). If the classical system is chaotic, then the trajectories of these two initial states will diverge exponentially fast. To be more precise, chaos is defined as an exponential divergence in two nearby trajectories, where one takes the joint limits in which the trajectories are evolved for an infinite amount of time, and the two initial points are infinitesimally close together. If we denote the distance between the trajectories as

$$\Delta(t) \equiv \sqrt{[x_1(t) - x_2(t)]^2 + [p_1(t) - p_2(t)]^2} \qquad (4.127)$$

(this is the distance in *phase space*, see below) then if the quantity

$$\lambda \equiv \lim_{\Delta(0) \to 0} \left\{ \lim_{t \to \infty} \left[\frac{\ln[\Delta(t)]}{t} \right] \right\}, \qquad (4.128)$$

is greater than zero, it follows that

$$\Delta(t) = \Delta(0) e^{\lambda t} \qquad (4.129)$$

for large t and small $\Delta(0)$. In this case the system is chaotic, and the exponent λ is called the *Lyapunov exponent*[5] [587, 688, 236, 502].

If all nearby initial states diverge exponentially, then a classical probability distribution over x and p will also broaden exponentially fast. Since a well-localized Gaussian quantum state evolves exactly as a well-localized classical probability density, we can conclude that the variances of the quantum state will also increase exponentially.[6] This contradicts the traditional argument that a well-localized classical object will remain in the Ehrenfest regime, and thus behave classically.

---

[5] In fact, for each degree of freedom of the system (each dimension of the phase space) there is a different stretching rate, and the directions of these stretching rates rotate in phase space during the evolution. Thus if one merely takes the phase space distance between two nearby trajectories, this distance will oscillate in addition to the divergence. If one averages over multiple trajectories, these oscillations can be eliminated, and in the long-time limit the maximal stretching rate dominates, which is the Lyapunov exponent.

[6] This argument is due to Zurek and Paz [724].

A second perplexing issue is that while classical mechanics is nonlinear, quantum mechanics is linear, and quantum systems with a bounded phase space have a discrete frequency spectrum.[7] Together these facts preclude quantum dynamics from exhibiting chaos. The modern analysis will also reveal that classical dynamics can emerge in much smaller systems than predicted by the traditional approach.

Visualizing a state in phase space: the Wigner function

When thinking about the QCT it is conceptually useful to think of the state of a dynamical system as a joint probability distribution over the coordinates and momenta. The full state of a classical system is given by a vector consisting of all the positions and momenta. The vector space in which this vector lives is the state space of the system. One can visualize the dynamics of a system by plotting its state in state space as it evolves: that is, by assigning an orthogonal axis to each of the coordinates and momenta. This visualization of the motion is called *phase space*. The state space, or phase space, of a single particle has only two dimensions, making it easy to display. As an example, the evolution of a harmonic oscillator is a circle in phase space, and this is true for every initial state. (Although note that the relative scaling of the two axes is meaningless, because x and p have different units.) It follows immediately that the dynamics of the time-evolution of the harmonic oscillator is merely a rotation in phase space.

If the positions and/or momenta of a classical system are not completely known, then its state is instead given by a probability distribution over the possible states, and thus a probability distribution in phase space. One can determine the evolution of this probability distribution from the equations of motion of the system [288, 609, 21]. For a single particle, evolving under the force $F(x) = -dV/dx$, the equation of motion for the probability distribution, $P(x,p)$, is

$$\frac{\partial P}{\partial t} = \left[-\left(\frac{p}{m}\right)\frac{\partial}{\partial x} - F(x)\frac{\partial}{\partial p} \right] P. \tag{4.130}$$

A probability density in phase space is the most general classical state, the classical equivalent of the quantum density matrix.

It is also possible to visualize the state of a quantum system as something very similar to a probability distribution in phase space. This phase space representation of a quantum state was devised by Wigner, and is called the *Wigner function*. Note that the phase space probability distribution for a classical particle, $P(x,p)$, has the property that $P(x) = \int_{-\infty}^{\infty} P(x,p)dp$ and $P(p) = \int_{-\infty}^{\infty} P(x,p)dx$, where $P(x)$ and $P(p)$ are the probability distributions for x and p, respectively. The Wigner function also has this important property. Nevertheless, it is not a real probability distribution because it can be negative. A theorem due to Hudson states that the only pure quantum state with a Wigner function that is non-negative over the whole phase space is the Gaussian [274].

---

[7] The term "bounded phase space" means that there is a finite region of space that the system can explore when it has a finite initial energy E.

The Wigner function is given by [252]

$$W(x,p) = \frac{1}{\pi\hbar} \int_{-\infty}^{\infty} \langle x-y|\rho|x+y\rangle \, e^{2ipy/\hbar} \, dy, \qquad (4.131)$$

where $\langle x'|\rho|x''\rangle$ is the density matrix in the position representation. The arguments x and p of the Wigner function define the quantum mechanical phase space. Just as in the classical case, the evolution of a quantum harmonic oscillator is merely rotation of the Wigner function in phase space. A comprehensive discussion of the Wigner function can be found in [252].

As an example of a Wigner function, consider a coherent state of a harmonic oscillator. This is defined as an eigenstate of the annihilation operator, and is given by $|\alpha\rangle = e^{-|\alpha|^2/2} \sum_{n=0}^{\infty} |n\rangle$ where the states $|n\rangle$ are the Fock states of the oscillator: $a^\dagger a|n\rangle = n|n\rangle$. The eigenvalue of the coherent state is α, and the expectation values of position and momentum for this state are given in Appendix D. If we define dimensionless position and momentum variables by $\tilde{x} = \sqrt{2m\omega/\hbar}\,x$ and $\tilde{p} = \sqrt{2/(\hbar m\omega)}\,p$, then we have $[\tilde{x},\tilde{p}] = 2i$, and $\mathrm{Re}[\alpha] = \langle\tilde{x}\rangle/2$ and $\mathrm{Im}[\alpha] = \langle\tilde{p}\rangle/2$. The variances and covariance of $\tilde{x}$ and $\tilde{p}$ for the coherent state are $V_{\tilde{x}} = V_{\tilde{p}} = 1$, and $\tilde{C} = 0$, where the covariance C is given by $C = \langle\tilde{x}\tilde{p}+\tilde{x}\tilde{p}\rangle/2 - \langle\tilde{x}\rangle\langle\tilde{p}\rangle$. The coherent state is, in fact, a Gaussian distribution in phase space, localized about the mean values $\langle\tilde{x}\rangle$ and $\langle\tilde{p}\rangle$. Specifically, the Wigner function for the coherent state is

$$W^{(\alpha)}(\tilde{x},\tilde{p}) = \frac{1}{2\pi} e^{-[(\tilde{x}-\langle\tilde{x}\rangle)^2+(\tilde{p}-\langle\tilde{p}\rangle)^2]/2} \qquad (4.132)$$

The Wigner function is a nice way to visualize the quantum-to-classical transition. If $W(x,p)$ remains well-localized in phase space, then we can interpret the mean values of position and momentum as the classical position and momentum, and their equations of motion as the classical ones. It is when the Wigner function spreads sufficiently to violate the inequality in Eq. (4.125) that quantum motion differs from classical. Note that since evolution in general involves the rotation of the Wigner function in phase space, if the variance of the momentum spreads in phase space, this will quickly translate into a spread in position.

Continuous measurements, localization, and classical trajectories

We now observe the following obvious fact: in order to discover the laws of classical motion, and in particular chaos, it was necessary to observe the trajectories of macroscopic systems. Given this trivial statement, and the benefit of hindsight, it is not surprising that one must include the measurement process to understand the emergence of classical trajectories from quantum mechanics. Further, since it is the *position* of macroscopic objects that is usually observed in practice, it is natural to consider a continuous measurement of position.

Ehrenfest's theorem tells us that a sufficiently well-localized wavefunction will obey classical dynamics. Since the measurement provides a continuous stream of information

about x, we can expect that under the right conditions it could keep the wavefunction localized within the Ehrenfest regime *indefinitely*. To explore this possibility further, recall that
in narrowing the position uncertainty, the measurement will in general spread the wavefunction in momentum, and this spread will feed back into the position through the normal
dynamical evolution. The fact that uncertainty in momentum and position are continually
converted into one another means that a continuous measurement of position is effectively
a continuous measurement of momentum too. A continuous position measurement is actually a continuous measurement of the full location of the state in phase space. This also
tells us that to keep the state localized in position the measurement must keep it localized
in all the canonical variables. Once we take into account the noise introduced by the measurement (see below), we will find that it is Heisenberg's uncertainty limit, $\sqrt{V_x V_p} \geq \hbar/2$,
that places the fundamental bound on how well the position can be localized such that the
mean of x obeys the smooth trajectories of classical mechanics. Note that $\hbar$ has units of the
product of position and momentum, and thus corresponds to an area in phase space.

From the above analysis we might expect a continuous measurement to induce classical
motion whenever the potential, and thus the phase space structure of the dynamics, varies
on a scale that is larger than the phase space area $\hbar/2$. But there is another condition
that must be satisfied to realize the smooth trajectories of classical dynamics: the noise
added to the motion of the means of x and p by the measurement must be negligible. This
noise is generated by the collapse of the wavefunction (that is, by the dW that appears
in the stochastic master equation), and is sometimes referred to as the "projection noise."
The stronger the measurement the more it is able to prevent the spreading induced by the
quantum dynamics, but the more projection noise it generates (up to the limit imposed by
the localization of the Wigner function). The size of the projection noise is determined
by the rate at which the measurement must reduce the variance of the position. If the
momentum spread is larger, then the continual conversion of momentum into position will
rapidly increase the position variance. To balance this spread the measurement will need
to be strong, and this will introduce significant projection noise. This is why both position
and momentum must be well localized for classical dynamics to emerge, and why it is
Heisenberg's uncertainty principle that sets the limit as to when this is possible.

When the phase space structure of the dynamics is sufficiently large compared to $\hbar/2$,
then there is a range of measurement strengths for which the localization and low-noise
requirements are met simultaneously, and the trajectories of smooth classical dynamics are
generated by the quantum evolution. As the scale of the phase space structure of a system
increases – loosely speaking, as the action increases – so does the range of measurement
strengths over which the two conditions are met. The classical limit is the limit in which
almost any measurement strength will generate the smooth trajectories of the equivalent
classical Hamiltonian.

To observe classical trajectories a system must be measured continuously, and we have
explained above how this measurement induces these trajectories. But it is not necessary
to make a measurement to induce the quantum-to-classical transition. As discussed in
Section 3.6, the effect of a thermal bath is that of a continuous measurement in which

the density matrix is averaged over all the possible measurement results. In showing that continuous measurement induces the trajectories of classical mechanics, we have therefore automatically shown that a thermal bath induces classical motion: the motion induced by the interaction with the bath is simply the average over a set of possible classical trajectories (with a little noise). This is exactly the same as that of an unobserved classical system that interacts with a thermal bath. The classical motion is the motion of the phase space probability distribution, and this is given by averaging over all the possible trajectories, taking into account the small amount of noise injected by the bath. A bath is therefore sufficient to induce fully classical behavior in a quantum system. In fact, the notion that thermal environments induced the quantum-to-classical transition was introduced before continuous measurement theory was applied to the problem, but without continuous measurement theory the analysis was much more difficult [724]. If one is not interested in trajectories then the measurement analysis above is simply a powerful mathematical device for establishing the effects of thermal baths on the motion of quantum systems.

So far we have explained the quantum-to-classical transition purely in qualitative terms. In lieu of presenting a fully quantitative version of the above analysis, we illustrate it below with a single quantitative example. This shows that even very tiny systems can make the transition from quantum to classical motion. References to articles that give quantitative analyses of the QCT can be found at the end of this chapter.

A numerical example: the Duffing oscillator

The emergence of classical chaos from quantum mechanics can be verified by performing numerical simulations. A well-studied example of a classically chaotic system is the "Duffing oscillator," a particle in a double-well potential driven by a sinusoidal force [587, p. 35]. The Hamiltonian for this system is

$$H = p^2/(2m) - ax^2 + bx^4 + \lambda x \cos(\Omega t), \qquad (4.133)$$

where m is the mass, a and b determine the precise shape of the double-well, and λ and Ω are, respectively, the strength and frequency of the sinusoidally varying, linear driving force. The action of the system (essentially the size of the phase space that it explores), is determined by these parameters, and we can change the area of the phase space with respect to $\hbar$ by changing them. However there is a more "canonical" way to scale the size of the phase space: fix all the parameters and change $\hbar$. While it seems strange at first, this method emerges naturally by changing to dimensionless (scaled) position and momentum. If we define scaled position and momentum $\tilde{x} = x/\sqrt{\hbar/(\kappa m\omega)}$, $\tilde{p} = p/\sqrt{m\omega\hbar/\kappa}$, then the Hamiltonian becomes

$$H = \tilde{p}^2/2 - A\tilde{x}^2 + B\tilde{x}^4 + \tilde{\lambda}\tilde{x}\cos(\tilde{\Omega}t), \qquad (4.134)$$

and $[\tilde{x}, \tilde{p}] = i\kappa$. In this scaling of the variables we have introduced the parameters ω and κ, and each plays a specific role in scaling the phase space of the system. The first has units of inverse time, and along with the mass, m, scales the momentum axis of phase space

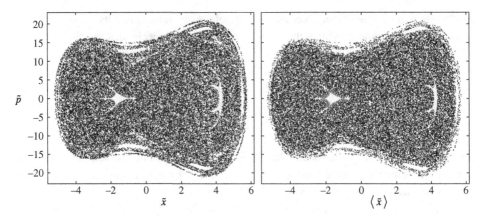

Figure 4.1 Two stroboscopic maps depicting the chaotic sea in the phase space structure of the driven Duffing oscillator. The one on the left is the classical dynamics. The one on the right is the quantum oscillator under a continuous measurement of the position, in a parameter regime in which this measurement induces the quantum-to-classical transition.

with respect to the position axis. It does not affect the dynamics, but changes the physical system to which the dynamics corresponds. The parameter κ scales the total area of the phase space with respect to $\hbar$. It is κ that determines whether the system is "big" enough to make the transition to classical dynamics.

For our example we choose $A = 10$, $B = 0.5$, $\tilde{\lambda} = 10$, $\tilde{\Omega} = 6.07$, and $\kappa = 10^{-5}$. We also measure the position continuously so that the master equation is given by Eq. (3.39) with $X = \tilde{x}$, and choose the measurement strength $\tilde{k} = 1/\kappa$. We simulate the evolution of the system, and compare a stroboscopic map of the evolution with that of the equivalent classical Duffing oscillator. A stroboscopic map is generated by plotting the location of the system in phase space, at intervals separated by one period of the driving. Maps that contain an apparently random scatter of points indicate chaos, whereas simple patterns indicate integrable dynamics In Fig. 4.1 we show the maps generated by the quantum and classical systems.

An example of the physical size of an observed quantum system that has the effectively classical dynamics depicted in Fig. 4.1 is as follows. If we choose the mass m to be that of a single gold atom (3.3×10^{-25} kg), and the frequency to be $\omega = 2\pi \times 10$ Hz, then the real position and velocity would be $x = 0.7$ mm $\times \tilde{x}$ and $v = p/m = 4.5$ cm/s $\times \tilde{p}$.

4.5 Irreversibility and the quantum measurement problem

The "quantum measurement problem" (QMP) is more of a philosophical rather than a physical problem, but no education in quantum measurement theory is quite complete without an understanding of it. Elucidating the QMP also helps to clarify why measurements are largely unnecessary in describing quantum behavior, even though they induce nonlinear non-unitary dynamics.

We first recall that all quantum measurements can be performed using a probe system and making a von Neumann measurement on the probe. A von Neumann measurement simply divides the probe system up into a set of mutually orthogonal states, and asserts that one of these states is the state that the probe is "really" in. This process in which a superposition of states is turned into a single state is called the "collapse of the wavefunction."

The fundamental laws of physics are specified as forces, and it is these forces that cause acceleration as specified by Newton's second law. In quantum mechanics the equation that determines the motion due to all forces is Schrödinger's equation, and the resulting dynamics is unitary evolution. The quantum measurement problem, originally elucidated by Schrödinger in his famous discussion involving a cat, is that the action of a measurement – the collapse of the wavefunction – does not correspond to the action of any force, and thus to any physical process. In this sense *there is no collapse.* Any set of mutually orthogonal subspaces defines a set of mutually exclusive possibilities, but there is no physical mechanism by which one of these is *chosen* as the one that we experience as reality. Measurements do not exist.

So if measurements are unreal, how do we decide when it is appropriate to pretend that a measurement has occurred, and select one of the orthogonal subspaces as the outcome? We explain this in two stages. First, the following process acts like a measurement, although it cannot guarantee that the measurement will not be undone. Consider a system in the state $|\psi\rangle = \sum_n c_n |n\rangle$, and allow the system to interact with a second system so that the joint state becomes

$$w = \sum_n c_n |n\rangle \otimes |n\rangle_{\text{sys2}}. \tag{4.135}$$

If we now trace out the second system, the state of the first system becomes the mixture

$$\rho = \sum_n |c_n|^2 |n\rangle\langle n|. \tag{4.136}$$

The interaction between the two systems, followed by the loss of access to the second system, turns the initial superposition into a mixture. Since a mixture of states is equivalent to the situation in which the system is really in one of the possible states $|n\rangle$, but we do not know which one, the interaction produces an effective collapse of the wavefunction, but without selecting one of the possible outcomes.

To be able to safely assume that a measurement has been performed, we need to be sure that it is not going to be undone. In the above scenario a second interaction could undo the correlation, and return the system to its initial superposition. It turns out that the effectively irreversible behavior of large quantum systems provides a physical process, in fact the only known process, that allows us to safely apply a measurement and pretend that a collapse has taken place. If we replace the second quantum system above with a large thermal bath, then as we have seen in this chapter, the interaction between the system and bath correlates the energy eigenstates of the system with those of the bath. Even though the system (and any observer) still has access to the bath, the evolution *never undoes* the correlation (at least for all practical purposes), and this is possible only because of the effectively irreversible motion. What is more, because of the complexity of the bath, no observer has enough access to the bath microstates to reverse the induced correlation. As a result we can safely assume that the correlation will never be undone, and apply a measurement that projects the system onto its energy eigenstates. When a microscopic quantum system is correlated with the macroscopically distinguishable states of a macroscopic system, then because of the unavoidable interaction of the macroscopic system with the environment, we can assume that a measurement of these macroscopic states, and thus of the microscopic system, has taken place. Thus it is the emergent irreversible behavior that leads to thermodynamics that provides the irreversibility necessary for the emergence of an apparent measurement process.

The evolution by which an effectively irreversible interaction with a large environment turns a superposition into a mixture is referred to as "decoherence." While decoherence effectively removes superpositions, and makes the application of measurement theory unambiguous, the quantum measurement problem remains in that there is no physical mechanism for collapse: (i) there is no collapse of the joint state of the system and environment, which ultimately remains in a superposition; (ii) there is no mechanism that selects a single outcome. Due to the process of decoherence the measurement problem is not a practical one, but is philosophically disturbing. While we are able to describe the dynamical behavior of the physical world, the lack of a physical mechanism for collapse makes it hard to understand why we experience reality the way we do.

There are various "interpretations" of quantum theory that try to reduce the dissonance between quantum measurement and everyday experience. These include the "many-worlds" [170, 146], "Bohmian" [70, 71, 674], and "epistemic" [568] interpretations such as "quantum Bayesianism" [111, 187, 365, 191]. Other authors have suggested that there may be a yet-undiscovered physical mechanism that induces collapse. In the author's opinion it is impossible to remove the measurement problem by interpretation, and the beauty, simplicity, and counter-intuitive nature of the departure of quantum theory from classical theories indicate that it is a fundamental aspect of reality. A deeper theory may enrich this new aspect of reality, but will not revoke it. Of course that is only the author's opinion.

On a more practical note, the non-existence of measurements makes it clear that the explicit use of measurement theory is not necessary to understand physical processes. All measurements do is assert that a particular set of orthogonal states are to be regarded as the mutually exclusive possibilities for the physical reality we will observe. Because of this we can conceptually replace measurements by selecting a set of orthogonal states for a probe system, and thinking of the observer as "being in" one of these probe states. A process in which an observer takes an action that depends on the result of a measurement can then be described by a joint unitary between the system and the probe, such that the evolution of the system depends on the state of the probe. This is shown explicitly in Section 1.5. Other processes such as post-selection are understood in a similar way (see Exercise 25).

The replacement of measurements with unitary interactions is not merely conceptual. Any real process that involves physically making a sequence of measurements can be replaced by a procedure that does not, with the possible exception of a single measurement at the end to specify the mutually exclusive outcomes. This does not mean that the two ways of implementing a process will always be interchangeable in terms of the practicality of implementation. This would depend on the relative difficulty of making measurements and performing other operations.

History and further reading

We gave references for further reading for some of the topics in this chapter as we went along. Here is some history and further reading for those for which we did not.

Maxwell's demon was introduced by Maxwell in 1872 [271]. In 1929 Szilard suggested a resolution of the paradox, and his construction is now called Szilard's engine [583]. But it was not until 1982 that the paradox was resolved when Bennett realized that the solution was provided by Landauer's erasure principle [46, 352]. It appears to have taken some time for it to be understood that Landauer's principle was a direct consequence of the reversibility of the laws of physics. The derivation of Landauer's principle from statistical mechanics, and thus from this reversibility, is developed in the following progression of papers [723, 558, 481, 285, 33, 529, 616, 402]. Our treatment here drew mainly from [285]. The most general treatment is that by Maroney in [402]. Further work on Maxwell's demon, and more generally on work extraction by, and the thermodynamics of, measurement and feedback control can be found in [108, 109, 332, 6, 497, 528, 299, 530, 268, 301, 582, 399, 269, 400, 139]. The fact that bare measurements do non-negative work on a system was derived in [304].

The idea that the environment induces the quantum-to-classical transition was introduced in the 1970s and 1980s [249, 721, 722, 317]. The explanation of the quantum-to-classical transition in terms of measurement theory is more recent. To the author's knowledge it was Chirikov in 1991 who first suggested that measurement may be involved in the emergence of classical chaos from quantum mechanics [122]. In 1994 Spiller and Ralph showed that classical chaos emerged when a damped open system was unraveled using the trajectories of the quantum-state diffusion model [475], equivalent to a joint measurement of position and momentum, although they did not specifically consider the quantum-to-classical transition [570]. Concurrently Zurek and Paz argued, by analyzing the dynamics of the phase space distribution, and without using measurement theory, that an environment that coupled to position would induce the quantum-to-classical transition [724]. Coupling to position is ubiquitous due to the presence of the electromagnetic field, which continually provides position information. Percival and co-workers then showed that the quantum-state diffusion model induced the transition, producing the trajectories of classical mechanics [537, 90, 476, 477, 475]. In 2000 Bhattacharya *et al.* showed that a continuous position measurement, equivalent to the coupling considered by Zurek and Paz, induced the trajectories of classical mechanics and the corresponding Lyapunov exponent [62, 63]. We drew from that work in Section 4.4. It has also been shown that the dynamics of continuously observed quantum systems become chaotic before they become classical, thus exhibiting "quantum chaos" [237]. Further work on the quantum-to-classical transition can be found in [553, 206, 207, 208, 173, 227, 171, 325, 178, 172, 174].

Exercises

1. Consider a system of N two-level systems, where the energy of the two levels is 0 and Δ, respectively. The energy eigenvalues of the whole system are given by $n\Delta$, where $n = 0, 1, 2, \ldots, N$. Show that the number of states with energy $n\Delta$ increases as the exponential of n. You will need to use Stirling's approximation, which is

$$N \approx N^N e^{-N} = e^{N(\ln N - 1)}, \quad N \gg 1, \tag{4.137}$$

and the combinatorial formula $N!/(n!(N-n)!)$. If you are not familiar with this formula, I recommend learning how it is derived.

2. Consider a large system consisting of N spin-1/2 systems, each of whose two levels has energy 0 and Δ, respectively. Now consider the space of states of the large system that have energy equal to $M\Delta$. If the state of the large system is confined to this space, but completely mixed on it, what is the density matrix of each of the spin-half systems? You will need to use the combinatorial formula $N!/(M!(N-M)!)$.

3. Solve Eq. (4.9) for β, when $E_j = \hbar\omega(j + 1/2)$, for $j = 0, 1, 2, \ldots, \infty$, and thus determine Ω as a function of E_{unv} for a universe composed of many harmonic oscillators.

4. An N-level quantum system is connected to a thermal bath at temperature T. The system starts in the Boltzmann state. We are going to reduce the entropy of the system by increasing its energy levels, and letting it re-thermalize to the new Boltzmann state. The thermalization takes place on a timescale $1/\gamma$. Raising the system's energy levels requires work. If we denote the energy levels by $\{E_n\}$, and the population of each level by P_n, then the amount of work done on the system when its energy levels are changed by $\{\Delta E_n\}$, and the populations are constant, is $\sum_n \Delta E_n P_n$. We have the choice of changing the energy level rapidly, so that the $\{P_n\}$ remain the same during the change, or slowly, so that as we change the energy levels the populations continually change to their new equilibrium values. (i) Which choice means we do the least work on the system for a given set of changes $\{\Delta E_n\}$, and what is this minimum energy? Now consider the reverse process in which we increase the entropy of the system by lowering the energy levels. In this process we extract work from the system. (ii) To extract the maximum work should we perform the change quickly (adiabatic) or slowly (isothermal), and why? The answers to the above questions are the reasons that the free energy $F = \langle E \rangle - TS(\rho)$, is the maximum amount of extractable work, at least for simple operations that merely change the energy levels.

5. The free energy, $F = \langle E \rangle - TS(\rho)$, gives the maximum amount of work that can be extracted from a microscopic system when the environment (a thermal bath) is at temperature T. The free energy can be thought of as the "real" energy, in the sense that it

is the energy that is actually available to be used. Now consider a measuring device that makes a measurement on a system S. The measuring device will in general have a system $\mathcal{M}$ that can interact with S, and a source of work that it can use. If the initial energy of the system is $\langle E \rangle_i$, and its energy after the measurement is $\langle E \rangle_f$, and the measurement reduces the system's entropy (on average) by ΔS, how much total work (and/or equivalently free energy) must the measuring device expend, as a minimum, to make the measurement?

6. Consider a Maxwell demon that wishes to make a measurement on a system in state ρ, and extract work from it. The demon's memory is the probe system that is used to make the measurement, and that stores the measurement result. The work that can be extracted by the demon is equal to the average reduction in the entropy of the system due to the measurement. Use Theorem 8 to show that, if the probe starts at the same temperature as the system, the measurement increases the average energy of the probe by at least $\Delta S T$, where ΔS is the average reduction in the entropy of the system. You need to know your thermodynamics for this one. Hint: consider the final state of the probe averaged over the measurement results. Theorem 8 lower-bounds the entropy increase of the probe between its initial state and this final state. The fact that a thermal equilibrium (Boltzmann) state is the state with the maximum entropy for a given average value of energy means that there is an equilibrium state that has the same entropy, and no more energy, than the final average state of the probe. Now you can integrate the entropy change between the initial state and this equilibrium state to determine a lower bound on the energy increase.

7. If Maxwell's demon has a memory with N orthogonal states, and starts in a thermal state at temperature T_D, what is the maximum amount of extra work that it can extract from an N-state system that is at a temperature $T > T_D$? Hint: use the laws of thermodynamics, and remember that the measurement process is merely a unitary interaction between the two systems.

8. Consider a universe consisting of a system and bath, in which the total energy of the universe is fixed. To simplify the analysis, assume that the joint states of the system and bath are confined to a degenerate subspace of the universe in which all states have energy E. Assume that for each system-state, $|j\rangle$, with a given energy, ε_j, there are many bath-states that have energy $\epsilon_j = E - \varepsilon_j$. Finally, let the number of bath-states with a given energy ϵ be proportional to $e^{\beta \epsilon}$. With these assumptions, the subspace of the universe with energy E is spanned by the basis of states $|j\rangle |m_j\rangle$, where $|j\rangle$ is the system-state with energy ε_j, and the $|m_j\rangle$ are all the states of the bath that have energy ϵ_j. Now take a random state of this subspace, namely,

$$|\psi\rangle = \frac{1}{\sqrt{\mathcal{N}}} \sum_{j,m_j} c_{j,m_j} \exp(i\theta_{j,m_j}) |j, m_j\rangle, \tag{4.138}$$

where the real numbers c_{j,m_j} are independent zero-mean Gaussian random numbers with unit variance. Show that if the number of states of the bath for each j, m_j, is large, then the state of the system, given by tracing over the bath, will almost certainly be equal to the Boltzmann state.

9. Show that the power spectra of x and p for the equations of motion given by Eqs. (4.54) are the same as those for the equations of motion given by Eqs. (4.53), to first order in the small parameter γ/ω. This means assuming that $\gamma \ll \omega$, writing the spectra in terms of γ/ω (usually by pulling out factors of ω), expanding sub-expressions where necessary as power series in γ/ω, dropping all powers of γ/ω higher than 1, and comparing the two spectra that result.

10. Research question: use the insight of typicality to derive the weak-coupling (Markovian) master equation. To do this assume that all the energy eigenstates of the universe are typical, and that they have a dense spectrum. The evolution of the system (being a small subsystem of the universe) is a result of the dephasing of all the complex phases in Eq. (4.47). This dephasing, and the fact that the system is weakly coupled to the bath, may be sufficient to obtain the Markovian master equation.

11. Show that Eq. (4.67) gives Eq. (4.62).

12. Write down the Redfield master equation for a three-level system whose Hamiltonian is H and interaction with the environment is via X, where

$$H = \begin{pmatrix} 0 & 0 & 0 \\ 0 & \Delta & 0 \\ 0 & 0 & \eta\Delta \end{pmatrix} \quad \text{and} \quad X = \begin{pmatrix} 0 & a & b \\ a & 0 & 0 \\ b & 0 & 0 \end{pmatrix}. \tag{4.139}$$

(i) Write this equation as an explicit set of differential equations for the elements of the density matrix. (ii) Determine the steady state for the diagonal density matrix elements, and show that this is the correct Boltzmann state.

13. A coherent state, $|\alpha\rangle$, is defined by $a|\alpha\rangle = \alpha|\alpha\rangle$. Derive the number-state (Fock-state) representation for a coherent state. Note that $a|n\rangle = \sqrt{n}|n-1\rangle$, where $|n\rangle$ is a Fock state satisfying $a^\dagger a|n\rangle = n|n\rangle$.

14. Solve for the steady state of the thermal master equation for a single qubit, given in Eqs. (5.71) and (5.72), and derive Eq. (5.73).

15. The Redfield thermal master equation for two uncoupled harmonic oscillators is simple to write down, since it is just the sum of the thermal master equations for each oscillator (Eq. 4.82). Now consider the two oscillators to be coupled via the linear interaction $H_1 = \hbar g x_1 x_2$, where $x_1 \propto a + a^\dagger$ is the position coordinate of oscillator 1 and $x_2 \propto b + b^\dagger$ is the same for oscillator 2. (i) Define two new annihilation operators, c and d that are linear combinations of a and b, such that when the Hamiltonian is written in terms of these operators the oscillators c and d are uncoupled. (ii) Write down the Redfield thermal master equation for the new oscillators. (iii) Substitute a and b into this master equation to obtain the master equation for the original oscillators. Note that the original oscillators may have different frequencies.

16. Consider a driven harmonic oscillator evolving under the zero-temperature quantum optics master equation,

$$\dot{\rho} = -i\omega[a^\dagger a, \rho] - iE[ae^{-i\omega t} + a^\dagger e^{i\omega t}, \rho] - (\gamma/2)\left(a^\dagger a\rho + \rho a^\dagger a - 2a\rho a^\dagger\right).$$

Show that the steady state solution is a coherent state $|\alpha e^{-i\omega t}\rangle$ (defined by $a|\beta\rangle = \beta|\beta\rangle$), and determine the value of α. Hint: first move into the interaction picture with respect to the Hamiltonian $\hbar\omega a^\dagger a$. This eliminates the oscillatory part of the dynamics, and removes all explicit time-dependence from the master equation. One way to obtain the result is to show that the evolution generated by the Hamiltonian $-i(a + a^\dagger)$ for a coherent state $|\alpha\rangle$, to first order in time, is just a shift of the coherent amplitude α.

17. Consider the quantum optics master equation, Eq. (4.82), for a harmonic oscillator connected to a thermal bath at temperature T. Show that if the Hamiltonian of the oscillator is replaced with the Hamiltonian $H = \hbar\chi(a^\dagger a)^2$, then the master equation cannot give the right thermal state for the oscillator. What is the steady state of the master equation with this new Hamiltonian? Hint: note that the eigenstates of the new Hamiltonian are the same as the simple oscillator, but the energy levels are different. Now eliminate the Hamiltonian of the oscillator by moving into the interaction picture.

18. Consider a harmonic oscillator evolving under the zero-temperature quantum optics master equation in the interaction picture:

$$\dot{\rho} = -(\gamma/2)\left(a^\dagger a\rho + \rho a^\dagger a - 2a\rho a^\dagger\right). \tag{4.140}$$

(i) Calculate the equations of motion for $\langle a + a^\dagger \rangle$, $-i\langle a - a^\dagger \rangle$, and $\langle a^\dagger a \rangle$, by using the fact that $\langle \dot{X} \rangle = \mathrm{Tr}[\rho X]$, for any operator X.

(ii) Include the terms in the master equation that describe the evolution when $T > 0$, and again determine the resulting equations of motion for $\langle a + a^\dagger \rangle$, and $\langle a^\dagger a \rangle$. Solve these equations to determine the steady-state values.

(iii) Add into the quantum optics master equation the classical driving force (for an optical resonator, this would be a laser incident on one of the mirrors) given in the interaction picture by the term

$$\dot{\rho} = -i\alpha[a + a^\dagger, \rho]. \tag{4.141}$$

Derive and solve the equations of motion for the above three expectation values in this case.

19. Here we will explore something called *algorithmic cooling* [546, 82, 547]. Consider three uncoupled qubits, each with energy levels E_0 and E_1, and all of them at thermal equilibrium at some temperature T. Label the qubits 1, 2, and 3. Now consider a unitary operation applied jointly to all three, that rearranges the joint energy eigenstates of the three qubits, thus rearranging their populations. The operation is fast in that the qubits do not have time to re-thermalize while it acts. (i) What is the most pure state in which

the unitary can prepare qubit 1? (ii) To what temperature does this state correspond? (iii) Now qubits 2 and 3 are allowed to re-thermalize to temperature T while the state of qubit 1 is preserved. Then another fast joint unitary is applied to the three qubits that once again rearranges their joint energy eigenstates. What is the coldest state that the unitary can now prepare for qubit 1?

20. In the Monte Carlo method for simulating a master equation using quantum jumps (Poisson noise), explain why we can use P_0 in place of P_1.

21. Derive the equation of motion for $\langle a^\dagger a \rangle$ for the simple Markovian version of the Brownian motion master equation described in Sec. 4.3.5, and determine the steady-state value.

22. Show that the simple Markovian version of the Brownian motion master equation, described in Section 4.3.5, has the correct thermal steady state for a harmonic oscillator.

23. Consider the correlation function of the thermal noise for a strongly damped oscillator, given in Eq. (4.120). When $\gamma \ll \omega$ the response function (transfer function) of the oscillator is sharply peaked at the oscillator's frequency ω_0, with a width $\sim \gamma$. The thermal noise is thus filtered through this response function, so that only values of ω in the range $[\omega_0 - \gamma, \omega_0 + \gamma]$ are important in determining the spectrum of oscillator dynamics. Show that in this case, if $kT \gg \hbar\omega_0$ then $(\omega/\omega_0)\coth[\hbar\omega/(2kT)] \approx \gamma 2kT/(\hbar\omega) \approx \gamma 2n_T \gg 1$, where n_T is defined in Eq. (4.116).

24. Use the Langevin equations for a strongly damped oscillator, given in Eqs. (4.118) and (4.119), to derive the spectrum of a position measurement for such an oscillator.

25. Derive the equations of motion for the diagonal elements of the density matrix, in the number (Fock) basis, for a harmonic oscillator whose evolution is given by the quantum optics master equation for a non-zero temperature T. Denote the diagonal density matrix element for state $|n\rangle$, $n = 0, 1, \ldots$, by p_n.

    (i) You can use the following method to obtain the steady-state solution for this infinite set of equations, a method called "detailed balance." Detailed balance uses the fact that if the flow of probability from density matrix element a to element b is matched ("balanced") with the corresponding reverse flow from b to a, and if this is true for *every* pair of dynamical variables (that is, if it is true "in detail"), the system will be in a steady state. The flow of probability from a to b is given by the contribution to the derivative of b that is proportional to a. Since all the transitions in the present case are between density matrix elements that differ by just one photon, matching the rates for each transition is not difficult. Show that enforcing the "detailed balance" condition gives p_{n+1} in terms of p_n. Use this to solve for all the p_n in terms of p_0, and then determine p_0 from the normalization $\sum_n p_n = 1$. Show that the resulting steady state is the correct Boltzmann state.

    (ii) Now do the same analysis for a nonlinear oscillator, whose Hamiltonian is the same as the harmonic oscillator, but with the extra term

$$H_{\text{nonlin}} = \hbar\kappa(a^\dagger a)^2. \tag{4.142}$$

    Show that in this case the steady state is not the correct Boltzmann state.

26. Consider a single particle in a pure state. Show that if its wavefunction is Gaussian, and its uncertainty in position and momentum is sufficiently small, then the means of x and p, and their variances and covariance, obey exactly the same equations as a classical system, whose state is given by a Gaussian joint probability density for x and p. See, e.g., [153, 63, 235]

27. Using the physical parameters given at the end of Section 4.4 for the Duffing oscillator, what is the driving force λ when $\tilde{\lambda} = 10$?

28. Here we consider what it means for a dynamical system to have "low noise." The evolution of the means of the position and momentum of a single particle, under a continuous measurement of position, is given by

$$d\langle x \rangle = \langle p \rangle \, dt/m + \sqrt{8\mu} V_x \, dW, \qquad (4.143)$$

$$d\langle p \rangle = \langle F(x) \rangle \, dt + \sqrt{8\mu} C_{xp} \, dW, \qquad (4.144)$$

where $F(x)$ is the force on the particle, μ is the measurement strength, V_x is the variance in position, $C_{xp} \equiv \langle xp + px \rangle /2 - \langle x \rangle \langle p \rangle$ is the covariance between the position and momentum, and dW is the Wiener noise increment (see Chapter 3). By "low noise" we mean that the paths traced out by $\langle x \rangle$ and $\langle p \rangle$ are relatively smooth. This means that on a timescale over which the motion of $\langle x \rangle$ (for example) is approximately a straight line, the change in x due to the noise should be small compared to that due to the deterministic motion.

By examining the equations of motion for $\langle x \rangle$ and $\langle p \rangle$, determine timescales Δt_x and Δt_p such that over a time $t \ll \Delta t_x$ the motion of $\langle x \rangle$ is approximately linear (has small curvature), and similarly for Δt_p. Use these timescales to show that the motion of $\langle x \rangle$ (respectively, $\langle p \rangle$) has low noise when

$$\sqrt{8\mu} V_x \ll \sqrt{\frac{|p^3|}{m|F|}}, \quad \text{respectively} \quad \sqrt{8\mu} C_{xp} \ll \sqrt{\frac{m|F^3|}{|p\partial_x F|}}. \qquad (4.145)$$

5

Quantum feedback control

5.1 Introduction

The notion of feedback control comes to us from the classical world of mechanical and electrical engineering. Machines designed to operate a certain way may deviate because of small errors in their physical construction, because of inherent instability, or because of external sources of noise. Such deviations can be controlled by monitoring a machine's behavior, and using this information to apply forces that periodically or continuously correct the motion. This is called *feedback control*, and the device that receives the measurement signal, and translates it into the appropriate correcting forces, is called the "feedback controller" (or "controller" for short). We will usually refer to the system to be controlled as the *primary*, and the system that acts as a feedback controller as the *auxiliary*. The device(s) that actually apply the forces to the system are often called the actuator(s), and the prescription by which the control forces are chosen based on the measurement results is variously called the control *algorithm*, *law*, *strategy*, or *protocol*. Here we exclusively use the terms *strategy* and *protocol*.

Feedback control is also known as *closed-loop* control, because the flow of information to the controller, and the flow back to the system via actuators, are thought of as forming a loop. As a historical note, Maxwell appears to have been the first to perform a mathematical analysis of a feedback control system. He studied "governors," mechanical devices that control the speed of an engine by using the centrifugal force generated by the engine to control the flow of fuel [413]. In fact the use of centrifugal governors goes back even further, as they were used to control the pressure between millstones in windmills in the 17th century [253].

In considering feedback control, it is useful to distinguish two ways in which a system may deviate from its intended motion: it may deviate in a way that can be predicted beforehand (that is, in a known way), or it can deviate in an unpredictable way, so that after a time its state becomes less certain from the point of view of the observer who wishes to control it. In the first case, measurements are not required to correct the motion. It is the second case that is most common, and in which feedback control is important. For classical systems, the observer's state-of-knowledge is given by a probability density over the state space, or *phase space*. If we think of this state-of-knowledge as having a mean and

232

variance, then these correspond, respectively, to the known and unknown parts of the deviation from the intended, or *target* state. Without measurement the observers can apply forces to bring the mean closer to the target state, but cannot do anything about the variance. It is the role of the measurement to reduce this variance. In reducing the variance, the measurement will in general also change the mean of the observer's state-of-knowledge (this new mean is, on average, closer to the true state of the system than the previous mean). It is the job of the controller to then bring the new mean closer to the target state.

In quantum mechanics the observer's state-of-knowledge is given by the density matrix, and we can characterize the observer's uncertainty by, for example, its von Nuemann entropy. When our purpose is to control the *state* of a quantum system, then our goal will always be to bring the system as close a possible to a given *pure* state. We will refer to the state we wish the system to be in as the *target state*, which may be time-dependent. Since mixing represents a lack of control over the state of a system, specifying a mixed target state is inconsistent with our definition of what it means to control the state of a system.

There are a number of interesting features that distinguish quantum from classical feedback control. The first is that measurements, in general, *change* the state of the system in a random way. This means that the measurement must be taken into account when designing control protocols, and must be tailored to obtain the best performance. We will also find that the motion induced by the measurement can itself be used to effect control. Because measurements must be tailored in order to optimally extract information from quantum systems (see Chapter 6), realizing optimal measurements is often non-trivial. This brings us to the second unique feature of quantum feedback: it is useful not only for controlling motion but also for engineering measurements that might otherwise be difficult to realize.

Feedback control can be implemented without using explicit measurements, whose results are processed by some digital computation, but instead by coupling the system to the auxiliary in a more direct, mechanical way. This is, in fact, how a centrifugal governor works. We can also do this to implement feedback control in quantum systems, and in this case there is a sharper distinction between feedback that uses measurements and feedback that involves only a Hamiltonian coupling between the primary and auxiliary. We will discuss both kinds of feedback and the relationship between them below.

A note on terminology: states-of-knowledge versus estimates

As a point of terminology, we note that the density matrix is the observer's full state-of-knowledge, and should not be confused with an "estimate" of the state of the system. In classical measurement theory, the probability density, $P(\mathbf{x})$, for the state, $\mathbf{x}$, of a system is not an estimate of the system state, it is the observer's full state-of-knowledge. The expectation value of $\mathbf{x}$, calculated using $P(\mathbf{x})$, is an example of an estimate of the true state of the system. For quantum systems the eigenstate corresponding to the largest eigenvalue of ρ is an estimate of the state of the system. More generally, an estimate is something that is approximate. If one uses an approximate method to obtain the density matrix from a measurement record, rather than integrating the SME, then the result is an estimate of the observer's state-of-knowledge. Averaging the measurement record over a fixed time-window gives an estimate of the expectation value of the measured observable.

Objectives

A classic goal, or objective, of feedback control is to move a dynamical system near to a given state, $|\psi\rangle$, or to ensure that its evolution closely follows some desired evolution, $|\psi(t)\rangle$. These two goals are very similar, since the second is merely the first in which the target state, $|\psi\rangle$, changes from one time to the next. Variations on these goals involve taking a system to a given state at a given time, keeping a system at a given state, or keeping the evolution close to a desired evolution for a finite time.

The above objectives are not the only ones. An important, and more ambitious, objective is to control the full dynamics of a system. This is distinct from controlling the state of the system, because in controlling the full dynamics we will in general want to map every initial state to a different final state. For quantum systems, when we want to control the full dynamics we usually also want this dynamics to be unitary. As we will explain below, feedback that uses measurements cannot do this in a manner in which the control protocol is independent of the initial state. To do so one can instead use coherent feedback, which will be defined below.

A third control objective is that of using feedback to construct measurements that may be difficult to realize in any other way. Feedback involves making a measurement, and then either applying forces to the system, or modifying the next measurement to be made on the system in a way that depends on measurement results. When we say that feedback can be used to construct a measurement, we are referring to the fact that any sequence of measurements and Hamiltonian evolution can always alternatively be viewed as a single measurement (see Exercise 15 in Chapter 1) – and it is the construction of this single measurement that we have in mind when we speak of constructing a measurement using feedback. Measurements constructed in this way are called "adaptive measurements." In this chapter we examine a feedback protocol that increases the rate at which the state of the measured system is purified. We discuss other examples of adaptive measurements in Chapter 6.

Constraints and optimality

In implementing feedback control, there will always be limitations to the forces that can be applied, or to the timescales upon which these forces can be adjusted. There are also limitations to the rate at which information can be extracted from a system. In addition to limitations, there may also be costs associated with implementing the control that one may wish to minimize, such as the total energy required. Control theory is often concerned with obtaining the most effective control, given the limitations, or devising a control protocol that performs adequately while minimizing the cost, however this is defined. This is the subject of *optimal control*. It is only for very few systems that one can obtain analytic solutions for optimal control protocols, and searching for optimal solutions numerically is often prohibitively expensive. As a result, most practical feedback protocols, however they are constructed, are not usually known to be optimal. In special cases in which the limitations and the systems are sufficiently simple, optimal protocols can sometimes be determined. If the only limitation is the size of the control force, then the optimal solution is usually to apply this force at maximum strength, but to switch the direction of the force from time to time. This kind of control is called "bang-bang" control due to the sounds made by a mechanical device switching between two extremes. When we investigate measurement-based feedback protocols in Sections 5.4 and 5.5.3, we will encounter some optimal protocols for simple systems.

Overview of this chapter

In the next two sections we describe the general structure of feedback control in quantum systems, and discuss various ways that it can be implemented. Following this we focus on feedback control via continuous measurements. We introduce the reader to two dynamical effects of continuous measurement that have a direct bearing on feedback control, and consider the problem of optimal control of a single qubit. Next we turn to the general problem of obtaining optimal feedback protocols, introduce the classic method of formulating this problem, and present the exact solutions for a class of control problems for linear classical and quantum systems. In the final section we consider a special case of the control of nonlinear systems, for which optimal protocols can be derived. We examine examples of coherent feedback control in Chapter 8.

5.2 Measurements versus coherent interactions

We can perform feedback control by making a sequence of measurements on a system, and modifying the Hamiltonian in response to the result obtained from each measurement. Even if we take the continuum limit, so that this sequence becomes a continuous measurement with a corresponding continuous modification of the Hamiltonian, we can describe a single "cycle" of this feedback control process by a single measurement followed by a single unitary operation:

$$\rho_j(t+\Delta t) = \frac{U_j A_j \rho(t) A_j^\dagger U_j^\dagger}{\operatorname{tr}\left[A_j^\dagger A_j\right]}. \tag{5.1}$$

Here $\rho(t)$ is the density matrix of the system at time t, the set of operators $\{A_j\}$ are the measurement operators, j labels the measurement result, and U_j is the action of the feedback applied to the system upon obtaining result j.

If the measurement and feedback are continuous, then all we have to do is replace Δt with dt, the measurement operators A_j are those for a continuous measurement performed over the time interval dt (described in Chapter 3), and the unitary may be written as $U_j = \exp\left(-iH_j dt/\hbar\right)$ for some control Hamiltonian H_j.

The performance of a single cycle of a feedback protocol for controlling the state of a system, being the probability that the system is in the target state at time $t+\Delta t$, is therefore given by

$$P = \langle\phi| \left[\sum_j P_j \rho_j(t+\Delta t)\right] |\phi\rangle = \sum_j \langle\phi|U_j A_j \rho(t) A_j^\dagger U_j^\dagger|\phi\rangle, \tag{5.2}$$

where we have averaged over all the possible outcomes j, and $|\phi\rangle$ is the target state.

In Section 1.5 we showed how a single feedback cycle can be implemented by using a second quantum system without making any measurements. The reader may wish to read that section at this point, although we review it briefly now. As we learned in Chapter 1, every measurement can be implemented by allowing the system to interact with a second (auxiliary) system, and then making a von Neumann measurement on the auxiliary. If we measure the auxiliary in the basis $\{|j\rangle\}$, and the resulting measurement on the primary has measurement operators $\{A_j\}$, then this means that before we make the measurement on the auxiliary, in the subspace for which the auxiliary is in state $|j\rangle$ the system has the (unnormalized) state $A_j \rho A_j^\dagger$. We can therefore perform the feedback by applying the following unitary operator to the combined system:

$$U_{\text{fb}} = \sum_{n=1}^{N} U_j \otimes |j\rangle\langle j|, \tag{5.3}$$

where U_j is a unitary that acts on the space of the system. For state $|j\rangle$ of the auxiliary the operation U_j is applied to the system. If we now throw away the auxiliary, which means taking the partial trace over it, then the final state of the system is

$$\rho(t+\Delta t) = \sum_j U_j A_j \rho(t) A_j^\dagger U_j^\dagger = \sum_j P_j \rho_j(t+\Delta t). \tag{5.4}$$

The probability that the system is in the target state, $|\phi\rangle$, is precisely that given above in Eq. (5.2). This measurement-free procedure for performing feedback is referred to as a coherent version (or coherent implementation) of a measurement-based feedback cycle.

The coherent version of feedback control involves only the application of a time-dependent Hamiltonian to the joint system consisting of the primary and auxiliary, and since the required time-dependence does not depend on any measurement results, it is independent of the initial state of the system. The application of a predetermined time-dependent Hamiltonian to a system is often referred to as "open-loop" control, since there is no feedback loop. But in this case, because the open-loop control is applied also to an auxiliary system, it actually implements feedback.

We now define *coherent feedback control* as any control scenario in which open-loop control is applied to two interacting systems, and in which the purpose is to achieve some control objective for only one of the systems. This scenario is depicted in Fig. 5.1. Since all measurements can be implemented using an auxiliary system, this coherent feedback (CF) scenario can implement in theory all feedback protocols that involve

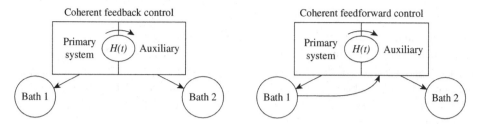

Figure 5.1 Here we depict the general definitions of coherent feedback and feedforward that we use in this text. Since coherent feedback subsumes measurement-based feedback, its definition covers all feedback scenarios. The arrows in the diagrams give the direction in which useful information flows. In feedback control information is in general lost to the baths, but no exploitable information comes from the baths. In feedforward the auxiliary system has access to Bath 1 and can extract useful information from it.

explicit measurements. It turns out, however, the coherent feedback is more general than measurement-based feedback because not all coherent protocols can be reproduced by those using measurements. We will explain why shortly, but first we pause to consider the fact that there is no explicit feedback loop in our definition of coherent feedback.

In classical control theory the process of feedback is invariably written as a loop. An arrow goes from the system to the controller, representing the information acquired by the controller about the system, and an arrow goes from the controller to the system representing the forces applied by the controller. It is understood that there is a causal relation between the two arrows, in that information must be acquired by the controller before it can apply forces based upon it. However, in the basic physical description of the interaction between a system and a controller, there is no loop: there is merely an interaction Hamiltonian. The system affects the controller, and vice versa. The existence of a loop is merely a statement about how feedback works; that any force applied to the system by the controller constitutes feedback only when it is based on some information that has been obtained about the system via the interaction. It is a statement about how the interaction Hamiltonian is used. In the Watt governor the absence of any basic causal loop is especially clear, since there is no explicit information gathering. The system and controller are merely two mechanical systems coupled by a constant Hamiltonian.

In discussing the redundancy of an explicit feedback loop in the definition of feedback control, it is worth contrasting feedback with feedforward. Unfortunately, there is more than one definition of feedforward used by the classical control community, but as far as the author is aware, in only one of these definitions is feedforward distinct from open-loop control and feedback control as far as the exploitation of information is concerned. Under this definition, feedforward control is a scenario in which the controller has access to the noise driving the primary before this noise affects it. Because of this it can reduce the entropy of the primary without measuring it. In Fig. 5.1 we depict the general scenario for coherent feedforward control, under the above definition of feedforward, although we do not discuss feedforward any further in this book.

Coherent feedback can outperform measurement-based feedback

The reason that measurement-based feedback (MF) is not as general as coherent feedback is that the unitary operator that corresponds to the feedback part of MF is restricted to the form given above for U_{fb}. This means that the unitary operator that implements the measurement and feedback steps must always be broken down into the action of two consecutive unitaries, the second of which must have the form of U_{fb}. We now give an example to show that this is not always optimal.

Consider the task of preparing a single qubit in a pure state, from an arbitrary mixed state. We have an auxiliary qubit already in a pure state, and we are allowed to use any interaction Hamiltonian subject to the constraint on the speed of evolution that the Hamiltonian can generate. The speed at which a Hamiltonian can transform one state into another that is orthogonal to it is bounded by the difference between its maximum and minimum eigenvalues. If we denote these eigenvalues by $\lambda_{max}(H)$ and $\lambda_{min}(H)$, then the maximum speed of evolution is determined by $\Delta\lambda(H) = \lambda_{max}(H) - \lambda_{min}(H)$. Our constraint is that $\Delta\lambda(H) \leq 2\mu$.

We will discuss evolution speed and minimum-time paths in Hilbert space in more detail in Section 8.1. For now we just need to know that the fastest way to prepare the primary qubit in a pure state is to apply the interaction Hamiltonian

$$H_I = \hbar\mu\,\sigma_x \otimes \sigma_x = \hbar\mu(|01\rangle\langle 10| + |10\rangle\langle 01|) \tag{5.5}$$

for a time of $\tau = \pi/(2\mu)$. If we start the auxiliary in state $|0\rangle$, then this interaction Hamiltonian performs one of two actions on the primary depending on its state. If the primary is in state $|0\rangle$ it does nothing, and if it is in state $|1\rangle$ it flips it to state $|0\rangle$. The result is that for any initial mixture of the states $|0\rangle$ and $|1\rangle$ the primary is prepared in the pure state $|0\rangle$.

There are two important points to note about the coherent feedback control protocol achieved by the interaction H_I. The first is that there is only a single outcome. This means that if we write the evolution using an operator-sum representation (Kraus representation), so that the final state is expressed as $\rho(\tau) = \sum_n A_n\rho(0)A_n^\dagger$ for some set of measurement operators $\{A_n\}$, then there is only one operator in the set, namely $A_1 = (|0\rangle + |1\rangle)\langle 0|$. The second is that the evolution cannot be broken into two parts, in which the second part has the form of U_{fb}.

It turns out that for a single qubit primary and auxiliary there is a feedback protocol that prepares the primary in the state $|0\rangle$, that is as fast as the one above, and that can be broken up into a measurement and feedback step (see Exercise 6). But for larger systems this is not the case. That is, if we wish to prepare an initially mixed N-state system in a pure state, using an N-state auxiliary, then the generalization of H_I for this case is faster than the fastest coherent implementation of an MF protocol. The reason for this is that in an MF protocol a basis of the auxiliary must first become fully correlated with a basis of the primary, so that for each basis state of the auxiliary the primary is reduced to a pure state. Only then can an operation of the form U_{fb} prepare the primary in a pure state (by transforming each basis state of the primary to the target state).

To implement the optimal coherent control protocol above, the interaction H_I must rotate the joint state $|10\rangle$ to the orthogonal state $|01\rangle$. It must therefore perform a $90°$ rotation in Hilbert space. The measurement-based version must first correlate the primary and auxiliary, and then perform the conditional rotations on the former. For a qubit each of these steps requires a $45°$ rotation, and thus takes no longer than the optimal protocol. However, for an N-state system the rotation angle required for each step is θ where $\cos(\theta) = 1/\sqrt{N}$.

We conclude from the above analysis that under a constraint on the interaction Hamiltonian, in general coherent feedback protocols can reduce the entropy of the primary faster than can measurement-based protocols. The faster a protocol can extract entropy from a system, the better it will perform in the steady state, other things being equal. The above analysis suggests, therefore, that for wide range of control problems coherent feedback protocols have the potential to outperform MF protocols.

Physical differences between coherent and measurement-based feedback

The dynamical differences between coherent and measurement-based feedback control are not the only important differences. The two implementations are quite different from a practical point of view. While both coherent and measurement-based feedback require that the system becomes correlated with another system, the key difference is that the latter requires *amplification*: to realize the measurement, a basis of the auxiliary must be correlated with a system whose states are macroscopically distinguishable. This process is typically much more technologically demanding than the coherent manipulation of mesoscopic systems, which is all that is required for coherent feedback. Measurement also comes with a blessing and a curse: the resulting classical signal can be processed in an essentially noise-free manner using classical computers. The downside is that the measurement and the processing, while sophisticated, can take considerably more time than the typical evolution-time of mesoscopic systems. It may be possible therefore to obtain the same level of processing in less time using coherent feedback with a sufficiently complex auxiliary. Nevertheless, the practical point of trade-off between coherent and measurement based control is not yet clear, especially because mesoscopic auxiliaries are subject to noise from their environments.

5.3 Explicit implementations of continuous-time feedback

5.3.1 Feedback via continuous measurements

In feedback implemented using continuous measurement, the Hamiltonian of the system can be modified as time goes by, based on the continuous stream of measurement results. The controller – in this case a classical computer – can choose the Hamiltonian at time t to be any function of the stream of measurement results up until that time. Since this stream of results from the initial time to time t is a continuous function, and the Hamiltonian is specified by a finite set of parameters, one speaks of the Hamiltonian as being a *functional* of the measurement record. Usually the Hamiltonian is a sum of terms, where the size of some (or all) of the terms can be varied by the controller. Thus $H = \sum_n \mu_n H_n$, where the real numbers μ_n are the *control parameters*.

In choosing the control Hamiltonian, it is actually clear that we do not need to consider all possible functionals of the measurement record, for the following reason. The density matrix of the system at time t constitutes the observer's complete state-of-knowledge of the system at that time. Along with a knowledge of the dynamics of the system, including sources of noise, it contains the full information about the future behavior of the system. In particular, it allows the observer to predict the probabilities of the outcomes of all future measurements on the system. Thus the density matrix, along with the specification of the system dynamics, constitutes all the information that the controller needs in order to determine the optimal choice for the Hamiltonian at time t. This means that the optimal Hamiltonian at time t can always be written as a function of the density matrix of the system at that time. This does not mean that determining the optimal Hamiltonian will be computationally feasible, methods for which we discuss in Section 5.5.

To determine the density matrix at time t, the controller must use the measurement record to solve the stochastic master equation for the system. The stochastic master equation, or SME, is therefore a prescription that tells the controller how to process the measurement record to obtain all the relevant information. The SME for the system, including feedback from a continuous measurement of an operator x, is therefore given by

$$d\rho = -(i/\hbar) \left[H_0 + \sum_n \mu_n(\rho) H_n, \rho \right] dt$$

$$- k[x, [x, \rho]] \, dt + \sqrt{2\eta k} \, (x\rho + \rho x - 2\langle x \rangle \rho) \, dW, \tag{5.6}$$

where H is now an explicit function of ρ. Recall from Chapter 3 that k is called the measurement strength, η is the measurement efficiency, and the measurement record, from which the observer determines dW, is given by $dr = \langle x \rangle \, dt + dW/\sqrt{8\eta k}$.

Processing the measurement record requires computational resources. If the dimension of the primary system is large, and the timescale on which the controller must respond is sufficiently fast, then the resources required to update ρ at each time-step can be prohibitive. This motivates developing feedback protocols that do not require the full knowledge of ρ. There are three main approaches to this problem that have been used to date. We consider the second two in the next section. The first approach is to fix a particular form for the density matrix – an *ansatz* – specified by a relatively small number of parameters, and that is general enough to provide a good approximation to the true density matrix throughout the evolution. In this case, one derives from the SME a set of (approximate) coupled differential equations for the parameters of the ansatz.

We will see in Section 5.5.2 that if we choose a Gaussian state as an ansatz we obtain the exact evolution for linear quantum systems under linear measurements. For nonlinear systems, a Gaussian ansatz has been shown to be effective for cooling an atom trapped in a standing-wave field inside an optical cavity [574, 575], and with a simple extension it is effective for feedback control of superpositions of two coherent states [305]. Gaussian ansatzes are explored in Exercises 7 and 8.

Wiseman–Milburn Markovian feedback

The second method of reducing the computational resources required for feedback control is much more drastic: one performs only some elementary processing of the measurement signal. In the simplest case, one attempts to control the system by performing no processing whatsoever, feeding the measurement record directly back to control the system. This was the form of feedback considered by Wiseman and Milburn when they introduced a consistent quantum treatment of feedback into quantum optics [681, 683, 668]. A number of effective control protocols can be obtained in this way. Feeding the unprocessed measurement signal back to the control Hamiltonian means making the control parameters μ_n directly proportional to the measurement signal. One must be careful in deriving the resulting stochastic master equation in this case, as the measurement signal is non-differentiable, and one must ensure that the feedback parameters change *after* the measurement noise has acted on the system.

To induce evolution of ρ that is proportional to the increment of the measurement record, dr, we must make the Hamiltonian proportional to dr/dt, since it is $d\rho/dt$ that is proportional to the Hamiltonian. Note also that since

$$\frac{dr}{dt} = \langle x \rangle + \frac{1}{\sqrt{8k}} \frac{dW}{dt}, \tag{5.7}$$

it is really dr/dt that is the measurement record, although we do not usually write it this way: dW/dt is white noise, an idealization of noise with very large, high-frequency fluctuations. This is essential because the measurement is continuous: the measurement result must contain no information about x as $dt \to 0$, so the standard deviation of the measurement error must become large as $dt \to 0$.

If we choose a single control Hamiltonian and make it proportional to dr/dt so that $H = H_0 + (dr/dt)F$, then the evolution of the system is given by the Wiseman–Milburn Markovian feedback SME:

$$d\rho = -\frac{i}{\hbar}[H_0 + H_{\text{fb}}, \rho]\,dt + 2k\left\{ \mathcal{D}\left(x - \frac{iF}{4\hbar k}\right) - \left(\frac{1-\eta}{\eta}\right)\mathcal{D}\left(\frac{iF}{4\hbar k}\right) \right\}\rho\,dt$$
$$+ \sqrt{2\eta k}\,\mathcal{H}\left(x - \frac{iF}{4\eta\hbar k}\right)\rho\,dW. \tag{5.8}$$

Here

$$H_{\text{fb}} = \frac{Fx + xF}{4}, \tag{5.9}$$

and we have followed Wiseman by defining the super-operators $\mathcal{H}(\cdot)$ and $\mathcal{D}(\cdot)$ as

$$\mathcal{H}(c)\rho \equiv c\rho + \rho c^\dagger - \text{Tr}[c + c^\dagger]\rho, \tag{5.10}$$

$$\mathcal{D}(c)\rho \equiv -\frac{1}{2}(c^\dagger c\rho + \rho c^\dagger c - 2c\rho c^\dagger), \tag{5.11}$$

for any operator c. Note that when c is Hermitian $\mathcal{D}(c)\rho = -[c,[c,\rho]]/2$. We give the derivation of the above SME in Appendix H. Of course, what we are usually interested in is the state of the system averaged over all the possible noise realizations and feedback operations. Recall from our discussion above that this is all we need if our measure of performance is linear in the state of the system. Averaging over the measurement noise dW is accomplished simply by setting $dW = 0$ in the SME, and the result is

$$\rho = -\frac{i}{\hbar}[H_0 + H_{\text{fb}}, \rho] + 2k\left\{\mathcal{D}\left(x - \frac{iF}{4\hbar k}\right) - \left(\frac{1-\eta}{\eta}\right)\mathcal{D}\left(\frac{iF}{4\hbar k}\right)\right\}\rho. \tag{5.12}$$

This master equation shows us that the effect of feeding back the measurement signal directly is to add a new term to the Hamiltonian, and to modify the Heisenberg noise term (the double commutator) by adding the feedback Hamiltonian to the measured observable, x. There is also an additional source of noise (the third term in Eq. (5.12)) if the measurement efficiency is not unity. Because Eq. (5.12) is a Markovian master equation, this kind of "direct" feedback was christened *Markovian feedback* by Wiseman. One can also feed back the measurement signal directly when one is detecting photons that leak out of a cavity (quantum jumps), or in homodyne detection. The master equations for both these cases may be found in [668].

Another simple way in which one can process the measurement record is to average it over some time-window, possibly with a chosen weighting function, in order to obtain an estimate for the measured observable. One can further differentiate this averaged signal to obtain an estimate of the derivative of the observable. A feedback protocol that uses this kind of processing is considered in [534].

5.3.2 Coherent feedback via unitary interactions

This kind of coherent feedback is conceptually the simplest. We couple the primary to the auxiliary via an interaction Hamiltonian, which we can write in general as the sum of products of a primary operator and an auxiliary resonator. If M_m are primary operators, and P_m are auxiliary operators, then the interaction Hamiltonian will be of the form $H_1(t) = \hbar \sum_m \lambda_m M_m(t) P_m(t)$. In general both systems will be coupled to their environments and will experience damping. The full evolution will therefore be given by the joint Hamiltonian of the two systems including the interaction $H_1(t)$, and Lindblad terms describing the irreversible motion.

5.3.3 Coherent feedback via one-way fields

So far we have introduced two ways to implement feedback that are physically rather different, one in which measurements are made on the system, and another that uses solely unitary evolution and involves a second "auxiliary" system. We now describe a third way to implement feedback, which we will call "field-mediated" coherent feedback. This feedback is coherent, in that no measurements are made, but the coupling between the primary system and the auxiliary is achieved by coupling both systems to traveling-wave fields. These fields introduce quantum noise into the system, since they are effectively zero-temperature thermal baths, and because of this the equations of motion for the system and auxiliary are no-longer unitary. In fact if we want to make a continuous measurement on a system we couple it to a traveling-wave field and continually measure this field as it propagates away from the system. In field-mediated coherent feedback, instead of making a measurement on the field, we allow this field to interact with the auxiliary system. In this sense the auxiliary system receives the continuous stream of information from the system, and processes this information so as to perform feedback on the system, instead of a classical computer processing the stream of measurement results. We can regard field-mediated feedback as being intermediate between unitary coherent feedback and measurement-based feedback, in that it is coherent in the sense that it avoids measurements, but is incoherent in that it involves interactions that carry information away from the system in the same way that continuous measurements do. Field-mediated coherent feedback is nevertheless commonly referred to simply as "coherent feedback."

To implement field-mediated coherent feedback we need to recall some things from Chapter 3. To make a continuous measurement on a quantum system we couple the system to the electromagnetic field, and measure the field that is generated by the system and propagates away from it. A simple example of this is an optical cavity, in which light is trapped between two parallel mirrors. In this case our system is a single mode of the trapped light, and the field is all the modes of light outside the cavity. Photons in the mode leak out and travel away from the cavity where we can detect them. More generally we will show in Chapter 7 that we can make continuous measurements of other systems by coupling them to a single mode in an electrical or optical cavity, and detecting the output light from the cavity. If we couple the system to a cavity mode in this way, but don't measure the output light, then the evolution of the system can be described by a Lindblad master equation, or by the input–output formalism (see Section 3.6 or Appendix F). The Lindblad operator that describes the loss of information from the system due to the "output channel" is the system observable via which it interacts with the cavity mode, and is the observable that would be measured if the output light were measured.

To implement the part of the feedback loop that provides the auxiliary with information about the primary, we couple the primary via one of its operators, L, to a mode of a superconducting or optical cavity. We also couple the auxiliary system, via one of its operators M, to a mode of a second cavity. We then take the output field from the first cavity and direct it into the second cavity. We can ensure that the field only propagates one way (from

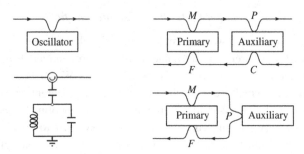

Figure 5.2 On the top left is a diagrammatic depiction of an oscillator coupled to a one-way quantum field that propagates to the right. This coupling provides a single input–output pair, and can be implemented in a superconducting circuit in the manner shown on the bottom left. Here an LC-oscillator is coupled to a transmission line via a capacitor and a counter-clockwise circulator. Superconducting circuits are discussed in detail in Chapter 7. On the top right is a diagram of the field-mediated feedback loop described by Eqs. (5.18) and (5.19). On the bottom right is the field-mediated feedback loop described by Eqs. (5.188) and (5.189).

the system to the auxiliary, and not vice versa) by using nifty devices called "isolators" or "circulators" (see Section 3.3). This configuration can be implemented in a compact way in an all-electrical circuit (see Chapters 7 and 8), and is called "cascading" two systems together, or a "cascade connection." This kind of coupling was first considered by Gardiner [198] and Carmichael [101], and Gardiner showed that it could be described in a very simple and intuitive way using the input–output formalism.

We now recall from Section 3.6 that if a system is coupled to an output channel by an operator M, then the evolution of an operator S of the system is given by the quantum Langevin equation

$$dS = \mathcal{L}_s S \, dt - \mathcal{K}(M)S \, dt + \sqrt{2}\left(\left[S, M^\dagger\right] db_{\text{in}} - db_{\text{in}}^\dagger [S, M]\right). \tag{5.13}$$

There are a number of terms here that need explaining. First, we have defined two linear superoperators that act on S. The first is

$$\mathcal{K}(M)S \equiv -2\mathcal{D}(M)S = M^\dagger M S + S M^\dagger M - 2M^\dagger S M, \tag{5.14}$$

in which M and S are any two operators. When M is Hermitian, $\mathcal{K}(M)S$ reduces to $[M, [M, S]]$. The second term in Eq. (5.13), that given by $\mathcal{K}(M)S$, is the usual Lindblad term in a master equation having a Markovian output channel given by the Lindblad operator M. The super-operator $\mathcal{L}_s$ gives the evolution of the system in the absence of the coupling to the field. If the system is not coupled to any other environment then its evolution is simply $\mathcal{L}_s S = -(i/\hbar)[S, H_s]$, where H_s is the system's Hamiltonian. If the system is coupled to an environment, then its evolution will also include one or more terms with the same form as $\mathcal{K}(M)S$. The operator ds_{in} is an infinitesimal increment of the field that is input to the system, and has the special commutation relation given in Section 3.6, Eq. (3.191). The output from the system is given by

$$s_{out}(t) = s_{in}(t) - \sqrt{2}M(t). \tag{5.15}$$

If we direct the output of this system (the primary) to the input of another system (the auxiliary), then all we have to do is to set the input field for the auxiliary equal to the output field for the primary. Let us say that the auxiliary is coupled to its output channel by an operator P, and denote the input and output field operators for the auxiliary by $c_{in}(t)$ and $c_{out}(t)$. The quantum Langevin equation for an operator A of the auxiliary is then

$$
\begin{aligned}
dA &= \mathcal{L}_a A\, dt - \mathcal{K}(P)A\, dt + \sqrt{2}\left(\left[A, P^\dagger\right] dc_{in} - dc_{in}^\dagger [A, P]\right), \\
&= \mathcal{L}_a A\, dt - \mathcal{K}(P)A\, dt - 2\left(\left[A, P^\dagger\right] M - M^\dagger [A, P]\right) dt \\
&\quad + \sqrt{2}\left(\left[A, P^\dagger\right] ds_{in} - ds_{in}^\dagger [A, P]\right).
\end{aligned} \tag{5.16}
$$

where in the second line we have connected the output of the primary to the input of the auxiliary by setting

$$c_{in}(t) = s_{out}(t) = s_{in}(t) - \sqrt{2}M(t). \tag{5.17}$$

The relationship between the primary and the auxiliary is asymmetric, in that the evolution of the auxiliary is influenced by the primary (the equation of motion for A contains M) but the primary is not influenced by the auxiliary. The latter is receiving information from the primary. The reason that we used the letters "M" and "P" to represent the two coupling operators above is because M is the operator of the system that is "measured" by the auxiliary, and the auxiliary operator P is used as a "probe" to measure this operator.

To complete the coherent feedback loop we must feed an output from the auxiliary into a new input for the primary. (We cannot use the old input, s_{in}, because when we consider the combined system as a single unit, there must always be at least one input and one output from this unit – the fields can only be traveling-wave fields with a continuum of modes if they can propagate in from, and out to, infinity.) We could connect the output $c_{out} = c_{in} - \sqrt{2}P$ back to the primary (the configuration depicted at the bottom right in Fig. 5.2), or we could provide the auxiliary with a new coupling to a new input–output pair, and feed the new output to the primary. We choose this second option, depicted at the top right in Fig. 5.2, because it gives us more flexibility.

In coupling the primary and auxiliary to new inputs and outputs, we need to define three more operators: the operators that couple each system to each input/output channel, and the new input field operator for the auxiliary (the input field operator for the primary will be set equal to this new output from the auxiliary). We denote the new coupling operator for the auxiliary by C (to indicate that it "controls" the force that is applied to the primary), and the new coupling operator for the system by F, to indicate that it is the "force" that is applied to the primary. Denoting the new input field to the auxiliary by a_{in}, the resulting

Langevin equations for the primary and auxiliary are

$$dS = \mathcal{L}_s S\, dt + 2\left(C^\dagger [S,F] - \left[S,F^\dagger\right]C\right) dt - [\mathcal{K}(M) + \mathcal{K}(F)]S\, dt$$
$$+ \sqrt{2}\left(\left[S,M^\dagger\right]ds_{in} - ds_{in}^\dagger[S,M]\right) + \sqrt{2}\left(\left[S,F^\dagger\right]da_{in} - da_{in}^\dagger[S,F]\right), \quad (5.18)$$

$$dA = \mathcal{L}_a A\, dt + 2\left(M^\dagger [A,P] - \left[A,P^\dagger\right]M\right) dt - [\mathcal{K}(P) + \mathcal{K}(C)]A\, dt$$
$$+ \sqrt{2}\left(\left[A,P^\dagger\right]ds_{in} - ds_{in}^\dagger[A,P]\right) + \sqrt{2}\left(\left[A,C^\dagger\right]da_{in} - da_{in}^\dagger[A,C]\right). \quad (5.19)$$

The second terms in each of the Langevin equations are the interactions between the primary and auxiliary induced by the fields. All subsequent terms merely give the usual dissipative/decohering evolutions of Lindblad output channels. It is instructive to rewrite the second terms as

$$C^\dagger [S,F] - \left[S,F^\dagger\right]C = -\frac{i}{\hbar}[S,H_1], \quad M^\dagger [A,P] - \left[A,P^\dagger\right]M = -\frac{i}{\hbar}[A,H_2],$$

where

$$H_1 = \hbar\left(iFC^\dagger + (iFC^\dagger)^\dagger\right), \qquad H_2 = \hbar\left(iPM^\dagger + (iPM^\dagger)^\dagger\right). \quad (5.20)$$

Here we have used the fact that $[S,C] = [A,M] = 0$. In this form we see that the motion induced in each system by the other is indeed Hamiltonian, but that each system feels a *different* interaction Hamiltonian. If the measured operator M and the control operator C are Hermitian then the action of the measurement and control is especially clear. In this case the evolution of the two systems can be written as

$$-\frac{i}{\hbar}[S,H_1] = -\frac{i}{\hbar}[S,iF + (iF)^\dagger]C, \qquad -\frac{i}{\hbar}[A,H_2] = -\frac{i}{\hbar}[A,iP + (iP)^\dagger]M.$$

This makes it clear that the "force" operator F drives the system, and the amount of this drive is determined by the eigenvalue of the "control" operator C. Conversely, the system operator M determines the amount by which the operator P drives the auxiliary. This is what imprints the value of M onto the auxiliary, thus realizing the measurement of M.

Note that the relationship between the primary and auxiliary in Eqs. (5.18) and (5.19) is symmetric, in that the primary could instead be regarded as the system that controls the auxiliary. In deriving these equations we have assumed that it takes no time for the field to travel between the two systems. If there were a large distance between the two systems, then M and P, for example, would be evaluated at different times in Eq. (5.19). In this case the action of the auxiliary on the system via the operator C would involve the information provided by M only at an earlier time. In fact, since we can think of the evolution of the auxiliary as the thing that processes the information about M, if this evolution takes an appreciable time, then the action of C on the system will effectively be delayed in the same way, since $C(t)$ will give the desired feedback action for the value of M at an earlier time.

The field-mediated coherent feedback equations, (5.18) and (5.19), are very general, in that the auxiliary could be arbitrarily complex, and classical (open-loop) controls could be used to make any or all of the operators of the two systems time-dependent. So long as we assume that the fields take no time to travel between the systems, we can calculate the master equation for the joint system directly from the quantum Langevin equations, which for nonlinear systems would be what one would have to solve in practice. Using the method in Section F.7 (see Exercise 13), and noting that all auxiliary operators commute with all primary operators when evaluated at the same time, t, the master equation is

$$\dot{\rho} = \mathcal{L}\rho - [\mathcal{K}(M) + \mathcal{K}(P) + \mathcal{K}(C) + \mathcal{K}(F)]\rho$$
$$+ [2F, \rho]C^{\dagger} - C[2F^{\dagger}, \rho] + [2P, \rho]M^{\dagger} - M[2P^{\dagger}, \rho], \tag{5.21}$$

where we have defined $\mathcal{L} = \mathcal{L}_s + \mathcal{L}_a$. If all the coupling operators are Hermitian this simplifies to

$$\dot{\rho} = \mathcal{L}\rho - [\mathcal{K}(M) + \mathcal{K}(P) + \mathcal{K}(C) + \mathcal{K}(F)]\rho - 2[C, [F, \rho]] - 2[M, [P, \rho]]. \tag{5.22}$$

The equations that describe field-mediated feedback control are not equivalent to those that describe unitary coherent feedback, which stems from the fact that in the former the coupling between the systems is mediated by an irreversible dynamics. The fields that couple the systems can usually be considered to be at zero-temperature, due to their high frequency. Because of this they do not introduce thermal noise into the system, and the quantum noise that they do introduce does not appear to limit the ability of this method to control the primary as compared with unitary coherent feedback. However, the relationship and relative advantages of field-mediated and unitary coherent feedback are not yet well understood. Recall that we obtained the above field-mediated feedback equations by creating a second output for the auxiliary to feed back to the system. If instead we feed back the output of the auxiliary that corresponds to the input receiving the signal from the system (so that the auxiliary has only one input and output) then we obtain somewhat different feedback equations. Deriving these is Exercise 11. We will analyze a concrete example of field-mediated coherent feedback in Section 8.3.2.

By connecting more than two systems together using cascade connections, and by using beam-splitters, it is possible to create complex networks. Network "elements" that add phase shifts or squeeze the fields can also be included. We will not discuss networks here, but further information about optical elements can be found in [342, 341], and discussions of linear networks can be found in [147, 222, 223, 453, 308].

5.3.4 Mixing one-way fields with unitary interactions: a coherent version of Markovian feedback

Naturally we can couple the primary system to an auxiliary using both unitary interactions and coupling via one-way fields. As one example of such a configuration, we show how a coherent feedback loop can be arranged to reproduce Wiseman–Milburn Markovian feedback. To do this we choose the auxiliary to be a harmonic oscillator, and couple the primary to it via a one-way field. The system is coupled via an operator $\sqrt{\kappa}M$, and the auxiliary is coupled via its annihilation operator a with damping rate γ. This effectively makes a continuous measurement on the primary and feeds the output into the auxiliary, in such a way that the annihilation operator a is driven by the measured operator M. Recall that to implement Markovian feedback, we need to apply a Hamiltonian, F, to the primary that is multiplied by the output of the continuous measurement. The trick used by Wiseman to do this is to give the auxiliary a large damping rate so that it relaxes on a timescale much faster than that upon which the state of the primary changes. This means that, to good approximation, the state of the auxiliary is determined at all times by the state of the primary. In this case the auxiliary is described as being "slaved" to the primary (see below). The auxiliary is then coupled to the primary via the interaction Hamiltonian

$$H_1 = \hbar g(a + a^\dagger)F, \tag{5.23}$$

so that the state of the auxiliary determines the force applied for the primary.

To analyze the above configuration, we first write down the quantum Langevin equations for the system and auxiliary, using the method described in the previous section. These are

$$dS = \mathcal{L}_s S\, dt - i(a + a^\dagger)[S, F]\, dt$$
$$- \mathcal{K}(M)S\, dt + \sqrt{2\kappa}\left(\left[S, M^\dagger\right] ds_{\text{in}} - ds_{\text{in}}^\dagger [S, M]\right), \tag{5.24}$$

$$dA = -i\omega\, dt - iF[A, a + a^\dagger]\, dt + 2\left(M^\dagger [A, P] - \left[A, P^\dagger\right] M\right) dt$$
$$- \mathcal{K}(P)A\, dt + \sqrt{2}\left(\left[A, P^\dagger\right] ds_{\text{in}} - ds_{\text{in}}^\dagger [A, P]\right), \tag{5.25}$$

where $P = \sqrt{\gamma/2}\,a$ and the super-operator $\mathcal{K}$ is defined in Eq. (5.14). We then determine the master equation corresponding to these Langevin equations using the method described in Exercise 13. This master equation is

$$\dot{\rho} = \mathcal{L}\rho - ig[(a + a^\dagger)F, \rho] - [\kappa\mathcal{K}(M) + (\gamma/2)\mathcal{K}(a)]\rho$$
$$+ \sqrt{2\kappa\gamma}([a, \rho]M^\dagger - M[a^\dagger, \rho]). \tag{5.26}$$

We now need to determine the steady state to which the auxiliary relaxes given the state of the primary. So long as this relaxation rate is fast compared to the motion of the primary, we can then replace the state of the auxiliary at all times by this steady-state. This technique is called *adiabatic elimination* because it eliminates the dynamics of the auxiliary. To proceed we write the equations of motion for each of the submatrices of the joint density matrix that correspond to the subspaces in which the auxiliary system is in one of its number-states. We denote these submatrices by $\rho_{jl} \equiv \langle j|\rho|l\rangle$, where $|j\rangle$ is the number state of the auxiliary with j excitations. Since the auxiliary is strongly damped we assume that it will be close to its ground state, $|0\rangle$, and so restrict ourselves to $l, j = 0, 1$. Note that the density matrix of the primary system, σ, is given by summing over the number states of the auxiliary; thus $\sigma = \rho_{00} + \rho_{11}$ if we ignore the number states with more than one excitation. The equations of motion for these submatrices are

$$\dot{\rho}_{00} = \{\mathcal{L}_s - \kappa \mathcal{K}(M)\}\rho_{00} + \gamma\rho_{11} + \sqrt{2\kappa\gamma}[M\rho_{01} + \text{H.c.}] + ig[F\rho_{10} - \text{H.c.}],$$
$$\dot{\rho}_{11} = \{\mathcal{L}_s - \kappa \mathcal{K}(M)\}\rho_{11} - \gamma\rho_{11} - \sqrt{2\kappa\gamma}[M\rho_{01} + \text{H.c.}] - ig[\rho_{10}F - \text{H.c.}],$$
$$\dot{\rho}_{01} = \{\mathcal{L}_s - \kappa \mathcal{K}(M)\}\rho_{01} - \frac{\gamma}{2}\rho_{01} + (\rho_{11} - \rho_{00})\left(\sqrt{2\kappa\gamma}M - igF\right), \tag{5.27}$$

where H.c. stands for the Hermitian conjugate of the preceding term. The equation of motion for the density matrix of the primary is then

$$\dot{\sigma} = \dot{\rho}_{00} + \dot{\rho}_{11} = \{\mathcal{L}_s - \kappa \mathcal{K}(M)\}\sigma + ig[F, \rho_{01} + \text{H.c.}]. \tag{5.28}$$

Now we observe from Eq. (5.27) that ρ_{01} is strongly damped at the rate $\gamma/2$. Setting $\dot{\rho}_{01}$ to zero, and dropping $[\mathcal{L}_s - \kappa \mathcal{K}(M)]\rho_{01}$ as insignificant compared to $\gamma\rho_{01}$, we find that the steady state is

$$\rho_{01}^{ss} = 2(\rho_{00} - \rho_{11})\left[\left(\frac{g}{\gamma}\right)iF - \sqrt{\frac{2\kappa}{\gamma}}M\right]. \tag{5.29}$$

Since we are assuming that γ is larger than the other dynamical timescales in the system, and assuming that F and M are on the order of unity, the parameters g/γ and $\sqrt{\kappa/\gamma}$ are small. We now note that ρ_{11} is also damped at the rate γ, and is therefore also smaller than ρ_{00} by a factor of the small parameters. We can therefore drop ρ_{11} in comparison to ρ_{00}, which means that $\sigma \approx \rho_{00}$ and

$$\rho_{01}^{ss} \approx 2\sigma\left[\left(\frac{g}{\gamma}\right)iF - \sqrt{\frac{2\kappa}{\gamma}}M\right]. \tag{5.30}$$

Substituting the expression for ρ_{01}^{ss} in for ρ_{01}, we obtain the master equation

$$\dot{\sigma} = \mathcal{L}_s\sigma - \kappa \mathcal{K}(M)\sigma - \frac{2g^2}{\gamma}[F, [F, \sigma]] - \sqrt{\frac{8g^2\kappa}{\gamma}}[F, M\sigma + \sigma M]. \tag{5.31}$$

If we set $g = \sqrt{\gamma}/(4\hbar\sqrt{2k})$ and $M = x$, then this master equation can be rearranged so that it is identical to that in Eq. (5.12) with $\eta = 1$.

In deriving the above master equation, we have dropped terms that are first-order in the small parameters $\sqrt{\kappa/\gamma}$ and g/γ. So the equation is only a good approximation when γ is very much larger than g and κ.

5.4 Feedback control via continuous measurements

In this section we will examine feedback control that is mediated by continuous measurements. It is not usually possible to obtain analytic solutions, or in fact solutions of any kind, for optimal continuous-measurement-based feedback protocols for nonlinear systems, quantum or classical. However, it is possible to gain insights into control with such feedback by examining key dynamical effects that continuous measurements can generate in quantum systems. With this approach in mind, we now consider the dynamical properties of measurements in the context of feedback control. We will discuss optimal control for linear systems in Section 5.5.2, and readers especially interested in linear systems may skip directly to that section. In discussing examples of continuous-time feedback the Bloch-sphere representation of a single qubit will be very useful. The definition of the Bloch sphere is given in Appendix D.

5.4.1 Rapid purification protocols

Recall that the essential function of measurement in a feedback loop is to reduce the entropy of the system. We will find in this section that the rate at which a measurement reduces the entropy can be altered by adjusting the eigenbasis of the observable that we measure (the "measurement basis"), and this can guide us in designing feedback strategies. In fact, we will find that in order to maximize the rate of entropy reduction, we will need to use feedback.

To begin, let us review what we know of the dynamics of a continuous measurement from Chapter 3. If we continuously measure the observable σ_z for a single qubit initially in a mixed state, then on average the purity of the state increases with time, and the state randomly approaches one of the two eigenstates of σ_z. The equation of motion for the qubit under this measurement is

$$d\rho = -k[\sigma_z, [\sigma_z, \rho]]\, dt + \sqrt{2k}(\sigma_z\rho + \rho\sigma_z - 2\langle\sigma_z\rangle\rho)\, dW$$
$$= -[\tilde{\sigma}_z, [\tilde{\sigma}_z, \rho]]\, dt + \sqrt{2}(\tilde{\sigma}_z\rho + \rho\tilde{\sigma}_z - 2\langle\tilde{\sigma}_z\rangle\rho)\, dW. \tag{5.32}$$

Here k is called the measurement strength, and it scales the average rate at which the purity increases with time. On the second line we have absorbed k into the measured observable by writing $\tilde{\sigma}_z = \sqrt{k}\sigma_z$. In doing so we see that k merely changes the eigenvalues of the measured operator. Thus the rate at which the purity increases, and equivalently the rate at which the measurement distinguishes between the two eigenstates, is determined by the difference between the eigenvalues of the two states. Since it is k that scales the purification rate, this rate is proportional to the square of the separation between the eigenvalues.

Consider measuring an observable of an N-dimensional system, whose eigenvalues are equally spaced. An example is the z-component of angular momentum, J_z, whose eigenvalues are $-j, -j+1, \ldots, j-1, j$, where $j = (N-1)/2$. If we start the system in a mixture of the eigenstates $|j\rangle$ and $|j-1\rangle$, then the difference between the eigenvalues of these states is

1. If instead we start in an equal mixture of $|j\rangle$ and $|-j\rangle$, this difference is $2j$. In the second case the measurement will therefore distinguish the two initial states $4j^2$ times as fast as in the first case.

The above observation suggests a way to increase the speed at which a measurement extracts information about a system. To see this, consider making a continuous measurement of J_z on an N-dimensional spin system, initially in the completely mixed state. As time proceeds, our state-of-knowledge – the probability density over the z-eigenstates – narrows about a single eigenstate. This means that the states with the largest probabilities have eigenvalues that are all close together. We can now increase the speed at which the system collapses to a single eigenstate by rearranging the eigenvalues of the measurement operator, so that the states with the highest probability have the eigenvalues that are furthest apart. Recall that instead of transforming the measurement operator, we can instead apply a unitary operator to the system. In this case we rearrange the eigenstates of the system so that those with the highest probabilities have the most well-separated eigenvalues. In fact, we can continually rearrange the eigenvalues as the measurement proceeds, so as to obtain the greatest increase in purity in each time-step. Such a "rapid-measurement" procedure is a continuous feedback process, but instead of being designed to control the system, it is designed to modify the measurement process. This kind of feedback control is called an *adaptive measurement*. We will see more adaptive measurements in this chapter, and in Chapter 6.

Since the above adaptive measurement involves only permuting the elements of a single basis, it can be performed on a classical system. Since we are measuring a single observable, the measurement is semiclassical (see Chapter 1), and thus the whole adaptive measurement works just as well on a classical system. Note that since we have a record of all the operations we have performed on the system during the measurement, once the measurement has projected the system onto a pure state, we can work backward and determine the initial state the system was in. Thus the adaptive measurement not only increases the rate at which the system is purified, but also the rate at which we obtain full information about the initial state. The maximum increase in speed that can be achieved, in the limit in which one requires a highly pure final state, is simply given by the ratio of the minimum difference between two eigenvalues, to the maximum possible difference between two eigenvalues. For an N-dimensional spin system, this maximum increase is $2j = N - 1$. This adaptive measurement is analyzed in detail in [131].

Rapid purification for a single qubit

We have seen that changing the measurement operator as a measurement proceeds can increase the rate at which information is extracted, but this effect is purely classical. There is, however, a much less intuitive, and purely quantum, effect that will allow us to increase the purification rate of a measurement even further. It turns out that the average amount by which a measurement purifies a system depends on the relationship between the eigenbasis of the measured observable, and that of the density matrix. To exploit this effect we must do more than merely permute the elements of a single basis, but rather continually change to a new basis as the measurement proceeds. To understand how it works, it is simplest to look at a measurement of a two-state system. We can visualize the whole process graphically if we use the "Bloch-sphere" representation, described in Appendix D.

Consider a single qubit whose density matrix, ρ, is diagonal in the z-basis, and whose Bloch vector, $\mathbf{a} = (a_x, a_y, a_z)$, points upward. Now let us make a continuous measurement of an observable whose eigenvectors lie in the xz-plane. In this case the measured observable is given by $M = \sin(\theta)\sigma_x + \cos(\theta)\sigma_z$, and the angle on the Bloch sphere between the measurement basis and the z-basis is θ. Substituting M into Eq. (3.39), writing the density matrix in terms of the Bloch vector, and setting $a_x = a_y = 0$, we obtain the infinitesimal change in the Bloch vector:

$$da_x = 4k\sin\theta\cos\theta\, a_z\, dt + \sqrt{8k}\sin\theta\, dW, \qquad (5.33)$$

$$da_y = 0, \qquad (5.34)$$

$$da_z = -4k\sin^2\theta\, a_z\, dt + \sqrt{8k}\cos\theta(1 - a_z^2)\, dW. \qquad (5.35)$$

To examine the rate at which we become more certain of the quantum state, we must use a measure of uncertainty. Since the form of the von Neumann entropy is algebraically complex, we will use instead the *impurity*, defined in Section 2.2.2. In terms of the Bloch vector elements, the impurity is $L = (1 - |\mathbf{a}|^2)/2$. Since the impurity is a nonlinear function of these elements, we must use the Ito-calculus relation, $dW^2 = dt$, to obtain its equation of motion. This is

$$dL = -8kL\left[1 - (1 - L)\cos^2\theta\right]dt + \sqrt{8k}\cos\theta\, L\sqrt{1 - 2L}\, dW. \qquad (5.36)$$

We note that to obtain this equation from Eqs. (5.33)–(5.35) we replaced a_z with $|a_z| = \sqrt{1 - 2L}$. This is fine because the probability distribution for dW is unchanged by the replacement $dW \to -dW$. The change in the impurity, averaged over all the measurement results, is then

$$\langle dL \rangle = -8k\left[1 - (1 - L)\cos^2\theta\right]L\, dt. \qquad (5.37)$$

So the question is, what value of θ gives the maximum average decrease in the impurity? Since $L \le 1/2$ it is clear by inspection that $|dL|$ is maximized when $\cos\theta = 0$ and thus $\theta = \pi/2$. Thus the maximum decrease in average impurity is obtained when the measurement direction $\mathbf{v}$ is *orthogonal* to the direction of the density matrix, $\mathbf{a}$. Another way to think of this is that $|\langle dL \rangle|$ is maximal when the basis of the measurement is maximally different from the eigenbasis of the density matrix. We will make precise the notion of "maximally different" below.

Recall now that in deriving the equation for $\langle dL \rangle$ above, we set $a_x = a_y = 0$, so that the eigenbasis of the density matrix is the σ_z basis. After a single time-step, dt, this will no longer be true if we measure in the σ_x basis. We can see how the eigenbasis of the density matrix changes under the σ_x measurement by setting $\theta = \pi/2$ in Eqs. (5.33)–(5.35).

We have

$$da_x = \sqrt{8k}\,dW, \tag{5.38}$$

$$da_y = 0, \tag{5.39}$$

$$da_z = -4ka_z\,dt. \tag{5.40}$$

This shows us that the direction of **a**, and thus the eigenbasis of the density matrix, changes in a random way. To get the maximum decrease in $\langle dL \rangle$ in the next time-step, we must therefore adjust the basis of the measured observable so that, in the Bloch-sphere representation, the basis states are orthogonal to the new density matrix. Equivalently, we can apply a unitary transformation to the system, to rotate the Bloch vector so that it once again points upward, and $a_x = a_y = 0$. This means applying a Hamiltonian to the system for an infinitesimal time-step dt, and since the direction of the rotation depends upon dW, the Hamiltonian is a *feedback* Hamiltonian.

To rotate the Bloch vector so that it points upward, we must rotate it in the xz-plane, and this means rotating it about the y-axis. We will describe the motion generated by an arbitrary 2D Hamiltonian in the next section. For now we note that the Hamiltonian

$$H(\mu) = \hbar\left(\frac{\mu}{2}\right)\sigma_y, \tag{5.41}$$

rotates the Bloch vector around the y-axis, at a constant angular velocity, in the clockwise direction as viewed in Fig. D.1. If this Hamiltonian acts for a time t, then the Bloch vector is rotated through an angle μt.

If we apply the above Hamiltonian for a time dt, it generates the following change in a_x and a_z:

$$da_x = \mu a_z\,dt, \tag{5.42}$$

$$da_z = -\mu a_x\,dt. \tag{5.43}$$

We need to choose the angle μ so that this rotation cancels the change in a_x that was generated by the measurement, returning a_x to zero. To cancel the change da_x given by Eq. (5.38), to first order in dt, we must therefore set

$$\mu(t) = -\left(\frac{\sqrt{8k}}{a_z(t)}\right)\frac{dW}{dt}. \tag{5.44}$$

This expression for μ is actually somewhat pathological. The Wiener process has infinitely rapid fluctuations, and its derivative, dW/dt, is therefore infinite. In practice we cannot make μ infinite, but if μ is large enough, the feedback Hamiltonian will be able to keep the Bloch vector pointing approximately north. It is also worth noting that in reality the measurement noise is not a true Wiener process, since it will a have a finite bandwidth – the

Wiener process is an idealization. As a result, μ need not be infinite to perfectly maintain $a_x = 0$, although it may need to be very large.

If we assume that the Hamiltonian feedback does maintain $a_x = 0$ during the measurement, which is the ideal case, then the evolution of the purity of the density matrix is very simple. In each time-step, a_z evolves under the measurement via $da_z = -4ka_z dt$, and then the Hamiltonian rotates the Bloch vector so that the new a_z is made equal to the new length of the Bloch vector. Thus at the end of each time-step, a_z is equal to the Bloch vector length, and so the evolution of a_z is in fact the evolution of this length. Now the length of the Bloch vector is $a \equiv |\mathbf{a}| = \sqrt{a_x^2 + a_z^2}$, and at the start of each time-step $a_x = 0$. So if the length at the start of a time-step is $a = a_z$, the length at the end of the time-step is

$$a + da = \sqrt{(da_x)^2 + (a_z + da_z)^2} = a_z \left[1 + 4k \left(\frac{1 - a_z^2}{a_z^2} \right) dt \right], \qquad (5.45)$$

and hence

$$da_z = da = 4k \left(\frac{1 - a_z^2}{a_z} \right) dt. \qquad (5.46)$$

The equation for da_z is now completely deterministic, so we can write it as

$$\dot{a}_z = 4k \left(\frac{1 - a_z^2}{a_z} \right). \qquad (5.47)$$

This equation looks rather complex, but if we now calculate the equation of motion for the impurity, $L = (1 - a_z^2)/2$, we obtain

$$dL = -a_z \dot{a}_z = -8kL. \qquad (5.48)$$

The impurity therefore decays exponentially:

$$L_{fb}(t) = e^{-8kt} L(0), \qquad (5.49)$$

where the subscript "fb" indicates that this is the evolution of the impurity under the feedback protocol.

Let us now compare the impurity obtained using the above feedback control protocol, with the average purity that results when we make a bare measurement. For simplicity, consider a qubit that is initially in the completely mixed state, and is continuously measured in the z-basis. The density matrix remains diagonal in the z-basis throughout the measurement, and the resulting stochastic evolution is solved in Section 3.2. The density matrix at time t is given by

$$\rho(t) = \frac{\exp\left[\sqrt{8kZ(t)} \sigma_z \right]}{\exp\left[\sqrt{8kZ(t)} \right] + \exp\left[-\sqrt{8kZ(t)} \right]}, \qquad (5.50)$$

where the probability density for $Z(t)$ is

$$P(Z,t) = \frac{1}{\sqrt{4\pi t}} \left(\exp\left[(Z + \sqrt{8k}t)^2 /(2t) \right] + \exp\left[(Z - \sqrt{8k}t)^2 /(2t) \right] \right). \tag{5.51}$$

Calculating the impurity of $\rho(t)$, and averaging over Z, we get

$$\langle L(t) \rangle = \frac{e^{-4kt}}{\sqrt{8\pi}t} \int_{-\infty}^{\infty} \frac{\exp\left[-z^2 /(2t) \right]}{\cosh\left(\sqrt{8k}z \right)} dz. \tag{5.52}$$

This integral does not have an analytic solution, but we can obtain an approximate solution when $t \gg 1/(8k)$, which is also the limit in which $\langle L(t) \rangle \ll 1$. To do so we note that the integral in Eq. (5.52) is independent of t when $t \gg 1/(8k)$. In this case, the impurity is approximately

$$\langle L(t) \rangle = \frac{e^{-4kt}}{\sqrt{8\pi}tC}, \tag{5.53}$$

where C is a constant.

Let us denote the time taken for the bare measurement to achieve a given average impurity by t_{bare}, and that taken by the feedback protocol to achieve the same impurity as t_{fb}. We can determine the ratio $r \equiv t_{\text{bare}}/t_{\text{fb}}$ by setting $L_{\text{fb}}(t_{\text{fb}}) = \langle L(t_{\text{bare}}) \rangle$, and the result is

$$\frac{1}{r} = \frac{t_{\text{fb}}}{t_{\text{bare}}} = \frac{1}{2} + \frac{\ln t_{\text{bare}}}{8k t_{\text{bare}}} + \frac{\ln(8\pi C)}{8k t_{\text{bare}}}. \tag{5.54}$$

From this it is clear that the "speed-up" factor afforded by the feedback protocol, r, tends to 2 as $t \to \infty$.

A subtlety in rapid purification

The "rapid-purification" protocol described above maximizes the average purity of the measured system at any given time, t. There is a converse quantity that one may wish to minimize, and that is the *average time* taken for the measurement to achieve a given level of purity. At first one might assume that for a given feedback protocol, optimizing one would be more or less equivalent to optimizing the other, but this is not the case at all. In fact, the joint distribution of measurement times and purities in a continuous measurement is such that the feedback protocol above, while maximizing the average purity at a given time, actually *lengthens* the average time taken to reach a given level of purity! This fact, shown by Wiseman and Ralph [684], means that if we fix a target level of purity, in the majority of cases the qubit under the bare measurement will cross this level of purity *before* the feedback protocol reaches this level. However, the stragglers have a sufficiently low purity that the purity averaged over all the cases will be lower at any given time than that achieved by the feedback. If one wants to prepare a single qubit in a pure state in the fastest time, then the bare measurement would be best. Conversely, if one wishes to prepare *every* qubit in an ensemble with a certain level of purity, then the feedback protocol derived above would be the better option.

Rapid purification for N-dimensional systems

We saw above that if we measure a single qubit in a basis that is at right angles to the eigenbasis of the density matrix, then we obtain the maxim average increase in the purity of the state. So how do we maximize this same quantity if we are measuring a higher-dimensional system? It turns out that this question does not have a simple answer, although various results are known.

For a single qubit, two bases that are at $90°$ are, in a sense, maximally different from one another. Further, the inner product between any element of one basis and any element of the other basis is $1/\sqrt{2}$. In this sense, the bases are "unbiased" with respect to each other, since each vector in one basis shows no "bias" in its overlap with the vectors of the other. We can immediately extend this definition of what it means to be "unbiased" to higher-dimensional systems. When two bases in an N-dimensional vector space are unbiased with respect to each other, then the overlap of every vector in one basis with every vector in the other is $1/\sqrt{N}$ [280, 689]. It is further known that if N is an integer power of a prime, then it is possible to find $N + 1$ bases that are all mutually unbiased with respect to one another [689]. (Numerical studies indicate that if N is not a power of a prime, then the maximum number of mutually unbiased bases is less than $N + 1$ [503].)

Combes and Wiseman have shown that if one measures an N-dimensional system in a basis that is unbiased with respect to the eigenbasis of the density matrix, then in the long-time limit, the speed-up that can be achieved in obtaining a target average purity is proportional to N^2. The protocol that realizes this speed-up requires using the right unbiased bases, the details of which are given in [132].

5.4.2 Control via measurement back-action

The rapid-projection protocol discussed above shows us that the amount by which a measurement purifies a system depends on the measurement basis, and if we align this basis to maximize the purification, we also maximize the noise from the measurement. This reveals some of the complexity of measurement-based feedback control, and in particular the likely difficulty of finding optimal feedback protocols. If the Hamiltonian that performs the feedback is fast enough, then the noise introduced by the measurement is unlikely to matter, since it can be countered by the Hamiltonian. But in general there will be a non-trivial trade-off between the rate of entropy reduction and the measurement-induced noise.

As we will see now, the problem of optimal feedback control in quantum systems is even more complex, because the dynamics induced by the measurement (the measurement "back-action") can be used itself to control the system. That is, the measurement does not only introduce noise into the system but can also generate an effective deterministic motion. There is therefore a complex interplay between the information extracted by the measurement, both the noise and deterministic motion that it generates, and the Hamiltonian feedback.

To understand how a continuous measurement can generate deterministic motion, we need to examine the effect of a *diffusion gradient*. By this we mean that the strength of the noise in a system varies in some way with the state of the system. As an example, consider a particle that moves on a circle, and whose position on the circle is denoted by θ. Let us say that the motion of the particle is driven only by noise, so that it is "purely diffusive." In this case the motion is given by

$$d\theta = \sqrt{f(\theta)}\,dW. \tag{5.55}$$

where f is an arbitrary function of θ. If f is independent of θ, then the diffusion is referred to as being constant. If $df/d\theta$ is non-zero, then the motion has diffusion gradients. To determine the effect of a diffusion gradient, rather than use the SDE, it is much simpler to use an equation that describes the evolution of the probability density for θ. Such an equation is called a *Fokker–Planck* equation, and it can be derived from the SDE. We will not show how to derive it here, but the reader can find this, as well as more details regarding Fokker–Planck equations in [288, 197].

The Fokker–Planck equation for θ is

$$\frac{\partial P}{\partial t} = \frac{\partial^2}{\partial \theta^2} \left[f(\theta) P \right]. \tag{5.56}$$

A useful fact, which is not difficult to derive, is that if a probability density, $Q(x)$, satisfies the differential equation

$$\frac{\partial Q}{\partial t} = -\frac{\partial}{\partial x} J(x), \tag{5.57}$$

then $J(x)$ gives the *probability current*. That is, $J(x)$ gives the net rate at which probability is flowing across the point x. For example, if the diffusion is constant, then by comparing Eqs. (5.57) and (5.56), we see that the probability current is

$$J(\theta) = -f \frac{\partial P}{\partial \theta}. \tag{5.58}$$

This tells us that probability will flow in the direction in which the probability density is decreasing, and thus from regions of high probability to low probability. This process evens out the probability density until it is flat, being the basic effect of diffusion.

So what happens if there is a diffusion gradient? Once again comparing Eqs. (5.57) and (5.56), this time allowing f to be an arbitrary function of θ, we have

$$J(\theta) = -\frac{\partial}{\partial \theta} \left[f(\theta) P \right] = -\frac{\partial f}{\partial \theta} P - f(\theta) \frac{\partial P}{\partial \theta}. \tag{5.59}$$

This time there is the simple diffusion term as before, proportional to $f(\theta)$, but there is an *additional* term. This new term is proportional to P, and this tells us that it describes deterministic motion of the particle. (To see this, note that if there is something at point θ that makes the particle move at a constant speed $d\theta/dt = v$, then the flow of probability at θ will be given by $vP(\theta)$. This is because the particle only moves to point $\theta + d\theta$ if it is actually at point θ.) The new term is also proportional to the *gradient* of the diffusion strength, $f(\theta)$. This tells us that there will be a net flow of probability in the direction in which the diffusion strength is decreasing; that is, from locations at which the diffusion is large, to locations at which it is small. This is the effectively deterministic motion generated by a diffusion gradient. The ultimate result of this effect is that, if the state space is finite, the steady-state probability density will be peaked around regions of low diffusion. (If the state space is infinite, then the diffusion will ultimately distribute the probability everywhere, and so the particle is not confined to regions of low diffusion.)

The above analysis suggests that diffusion gradients can be used to control a system. So long as the state space is finite, one can confine a system close to some state by making it the region of lowest diffusion. Further, we already know that continuous measurements generate diffusion, and this diffusion is proportional to the strength of the measurement (among other things). We can therefore generate an effective diffusion gradient by changing the measurement strength depending on where the system is in state space. We can use this to confine the system to a given state, by turning off the measurement only when the system is close to that state.

A "back-action" control protocol for a single qubit

It is simple to construct a feedback protocol to control a single qubit using a diffusion gradient. To explain how this works, we refer to the diagram on the left in Fig. 5.3. Let us say that we wish to stabilize the qubit in the state $|0\rangle$, at the bottom of the Bloch sphere. The black arrow at angle θ to the vertical axis in the figure represents the Bloch vector of the state of the system at the current time, t. To move this state to $|0\rangle$ we want to generate diffusion along the shortest line that connects it to $|0\rangle$, being a longitudinal great circle. We induce diffusion along this great circle by making a measurement in a basis that lies along a direction orthogonal to the Bloch vector. While the resulting diffusion kicks the state in both directions, depicted in Fig. 5.3 by the curved dashed line, we can create a gradient in the diffusion by increasing the strength of the measurement with θ. This gradient "pushes" the state toward $\theta = 0$, and thus the south pole.

If we want to pin the state to $|0\rangle$, then we should turn off the diffusion, and thus the measurement, at $\theta = 0$. This is fine unless there are other sources of noise and decoherence that mix the state. In this case, if we turn off the measurement, the system is left unprotected against this noise, so that the Bloch vector will decrease in length and control is lost. We can prevent this by making another, simultaneous measurement whose basis is aligned with the Bloch vector. This causes no diffusion, but nevertheless continues to purify the state even when the orthogonal measurement is switched off. There is no need to vary the strength of this measurement, so we fix it at some maximum value μ.

For the first measurement, we will choose the measurement strength to be $k = \alpha\theta^2$ for some fixed α. If we include dephasing noise at rate β (dephasing noise is defined below

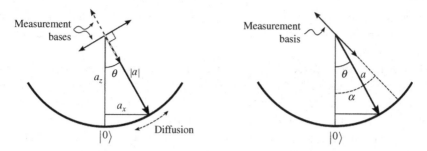

Figure 5.3 The diagram on the left-hand side depicts the feedback protocol described in Section 5.4.2, and that on the right the protocol in Section 5.4.3.

in Eq. (5.68)), then the evolution of the Bloch vector under the above feedback protocol is given (see Exercise 16) by

$$d\theta = \sqrt{8\alpha}\left(\frac{\theta}{|a|}\right)dW,\tag{5.60}$$

$$da = (1-a^2)\left(\left[\frac{4\alpha\theta^2}{a}\right]dt + \sqrt{8\mu}\,dV\right) - 8\beta a\,dt,\tag{5.61}$$

where $a(t)$ is the component of the Bloch vector *along the direction* of the Bloch vector. Thus the length of the Bloch vector is $|a|$, and $a \in [-1,1]$. (Note that we need only two variables to describe the motion, since the state stays on a single longitudinal great circle.) The angle θ is cyclic, and the natural interval to use for θ is $[-\pi,\pi]$. A subtlety here is that each qubit state on the great circle is actually labeled twice by the variables θ and a, even when we reduce θ to the interval $[-\pi,\pi]$. This is because the transformation $a \to -a$ is the same as $\theta \to \theta + \pi$.

If there is no decoherence ($\beta = 0$), then an initially pure state stays pure ($a = 1$), the second measurement has no effect, and the equation of motion for θ simplifies to

$$d\theta = \sqrt{8\alpha}\theta\,dW.\tag{5.62}$$

In this simple case, since the diffusion drops to zero at $\theta = 0$, the steady-state solution for the probability density over θ, $P(\theta)$, is a δ-function centered at $\theta = 0$. Another way to say this is that there is no steady-state solution: as $t \to \infty$, any initial probability density continues to narrow without bound about $\theta = 0$. Solving Eq. (5.62) numerically, one finds that the second moment of θ decays – asymptotically – as an exponential:

$$\langle \theta^2(t)\rangle \propto e^{-\eta kt}, \quad \text{for } t \gg k,\tag{5.63}$$

where $\eta \approx 1.23$. This shows that the control protocol, in controlling θ, acts on the timescale of the measurement strength, k. Recall that, loosely speaking, the measurement strength gives the timescale on which the variance of the observed quantity is reduced, and this reduction is also exponential. This tells us that measurements will allow one to control quantum systems *on the same timescale* as that on which they provide information about the value they measure.

When there is decoherence, the steady state of both θ and a is non-zero, and in this case an important question is the purity that the protocol can achieve in the steady state. In the regime of good control, in which the steady-state value of a is close to unity, it is not difficult to obtain an approximate answer to this question. When a deviates only a little from unity, the equation of motion for θ, Eq. (5.60), is close to Eq. (5.62), and so we can assume that the diffusion gradient is effective at confining θ to zero in the steady state. Substituting $\theta = 0$ into the equation of motion for a gives us a stochastic equation for a single variable, and the steady-state probability distribution for a can then be obtained using the equivalent Fokker–Planck equation (this method is described in [288, 197]). The approximate steady-state probability distribution for a is

$$P_{ss}(a) = \frac{\exp\left[-\frac{\beta}{\mu(1-a^2)}\right]}{\mathcal{N}(1-a^2)^2}, \tag{5.64}$$

where $\mathcal{N}$ is a normalization constant, to be chosen so that $\int_{-1}^{1} P_{ss}(a)\,da = 1$.

5.4.3 Near-optimal feedback control for a single qubit?

The analysis in the previous subsections has shown us that measurement has a non-trivial effect on the dynamics of a quantum system, and that this effect is likely to be important in determining optimal feedback protocols: by changing the basis of the measurement we can change the rate of purification, as well as the diffusion in state space which influences the deterministic motion. And both of these aspects of the motion are relevant to the problem of control. We now consider controlling a single qubit via a single continuous measurement, in which we are free to vary the measurement basis, the measurement strength, and the Hamiltonian of the qubit. The analysis in this section is drawn from [27].

All Hamiltonians for a single qubit can be written as

$$H = \hbar(\mu/2)\mathbf{n} \cdot \boldsymbol{\sigma}, \tag{5.65}$$

where $\boldsymbol{\sigma} \equiv (\sigma_x, \sigma_y, \sigma_z)$ and $\mathbf{n}$ is a three-dimensional unit vector. The axis about which the Hamiltonian rotates the qubit is given by $\mathbf{n}$, and the angular speed of the rotation is given by μ. We can also write the equation of motion given by $H_{\mathbf{n}}$ in terms of the elements of the Bloch vector, by using the handy relation

$$(\mathbf{m} \cdot \boldsymbol{\sigma})(\mathbf{n} \cdot \boldsymbol{\sigma}) = (\mathbf{m} \cdot \mathbf{n})I + i\boldsymbol{\sigma} \cdot (\mathbf{m} \times \mathbf{n}). \tag{5.66}$$

This equation of motion is

$$\dot{\mathbf{a}} = \mu\mathbf{n} \times \mathbf{a}. \tag{5.67}$$

We will take as the constraints on our control protocol that (i) the measurement strength, k is bounded, so that $k \leq k_{\mathrm{max}}$, and (ii) the angular speed μ is bounded, so that $\mu \leq \omega$. We can alternatively write this second constraint as $\sqrt{\mathrm{Tr}[H^2]} \leq \hbar\omega/\sqrt{2}$.

The first thing to note about this control problem is that it is very complex. The density matrix of the system is defined by three real parameters, as are the measurement and the Hamiltonian. This means that a feedback protocol is defined by a function with three inputs (the density matrix) and six outputs (those that define the choice of measurement and Hamiltonian). This presents a daunting task, even for numerical optimization via the Hamilton–Jacobi–Bellman (HJB) equation (see Section 5.5.1).

It turns out that most common noise sources for single qubits are symmetric about the z-axis (where the eigenbasis of σ_z is defined as the energy basis). If we restrict ourselves to this kind of noise then we can reduce the complexity of the control problem somewhat. Shortly we will discuss how we can simplify the problem further, at least approximately. But first we list a number of common noise models for qubits.

Common types of noise

There are a number of noise models that are good approximations to real noise in a variety of two-level systems. These go by the names of *dephasing*, *damping* (also known as *decay*), *thermal noise*, and *depolarizing noise*. The first of these destroys coherence between the σ_z eigenstates, turning an initially pure state into a mixture of the σ_z eigenstates. As a result it maintains the value of a_z, but reduces the values of a_x and a_y over time. The master equation describing dephasing is

$$\dot{\rho} = -(\gamma/2)[\sigma_z, [\sigma_z, \rho]] = -\gamma(\rho - \sigma_z \rho \sigma_z) \quad \text{(dephasing).} \tag{5.68}$$

Note that this is the master equation for a continuous measurement of σ_z, in which the measurement record is thrown away. We can rewrite this master equation as an equation of motion for the elements of the Bloch sphere vector **a**. In this representation the dephasing master equation is simply

$$\dot{a}_x = -2\gamma a_x \tag{5.69}$$

$$\dot{a}_y = -2\gamma a_y. \tag{5.70}$$

The master equation for thermalization is

$$\dot{\rho} = \gamma[1+\tilde{n}]\mathcal{D}(\sigma_-)\rho + \gamma\tilde{n}\mathcal{D}(\sigma_+)\rho \quad \text{(thermal noise),} \tag{5.71}$$

where the super-operator $\mathcal{D}(\cdot)$ is defined in Eq. (5.11). The rate-constant γ is referred to as the damping rate. The dimensionless constant $\tilde{n}$ depends upon the temperature, and is given by

$$\tilde{n} = \frac{1}{\exp\left[\hbar\Delta/(k_{\mathrm{B}}T)\right] - 1}, \tag{5.72}$$

where $\hbar\Delta$ is the energy difference between the ground and excited states of the qubit, T is the temperature of the bath, and k_B is Boltzmann's constant. The population of the excited state when the qubit is at the temperature of the bath (that is, the steady state of the above master equation), is given by

$$P_T = \frac{\tilde{n}}{1+2\tilde{n}} = \frac{1}{\exp\left[\hbar\Delta/(k_B T)\right]+1}. \tag{5.73}$$

The equivalent equations of motion for the Bloch vector are

$$\dot{a}_x = -(\gamma/2)a_x, \tag{5.74}$$

$$\dot{a}_y = -(\gamma/2)a_y, \tag{5.75}$$

$$\dot{a}_z = -\gamma(a_z+1-2\tilde{n}). \tag{5.76}$$

Damping is merely thermal noise with the temperature set to zero ($\tilde{n}=0$).

Now finally for depolarizing noise. This is just dephasing, but with equal amounts in the x, y, and z directions. Depolarizing noise therefore reduces all the elements of the Bloch vector at the same rate. The master equation for this case is

$$\dot{\rho} = -(\gamma/4)\left([\sigma_x,[\sigma_x,\rho]]+[\sigma_y,[\sigma_y,\rho]]+[\sigma_z,[\sigma_z,\rho]]\right) \quad \text{(depolarizing)}. \tag{5.77}$$

The equivalent equations of motion for the Bloch vector are

$$\dot{a}_j = -2\gamma a_j, \quad \text{for } j=x,y,z. \tag{5.78}$$

Reducing the complexity of the control problem

All the noise models discussed above are symmetric under rotations of the qubit about the z-axis. Since we allow our control protocol to use any Hamiltonian and to measure along any axis, if our objective is also invariant under z-rotations then the performance of our protocol will be also. In this case the angle ϕ on the Bloch sphere (defined in Appendix D) is irrelevant, and the state of the system can be parametrized by the length of the Bloch vector, a, and an angle $\tilde{\theta}$ between the Bloch vector and the state $|1\rangle$. The objective we choose here is that of maximizing the probability that the system is in the ground state, $|0\rangle$, and this has the above symmetry. We thus wish to maximize the steady-state value of $P = \langle 0|\rho|0\rangle$, where ρ is the state of the system. We will also define the angle $\theta = \pi - \tilde{\theta}$, which is the angle between the Bloch vector and the state $|0\rangle$. With this definition $P = (1+a\cos\theta)/2$.

We now present a number of arguments, each of which suggests that we can eliminate one or more parameters from the control problem, while exactly or approximately preserving the optimality of the protocol. Eliminating these parameters greatly simplifies the control problem.

The first argument is that, since we can measure in any basis and apply any Hamiltonian, it is reasonable to expect that we are always in a better position for the purposes of future control when the Bloch vector is closer to the state $|0\rangle$, given that its length is fixed. While intuitively obvious, this statement has not been rigorously proved, to the author's knowledge. Nevertheless, this suggests that we should choose the Hamiltonian at each time-step to rotate the state toward the target at the maximum possible speed, since this achieves the closest state to the target at the end of the time-step, over all choices of the Hamiltonian. This means setting $\mu = \omega$, and choosing the axis of rotation so that the Bloch vector travels on a great circle. Taking this choice for the Hamiltonian simplifies the control problem, which is now reduced to determining the parameters that describe the measurement.

The second argument is that, since we can measure in any basis, it is always best to measure at the maximum strength. The reason is that we can always choose to measure in the eigenbasis of the density matrix, a measurement that is effectively classical. As such it reduces the entropy of the system without introducing any quantum back-action, and as such does not change the direction of the Bloch vector. Since our goal is to achieve a state that is as pure as possible, this choice of measurement produces a benefit without any detriment to the future control. This argument suggests that we can set $k = k_{\text{max}}$ and still maintain optimality, leaving us only with the basis of the measurement undecided.

The third argument concerns the basis of the measurement. We note that the Hamiltonian we have chosen above preserves the value of ϕ. That is, if the Bloch vector lies in the xz-plane, then the Hamiltonian does not take it out of this plane, since it always rotates the Bloch vector in the direction of the south pole. We now ask whether it is beneficial to chose a measurement that changes ϕ. Of course, ϕ is irrelevant from the point of view of the performance measure, but a measurement that changes ϕ will have a different effect on θ and a than one that does not. We are most interested in the regime of good control, in which k_{max} and ω are large enough that the feedback protocol can keep $1 - P \ll 1$, and thus $1 - a \ll 1$ and $\theta \ll 1$. For small θ we note that the difference between a measurement that preserves ϕ and one that does not is second-order in θ. Thus in the regime of good control, making a measurement that changes ϕ can only have a minor effect on the performance. Restricting ourselves to measurements that preserve the value of ϕ, since neither the Hamiltonian nor the measurement changes ϕ it is a constant of the motion. Since all values of ϕ are equivalent due to the symmetry of the noise and the measure of performance, we can set $\phi = 0$ and consider motion restricted to the xz-plane.

The measurement basis can now be described by a single angle α; we specify the Bloch vectors of these two states as the vectors that lie in the xz-plane and along the line that is at an angle α to the target state $|0\rangle$ (the south pole). This configuration is depicted on the right in Fig. 5.3. If $\alpha = \theta$ then the measurement is "aligned" with the state, and thus is in the eigenbasis of the density matrix. In this case the measurement causes no diffusion in θ. The evolution of the density matrix is therefore given by

$$d\rho = -(i/\hbar)[H(t), \rho]\,dt + \sqrt{2k_{\text{max}}}(\sigma_\alpha \rho + \rho \sigma_\alpha - 2\langle \sigma_\alpha \rangle \rho)\,dW, \qquad (5.79)$$

where $\rho = (I + a\sin(\theta)\sigma_x - a\cos(\theta)\sigma_z)/2$ and

$$\sigma_\alpha = \sin(\alpha)\sigma_x - \cos(\alpha)\sigma_z. \tag{5.80}$$

The fourth and final argument is that in the regime of good control the measurement basis, defined by α, need not depend upon the length of the Bloch vector, a. This insight comes from examining the equations of motion for a and θ from a measurement at the angle α. The simplest way to derive these equations is to have the Bloch vector point directly upward, so that $a = a_z$ and $a_x = a_y = 0$, and make a measurement at an angle ψ to the z-axis. We write ρ in terms of the Bloch vector, and substitute this in Eq. (5.79) to derive the equations of motion for a_x and a_z. From these we use Ito's rule to obtain the equations of motion for $a = \sqrt{a_x^2 + a_z^2}$ and $\theta = \tan(a_x/a_z)$, which are

$$d\theta = 2k\sin(2\psi)\left(3 - \frac{2}{a^2}\right)dt + \sqrt{8k}\sin(\psi)\left(\frac{1}{a}\right)dW, \tag{5.81}$$

$$da = 4k\sin^2(\psi)\left(a - \frac{2}{a}\right)dt + \sqrt{8k}\cos(\psi)(1 - a^2)dW. \tag{5.82}$$

When a is close to unity, the regime of good control, we can expand these equations as a power series in $\Delta = 1 - a$. Keeping terms up to first order in Δ we have

$$d\theta = 2k\sin(2\psi)(1 - 4\Delta)dt + \sqrt{8k}\sin(\psi)(1 + \Delta)dW, \tag{5.83}$$

$$da = -4k\sin^2(\psi)(1 + 3\Delta)dt + 2\Delta\sqrt{8k}\cos(\psi)dW. \tag{5.84}$$

From these equations we see immediately that the motion of θ is independent of Δ to leading order. The length of the Bloch vector therefore has little effect on the dynamics, and thus the control, of θ in the regime of good control.

Examining the equation of motion for a we see that the deterministic part of this equation (the term multiplying dt) is also independent of a to leading order, but this is not true of the stochastic part (the term multiplying dW). The fact that the stochastic part is proportional to Δ is precisely the diffusion gradient induced by the measurement, and by which the measurement increases the length of the Bloch vector (makes the state more pure). The important fact for our purposes is that as far as this diffusion gradient is concerned, making ψ dependent on Δ has the same action as changing the measurement strength. On physical grounds it is apparent that modulating the measurement strength (that is, reducing it below its maximal value) is not useful as it cannot increase the rate at which the measurement purifies the state. A more quantitative argument is as follows. If we choose ψ as a function of Δ, so that the stochastic term is proportional to a higher power of Δ, then we will reduce the diffusion gradient, thus effectively reducing measurement strength. This argument does not apply for powers of Δ that are less than unity, but numerical simulations show that if we replace Δ with $\sqrt{\Delta}$, the rate of purification is reduced.[1]

---

[1] The author performed these simulations while working on [300], although they do not appear in that article.

With the above simplifications the state ρ is defined by two parameters, a and θ, and the feedback protocol is completely specified by a function $\alpha = f(\theta)$ that tells us how to chose the measurement angle based on the location of the Bloch vector. Finding the optimal $f(\theta)$ is a task that is feasible on a parallel computer. Deferring a discussion of numerical optimization to Section 8.1, we present here the results of applying this technique to $f(\theta)$.

Since we are interested in the regime of good control, we assume that $f(\theta)$ can be expanded in a power series in θ, and truncate this series to first order so that $f = c_0 + c_1\theta$. Since the performance is not affected by an overall scaling of time, we can measure all the rate constants in terms of one of them, and we choose this to be the measurement strength k_{max}. Choosing the noise to be thermal we set the damping rate to be $\gamma = 0.1k$ and set $\tilde{n} = 0.1$. With this choice of noise we expect to be in the regime of good control since $k_{max} \gg \gamma$.

We wish to determine how the optimal choices for c_0 and c_1 depend on the speed of the feedback Hamiltonian, given by ω. Numerical optimization reveals that, regardless of the value of c_1, the optimum value of c_0 is

$$c_0 = \begin{cases} 0 & \omega \lesssim 45k \\ \pi/2 & \omega \gtrsim 45k \end{cases} \quad \text{for } \gamma = 0.1, \tilde{n} = 0.1. \tag{5.85}$$

The dependence of c_1 on θ is more complex. When there is no feedback Hamiltonian ($\omega = 0$), then the optimal value is $c_1 = -0.5$. As the speed of the feedback Hamiltonian increases, the optimal value of c_1 decreases approximately exponentially to stabilize at a value of ≈ -0.7. The rate of decay of the exponential is ≈ 0.5. Of course these values might well change with the noise strength. Above $\omega \approx 30k$ the value of c_1 no longer has a significant effect on the performance, and can be set to zero.

The above results for the optimal values of c_0 and c_1 can be largely understood from our analysis of rapid purification and measurement-induced diffusion above. The fact that the induced diffusion increases with the average rate of purification explains why we should align the measurement with the Bloch vector as it approaches the target when ω is small, and why we should measure at right angles when ω is large. Since the diffusion is detrimental when we are at the target state, the protocol can only make use of the increased purification rate when the feedback is strong enough to counteract the diffusion. That c_1 is non-zero when $c_0 = 0$ is explained by the fact that this creates a diffusion gradient that pushes the Bloch vector toward the target state. What is not explained by these two effects is why c_1 is negative. This is likely due to the quantum Zeno effect (see Exercises 22 and 23) in which the measurement tends to drag the Bloch vector toward the measurement axis. Because of this a negative value of c_1 will drag the Bloch vector in the direction of the target state. The use of the Zeno effect to drag quantum states is explored in [470, 563].

In Fig. 5.4 we plot the performance as a function of c_0 and ω, using the optimal value of c_1 when $c_0 = 0$. With this choice for c_1 the plot shows the optimal performance of our simplified protocol along the lines where the value of c_1 is optimal, and thus the performance of the optimal protocol, which is denoted by the black line.

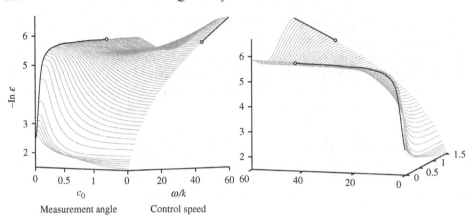

Figure 5.4 Performance of a feedback control protocol for a single qubit that is expected to be approximately optimal in the regime of good control. The performance is measured by the steady-state probability that the system is in the target state, P_{ss}. The error-probability is $\varepsilon = 1 - P_{ss}$, and the quantity plotted here is $-\ln(\varepsilon)$. The optimal protocol is given by the black line, and has a discontinuity at $\omega \approx 45k$, where k is the measurement strength.

5.4.4 Summary

We discussed rapid purification and control via measurement back-action to provide insight into quantum feedback control, and also to make clear its complexity. Because quantum measurement induces noisy dynamics in the measured system, and this dynamics has a non-trivial effect on the average motion as well as the entropy extraction, it is not at all clear in general how to choose the measurement so as to optimize a control objective. To determine an optimal control protocol for an arbitrary problem one would need to solve the Hamilton–Jacobi–Bellman equation (to be introduced below). But even for a single qubit such a task requires tremendous numerical resources. It is therefore worthwhile to investigate a variety of protocols, and compare their performance, so as to come up with "rules of thumb" that work well for certain kinds of problems and in certain regimes. Such rules of thumb would constitute a toolbox for use in quantum engineering.

In the final section of this chapter we present systematic methods for finding optimal control protocols. For special cases, and certain specific regimes for more general systems, it is possible to use these methods to determine truly optimal classical and quantum continuous-time feedback protocols.

5.5 Optimization

A key question in feedback control, or in fact any kind of control, is what is the best control that can be achieved given fixed resources, or conversely how to minimize the resources while achieving a given performance. The term "optimal" or "optimization" refers to performing a maximization or minimization under some constraints. The resources in question may be, for example, the energy used, the size of the feedback forces applied, and/or the timescale upon which the controller must respond. One further approach to optimization is to minimize a weighted sum of the resources used and the badness of the control (see, for example, "LQG control" which is discussed below), but in the author's opinion this tends to be less well motivated as the meaning of the results is less clear.

Maximizing performance for feedback control of a nonlinear system is usually impossible, but there are some cases in which it can be done. This is typically only possible when there is no constraint on the speed at which the control can be varied (so that it can be switched on or off instantaneously). In this case the optimal control can take a simple form that involves the control parameter (e.g., the applied force) switching one or more times between its two extreme values. This is often referred to as "bang-bang" control, after the sound that a mechanical system might make when it switches rapidly. A tool that can be useful in finding optimal control protocols, both for open-loop and feedback control, is the Hamilton–Jacobi-Bellman (HJB) equation. Here we introduce this equation, and describe how it can be used to determine whether or not a given control strategy is optimal, or as a numerical tool to search for optimal strategies. For deterministic dynamics, there is an alternative method that transforms the problem of finding an optimal strategy into that of solving a set of differential equations with constraints. We give a brief introduction to this method, which is called Pontryagin's maximum principle.

5.5.1 *Bellman's equation and the HJB equation*

We feel that it is simplest to consider first a dynamical system that has discrete time-steps. If the state of the system is x, and we label the time-steps by an integer n, then the dynamical system is described by an equation

$$x_{n+1} = f(x_n, n, u), \tag{5.86}$$

where f is some function. Here u is a parameter in the evolution function f, that can be modified as time goes by to control the system. Thus a feedback control strategy for this discrete-time system would specify u as a function of x and n. Of course x and u may be vectors.

Now let us say that the goal of our feedback control is to minimize a "total cost," D, and this total cost is the sum of a cost incurred at each time-step. The cost at each time-step, d, may be a function of the state of the system, the time, and the control applied at that time, so we can write $d(x, n, u)$. In the feedback control examples that we have considered so far, our cost was only a function of the state of the system at the final time. But if we wanted to minimize the energy expended by the controller, then the cost at each time-step would be a function of the control that is applied, namely the parameter (or set of parameters) u.

If the control is to be applied from a starting time $n = 0$, to a final time $n = N$, then the total cost is given by

$$D = \sum_{n=0}^{N} d(x_n, n, u[x_n, n]) \tag{5.87}$$

and depends on the control strategy u. While we have written the value of the control parameter u as a function of the state at the current time, it can in fact be a function of the states of the system at all previous times. To indicate this we could write $u[\{x_m : m \leq n\}, n]$, but this is rather cumbersome notation, so we will use $u[x_n, n]$, or simply u, as shorthand for this long expression. Since the function $u[x_n, n]$ defines the entire control strategy, we will also use u by itself to denote a given control strategy. Since D is a function of the control strategy, we could write $D(u)$ or D_u to denote this.

As an example, if our goal were to keep the state of the system close to x_{targ}, then our cost might be

$$d(x_n, n, u[x_n, n]) = d(x_n) = (x_n - x_{\text{targ}})^2. \tag{5.88}$$

Note that even though this cost does not explicitly depend on the value of the control at each time, it still depends on the control strategy, for this affects the evolution, and thus the values x_n, via the dynamical equation.

It is worth noting that since our dynamical update function f depends on the state of the system only at the current time, and not on the state at any earlier time, the optimal control at time n will also depend only on the state at time n. This is because all information regarding the future behavior of the system can be obtained from the present state x_n. It is for this reason that the optimal control strategy for a quantum system will depend only on the current state-of-knowledge, $\rho(t)$, and not on the state at early times. This is why, once we have processed the measurement record by using the stochastic master equation to obtain $\rho(t)$, the measurement record can be discarded. If instead of writing the control at time t, $u(t)$, in terms of $\rho(t)$, we write it directly in terms of the measurement record, then the optimal strategy will in general depend on the entire measurement record up to the current time.

The cost-to-go

We now introduce a function that will be very important for much of what follows. It is called the "cost-to-go," or the "cost function," and is the cost that will be incurred from the present time, m, until the final time, N, given that the current state of the system is $x_m = x$, and given a control strategy u. We will denote the cost-to-go by

$$C_u(x, m) = \sum_{n=m}^{N} d(x_n, n, u[x_n, n]). \tag{5.89}$$

Here the subscript indicates that the cost-to-go depends on the control strategy. The cost-to-go is the total cost that would be incurred if you started the system in state x at time m. So an optimal strategy, u_{opt}, is therefore defined by

$$C_{u_{opt}}(x,m) = \min_u [C_u(x,m)], \quad \forall x,m. \tag{5.90}$$

That is, an optimal feedback strategy is the one that gives the lowest cost over all the possible strategies, no matter at what time you start the system, or in what state you start the system.

We now make what is actually a simple observation, but one that is quite powerful. Imagine that we have found the optimal control strategy for the last time-step (that is, for going from $n = N - 1$ to $n = N$), and we have found this for all the possible values of x_{N-1}. The optimal strategy at time $N - 2$ will involve choosing some control u_{N-2} at that time, and then at time $N - 1$ *using the optimal strategy already obtained for that time-step*. This is because the total cost is the sum of the costs at each time-step, and the cost at each time-step, d_n, only depends on the values chosen at that time-step. Hence once we have found the control than minimizes the cost for the last time-step (which is $d_{N-1} + d_N$), that control is the right one to use at the last time-step, no matter from what initial time we started controlling the system.[2]

Given the above observation, the procedure for finding the optimal control at time $N - 2$, if we already have the optimal control for the final time-step, is as follows. We need to choose u_{N-2} to minimize the value of $C(x_{N-2},N - 2) = d_{N-2} + d_{N-1} + d_N$. We already know how to choose u_{N-1} to minimize $C(x_{N-1},N - 1) = d_{N-1} + d_N$, given x_{N-1}. Now the value of x_{N-1}, and thus the value of $C(x_{N-1},N - 1)$, will depend on our choice for u_{N-2}. So to determine u_{N-2} we have to minimize $d_{N-2} + C(x_{N-1},N - 1)$, given that we already know how to make $C(x_{N-1},N - 1) = C_{u_{opt}}(x_{N-1},N - 1)$. So the minimal value of $C(x_{N-2},N - 2)$ is given by

$$C_{u_{opt}}(x_{N-2},N - 2) = \min_{u_{N-2}} \left[d(x_{N-2},N - 2, u_{N-2}) + C_{u_{opt}}(x_{N-1},N - 1) \right]. \tag{5.91}$$

In more compact notation, this is

$$C_{N-2}^{opt} = \min_{u_{N-2}} \left[d_{N-2} + C_{N-1}^{opt} \right]. \tag{5.92}$$

---

[2] This argument can be viewed as an application of "the principle of optimality". This "principle" is the rather obvious fact that if one wants to find the minimum element of a set, one can do it in two parts. First divide the set into subsets and find the minimum element of each subset. Calling the set of these minima S, we can now find the minimum of the original set by finding the minimum element of S. It follows from the principle of optimality that the shortest path between two points also gives the shortest path between any other two points on the path. The problem of finding the minimal-cost protocol is exactly analogous to that of finding a minimal path.

More generally if we know the optimal control to apply from time $k + 1$ onward, then the value of the best cost-to-go at time k is given by

$$C_k^{\text{opt}} = \min_{u_k} \left[d_k + C_{k+1}^{\text{opt}} \right]. \tag{5.93}$$

The above equation is a central one in optimal control theory, and is called variously *Bellman's equation*, the *dynamic optimality equation*, or the *dynamic programming equation*. It gives a systematic procedure for finding an optimal control by stepping backward through time. For sufficiently simple control problems, it can be solved directly, and we give an example below. It can also be used to derive the HJB equation, which we will turn to in the next section.

In our derivation of Bellman's equation, we have assumed that the dynamical system is deterministic. However, the same reasoning works unchanged if we have a stochastic system, and we wish to minimize the *average* value of the total cost, where the average is taken over all possible noise realizations. In the case of a stochastic system, at each time-step the state of the system changes in a random way, so that there is more than one possible value for x_{n+1} given x_n. The cost-to-go is now the *average* cost-to-go, defined by

$$C_u(x, m) = \left\langle \sum_{n=m}^{N} d(x_n, n, u[x_n, n]) \right\rangle \tag{5.94}$$

where the average is over all possible future trajectories, given that the system starts in state x_m. If you now think through the process of stepping back through time, as we did above for the deterministic system, you will see that Bellman's equation holds just as well for the average cost-to-go.

Solving Bellman's equation: a simple example

To get a better understanding of the optimality equation and its properties, let us solve it for a simple case. Imagine that someone has an amount of money invested in the bank, but they are not allowed to withdraw any of the money, only to spend the interest each year. If they want they can deposit more money into the account, but they can still only spend the interest. The question we ask is, if they want to have the maximum amount of money to spend over a fixed number of years, should they reinvest some of their yearly interest payments, or should they spend all of them?

We now formulate the problem using Bellman's equation. Each time-step is one year, so time n will represent the end of the nth year. Let us say that the fixed time period is N years. If each year the interest is a fraction p of the total money invested, and there is X_1 money in the account in the first year, then at time $n = 1$ we receive the interest pX_1. If we reinvest a fraction s of this interest, then in the next year the payment will be $pX_2 = p(X_1 + spX_1) = (p + sp^2)X_1$. In this case we want to maximize a quantity rather than minimize it, so while we will still denote this quantity by C, we will refer to it as a *reward* rather than a *cost*. Apart from this difference, the problem has exactly the dynamical form we have been considering above. There is a value, X, that may change at each time-step, and the total reward is the sum of a reward at each time-step. There is a control parameter that the person can chose at each time-step, s, and this changes the way X evolves. The reward at time-step n is $d(X_n, s_n) = (1-s)pX_n$, and so depends on the state X_n as well as the control s_n. Bellman's equation for this problem is therefore

$$C^{\text{opt}}(k, X_k) = \max_{s_k} \left[d(k, X_k, s_k) + C^{\text{opt}}(k+1, (1+ps)X_k) \right]$$

$$= \max_{s_k} \left[(1-s_k)pX_k + C^{\text{opt}}(k+1, (1+ps_k)X_k) \right]. \quad (5.95)$$

Now let us step back from the final time. At the final time-step, when we receive the final interest payment, it is clear that we should spend all of it. Thus the optimal strategy at time N is $s_N = 0$. So since the total amount of money invested in year N is X_N, we have $C(N, X_N) = pX_N$. Now stepping back to time $N-1$, Bellman's equation gives

$$C^{\text{opt}}(N-1, X_{N-1}) = \max_s \left[(1-s)pX_{N-1} + C^{\text{opt}}(N, (1+sp)X_{N-1}) \right]$$

$$= \max_s \left[(1-s)pX_{N-1} + p(1+ps)X_{N-1} \right]$$

$$= \max_s \left[(2p + (p-1)ps \right] X_{N-1} = 2pX_{N-1}, \quad (5.96)$$

where we have written s as shorthand for s_{N-1}, and used the fact that the final amount invested is $X_N = (1+ps)X_{N-1}$. Since $p < 1$, it is obvious that $s = 0$ gives the maximum. It is clear why the total payment is $2pX_{N-1}$ if we re-invest nothing at the last two time points: each payment is simply p times what is in the bank at time $N-1$, and there are two payments.

If we now step back again, we find that

$$C^{\text{opt}}(N-2, X_{N-2}) = \max_s \left[(1-s)pX_{N-2} + C^{\text{opt}}(N-1, (1+sp)X_{N-2}) \right]$$

$$= \max_s \left[(1-s)pX_{N-2} + (1+sp)(2p)X_{N-2} \right]$$

$$= \max_s \left[3p + (2p-1)ps \right] X_{N-2}. \quad (5.97)$$

This time it is optimal to invest nothing if $2p < 1$. If this is the case, and we invest nothing, then the total reward is $3pX_{N-2}$. We see a pattern. If we step back to time $N-n$, and each time we choose to reinvest nothing, then the total payment for the last $n+1$ steps will be $(n+1)pX_{N-n}$. At time-step $N-n$ it will be optimal to reinvest nothing so long as $np < 1$.

So what happens when we have stepped back to the point where $np > 1$? In this case we have

$$C^{\text{opt}}(N-n, X_{N-n}) = \max_s \left[(1-s)pX_{N-n} + C^{\text{opt}}(N-n+1, (1+sp)X_{N-n}) \right]$$

$$= \max_s \left[(n+1)p + (np-1)ps \right] X_{N-n} \quad (5.98)$$

with $np > 1$. Now the term containing s makes a *positive* contribution to the total reward, so the optimal choice is $s = 1$. This time the reward is

$$C^{\mathrm{opt}}(N-n, X_{N-n}) = np(1+p)X_{N-n}. \tag{5.99}$$

When we step back again, since the reward this time is greater than the reward last time (that is, $np[1 + p] > np$), it will *still* be optimal to choose $s = 1$, and this continues right back to the start (since the reward can never decrease as we step back).

So now we have the complete optimal control strategy. We invest everything for the first few years, and at a certain point we switch to spending all of the interest payment each year. The point at which we switch is determined by the interest rate p. If m is the integer such that $mp > 1 > (m-1)p$, then we reinvest all of the payments for the first $N - m$ years, and spend all of the payments for the last m years.

In the above simple control problem, we were able to obtain a full solution analytically. This solution involves setting the control parameter, s, equal to its maximum value for some of the time, and then switching it to its minimum value. The optimal control strategy is therefore a "bang-bang" strategy, in that s takes only its extreme values. In the next section we will consider controlling systems that evolve continuously in time. In this case the only way to implement a "bang-bang" strategy, in which the control parameter takes only its extreme values, is to switch the control *instantaneously* between these values. Thus while bang-bang strategies are the simplest to find, they will only be optimal solutions to Bellman's equation if there are no constraints placed on the *rate* at which the control parameter(s) can be varied. In practice there is always a constraint on this rate, and thus bang-bang control can be viewed as an idealization. Still, obtaining such optimal control strategies provides insight into a given control problem. One can always approximate a bang-bang strategy using a strategy with a finite rate.

The Hamilton–Jacobi–Bellman equation

The Hamilton–Jacobi–Bellman (HJB) equation is a differential equation that is the continuous-time version of Bellman's equation. Its name derives from its similarity with the Hamilton–Jacobi equation that governs the solutions of classical Hamiltonian dynamical systems [21, 587]. To obtain the HJB equation, we consider a dynamical system given by

$$d\mathbf{x} = \mathbf{f}(\mathbf{x}, t, \mathbf{u}) + \mathbf{g}(\mathbf{x}, t, \mathbf{u}) \, dW, \tag{5.100}$$

where dW is the Wiener increment of Gaussian noise. In analogy with the cost defined above for discrete-time dynamics, we can define the cost-to-go as

$$C_{\mathbf{u}}(\mathbf{x}, t, T) = \int_t^T J\left[\mathbf{x}(t'), t', \mathbf{u}(t')\right] dt' + K[\mathbf{x}(T), T]. \tag{5.101}$$

The function K is included to allow the cost to contain a non-zero contribution from the state at the final time.

The continuous-time version of Bellman's equation, Eq. (5.94), is then

$$C_{\text{opt}}(\mathbf{x}, t, T) = \min_{\mathbf{u}} \left\langle C_{\mathbf{u}}(\mathbf{x}, t, s) + C_{\text{opt}}(\mathbf{x}(s), s, T) \right\rangle. \tag{5.102}$$

To turn this into a differential equation, we observe that if we choose $s = t + dt$, then Bellman's equation gives us $C_{\text{opt}}(\mathbf{x}(t), t, T)$ in terms of $C_{\text{opt}}(\mathbf{x}(t+dt), t+dt, T)$, and is therefore an equation of motion for C_{opt}. All we have to do is to rewrite it in differential form. To do this we first expand $C_{\text{opt}}(\mathbf{x}(t+dt), t+dt, T)$ in a power series about t. Using C_{opt} as shorthand for $C_{\text{opt}}(\mathbf{x}(t), t, T)$, we obtain

$$C_{\text{opt}}(\mathbf{x}(t+dt), t+dt, T) = C_{\text{opt}} + \alpha\, dt + \mathbf{b} \cdot d\mathbf{x} + d\mathbf{x}^t \frac{\Gamma}{2}\, d\mathbf{x}$$

$$= C_{\text{opt}} + \alpha\, dt + \mathbf{b} \cdot (\mathbf{f}\,dt + \mathbf{g}\,dW) + \mathbf{g}^t \frac{\Gamma}{2}\, \mathbf{g}\,dt. \tag{5.103}$$

In the above equation, we have defined the scalar α, vector $\mathbf{b}$, and matrix Γ as

$$\alpha = \frac{\partial C_{\text{opt}}}{\partial t}, \quad \mathbf{b} = \frac{\partial C_{\text{opt}}}{\partial \mathbf{x}}, \quad \Gamma = \frac{\partial^2 C_{\text{opt}}}{\partial \mathbf{x}^2}, \tag{5.104}$$

and $d\mathbf{x} = \mathbf{x}(t+dt) - \mathbf{x}(t)$. Note that $\partial^2 C_{\text{opt}}/\partial \mathbf{x}^2$ denotes the matrix of second derivatives of C_{opt}, usually referred to as the *Hessian*. To obtain the differential equation, we need the power series expansion only to first-order in dt. But to do so we need to keep the term that is second-order in $d\mathbf{x}$, because the equation of motion for $\mathbf{x}$ contains dW, and $dW^2 = dt$ [288]. Setting $s = t+dt$ in Bellman's equation, substituting in the above expansion, and noting that $\langle dW \rangle = 0$ and $\langle C_{\text{opt}} \rangle = C_{\text{opt}}$, we have

$$\min_{\mathbf{u}} \left[J + \frac{\partial C_{\text{opt}}}{\partial t} + \frac{\partial C_{\text{opt}}}{\partial \mathbf{x}} \cdot \mathbf{f} + \left(\frac{1}{2}\right) \mathbf{g}^t \frac{\partial^2 C_{\text{opt}}}{\partial \mathbf{x}^2}\, \mathbf{g} \right] = 0. \tag{5.105}$$

This is the HJB equation. Although we have not derived it rigorously, by its construction from Bellman's equation, if C satisfies this equation it is the optimal (minimal) C, and thus any control strategy $\mathbf{u}$ that gives a C that satisfies the HJB equation is an optimal strategy. Note that if the motion of the system is deterministic, then the HJB equation simplifies by setting $\mathbf{g} = 0$. We now show how this equation can be used to advantage.

Using the HJB equation I: the verification theorem

The HJB equation can be used to determine if a given control strategy – obtained for example using an intuitive guess – is optimal. If a given problem possesses an optimal strategy in which the control parameters take their extreme values, then this strategy can often be guessed. If this is the case, and if the equations of motion are sufficiently simple that they can be solved for this strategy, and thus the cost function C determined, then the HJB equation provides a greatly simplified way to determine if the strategy is optimal.

First note that at $t = T$, the optimal cost function C must satisfy $C(\mathbf{x}, T) = K(\mathbf{x}, T)$. So to determine if a guessed strategy, $\mathbf{u}_{\text{guess}}(\mathbf{x}, t)$, is an optimal strategy we must solve the equations of motion to determine $C(\mathbf{x}, t, \mathbf{u}_{\text{guess}})$, and check that

$$C_{\text{g}} \equiv C(\mathbf{x}, T, \mathbf{u}_{\text{guess}}) = K(\mathbf{x}, T). \tag{5.106}$$

Now assume that our guess for the control strategy is indeed an optimal strategy. If we substitute C_{g} into the HJB equation, then since the control $\mathbf{u}$ has already been optimized for C_{g}, the minimization over $\mathbf{u}$ now only applies to the value of the control that appears in $\mathbf{f}$, $\mathbf{g}$, and J. To make this especially clear in our notation, we will denote the control values that need to be minimized by $\mathbf{v}$ instead of $\mathbf{u}$. Since the term $\partial C_{\text{g}}/\partial t$ no longer has anything to minimize, we can remove it from the minimization, and the HJB equation becomes

$$\frac{\partial C_{\text{g}}}{\partial t} = -\min_{\mathbf{v}} \left[J_{\mathbf{v}} + \frac{\partial C_{\text{g}}}{\partial \mathbf{x}} \cdot \mathbf{f}_{\mathbf{v}} + \left(\frac{1}{2} \right) \mathbf{g}_{\mathbf{v}}^{\text{t}} \frac{\partial^2 C_{\text{g}}}{\partial \mathbf{x}^2} \mathbf{g}_{\mathbf{v}} \right], \tag{5.107}$$

where

$$C_{\text{g}} = C(\mathbf{x}, t, \mathbf{u}_{\text{guess}}), \quad \mathbf{f}_{\mathbf{v}} = \mathbf{f}(\mathbf{x}, t, \mathbf{v}(\mathbf{x}, t)), \tag{5.108}$$

$$\mathbf{g}_{\mathbf{v}} = \mathbf{g}(\mathbf{x}, t, \mathbf{v}(\mathbf{x}, t)), \quad J_{\mathbf{v}} = J(\mathbf{x}, t, \mathbf{v}(\mathbf{x}, t)). \tag{5.109}$$

So to determine if $\mathbf{u}_{\text{guess}}$ is an optimal strategy, we need to minimize the RHS over $\mathbf{v}(\mathbf{x}, t)$. And now here is the great simplification: because the HJB equation must be true for *every* value of x and t, the minimization can be done *separately* for each pair $(\mathbf{x}, t)$. This makes it feasible numerically, and often the minimum can even be found analytically.

Let us assume that we have found the values of the control parameters $\mathbf{v}(\mathbf{x}, t)$, that minimize the RHS for each t, and denote these by $\mathbf{v}_{\text{min}}(\mathbf{x}, t)$. If $\mathbf{v}_{\text{min}}(\mathbf{x}, t) = \mathbf{u}_{\text{guess}}$, then our guess is an optimal solution. In symbols,

$$\mathbf{v}_{\text{min}}(\mathbf{x}, t) = \mathbf{u}_{\text{guess}}(\mathbf{x}, t) \quad \Rightarrow \quad \mathbf{u}_{\text{guess}} = \mathbf{u}_{\text{opt}}. \tag{5.110}$$

The above procedure sums up the "verification theorem": this theorem states that the above procedure is a way to verify that a strategy is optimal. Note, however, that to use this procedure, the cost function C_{g} must be differentiable once with respect to t, and twice with respect to $\mathbf{x}$; if this is not the case, we cannot evaluate the derivatives of C_{g} that appear in the HJB equation. This is a serious problem because, as we mentioned above, solutions that can be guessed are often bang-bang strategies, in which the control changes discontinuously. Fortunately, there is a way around this, which we describe below.

The HJB equation for minimum-average-time problems

All cost functions we have considered so far are functions of the dynamical variables. But there is a natural control problem that is not covered by these cost functions: that of minimizing the average time it takes to achieve a desired result. It turns out that we can also derive an HJB equation for this kind of cost function (although it is a little different from the one above) and use it to verify optimal solutions in the same way.

To construct a minimal-time cost we first decide upon a function $\tau(\mathbf{x}(t),t)$. The cost is defined as the average time it takes until $\tau(\mathbf{x}(t),t)$ reaches a chosen value τ_c, called the *threshold*. The cost function, or cost-to-go $C(\mathbf{y},t)$, is the average time still remaining until $\tau(\mathbf{x}(t),t)$ reaches the threshold, given that the current time is t and the current state of the system is $\mathbf{y}$. Thus if we define $\mathbf{x}_c$ as a state for which τ is equal to τ_c, it follows that $C(\mathbf{x}_c,t)=0$.

With the above cost function the HJB equation is [181]

$$\frac{\partial C_g}{\partial t} = -\min_{\mathbf{v}} \left[\frac{\partial C_g}{\partial \mathbf{x}} \cdot \mathbf{f_v} + \left(\frac{1}{2}\right) \mathbf{g_v^t} \frac{\partial^2 C_g}{\partial \mathbf{x}^2} \mathbf{g_v} \right] - 1. \tag{5.111}$$

Note that this HJB equation does not contain either the function $\tau(\mathbf{x}(t),t)$, or the threshold value τ_c. These have already played their role by determining C. Note also that the terminal condition on the time-optimal HJB equation is $C(\mathbf{x}_c,t)=0$. This is automatically satisfied by the cost function since it is part of its definition.

A simple example of using the verification theorem

The fact that the HJB equation allows one to determine whether a control strategy is optimal merely by checking some derivative inequalities may seem too good to be true – is there a catch? There are two aspects of this procedure that could be thought of as the catches. The first is that the inequality one must check is itself an optimization problem – the beauty of the HJB equation is that it converts an optimization over a dynamical process to a set of simpler optimizations at single points in time. The second catch is that, in order to determine the optimality of a control strategy, the method must be able to explore all points in the system's state space for the chosen control strategy. This means that to apply the verification procedure you must first solve the evolution of the system, under the control strategy, for *every possible initial condition*, not merely one or two initial conditions that may happen to be of interest.

As a simple example we use the verification theorem to show that the control strategy for rapid purification of a qubit presented in Section 5.4.1 is optimal. Recall that the objective is to minimize the average impurity of the qubit at a future time T. The cost function is therefore $C = \langle L(T) \rangle$, where L is the impurity. We have a fixed measurement strength k, but we can choose the angle θ on the Bloch sphere between the basis of the measured observable and the Bloch vector. Note that we have complete freedom in choosing the value of θ at each time t only if we can apply a control Hamiltonian, H, that is infinitely fast. This is a good approximation to a real control situation only if $\text{Tr}[H^2] \gg \hbar^2 k^2$. For a measurement with strength k and angle θ the evolution of the impurity is given by

$$dL = -8kL\left[1-(1-L)\cos^2\theta\right] dt + \sqrt{8k}\cos\theta L\sqrt{1-2L}\, dW. \tag{5.112}$$

The control strategy presented in Section 5.4.1 is to choose $\cos\theta = 0$ for all t. The resulting evolution of the impurity is $L = L(0)e^{-8kt}$. Since the cost is simply the impurity at time T, the cost-to-go at time t, given that the impurity at that time is L, is given by

$$C(L,t) = L(T) = Le^{-8k(T-t)}. \tag{5.113}$$

Since there is only one dynamical variable, and since $J = 0$, the HJB equation for this problem is

$$\frac{\partial C}{\partial t} = -\min_\theta \left[f(L,\theta)\frac{\partial C}{\partial L} + \frac{1}{2}[g(L,\theta)]^2 \frac{\partial^2 C}{\partial L^2} \right], \tag{5.114}$$

where

$$f = -8kL\left[1 - (1-L)\cos^2\theta\right], \tag{5.115}$$

$$g = \sqrt{8k}\cos\theta L\sqrt{1-2L}. \tag{5.116}$$

It is important to note that in obtaining the cost function, $C(L,T)$, the values of the control variables (in this case θ) have been *fixed* by the control protocol. Thus in the minimization it is only the value of θ that appears in the functions $f(L,\theta)$ and $g(L,\theta)$ that takes part in the minimization. To satisfy the HJB equation the value of the control that minimizes the right-hand side for each t must be the same control as used in the protocol. Substituting C, f, and g into the HJB equation we obtain

$$L = -\min_\theta \left[-L\left(1 - (1-L)\cos^2\theta\right) \right], \tag{5.117}$$

and this is clearly minimized by $\theta = \pi/2$. Since the cost function satisfies the final condition (in this case by construction), the protocol is optimal.

Testing discontinuous control strategies

It turns out that there is a straightforward way to use the HJB equation to test whether a *piecewise-continuous* control strategy is optimal. A control strategy is piecewise continuous if it is discontinuous only at a finite set of points. By examining the relationship between the control strategy and cost function, we see that as long as the former is continuous at the final time, the cost function will be continuous. In this case, the effect of the discontinuous points in the control strategy is to make the *derivatives* of the cost function undefined at these points. So long as the cost function is continuous, we can verify that a piecewise-continuous control strategy is optimal by checking that it satisfies the HJB equation where its derivatives are well defined, and then by checking some additional conditions at the set of points where the derivatives are undefined.

To explain how to check the conditions at the points where the derivatives of C_g are undefined, we need to define some things. Let us denote a point at which the derivatives are undefined by $t = \tau$ and $\mathbf{x}(\tau) \equiv \mathbf{d}$. The cost function is given by analytic functions on either side of $(t, \mathbf{x}(t)) = (\tau, \mathbf{x}(\tau)) = (\tau, \mathbf{d})$, and so we can write it as

$$C_g(\mathbf{x}, t) = \begin{cases} C^+(\mathbf{x}(t), t) & \text{for } t \geq \tau \\ C^-(\mathbf{x}(t), t) & \text{for } t \leq \tau \end{cases} \tag{5.118}$$

with $C(\mathbf{d}, \tau)^+ = C(\mathbf{d}, \tau)^-$.

Let q denote a real number, $\mathbf{p}$ a real vector, and Q a real matrix, where $\mathbf{p}$ and Q have the dimension of $\mathbf{x}$. For each point at which C_g is discontinuous, we determine the set of triples $(q, \mathbf{p}, Q)$ for which

$$\partial_t C^+ \Delta t + \partial_{\mathbf{x}} C^+ \cdot \Delta \mathbf{x} + \Delta \mathbf{x}^{\mathrm{T}} \partial_{\mathbf{x}}^2 C^+ \Delta \mathbf{x} \leq q \Delta t + \mathbf{p} \cdot \Delta \mathbf{x} + \Delta \mathbf{x}^{\mathrm{T}} Q \Delta \mathbf{x} \tag{5.119}$$

and

$$\partial_t C^- \Delta t + \partial_{\mathbf{x}} C^- \cdot \Delta \mathbf{x} + \Delta \mathbf{x}^{\mathrm{T}} \partial_{\mathbf{x}}^2 C^- \Delta \mathbf{x} \leq q \Delta t + \mathbf{p} \cdot \Delta \mathbf{x} + \Delta \mathbf{x}^{\mathrm{T}} Q \Delta \mathbf{x}, \tag{5.120}$$

where the number $\partial_t C^-$, the vector $\partial_{\mathbf{x}} C^-$, and the matrix $\partial_{\mathbf{x}}^2 C^-$ are

$$\partial_t C^{\pm} = \left. \frac{\partial C^{\pm}}{\partial t} \right|_{(\mathbf{d}, \tau)}, \qquad \partial_{\mathbf{x}} C^{\pm} = \left. \frac{\partial C^{\pm}}{\partial \mathbf{x}} \right|_{(\mathbf{d}, \tau)}, \qquad \partial_{\mathbf{x}} C^{\pm} = \left. \frac{\partial^2 C^{\pm}}{\partial \mathbf{x}^2} \right|_{(\mathbf{d}, \tau)}$$

and

$$\Delta t = t - \tau \quad \text{and} \quad \Delta \mathbf{x} = \mathbf{x} - \mathbf{d}. \tag{5.121}$$

This set of tuples is called the "superderivative" and we denote it by D^+. We also determine the set of tuples for which

$$\partial_t C^+ \Delta t + \partial_{\mathbf{x}} C^+ \cdot \Delta \mathbf{x} + \Delta \mathbf{x}^{\mathrm{T}} \partial_{\mathbf{x}}^2 C^+ \Delta \mathbf{x} \geq q \Delta t + \mathbf{p} \cdot \Delta \mathbf{x} + \Delta \mathbf{x}^{\mathrm{T}} Q \Delta \mathbf{x} \tag{5.122}$$

and

$$\partial_t C^- \Delta t + \partial_{\mathbf{x}} C^- \cdot \Delta \mathbf{x} + \Delta \mathbf{x}^{\mathrm{T}} \partial_{\mathbf{x}}^2 C^- \Delta \mathbf{x} \geq q \Delta t + \mathbf{p} \cdot \Delta \mathbf{x} + \Delta \mathbf{x}^{\mathrm{T}} Q \Delta \mathbf{x}, \tag{5.123}$$

and this set is called the "subderivative," denoted by D^-. (For the purposes of satisfying the above inequalities, Δt and $\Delta \mathbf{x}$ are arbitrarily small. Thus if the term proportional to $\delta \mathbf{x}$ on the left-hand side is not zero, it dominates over the second-order term. In this case the matrix Q is unnecessary and we can set it equal to zero.)

The first thing we must check is that

$$G(\mathbf{d}, f(\mathbf{d}), \mathbf{p}, Q) \geq 0 \quad \text{whenever } (q, \mathbf{p}, Q) \in D^+, \tag{5.124}$$

$$G(\mathbf{d}, f(\mathbf{d}), \mathbf{p}, Q) \leq 0 \quad \text{whenever } (q, \mathbf{p}, Q) \in D^-. \tag{5.125}$$

The second thing we must check is that the superderivative is not empty at any of the points of discontinuity.

The third condition we must check is that at each point of discontinuity there exists at least one element of the superderivative for which

$$q = -J - \mathbf{p} \cdot \mathbf{f} - \left(\frac{1}{2}\right) \mathbf{g}^t Q \mathbf{g}, \tag{5.126}$$

where naturally

$$\mathbf{f} = \mathbf{f}(\mathbf{d}, \tau, \mathbf{u}^{\text{guess}}(\mathbf{d}, \tau)), \tag{5.127}$$

$$\mathbf{g} = \mathbf{g}(\mathbf{d}, \tau, \mathbf{u}^{\text{guess}}(\mathbf{d}, \tau)), \tag{5.128}$$

$$J = J(\mathbf{d}, \tau, \mathbf{u}^{\text{guess}}(\mathbf{d}, \tau)). \tag{5.129}$$

If these three conditions are satisfied, then the piecewise-continuous strategy $\mathbf{u}^{\text{guess}}$ is optimal.

Viscosity solutions of differential equations

It is not at all obvious that the procedure described above for checking if a piecewise-continuous control strategy is optimal is the right one. While the proof that it is is beyond our scope here, we can at least make its structure more understandable. The above procedure is derived from the concept of a *viscosity solution* to a differential equation. A viscosity solution extends the notion of a solution to a differential equation to include functions for which the derivative is not defined at every point. At a point $\mathbf{z} = \mathbf{d}$ at which a function $f(\mathbf{z})$ has a discontinuity in its derivative, the *superderivative* (or *superjet*) is defined as being the set of all tuples $(\mathbf{p}, Q)$ that satisfy

$$\partial_z f^{\pm} \Delta \mathbf{z} + \partial_z^2 f^{\pm} (\Delta \mathbf{z})^2 \leq p \Delta \mathbf{z} + Q (\Delta \mathbf{z})^2 \tag{5.130}$$

in the limit as $\Delta \mathbf{z} \to 0$. Here naturally $\Delta \mathbf{z} = \mathbf{z} - \mathbf{d}, f^+$ is the function on the positive side of the point $\mathbf{z} = \mathbf{d}$, and f^- is the function on the negative side of $\mathbf{z} = \mathbf{d}$. Note that the condition given by Eq. (5.130) is equivalent to

$$\mathbf{p} \geq \partial_z f^+ \tag{5.131}$$

if $\partial_z f^+ \neq 0$, and

$$Q \geq \partial_z^2 f^+ \tag{5.132}$$

otherwise, and the same for f^-. The *subderivative* set of tuples $(\mathbf{p}, Q)$ is defined in the same way, but with the inequality reversed. Note that in our discussion of the HJB equation above, we did not need to include the second derivative of t in our superderivative because the HJB equation does not contain it.

We can write a second-order differential equation for f as

$$G(x, f, \partial_z f, \partial_z^2 f) = 0 \tag{5.133}$$

for some "function" G. The function f is said to be a *viscosity solution* of this differential equation if whenever the tuple $(\mathbf{p}, Q)$ is in the superderivative set, then

$$G(\mathbf{d}, f(\mathbf{d}), \mathbf{p}, Q) \geq 0, \tag{5.134}$$

and whenever $(\mathbf{p}, Q)$ is in the subderivative set then

$$G(\mathbf{d}, f(\mathbf{d}), \mathbf{p}, Q) \leq 0. \tag{5.135}$$

It turns out that the concept of a viscosity solution is the right one for formulating the HJB equation for piecewise-continuous control strategies. The procedure we described above for checking whether such a control strategy was optimal is essentially that for checking that the cost function C_g is a viscosity solution to the HJB equation.

Using the HJB equation II: numerical searches for optimal strategies

The HJB equation can also be used to aid a numerical search for an optimal protocol. To do so we exploit the fact that the cost-to-go, $C(\mathbf{x}, t, \mathbf{u})$, satisfies the differential equation

$$\frac{\partial C_g}{\partial t} = -\left[J_u + \frac{\partial C_g}{\partial \mathbf{x}} \cdot \mathbf{f_u} + \left(\frac{1}{2}\right) \mathbf{g}_u^{\mathsf{t}} \frac{\partial^2 C_g}{\partial \mathbf{x}^2} \mathbf{g_u} \right], \tag{5.136}$$

for *every* protocol. But the expression in the square brackets, which we will refer to as G, will only be minimized if the protocol is optimal. We can therefore perform a search in the space of protocols (all functions $\mathbf{u}(\mathbf{x}, t)$) to find protocols that minimize G. Since $\mathbf{u}$ is a function of both time and the state, the search space can be very large. It is nevertheless possible to simplify the search by starting close to the the final time, T. That is, we perform the optimization for the time-interval $[T - \Delta t, T]$ for every state that the system could be in at the start of this interval. Once we know the optimal protocol for each state at time $T - \Delta t$, we can step back and perform an optimization for the interval $[T - 2\Delta t, T]$, once again for every state of the system at the start of the interval. This time we only have to search for the optimal control on the part of the interval that we have not already solved, because whatever the state is at time $T - \Delta t$ we just use the optimal protocol we have already found on the interval $[T - \Delta t, T]$. Once we have found the set of optimal protocols for the interval $[T - \Delta t, T]$, we iterate the above procedure by repeatedly stepping backward in time. Because of the form of the cost, either that in Eq. (5.101) or the minimum-average-time cost, this backward-in-time procedure is guaranteed to find a globally optimal protocol.

If $\mathbf{v}$ is not a function of time but only of the state, $\mathbf{x}$, then the task is significantly easier. Not only is the search space reduced, but it usually means that if the minimization in the HJB equation is fulfilled at a single time, it is fulfilled at all times. Further, if one can obtain an analytic solution to the equations of motion for every $\mathbf{v}$, then one can obtain an analytic expression for the right-hand side in terms of $\mathbf{v}$. In this case numerical minimization is likely to be easy.

Pontryagin's maximum principle

This "principle" is a theorem which can be useful for finding optimal controls for deterministic systems. The HJB equation gives necessary and sufficient conditions for optimality, but to do so it requires that the evolution of the system be obtained for every initial condition. Pontryagin's maximum principle gives merely necessary conditions for optimality, but because of this it requires only that the trajectory in question be analyzed.

Consider the deterministic dynamical system defined by

$$\dot{\mathbf{x}} = \mathbf{f}(\mathbf{x}, t, \mathbf{u}), \tag{5.137}$$

where the control $\mathbf{u}(t)$ is any function(al) of the evolution $\mathbf{x}(t')$ for $t' \in [0, t]$, and the cost is given by Eq. (5.101). Note that if we do not want to constrain the control problem to ending at a fixed time T, then we simply allow T to be an undetermined variable that appears in the cost function, and that will be determined by minimizing the cost. One can also demand that the control take the system to one of a specified set of final states. Pontryagin's theorem allows this set to be any hyperplane in the vector space, $\mathcal{S}$, of the system states $\mathbf{x}$. We specify the hyperplane as a point, $\mathbf{p}$, to which is added any vector in a vector space $\mathcal{V}$ that is a subspace of $\mathcal{S}$. If we do demand that the control take the system to a specified hyperplane of final states, then with constraints on the control function $\mathbf{u}$ there may be no control function that gives a solution. In this case Pontryagin's theorem applies *if* there is a solution.

Theorem 13 (Pontryagin) Define the function

$$H(\mathbf{x}, t, \mathbf{u}, \boldsymbol{\lambda}) = \mathbf{f}(\mathbf{x}, t, \mathbf{u}) \cdot \boldsymbol{\lambda} + J(\mathbf{x}, t, T) \tag{5.138}$$

where we have introduced a new vector-valued function of time, $\boldsymbol{\lambda}(t)$, with the same dimension as $\mathbf{x}$. Introducing another new function of time $\mu(t)$, the following statements hold:

(i) The optimal control $\mathbf{u}^*(t)$ is given by setting

$$H(\mathbf{x}(t), t, \mathbf{u}^*(t), \boldsymbol{\lambda}) = -\mu(t), \tag{5.139}$$

and the maximum value of H is $-\mu(t)$.

(ii) The function of time $\boldsymbol{\lambda}$ is determined by the differential equation

$$\dot{\boldsymbol{\lambda}} = -\frac{\partial H}{\partial \mathbf{x}} = -\frac{\partial \mathbf{f}}{\partial \mathbf{x}} \boldsymbol{\lambda} + \frac{\partial J}{\partial \mathbf{x}}. \tag{5.140}$$

(iii) The function of time μ is determined by the differential equation

$$\dot{\mu} = -\frac{\partial H}{\partial t} = -\frac{\partial \mathbf{f}}{\partial t} \cdot \boldsymbol{\lambda} + \frac{\partial J}{\partial t}. \tag{5.141}$$

(iv) If the final time T is unconstrained (see above), then

$$\mu(T) = -\frac{\partial}{\partial T} K(x,T).$$ (5.142)

Note that if $\partial \mathbf{f}/\partial t = \partial J/\partial t = \partial K/\partial T = 0$, then (iii) and (iv) imply that $\mu(t) = 0$ for all t (when T is unconstrained).

(v) For a specified hyperplane $\mathbf{p} + \mathcal{V}$ of allowed final states (see above),

$$\left(\boldsymbol{\lambda}(T) - \frac{\partial K}{\partial \mathbf{x}} \right) \cdot \mathbf{y} = 0$$ (5.143)

for all $\mathbf{y} \in \mathcal{V}$.

As a simple example we apply Pontryagin's theorem to the problem considered above, that of bringing a particle in one dimension to rest at a fixed location in the shortest time. The state of the particle is the position x and velocity v, so $\mathbf{x} = (x_1, x_2)^{\mathrm{T}} = (x, v)^{\mathrm{T}}$. The control is a force applied to the particle, which we we write as an acceleration. The constraint is a bound on the maximum acceleration that we can apply. So $\dot{v} = u(t)$, and $-a \le u(t) \le a$. The equation of motion is

$$\dot{\mathbf{x}} = \begin{pmatrix} f_1 \\ f_2 \end{pmatrix} = \begin{pmatrix} v \\ u \end{pmatrix}.$$ (5.144)

Since we want to bring the particle to rest at the origin, the set of allowed final states is just the point $\mathbf{x}(T) = (0,0)^{\mathrm{T}}$. Because it is the time T that we wish to minimize, we need to define the cost so that it is proportional to T (at the initial time $t = 0$). This is done easily by setting $J = 1$ and $K = 0$ so that the cost-to-go at time t is

$$C_{\mathbf{u}}(\mathbf{x},t,T) = \int_t^T J\, dt' + K[\mathbf{x}(T),T] = \int_t^T dt' = T - t.$$ (5.145)

To apply Pontryagin's theorem to this problem, we first note that since the time T is unconstrained, and $\partial \mathbf{f}/\partial t = \partial J/\partial t = \partial K/\partial T = 0$, statements (iii) and (iv) tell us that $\mu = 0$. We now write down Pontryagin's function, which is

$$H = \lambda_1 v + \lambda_2 u - 1,$$ (5.146)

and note that under the optimal control H is maximized, and is equal to $\mu = 0$. From this we immediately obtain two facts: (i) The value of u that maximizes H depends on the sign of λ_2. If $\lambda_2 > 0$ then $u = a$, and if $\lambda_2 < 0$ then $u = -a$. (ii) At $t = T$ the velocity v is required to be zero, so Eq. (5.146) gives us $|\lambda_2(T)| = 1/a$.

We now know that the optimal control is equal to one of its extreme values at every time (this is usually the case when we constrain the value of the controls but not their time-derivatives), but we don't know when it should be equal to which extreme. To determine this we need to solve the equations of motion for λ_1 and λ_2. Since $f_1 = v$ and $f_2 = u$, these equations are

$$\begin{pmatrix} \dot{\lambda}_1 \\ \dot{\lambda}_2 \end{pmatrix} = \begin{pmatrix} \partial f_1/\partial x_1 & \partial f_2/\partial x_1 \\ \partial f_1/\partial x_2 & \partial f_2/\partial x_2 \end{pmatrix} \begin{pmatrix} \lambda_1 \\ \lambda_2 \end{pmatrix} = \begin{pmatrix} 0 \\ \lambda_1 \end{pmatrix}. \tag{5.147}$$

Thus λ_1 is constant ($\lambda_1 = \lambda_1(0)$), and the solution for λ_2 is $\lambda_2(t) = \lambda_2(0) + \lambda_1 t$. For convenience in what follows we will define $\lambda(0) = -\alpha$. We have information about the final value of λ_2, but not the initial value, so we can work backward to get the initial values. Of course we actually want to know the control as a function of the initial state, $\mathbf{x}(0)$. But instead we know the value of the control as a function of the values of λ_2 and λ_1. So we first determine the values of the control given the initial values of the λs, and the value of the control determines the evolution of the state. Since we know the final state we can then determine the initial state as a function of the control, and invert this relationship to obtain our answer.

So let us consider the case in which $\lambda_2(T) = -1/a$. Since the time-dependence of λ_2 is a straight line, we have two cases: either λ_2 starts off negative and remains negative, or it starts off positive and crosses zero (just once) to become negative. Substituting $\lambda_2(T)$ into the solution for $\lambda_2(t)$ gives us $\lambda_2(0) = -1/a + \alpha T$. Thus if $\alpha \leq 1/(aT)$ then λ_2 is always negative, and so u is always equal to $-a$. Otherwise λ_2 starts out positive and turns negative at $\tau = T - 1/(a\alpha)$. Now we know the value of the control for both cases we can determine the evolution of the particle in both cases. If λ_2 starts positive, the control is $u(t) = a$ for $t \in [0, T - 1/(a\alpha)]$, and $u(t) = -a$ for the rest of the time. The initial state of the particle is then $x(0) = 1/(a\alpha^2) - 2T/\alpha + aT^2/2$ and $v(0) = 2/\alpha - aT$. We can therefore solve for T and α in terms of $x(0)$ and $v(0)$, and thus obtain the control protocol (which is determined by the total time T and the switching time τ). The case in which $\lambda_2(T) = 1/a$ is essentially identical apart from a reversal of the signs of all the signed quantities. This reflects the fact that the control for $(x(0), v(0))$ is the same as that for $(-x(0), -v(0))$, except that the direction of the acceleration is reversed.

5.5.2 Optimal control for linear quantum systems

The equation of motion of quantum mechanics, Schrödinger's equation, is a linear differential equation. This means that a linear combination of any two solutions is also a solution. If this were not the case then it could only evolve probability distributions if it were stochastic. But this linearity is not the linearity we are referring to in the title of this section. By a "linear quantum system," we mean a system for which the Heisenberg equations of motion for a canonical set of coordinates are linear. The term "canonical coordinates" comes from classical Hamiltonian mechanics, and in the quantum context means any set of coordinates that can be divided into pairs satisfying the commutation relations of position and momentum. Thus the set of coordinates $\{x_j, p_k\}$ with commutation relations $[x_j, p_k] = i\hbar\delta_{jk}$, with all other commutators equal to zero, are a canonical set. It turns out to be useful to define a (real) matrix that characterizes the canonical commutation relations. We first define a vector containing all the coordinates:

$$\mathbf{v} = (v_1, \ldots, v_{2N}) \equiv (x_1, p_1, x_2, p_2, \ldots, x_N, p_N), \tag{5.148}$$

and then define a matrix, Σ, whose elements are given by $\Sigma_{j,k} = [v_j, v_k]/(i\hbar)$ (we divide the commutator by $i\hbar$ to make the matrix real and dimensionless). The matrix Σ is given by

$$\Sigma = I_N \otimes \begin{pmatrix} 0 & 1 \\ -1 & 0 \end{pmatrix}, \tag{5.149}$$

and is an example of a *symplectic* matrix. Here I_N denotes the N-dimensional identity matrix, and the tensor product – denoted by "$\otimes$" – is defined in Appendix A. A matrix S is symplectic if it has an even dimension, and satisfies $S^T \Omega S = \Omega$, where Ω is an invertible skew-symmetric matrix.[3] Symplectic matrices are important in classical mechanics, and appear here because the dynamics of linear quantum systems, as we shall discuss below, is the same as that of linear classical systems driven by the right amount of Gaussian noise. Note that the action of Σ on the vector $\mathbf{v}$ is very simple: it converts each of the x_i operators to the negative of its corresponding momentum operator, p_i, and all the p_i operators to their respective position operators. This reflects the fact that if x_i appears by itself in the Hamiltonian, it generates a negative term in the equation of motion for p_i. Conversely, a p_i in the Hamiltonian generates a positive term in the equation for x_i. This is also true classically.

From the canonical commutation relations, and the fact that the Heisenberg equation of motion for any operator a is given by $\dot{a} = -(i/\hbar)[a, H]$, it follows that the equations of motion for a set of canonical coordinates are linear if and only if the Hamiltonian is no more than quadratic in these coordinates. The most general Hamiltonian quadratic in $\{x_j, p_k\}$ may be written as

$$H = \frac{1}{2}\mathbf{v}^T A \mathbf{v} + \mathbf{v}\Sigma B\mathbf{c}, \tag{5.150}$$

where A is real and symmetric, B is any real $2N \times K$ matrix, and $\mathbf{c}$ is a real vector of dimension K. The resulting equation of motion is

$$\frac{d}{dt}\mathbf{v} = A\mathbf{v} + B\mathbf{c}. \tag{5.151}$$

This equation is also an example of the usefulness of the matrix Σ: it transforms between the matrix B that appears in the equations of motion and the matrix in the Hamiltonian that generates them.

---

[3] A matrix A is skew-symmetric, a.k.a anti-symmetric, if and only if $A^t = -A$.

If M and B are constant, and $\mathbf{c}$ is time-dependent, then the solution to this linear differential equation is

$$\mathbf{v}(t) = e^{At}\mathbf{v}(0) + \int_0^t e^{A(t-s)}B\mathbf{c}(s)ds. \tag{5.152}$$

It turns out that if the initial state of a linear quantum system is Gaussian in all the canonical coordinates, then the linear dynamics keeps it Gaussian, and this leads to closed equations of motion for the means and (co-)variances of the coordinates. The variance of an observable a is defined by $C_{aa} = \langle a^2\rangle - \langle a\rangle^2$, and similarly the covariance between two observables a and b is defined by $C_{ab} = (1/2)\langle ab + ba\rangle - \langle a\rangle\langle b\rangle$. The *covariance matrix*, C, that has as its elements all the variances and covariances of the dynamical variables, is defined by

$$C = \langle \mathbf{v}\mathbf{v}^{\mathrm{T}}\rangle - \langle \mathbf{v}\rangle\langle \mathbf{v}\rangle^{\mathrm{T}}. \tag{5.153}$$

It is useful to note that the covariance matrix for a general Gaussian state must satisfy

$$C + \frac{i\hbar}{2}\Sigma \geq 0. \tag{5.154}$$

The meaning of this matrix inequality is that the matrix expression on the left-hand side has no negative eigenvalues.

The equation of motion for the means is simply the equation for the coordinates, thus

$$\frac{d}{dt}\langle \mathbf{v}\rangle = A\langle \mathbf{v}\rangle + B\mathbf{c}, \tag{5.155}$$

and the equation of motion for the covariance matrix is

$$\frac{d}{dt}C = (\Sigma A)C + C(\Sigma A)^{\mathrm{T}}. \tag{5.156}$$

The evolution of Gaussian states in linear systems is therefore simple to solve, since only the means and variances need to be calculated, and the equations of motion for these are linear. But there is something else about linear quantum systems that makes them special. The dynamics of a Gaussian state for a linear quantum system is exactly the same as that for the equivalent classical system, so long as the "state" of the latter is the same Gaussian probability distribution but now for the classical coordinates. (The general expression for a Gaussian probability distribution with N variables is given in Appendix G.) Here when we say the "equivalent classical system," we mean a classical system in which the Hamiltonian is just the quantum Hamiltonian with the quantum (operator) coordinates replaced by the classical coordinates. So that means that Eqs. (5.150) through (5.155), along with Eq. (5.156), are also valid for classical systems.

For the purposes of feedback control, we need to include continuous measurements (or, equivalently, input and output fields) in the dynamics of our linear system. Recall from

Chapter 3 that the evolution of the density matrix, under a measurement of a single observable x, is given by a stochastic master equation (SME). It turns out that if the measured observable is a linear combination of the canonical coordinates, and the initial state is Gaussian, then the SME preserves the Gaussian nature of the state, just as does linear dynamics. We will refer to measurements of linear combinations of the canonical coordinates as *linear measurements*.

The difference between linear classical measurements and their quantum counterparts is that the latter introduce noise into the system so as to preserve Heisenberg's uncertainty principle. Nevertheless, it turns out that we can reproduce the dynamics of a linear quantum system under a linear measurement, by making the equivalent measurement on a classical system, and driving the classical system with the right amount of noise. Thus for every linearly measured linear quantum system in a Gaussian state, we can find a noisy, measured, linear classical system for which the Kushner–Stratonovich equation gives the same evolution for the observer's classical state-of-knowledge as does the quantum SME.

To see the equivalence of measured linear quantum and classical systems, we first write down the equations for the means and variances of a measured linear quantum system. Let us measure a set of observables $\{m_i\}$ given as the vector $\mathbf{m} = (m_1,\ldots,m_L) = M\mathbf{v}$, where the m_i are chosen to have units of position. Further, if we measure each of the m_i with efficiency η_i, then the set of L measurement records is given by

$$\mathbf{dy} = \sqrt{8\Upsilon}M\langle\mathbf{v}\rangle\,dt + \mathbf{dW}, \tag{5.157}$$

where $\mathbf{dW} = (dW_1,\ldots,dW_L)$ is a vector of L mutually independent Wiener noises, and Υ is a diagonal matrix that gives the efficiencies of the measurements:

$$\Upsilon = \begin{pmatrix} \eta_1 & & 0 \\ & \ddots & \\ 0 & & \eta_L \end{pmatrix}. \tag{5.158}$$

Note that the elements of M determine the strength with which each observable m_i is measured. In particular, the elements of M are proportional to the square-roots of the measurement strengths. The evolution of the means and variances of the observer's state-of-knowledge is given by

$$\langle\dot{\mathbf{v}}\rangle = A\langle\mathbf{v}\rangle + B\mathbf{c} + CM^{\mathrm{T}}\sqrt{8\Upsilon}\,\mathbf{dW} \tag{5.159}$$

$$\dot{C} = (\Sigma A)C + C(\Sigma A)^{\mathrm{T}} + \hbar^2 D - 8CM^{\mathrm{T}}\Upsilon MC. \tag{5.160}$$

Here it is the matrix D that gives the increase in the variances from the back-action noise of the measurement (the noise that preserves the uncertainty principle), and it is the final term in Eq. (5.160) that gives the continual reduction in the variances of the observables due to the information provided by the measurement. The matrix D is given by

$$D = 2\Sigma M^{\mathrm{T}} M\Sigma^{\mathrm{T}}. \tag{5.161}$$

Note that the symplectic matrix Σ appears here, where it reflects the fact that a measurement of x_i generates noise affecting not itself but its momentum variable p_i (and vice versa). The equation of motion for the variances is an example of a *matrix Riccati differential equation* [127, 508, 469]. Note that the evolution of the covariances does not depend on the measurement results at all, only on the fact that we are making a measurement! This is a unique feature of Gaussian measurements made on quantities that have a Gaussian probability distribution: in this case the reduction in the variance is the same for every measurement result.

The equations we have presented above are easily reproduced by a linear classical system subjected to a continuous linear measurement. In particular, if we drive a classical system with a set of independent Wiener noises denoted by $\mathbf{dY}$, so that the equations of motion are

$$\mathbf{dv} = A\mathbf{v}\,dt + B\mathbf{c}\,dt + \hbar\sqrt{D}\,\mathbf{dY}, \qquad (5.162)$$

then this noise adds precisely the term $\hbar^2 D$ to the equations of motion for the covariance matrix. If we now make a set of continuous measurements on this system, measuring the observables m_i as in the quantum case, then the equations for the means and variances given by the Kushner–Stratonovich equation are exactly those above in Eqs. (5.159) and (5.160).

If we add feedback, we can keep the classical and quantum dynamics the same by ensuring that the feedback keeps the dynamics linear. If one implements feedback by driving the system with a linear combination of the means, then the linearity is preserved. So the most general kind of linear feedback that we can apply will give us the quantum dynamical equations

$$\frac{d}{dt}\langle\mathbf{v}\rangle = A\langle\mathbf{v}\rangle + B\mathbf{c} + E\langle\mathbf{v}\rangle = (A+E)\langle\mathbf{v}\rangle + B\mathbf{c}, \qquad (5.163)$$

where E is selected by the feedback. It is usual to split the E matrix into two parts, $E = FG$, and write the equations of motion as

$$\frac{d}{dt}\langle\mathbf{v}\rangle = A\langle\mathbf{v}\rangle + B\mathbf{c} + F\mathbf{u}, \qquad (5.164)$$

where the vector $\mathbf{u}$ is

$$\mathbf{u} = G\langle\mathbf{v}\rangle. \qquad (5.165)$$

The elements of $\mathbf{u}$ are called the "control inputs." The reason for dividing up the feedback in this way is that it allows us to fix F, placing a restriction on the control inputs, while trying to find the optimal G. The matrix G is often called the "feedback gain."

Linear, quadratic, Gaussian (LQG) control

We have spent some time showing how linear quantum dynamics is equivalent to classical dynamics. We have done so because this fact allows classical control protocols for linear systems to be translated directly to their quantum counterparts. For linear classical systems there are two kinds of optimal control problems that have analytic solutions. The first, which we consider now, is called "LQG" control, which is short for "linear system, quadratic cost, and Gaussian noise." The second analytically solvable control problem is called "linear, exponential, Gaussian" (LEG) control, and we will return to this briefly below.

In LQG control the objective is to minimize a quadratic function of the dynamical variables and the control inputs. Specifically, the objective is to minimize, over some specified time, the time-integral of the expectation value of

$$f = \mathbf{v}^{\mathrm{T}}Q\mathbf{v} + \mathbf{u}^{\mathrm{T}}R\mathbf{u}, \qquad (5.166)$$

where Q and R are symmetric matrices. In addition, one may also wish to minimize the first term in f at the final time. The justification for this quadratic objective is that energy is a quadratic function of the canonical coordinates, and so the objective allows one to minimize some weighted average of the energy of the system – or its mean-square deviations from some desired motion – along with the energy used to control the system (the energy of the control inputs). No doubt the real reason that LQG is standard fare in control theory is because it is one of the few control problems with a fully analytic solution. Most realistic control problems are nonlinear, and as a result optimal control protocols are virtually impossible to obtain. Because of this there is a large gap between control engineers who build real systems, and those control theorists whose interest is in proving rigorous theorems about such things as whether a control protocol converges in the infinite-time limit. The former worry little about optimality or mathematically rigorous proofs of performance, as they are mainly interested in designing intuitive methods that work effectively.

The weakness of LQG control is that given the arbitrariness of the relative weighting between the energy of the system and that of the control inputs, it is unclear what physical resource is actually being optimized. Thus LQG control does not obtain the maximal performance under a single well-motivated constraint. An example of an optimization of which the meaning is much clearer is that in which the mean-square deviations of the system are minimized, given that the control inputs are limited to some fixed range of values. Problems like this often have solutions that involve switching the inputs instantaneously from their maximum to their minimum values, and these are called "bang-bang" protocols. Note that bang-bang protocols are examples of piecewise-constant protocols. The problem is that for these "bounded controls" protocols, one cannot specify the problem without relaxing the linearity restriction (think about the range of the inputs for the linear system when G is fixed). If one does restrict the dynamics to be linear, and places a bound on G, then one can find optimal solutions, although the motivation is less clear. With the above caveats, LQG control is nevertheless of some interest, and so we now give the solution. This solution can be obtained by using the HJB equation [414, 658].

Consider a noisy linear classical system with feedback that has the underlying (pre-feedback) dynamics given by Eq. (5.162), and on which is made the set of continuous measurements given in Eq. (5.157). It turns out that the feedback that minimizes the cost J for such a linear system is also a linear function of the state variables. That is, the optimal feedback protocol is precisely the linear form of feedback that we introduced above. Under the optimal feedback the system therefore *remains linear* [414, 658]. Since the matrix F is fixed as part of the specification of the optimal control problem, the question of determining the optimal feedback protocol is simply that of determining G.

When the linear feedback specified by Eqs. (5.163)–(5.165) is applied to the system, then the state-of-knowledge of the observer evolves via

$$\langle \dot{\mathbf{v}} \rangle = (A + E)\langle \mathbf{v} \rangle + B\mathbf{c} + CM^{\mathrm{T}}\sqrt{8\Upsilon}\, \mathbf{dW}, \tag{5.167}$$

$$\dot{C} = \Sigma AC + C(\Sigma A)^{\mathrm{T}} + \hbar^2 D - 8CM^{\mathrm{T}}\Upsilon MC, \tag{5.168}$$

with $E\langle \mathbf{v} \rangle = FG\langle \mathbf{v} \rangle = F\mathbf{u}$. Note that E does not appear in the equation of motion for the covariance matrix for the same reason that B does not appear in it.

If the objective is to minimize the cost

$$J = \int_0^T \left[\langle \mathbf{v}^{\mathrm{T}}(t)Q\mathbf{v}(t) \rangle + \langle \mathbf{u}^{\mathrm{T}}(t)R\mathbf{u}(t) \rangle \right] dt + \langle \mathbf{v}^{\mathrm{T}}(T)P\mathbf{v}(T) \rangle, \tag{5.169}$$

then the optimal feedback – that is, the optimal choice for G – is obtained by solving the following matrix Riccati differential equation for a matrix S:

$$\dot{S} = -A^{\mathrm{T}}S - SA + SFR^{-1}F^{\mathrm{T}}S - Q \quad \text{with} \ S(T) = P. \tag{5.170}$$

Note that P gives the final-time boundary condition for S. The optimal choice for the "feedback gain," G, is then

$$G(t) = -R^{-1}F^{\mathrm{T}}S(t). \tag{5.171}$$

If all the matrices defining the dynamics of the system are constant (that is, do not depend on time), as well as those that define the cost function J, then the LQG problem has a much simpler steady-state solution in which G is constant. (The steady-state solution is obtained by taking $T \to \infty$ in the definition of the cost function, in which case the term given by the matrix P can be neglected.) To obtain the steady-state solution for G, all we have to do is obtain the steady-state solution for S by solving

$$0 = -A^{\mathrm{T}}S_{\mathrm{ss}} - S_{\mathrm{ss}}A + S_{\mathrm{ss}}FR^{-1}F^{\mathrm{T}}S_{\mathrm{ss}} - Q, \tag{5.172}$$

and then

$$G_{\mathrm{ss}} = -R^{-1}F^{\mathrm{T}}S_{\mathrm{ss}}. \tag{5.173}$$

Optimizing the measurement

We have used the results from classical control theory to obtain optimal control protocols for linear quantum systems. But note that for quantum systems some of the noise comes from the measurement process itself. The reason we could use LQG theory "as-is" was that we assumed that the measurement was fixed, and thus the noise on the system was fixed too. However, if we have freedom to choose what we measure, and how strongly we measure it, then we can only obtain the *true quantum LQG optimal control* by optimizing also over the measurement that we make.

The measurement is defined by the matrix M, along with the set of efficiencies given by the matrix Υ. It is clear that perfectly efficient measurements (defined by $\Upsilon = I$) are always optimal for feedback control, because the back-action noise is independent of efficiency, and this maximizes the controller's information. To find the optimal choice for M, we first determine the optimal control $G(t)$ for each M, and from this calculate the minimum value of the cost as a function of M. Then we have to minimize this function. For simple cases the entire calculation can be done analytically, but in general it must be done using a numerical search method. Numerical search methods for optimization are guaranteed to find the true minimum value of a function only if it does possess other minima that are not as small. Such additional minima are referred to as "local minima," and numerical methods may find these instead of the true "global" minimum. To the author's knowledge it is not known whether the optimal cost as a function of the measurement matrix has local minima. However, there is a special case in which the optimization problem does not have local minima. This is when the interaction with the fields that mediates the continuous measurement is fixed (that is, the observable(s) being measured are fixed) and one wants to determine the *unraveling* that provides the best control [154, 677] (see Section 3.4 for the definition of an unraveling). References on numerical optimization can be found in the "History and further reading" section of Chapter 8.

Linear, exponential, Gaussian control

There is a second optimal control problem for linear systems that can be fully solved. This is called "linear, exponential, Gaussian" (LEG) control, or alternatively *risk-sensitive* control. This differs from LQG control only in that instead of minimizing the expectation value of the quadratic cost f given in Eq. (5.166), one minimizes

$$g = -\ln\left[\left\langle e^{-\alpha f}\right\rangle\right]. \tag{5.174}$$

This form of the cost is referred to as risk-sensitive because it weights large deviations from the target state either more heavily (α negative) so that the control is "risk-averse," or less heavily (α positive) so that the control is risk-preferring. LEG control satisfies a separation principle with modified estimates that depend on the risk parameter α, and has certainty equivalence. Further details regarding LEG control for classical systems can be found in references [656, 657]. LEG control has been generalized to quantum systems by James [307].

"Robust" control of linear systems

Optimal control is about minimizing the resources, or maximizing the performance, of a control protocol. Robust control, on the other hand, is about ensuring that the protocol does not perform too badly if the parameters that define the system are not those that we initially envisaged when designing the protocol. The term "robust" refers to the robustness of the performance against deviations in the system specifications. We will not discuss robust control in this book, but for linear systems it turns out that analytic solutions for robust control protocols can be obtained. Details can be found in [717, 716].

Certainty equivalence, state-estimates, and the separation "principle"

There are a few terms that are often used in the control theory literature that we feel are worth mentioning. Recall that in Section 5.3 we described the difference between the observer's full *state-of-knowledge* as apposed to an *estimate* of the system's state or of some observable. Recall from Chapter 1 that the observer's state-of-knowledge, when combined with knowledge of the system's dynamics, contains complete information about the probabilities of the outcomes of all future measurements on the system. Thus the optimal control protocol for any control problem can *always* be realized by choosing the controls to be functions only of the state-of-knowledge. But this need not be true for estimates of the system's state or observables; the optimal control protocol may not be a function of any given set of estimates.

Control problems for which the optimal protocol can be obtained by separately obtaining a set of estimates, and then optimizing over all control protocols for which the controls are functions of these estimates, are said to obey the *separation principle*. This refers to the fact that the problem of obtaining the estimates can be separated from that of optimizing the control protocol.

Consider a control problem for a noisy system, but in which the observer has such good measurements that she always knows the state of the system at each time. In this case, the controls for the optimal control protocol will always be a function of the true state of the system, since this state is always available. Consider again the same control problem for the same system, but now the observer does not have full knowledge of the state, and thus only has an estimate. If the controls for the optimal protocol for the latter problem can be obtained by taking the optimal protocol for the former problem and simply replacing the true state by the estimate of the state in the functions that determine the controls, then the latter problem is referred to as having *certainty equivalence*. In fact, the classical LQG control problem has certainty equivalence. This can be seen immediately from the fact that the controls do not depend on the variances, but only on the means (these being the estimates of the true state): if we increase the measurement strength so that the variances go to zero, the means becomes the true state of the system, and the optimal controls do not change.

5.5.3 Optimal control for nonlinear quantum systems

In Section 5.5.1 we showed that the protocol for rapid purification in Section 5.4.1 was optimal. This protocol is only strictly applicable when the control operations are instantaneous, and is therefore a good approximation to a real feedback protocol only when the control speed is much faster than the measurement rate. In this regime, which we will refer to as *strong feedback*, continuous-time feedback protocols simplify and it is possible to obtain optimal protocols.

An important regime that can also provide simplifications is that of "good control," defined as the regime in which the control protocol is able to keep the system close to target state. This means that the measurement strength and/or the rate of the control Hamiltonian is much greater than the rate of disturbance introduced by the noise. In Section 8.3.3 we will examine optimal coherent feedback protocols in the regime of good control. Here we will consider continuous-measurement-based feedback control in the regime of both good control and strong feedback, and show that some optimality results can be derived.

Optimal protocols in the regime of good control and strong feedback

Consider the following control problem: Maximize the probability that the system is in the state $|\psi\rangle$ at every future time T. Since we can perform any unitary operation in zero time (the regime of strong feedback), to achieve this goal we need only to maximize the maximum eigenvalue of the density matrix at each future time; it doesn't matter what the eigenstate that corresponds to this eigenvalue is, because we can instantly change this eigenstate to $|\psi\rangle$. The control problem therefore concerns only the eigenvalues of the density matrix and their dynamics under continuous measurement. To achieve the objective we have at our disposal a continuous measurement of a fixed operator X with maximum measurement strength k, and can perform any Hamiltonian control. Since the Hamiltonian control takes no time, we are able to effectively measure any operator of the form $X = UYU^\dagger$ where U can be any unitary and Y is some fixed observable. As usual we will denote the system density matrix by ρ, and the eigenvalues and eigenvectors of the density matrix by λ_j and $|j\rangle$, respectively. We will label the eigenvalues in decreasing order so that λ_0 is always the largest eigenvalue. If we assume that we are in the regime of good control, then $\lambda_0 = 1 - \Delta$ with $\Delta \ll 1$.

Since we need only consider the evolution of the eigenvalues of the density matrix, it is useful to derive the equations of motion for these from the stochastic master equation (Eq. 3.38). Deriving these equations is the subject of Exercise 26, and can be done using time-independent perturbation theory. The result is

$$d\lambda_n = 8k\,dt \left[\sum_{l \neq n} \frac{\lambda_l}{\lambda_n - \lambda_l} |X_{nl}|^2 \right] \lambda_n + \sqrt{8k}\,dW \left(X_{nn} - \sum_l \lambda_l X_{ll} \right) \lambda_n, \qquad (5.175)$$

where $X_{jk} = \langle j|X|k\rangle$ are the matrix elements of X in the eigenbasis of ρ. It is important to note that since the eigenvectors of ρ *change with time*, the matrix elements X_{jk} will only be constant if X changes with time so that its matrix elements remain fixed with respect to the eigenbasis of ρ.

To exploit the fact that Δ is small (the regime of good control), we derive the equation of motion for Δ and simplify it by dropping terms that are second-order and higher in Δ. To do this we note that to first order in Δ we have $\Delta = (1 - \text{Tr}[\rho^2])/2$. From the stochastic master equation we calculate the derivative of ρ^2, and expand it to first-order in Δ. The resulting equation of motion for Δ is

$$d\Delta = -8k \sum_{j\neq 0} \lambda_j |X_{j0}|^2 \, dt - \sqrt{8k} \left[\Delta X_{00} - \sum_{j\neq 0} \lambda_j X_{jj} \right] dW. \tag{5.176}$$

We now consider optimizing the average reduction of Δ in each time-step separately. We refer to this strategy as a locally optimal protocol because it is optimal from one infinitesimal time-step to the next, but may not be optimal across larger time-differences. The average reduction in Δ is given by taking the average on both sides of Eq. (5.176), and is $d\langle \Delta \rangle = -8k \sum_{j\neq 0} \lambda_j |X_{j0}|^2 \, dt$. The following theorem shows how to chose U to maximize $d\langle \Delta \rangle$.

Theorem 14 Let X and U be N-dimensional operators, respectively Hermitian and unitary, and ρ an N-dimensional density matrix. Without loss of generality we choose X and ρ to be diagonal. The set of numbers $\{\lambda_i\}$ are the eigenvalues of ρ labeled in decreasing order. We arrange the eigenvalues of X so that the two extreme values are in the first 2-by-2 block. (If the extreme eigenvalues of X are degenerate, then one has a choice as to which eigenvectors to select to correspond to the extreme eigenvalues.) The same block of ρ contains the two largest eigenvalues λ_0 and λ_1. The maximum of $F(U) \equiv \sum_{i\neq 0} \lambda_i |\langle i| UXU^\dagger |0\rangle|^2 \equiv \sum_{i\neq 0} \lambda_i |X_{i0}^u|^2$ is achieved if

$$U = U_{opt} = U_2^u \oplus V, \tag{5.177}$$

where U_2^u is any 2-by-2 unitary unbiased w.r.t the basis $\{(1,0),(0,1)\}$:

$$U_2^u = \frac{e^{i\phi}}{\sqrt{2}} \begin{pmatrix} e^{i\theta_1} & -e^{i\theta_2} \\ e^{-i\theta_2} & e^{-i\theta_1} \end{pmatrix}, \tag{5.178}$$

and V is any unitary with dimension $N-2$.

Proof A constructive proof of this theorem, using the functional derivative, is given in [554]. Here we present a shorter proof, also given in [554]. We first obtain an upper bound on $F(U)$. This is

$$F(U) \le \lambda_1 \sum_{i=1}^{N-1} |X_{i0}^u|^2 = \lambda_1 \left[\sum_{i=0}^{N-1} |X_{i0}^u|^2 - |X_{00}^u|^2 \right]$$

$$= \lambda_1 \text{Var}\left(X, U^\dagger |0 > \right) \le \lambda_1 \frac{(x_{max} - x_{min})^2}{4}. \tag{5.179}$$

Here $\text{Var}(X, |\psi\rangle)$ denotes the variance of X in the state $|\psi\rangle$, and x_{min} and x_{max} are the extreme eigenvalues of X. The first inequality is immediate, and the last is well known [192, theorem 2]. Since U_{opt} saturates this bound, it achieves the maximum. That only unitaries of the above form achieve the maximum is simplest to show when the eigenvalues of ρ are non-degenerate: to saturate the first inequality one must restrict U to the subspace spanned by $\{|0\rangle, |1\rangle\}$, and to achieve the last, U must be unbiased w.r.t to the eigenbasis. When the eigenvalues of ρ are degenerate, a careful analysis shows that this remains true [554] $\qquad \square$

This theorem shows us immediately that for $N = 2$ and $N = 3$ there is only one locally optimal protocol, because in these cases the operator V in the theorem is trivial. For $N \geq 4$ there are many possible locally optimal protocols, and some will be better than others because the choice of V at time t will in general affect the value of $d\langle \Delta \rangle$ at later times. Optimizing V is non-trivial, as the equations of motion for the eigenvalues of the density matrix are nonlinear. For $N = 4$ this optimization is not difficult, and is the subject of Exercise 28.

The locally optimal protocol for $N = 2$ corresponds to the optimal protocol for rapid purification in Section 5.4.1, and in Exercise 21 you can show that it is also the optimal protocol for the control problem considered here. We now use the verification theorem described in Section 5.5.1 to show that the locally optimal protocol for $N = 3$, for the case in which Y is proportional to J_z, is also a fully optimal protocol.

We must first determine the equation of motion for the cost function under the proposed protocol. The optimal X is given by $X_{01} = 1 = -X_{10}$ with all other elements equal to zero. Note that since $\mathrm{Tr}[\rho] = 1$ only two of the three eigenvalues are independent dynamical variables. The resulting equations of motion for these eigenvalues are

$$\dot{\lambda}_1 = -8k\lambda_1, \qquad \dot{\lambda}_2 = 0. \tag{5.180}$$

Recall that by definition $\lambda_1 \geq \lambda_2$. But observe that if $\lambda_1 = \lambda_2$ then λ_1 will be *smaller* than λ_2 after a time-step δt. This means that we must apply a unitary to swap λ_1 and λ_2 before the next time-step. Two time-steps later we have to swap them back. In the limit in which this swapping is done continuously, the rate of reduction of λ_1 is *split equally* between λ_1 and λ_2, so they both decay at the rate $-4k$. The equations of motion are thus

$$\dot{\lambda}_1 = -4k\lambda_1, \qquad \dot{\lambda}_2 = -4k\lambda_2. \tag{5.181}$$

So the optimal feedback protocol has two "phases." When $\lambda_1 > \lambda_2$ the evolution of the eigenvalues is $\lambda_1(t) = e^{-8kt}\lambda_1(0)$ and $\lambda_2(t) = \lambda_2(0)$. When $\lambda_1(t)$ hits the value of $\lambda_2(0)$, which happens at time $\tau = \ln[\lambda_1(0)/\lambda_2(0)]/(8k)$, the evolution changes and the resulting evolution is

$$\lambda_1(t) = \begin{cases} e^{-8kt}\lambda_1(0) & t \leq \tau \\ e^{-8k\tau}e^{-4k(t-\tau)}\lambda_1(0) & t > \tau \end{cases}, \quad \lambda_2(t) = \begin{cases} \lambda_2(0) & t < \tau \\ e^{-4k(t-\tau)}\lambda_2(0) & t \geq \tau. \end{cases} \tag{5.182}$$

The cost is the quantity that the protocol is supposed to minimize, and this is $\Delta(T) = \lambda_1(T) + \lambda_2(T)$. The cost function is the value of the cost as a function of the current time, t, and the current values of the dynamical variables, which we will denote simply by λ_1

and λ_2. With these definitions the switching time is given by $t + \ln[\lambda_1/\lambda_2]/(8k)$, and using Eqs. (5.182) the cost function becomes

$$C(t, \lambda_1, \lambda_2) = \begin{cases} e^{-8k(T-t)}\lambda_1 + \lambda_2 & t \geq t_c, \\ \sqrt{\lambda_1 \lambda_2} e^{-4k(T-t)} & t < t_c, \end{cases} \tag{5.183}$$

where $t_c = T - \ln(\lambda_1/\lambda_2)/(8k)$. We can now apply the verification theorem that uses viscosity solutions, described in Section 5.5.1. We first check that when t is in each of the regions $[t_c, T]$ and $[0, t_c]$ the cost function satisfies the HJB equation given in Eq. (5.107). For the first region the HJB equation becomes

$$\frac{\partial C}{\partial t} = -8 \min_{U} \left\{ -\lambda_1 |X_{10}|^2 \alpha(t) - \lambda_2 |X_{20}|^2 - \left(\frac{\lambda_1 \lambda_2 \beta(t)}{\lambda_1 - \lambda_2} \right) |X_{21}|^2 \right\}.$$

with $\alpha(t) = e^{-8a^2(T-t)}$ and $\beta(t) = 1 - \alpha(t)$. In this region t is such that $\lambda_1 e^{-8a^2 t} \geq \lambda_2$, and because of this we can recast the problem as the maximization of

$$F(U) = \eta |X_{10}|^2 + |X_{20}|^2 + \xi |X_{21}|^2, \tag{5.184}$$

where

$$\eta = \frac{\lambda_1}{\lambda_2} e^{-8a^2(T-t)} > 1 > \xi = \eta \left[\frac{e^{8a^2(T-t)} - 1}{\lambda_1/\lambda_2 - 1} \right]. \tag{5.185}$$

Performing this maximization numerically we find that the HJB equation is

$$\frac{\partial C}{\partial t} = 8k\lambda_1 e^{-8k(T-t)}. \tag{5.186}$$

By calculating the time-derivative of C in the interval $[t_c, T]$ we see that Eq. (5.186) is true, and thus C satisfies the HJB equation. The protocol is therefore optimal when $T < \tau$.

Calculating the HJB equation for the second interval we find that it is

$$\frac{\partial C}{\partial t} = -\min_{U} \left\{ -(X_{11} - X_{22})^2 - 4 \left(|X_{10}|^2 + |X_{20}|^2 + |X_{21}|^2 \right) \right\} C.$$

Again performing the optimization numerically we find that

$$\frac{\partial C}{\partial t} = 4kC, \tag{5.187}$$

and this is indeed satisfied by C on the interval $[0, t_c]$.

The last thing to do to confirm that the protocol is optimal is to check the super- and sub-derivatives at the point at which the derivatives are undefined, $t = t_c$. We leave this as Exercise 29.

History and further reading

The continuous-time feedback control of quantum systems, using the fully developed theory of continuous quantum measurements, was first considered by Belavkin [41, 42, 43] and by Wiseman and Milburn [681, 668, 682, 672], more or less concurrently. Since Belavkin had derived the stochastic master equation (SME) as the quantum analogue of the Kushner–Stratonovich equation, it was natural for him to consider the connection between quantum and classical feedback control at the level of controlling observables such a position and momentum. He therefore elucidated the fact that the quantum equations for linear systems had the same form as the Kalman filter, and so the classical LQG methods could be applied. Belavkin did not apply his methods to concrete physical situations, however, and his highly mathematical style appears to have prevented his work from being digested by the physics community.

Wiseman and Milburn (WM) took a very different approach. While LQG control involves processing (filtering) the measurement record to obtain a real-time estimate of an observable, WM instead considered applying a force (feedback Hamiltonian) to a system that was proportional to the raw measurement record. They realized that if they did that, they could average over the action of the feedback and obtain a master equation that was still linear and Markovian, and could therefore be solved analytically for a range of problems. They showed that their Markovian feedback could be used to obtain a number of interesting effects in quantum optics [683, 671].

In 1998 two sets of authors, again concurrently, rediscovered the connection between the SME for linear systems and the Kalman filter, and showed that LQG control theory could be applied directly to linear quantum systems under continuous measurements of the canonical coordinates (Yanagisawa and Kimura [698] and Doherty and Jacobs [153]). The general formulation of continuous-time measurement-based feedback control for linear systems, presented in Section 5.5.2 was elucidated by Wiseman and Doherty in [677]. A general formulation of Wiseman–Milburn Markovian feedback was developed in [119].

In the last decade quite a number of theoretical papers have been written about continuous-time feedback control via continuous measurements. Various topics studied include quantum state-preparation and preservation [683, 398, 153, 155, 156, 600, 524, 574, 578, 499, 526, 577, 396, 575, 603, 663, 102, 694, 646, 713, 305], metrology [157, 670, 678, 54, 55, 579], quantum error-correction [4, 5, 534, 114], and explorations of optimality [291, 554]. Work with a mathematical, control-theoretic emphasis has focused on exact mathematical results on stability and convergence [628, 7, 629, 426, 8, 626, 496, 606], on optimality [307], on structural properties of quantum noise [695, 607], and on the exploration of different algorithms for estimation and control [424, 425, 158, 692, 693, 117]. Rapid state-reduction was introduced in [284], and further work on this subject can be found in [130, 318, 228, 684, 251, 675, 132, 593, 525, 369]. The protocol discussed in Section 5.4.2 for controlling systems with measurements only was drawn from [300]. Another approach to controlling systems solely via measurements is given by Ashhab and Nori in [22].

Continuous measurement-based quantum feedback control, including adaptive measurement, was first demonstrated with optical systems and trapped single atoms [567, 431, 179, 19, 95, 250, 347, 340, 541, 209, 536, 85]. Monitoring and feedback control of quantum superconducting systems only became possible with the development of ultra-low-noise amplifiers [703, 637, 638, 517]. Real-time monitoring of electron and nuclear spins has also been demonstrated using optical read-out [444, 622].

Feedback control of quantum systems without the use of measurements, now referred to as coherent feedback, was first considered by Wiseman and Milburn. They showed that their measurement-based Markovian feedback control could be reproduced by coupling the system to another quantum system via a one-way quantum field. They also showed that there were control objectives that could be achieved by using coherent feedback that could not be achieved by measurement-based feedback because of the action of the measurement [682].

In 2000 Lloyd introduced the idea of quantum feedback control without measurements in a very different setting to that of WM [378, 442]. He considered a quantum system that acted as a feedback controller for another quantum system, by exploiting only unitary interactions between the two. Realizations of this form of feedback in quantum optics were devised by Vitali *et al.* [641]. In 2003 Yanagisawa and Kimura, and a little later James and collaborators, began to analyze coherent feedback in the paradigm considered by Wiseman and Milburn, from a control theoretical point of view. These authors have produced a considerable body of work codifying and understanding the structure of linear coherent feedback networks [699, 700, 147, 452, 453, 222, 223, 308, 711, 381, 555]. In 2009 Nurdin, James, and Petersen [452] found a numerical example in which linear coherent feedback networks could outperform linear measurement-based feedback, which stimulated interest in the physics community in the relationship between measurement-based and coherent feedback. Hamerly and Mabuchi showed that time-independent linear coherent feedback significantly outperforms linear measurement-based cooling for harmonic oscillators [240, 241]. We analyze coherent and measurement-based feedback cooling of harmonic oscillators in Chapter 8. Our discussion of the potential superiority of coherent feedback was drawn from [649]. Applications of field-mediated coherent feedback have been considered in [329, 385, 691, 714, 240, 241, 715]. Examples of coherent feedback have been realized in optics [384], in superconducting circuits [328, 330], and for cooling mechanical resonators in nano-electromechanical and optomechanical systems [437, 336, 542, 543, 112, 597].

The input–output formulation of open systems is used extensively in continuous-time coherent feedback networks to describe cascade (one-way field-mediated) connections between quantum systems. Input-output theory was developed by Collet and Gardiner [128, 199], and its application to describing cascade connections was introduced by Gardiner [198]. The input–output formalism has also been extended to handle non-Markovian couplings to the fields [715]. Further details about optics and the optical elements used to construct field-mediated optical networks can be found in, e.g., [327, 341].

Classical LQG and risk-sensitive control of linear classical systems are described in, e.g., [414, 658, 656] and robust control of these systems is discussed in [717, 716]. Further details regarding the HJB equation and Pontriagin's maximum principle can be found in standard textbooks on optimal control theory, such as Kirk [335]. The classic verification theorem for differentiable functions can be found in, e.g., [181], and the verification theorem for viscosity solutions was proved in [224, 718].

Exercises

1. Show that given the freedom to choose any feedback unitary operators U_j, feedback can be used to turn any measurement into a bare measurement.
2. A (measurement-based) feedback process consists of a sequence of measurement and feedback steps. If one is not concerned with the time taken, a feedback protocol in which one changes the measurement basis from step to step can instead be realized by applying the same measurement each time, but applying unitary feedback operations to the system in between the measurements. Show that this is true.
3. Consider a feedback process in which a measurement described by the operators $\{A_j\}$ is made on a system in the initial state ρ. Then a unitary operator U_j is applied to the system upon obtaining the result j. The purpose of the feedback is to minimize the entropy of the final state, $S_j = \text{Tr}[U_j A_j \rho A_j^\dagger U_j^\dagger]$, where this entropy is averaged over the measurement outcomes. (i) How does this entropy depend on the unitary operators U_j? (ii) If the goal is instead to minimize the entropy of the final state, where it is the final state that is averaged over the measurement outcomes, how does one choose the unitaries U_j to minimize this entropy? Hint: the results in Chapter 2 imply that the entropy of the sum of two density matrices is minimal when the density matrices have the same eigenbasis.
4. Consider a unitary operator U_swap that will swap the states of two systems, for any initial pure-states. Show that this operator will also swap the states of the two systems if they are in mixed states. Conversely, show that any unitary operator that will transfer a completely mixed state from one system to another with the same dimension will also swap any initial pure-states between two systems with the same dimension.
5. Determine the explicit coherent-feedback correlation and actuation steps (that is, the unitary operators that implement them, and the Hamiltonians that generate these unitary operators) required to swap the states of two N-level quantum systems. Hint: a swap operation transforms each joint state $|n\rangle|m\rangle$ to $|m\rangle|n\rangle$.
6. If we want to remove all the entropy from a qubit, we can perform a swap operation between the qubit and another qubit that has been prepared in the ground state. Assume that there is a maximum speed at which we can transform from one state to another, which may be thought of as a maximum speed of rotation in Hilbert space. The swap operation between two qubits requires transforming the state $|01\rangle$ to $|10\rangle$ (and vice versa). This requires a 90° rotation in Hilbert space. Now consider how we might

remove the entropy from the qubit by using a correlation and actuation (measurement-like) operation. To do so we must first fully correlate the two qubits, so that, for each state of the second qubit, we can then apply a unitary to the first qubit to leave it in the ground state. The task is to obtain a correlation/actuation method that takes no longer than the swap, under the restriction on the speed of rotation in Hilbert space. Hint: if we start with a qubit in a single state, we can obtain one of two orthogonal states by choosing to either (i) rotate the initial state by 45° in one direction, or (ii) rotate it by 45° in the other direction.

7. If one makes a continuous measurement of the position of a single particle that is initially in a Gaussian state, then the state remains Gaussian. Apply an efficient continuous measurement to the general pure Gaussian state given in Section G.2, and derive equations of motion for the parameters of the Gaussian state.

8. If one makes a continuous measurement of the position of a single particle that is initially in a Gaussian state, then the state remains Gaussian. In the previous exercise we showed this to be true for a pure Gaussian state. One way to show that it is true for a general Gaussian state is to start from the stochastic master equation for a continuous measurement of position, and derive the equations of motion for the means and variances of x and p using $d\langle c \rangle = \text{Tr}[(d\rho)c]$, where c is an arbitrary operator. To do this one must remember that the covariance of x and p must be symmetrized, so that $C_{xp} = \langle xp + px \rangle/2 - \langle x \rangle \langle p \rangle$. The third cumulants of x and p appear in the equations of motion for the variances and covariance, and the assumption that the state is Gaussian means setting these cumulants to zero. The result is a closed set of equations for the means and variances. Use the above method to calculate $d\langle x \rangle$ and $dV_x = d\langle x^2 \rangle - 2\langle x \rangle d\langle x \rangle + (d\langle x \rangle)^2$.

9. Consider one optical cavity mode damped at the rate γ, whose output is fed to the single input of a second optical cavity damped at the rate κ. The output of the second cavity is then fed into a second input to the first cavity. Determine the equations of motion for the annihilation operators of both cavities, and show that the interaction between them is that given by the Hamiltonian $H = -i\sqrt{\gamma\kappa}(ab^\dagger - ba^\dagger)$, where a and b are the annihilation operators of the respective cavities.

10. Calculate the power spectrum of the second (free) output from the first optical cavity in Exercise 9.

11. Consider a field-mediated feedback loop between a primary and auxiliary using the same operator definitions as in Eqs. (5.18) and (5.19). (i) Show that the configuration depicted at the bottom-right in Fig. 5.2 is described by the Langevin equations

$$dS = \left\{ \mathcal{L}_s S - 2 \left\{ \left[S, F^\dagger \right] (P+M) - (P+M)^\dagger [S,F] \right\} - [\mathcal{K}(M) + \mathcal{K}(F)]S \right\} dt$$
$$+ \sqrt{2} \left(\left[S,(M+F)^\dagger \right] ds_{\text{in}} - ds_{\text{in}}^\dagger [S,(M+F)] \right), \qquad (5.188)$$

$$dA = \left\{ \mathcal{L}_a A - 2 \left(\left[A, P^\dagger \right] M - M^\dagger [A,P] \right) - \mathcal{K}(P)A \right\} dt$$
$$+ \sqrt{2} \left(\left[A, P^\dagger \right] ds_{\text{in}} - ds_{\text{in}}^\dagger [A,P] \right). \qquad (5.189)$$

(ii) Show that these equations can be rewritten as

$$dS = \left\{ \mathcal{L}_s S - 2i \left[S, iFP^\dagger + (iFP^\dagger)^\dagger \right] - \mathcal{K}(M+F)S \right\} dt$$
$$+ \sqrt{2} \left(\left[S, (M+F)^\dagger \right] ds_{\text{in}} - ds_{\text{in}}^\dagger [S, (M+F)] \right), \tag{5.190}$$

$$dA = \left\{ \mathcal{L}_a A - 2i \left[A, iPM^\dagger + (iPM^\dagger)^\dagger \right] - \mathcal{K}(P)A \right\} dt$$
$$+ \sqrt{2} \left(\left[A, P^\dagger \right] ds_{\text{in}} - ds_{\text{in}}^\dagger [A, P] \right). \tag{5.191}$$

(iii) How should we choose the operator F so that the equation of motion for the system operator S has no quantum noise terms?

12. Consider the master equation given by Eq. (5.22). What happens to the terms of the RHS when we set $F = -C$ and $P = -M$?

13. Deriving the master equation (5.21) from the quantum Langevin equations (5.18) and (5.19) is not entirely straightforward, due to the fact that each of the Langevin equations describes the evolution of the operators only for one of the subsystems, not for both. The master equation must be valid for any operator of the combined system. To obtain the master equation we first write down a single Langevin equation for an operator X that will reduce either to Eq. (5.18) or Eq. (5.19), depending on whether we set X equal to an operator of the primary system or the auxiliary system. This is not difficult: by inspecting Eqs. (5.18) and (5.19), we see that the terms in (5.18) vanish when we replace the system operator S with an auxiliary operator A, and vice versa. We can therefore construct the needed Langevin equation for X by including all the terms from both (5.18) and (5.19). Your tasks are (i) to write down the Langevin equation for an operator of the combined system, and (ii) to use this equation, and the method described in Section F.7 to derive Eq. (5.21).

14. Consider a single qubit in the state

$$\rho = U \begin{pmatrix} p & 0 \\ 0 & 1-p \end{pmatrix} U^\dagger, \qquad U = \begin{pmatrix} \cos(\theta/2) & \sin(\theta/2) \\ -\sin(\theta/2) & \cos(\theta/2) \end{pmatrix}, \tag{5.192}$$

where ρ and U are written in the σ_z basis. Here U rotates the σ_z eigenstates by angle θ on the Bloch sphere toward the σ_x basis. When $\theta = \pi/2$ the σ_z eigenstates are transformed to σ_x eigenstates. The diagonal elements of ρ give the probability distribution for the system to be in either of the σ_z eigenstates. We will refer to these diagonal elements as the z-distribution.

(i) For what value of θ is the entropy of the z-distribution greatest? (ii) Now consider the measurement given by the operators $A_\pm = \sqrt{1-k}|\pm\rangle\langle\pm| + \sqrt{k}|\mp\rangle\langle\mp|$ where $|\pm\rangle$ are the eigenstates of σ_z. Make this measurement on the following three states: (1) the state ρ with $\theta = 0$; (2) the state ρ with $\theta = \pi/2$; 3) the state in (2) but with the off-diagonal elements set to zero. How does ΔI_{sys} compare in the three cases? (iii) How do the answers to question (ii) relate to the rapid purification protocol for a single qubit described in Section 5.4.1?

15. Consider the feedback protocol in Section 5.4.3, and determine its performance in the steady state when the measurement has efficiency η.

16. Derive Eqs. (5.60) and Eqs. (5.61). This means deriving the equations of motion for the Bloch vector length, $a \equiv |\mathbf{a}|$, and the angle $\theta = \tan^{-1}(a_x/a_z)$ under a continuous measurement in the σ_x basis, and another in the σ_z basis. To do this it is simplest to align the z-axis with the Bloch vector so that $a = a_z$ and $a_x = a_y = 0$, as is done in the derivation of Eqs. (5.38)–(5.40). One then uses Ito calculus to change variables to a and θ.

17. Consider a deterministic linear dynamical system given by $\dot{\mathbf{x}} = A(\mathbf{u})\mathbf{x}$, where A is a time-independent matrix whose elements are the control parameters $\mathbf{u}$. The cost to be minimized is $C = \mathbf{b}^{\mathrm{T}}\mathbf{x}(\tau)$, for some fixed vector $\mathbf{b}$ and the final time τ, where the superscript "T" denotes transpose. Show that in this case the HJB equation for the cost function becomes

$$\mathbf{b}^{\mathrm{T}} e^{A(\mathbf{u})(\tau - t)} A(\mathbf{u})\mathbf{x} = \min_{\mathbf{v}} \left[\mathbf{b}^{\mathrm{T}} e^{A(\mathbf{u})(\tau - t)} A(\mathbf{v})\mathbf{x} \right], \quad 0 \le t \le T. \tag{5.193}$$

Note that if the elements of A can be varied in time, but the guess for the optimal controls is constant in time, the HJB equation *still* has this simple form, with the only change that $\mathbf{v}$ can have a different value at each time:

$$\mathbf{b}^{\mathrm{T}} e^{A(\mathbf{u})(\tau - t)} A(\mathbf{u})\mathbf{x} = \min_{\mathbf{v}(t)} \left[\mathbf{b}^{\mathrm{T}} e^{A(\mathbf{u})(\tau - t)} A(\mathbf{v}(t))\mathbf{x} \right], \quad 0 \le t \le T. \tag{5.194}$$

18. Consider the problem of bringing a particle moving on a line to rest at the origin in the shortest time, given that a maximum acceleration of a can be applied. We solved this problem in the text using Pontryagin's theorem. Now consider guessing a solution and using the verification theorem. It is fairly obvious that the solution to the problem is to accelerate the particle toward the origin as fast as possible, until the point at which we must decelerate it as fast as possible in order to avoid an overshoot. (i) Given this control strategy, work out the evolution of the particle for all initial conditions, the total time to reach the origin, T, and the time at which the control is switched from acceleration to deceleration. (ii) Use the HJB equation in the form given by Eq. (5.111) to verify that this is indeed the optimal control protocol.

19. Derive the HJB equation given in Eq. (5.111).

20. Consider the problem of using a continuous measurement, combined with feedback, to prepare an initially mixed qubit in a given pure state, with a specified minimum probability p, in the shortest time (on average). We simplify the problem by assuming that we can apply any unitary operation instantly (the regime of strong feedback). In this case the problem reduces to that of making one of the eigenvalues of the density matrix at least as large as p in the shortest time. The constraints are that (i) we must make a continuous measurement with measurement strength no greater than k; (ii) at each time we can measure any operator of the form $U\sigma_z U^{\dagger}$ for any unitary U. Use the HJB equation for minimum-average-time problems to show that the optimal

protocol is to measure in the eigenbasis of the density matrix. This is the Wiseman–Ralph protocol [684, 675].

21. Consider the problem of using a continuous measurement, combined with feedback, to maximize, on average, the maximum eigenvalue of the density matrix at a fixed future time T. We assume that we are in the regime of strong feedback, so that any unitary operation can be applied instantly. The constraints are that (i) we must make a continuous measurement with measurement strength no greater than k; (ii) at each time we can measure any operator of the form $U\sigma_z U^\dagger$ for any unitary U. Use the verification theorem to show that the optimal protocol is to keep the measurement basis at an angle of $90°$ to the eigenbasis of the density matrix, where the angle is measured on the Bloch sphere. This is the same protocol as the one presented in Section 5.4.1 for rapid purification of a qubit, and the goal described here is essentially the same, but with a different measure of performance. The measure used in this exercise is, in fact, more well-motivated physically, because it corresponds to something that can be measured.

22. Consider a quantum harmonic oscillator subjected to a continuous measurement of its position. (i) Use Eqs. (5.159) and (5.160) to write down the equations of motion for the means and covariances of the position and momentum of the oscillator. (ii) Use Eq. (5.162) to write down the equations of motion for a noisy classical oscillator, and the corresponding classical continuous measurement, which are together equivalent to the measured quantum oscillator.

23. Consider a free particle with mass m subjected to a continuous measurement of its position with measurement strength k. We will use Eqs. (5.172) and (5.173) to obtain the optimal LQG protocol for minimizing the steady-state energy of the particle, given that we can apply a linear force to it. (i) Denoting the "coordinate" vector by $(x,p)^{\mathrm{T}}$, determine the matrices A, B, and M for the oscillator. (ii) Since the feedback involves applying a force, and since the matrix G is arbitrary, we can set the matrix F to be

$$F = \begin{pmatrix} 0 & 0 \\ 0 & 1 \end{pmatrix}. \tag{5.195}$$

Since we wish to minimize the energy of the oscillator we can choose

$$Q = \frac{1}{2}\begin{pmatrix} 1/m & 0 \\ 0 & m\omega^2 \end{pmatrix}, \quad R = \alpha^2 Q, \tag{5.196}$$

where α is a constant with units of time that sets the relative weighting of Q and R. Determine S_{ss} and G_{ss}. (iii) Calculate the steady-state energy of the oscillator under the feedback protocol. (iv) Find the value of the measurement strength k that gives the minimum steady-state energy.

24. Consider a quantum harmonic oscillator with frequency ω and mass m subjected to a continuous measurement of its position with measurement strength k. We will use Eqs. (5.172) and (5.173) to obtain the optimal LQG protocol for minimizing the steady-state energy of the oscillator, given that we can apply a linear force to it. (i) Denoting the

coordinate vector by $(x,p)^\mathsf{T}$, determine the matrices A, B, and M for the oscillator. (ii) The matrices F, Q, and R are given by Eqs. (5.195) and (5.196) above. Determine S_{ss} and G_{ss}. (iii) Calculate the steady-state energy of the particle under the feedback protocol. (iv) Find the value of the measurement strength k that gives the minimum steady-state energy. (v) What is the minimal energy found in (iv) when $\eta \to 0$?

25. Consider a random variable f. Show that if the variance of f, V_f, is much less than $1/\alpha^2$, then $g = -\ln[\langle\exp(-\alpha f)\rangle]$ is given approximately by

$$g \approx \alpha \left(\langle f \rangle - \frac{\alpha}{2} V_f \right). \tag{5.197}$$

This means that g contains information about the variance of f, and thus is sensitive to the "risk" involved in relying on the value of f.

26. Given the stochastic master equation, Eq. (3.38), determine the resulting stochastic equations of motion for the eigenvalues of ρ. You can do this by using time-independent perturbation theory (TIPT) in the following way. The stochastic master equation gives an expression for $d\rho$, where $\rho(t+dt) = \rho(t) + d\rho$. Since ρ is a matrix, this expression has the usual TIPT form $H' = H + \lambda V$, where $\rho(t+dt)$ takes the place of H'. TIPT therefore shows us how to determine the eigenvalues of $\rho(t+dt)$ in terms of those of $\rho(t)$, in powers of $d\rho$, which in turn translates to powers of dt and dW. The equation of motion for the eigenvalues is given by expanding only to the first power of dt. However, since $(dW)^2 = dt$ (the Ito rule), we must calculate the perturbation expansion to second order in $d\rho$, and then keep all terms that are first-order in dt. The recursion relations for TIPT are given in Appendix G. Note that this procedure gives the exact equations of motion for the eigenvalues, not merely an approximation to them.

27. Derive Eq. (5.176).

28. Consider the locally optimal protocol described by Theorem 14 applied to a four-dimensional system. To find the best locally optimal protocol we need to determine the operator V, which in this case is two-dimensional. It is clear that the best thing we can do is to increase the second-smallest eigenvalue of ρ, λ_2, at the fastest average rate, and we know from the rapid-purification protocol in Section 5.4.1 that the way to do this is to chose V so as to measure in a basis unbiased with respect to the eigenstates $|3\rangle$ and $|4\rangle$. Taking the measured observable to be $Y = 2J_z$, so that $(Y_{00}, Y_{11}, Y_{22}, Y_{33}) = (3, 1, -1, -3)$, determine the form of X for this choice of V (taking V to be real), and show that the resulting equations of motion of the eigenvectors of ρ are

$$\dot{\lambda}_1 = \dot{\Delta} = -24k\lambda_1, \tag{5.198}$$

$$\dot{\lambda}_2 = -\dot{\lambda}_3 = 8k\lambda_2\lambda_3/(\lambda_2 - \lambda_3), \tag{5.199}$$

29. Show that the super- and sub-derivatives of the cost function in Eq. (5.183) match at the point $t = t_c$.

6

Metrology

While the term *metrology* refers to measurement techniques in general, it is used more specifically to mean the study of techniques to measure classical quantities. Metrology is concerned with how to make measurements as accurately as possible, and to quantify how accurate a measurement technique is. Precise measurements of quantities such as frequency and mass are important for establishing and maintaining standard units. Other effectively classical quantities that one may wish to measure are acceleration and the strengths of magnetic or electric fields.

Metrology is concerned not only with measurements of single fixed quantities but also with quantities that vary in time, usually called signals or waveforms. Since all signals have some maximum rate at which they change, they can be discretized by sampling with sufficient frequency. Thus all signals can be regarded as finite sets of values to be measured. Nevertheless, the fact that these values arrive in a sequence with a given duration places practical limitations on their measurement. Optimal methods for measuring signals are important in communication, and in many scientific applications where data is time-varying and noisy.

In determining the value of a classical quantity, after processing the measurement results we usually need to arrive at a single "estimate" for the value. If we use standard (Bayesian) measurement theory, then we first obtain a probability distribution for the possible values, and from this we can choose our estimate. The meanings of the terms "metrology" and "estimation" therefore overlap. Since we are concerned in this book with quantum systems and quantum measurements, the metrology we consider here is that in which quantum systems are used as the probes for measuring classical quantities. In fact, for measurements involving light, the discrete nature of photons is essential for determining the limits to precision. When using quantum systems as probes, the classical value to be determined appears as a parameter in the Hamiltonian of the system, and hence metrology is often referred to as *parameter estimation*.

Our treatment of metrology here is fairly brief. Its purpose is to introduce some of the key concepts, terminology, and topics of metrology, so as to give the reader a base from which to explore the literature. The first part of this chapter deals with the measurement of single-valued quantities, while the second part discusses the estimation of signals. In the latter case we must consider how to handle the dynamics of the probe system during the influence of the signal, and the back-action that results from this dynamics.

6.1 Metrology of single quantities

6.1.1 The Cramér–Rao bound

Let us recall briefly classical measurement theory, discussed in Section 1.2: when we make a measurement that gives result y, to determine the value of a quantity x, the measurement is characterized by the likelihood function, $P(y|x)$. This function is the probability distribution for the measurement result y given that the value of the quantity is x. Once we have obtained the measurement result, Bayes' theorem tells us how to take this information into account to determine our new probability distribution (our state-of-knowledge) for x.

What we did not discuss in Section 1.2 is that, once we have the probability distribution for x, we must usually choose a single value to use as our result for x. This value is called our *estimate* for x, and we will denote it by $\tilde{x}$. As examples, we could choose $\tilde{x}$ to be the value of x at which the posterior probability distribution for x is maximum, or the average value of x given by this distribution. The rule that we use to choose $\tilde{x}$, as a function of the measurement result, y, is called the *estimator*. Ideally we want to choose this estimator so that $\tilde{x}$ deviates from x as little as possible, on average. To measure the performance of an estimator, we should average the mean-square error over all values of y and all values of x. However, in the interests of avoiding the prior, it is common to consider the mean-square error for each value of x separately, even though the estimator cannot depend on x because the true value of x is not known to the person making the measurements. This avoids the use of the prior, but will strictly give optimal results (that is, achieve the lowest error) only when the prior is unity.

If we analyze estimators by considering how they perform for each value of x, then it is useful to define an *unbiased* estimator as one for which the error, $\epsilon(y,x) = \tilde{x}(y) - x$, gives zero when averaged over all possible values of y (for each value of x). We can evaluate the performance of an unbiased estimator by averaging the square of the error over all values of y. This mean-square error, ϵ_{ms}, is

$$\epsilon_{\text{ms}}(x) = \int [\tilde{x}(y) - x]^2 P(y|x) \, dy. \tag{6.1}$$

Note that whatever the minimum possible value for ϵ_{ms} (for a given value of x), this minimum is determined entirely by the likelihood function.

The Cramér–Rao bound states that whatever estimator function we use,

$$\epsilon_{\text{ms}}(x) \geq \frac{1}{\mathcal{I}(x)}, \tag{6.2}$$

where $\mathcal{I}(x)$ is called the *Fisher information*, and is given by

$$\mathcal{I}(x) = \int \left(\frac{\partial}{\partial x} \ln [P(y|x)] \right)^2 P(y|x) \, dy = \int \left(\frac{\partial^2}{\partial x^2} \ln [P(y|x)] \right) P(y|x) \, dy, \tag{6.3}$$

where the second equality follows from the fact that $P(y|x)$ is a probability distribution. It turns out that the expectation value of $\partial_x(\ln[P(y|x)])$ over $P(y|x)$ is zero, and thus the Fisher information is the variance of $\partial_x(\ln[P(y|x)])$ over $P(y|x)$. (Here we are using ∂_x as shorthand for $\partial/\partial x$.) Since the minimum mean-square error is bounded by $1/\mathcal{I}(x)$, the Fisher information characterizes how much the measurement tells us about x. The Cramér–Rao bound is useful because it allows us to determine the limits to the precision that can be achieved by a given measurement (assuming that the prior function is relatively flat).

In many cases we make more than one measurement to obtain our estimate for x, and thus the likelihood function is the product of the likelihood functions for each measurement. If we make m identical measurements, with measurement results $y_1,\ldots,y_n$, then $P(y|x) = \prod_{j=1}^{m} P(y_j|x)$. Substituting this expression for $P(y|x)$ into the right-most expression in Eq. (6.3), we obtain

$$\mathcal{I}(x) = \sum_{j=1}^{m} \mathcal{I}_j(x) = m\mathcal{I}_1(x), \tag{6.4}$$

where $\mathcal{I}_j(x)$ is the Fisher information for the jth likelihood function. The maximum precision, as defined by the minimum root-mean-square error, $\epsilon_{\mathrm{rms}}(x) = \sqrt{\epsilon_{\mathrm{ms}}(x)}$, scales as $1/\sqrt{m}$.

The proof of the Cramér–Rao bound is quite short. If the estimator is unbiased, then

$$\langle \epsilon(x) \rangle = \int \epsilon(y,x)P(y|x)\,dy = \int [\tilde{x}(y) - x]P(y|x)\,dy = 0. \tag{6.5}$$

Differentiating $\langle \epsilon(x) \rangle$ with respect to x, and rearranging gives us

$$\int \epsilon(y,x)\frac{\partial}{\partial x}P(y|x)\,dy = \int P(y|x)\,dy = 1. \tag{6.6}$$

Now there are just two more steps. We note that $\partial_x P = P\partial_x \ln[P]$, and substituting this into the above equation we can write the result as

$$\int \left(\epsilon(y,x)\sqrt{P(y|x)} \right) \left(\sqrt{P(y|x)}\frac{\partial}{\partial x}\ln[P(y|x)] \right) dy = 1. \tag{6.7}$$

We finally use the Cauchy–Schwartz inequality, $(\int A^2\,dy)(\int B^2\,dy) \geq (\int AB\,dy)^2$, and this gives us

$$\int [\epsilon(y,x)]^2 P(y|x)\,dy \times \int \left(\frac{\partial}{\partial x}\ln[P(y|x)] \right)^2 P(y|x)\,dy = \epsilon_{\mathrm{ms}}(x)\mathcal{I}(x) \geq 1. \tag{6.8}$$

6.1.2 Optimizing the Cramér–Rao bound

The Cramér–Rao bound applies to any parameter estimation problem, and thus to quantum mechanics. The role of quantum mechanics is merely to determine the conditional probability $P(y|x)$. Let us say that we make a measurement with measurement operators $\{A_y\}$

on a quantum system to determine some parameter x. We can obtain information about x so long as the density matrix of the system depends upon x. Writing this density matrix as ρ_x, the conditional probability is given by $P(y|x) = \text{Tr}[E_y \rho_x]$ with $E_y \equiv A_y^\dagger A_y$. By defining an operator L_x that satisfies

$$\frac{\partial \rho_x}{\partial x} = \frac{L_x \rho_x + \rho_x L_x}{2},\tag{6.9}$$

we can write the Fisher information in terms of ρ_x in a compact way. The operator L_x is called the *symmetric logarithmic derivative*, and satisfies $L_x = L_x^\dagger$. Substituting the quantum mechanical expression for the conditional probability into the Fisher information, and writing the result in terms of L_x, we obtain

$$\mathcal{I}(x) = \int \left(\frac{\partial}{\partial x} \ln[P(y|x)]\right)^2 P(y|x)\, dy = \int \left(\frac{\partial}{\partial x} P(y|x)\right)^2 \frac{1}{P(y|x)}\, dy$$

$$= \int \frac{\left(\text{Re}\left[\text{Tr}[E_y \rho_x L_x]\right]\right)^2}{\text{Tr}[E_y \rho_x]}\, dy.\tag{6.10}$$

This expression for the Fisher information explicitly includes the measurement. A natural question to ask is: given the influence of the parameter x upon the system, encapsulated by ρ_x, what measurement extracts the most information about x? It turns out that by using some clever tricks we can explicitly maximize the Fisher information over all measurements.

The proof begins with obtaining an upper bound on the Fisher information. We have

$$\frac{\left(\text{Re}\left[\text{Tr}[E_y \rho_x L_x]\right]\right)^2}{\text{Tr}[E_y \rho_x]} \leq \left|\frac{\text{Tr}[E_y \rho_x L_x]}{\sqrt{\text{Tr}[E_y \rho_x]}}\right|^2 \leq \text{Tr}\left[E_y L_x \rho_x L_x\right].\tag{6.11}$$

The first inequality is obvious, but the second is not at all. To obtain this inequality we must cleverly rearrange the second expression, factoring the operators into square roots and using the cyclic property of the trace, to give

$$\left|\frac{\text{Tr}[E_y \rho_x L_x]}{\sqrt{\text{Tr}[E_y \rho_x]}}\right|^2 = \left|\text{Tr}\left[\left(\frac{\sqrt{E_y}\sqrt{\rho_x}}{\sqrt{\text{Tr}[E_y \rho_x]}}\right)\sqrt{\rho_x} L_x \sqrt{E_y}\right]\right|^2.\tag{6.12}$$

It is now possible to apply the Cauchy–Schwartz inequality in the form $|\text{Tr}[B^\dagger C]|^2 \leq \text{Tr}[B^\dagger B]\text{Tr}[C^\dagger C]$, with $C = \sqrt{\rho_x} L_x \sqrt{E_y}$, to obtain the second inequality in Eq. (6.11).

Using the above inequalities in the expression for the Fisher information we have

$$\mathcal{I}(x) \leq \int \text{Tr}\left[E_y L_x \rho_x L_x\right] dy = \text{Tr}\left[\left(\int A_y^\dagger A_y\, dy\right) L_x \rho_x L_x\right] = \text{Tr}[L_x \rho_x L_x],\tag{6.13}$$

where we have used the basic relation of the measurement operators, $\int A_y^\dagger A_y\, dy = I$. The upper bound on the Fisher information, $\text{Tr}[L_x \rho_x L_x] = \text{Tr}\left[\rho_x L_x^2\right]$ is referred to as the *quantum Fisher information*, and the Cramér–Rao bound with the Fisher information replaced with this quantity is called the quantum Cramér–Rao bound.

By inspecting the conditions for equality in the inequalities used to obtain the above upper bound, it is possible to determine a measurement that achieves the maximum. The measurement operators for this optimal measurement are rank-one projectors, each of which projects onto an eigenstate of the Hermitian operator L_x. Of course, there is no guarantee that this measurement will be physically feasible for any given scenario. Further information regarding the quantum Fisher information and optimal estimators that achieve the bound, given the optimal measurement, can be found in [87, 88, 466]. In the limit of many measurements, the maximum of the likelihood function is an optimal estimator.

6.1.3 Resources and limits to precision

In pushing the limits of precision measurements, a question naturally arises as to the resources required to achieve a given precision. Consider, for example, the measurement of the phase of a laser. This is of practical importance because many measuring techniques involve a physical interaction in which the quantity to be measured induces a phase shift in a laser beam, which is then measured. For simplicity, imagine that the laser light is trapped in a single mode of an optical cavity, and that we will make a measurement on this mode. (A continuous measurement of the phase of a traveling-wave laser can be regarded, approximately, as a sequence of measurements, each performed on a newly minted mode.) Since the laser is in a coherent state $|\alpha\rangle = |Ae^{i\theta}\rangle$, the uncertainty in the phase is given by calculating the standard deviation of the Pegg–Barnett phase operator, Φ, in this state. (The phase operator Φ is defined in Appendix D, as are coherent states.) But the coherent state is a blob centered at phase θ and amplitude A. Because of this, for $\alpha \gg 1$ the standard deviation of Φ is approximately that of the phase-quadrature, $Y = -i(a - a^\dagger)/2$, divided by A. (You can show this in Exercise 3.) The standard deviation of Y is unity, and the energy of the state is $\langle E \rangle = \hbar\omega A^2$, so the error in the measurement of the phase, being the standard deviation of the phase operator, is

$$\Delta\theta \approx \frac{\hbar\omega}{\sqrt{E}} = \frac{1}{\sqrt{n}}, \quad n \gg 1, \tag{6.14}$$

where n is the average number of photons. If one wishes to detect small phase shifts about a known phase ϕ, then homodyne detection, discussed in Section 3.4, can be used to achieve the above accuracy.

Now consider what happens if we make m independent measurements like the one above, and average the answers to obtain our final result. When we add independent random variables together, the variance of the result is the sum of the variances of the variables. Because of this, the error in the final result, being the average of m independent measurements, is proportional to $1/\sqrt{m}$. This scaling with the number of independent measurements is a fundamental property of combining independent information. But the scaling with n, the number of photons, is not fundamental, even though it has the same form. It is merely due to the fact that we chose a coherent state as the state of the mode on which the phase shift is imprinted.

Let us assume that the phase-shift we wish to measure could be any value in the range $[0, 2\pi)$. If the mode is allowed to contain up to n photons, then a measurement on the mode can provide exactly $n + 1$ distinguishable outcomes. The maximum possible precision is therefore the width of the total range of values of θ, divided by $n+1$, being $\Delta\theta_{min} = 2\pi/(n+1)$. But for large n this is a much lower error than we can achieve with the coherent state. This tells us that there should be better initial states that we can choose to measure phase. The problem with the coherent state is that it already contains more phase uncertainty than is necessary. If we start the mode in an eigenstate of the Pegg–Barnett phase operator, of which there are naturally $n + 1$ for a system with at most n photons, then the phase of the initial mode is as certain as possible. When we apply the phase shift we would then be able to determine it with accuracy $\Delta\theta_{min}$ if we were able to make a von Neumann measurement in the basis of the phase operator. In fact it is very difficult to construct a measurement of the phase operator, and no practical methods exist for large numbers of photons. Methods to determine phase therefore always measure some other observable.

Coherent states are regarded as classical states because they are arguably the closest thing to a classical field in terms of their properties, and are the easiest pure states to prepare. To obtain a phase measurement error proportional to $1/n$, rather than $1/\sqrt{n}$, we have to work much harder to create the necessary quantum states. The error of $1/\sqrt{n}$ is referred to as the "standard quantum limit" for metrology, because it is straightforward to obtain, either by using coherent states or by making multiple independent measurements. The error proportional to $1/n$ is called the "Heisenberg limit." This is because the phase operator is conjugate to the photon-number operator, and so Heisenberg's uncertainty principle tells us that the minimum error in phase is $\sim 1/n$ given that the spread in photon number is $\sim n$. The term "quantum metrology" refers to the use of states with non-classical properties to achieve higher precision when measuring classical quantities.

There are two further, and related contexts in which the precision of a measurement scales as the inverse of a countable resource. In these contexts this scaling is also referred to as the Heisenberg limit, although the motivation for this terminology is not as clear. Recall that the classical quantities we wish to measure are imprinted on our quantum system by some physical process. This means that a Hamiltonian $H = \hbar\lambda G$ is applied to the system, where λ is the quantity to be measured. For example, if the physical process causes a phase shift, then G is the photon-number operator. If instead of applying the physical process to the probe system just once, we apply it N times, then this creates N times the effect on the system. This means that the system is N times more sensitive to a change in the value of λ. Thus the precision to which a change in λ can be measured scales as $1/N$.

If we consider using N probe systems for our measurement, and subjecting each to the influence of the parameter λ, then it turns out that we can achieve the same $1/N$ increase in precision as above, but doing so requires that we prepare highly non-classical states. To see this consider the use of a qubit to probe a parameter that acts on the qubit as $H = \hbar\lambda\sigma_z$. If we start the qubit in the σ_x eigenstate $|+\rangle \equiv (|0\rangle + |1\rangle)/\sqrt{2}$, then the action of H changes the state to $|\psi(\phi)\rangle = (|0\rangle + e^{-i\phi}|1\rangle)/\sqrt{2}$ where $\phi = 2\lambda t$. The rotational phase ϕ is imprinted on the qubit, and information about it can now be extracted by measurement. If we apply the phase shift N times, then the resulting state is, naturally, $|\psi(N\phi)\rangle$. Now consider starting N qubits in the joint state

$$|\psi_N\rangle = \frac{|0\rangle_1|0\rangle_2\cdots|0\rangle_N + |1\rangle_1|1\rangle_2\cdots|1\rangle_N}{\sqrt{2}} \equiv \frac{|00\cdots0\rangle + |11\cdots1\rangle}{\sqrt{2}}, \tag{6.15}$$

where $|x\rangle_j$ denotes state $|x\rangle$ of the jth qubit. If we apply the action of λ to each qubit, then the joint Hamiltonian is $H_{\text{tot}} = \hbar\lambda\sum_j \sigma_z^{(j)}$, where $\sigma_z^{(j)}$ is the σ_z operator for the jth qubit. The action of H_{tot} on $|\psi_N\rangle$ is

$$H_{\text{tot}}|\psi_N\rangle = \frac{e^{iN\lambda t}|00\cdots0\rangle + e^{-iN\lambda t}|11\cdots1\rangle}{\sqrt{2}} = \frac{|00\cdots0\rangle + e^{-iN\phi}|11\cdots1\rangle}{\sqrt{2}}. \tag{6.16}$$

We have therefore achieved the same influence, a phase of $N\phi$ between two orthogonal states, as if we had applied the phase shift N times to a single qubit.

Now let us think about the use of N probe systems a little further. If we have N systems each with n states, then the total number of states is $n_{\text{tot}} = n^N$. If we could use all these states as if they were the states of a single system, then we could achieve a precision proportional to $1/n_{\text{tot}}$, being the Heisenberg limit for the combined system. This would be much more precise than the scaling of $1/N$ achieved in the method above. The reason that the above method can at most increase the precision by $1/N$ is that the phase shift is applied separately to each qubit. That is, the Hamiltonian that gives the action of λ on the joint system is restricted to be a sum of Hamiltonians each of which acts only on a single system. In order to take advantage of all the states in the combined state space we would need a Hamiltonian with terms that were products of operators from all the systems. Such products, referred to as N-body interactions because they involve all N subsystems at once, do not occur naturally.

Scenarios in which a parameter influences N systems separately are the most readily available. Nevertheless there are certainly situations in which a parameter controls all two-body interactions between a set of N systems. If the operator G contains all k-body interactions between N systems, then the precision with which λ can be measured scales as $1/N^k$. We will not discuss this further here, but the details can be found in [72, 73, 74].

6.1.4 Adaptive measurements

An adaptive measurement is a sequence of measurements in which each measurement is chosen based on the results of the previous measurements. A continuous measurement can therefore also be adaptive, where some aspect of the measurement is changed with time. An adaptive measurement is an example of a feedback process, which is the subject of Chapter 5. There are a number of reasons for making adaptive measurements. Since quantum measurements will in general disturb the system they are measuring, a special kind of measurement may be required to obtain the most information about a given parameter. The problem is that many measurements are impossible or impractical to realize, and some measurements that would otherwise be impractical can be realized by using an adaptive measurement process. We explore examples of these kinds of adaptive measurements in the exercises at the end of this chapter.

Another motivation for making adaptive measurements is that of controlling quantum systems using feedback. Because measurements can have a non-trivial dynamical effect, it is possible to realize feedback control by changing the measurement with time instead of changing the Hamiltonian of the system. In Chapter 5 we present in detail two examples of adaptive measurements. In Section 5.4.2 we examine an adaptive measurement that implements continuous-time feedback control of a single qubit. In Section 5.4.1 we examine a continuous-time adaptive measurement that increases the speed at which the measured system is projected onto a pure state.

A third situation in which one can use an adaptive measurement to advantage is where a parameter is to be estimated from a measurement on many copies of an identical state. Each state contains the same information about the parameter, and we extract this information by measuring one after the other. As we measure each of the states, in obtaining information about the parameter we also obtain information about the state. The amount of information that each measurement obtains about the parameter depends on the relationship between the measurement and the state. Because of this, as our knowledge of the state increases, we can change the measurement we are making to increase the rate at which information is obtained about the parameter.

As a simple example consider making a measurement on a qubit in the state

$$|\psi(\phi)\rangle = \frac{|0\rangle + e^{i\phi}|1\rangle}{\sqrt{2}}. \tag{6.17}$$

If we make a two-outcome measurement with the measurement operators

$$A_0 = |\psi(\theta)\rangle\langle\psi(\theta)|, \tag{6.18}$$

$$A_1 = |\chi(\theta)\rangle\langle\chi(\theta)|, \tag{6.19}$$

with $\langle\psi(\theta)|\chi(\theta)\rangle = 0$, then the amount of information we obtain about ϕ, on average, depends on the relationship between the true value of ϕ and the value we choose for θ. The average amount of information we obtain from a single measurement with a given value of θ depends on the current distribution of ϕ, and this distribution changes upon each measurement. If the distribution of ϕ is flat on the full interval $[0, 2\pi]$, then all values of θ are equally good. But as this distribution narrows, some choices of θ are better than others. As an example, if the distribution for ϕ is sharply peaked about the value $\hat{\phi}$, then in trying to determine ϕ we are essentially trying to distinguish between a bunch of quantum states that are very close together. If we make the above measurement with $\theta = \phi$ then the outcome is only weakly dependent on the value of ϕ, whereas if we choose $\theta = \hat{\phi} + \pi/2$, then the outcome depends on ϕ more strongly. In fact, from Exercise 19 in Chapter 1 we can expect the choice $\theta = \hat{\phi} + \pi/2$ to be optimal. You can explore the likelihood function for this estimation problem in Exercise 7.

A situation in which we must determine a set of parameters by measuring many identical states is that in which we want to determine the state itself. In this case the parameters consist of all the independent elements of the density matrix describing the state. The process of determining a quantum state is called "quantum-state tomography." This process can also benefit from adaptive measurements because the likelihood function has a non-trivial dependence on the basis of the measurement.

We note that to perform quantum-state tomography it is not sufficient to measure in a single basis, as this provides information only about the diagonal elements of the density matrix in that basis. It is possible to determine a density matrix by repeatedly making a single measurement, but to do so at least some of the measurement operators must not commute with each other. An example of such a measurement for a single qubit is one in which the measurement operators are $A_j = |\psi_j\rangle\langle\psi_j|$, with

$$|\psi_0\rangle = |1\rangle, \qquad\qquad |\psi_1\rangle = \sqrt{1/3}\,|1\rangle + \sqrt{2/3}\,|0\rangle,$$
$$|\psi_2\rangle = \sqrt{1/3}\,|1\rangle + \sqrt{2/3}\,e^{i2\pi/3}|0\rangle, \quad |\psi_3\rangle = \sqrt{1/3}\,|1\rangle + \sqrt{2/3}\,e^{-i2\pi/3}|0\rangle.$$

In this case the operators A_j project the system onto states that form a symmetrical pattern on the Bloch sphere: they lie at the points of a tetrahedron. Because of this, the measurement is referred to as being "symmetric and informationally complete" (SIC), and this notion can be generalized to systems with higher dimensions [511].

6.2 Metrology of signals

A wide variety of situations require one to determine a *signal*, defined as some quantity that changes with time. To do this we must have a system we can measure that is sensitive to the signal. This system acts as an antenna, and we will likely need to measure the system repeatedly, or continuously, to infer the signal from the resulting dynamics. Such a detection procedure raises immediately the question of Heisenberg's uncertainty principle. In trying to track the motion of a system we are liable to add noise to it. We might, for example, use an oscillator as a probe to measure a time-varying force. If we do this by making a sequence of position measurements on the oscillator, the resulting back-action noise on the momentum leads to what is referred to as the *standard quantum limit* for force detection [84, 110, 462]. In fact, it is possible to cancel the effect of the back-action noise, with the result that quantum mechanics places no additional limitation on the metrology of signals. A number of ways to cancel the back-action noise in the estimation of signals have been suggested [75, 306, 380, 333, 331], with the first due to Unruh [620]. But a revolution in our understanding of the metrology of signals came with the work of Tsang and Caves in 2012, who, in discovering a new noise-cancelation technique, realized that *quantum-mechanics-free* subsystems could be engineered within quantum systems. Here we first discuss the concept of a quantum-mechanics-free subsystem, and then consider how such a system can be implemented for use in detecting a time-varying force. This example has applications to gravitational-wave detection [417, 418, 169] and accelerometers.

6.2.1 Quantum-mechanics-free subsystems

The observation of Tsang and Caves was that, for quantum systems with continuous dynamical variables, a joint system can be constructed in which a subsystem has conjugate position and momentum variables that commute, and for which the dynamics are non-trivial, thus realizing an effectively classical system within a quantum system. All the observables of this system are non-demolition observables. To understand how this works, the formalism of Hamiltonian mechanics is useful.

Consider two systems, each with N degrees of freedom. We will call the positions and momenta of the first system $\{q_j\}$ and $\{p_j\}$, respectively, and those of the second system $\{Q_j\}$ and $\{P_j\}$. The commutation relations for these coordinates are, naturally, $[q_j, p_k] = [Q_j, P_k] = i\hbar\delta_{jk}$. We now observe that since q_j commutes with P_j, each of these pairs could potentially be the position and momentum of a classical system. But for this to be true they would need to be coupled to each other in the right way. Now consider what happens if the Hamiltonian of the two systems has the form

$$H = \frac{1}{2}\sum_j p_j F_j + F_j p_j - (Q_j G_j + G_j Q_j) + f(\{q_n\}, \{P_n\}, t), \tag{6.20}$$

where $F_j = F_j(q_j, P_j, t)$ and $G_j = G_j(q_j, P_j, t)$. Since q_j commutes with everything but p_j, and P_j commutes with everything but Q_j, the resulting equations of motion for q_j and P_j are

$$\dot{q}_j = \frac{-i}{\hbar}[q_j, H] = \frac{\partial H}{\partial p_j} = F_j(q_j, P_j, t), \tag{6.21}$$

$$\dot{P}_j = \frac{-i}{\hbar}[P_j, H] = -\frac{\partial H}{\partial Q_j} = G_j(q_j, P_j, t). \tag{6.22}$$

Further, if we can arrange F_j and G_j so that there is a function $H_{\text{sub}}(q_j, P_j, t)$ such that

$$\dot{q}_j = F_j(q_j, P_j, t) = \frac{\partial H_{\text{sub}}}{\partial P_j}, \tag{6.23}$$

$$\dot{P}_j = G_j(q_j, P_j, t) = -\frac{\partial H_{\text{sub}}}{\partial q_j}, \tag{6.24}$$

then q_j and P_j would have the classical dynamics of the Hamiltonian H_{sub}.

Now let us see how we might engineer two quantum systems to realize a quantum-mechanics-free harmonic oscillator. Denoting the coordinates of our two systems by (q, p) and (Q, P), respectively, we need $F = P/m$ and $G = -m\omega^2 q$, where as usual m and ω are the mass and frequency of our desired oscillator. The joint Hamiltonian we need is therefore

$$H = \frac{pP}{m} + m\omega^2 qQ. \tag{6.25}$$

If we now transform to new coordinates

$$q' = \frac{1}{2}(q+2Q), \quad Q' = \frac{1}{2}(q-2Q), \tag{6.26}$$

$$p' = \frac{1}{2}(2p+P), \quad P' = \frac{1}{2}(2p-P), \tag{6.27}$$

then this Hamiltonian becomes

$$H = \frac{p'^2}{2m} + \frac{1}{2}m\omega^2 q'^2 + \frac{P'^2}{2(-m)} + \frac{1}{2}(-m)\omega^2 Q'^2. \tag{6.28}$$

This is a much more familiar, and rather interesting Hamiltonian. It consists of two uncou-pled harmonic oscillators, with the twist that the second one has negative mass. Negative mass means that the Hamiltonian of the oscillator is merely the negative of that of a normal oscillator. This means that the energy levels are the negative of the usual energy levels, and therefore extend down instead of up: when we add a "phonon" to a negative-mass oscilla-tor the energy of the oscillator *decreases* by $\hbar\omega$. Since the evolution of a quantum system is given by $U(t) = \exp(-iHt/\hbar)$, the evolution of the negative-mass oscillator is the time-reversed evolution of the normal oscillator. If we think of the normal oscillator as rotating clockwise in phase space, then the negative-mass oscillator rotates counter-clockwise.

Inverting the transformation in Eqs. (6.26) and (6.27), the quantum-mechanics-free subsystem consists of the coordinates

$$q = \frac{1}{2}(q'+Q'), \quad P = \frac{1}{2}(p'-P'), \tag{6.29}$$

and has the effective Hamiltonian

$$H_{\text{sub}} = \frac{p^2}{2m} + \frac{1}{2}m\omega^2 q^2. \tag{6.30}$$

So how is it that this subsystem acts as if q and P commute with themselves from one time to another, and in particular that a measurement of Q introduces no back-action noise into the dynamics? This can be understood quite simply. A measurement of the oscilla-tor position q is a measurement of the sum of the real coordinates q' and Q', and thus introduces the same back-action into p' and P'. There is no effect on P because it is the difference of the real momenta, and so the contributions from both cancel. But this is not enough to cancel the back-action that will then feed back into q: since q is the sum of the two positions, and the same noise is introduced into their corresponding momenta, if both oscillators were identical the same noise would be introduced into both positions, and thus into q. But the dynamics of both oscillators is not the same. The second oscillator has time-reversed motion, and thus the effect of the momentum on the position is the *opposite* of that of the first oscillator. Thus the noise on Q' is the opposite of that on q', and the two cancel leaving Q noise-free. Simple but brilliant. In fact, quantum-mechanics-free (QMF) subsystems were discovered by asking how quantum noise might be made to cancel in a network of systems [613]. Of course, to realize this QMF oscillator we must work out how to realize an oscillator that effectively has negative mass, and we consider how to do this next.

6.2.2 Oscillator-mediated force detection

The reason we want to realize a QMF oscillator, introduced above, is to use it as a back-action-free antenna to detect a time-varying force. Assume for the sake of argument that we have a positive and a negative oscillator with identical frequencies. If their respective coordinates are (q',p') and (Q',P') then the QMF oscillator they create has the coordinates (q,P) given in Eq. (6.29). If we apply a force $F(t)$ only to the positive oscillator then this force also drives the QMF oscillator, since

$$\dot{P} = \frac{1}{2}(\dot{p}' - \dot{P}') = -m\omega^2 q + F(t). \tag{6.31}$$

If we then make a joint measurement of the positions of both oscillators, so as to measure the QMF coordinate q, then we detect the motion of the QMF oscillator induced by the force, and there is no back-action from the measurement. In what follows we draw heavily upon Sections 7.2, 7.5, 7.6, and 7.7.3, so you may wish to study those sections first.

It turns out that there is a way to engineer a negative-mass oscillator. If we have two oscillators whose respective frequencies are $\Omega + \omega$ and $\Omega - \omega$, then their joint Hamiltonian is

$$H = \hbar(\Omega+\omega)a^{\dagger}a + \hbar(\Omega-\omega)b^{\dagger}b = \hbar\Omega(a^{\dagger}a+b^{\dagger}b) + \hbar\omega a^{\dagger}a - \hbar\omega b^{\dagger}b. \tag{6.32}$$

If we now move into the interaction picture with respect to the Hamiltonian $\hbar\Omega(a^{\dagger}a+b^{\dagger}b)$, then the Hamiltonian that evolves the states is

$$H_{\rm I} = \hbar\omega a^{\dagger}a - \hbar\omega b^{\dagger}b, \tag{6.33}$$

and this is exactly the Hamiltonian in Eq. (6.28) including the negative-mass oscillator. The question now is how to effectively remove the offset frequency Ω from two positive mass oscillators, at least as far as our force-detection procedure is concerned. We can do this by using the technique of frequency conversion discussed in Section 7.2.

We start by considering the technique for measuring the position of a mechanical resonator by using an optical or electrical resonator described in Section 7.7.3. Here we will speak of this probe resonator as being an optical cavity, but the analysis applies equally well to an electrical resonator. We recall from Section 7.6 that the mechanical resonator forms one of the end-mirrors of the probe cavity, and so the motion of the mechanical resonator imprints a phase-shift onto the light in the cavity. The probe cavity is driven with a laser, and the light that is output from the cavity is measured using homodyne detection.

Here we will strongly drive the cavity so as to convert the nonlinear interaction between the cavity and the mechanical resonator, given by Eq. (7.80), into a linear interaction using the method described in Section 7.5.

The frequency of the probe cavity is Ω, and that of the mechanical resonator is ω. If we set the frequency of the laser driving the probe cavity equal to that of the cavity, then in the Schrödinger picture the effective Hamiltonian is

$$H_1 = \Omega a^\dagger a + \lambda(a e^{i\Omega t} + a^\dagger e^{-i\Omega t})x_b + \omega b^\dagger b, \qquad (6.34)$$

where $x_b = b + b^\dagger$ is the scaled position of the mechanical resonator, and λ is the strength of the interaction between the two resonators. Note that the interaction between the probe and the mechanical resonator is modulated at the frequency Ω. This removes the frequency of the probe cavity from the point of view of the interaction. As far as the mechanical resonator is concerned it is interacting with a "oscillator" with zero frequency. This is made clear by moving into the interaction picture with respect to the probe Hamiltonian $H_0 = \Omega a^\dagger a$.

To create the back-action-free measurement, we need to couple the probe cavity to another oscillator that has an effectively negative frequency. Let us see what happens if we couple the probe cavity to another optical mode that has frequency $\Omega - \omega$. The Hamiltonian of all three oscillators is then

$$H_2 = \hbar\Omega a^\dagger a + \hbar(\Omega - \omega)c^\dagger c + \hbar\omega b^\dagger b + \hbar\lambda(a e^{i\Omega t} + a^\dagger e^{-i\Omega t})x_b + \hbar\mu(a + a^\dagger)x_c,$$

where $x_c \equiv c + c^\dagger$ and μ is the strength of the interaction between the probe and auxiliary. While the linear interaction with the mechanical resonator is modulated at frequency Ω, the interaction with the auxiliary is not modulated at all. This is the crucial difference that allows the auxiliary to act as the negative twin for the mechanical resonator. To see this we move into the interaction picture with respect to the Hamiltonian $H = \hbar\Omega(a^\dagger a + c^\dagger c)$ and make the rotating-wave approximation (see Sections 7.2 and G.4). The resulting interaction picture version of H_2 is

$$H_I = \hbar\lambda(a + a^\dagger)(x_b + x_c) - \hbar\omega c^\dagger c + \hbar\omega b^\dagger b. \qquad (6.35)$$

where we have set $\mu = \lambda$. This shows us that from the point of view of the probe, it is interacting with, and thus measuring, the sum of the positions of a positive and negative pair of oscillators. The homodyne measurement, since it is made by interfering the output light with a laser at frequency Ω, demodulates the output at this frequency. This means that it measures the changes in the output light that are due to the interaction Hamiltonian H_I. Another way to think of this is that homodyning measures the motion in the rotating frame corresponding to the interaction picture.

The above measurement scheme results in a measurement that can be described purely classically – it is a classical continuous measurement of a classic oscillator, driven by a classical force. The size of the measurement noise, and thus the measurement strength, is

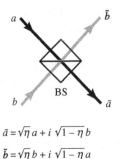

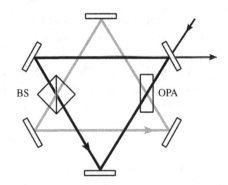

$$\tilde{a} = \sqrt{\eta}\, a + i\,\sqrt{1-\eta}\, b$$

$$\tilde{b} = \sqrt{\eta}\, b + i\,\sqrt{1-\eta}\, a$$

Figure 6.1 On the left we depict a beam-splitter with input modes a and b, and show the transformation between its input modes and its output modes $\tilde{a}$ and $\tilde{b}$. Note that the symbol we use to depict a beam-splitter in this chapter is different from that used in the other chapters. On the right we show two ring cavities, whose respective modes are coupled together by a beam-splitter and an optical parametric amplifier.

determined by the parameters of the probe cavity and the interaction strength between the probe and the mechanical resonator.

Engineering a linear interaction between optical modes

One way to construct a linear interaction between two optical modes is the scheme depicted on the right in Fig. 6.1. Here each mode is the counter-clockwise traveling-wave mode of a ring cavity formed by three mirrors. The probe mode can be driven by a laser through one of the mirrors that is partially transmitting (depicted in the diagram at the top right) via which the output is also detected. The two modes are coupled by two optical elements. On the left is a beam-splitter, and on the right is an optical parametric amplifier (OPA). The beam-splitter transforms the mode operators of the two modes as shown in the diagram on the left in Fig 6.1. This creates an effective Hamiltonian in the following way. If we make the transitivity η of the beam-splitter very small ($\eta \ll 1$), then on each round trip through the ring cavity the transformation applied to the two modes by the beam-splitter is infinitesimal. Since the round-trip time is very short, this infinitesimal transformation is applied many-times on the timescale of the rest of the dynamics, and thus acts as if it is generated continually by a Hamiltonian of the form

$$H_{\mathrm{BS}} = \hbar\mu(cb^{\dagger} + bc^{\dagger}). \tag{6.36}$$

Deriving this Hamiltonian is the subject of Exercise 12.

The optical parametric amplifier is a nonlinear optical crystal that couples the two modes via the Hamiltonian

$$H_{\mathrm{OPA}} = \hbar\mu(cb + c^{\dagger}b^{\dagger}). \tag{6.37}$$

It creates and destroys photons in pairs, one in each mode. Adding together the Hamiltonians generated by the beam-splitter and the OPA gives a general linear interaction.

History and further reading

Classical estimation theory is discussed in detail in the books by van Trees [627] and Levy [368]. These consider the estimation of both single quantities and signals. The quantum version of the Cramér–Rao bound, which involves optimizing the classical Fisher information over all quantum measurements, was obtained originally by Helstrom [248] and Holevo [259]. The method of deriving the bound by first deriving the classical Cramér–Rao bound and then performing the measurement optimization was introduced by Braunstein and Caves [87], via which they determined the optimal measurement. Further discussions of the quantum Cramér–Rao bound can be found in [88, 466]

There is a considerable literature concerned with schemes to beat the standard quantum limit for measurements of optical phase (see, e.g., [705, 262, 159, 152, 160, 56]). Reference [56] discusses the equivalence between repeated application of a phase shift, and the application of a single phase shift to a larger system. Regarding the difficulty of realizing measurements of canonical phase (the Pegg–Barnett phase operator [471, 472]), Pregnell and Pegg have shown that such a measurement can in theory be realized probabilistically, although not necessarily in a practical way [492]. Berry and Wiseman have shown that the canonical phase measurement may be approximated rather well for the purposes of phase estimation by an adaptive measurement that counts photons at the two outputs of an interferometer [54]. Experiments in which phase measurements are realized beyond the standard quantum limit are discussed in [250, 690]. A quantum adaptive measurement of optical phase was realized in [19]. Further examples of adaptive measurements can be found in [456, 316, 670, 287].

The scaling of precision for metrology that uses many-body Hamiltonians is discussed in [72, 73, 74].

In Section 7.7.1 we derive the standard quantum limit (or "SQL") for position measurement of an oscillator, Eq. (7.85), also known as the SQL for force detection, which arises from the relationship between the measurement noise and the resulting back-action noise. Various analyses of this standard quantum limit can be found in [110, 105, 107, 462]. We gave a number of references to work on canceling back-action noise and thus beating the SQL at the start of Section 6.2. Once you have determined the noise in a given detection scheme – that is, you have fixed the probe system and the measurement – then there is no further quantum mechanics in the signal estimation problem. A wealth of techniques for estimating signals within a noisy data-stream can be found in [627, 368]. Some recent theoretical work on signal detection and estimation with quantum systems is given in [612, 615, 614], and recent experiments are described in [655, 281].

Exercises

1. (i) Calculate the Fisher information for a measurement with a Gaussian likelihood function. Now assume the prior is flat. Consider the estimator in which $\tilde{x}$ is chosen to be the value of x at which the likelihood function is maximum. (ii) Is this estimator unbiased? (iii) Does it reach the Cramér–Rao bound?

2. (i) Show that

$$\int \left(\frac{\partial}{\partial x} \ln [P(y|x)] \right) P(y|x)\, dy = 0. \tag{6.38}$$

Hint: use integration by parts, and remember that $P(y|x)$ is the probability distribution for y, and thus has certain properties. (ii) Show that the second equality in Eq. (6.3) is true.

3. For a mode (a harmonic oscillator) the classical quadrature variables x and y (proportional to position and momentum for a mechanical oscillator) correspond, respectively, to the quantum operators $X = (a + a^\dagger)/2$ and $Y = -i(a - a^\dagger)/2$. Take as our definition of amplitude and phase the variables $A = |x + iy|$ and $\theta = \arg(x + iy)$. If the system is in a coherent state $|\alpha\rangle$, then it has a probability distribution for x and y with means $\text{Re}[\alpha]$ and $\text{Im}[\alpha]$, and variances both equal to unity. (i) Use the relationship between the phase and the quadrature variables to show that for $|\alpha| \gg 1$, a change in the θ direction of distance d is equal to change in θ of $d/|\alpha|$. Use this, and the above information, to determine approximately the standard deviation of θ in the state $|\alpha\rangle$ for $\alpha \gg 1$. (ii) Calculate the standard deviation of the canonical (Pegg–Barnett) phase operator in a coherent state, for $|\alpha| \gg 1$.

4. Here you can analyze the Mach–Zehnder interferometer (MZI). (The MZI is mathematically equivalent to a Michelson interferometer, both of which are described in [160].) Consider a single mode of an optical beam, described by the mode (annihilation) operator a, incident on an MZI, a configuration of two beam-splitters and two mirrors as shown in Fig. 6.2. The mode is split into two paths by the initial beam-splitter, and these two paths are recombined at the second beam-splitter. Photo-detectors measure the intensities of the two output modes, whose respective outputs are subtracted. The interferometer measures a phase shift ϕ applied to one of the optical paths. Note that this scenario is essentially homodyne detection, discussed in Section 3.4. The mode with annihilation operator b that enters the second input port of beam-splitter 1 is in the vacuum state. In Fig. 6.1 we show how the beam-splitters transform their input modes into their output modes, where we have assumed a $\pi/2$ phase change upon reflection. We also assume a $\pi/2$ phase change at each mirror. The operator that applies the phase shift ϕ is $\exp(-i\phi a^\dagger a)$. (i) Assume that there is a coherent state $|\alpha\rangle$ in mode a (coherent states are defined in Appendix D). Determine the state of modes c and d. To do this, you have to first write the coherent state in

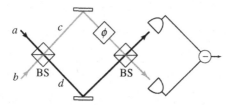

Figure 6.2 The Mach–Zehnder interferometer used in Exercises 4 and 6.

terms of the action of the mode operators on the vacuum state, and then use the beam-splitter transformation relations on the mode operators. The coherent state is given by $|\alpha\rangle = \exp(\alpha a^\dagger - \alpha^* a)|0\rangle$. (ii) Show that if the phase shift applied to mode c is zero, then one of the output modes is dark (no photons) and the other is bright. (iii) Show that the output signal, the difference of the signals from the two photo-detectors, is equal to $I\cos(\phi)$, where I is the energy in the input mode a.

5. Consider the beam-splitter depicted in Fig. 6.1. If there is one photon in each of the input ports, what is the state of the output ports? Recall that to determine the output states, you have to write the input states in terms of the action of the mode operators on the vacuum, and then apply the beam-splitter transformation relations to the mode operators. (The fact that both output ports never get one photon each is called the Hong–Ou–Mandel effect [263].)

6. Consider two optical modes in the joint state $|\psi_{NOON}\rangle \equiv (|N\rangle|0\rangle + |0\rangle|N\rangle)/\sqrt{2}$. First apply a phase shift ϕ to one mode, generated by the operator $\exp(-i\phi a^\dagger a)$, and then feed the two modes respectively into the two inputs of a beam-splitter, and measure the two outputs with photo-detectors (as in the measurement scheme for the Mach–Zehnder interferometer in Fig. 6.2). Show that the difference of the signals from the photodetectors is proportional to $\cos(N\phi)$. The state $|\psi_{NOON}\rangle$, called a "noon state" [160], is thus much more sensitive to a phase shift than a coherent state input to a Mach–Zehnder interferometer, and it achieves the Heisenberg limit. However, noon states are not easy to produce.

7. Consider a single qubit prepared in the state $|\psi(\phi)\rangle = |0\rangle + e^{-i\phi}|1\rangle$ for some value of ϕ, where ϕ has the probability distribution $P(\phi)$. We make a measurement on the qubit with measurement operators

$$A_0(\theta) = |\psi(\theta)\rangle\langle\psi(\theta)|, \tag{6.39}$$

$$A_1(\theta) = |\chi(\theta)\rangle\langle\chi(\theta)|, \tag{6.40}$$

where $|\chi(\theta)\rangle = |0\rangle - e^{-i\phi}|1\rangle$. (i) If we denote the measurement results by the variable a, so that a is 0 or 1, what is the likelihood function $P(a|\phi)$ for this measurement? (ii) Write down an expression for the average change in the entropy of the observer's state-of-knowledge of ϕ when he or she makes the measurement. (iii) If the probability distribution for ϕ is

$$P(\phi) = \frac{e^{-\phi^2/b}}{\sqrt{\pi b}} \qquad (6.41)$$

with $b \ll 1$, what choice for θ maximizes the average change in entropy of the observer's state-of-knowledge of ϕ?

8. Consider that a single bit of information has been encoded in a qubit, using the two states $|a(\phi)\rangle = |0\rangle + e^{-i\phi}|1\rangle$ and $|b(\phi)\rangle = |0\rangle - e^{-i\phi}|1\rangle$, but we don't know what value of ϕ was used, except that it is in the range $[0, \pi]$. If we have many of these qubits, all prepared in the same initial state, consider how you could determine the encoding basis, and the encoded information, by making a von Neumann measurement on each of the qubits in turn. Your procedure might be adaptive. Now consider the entire joint state of the N qubits. What von Neuman measurement would you make on all the qubits at once to determine the encoding and the encoded information?

9. For this exercise you will need to have read Sections 3.2 and 3.4. The solution to the linear SSE for homodyne detection of a general quadrature $X_\theta = ae^{-i\theta} + a^\dagger e^{i\theta}$ is given in Exercise 18 of Chapter 3. Assume that the mode is in a coherent state $|\alpha\rangle$, with $\alpha = Ae^{i\phi}$, and that the purpose of making the homodyne measurement is to determine the phase, ϕ. If we want to determine a small phase shift about a known phase ϕ_0, so that $\phi = \phi_0 + \Delta\phi$, then we can do this accurately with homodyne detection. Homodyne detection is a good approximation to a measurement of the phase ϕ if we set $\theta = \phi + \pi/2$. This is made clear by the relationship between the quadratures and the amplitude and phase in Fig. D.2 of Appendix D. The amplitude and phase are essentially polar coordinates, whereas the quadrature X_θ is a linear combination of the Cartesian coordinates. (i) Assume that $\phi_0 = 0$ and the distribution for $\Delta\phi$ is Gaussian with mean zero and standard deviation $\sigma \ll 1$. Choose $\theta = \pi/2$, and substitute this into the solution for homodyne detection in Exercise 18 in Chapter 3. Because the value of θ does not depend in any way on the measurement record, S is deterministic, and so the outcome of the homodyne measurement is labeled only by the random variable R. (i) Determine the likelihood function $P(R|\Delta\phi)$ at time t. (ii) What is the likelihood function as $t \to \infty$? (iii) What happens to the likelihood function if we choose a different value of θ?

10. For this exercise you will need to have read Sections 3.2 and 3.4. The solution to the linear SSE for homodyne detection of a general quadrature $X_\theta = ae^{-i\theta} + a^\dagger e^{i\theta}$ is given in Exercise 18 of Chapter 3. As explained in Exercise 9 above, homodyne measurement is a good approximation to a measurement of phase if θ is chosen to be at $90°$ to the phase. If instead θ is equal to the phase, then we obtain a measurement of the amplitude. But if the phase is completely unknown to start with, we cannot know what quadrature to choose to obtain the most information about the phase. Nevertheless, we can use an adaptive measurement procedure to obtain a better phase measurement than is possible otherwise. The idea is to start the homodyne measurement by picking

an arbitrary quadrature, and to change the quadrature being measured as information about the phase is obtained, so as to improve the phase measurement as time goes by. This is Wiseman's adaptive phase measurement, described in [670], and the following problem is solved in that paper.

To implement the adaptive phase measurement, we use the solution for homodyne detection given in Exercise 18 in Chapter 3. We choose $\theta(t)$ to be equal to our estimate of ϕ as the measurement proceeds. Because our estimate will depend on the measurement record, the variable $S(t)$ is a stochastic integral, and thus a random variable. The measurement result at time t is thus labeled by both S and R, and so the likelihood function is $P(S, R|\phi)$. For a general initial state the evolution for the adaptive measurement protocol does not have an analytic solution. But it does when the initial state has only zero or one photon, and is given by $|\psi(0)\rangle = (\sqrt{p}|0\rangle + \sqrt{1-p}e^{i\phi}|1\rangle)/\sqrt{2}$. The task is to measure the phase ϕ. (i) Show that in this case the likelihood function is $P(R, S|\phi) = \text{Re}[R^* e^{i\phi}] + c$ where c is a constant. (Hint: since the field is a two-state system, you can replace a with $\sigma = |0\rangle\langle1|$.) Thus the maximum likelihood estimate for ϕ is $\hat{\phi} = \arg[R]$. (ii) You can now substitute this estimate for ϕ for θ in the expression for R, given in Eq. (3.270), and solve the resulting stochastic differential equation for R. (iii) Show that in the limit $t \to \infty$, $R = e^{i\varphi}$ where φ is a random variable. (iv) From this solution for the adaptive measurement, determine the measurement operators for the measurement when $t \to \infty$. These are all projectors.

11. Consider two oscillators, one with positive mass and one with negative mass, as described by the Hamiltonian in Eq. (6.28). By considering two small consecutive time-steps of size Δt, show that regardless of the initial state of either oscillator, the motion of the coordinate $q = (q' + Q')/2$ is unaffected by a force F if this force is applied to both oscillators simultaneously.

12. Here we consider how the beam-splitter that couples the two ring-cavities in Fig. 6.1 creates the effective Hamiltonian in Eq. (6.36). The beam-splitter transformation is given in Fig. 6.1, and we see from the configuration of the ring-cavities that when the light in mode a hits the beam-splitter, the light that goes back into mode a is the light that goes into mode $\tilde{a}$ in the diagram of the beam-splitter. Thus the beam-splitter transforms mode a as $a \to \sqrt{\eta} + i\sqrt{1-\eta}b$. This transformation is applied to the mode a on each round trip through the beam-splitter. (i) Consider an arbitrary Hamiltonian acting on the mode a for a short amount of time, and write down the change in a induced to first order in time. (ii) Set $\eta \ll 1$, and set the transformation of a obtained in part (i) equal to the transformation generated by the beam-splitter to determine the Hamiltonian that generates the beam-splitter transformation.

13. Use the method in Section 7.5 to derive the Hamiltonian in Eq. (6.34). You will need to use the displacement picture.

14. Consider a classical damped oscillator driven by a sinusoidal force with a fixed amplitude, so that the oscillator reaches a steady state. Make a continuous classical measurement of the position of the oscillator. Multiply the output of the measurement

by the oscillator frequency (this demodulates the signal) and average the result for a time T. Use this average (multiplied by a suitable constant) as an estimate of the amplitude of the applied force. Assume that the time T is long compared to a single period of the oscillator. What is the error in the estimate of the force?

7

Quantum mesoscopic systems I:
circuits and measurements

Superconducting circuits, photonic cavities, and tiny mechanical oscillators are versatile systems in which quantum dynamics can be observed and controlled. These various technologies can be combined, and the resulting "circuits" are often referred to as *nano-electromechanical* or *optomechanical* systems. In the context of quantum physics, such devices are referred to as "mesoscopic" because they contain a large number of atoms. This terminology distinguishes them from the "microscopic" systems of atom-optics, such as single ions [364] and atoms [545, 233, 26], and the spins of individual nuclei [412]. Nano-engineered systems provide an especially powerful arena for exploring quantum measurement and control because of their potential to realize complex circuits. In this chapter we explain how to treat these systems and describe in detail a common method used to measure them.

7.1 Superconducting circuits

The electrical circuits that we describe in this chapter are quite different from those that use normal conductors, such as copper wire at room temperature. Normal conductors are subject to strong dissipation and decoherence from the environment. As a result the dynamical degrees of freedom of the circuits, the currents and voltages, behave as classical variables. In superconducting circuits, the resistance, and thus the dissipation, is so small that the currents and voltages behave as quantum mechanical degrees of freedom. Of course, the current in a wire is actually composed of many elementary particles that each obey quantum mechanics. Why then should the current act like a single quantum degree of freedom? A mechanical analogy is useful here. Recall that even though a macroscopic pendulum is composed of many particles, when one talks about the motion of the pendulum one means the motion of the *center-of-mass* of its constituent particles. Since the center of mass is merely a linear combination of the coordinates of each particle, it is itself a valid quantum mechanical coordinate with a conjugate momentum. In a similar way the total current in a superconducting wire is a collective degree of freedom that obeys quantum mechanics.

The standard method for determining the quantum dynamics of current and voltage in a circuit is phenomenological. Rather than starting with the microscopic dynamics of the electrons in the circuit, one instead determines the circuit's classical equations of motion.

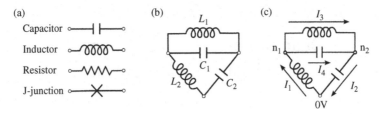

Figure 7.1 (a) The diagrams for four circuit elements. (b) An example of a simple circuit with no resistance. (c) The circuit in (b), with various annotations: (i) positive directions are defined for the currents in each link; (ii) the ground node is defined, and the active nodes are labeled n_1 and n_2; (iii) The links of the "spanning tree" are shown thick, while the others are thin.

These equations are second order like the equations of classical mechanics. We then search for a way to derive the equations of motion from a Hamiltonian, and demand that this Hamiltonian gives the overall energy of the circuit. We find that this is possible, and that the canonical coordinates and momenta in this description are not the voltages and currents, but the integrals of these with time. The time-integral of voltage is referred to as *flux*,[1] and in the formulation we will use it is the canonical coordinate. The canonical momentum is the time-integral of current, and thus has units of charge. The canonical momenta correspond to the physical charge at certain locations in the circuit, but the canonical coordinates (the fluxes) do not have a similar interpretation. Finally we perform the standard "quantization" procedure in which we give the canonical coordinates and momenta operators with the Heisenberg commutation relations. There is no guarantee that such a procedure will give the correct quantum dynamics of the circuit variables, since we have not started from a microscopic QED description, but experiments continue to confirm that it does so.

Classical equations of motion for a circuit

A circuit consists of a number of *circuit elements*, each of which has two *terminals* that can be connected to the terminals of other elements [3]. A circuit is therefore a *network*: a set of points, called nodes (the *terminals*), connected by lines, or "links" (the *circuit elements*). Figure 7.1 shows the symbols used to represent four common circuit elements, along with an example of a circuit. The circuit is a network with four links and three nodes. We note that in circuit theory the links are usually called branches. Circuits also have *loops*. A loop is a series of connected links that form a polygon. The smallest loops are those whose polygon is not intersected by any links, and these are referred to as the *irreducible* loops. The circuit in Fig.7.1 has three loops and two irreducible loops.

---

[1] The term "flux" derives from Faraday's law. The voltage around a conducting loop is given by $V = d\Phi/dt$, where Φ is the magnetic flux through the loop. Thus the anti-derivative of the voltage has units of flux, even though in our case the voltage in question is not necessarily around a loop, but may be the voltage difference between two nodes of a circuit. Of course, when we sum the voltages around any loop of a circuit, the sum of the corresponding "fluxes" for each voltage will be equal to the total magnetic flux through the loop because of Faraday's law. This is why it is sensible to refer to the time-integral of each voltage as "flux."

Each link in a circuit will in general have a different current flowing through it and a different voltage across it. But these currents and voltages are not independent dynamical variables because of Kirchhoff's laws [3]: (i) the sum of the currents flowing into a node is zero (charge does not build up at the nodes, although it may do so in the circuit elements), and (ii) the sum of the voltages around a loop is equal to the rate of change of magnetic flux through the loop (Faraday's law). Further, each circuit element enforces a specific relationship between the voltage across it and the current through it; this enforcement is in fact the only role that the circuit elements play in determining the dynamics.

We will first explain how to obtain the classical equations of motion of a circuit, and then construct a Lagrangian that gives these equations. The Hamiltonian is then obtained from the Lagrangian. As usual this Hamiltonian governs both the classical and quantum dynamics. We go through this procedure to aid understanding, but it is important to note that when deriving the Hamiltonian for a circuit we do not need to derive the equations of motion prior to obtaining the Lagrangian. As we will see, there is a very simple procedure for obtaining the Lagrangian directly from the circuit, and we summarize this below.

Classical circuit theory provides a systematic way to obtain the equations of motion for the independent dynamical variables of any circuit [143, 3]. This procedure is as follows.

Procedure for obtaining the equations of motion (long method)

1. Define a sign convention for the direction of each current in each link.
2. Choose one of the nodes and define its voltage to be zero (ground).
3. Label the other nodes in the circuit. These are called the active nodes, and each one will give us one dynamical equation. We will discuss the total number of independent dynamical variables below.
4. For each active node select a path that connects it to the ground node (a path is a number of connected links). The resulting set of paths is called the *spanning tree*. The spanning tree must be such that, for each active node, there is only one way to travel from it to the ground node by traversing the tree. In Fig. 7.1c we show a possible choice for the ground node and the spanning tree.

   Note: the voltage at an active node is the sum of the voltages across each link in the path that connects it to the ground node (in this sum, a "link voltage" may have a minus sign depending on the direction chosen for the current in the link). Since the fluxes are simply the time-integrals of the voltages, the same relationship holds between the fluxes for the active nodes and the fluxes for the links. The links that are not part of the spanning tree form irreducible loops with the spanning-tree links. Recall that the sum of the fluxes around a loop is equal to the magnetic flux through the loop. The result of this is that we can write all the link fluxes as linear combinations of the fluxes for the active nodes and the fluxes through the irreducible loops. While this may all seem rather complex now, it should be much clearer when we apply it in the example below.

5. For the next step in our systematic procedure, we need to know the relationship between the current and voltage for each link (that is, for each circuit element). These rules are

Table 7.1. *The relationship between voltage and current for various circuit elements. Capacitance is denoted by C, inductance by L, and flux by ϕ. Here flux is merely defined as the time-integral of voltage: $V = \dot{\phi}$.*

Element	Voltage relation	Flux relation
Capacitor	$I = C\dot{V}$	$I = C\ddot{\phi}$
Inductor	$V = L\dot{I}$	$I = \phi/L$
J-junction		$I = I_0 \sin(2\pi\phi/\phi_0)$

given in Table 7.1. We have included the Josephson junction in this table, but will not describe it in detail until we use it later. We can now carry out this step: obtain an equation from each active node by setting the currents flowing into the node from the inductors equal to those flowing out of the node into the capacitors. The reason we place the inductors and capacitors on separate sides of the equation is that it will make the connection with the Lagrangian easier to see later on.

6. Replace the currents in these equations with their corresponding fluxes by using the rules for each circuit element. Each of these fluxes is the flux for a single link.

7. Replace the link fluxes by their expressions in terms of the fluxes for the active nodes and the fluxes through the irreducible loops.

8. Finally, define a flux for each active node to be the time-integral of the active-node voltage, and thus replace each active-node voltage by the time-derivative of its active-node flux.

An example: the circuit in Figure 7.1

As an example of the above procedure, let us derive the equations of motion for the circuit in Fig. 7.1. Using the directions for the currents defined there, Kirchhoff's current equations for the active nodes, placing all the capacitor currents on the left-hand side, are

$$
\begin{aligned}
\text{Node 1:} \quad I_4 &= I_1 - I_3, \\
\text{Node 2:} \quad I_2 - I_4 &= I_3.
\end{aligned}
\tag{7.1}
$$

Now using the rules for capacitors and inductors in Table 7.1 to replace the link currents with the link fluxes, we obtain

$$
\begin{aligned}
\text{Node 1:} \quad C_1\ddot{\tilde{\phi}}_4 &= \tilde{\phi}_1/L_2 - \tilde{\phi}_3/L_1, \\
\text{Node 2:} \quad C_2\ddot{\tilde{\phi}}_2 - C_1\ddot{\tilde{\phi}}_4 &= \tilde{\phi}_3/L_1,
\end{aligned}
\tag{7.2}
$$

where $\tilde{\phi}_i$ is the flux for the ith link. By traversing the paths that connect the active nodes to the ground node, we find the relationships between the active-node fluxes, ϕ_1 and ϕ_2, and the link fluxes. These are

$$\tilde{\phi}_1 = -\phi_1, \tag{7.3}$$

$$\tilde{\phi}_2 = \phi_2. \tag{7.4}$$

By summing the voltages around the two irreducible loops, we find that

$$
\begin{aligned}
\text{Top loop:} \quad & \tilde{\phi}_4 = \Phi_B - \tilde{\phi}_1 - \tilde{\phi}_2 = \Phi_B + \phi_1 - \phi_2, \\
\text{Bottom loop:} \quad & \tilde{\phi}_3 = \Phi_A + \tilde{\phi}_4 = \Phi_A + \Phi_B + \phi_1 - \phi_2.
\end{aligned} \tag{7.5}
$$

where Φ_A and Φ_B are the applied magnetic fluxes through the top and bottom loops of the circuit, respectively.

We finally substitute the active-node fluxes ϕ_1 and ϕ_2 into the equations of motion (7.2). If we assume that the external fluxes Φ_A and Φ_B are constant, then we have

$$C_1(\ddot{\phi}_1 - \ddot{\phi}_2) = \frac{\phi_2 - \phi_1}{L_1} - \frac{\phi_1}{L_2} - F, \tag{7.6}$$

$$C_2\ddot{\phi}_2 - C_1(\ddot{\phi}_1 - \ddot{\phi}_2) = -\frac{\phi_2 - \phi_1}{L_1} + F, \tag{7.7}$$

where we have defined the constant "force" $F = (\Phi_A + \Phi_B)/L_1$. Note that these differential equations are second-order, with no first derivatives, and thus have the same form as Newton's equations of motion.

The circuit Lagrangian and Hamiltonian

To determine the quantum mechanical coordinates that satisfy the canonical commutation relations, we have to obtain the Hamiltonian for the circuit. To do this we need to first find a Lagrangian for the circuit, which is a function of the coordinates ϕ_i and their derivatives $\dot{\phi}_i$, such that the equations of motion are given by Lagrange's equations. Denoting the Lagrangian by $\mathcal{L}$, these equations are

$$\frac{d}{dt}\left(\frac{\partial\mathcal{L}}{\partial\dot{\phi}_i}\right) = \frac{\partial\mathcal{L}}{\partial\phi_i}. \tag{7.8}$$

We see that the equations of motion above, Eqs. (7.6) and (7.7), have the form of Newton's equations in which the capacitances play the role of masses, and the inductors play the role of forces. We recall now that the Lagrangian is the difference of the kinetic and potential energies, and with the above identifications, the electric circuit version would have to be the difference between the total capacitor energies and the total inductor energies. We can write these down from the circuit diagram, using the fact that a capacitor with capacitance C and flux ϕ has energy $C\dot{\phi}/2$, and similarly an inductance L has energy $\phi/(2L)$. Writing the capacitive and inductive energies for the circuit in terms of the active-node fluxes, we find that the Lagrangian is

$$\mathcal{L} = \frac{C_2 \dot{\phi}_2^2}{2} + \frac{C_1(\dot{\phi}_1 - \dot{\phi}_2)^2}{2} - \left[\frac{\phi_1^2}{2L_2} + \frac{(\Phi_A + \Phi_B + \phi_1 - \phi_2)^2}{2L_1} \right]. \tag{7.9}$$

Indeed, this Lagrangian gives the correct equations of motion. The rules of Hamiltonian mechanics now tell us that the Hamiltonian is given by $H = \sum_i \dot{\phi}_i p_i - \mathcal{L}$, where the p_i are the canonical momenta defined by $p_i \equiv \partial \mathcal{L}/\partial \dot{\phi}_i$. In fact, for the above circuit, and all circuits we will be concerned with, the kinetic (capacitive) part of the Lagrangian is linear. This means that the Lagrangian can be written as

$$\mathcal{L} = \tfrac{1}{2} \underline{\dot{\phi}}^\mathrm{T} C \underline{\dot{\phi}} + \mathbf{a}^\mathrm{T} \underline{\dot{\phi}} - E_\mathrm{v}(\underline{\phi}), \tag{7.10}$$

where $\underline{\phi} = (\phi_1, \phi_2, \ldots, \phi_N)$, C is a matrix of capacitances, $\mathbf{a}$ is a real vector (we will see shortly that $\mathbf{a}$ arises from voltage sources), and E_v is the part of the circuit energy that depends on the coordinates only. In this case the canonical momenta are given by

$$\mathbf{p} = C\underline{\dot{\phi}} - \mathbf{a}, \tag{7.11}$$

where $p = (p_1, p_2, \ldots, p_N)$, and the Hamiltonian is

$$H = \tfrac{1}{2}(\mathbf{p} + \mathbf{a})^\mathrm{T} C^{-1}(\mathbf{p} + \mathbf{a}) + E_\mathrm{v}(\underline{\phi}). \tag{7.12}$$

Note that the canonical momentum for a given node is only well defined if there is at least one capacitor connected to it. But since all circuit elements always have some "parasitic" capacitance in parallel with them, this causes no problem.

Using Eq. (7.11) we find that the momenta for our circuit are

$$p_1 = C_1(\dot{\phi}_1 - \dot{\phi}_2), \tag{7.13}$$

$$p_2 = C_2\dot{\phi}_2 - C_1(\dot{\phi}_1 - \dot{\phi}_2), \tag{7.14}$$

and the Hamiltonian is

$$H = \frac{p_1^2}{2C_1} + \frac{(p_1 + p_2)^2}{2C_2} + \left[\frac{\phi_1^2}{2L_2} + \frac{(\Phi_A + \Phi_B + \phi_1 - \phi_2)^2}{2L_1} \right]. \tag{7.15}$$

This is also the quantum Hamiltonian for the superconducting circuit, where the commutation relations between the operators ϕ_j and p_k are

$$[\phi_j, p_k] = i\hbar \delta_{jk}. \tag{7.16}$$

Note that the canonical momenta are linear combinations of the charges on the capacitors in the circuit, since the relationship between charge Q, voltage V, and flux ϕ across a capacitor C is $Q = CV = C\dot{\phi}$

Including voltage and current sources

We can include current and voltage sources as links in which either the current through the link is fixed or the voltage across the link is fixed. It is not obvious at first how to include these in the Lagrangian. It becomes clear, though, if we place the voltage source in *series* with a capacitor, and the current source in *parallel* with an inductor. We now consider treating the voltage and capacitor as a single link, and similarly the current and inductor. To write down the Lagrangian, we need the energy of the link, in terms of the flux across the link (or its derivative). For a capacitor by itself, the relationship is $E = C\dot{\phi}^2/2$. But if we include a voltage source V in the link, then the total voltage across the link is no longer the voltage across the capacitor. If the total link voltage is $V_{\text{link}} = \dot{\phi}$, then the voltage across the capacitor is $V_{\text{cap}} = V_{\text{link}} - V$ and so $\dot{\phi}_{\text{cap}} = \dot{\phi} - V$. The energy stored in the capacitor is therefore $E = C\dot{\phi}_{\text{cap}}^2/2 = C(\dot{\phi} - V)^2/2$. A similar line of reasoning for an inductor in parallel with a current source I shows that the effective energy of the link, in terms of the total flux across the link, is $E = (\phi + IL)^2/(2L)$.

If we need to add a current source that is not in parallel with an inductor, we can do this by first including the current source together with the parallel inductance in the circuit, calculating the Lagrangian, and then taking the limit as the inductance, L, tends to infinity. In this limit the inductance is effectively an open circuit, leaving the current source by itself. Since the current source and the inductance together contribute the term $-(\phi + IL)^2/(2L)$ to the Lagrangian, the current source alone contributes the term

$$\lim_{L \to \infty} -\frac{(\phi + IL)^2}{2L} = -I\phi - \lim_{L \to \infty} \left[\frac{\phi^2}{2L} + \frac{I^2 L}{2} \right] \equiv -I\phi, \tag{7.17}$$

where the final equivalence is due to the fact that adding a constant to the Lagrangian has no effect on the equations of motion.

We also note that a voltage source in parallel with a capacitor merely eliminates the capacitor, since the voltage source fixes the capacitor's energy.

7.1.1 Procedure for obtaining the circuit Lagrangian (short method)

To summarize what we now know, to write down the Lagrangian for a circuit we follow these steps:

1. Add up all the energies of the capacitor links as a function of the fluxes across the links. Each voltage source must be included as part of one of these capacitor links, as described above.
2. Add up all the energies of the inductor links, in terms of the fluxes across the links, including the contributions of any current sources.
3. The Lagrangian is the total capacitor energy minus the inductor energy.
4. For each irreducible loop, the sum of the fluxes around the loop must add up to the externally applied magnetic flux through the loop (which may be zero). Thus each loop gives us a constraint equation that will eliminate one flux variable.

5. Substitute these constraint equations into the Lagrangian so that the Lagrangian is left
   only with a set of independent fluxes and their derivatives.

The Hamiltonian is then obtained from the Lagrangian using the usual method of classical mechanics, as described above.

The number of dynamical variables in a circuit

Every capacitance in a circuit contributes one dynamical degree of freedom. Since the total flux added up around each loop is set equal to the magnetic flux through the loop, each loop imposes a constraint that reduces the number of dynamical variables by one. So the total number of dynamical variables in a circuit can be read off directly: it is the number of capacitances minus the number of irreducible loops. To obtain sensible dynamical equations a circuit must have enough capacitors. This is equivalent to the requirement in classical mechanics that each particle acted on by a force must have a mass.

It is worth discussing the constraint imposed by the loops a little more. Recall that this constraint is Faraday's law: the sum of all the voltages around a loop, V_{tot}, is the negative of the rate of change of the total magnetic flux through the loop, $d\Phi_{tot}/dt$. Note that Φ_{tot} is the sum of the externally applied flux and the flux generated by the current in the loop. However, we actually apply this rule to the integral of both these quantities: $\int V_{tot} + \Phi_{tot} = 0$. Because of this there is a constant of integration that is undetermined, so we should really write $\int V_{tot}(t) + \Phi_{tot}(t) = c$. In fact, it turns out that in a superconducting loop c is quantized. This is due to the relationship between the flux and the phase of the wavefunction for the current. The quantization is imposed by the fact that the phase can only change by an integer multiple of 2π around the loop. The result is that $\int V_{tot}(t) + \Phi_{tot}(t) = n\phi_0$, where $\phi_0 = h/(2e)$ and is called the "flux quantum." For the loops in our circuits, n is set, in a non-deterministic way, by the conditions chosen when the superconductor is brought into its superconducting phase. We usually assume that $n = 0$, and this is why we use the constraint $\int V_{tot} + \Phi_{tot} = 0$.

Faraday's law has further implications for the magnetic flux through a loop in which there are no capacitances. In this case there can be no net voltage around the loop (a voltage would cause an infinite current), and so the total magnetic flux through the loop cannot change. Any magnetic flux that passes through such a loop when it becomes superconducting is therefore trapped. Further, since there is no voltage contribution to the flux, the total flux is the magnetic flux, and so the latter is quantized.

7.2 Resonance and the rotating-wave approximation

We now pause to discuss the phenomenon of resonance, which we will need more than once in what follows. Resonance has to do with the response of a system to an applied force, or to the exchange of energy between two systems. Consider a quantum system with Hamiltonian H. If this system is prepared in an eigenstate of H it will remain in that state. If we apply an additional force to the system – that is, change the Hamiltonian – we

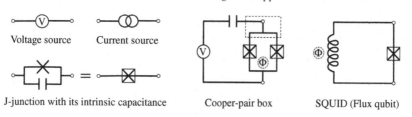

Figure 7.2 Displayed here are the circuit symbols for voltage and current sources, and the compact symbol for the Josephson junction (JJ) combined with its own capacitance. Also shown are the circuit diagrams for the "split" Cooper-pair box, and a loop with a JJ, also known as a "Superconducting quantum-interference device" (SQUID). In both circuits the symbol Φ denotes a magnetic flux through the loop. Note that there is no difference between the flux through a loop and that through an inductor in the loop.

can cause one eigenstate of the original Hamiltonian to evolve to another. It is useful to think of two ways in which we can apply a force. If we apply a constant force, then we could think of this as deforming the system. The force changes the Hamiltonian producing new eigenstates that are linear combinations of the original eigenstates. Let us consider just two of these original eigenstates, and call them $|E_1\rangle$ and $|E_2\rangle$. An applied force that "couples" these states together adds a matrix element to H that connects the two states: $\langle E_1|H|E_2\rangle = V$. To induce evolution that takes a system initially in state $|E_1\rangle$ and puts a significant population in state $|E_2\rangle$ at some future time, V must be an appreciable fraction of the energy difference between the states, $\Delta E = |E_1 - E_2|$. If $V \ll \Delta E$, then the force has little effect, and the amount of population in state $|E_2\rangle$ at any future time will remain small. These facts can be verified easily by diagonalizing a 2×2 matrix. The rule can be summarized loosely by saying that to induce "transitions" between states with an energy "gap" ΔE, a constant force must change the energy of the system by approximately ΔE.

There is another way to apply a force to a system to induce it to move from one energy eigenstate to another, and this requires a much smaller force. To do so one applies a time-dependent force oscillating as a sine wave with angular frequency $\omega = \Delta E/\hbar$. The matrix element connecting states $|E_1\rangle$ and $|E_2\rangle$ is now $\langle E_1|H|E_2\rangle = V\cos(\omega t)$. This oscillating force will cause a complete swap of population between $|E_1\rangle$ and $|E_2\rangle$, and do this in a time $t = \pi\hbar/(2V)$, regardless of the size of V. This is the phenomenon of resonance, of which a special case is the application of an oscillating force to a harmonic oscillator. The more the frequency of the applied force differs from ΔE, the less effective the force is at transferring population between the two energy levels, and thus the less effective it is at adding energy to or subtracting energy from the system.

Resonance is understood most easily by using the interaction picture. Consider the example of a Harmonic oscillator with frequency ω subjected to a force oscillating at ω. The Hamiltonian is

$$H = \hbar\omega a^\dagger a + \hbar Fx\cos(\omega t) = \hbar\omega a^\dagger a + \hbar 2g\cos(\omega t)(a + a^\dagger), \qquad (7.18)$$

where $g = \sqrt{\hbar/(8m\omega)}F$, and m is the mass of the oscillator. If moving to the interaction picture means evolving the operators using the oscillator Hamiltonian, $H_0 = \hbar\omega a^\dagger a$. The resulting Hamiltonian that generates the evolution of the states is

$$H = \hbar 2g\cos(\omega t)(ae^{-i\omega t} + a^\dagger e^{i\omega t})$$
$$= \hbar g(a + a^\dagger) + \hbar g(ae^{-2i\omega t} + a^\dagger e^{i2\omega t}). \qquad (7.19)$$

Now consider what happens if the size of the "force," g, is much smaller than ω. The timescale at which the states evolve is given by g, and so if $\omega \gg g$ the oscillating factors $e^{\pm 2i\omega t}$ in the Hamiltonian will oscillate sufficiently fast that they will cancel themselves out, to a very good approximation. The Hamiltonian is then effectively given by

$$H = \hbar g(a + a^\dagger). \qquad (7.20)$$

The oscillating terms that we dropped to obtain this Hamiltonian are referred to as the "off-resonant" terms, and dropping them is called the *rotating-wave approximation*, or RWA. Further details about the RWA are given in Appendix G. Basically, the RWA involves dropping any terms in the Hamiltonian that oscillate fast compared to the magnitude of all the terms.

We see from the Hamiltonian in Eq. (7.20) that the evolution is dependent only on the size of the force, g, and no longer on the energy gaps between the levels of the harmonic oscillator, $\Delta E = \hbar\omega$. Hence our claim above that a resonant force will induce transitions between energy levels no matter how small it is. If g is not much smaller than ω, then the force will still cause transitions but the motion is more easily analyzed using coherent states. If the frequency of the driving force is not close to that of the oscillator, then all the terms in the interaction Hamiltonian oscillate, and the effect of the force is reduced.

The reason that oscillating forces can inject energy into a system when stationary forces cannot can be understood more readily by examining an interaction between two quantum systems. A weak interaction between two systems will induce energy exchange between them if the energy gaps of one system are similar in size to the energy gaps of the other. In this case one system can drop an energy level while the other gains an energy level, with energy being conserved in the process. It is instructive to examine this process in the interaction picture. Consider two harmonic oscillators with frequencies ω_A and ω_B, interacting via a force proportional to the distance between them. The Hamiltonian is

$$H = \hbar\tilde{\omega}_A a^\dagger a - \hbar f(x_A - x_B)^2 + \hbar\tilde{\omega}_B b^\dagger b$$
$$= \hbar\omega_A a^\dagger a + 2\hbar f x_A x_B + \hbar\omega_B b^\dagger b, \qquad (7.21)$$

where we have absorbed the frequency shift induced in the oscillators into the frequencies ω_A and ω_B. Transforming to the interaction picture (moving the evolution of the individual oscillator Hamiltonians to the operators), we have

$$H = \hbar g(b^\dagger a e^{-i\Delta t} + a^\dagger b e^{i\Delta t}) + \hbar g(ab e^{-i(\omega_A + \omega_B)t} + a^\dagger b^\dagger e^{i(\omega_A + \omega_B)t})$$

$$\approx \hbar g(b^\dagger a e^{-i\Delta t} + a^\dagger b e^{i\Delta t}), \quad g \ll (\omega_A, \omega_B), \tag{7.22}$$

where $\Delta = \omega_A - \omega_B$ is the difference between the two frequencies. When the frequencies of the two oscillators are equal, then $\Delta = 0$, and the interaction obtains the maximum rate of energy exchange between the systems. Further, we see that the two terms in the interaction Hamiltonian, $b^\dagger a$ and $a^\dagger b$, describe the simultaneous loss of one quantum of energy from one oscillator and the gain of one quantum of energy to other. If the frequencies are different, then the action of these terms no longer conserves energy, because the quanta of the two oscillators have different energies. In this case the terms in the Hamiltonian oscillate at the difference frequency Δ, and this suppresses the energy exchange.

Examining the resonance effect for the applied force and for the coupled oscillators, we see that it is essentially the same in both cases. In the latter the force is replaced by the oscillating annihilation and creation operators. The classical force is in fact the classical equivalent of a mode of the electromagnetic field, or some other oscillating field. The fact that the force is oscillating means that it is a wave, and waves carry energy. An oscillating force is therefore a source of energy, and it injects energy into a quantum oscillator by virtue of the fact that it is a classical description of what is really a quantum oscillator.

There are two effects due to resonance that we will need below. The first is the fact that if a system has many different energy gaps, then a force applied at the right frequency is able to induce transitions between just one of these gaps. This will happen so long as the force is sufficiently weak, and the gaps are sufficiently different. The second effect is the use of a classical oscillating force to perform frequency conversion.

Frequency conversion

We saw above that two coupled oscillators with different frequencies will not exchange energy unless the size of the coupling term is similar to the frequency difference. But we can use resonance to induce them to exchange energy in the same way that we used an oscillating force to inject energy into a resonator: we can modulate the size of the interaction between the oscillators at the frequency difference. If we do this the Hamiltonian becomes

$$H = \hbar \omega_A a^\dagger a + 2\hbar f \cos(\Delta t) x_A x_B + \hbar \omega_B b^\dagger b, \tag{7.23}$$

where we have used the symbols as defined in Eq. (7.22). In the interaction picture this becomes

$$H \approx \hbar g(b^\dagger a + a^\dagger b) + \hbar g(b^\dagger a e^{-i2\Delta t} + a^\dagger b e^{i2\Delta t}), \tag{7.24}$$

where we have dropped the terms that oscillate at least as fast as ω_A and ω_B, assuming that Δ is less than each of them. When $g \ll \Delta$ we can also drop the terms that oscillate at $\pm\Delta$. The time-independent term in the Hamiltonian describes the exchange of quanta for resonators that are resonant. By modulating the interaction strength we have made the resonators appear to each other to be on resonance. This is the classical equivalent of adding a third quantum oscillator that has quanta of just the right energy to make up the energy lost or gained when the two original oscillators exchange quanta. This method of converting between different frequencies is invaluable in constructing mesoscopic circuits containing multiple systems.

7.3 Superconducting harmonic oscillators

LC-resonators

The simplest circuit, consisting of an inductor and a capacitor, is also a very useful one. It is a harmonic oscillator, and often referred to as an "LC" oscillator. If we follow the procedure above, we find that the Hamiltonian for this circuit is

$$H = \frac{Q^2}{2C} + \frac{\phi^2}{2L}, \quad [\phi, Q] = i\hbar. \tag{7.25}$$

Here C is the capacitance, L is the inductance, ϕ is the flux, and we have called the canonical momentum Q, since it is the charge on the capacitor. We have also set the magnetic flux through the single loop of the circuit to be zero.

The angular frequency of the resonator is $\omega = 1/\sqrt{LC}$. The annihilation operator for the oscillator is

$$a = \sqrt{\frac{C}{2\hbar L}}\phi - i\sqrt{\frac{L}{2\hbar C}}Q, \tag{7.26}$$

where $[a, a^\dagger] = 1$ and

$$\phi = \sqrt{\frac{\hbar L}{2C}}(a + a^\dagger), \quad Q = -i\sqrt{\frac{\hbar C}{2L}}(a - a^\dagger). \tag{7.27}$$

Superconducting "microwave cavity" resonators

The superconducting LC-oscillator described above is referred to as a "lumped element" device, since its electromagnetic oscillation can be described as due to two individual circuit elements. The opposite of this is an expanse of superconductor throughout which the electric field is free to vary. While an LC-oscillator has just a single oscillation frequency, a piece of superconductor can support multiple waves with different wavelengths and frequencies, and so is often referred to as a "cavity." An example is a strip of superconductor on a flat substrate, depicted in Fig. 7.3. The ends of the strip are reflecting boundaries, so that the strip acts a bit like a guitar string, supporting a discrete set of standing waves. The waves in the strip couple, via a capacitive interaction, to the superconducting lines that terminate near to each end . These lines are usually sufficiently long that they support an effective continuum of traveling waves, in which case they are called *transmission lines*. Signals can be fed into the strip via these lines, and signals that travel along the lines from

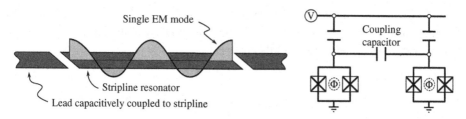

Figure 7.3 On the left is shown a simple layout for a "stripline" resonator. The strip of superconductor in the center supports a set of discrete standing-wave modes of the electromagnetic field. The modes of this resonator are coupled capacitively to two leads, one on each end. On the right we show a circuit diagram for two Cooper-pair boxes coupled capacitively.

the strip can be measured. The strip must be long enough to support a wave with the desired frequency.

Electrical oscillators that support standing waves along a single spatial direction are referred to variously as *stripline, coplanar wave-guide,* or *transmission-line* resonators. They are also called one-dimensional *superconducting microwave cavities.* To describe these cavities in a circuit, one can use a model also typically used to describe transmission lines. In this model the strip is represented by a chain of many identical inductors connected in series, and each inductor is also connected to ground via a capacitor. The chain of the inductors is terminated at each end with a capacitance. If one takes the limit in which there are many inductors per-unit-distance, the result is a continuously varying field. If the inductance per-unit length is $\mathcal{L}$, and the capacitance per unit length is $\mathcal{C}$, then the speed of the waves is $v = 1/\sqrt{\mathcal{L}\mathcal{C}}$. The frequencies of the modes supported by the cavity are $\omega_n = n\pi v/L$, where n is a positive integer and L is the length of the cavity [67, 488]. A wealth of information about microwave cavities can be found in [488].

The coupling of the superconducting cavity to the transmission lines is exactly analogous to the coupling of an optical cavity to the outside world via transmission through its end-mirrors. This is because transmission lines support a continuum of traveling modes in the same way that free space does. Thus the coupling of the former to a transmission line, as well as that of superconducting LC oscillators, is described by the theory of damping developed in Chapter 4, and the input–output theory described in Chapter 3 and Appendix F. We can also use the homodyne detection theory of Chapter 3 for measurements on superconducting oscillators. This will be discussed further in Section 7.7.3 below.

Superconducting strips are not the only way to create cavities for microwaves. A genuine three-dimensional cavity can be made by hollowing out a small region inside a solid block of superconductor. If we make two small holes in the superconductor that connect the cavity to the outside, we can insert into these holes cables that contain superconducting leads, so that the leads terminate at the sides of the cavity. The microwaves in the transmission lines couple evanescently to the modes supported by the cavity. A superconducting qubit can then be coupled to the cavity modes merely by placing the qubit circuit, fabricated on a tiny wafer, inside the cavity [463], [517]. We will discuss the coupling of superconducting qubits to superconducting cavities further below.

7.4 Superconducting nonlinear oscillators and qubits

To create two-level quantum systems using superconducting circuits, it is enough to create a nonlinear resonator with a sufficiently strong nonlinearity. The reason for this is the phenomenon of *resonance*, discussed above. The difference between linear (harmonic) resonators and nonlinear resonators is that in the latter the energy gaps between adjacent energy levels are in general all different. Recall that to couple together energy levels separated by a gap ΔE, we need to apply a force oscillating at a frequency $\omega = \Delta E / \hbar$. What is useful to use here is that a force with frequency ω will couple levels with the gap ΔE, but will not couple levels for which the gap is sufficiently different. The force is referred to as being "on resonance" with the gap or transition ΔE, and "off-resonant" with the other transitions. Since a linear force gives a term in the Hamiltonian proportional to a and $a^\dagger$, it only couples adjacent oscillator states together. So by modulating the force at the right frequency, we can couple together only the lowest two levels of the oscillator, and restrict the dynamics to this two-dimensional subspace. In Appendix G we show quantitatively how a transition between two levels is suppressed by a mismatch between the frequency of the coupling and the energy gap of the transition.

We now introduce the Josephson junction, a remarkable superconducting circuit element that is highly nonlinear. We can use this to create nonlinear oscillators and effective qubits.

7.4.1 The Josephson junction

The Josephson junction, or "JJ" for short, is a thin wall made of normal conductor that separates two lumps of superconductor. The BCS theory of superconductivity tells us that the current flowing in a superconductor is made up of particles called "Cooper-pairs," where each Cooper-pair is a pair of electrons acting as if they were a single particle. While electrons are fermions, the Cooper-pairs are bosons. The Cooper-pairs that form the superconducting current are all in the same quantum state, meaning that they all have the same spatial wavefunction. In fact this is an effective wavefunction derived from the order parameter in the BCS theory [2], but that is well beyond our needs here. Many of the properties of superconductors can be derived by considering the dynamics of the effective wavefunction, called the supercurrent wavefunction.

Josephson realized that the supercurrent wavefunctions in each of the two superconducting lumps should penetrate into the wall between them in the same way that the wavefunction of a particle penetrates into a potential barrier. Because of this penetration, the wavefunctions of the two superconductors overlap inside the wall, and this leads to a coupling between the respective supercurrents. What is remarkable is that this coupling creates a strongly nonlinear dynamics for the flux across the junction: the supercurrent that flows through the junction, I, is related to the flux across the junction, ϕ, by

$$I = I_0 \sin(2\pi\phi/\phi_0). \tag{7.28}$$

Here $\phi_0 = h/(2e)$ is a fundamental constant, referred to as the "flux quantum" (see above), and I_0 is determined by the junction geometry. As we will see, unlike in capacitors and inductors the quantum operator representing the flux ϕ across the junction is a cyclic variable, taking values in the range $[0, \phi_0)$. We will also see that the highly nonlinear relationship between I and ϕ is due directly to the quantum mechanical nature of the supercurrent wavefunction.

The Josephson-junction relation, Eq. (7.28), is actually quite simple to derive from the supercurrent wavefunction, as is the cyclic nature of the flux, and we do this now. For simplicity we assume that the two supercurrent wavefunctions, one in each lump of super-conductor, are close to plane waves and have essentially equal charge densities. We can write their values at the wall as $\psi_j = A_j e^{-i\theta_j}$, where $j = 1, 2$ labels the two lumps. Inside the conducting wall the wavefunctions decay exponentially with same decay constant γ. So in the wall the total wavefunction is

$$\psi(x) = A_1 e^{-\gamma(x+d)} e^{-i\theta_1} + A_2 e^{\gamma(x-d)} e^{-i\theta_2}, \quad x \in [-d, d] \tag{7.29}$$

where $2d$ is the thickness of the wall and x is the position inside the wall, with $x = 0$ being the center of the wall. We can now calculate the current through the wall, since it is proportional to the probability current for the wavefunction ψ. Using the expression for the probability current, J, we have

$$J = \frac{\hbar}{m_c} \mathrm{Im} \left[\psi^* \frac{\partial}{\partial x} \psi \right] = \frac{2\gamma \hbar A^2}{m_c} \sin(\theta_2 - \theta_1) \equiv J_0 \sin(\theta), \tag{7.30}$$

where m_c is the mass of a Cooper-pair, and we have defined $\theta \equiv \theta_2 - \theta_1$. The current through the junction is thus proportional to the sine of the phase difference between the two supercurrent wavefunctions.

We now note that the energy of the Cooper-pairs is, by the usual rules of quantum mechanics, equal to $\hbar$ times the oscillation frequency of their wavefunction. Since the oscillation frequency of ψ_j is simply $\omega_j = d\theta_j/dt$, the energy difference between the Cooper-pairs on either side of the wall is

$$\Delta E = \hbar \frac{d}{dt}(\theta_1 - \theta_2) = -\hbar\dot{\theta}. \tag{7.31}$$

But since the Cooper-pairs have charge $-2e$, and they flow across the wall, their energy difference between the two sides of the wall must be equal to the voltage difference across the wall, V, multiplied by $-2e$. Thus $V = -\Delta E/(2e) = \hbar/(2e)\dot{\theta}$. Now of course the flux across the junction is defined by $\dot{\phi} = V$. In view of Eq. (7.31), this means that the flux is essentially the phase difference multiplied by $\hbar/(2e) = \phi_0/(2\pi)$, where $\phi_0 \equiv h/(2e)$ is the flux quantum. However, since the flux need not be confined to the range $[0, 2\pi)$, the relationship is technically

$$\theta = 2\pi \left[\frac{\phi}{\phi_0} \mod 1 \right]. \tag{7.32}$$

This last relation, when substituted into Eq. (7.30), gives immediately the Josephson-junction relation between current and flux, Eq. (7.28). Now, while the flux need not be confined to the range of θ, it is the latter that is quantum mechanically the more fundametal quantity, since it is the phase of the wavefunction. Because of this, when quantizing the Hamiltonian of the Josephson junction, we take the phase difference θ to be the canonical quantum degree of freedom, rather than the flux. This makes the Hamiltonian of the Josephson junction quite distinct from that of LC circuits, because this phase is a cyclic variable.

For the purpose of writing down the Hamiltonians of circuits with Josephson junctions (JJs), it is the expression for the energy of the JJ that we need. We can calculate this energy at time t by integrating the power over time, assuming that at $t = -\infty$ all fields are zero. The energy is

$$E_{JJ} = \int_{-\infty}^{t} I(s)V(s)\,ds = I_0 \int_{-\infty}^{t} \sin\left[\frac{2\pi\phi}{\phi_0} \right] \frac{d\phi}{ds}\,ds = -E_J \cos\left[\frac{2\pi\phi}{\phi_0} \right], \tag{7.33}$$

where the *Josephson energy* is defined by

$$E_J \equiv \phi_0 I_0/(2\pi). \tag{7.34}$$

We can now include JJs as elements in our circuits using Eq. (7.28) or Eq. (7.33). However, every real JJ also has a capacitance. So when we include a JJ in a circuit, we must always place it in parallel with this capacitance. We will refer to a JJ without a parallel capacitance as a *bare* JJ. For convenience we define a circuit symbol that represents the parallel combination of a capacitance and a bare JJ. The circuit symbol for a bare JJ is shown in Fig. 7.1, and that for a real JJ is shown in Fig. 7.2.

The Hamiltonian of an isolated Josephson junction

Since a real JJ has its own capacitance, it forms a circuit all by itself, consisting of two links (a capacitor and a bare JJ) and two nodes. Recall from the previous section that it is the phase difference, θ, given by Eq. (7.32), that we regard as the canonical quantum coordinate. But to keep with units of flux, we will multiply the phase by $\phi_0/2\pi$, and use the resulting *cyclic flux*, $\phi_J \equiv \phi_0\theta/(2\pi)$, as our canonical coordinate.

Using the expression for the energy of the JJ, Eq. (7.33), the Lagrangian for the isolated JJ is

$$\mathcal{L}_{JJ} = \frac{C_J\dot{\phi}_J^2}{2} + E_J \cos\left(\frac{2\pi\phi_J}{\phi_0} \right). \tag{7.35}$$

The canonical momentum is

$$Q_J = \frac{\partial L_{JJ}}{\partial \dot\phi_J} = C_J \dot\phi_J, \tag{7.36}$$

and the Hamiltonian is

$$H_{JJ} = \frac{Q_J^2}{2C_J} - E_J \cos\left(\frac{2\pi\phi_J}{\phi_0}\right). \tag{7.37}$$

We now apply the canonical commutation relations, $[\phi_J, Q_J] = i\hbar$, but here there is a twist. Recall that if the canonical variable ϕ_J takes all values on the real line, then we can construct an operator Q_J that satisfies the commutation relation, and the eigenvalues of this operator also take all values on the real line. The relationship that is required between the operators ϕ_J and Q_J to satisfy the commutation relation is the Fourier transform. But in the present case the eigenvalues of the operator ϕ_J do not take all values on the real line, but are confined instead to the interval $[0, \phi_0)$. It turns out that we can still obtain an operator Q_J that satisfies the canonical commutation relation, but because ϕ_J is confined to a finite interval, we must use the Fourier series instead of the Fourier transform. The result is that the set of eigenvalues of Q_J is now discrete. These eigenvalues are $n(\hbar/\phi_0) = 2en$ where n is any integer. Since the eigenvalues have units of charge, and each Cooper-pair has a charge of $-2e$, we see that n can be interpreted as corresponding to a number of Cooper-pairs. The operators ϕ_J and Q_J are given by

$$\phi_J = \left(\frac{\phi}{2\pi}\right)\hat\theta_J, \quad \text{where } \hat\theta_J|\theta\rangle = \theta|\theta\rangle, \quad \theta \in [0, 2\pi), \tag{7.38}$$

$$Q_J = -2eN_J, \quad \text{where } N_J|n\rangle = n|n\rangle, \quad n = 0, \pm1, \pm2, \dots \tag{7.39}$$

The relationship between the number and phase operators is

$$|\theta\rangle = \sum_{n=-\infty}^{\infty} e^{i\theta n}|n\rangle, \quad |n\rangle = \frac{1}{2\pi}\int_0^\infty e^{-i\theta n}|\theta\rangle\, d\theta, \tag{7.40}$$

and their commutator is $[\hat\theta_J, N_J] = i$. Because of this commutation relation, the operators each generate a shift in the value of the other. That is,

$$\exp(-im\hat\theta_J)|n\rangle = |n+m\rangle, \tag{7.41}$$
$$\exp(-i\delta N_J)|\theta\rangle = |\theta+\delta\rangle, \tag{7.42}$$

The first relation means that we can write $\exp(-i\hat\theta_J) = \sum_n |n+1\rangle\langle n|$. We can use this relation to rewrite the nonlinear term in the Hamiltonian, Eq. (7.37). In particular,

$$\cos(\hat\theta) = \frac{1}{2}\left[e^{-i\hat\theta_J} + e^{i\hat\theta_J}\right] = \sum_{n=-\infty}^{\infty} |n+1\rangle\langle n| + |n\rangle\langle n+1|. \tag{7.43}$$

This expression makes the physical interpretation for the nonlinear term clear: it induces transitions that change the number of Cooper-pairs by plus or minus one. Since the nonlinear term comes from the junction, this makes the interpretation for the number operator clear. Since Cooper-pairs can tunnel through the junction, we can expect the nonlinear term to describe this tunneling. Since this term involves changing n by one, n must represent the number of Cooper-pairs on one side of the junction.

In terms of the number and phase operators, the Hamiltonian for the isolated JJ is

$$H_{JJ} = \left(\frac{(2e)^2}{2C_J} \right) N_J^2 - E_J \cos(\hat{\theta}). \tag{7.44}$$

While this is a neat theoretical description, real JJs presently have physical imperfections that need to be taken into account. One of these is that real capacitors always have some residual bias charge even at zero voltage, due, for example, to a difference between the work functions of the capacitor's plates. Since the N_J represents the number of Cooper-pairs on one side of the junction, it also represents the charge on the capacitor. A residual bias charge therefore enters the Hamiltonian as an additional charge added to the value of N_J. If we denote the bias charge by Q_C, and define the dimensionless quantity $N_c \equiv Q_C/(2e)$, then the Hamiltonian becomes

$$H_{JJ} = \frac{(2e)^2}{2C_J}(N_J - N_c)^2 - E_J \cos(\hat{\theta}). \tag{7.45}$$

All three parameters in the Hamiltonian, C_J, N_c, and E_J are subject to fluctuations in time, and this is an important source of noise for circuits with Josephson junctions.

7.4.2 The Cooper-pair box and the transmon

A circuit for a "Cooper-pair box" (CPB), also known as a "charge qubit," is shown in Fig. 7.2. The circuit shown is in fact a "split" CPB, which has two JJs in parallel, with an external flux applied through the loop that they form. The qualifier "split" refers to the fact that this circuit is obtained from the original CPB circuit by replacing a single JJ with two parallel JJs. The reason for the applied flux will become clear below. The CPB can be thought of as a superconducting "island." This island is the piece of the circuit enclosed by the dashed box in Fig. 7.2. To get to the island from the voltage source you either have to cross the capacitor on one side, or one of the JJs on the other. One can speak of the island as "being formed by a capacitor and two JJs."

Using the procedure in Section 7.1.1, the Lagrangian for the split CPB is

$$\mathcal{L} = \frac{C(\dot{\phi} - V)^2}{2} + \frac{C_{J1}\dot{\phi}_{c1}^2}{2} + \frac{C_{J2}\dot{\phi}_{c1}^2}{2} + E_1 \cos\theta_1 + E_2 \cos\theta_2, \tag{7.46}$$

where C is the capacitor connected to the voltage source, and C_{J1} and C_{J2} are the capacitances of the two JJs. Note that we have absorbed the offset charge of the capacitor into the definition of V (cf. Eq. (7.45)). To apply the constraints we recall that the flux ϕ is not the flux across the capacitor, but the sum of the fluxes across the voltage source and the capacitor. The constraints for the JJs are $\dot{\phi}_{c1} = -(\phi_0/2\pi)\dot{\theta}_1$ and $\dot{\phi}_{c2} = -(\phi_0/2\pi)\dot{\theta}_2$, and the constraints for the two loops are $\dot{\phi} = -(\phi_0/2\pi)\dot{\theta}_1$ and $\theta_1 + \theta_2 = 2\pi\,(\Phi/\phi_0 \bmod 1)$. Defining $\varphi = \phi$ and $\theta = 2\pi\,(\varphi/\phi_0 \bmod 1)$ the Lagrangian is

$$\mathcal{L} = \frac{C_\Sigma(\dot{\varphi} - \eta V)^2}{2} + E_1\cos\theta + E_2\cos\left(\frac{2\pi\,\Phi}{\phi_0} - \theta\right). \tag{7.47}$$

where $C_\Sigma = C + C_{J1} + C_{J2}$ and $\eta = C/C_\Sigma$. We now make the change of variable $\varphi \to \varphi + \Phi/2$ and use the trigonometric identity $\cos(A \pm B) = \cos A \cos B \pm \sin A \sin B$ to rewrite the resulting expression as

$$\mathcal{L} = \frac{C_\Sigma(\dot{\varphi} - \eta V)^2}{2} + E\cos\theta + E'\sin\theta, \tag{7.48}$$

where

$$E = \cos(\pi\,\Phi/\phi_0)(E_1 + E_2), \tag{7.49}$$

$$E' = \sin(\pi\,\Phi/\phi_0)(E_1 - E_2). \tag{7.50}$$

We can further rewrite this as

$$\mathcal{L} = \frac{C_\Sigma(\dot{\varphi} - \eta V)^2}{2} + \tilde{E}\cos(\theta + \delta\theta), \tag{7.51}$$

where

$$\tilde{E}^2 = E_1^2 + E_2^2 + 2E_1 E_2\cos(2\pi\,\Phi/\phi_0), \tag{7.52}$$

$$\tan(\delta\theta) = -\left(\frac{E_1 - E_2}{E_1 + E_2}\right)\tan(\pi\,\Phi/\phi_0). \tag{7.53}$$

We have obtained a Lagrangian with a cosine potential in which the height of the potential, as well as the offset phase $\delta\theta$, can be changed by changing the applied magnetic flux. In fact, the Lagrangian is the same Lagrangian we would obtain by replacing the two JJs with a single JJ, but in that case the height of the potential would be fixed by the Josephson energy of the junction.

Calculating the canonical momenta, Q, we find that the Hamiltonian is

$$H_{\mathrm{JJ}} = \frac{(Q + CV)^2}{2C_\Sigma} - \tilde{E}\cos\left[\frac{2\pi}{\phi_0}(\varphi + \delta\varphi)\right], \tag{7.54}$$

where $\delta\varphi = \phi_0\delta\theta/(2\pi)$. Note that all the voltage fluxes that appear as coordinates (as opposed to velocities) in the Lagrangian are cyclic, so Q has discrete energy levels, just as it does for the isolated JJ. Changing variables to the Cooper-pair number-operator N_{J}, as we did for the isolated JJ, the Hamiltonian is

$$H_{\mathrm{JJ}} = E_{\mathrm{c}}(N_{\mathrm{J}} - N_{\mathrm{v}})^2 - \tilde{E}\cos(\varphi + \delta\varphi), \tag{7.55}$$

where $E_c \equiv (2e)^2/(2C_\Sigma)$ and $N_v = CV/(2e)$.

To use the CPB as a single qubit, we need two states whose energy difference is small compared to the distance to nearby states. It turns out that this is true for the lowest two energy eigenstates when $E_c \gg \tilde{E}$. Further, when $N_{\mathrm{mod}} \equiv (N_v \bmod 1)$ is close to zero, these states are close to the eigenstates of the charge N_J with eigenvalues N_v and $N_v + 1$. Conversely, when N_{mod} is close to $1/2$, the lowest two energy states are nearly equal superpositions of the same charge states. (Note that if $N_v = n + 1/2$ the eigenstates of charge with eigenvalues n and $n+1$ are degenerate.)

Let us denote the charge eigenstates with eigenvalues N_v and $N_v + 1$ as $|0\rangle$ and $|1\rangle$ respectively. In view of the above results, when we restrict the Hamiltonian to the lowest two energy eigenstates, we can write it in the charge basis $\{|0\rangle, |1\rangle\}$ as

$$H_{\mathrm{qubit}} = \hbar\Delta\sigma_z + \hbar\lambda\sigma_x, \tag{7.56}$$

where σ_z and σ_x are the usual Pauli matrices. When $E_c \gg \tilde{E}$ the parameters are [639]

$$\Delta \approx E_c(1/2 - N_{\mathrm{mod}}), \qquad \lambda \approx \tilde{E}/2, \tag{7.57}$$

where $N_{\mathrm{mod}} \in [0, 1/2]$ and can be externally varied.

Various designs for CPBs can be found in [96, 560, 391].

The transmon

The so-called *transmon* is very similar to the Cooper-pair box. The only difference in the circuit is that an extra capacitor may be connected in parallel with the JJs [339, 76]. The resulting Hamiltonian has exactly the same form as that of the CPB. The sole purpose of the extra capacitor is to increase the capacitance C_Σ in the Hamiltonian, and thus reduce the value of E_c. The difference between the CPB and the transmon is in the choice of the relative sizes of E_c and $\tilde{E}$. In the CPB $E_c \gg \tilde{E}$, whereas for the transmon this relationship is the other way around: $E_c \ll \tilde{E}$. This has the effect of reducing somewhat the differences in the energy gaps between adjacent levels (that is, making the system more like a harmonic oscillator, reducing the "anharmonicity"), but has the great benefit of making the resulting qubit much less sensitive to charge noise. Noise is a problem for all superconducting qubits, and for CPBs the problem is the sensitivity to fluctuations of charge in the environment. CPBs are especially sensitive to charge noise because their energy states are more localized in charge. This means that environmental charge fluctuations affect their energy levels, and thus induce dephasing noise into the dynamics. The rate at which charge varies with energy is referred to as the "charge dispersion." By reducing the relative contribution of the charge to the energy, the charge sensitivity is greatly reduced. While the states of the transmon cannot be measured using a charge sensor as they can for the CPB, both kinds of qubits can be measured by using an oscillator. We discuss this further below.

Coupling CPBs/transmons together

Cooper-pair boxes (or transmons) can be coupled together by linking their circuits with a capacitor, a configuration shown in Fig. 7.3. This results in a coupling Hamiltonian of the form $H \propto \sigma_z^{(1)}\sigma_z^{(2)}$ (here the superscripts label the qubits). If one wants to turn the coupling on and off then a capacitive coupling is not so practical, as the introduction of a switch opens the qubit to noise from the circuit that controls the switch. One way around this problem is to couple the CPBs together indirectly by coupling each to the same resonator. If we set the resonator to be sufficiently far off resonance with the qubits then the resonator can be eliminated, and the resulting effective interaction between the qubits has the form $H \propto \sigma_y^{(1)}\sigma_y^{(2)}$. It turns out that this interaction can be turned off and on by changing the Josephson energies of the qubits via the flux applied to each [391]. We show how to couple CPBs to resonators in the next section.

7.4.3 Coupling qubits to resonators

Later in this chapter we will consider making measurements on mesoscopic systems by coupling them to a cavity or oscillator. One can usually couple two electrical devices together by linking their respective circuits together via a capacitor or mutual inductance. Another way to do it is to replace the voltage source in one of the circuits with the same voltage source in series with the second device. This effectively adds the voltage across the second device to the voltage that appears in the Hamiltonian of the first. We give an example of this method in Fig. 7.4, where we couple a Cooper-pair box (CPB) to an LC-oscillator. The resulting coupling Hamiltonian can also be obtained in coupling to a stripline or 3D-cavity resonator, and thus the LC-resonator serves as a good circuit model for these distributed resonators.

To obtain the Hamiltonian for the CPB/LC-resonator in Fig. 7.4, we add the contribution of the inductor and capacitor in the LC-resonator to the Lagrangian of the original CPB, given in Eq. (7.51). Only one of the constraints in the original CPB circuit is modified, the

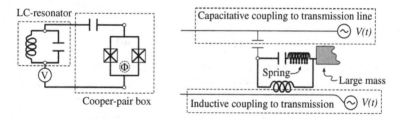

Figure 7.4 The circuit on the left is a Cooper-pair box (CPB) coupled to an LC-resonator. The circuit on the right is a mechanical oscillator coupled to an LC-resonator. Here the mechanical oscillator forms one plate of the capacitor in the LC-resonator. This plate oscillates on a spring that is anchored to a large mass. In the dashed boxes we show how the LC-resonator can be coupled to a transmission line either capacitively or inductively. The transmission line can be used to apply an input signal, $V(t)$, and measure an output signal.

one coming from the loop containing the voltage source. The new version of this constraint is $\dot{\phi} + \dot{\phi}_{res} = -(\phi_0/2\pi)\dot{\theta}_1$ where $\dot{\phi}_{res}$ is the flux across the resonator's inductor. Using this to write ϕ in terms of ϕ_{res} and θ_1, and defining $\theta = \theta_1$ and $\dot{\varphi} = -(\phi_0/2\pi)\dot{\theta}$, the Lagrangian is

$$\mathcal{L} = \frac{C_\Sigma(\dot{\varphi} - \dot{\phi}_{res} - \eta V)^2}{2} + \tilde{E}\cos(\theta + \delta\theta) + \frac{C_{res}(\dot{\phi}_{res})^2}{2} + \frac{\phi_{res}^2}{2L_{res}}, \tag{7.58}$$

where C_Σ and η are defined just below Eq. (7.46), and $\delta\theta$ is defined in Eq. (7.53). Using the general formula given in Eq. (7.12) we obtain the Hamiltonian

$$H = H_{CPB} + gQ_{res}N_J + \frac{(Q_{res} - Q_0)^2}{2C_{ser}} + \frac{\phi_{res}^2}{2L_{res}}, \tag{7.59}$$

where the CPB Hamiltonian is

$$H_{CPB} = E_c(N_J - rN_v)^2 - \tilde{E}\cos(\varphi + \delta\varphi), \tag{7.60}$$

the resonator momentum conjugate to ϕ_{res} is denoted by Q_{res},

$$C_{ser} = \frac{C_\Sigma C_{res}}{C_\Sigma + C_{res}}, \quad r = \frac{C_{ser}}{C_\Sigma}, \quad Q_0 = rCV, \quad g = \frac{2e}{C_{res}}, \tag{7.61}$$

and the other variables are defined above in Section 7.4.2. We see that the coupling term between the CPB and the resonator is the product of their respective momenta.

We now want to know the form that the coupling takes when the Hamiltonian is restricted to the lowest two energy eigenstates to form a single qubit. We note that we can absorb the coupling term into the quadratic term in the CPB Hamiltonian, which results in the replacement $rN_v \rightarrow rN_v + gQ_{res}/(2E_c)$. The Hamiltonian with the coupling term is therefore given by replacing N_v with $rN_v + gQ_{res}/(2E_c)$ in Eq. (7.56). The resulting coupling between the qubit and LC-resonator is thus

$$H = \hbar\Delta\sigma_z + \hbar\lambda\sigma_x + \frac{g}{2}\sigma_z Q_{res} + \frac{(Q_{res} - Q_0)^2}{2C_{ser}} + \frac{\phi_{res}^2}{2L_{res}}. \tag{7.62}$$

7.4.4 The RF-SQUID and flux qubits

A circuit consisting of a loop in which a single JJ is connected to an inductor is traditionally called an RF-SQUID. The acronym SQUID stands for "superconducting quantum interference device", and refers to circuits containing JJs that were originally used for measuring tiny magnetic fields. The prefix "RF" in the name for this particular circuit comes from the way in which it detects these fields [337]. But this history does not really concern us here.

The RF-SQUID circuit is depicted in Fig. 7.2. The inductor in the circuit has an external flux Φ applied to it. Using the procedure in Section 7.1.1 the Lagrangian for the circuit is the Lagrangian for the isolated JJ, Eq. (7.35), minus the energy of the inductor. Note that the external magnetic flux through the inductor simply contributes to the total external flux through the loop that contains the inductor: there is no need to distinguish between magnetic flux applied to the inductor and that applied to the loop. The Lagrangian is therefore

$$\mathcal{L} = \frac{C_J \dot{\phi}_c^2}{2} + E_J \cos\left(\frac{2\pi \phi_J}{\phi_0}\right) - \frac{\phi_l^2}{2L}, \tag{7.63}$$

where ϕ_c is the capacitor flux, and ϕ_l is the inductor flux. Since we have three elements in parallel (the inductor, the JJ, and the JJs capacitance), we have two loops, and thus two constraints. We must also choose which element is in the "middle" dividing the two loops, although this choice does not effect the result. Choosing the capacitor we note that the loop between the capacitor and the bare junction has no external flux, but the loop between the capacitor and the inductor has external flux Φ. So the loop constraints are $\phi_J = -\phi_c$ mod ϕ_0, and $\phi_c = -\phi_l - \Phi$. Substituting in ϕ_l for the other two fluxes, and setting $\Phi = 0$, the Lagrangian becomes

$$\mathcal{L} = \frac{C_J \dot{\phi}_l^2}{2} + E_J \cos\left[\frac{2\pi}{\phi_0}(\phi_l + \Phi)\right] - \frac{\phi_l^2}{2L}. \tag{7.64}$$

Note that although the flux that appears in the cosine term is cyclic, the flux that appears in the inductive term is not. While we can replace the cyclic flux with the continuous flux ϕ_l, we cannot do the reverse – cyclic variables can only appear in periodic functions, since they would generate unphysical discontinuities anywhere else. We take this to imply that the canonical momentum must be conjugate to the continuous variable, and not the cyclic variable. As a result, the addition of the inductor changes the canonical momentum from a discrete variable to a continuous one.

Calculating the conjugate momentum using the Lagrangian in Eq. (7.64), the Hamiltonian becomes

$$H = \frac{q^2}{2C_J} - E_J \cos\left[\frac{2\pi}{\phi_0}(\phi + \Phi)\right] + \frac{\phi^2}{2L}, \tag{7.65}$$

where the canonical variables are $\phi \equiv \phi_l$ and q. The system is a highly nonlinear oscillator, with a cosine term added to the quadratic potential.

We introduced above the Josephson energy $E_J \equiv \phi_0 I_0/(2\pi)$. This parameter gives the height of the cosine potential. Another parameter that is also useful characterizes the wells of the cosine potential. If we expand the cosine in a power series about a minimum (e.g., $2\pi(\phi + \Phi)/\phi_0 \equiv \theta = \pi$), then the second-order term (the parabolic shape) is equal to $E_J \theta^2 (2\pi)^2/(2\phi_0^2) \equiv \theta^2/2L_J$. The new parameter defined by $L_J \equiv \phi_0/(2\pi I_0)$ is called the Josephson inductance. The meaning of L_J is as follows: the parabolic potential that most closely matches the wells of the cosine is the potential due to an inductance equal to L_J.

To form a qubit the parameters are chosen to create a double-well potential, by setting $L_J = L$ and choosing Φ so that the peak of the cosine is in the center of the parabolic potential. If we restrict the Hamiltonian to the two lowest energy levels in the double well, and define the symmetric and anti-symmetric combinations of these two energy levels to be the eigenstates of the operator σ_z, then the Hamiltonian takes the form [144]

$$H_{\text{qubit}} = \hbar\Delta\sigma_z + \hbar\lambda\sigma_x. \tag{7.66}$$

The expressions for the parameters Δ and λ are fairly complex, and contain constants that must be determined numerically. The coupling between the levels has the form

$$\lambda \propto (1 + \Phi/\phi_0)^{-1}, \tag{7.67}$$

and can therefore be controlled using the external flux. These qubits are commonly referred to as "flux qubits." The simple design for flux qubits presented above can be greatly improved to make them less sensitive to noise [429, 68, 701], in a manner analogous to that in which the transmon is obtained from the CPB [339]. These more advanced designs use multiple JJs to obtain a single flux qubit. Further details regarding the original flux qubit can be found in [98, 594], and more advanced designs can be found in [429, 457, 68, 701, 720, 176].

Coupling flux qubits together

Flux qubits can be coupled together directly by using a mutual inductance. In the simplest case a second inductor is included in the SQUID loop of each qubit, arranged so that there is a mutual inductance between them. The direct inductive coupling used in [429] gives an interaction Hamiltonian of the form $H \propto \sigma_z\sigma_z$. Like charge qubits, flux qubits can be coupled together by using an LC-resonator as an intermediary, for example by coupling two qubits inductively to the same resonator. Adiabatic elimination of the resonator gives an interaction of the form $H \propto \sigma_y\sigma_y$ [392]. Further methods for coupling flux qubits together can be found in [226, 59, 450, 624].

7.5 Electromechanical systems

Mechanical resonators

Micro- and nano-mechanical resonators are tiny pieces of a solid that are connected to a much larger object but have enough freedom to vibrate with a discrete set of frequencies. They can take the form of wires, bridges, flaps (cantilevers), or membranes. Any piece of solid matter has many ways in which it can be distorted, and thus many degrees of freedom that can oscillate harmonically at a set of frequencies. The oscillation at each frequency is called a mode. These modes are simple to describe quantum mechanically, because they are just the mechanical harmonic oscillators introduced in undergraduate classes. For nano-resonators it is the task of enumerating the modes and determining their frequencies in which the complexity lies. This task requires the continuum mechanics of stress and strain, but we will not be concerned with these details here. Nano-resonators can be fabricated using etching techniques applied to layered structures (the same methods used to fabricate micro-electronic and superconducting circuits). Carbon nano-tubes can also be used as nano-resonators. An excellent overview of micro- and nano-resonators is given in [486].

To do interesting things with nano-resonators we need to couple them to other systems, such as superconducting circuits or optics. We will discuss interfacing resonators with optics in the next section. Since the dynamical degree of freedom of a nano-resonator is position, to couple them to superconducting circuits we look for circuit elements whose properties depend in some way on position (geometry). In fact geometry is crucial for both capacitors and inductors: for capacitors it is the distance between the plates, and for inductors it is the area of the loop that they enclose. If we arrange the resonator so that it forms one of the plates of a capacitor, then the movement of the resonator will change the capacitance, coupling the motion to the dynamics of the circuit. This is called a *capacitive* coupling. We could alternatively make the resonator part of an inductive loop, so that when the resonator moves it changes the area of the loop, and thus the inductance. This would be an *inductive* coupling. Both capacitive and inductive couplings have been realized in experiments. A third method is to make the resonator out of a piezoelectric material, so that when it oscillates the stretching and compressing creates a corresponding voltage across it. In this case the resonator becomes a voltage source which can be inserted into a circuit.

We will derive the Hamiltonian for capacitive coupling here. References that analyze an inductive coupling are given at the end of the chapter. We first note that a capacitor is simply two lumps of superconductor with a gap between them so that no current can flow. Further, a lump of superconductor connected to a voltage source is usually called an *electrode* or *gate*. To create a capacitor out of a resonator, we just place an electrode on the resonator (for example, coat one of its surfaces with a superconductor, and connect it to a voltage source) and position another electrode nearby. In Fig. 7.4 we depict an LC circuit in which a resonator is coupled to the capacitor in this way. Since the LC circuit is itself a harmonic oscillator, the result is two coupled oscillators, one mechanical and one superconducting. We now wish to determine the Hamiltonian of these coupled oscillators.

First we work out how the capacitance depends on the position of the resonator. The capacitance between two electrodes is inversely proportional to the distance, d, between the electrodes. If the oscillator is one of the electrodes, then this distance is $d = d_0 + \xi x_m$, where x_m is the extension of the resonator away from its equilibrium position, and ξ is some dimensionless constant less than or equal to unity. If we denote the value of the capacitance when $x_m = 0$ as C_0 then $C = C_0/(1 + \xi x_m/d_0)$. The frequency of the LC-oscillator is therefore

$$\omega = \frac{1}{\sqrt{LC}} \simeq \omega_0 \left[1 + \frac{1}{2} \left(\frac{\xi}{d_0} \right) x_m + \frac{3}{8} \left(\frac{\xi}{d_0} \right)^2 x_m^2 + \cdots \right], \qquad (7.68)$$

where $\omega_0 = 1/\sqrt{LC_0}$. We have expanded the expression for ω in powers of x_m/d_0 because the amplitude of oscillation of the resonator is typically *very* much smaller than the equilibrium gap d_0. This shows us that if the nano-resonator were classical, and had extension x_m, the Hamiltonian of the LC-oscillator would be

$$H_{LC} = \hbar \omega_0 \left[1 + \left(\frac{\xi}{2d_0} \right) x_m \right] \left(a^\dagger a + \tfrac{1}{2} \right), \qquad (7.69)$$

to first-order in x_m/d_0. Of course, we need to describe the resonator quantum mechanically. To do so we argue that given Eq. (7.69), if the resonator is in the eigenstate of its position operator, x, with eigenvalue x_m, then the evolution of the LC-oscillator would be given by the Hamiltonian H_{LC} above. This will be the case for every eigenvalue of x if we take the interaction term between the two oscillators to be given by H_{LC} with x_m replaced by the quantum operator for the resonators position, x. We also have $x = (b + b^\dagger)\sqrt{\hbar/(2m\omega_m)}$, where b, ω_m, and m are the mechanical resonator's annihilation operator, frequency, and mass, respectively. Putting this expression for x into H_{LC}, and adding-in the Hamiltonian of the mechanical resonator, the joint Hamiltonian of the two resonators becomes

$$H = \hbar\omega_m b^\dagger b + \hbar g (b + b^\dagger) a^\dagger a + \hbar \frac{g}{2}(b + b^\dagger) + \hbar\omega_0 a^\dagger a, \qquad (7.70)$$

with

$$g = \omega_0 \frac{\xi \Delta x_m}{2d_0}, \qquad (7.71)$$

and $\Delta x_m \equiv \sqrt{\hbar/(2m\omega_m)}$ is the standard deviation of the resonator's position when it is in the ground state. This quantity is also referred to as the "zero point" uncertainty in the resonator's position. The interaction term $\hbar\frac{g}{2}(b + b^\dagger)$ is often dropped in analyzing electromechanical systems, and we will do so too, because all it usually does is shift the equilibrium point of the mechanical resonator. Note that our derivation of the Hamiltonian, Eq. (7.70), is phenomenological (for want of a better term), since we have not derived the interaction between the two oscillators microscopically.

Coupling a superconducting qubit to a mechanical resonator

A simple way to couple a superconducting qubit to a mechanical resonator is to place the resonator next to the island of a Cooper-pair box (CPB), and put a conductor on the resonator. If we apply a voltage to the conducting "gate" on the resonator, then there is a capacitive force between this gate and the island of the CPB. This capacitance depends on the distance between the gate and the island, and is therefore proportional to the position of the resonator as described just above Eq. (7.68). In doing this we are actually just changing the capacitance C in the CPB circuit (shown in Fig. 7.2). From Eq. (7.55) we see that changing C changes N_v, and so the resulting interaction has the same form as the coupling between a CPB and a superconducting resonator (see the discussion just above Eq. (7.62)). The capacitative interaction is

$$H_I = \hbar g \sigma_z x, \qquad (7.72)$$

where x is the position of the resonator, and g is some coupling constant. Determining this coupling constant is Exercise 11.

Parametric coupling versus linear coupling

The interaction between the mechanical resonator and the LC circuit is especially interesting because it is nonlinear. In fact, the joint evolution of the two resonators can generate the evolution of a Kerr nonlinearity for the LC-oscillator (see Exercise 9), as well as other interesting behavior. The coupling term $g(b + b^\dagger)a^\dagger a$ is often called *parametric coupling* because (i) the frequency ω_0 is a "parameter" in the Hamiltonian of the LC-oscillator, and (ii) the coupling to the nano-resonator appears as a shift in this frequency. Similarly, if one modulates the frequency of an oscillator periodically with time, it is referred to as *parametric driving*. It turns out that such driving can be used to amplify a signal, a technique called *parametric amplification* (see Exercise 19).

Linear driving, on the other hand, is the application of a force $F(t)$ to a mechanical system, in which the force does not vary with position but may change with time. The potential energy corresponding to such a force is $V(x) = -F(t)x$, and so is linear in x. This kind of driving is called linear driving for this reason. The equivalent of linear driving for an electrical resonator is the insertion of a time-dependent voltage or current source into the circuit, and for an optical cavity it is the injection of a coherent beam of light through one of the cavity end-mirrors.

For a superconducting LC-resonator the Hamiltonian that results from applying an oscillating voltage in series with the capacitor can be determined using the methods in Section 7.1. If the voltage is given by $V(t) = V_0 \cos(\omega t)$ then the Hamiltonian is

$$H = \hbar\omega a^\dagger a - i\hbar \left(\frac{V_0}{\sqrt{2\hbar CL}} \right)(a - a^\dagger)\cos(\omega t)$$

$$\approx \hbar\omega a^\dagger a + \hbar(E^* a e^{i\omega t} + E a^\dagger e^{-i\omega t}), \tag{7.73}$$

where $E = (V_0/\sqrt{2\hbar CL})e^{i\pi/2}$ and in the second line we have made the rotating-wave approximation (RWA) described above. For a mechanical resonator the application of an oscillating force is described by the term $H_{\text{force}} = -F\cos(\omega t)x = \hbar F/\sqrt{2\hbar m\omega}\cos(\omega t)$ $(a+a)^\dagger$, and thus $E = F/\sqrt{2\hbar m\omega}$. For an optical cavity driven by a laser incident on one of the end-mirrors, $|E| = \sqrt{P\gamma/(\hbar\omega)}$, where P is the power of the laser and γ is the damping rate of the cavity. The phase of E is given by the phase of the laser. We note that the RWA is an excellent approximation for optical cavities, since optical frequencies are extremely high, and it is usually a very good approximation for superconducting oscillators such as LC circuits. Whether or not it is good for mechanical resonators depends on the relative sizes of ω_0 and E.

With the linear driving given by the last line of Eq. (7.73), if the resonator is weakly damped at the rate γ, in which the damping is given by the Lindblad master equation, Eq. (4.82), then the steady state is the coherent state $|\alpha\rangle$ with $\alpha = 2E/(\gamma)$.

Linear coupling is defined as coupling in which the interaction Hamiltonian is no more than quadratic in the creation and annihilation operators. Since an interaction Hamiltonian is the product of two Hermitian operators, one for each system, a linear coupling is restricted to the form

$$H = g(ae^{-i\theta} + a^\dagger e^{i\theta})(be^{-i\phi} + b^\dagger e^{i\phi}), \tag{7.74}$$

where g, ϕ, and θ may be time-dependent. For a given physical system, it may be the case that couplings are easy to realize for only one specific linear combination of x and p (one value of θ and ϕ). For example, if two mechanical resonators apply a mutual force to each other that is proportional to the distance between them, then the coupling is the product of their respective position operators. One cannot easily obtain a coupling that is the product of their momenta.

Obtaining a linear coupling from a parametric coupling

We can turn the parametric interaction between a mechanical oscillator and an LC circuit into one that is almost linear by driving the latter with a voltage signal oscillating at some frequency $\omega = \omega_0 + \Delta$. Here Δ is referred to as the "detuning" (of the driving signal with respect to the oscillation frequency). With this linear driving, the Hamiltonian for the two oscillators under the rotating-wave approximation becomes

$$H = \hbar\omega_m b^\dagger b + \hbar(\omega_0 + gX)a^\dagger a + \hbar(E^* a e^{i\omega t} + E a^\dagger e^{-i\omega t}), \tag{7.75}$$

where $E = |E|e^{i\phi}$ is determined by the driving field as described above, and $X \equiv a + a^\dagger$.

Recall now that the Heisenberg picture is obtained from the Schrödinger picture by applying a (time-dependent) unitary transformation to the operators, and the inverse transformation to the state of the system, so that the state remains fixed in time and the operators evolve instead. One can also use an "interaction picture," in which one transforms the harmonic motion of the LC oscillators to the operators, but keeps the rest of the Hamiltonian to evolve the system state. If we move into this interaction picture, the above Hamiltonian becomes

$$H = \hbar\omega_m b^\dagger b + \hbar(\Delta + gX)a^\dagger a + \hbar(E^* a + E a^\dagger). \tag{7.76}$$

More generally, one can apply any fixed (time-independent) unitary transformation to the operators and state to obtain a different "picture." We now apply a transformation of this type to the LC-oscillator. We choose the unitary

$$U = e^{\alpha a - \alpha^* a^\dagger} \tag{7.77}$$

that shifts the annihilation operator by an amount α, so that $a \to a + \alpha$. Choosing $\alpha = -E/\Delta$, we find that in this "displacement picture" the Hamiltonian becomes

$$H = \hbar\omega_m b^\dagger b + \hbar g|\alpha|^2(b + b^\dagger) + \hbar\Delta a^\dagger a$$
$$- \hbar g(b + b^\dagger)(\alpha^* a + \alpha a^\dagger) + \hbar g(b + b^\dagger)a^\dagger a. \tag{7.78}$$

The displacement transformation has canceled the driving of the LC-oscillator, created an effective force on the mechanical resonator with amplitude $g|\alpha|^2$ for the nano-resonator, and an effective linear interaction. If $|\alpha|^2 \gg \langle a^\dagger a \rangle$, then the linear interaction dominates over the nonlinear one, to give a nearly linear interaction. The effect of the force is merely to shift the equilibrium point of the mechanical resonator, and because of this we can eliminate this force by using a shifted version of the annihilation operator. Transforming

$b \rightarrow b - g|\alpha|^2/\omega$, and dropping the nonlinear term, the Hamiltonian becomes

$$H = \hbar\omega_m b^\dagger b + \hbar\Delta a^\dagger a - \hbar\lambda(b + b^\dagger)(ae^{-i\theta} + a^\dagger e^{i\theta}), \qquad (7.79)$$

where we have written $\alpha = |\alpha|e^{i\theta}$ and defined the linear coupling constant $\lambda = g|\alpha|$.

7.6 Optomechanical systems

Fabry–Perot cavities

An "optomechanical" system is one in which a mechanical degree of freedom interacts with one or more modes of a light field trapped in an optical cavity. The basic operation of an optical cavity is described in Section 3.3, but we recap a little here. A traditional standing-wave optical cavity, also called a Fabry–Perot cavity, consists of two parallel mirrors between which a beam of light can be trapped as a standing wave. Since no real mirrors are perfectly reflecting, the light in an optical cavity will eventually dissipate by leaking out through the mirrors. The light that is output from a cavity in this way can be detected using photon-counters or photodiodes. The distance between the mirrors is referred to as the length of the cavity. If this length is L, then standing-wave modes of frequency $\omega_n = c/(2L)$ can be trapped by the cavity, where c is the speed of light. Each cavity mode is described by an annihilation operator a_n, where the operator $N = a_n^\dagger a_n$ corresponds to the number of photons in the mode. The Hamiltonian for mode n is $H_n = \hbar\omega_n(a_n^\dagger a_n + 1/2)$, and one often drops the zero-point energy $\hbar\omega/2$ (as we do in most of our treatments of harmonic oscillators) because it does not affect the dynamics. A single cavity mode is therefore a single quantum harmonic oscillator. Damping of the mode via one of the cavity mirrors (or "end-mirrors") is described well by the quantum optical master equation given in Section 4.3.1. As with all other oscillators, the *quality factor*, Q, of an optical mode is defined as the frequency of the mode divided by the damping rate. A cavity that possesses high-quality (low-loss) modes is often referred to as a "high-Q" cavity. The definitions of various quantum states and operators for cavity modes are given in Appendix D.

If we construct an optical cavity in which one of the end-mirrors is free to oscillate, as in Fig. 7.5, the result is an optomechanical system: the light pushes the mirror as it bounces off it, and the movement of the mirror shifts the phase of the reflected light. The interaction between the center-of-mass of the mirror and a single mode of the light can be described very simply using a phenomenological Hamiltonian. We first note that the frequency of the mode is given by $c/(2L)$, and a shift of length x in the position of the mirror changes L to $L + x$. If this change in L is slow compared to the oscillation frequency, then the adiabatic approximation tells us that the frequency of the mode should effectively follow the change in the cavity length. Replacing the number x by the quantum operator for the position of the mirror, and expanding the expression for the frequency to first-order in x/L,

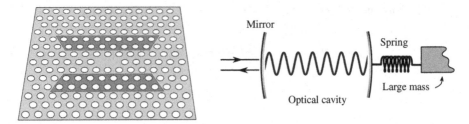

Figure 7.5 Here we show two ways to create optomechanical systems. On the left is an example of a photonic crystal, a thin transparent sheet of material in which a regular array of holes (white circles) is drilled. Light can only propagate in the region with two missing holes, which thus forms a high-Q optical cavity. The shaded region, if removed, would create a thin bridge forming a mechanical resonator with which the cavity modes will interact. On the right is an optical cavity in which one of the mirrors is a mechanical oscillator.

the Hamiltonian for the optical mode and the oscillating mirror becomes

$$H = \hbar\omega_m b^\dagger b + \hbar g(b + b^\dagger)a^\dagger a + \hbar \frac{g}{2}(b + b^\dagger) + \hbar\omega_0 a^\dagger a, \qquad (7.80)$$

where ω_m is the frequency of the mechanical oscillator, ω_0 that of the optical mode, and $g = (\omega_0/L)\sqrt{\hbar/(2m\omega_m)}$ is the rate of the coupling between them. Note that this coupling, referred to as the *optomechanical coupling Hamiltonian* is essentially the same as that between a mechanical resonator and an LC-oscillator or a superconducting stripline resonator.

Membrane-in-the-middle cavities

The method described above is not the only way to couple a mechanical oscillator to an optical mode. A very different coupling can be obtained by placing a thin partially reflecting mirror inside an optical cavity to split it into two subcavities. This mirror may be a transparent membrane that is partially reflecting merely because of its refractive index. In this case we allow the membrane to vibrate instead of one of the end-mirrors. It turns out that the frequencies of each mode of the double cavity now depend on the position of the membrane with respect to the nodes and antinodes of the mode. Denoting the frequency change by $\Delta\omega$ this relationship is $\Delta\omega(x) = (c/L)\cos^{-1}[r\cos(4\pi x/\lambda)]$, where x is the position of the membrane, L is the length of the cavity, r is the amplitude reflectivity of the membrane, and λ is the wavelength of the cavity mode [309]. This rather complex relationship is interesting because it shows that if we place the membrane at an extremum of $\Delta\omega(x)$, the mode frequency will depend *quadratically* on x. So if the membrane is free to oscillate, and placed in the cavity at an extremum of $\Delta\omega(x)$, the coupling between the position of the membrane and the cavity mode is given by

$$H = \hbar g x^2 a^\dagger a \qquad (7.81)$$

for some coupling constant g. This interaction generates a very different class of dynamics than is possible with an oscillating end-mirror.

Photonic crystal cavities

A remarkable way to create an ultra-small, ultra-high-Q optical cavity is to use something called a *photonic crystal* [491]. Recall that a real crystal is a substance in which the atoms are arranged in a regular pattern. A photonic crystal is a transparent material in which a regular crystal-like lattice of dots has been created. The dots have a different refractive index from the material itself. A photonic crystal is thus a regular pattern formed by a varying refractive index. The effect of the pattern is that when light tries to pass through the substance it is subject to diffraction. This is exactly analogous to the diffraction of electrons through a real crystal, which was the effect that allowed Davisson and Germer to reveal the wave nature of electrons. The term "photonic crystal" refers to the fact that it is a crystal structure designed to refract photons. To induce strong refraction, the periodicity of the lattice must be close to the wavelength of the light. So for optical light the dots need to be about half a micron in size, whereas for microwaves this length can be centimeters. Creating a three-dimensional photonic crystal is not easy, but it is much easier to make one that is two-dimensional. In this case one forms a thin sheet of transparent material. The material is chosen so that light that is injected into one edge of the sheet will flow only in the plane of the sheet, being trapped by total internal reflection. One then drills a lattice of holes in the sheet, which creates the photonic crystal. A two-dimensional photonic crystal is depicted in Fig. 7.5.

Because of the multiple refractions/reflections caused by the lattice structure, known as distributed Bragg reflection (DBR), there are many wavelengths that cannot propagate inside a photonic crystal. It is this effect that allows one to trap light very effectively in a small region of the crystal sheet. To do this one merely has to remove one of the holes. Light that cannot propagate in the rest of the sheet can do so in the vicinity of the missing hole, and can thus be trapped in that tiny region. This region is called a *photonic cavity*, and is shown in Fig. 7.5. If one makes a path of missing dots, then light will propagate only along the path, even if the path changes direction, providing a photonic wave-guide. By having a photonic wave-guide pass near a photonic cavity, light will couple from the guide into the cavity, and vice versa.

A photonic cavity can be coupled to a mechanical resonator by cutting two square holes out of the crystal sheet, to form a thin strip, or "bridge" between them. If one puts a photonic cavity on the bridge, then when the bridge vibrates it changes the shape of the photonic cavity, and thus realizes an optomechanical coupling between the cavity and the mechanical modes of the bridge. A depiction of this kind of opto-mechanical system, first introduced by Painter and collabarotors [527, 112] is shown in Fig. 7.5.

Table 7.2. *The relationship between the amplifier parameters A and G, and measurement strength k and efficiency η, for a position measurement of a harmonic oscillator (a linear phase preserving measurement). The oscillator damping rate is γ, and the ground-state position variance is V_x.*

Amplifier parameters	Measurement parameters
A (noise added)	$1/(4\eta k) + (4V_x)^2 k/\gamma^2$
G (gain)	$4/\gamma$

7.7 Measuring mesoscopic systems

This section is intended to serve theorists who are familiar with the continuous measurement theory of Chapter 3 and want to apply this to real systems. It is also intended to serve experimentalists who are familiar with real measurements, and want to link these with analyses in the theoretical literature. The language used by experimentalists is usually that of amplifiers, involving noise power and signal-to-noise ratio, and this is quite different from the language of Chapter 3. Fortunately, there are only two parameters to translate: the measurement strength, k, and the measurement efficiency, η. We begin by describing the language of amplifiers and how it relates to the language of continuous measurements. We then give a concise description of the meaning of the two theoretical parameters k and η in terms of noise power, and present in Table 7.2 the explicit mapping between these parameters and those of amplifiers, for a continuous measurement of the position of a harmonic oscillator.

7.7.1 Amplifiers and continuous measurements

The language of amplifiers is natural for describing continuous quantum measurements for the following reason. To measure an observable of a quantum system that contains only a few quanta, the value of the observable must be turned into a classical signal. To see that this classical signal must have a much larger energy than the observable, recall that a classical number is something with negligible uncertainty. To store such a number physically we must use *many* quanta, so that the uncertainty of a single quantum is negligible compared to the value of the number.

Any real measuring device must amplify the very tiny variations of an observable of the system so that it can produce a classical measurement result. When we discussed continuous measurement theory in Chapter 3 we glossed over this fact. We took it for granted that the measuring device produced a classical number, dy, proportional to the expectation value of the mesoscopic observable. For optical systems one of two procedures is used to make measurements. For very weak signals in which it is necessary to detect individual photons, a photon-counter is used, and it is the photon-counter that performs the amplification. It turns the effect of a single photon into an electrical pulse large enough to be clearly detected above background noise. But photon-counters have traditionally been highly inefficient, as well as subject to "dark counts" in which they register a photon when none is there. There are "superconducting nanowire" [441] detectors that have much higher efficiency, but they presently suffer from a long "dead time" (the minimum time-difference between which individual photons can be resolved). At the time of writing there are no fast single-photon detectors with $\eta \approx 1$.

Homodyne detection works by interfering the beam to be measured with another beam whose intensity can be arbitrarily high. Because of this, single-photon detectors are not required, and the second kind of detectors for measuring optical beams can be used. These detectors, of which photo-diodes are an example, cannot detect individual photons, but can detect the energy in a sufficiently intense beam with high efficiency. Homodyne detection of optical beams is thus something of an exception, in that it does not require an amplification stage.

If we use homodyne detection to measure the microwave signals of superconducting systems, then we do have to use an amplifier. Building an amplifier that can amplify such small signals without swamping them with noise is the key experimental challenge in performing any quantum measurement. If the amplifier adds a ton of noise it produces a highly inefficient measurement ($\eta \ll 1$). Due to this inefficiency a much longer time is required to obtain an accurate result, and unless the measurement is a non-demolition measurement this will result in more back-action noise in the system. Real-time feedback control becomes impossible when measurements are highly inefficient, because the rate of information extraction is too slow and the back-action noise is comparatively large.

When experimentalists can make only highly inefficient measurements they must repeat the measurement many times, preparing the system in the same initial state each time. When the results of the measurement are averaged one obtains an accurate measurement of the initial preparation, because this preparation is the same across all repetitions. But this process does not reveal the state of any single system after the measurement, since this state varies from repetition to repetition, and each measurement is very noisy.

Noise and noise power

In Chapter 3 we initially expressed the noise on the measurement record as a Gaussian random increment in each time-step dt. Here we wish to describe this noise in terms of its power spectral density, or *power spectrum*. We introduced and discussed the power spectrum in Section 3.1.4. We summarize that section now to connect the key concepts to measurements and amplification.

The power of a signal $f(t)$ is defined as the time-average of its mean-square, $\langle f(t) \rangle$. Thus the power need not have units of power; the name derives from the fact that the power of physical waves, such as light, is proportional to the square of their amplitude. What the power does do is characterize the variance of the noise in a signal. This is because noise virtually always has mean zero, and so the mean-square is also the variance.

It is useful to know how the power (the variance in the noise) is distributed across the frequency spectrum. The function that gives this distribution is called the power spectral density, or *power spectrum* for short. In fact, the power spectrum, denoted by $S(\omega)$, is defined both for positive and negative frequencies, and is a symmetric function of frequency. As a result the total power-per-unit-frequency at frequency ω is *twice* the value of the power spectrum at ω. Since the "power-per-unit-frequency" is rather a mouthful, we will refer to the power-per-unit-frequency at ω simply as the *noise power*, or even just the *noise*, and denote it by $P(\omega) = 2S(\omega)$

The noise power is useful because a signal that we may want to detect within the noise is usually restricted to a band of frequencies, and may even be the amplitude of a single sine wave (a single frequency). To determine the amplitude of a single sine wave within a noisy signal, one multiplies the signal by a sine wave at the same frequency (referred to as *modulating* the signal) and then averages the result over time. The error in the resulting estimate of the amplitude is proportional to the value of the power spectrum evaluated at that frequency.

The measurement record can be written as $r(t) = \langle x(t) \rangle + \zeta(t)/\sqrt{8k}$, where k is the measurement strength, and $\zeta(t)$ is white noise. White noise has a constant power-per-unit frequency, and $\zeta(t)$ is defined so the power spectrum is unity. Thus the noise power on the measurement record that comes from $\zeta(t)$, is

$$P(\omega) = 2S(\omega) = 1/4k. \tag{7.82}$$

To connect the parameters k and η to the language of amplifiers, we also need to consider the "back-action" noise from the measurement when the system being measured is a harmonic oscillator. If we measure a deterministic classical oscillator, then the measurement noise will be the only noise in the measurement signal. But if we measure a quantum oscillator, then the oscillator is driven by this back-action noise. Recall that the origin of this noise is just Heisenberg's uncertainty principle that the position and momentum cannot both have zero uncertainty. Since in a harmonic oscillator the momentum affects the position, and vice versa over time, under a continuous measurement of position both the position and momentum remain uncertain. This uncertainty means that each time a measurement is made, the resulting change in the wavefunction causes the mean position and momentum to change randomly. As a result the measurement signal contains not only the measurement noise, but also this back-action or "projection" noise due to the quantum nature of the oscillator.

We can completely solve the dynamics of a continuously measured harmonic oscillator, and determine the spectrum of the measurement signal, $r(t)$. In this case we do not drive the harmonic oscillator, but merely measure its position. Thus if we were making the equivalent measurement on a classical oscillator, the oscillator would be perfectly still, and the only noise on the measurement signal would be the measurement noise. For a quantum oscillator there are two additional sources of noise. The first is the position and momentum uncertainty of the ground state: when we measure the position we will observe fluctuations at least as large as the ground-state position uncertainty. The second additional source of noise is the back-action noise from the measurement. This white noise drives the oscillator and causes the position to fluctuate.

The dynamics of a weakly damped harmonic oscillator under a continuous position measurement is given in Eq. (3.237), and we repeat it here for convenience:

$$d\rho = -i[\omega_0 a^\dagger a, \rho]\,dt - k[x,[x,\rho]]\,dt + \sqrt{2k}(x\rho + \rho x - 2\langle x\rangle\rho)\,dW,$$
$$-\frac{\gamma}{2}\Big[(n_T+1)(a^\dagger a\rho + \rho a^\dagger a - 2a\rho a^\dagger) - n_T(aa^\dagger\rho + \rho aa^\dagger - 2a^\dagger\rho a)\Big]\,dt,$$

in which ω_0 is the frequency of the oscillator, k is the measurement strength, $\gamma \ll \omega_0$ is the damping rate, and $n_T = 1/(e^{\hbar\omega_0/kT} - 1)$ is the average number of phonons in the resonator at temperature T.

In Sections 3.1.4 and 3.7.2 we calculated the power spectrum of $r(t)$, and found it to be

$$S_r(\omega) = \frac{1}{8k} + \frac{V_x\gamma(2n_T+1) + 4V_x^2 k}{2(|\omega| - \omega_0)^2 + \gamma^2/2}, \tag{7.83}$$

where $V_x = \sqrt{\hbar/(2m\omega_0)}$ is the variance of the position for the oscillator's ground state.

Imagine that the oscillator is now driven by an oscillating force (which we have not included above), and we wish to determine the steady-state amplitude of the resulting oscillations. These oscillations would appear in the signal $\langle x(t)\rangle$, and we could detect them by modulating the signal at the oscillator's frequency and averaging the result over time. The error in our estimate of this amplitude is proportional to the square root of the power spectrum of the noise, $S_r(\omega)$, at the oscillator's frequency. The power spectrum evaluated at $\omega = \pm\omega_0$ therefore tells us how much noise there will be on a measurement of the oscillator's amplitude. We will now see that the quantum back-action noise imposes a limit on how small this noise can be for a continuous position measurement.

If we sum the values of the power spectrum at $\omega = \pm\omega_0$, and set the temperature to zero ($n_T = 0$), we have

$$P(\omega_0) = S_r(\omega_0) + S_r(-\omega_0) = \frac{1}{4k} + \frac{4V_x}{\gamma} + \frac{(4V_x)^2 k}{\gamma^2}. \tag{7.84}$$

By examining the origin of the three terms that contribute to the noise, we see that the middle term comes from the quantum fluctuations of the position in the oscillator's ground state, and is therefore intrinsic to the oscillator. The other two terms depend on k, and are therefore due to the measurement. The first term is the measurement noise, also called the "shot noise" or the "imprecision noise" of the measurement. It characterizes how rapidly information is being extracted about position – the smaller the imprecision noise, the faster we obtain information about x. The last term comes from the quantum back-action of the measurement; the faster we extract information about x, the faster we feed noise into the momentum.

Since the imprecision noise decreases with k, and the back-action noise increases, there is a "sweet spot": a value for k in which the total noise reaches a minimum. Taking the derivative of the total noise with respect to k, we find that this sweet spot is $k = \gamma/(8V_x)$. Substituting this value for k into Eq. (7.84) above we have

$$\min_k P(\omega_0) = \frac{2V_x}{\gamma} + \frac{4V_x}{\gamma} + \frac{2V_x}{\gamma} = \frac{8V_x}{\gamma}. \tag{7.85}$$

The minimum noise in the position measurement is exactly *twice* the intrinsic fluctuations of the position. This minimum noise level is called the "standard quantum limit" (or "SQL") for a measurement of position.

The phonon/photon as a unit of noise

We now include a non-zero temperature in the expression for the noise, Eq. (7.85), and rearrange it a little. The result is

$$P(\omega_0) = \frac{8V_x}{\gamma} \left[\frac{\gamma}{32V_x k} + \left(n_T + \frac{1}{2} \right) + \frac{2V_x k}{\gamma} \right]. \tag{7.86}$$

Since n_T is the average number of phonons in the oscillator, we see that one phonon contributes an amount $8V_x/\gamma$ to the position noise. Because of this the ground-state fluctuations of the resonator are often referred to as having "half a phonon" of noise. Further, if we set $k = \gamma/(8V_x)$ so as to minimize the noise from the measurement, we have

$$\min_k P(\omega_0) = \frac{8V_x}{\gamma} \left[\frac{1}{4} + \left(n_T + \frac{1}{2} \right) + \frac{1}{4} \right]. \tag{7.87}$$

The minimum noise added by the measurement is therefore also "half a phonon." If the oscillator is an optical mode or electrical oscillator we replace "phonon" with "photon."

Inefficiency: measurements with extra noise

We can easily extend the above results to include inefficiency in the measurement. Inefficiency increases the imprecision noise without changing the interaction with the system, and so the size of the back-action remains the same. Inefficiency simply means that additional classical noise has been added to the measurement signal after the information has been obtained from the system, but before the observer has had a chance to record it. Inefficient measurements therefore produce more back-action for a given amount of information: they get less bang for their buck. The total noise at ω_0 for an inefficient measurement of the oscillator is thus

$$P(\omega_0) = \frac{1}{4k} + P_{\text{xtr}} + \frac{8V_x}{\gamma} \left(n_T + \frac{1}{2} \right) + \frac{(4V_x)^2 k}{\gamma^2}, \tag{7.88}$$

where we have denoted the additional imprecision noise power by P_{xtr}. The inefficiency of the measurement, denoted by η, is defined by writing the total imprecision noise as $1/(4\eta k)$. The resulting relationship between P_{xtr} and η is

$$P_{\text{xtr}} = \frac{1-\eta}{4k\eta}, \qquad \eta = \frac{1}{1+4kP_{\text{xtr}}}. \tag{7.89}$$

Using the above definition of the noise added by one photon, we can also express the extra measurement noise in units of photons. The number of photons of extra noise is

$$n_{\text{xtr}} = \gamma P_{\text{xtr}}/(8V_x). \tag{7.90}$$

If we choose the measurement strength k to minimize the noise ($k = \gamma/(8V_x)$), then $n_{\text{xtr}} = kP_{\text{xtr}}$ and the measurement efficiency is

$$\eta = \frac{1}{1+4n_{\text{xtr}}}. \tag{7.91}$$

Quantum measurements that are efficient, so that $\eta = 1$, are described as being "quantum-limited." The ultimate goal of any quantum measurement technique for mesoscopic systems is to realize a quantum-limited measurement.

Noise added by an amplifier, and the standard quantum limit

The gain of an amplifier, denoted by G, is defined as the factor by which the amplifier increases the power of the signal. The amplitude of the signal is therefore multiplied by $\sqrt{G}$. Amplifiers are not perfect, and may add noise to the signal in the process of amplifying it. The noise added by an amplifier is defined as the additional noise in the output. If the input noise power is S, and the gain of the amplifier is G, then a perfect amplifier would give an output signal with noise power GS. If the noise power added by the amplifier is A, the output will have noise power $GS + A$.

As discussed above, all quantum measurements on mesoscopic systems require an amplifier to turn very tiny signals at the level of a single quantum of energy into classical numbers. This means that to obtain our classical measurement signal $r(t)$, there is necessarily an intrinsic amplification process, mediated by the interaction with the measured system, that turns the effect of the system on the probe into a well-defined number. The noise power for the measurement signal calculated above, $P(\omega_0)$, can therefore be regarded as the output of an amplifier. Above we found that when we chose the measurement strength, k, to minimize the noise power, the total noise power was *twice* the noise power of the quantum fluctuations of x, the observable being measured. The minimum noise over and above the fluctuations of x was equivalent to that added by half a phonon (half a quantum of energy in the oscillator). In the language of amplifiers, this half a quantum of added noise is regarded as noise added by the amplifier.

If we regard the position of a quantum harmonic oscillator as a signal that we wish to turn into a classical number, then the resulting classical signal will have a minimum of half a quantum of additional noise. This is referred to as the quantum limit to amplification, which is also the standard quantum limit introduced above. One way to derive this limit is to consider an amplification process in which the position and momentum of a harmonic oscillator are both amplified and stored in a second oscillator. If we denote the annihilation operator of the first oscillator by a, and the second by b, then this amplification process means that $b = \sqrt{G}a$. If G is very large then the amplitude of the second oscillator is essentially a classical number, since it has many quanta. One can show that the uncertainty in the resulting state of the second oscillator (the amplified signal) has added noise of at least G times the contribution of half a quantum (this is Exercise 18). This quantum limit to amplification comes from the fact that one is trying to determine both x and p at the same time. In our position measurement, since we are attempting to track the position over time, we are similarly obtaining information about the momentum as well as the position, and that is why this measurement has the same quantum limit.

Amplification processes in which information about both position and momentum is simultaneously turned into a classical signal provide information about both the amplitude and phase of the oscillation. Such processes are often referred to as "phase preserving" or "phase insensitive" amplifiers. We prefer the former term, as we feel it is less confusing. If one does not care about the phase of the oscillation, and only wants information about the amplitude (or vice versa), then it is possible to make a measurement (perform an amplification) without the additional noise. Amplification processes like this are called "phase non-preserving" or "phase sensitive" amplifiers. The quantum limit to amplification applies only to phase preserving amplifiers.

The noise temperature of an amplifier

The "noise temperature" of an amplifier is merely a way of characterizing the amount of noise added to the signal by an amplifier. But to understand this terminology we must first understand why noise is considered to have a *temperature*. This notion comes from the fact that if one measures the voltage across an otherwise unconnected resistor, one will obtain white noise with noise power $P(\omega) = Tk_{\mathrm{B}}$, where T is the temperature of the resistor and k_{B} is Boltzmann's constant. This noise is called *Jonson–Nyquist* noise, after Jonson who first observed it [315], and Nyquist who explained its origin theoretically [454]. As a result of Jonson–Nyquist noise, a "temperature" T can be associated with any noise source via the formula $T = P(\omega)/k_{\mathrm{B}}$.

We can now define the noise temperature of an *amplifier*. Let us say that our amplifier adds noise of power A to the output signal. To define the noise temperature of the amplifier we consider how much noise we would have to add to the input signal of a perfect amplifier, to get the same noise added at the output. If the gain of the amplifier is G, then we must add noise with power A/G to the input. The noise temperature of the amplifier, T_{amp} is the temperature of the noise added at the input: $T_{\mathrm{amp}} = A/(Gk_{\mathrm{B}})$.

7.7.2 Translating between experiment and theory

To summarize the above discussion, a continuous measurement of a quantum system can be thought of as an amplification process that produces a classical output signal. We denote the gain of the amplifier by G, the noise power added by the amplifier to the output signal as A, and the noise temperature of the amplifier by $T_{\mathrm{amp}} \equiv A/(Gk_{\mathrm{B}})$. The gain and added noise may be a result of multiple stages of amplification, but we are only concerned here with the overall gain and added noise. It is crucial to remember that the added noise A is defined as *any noise that would not appear* in the dynamics of the observable being measured, if it were not being measured.

One can alternatively use the language developed by quantum measurement theorists to describe a continuous measurement. In this case the measurement signal is written as the quantity being measured, whose dynamics *includes* the influence of the noise on the system, plus white noise with noise power equal to $1/(4\eta k)$, where k is the measurement strength, and $\eta \in [0,1]$ is the measurement efficiency. What distinguishes k from η is that the effect of the measurement on the dynamics of the measured system (the back-action noise) is due to k alone, and not to η.

In the amplifier language, the definition of the noise added, A, includes the back-action noise of the measurement, while in the continuous-measurement language the definition of the imprecision noise, and thus the measurement efficiency, η, does not. Because of this, and because the effect of the back-action noise depends upon the system Hamiltonian and the measured observable, one cannot give a single formula to translate between the parameters G and A, and k and η. For a phase-preserving measurement of the position or momentum of a harmonic oscillator, the effect of the back-action noise can be determined analytically, and is given above. (A phase-preserving measurement is one that obtains equal information about both position and momentum. A continuous measurement of either position or momentum is phase-preserving in the steady state.) For this case we can determine the relationship between the two sets of parameters easily by writing Eq. (7.88), which gives the noise power in terms of k and η, in terms of G and A instead. This version of Eq. (7.88) is

$$P(\omega_0) = G(2V_x)\left(n_T + \frac{1}{2}\right) + A. \tag{7.92}$$

We give the resulting mapping between the parameters in Table 7.2.

7.7.3 Implementing a continuous measurement

In Chapter 3 we derived a theory for continuous quantum measurement, and we showed above how this theory relates to the language of amplifiers. But the only real measurement technique that we have discussed so far is that of homodyne detection of the light from an optical cavity. Here we take this further and analyze a real measurement scenario for a mesoscopic system. While there are many ways to make measurements, we

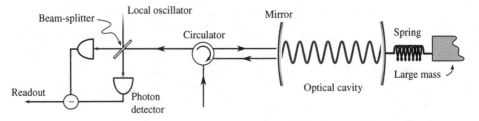

Figure 7.6 A configuration for measuring the position of an oscillator using optics. Here the oscillator is the mirror on the right-hand side of the optical cavity, which is attached to a spring. The phase
of the light that is output from the cavity is measured using the technique of homodyne detection,
discussed in Section 3.4.

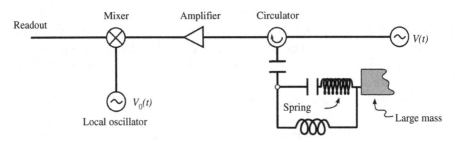

Figure 7.7 A configuration for measuring the position of an oscillator using a superconducting electromechanical circuit. The oscillator is one plate of a capacitor that is attached to a spring. The
electrical signal from the LC circuit is coupled to the same line on which the input signal drives the
circuit, but the circulator ensures that this output travels only to the left. It is then amplified, often
using two or three stages of amplification, before the phase of the signal is measured using homodyne
detection. To perform this measurement the output signal is interfered ("mixed") with another signal
of constant amplitude, called the "local oscillator."

discuss in detail here a technique that is widely used because it achieves the highest efficiency, at least with present experimental technology. This technique, which we will refer
to as an "oscillator-mediated" measurement, uses homodyne detection; it can be implemented with both superconducting and optomechanical systems; and the same theoretical
treatment applies to both.

An oscillator-mediated measurement in depicted in Figs. 7.6 and 7.7 for the concrete
example of measuring the position of a mechanical oscillator. Figure 7.6 depicts a measurement mediated by a single mode of an optical cavity, and Fig. 7.7 a measurement made
via a superconducting LC-oscillator. We now explain the various elements of the technique
in these two cases. The system to be measured, which we will call the target, is coupled
to a single mode of a superconducting resonator or optical cavity. The target might be a
superconducting qubit or a mechanical oscillator, for example. When an optical cavity is
used, the output light is measured via homodyne detection. This involves "mixing" (adding
together) the light from the cavity with a laser beam using a beam-splitter, and the intensity
of the resulting beam is measured. For a superconducting oscillator, an equivalent homodyning procedure is performed. The oscillator is coupled to an electrical circuit that acts as

a transmission line with many modes. The voltage signal output into this transmission line is then mixed with a second voltage signal, and the total voltage is measured. The primary difference between optical and superconducting systems is that the signal from the latter must be amplified before it can be reliably mixed and measured. Creating an amplifier with the exceedingly low noise required to make even reasonably efficient measurements was a great challenge. At the time of writing, such amplifiers have been developed only recently. These are non-degenerate parametric amplifiers that use a ring of Josephson junctions as their key element [51, 50, 1, 637, 638]. In recent experiments that achieve near-quantum-limited measurements in superconducting circuits, the output of the oscillator or cavity mode is amplified with one of these parametric amplifiers, and is then amplified again with a high-electron-mobility transistor (HEMT) amplifier before being fed into electronics at room temperature [51, 637, 638]. The HEMT amplifier cannot be used alone, as its noise temperature is too high.

From now on we will usually refer to the system mediating the measurement as a cavity, but the analysis applies equally well to a measurement mediated by any harmonic oscillator. A measurement made on the cavity realizes a measurement of the target because of the target's effect on the cavity. An example in which this process is especially clear is when the target is a mechanical resonator. In this case the interaction Hamiltonian is $H_1 = gxa^\dagger a$ where x is the mechanical position and $a^\dagger a$ is the number operator for the cavity mode (see Eq. (7.80)). Since the number operator generates a phase shift of the light in the mode, the interaction produces a phase shift of the cavity light that is proportional to the position of the mechanical oscillator. Homodyne detection of the phase-quadrature of the light therefore measures the position of the oscillator. If the interaction were instead proportional to $x(a + a^\dagger)$, then since the amplitude quadrature $a + a^\dagger$ generates a shift in the phase quadrature of the cavity, the same homodyne measurement would measure the mechanical position. If the interaction were proportional to the phase quadrature of the cavity mode, then homodyne detection of the amplitude quadrature would be the appropriate measurement.

Our purpose in the remainder of this section is to show that the indirect oscillator-mediated measurement provides a continuous Gaussian measurement of the target. In particular it realizes a continuous measurement of the target operator that appears in the interaction with the cavity. We show that the stochastic master equation that describes this continuous measurement can be derived from the interaction with, and measurement of, the superconducting/optical resonator. This derivation allows us to determine the strength of the measurement of the target as a function of the rate of the cavity/target interaction and the damping rate of the cavity.

For now we will assume that the interaction with the target has the form $H_1 = Aa^\dagger a$. We will consider the modifications required for a linear interaction below. The Hamiltonian for the cavity and target is

$$H = H_{\text{tg}} + \hbar g A a^\dagger a + i\hbar E(a e^{i\omega t} - a^\dagger e^{-i\omega t}) + \hbar \omega a^\dagger a, \tag{7.93}$$

where H_{tg} is the Hamiltonian for the target, g is the rate of the coupling between the two systems, and E is the strength of the laser driving the cavity. We now transform to the interaction picture with respect to the Hamiltonian of the cavity mode, and the Hamiltonian for the remaining evolution becomes

$$H = H_{\text{tg}} + \hbar g A a^\dagger a + i\hbar E(a - a^\dagger). \tag{7.94}$$

The interaction picture simplifies the dynamics by eliminating the simple oscillation of the cavity mode at the optical or microwave frequency. The cavity is also damped via transmission through one of its mirrors, and the output light is measured. The master equation describing the two systems is therefore

$$\dot{\rho} = -\frac{i}{\hbar}[H, \rho] - \frac{\kappa}{2}(a^\dagger a\rho + \rho a^\dagger a - 2a\rho a^\dagger), \tag{7.95}$$

where κ is the damping rate of the cavity.

We now note that in the absence of the interaction with the target the cavity steady state will be the coherent state $|\alpha\rangle$ with $\alpha = 2E/\kappa$ (see Exercise 16). In view of this we transform to a new "picture" by applying a displacement operator that shifts the coherent state $|\alpha\rangle$ to the vacuum state. This operator is $D(-\alpha)$, where $D(\alpha) = \exp(\alpha a^\dagger - \alpha^* a)$. To obtain the new picture we apply $D(-\alpha)$ to the cavity state, and its inverse to the operators. This transforms the annihilation operator as $a \to a + \alpha$, and so the Hamiltonian in the new picture is

$$\begin{aligned} H &= H_{\text{tg}} + \hbar g A(a^\dagger + \alpha)(a + \alpha) + i\hbar E(a - a^\dagger) \\ &\approx H_{\text{tg}} + \hbar g A[|\alpha|^2 + \alpha(a + a^\dagger)] + i\hbar E(a - a^\dagger), \end{aligned} \tag{7.96}$$

and the master equation becomes

$$\dot{\rho} = -\frac{i}{\hbar}[H, \rho] - \frac{\kappa}{2}(a^\dagger a\rho + \rho a^\dagger a - 2a\rho a^\dagger) - E[a - a^\dagger, \rho]. \tag{7.97}$$

We have dropped the term $\hbar g A a^\dagger a$ from the Hamiltonian, under the assumption that $\alpha \langle a + a^\dagger\rangle \gg a^\dagger a$. Note that while the driving term still appears in the Hamiltonian, its effect is now canceled by the last term in the master equation, which comes from the original damping terms. As a result the steady state of the cavity, in the absence of the interaction with the target, is now the vacuum state.

Since our purpose is to measure the target, not the cavity, we would like to obtain a description of the measurement process that makes it clear that we are performing a measurement of the observable A of the target. Note that the measurement of the cavity phase will faithfully measure A if the phase of the cavity at each time is determined by the value of A. If the cavity is damped at a rate that is much faster than the rate at which A changes the cavity state, then the cavity will reach a steady state, and this steady state will quickly adjust to changes in A. In this case the state of the cavity will faithfully follow the value of A.

Further, when the cavity damping is fast compared to the effect of the mechanical resonator, the energy in the resonator remains close to the vacuum, at least in the displacement picture. This means that we can approximate the state of the cavity, during the measurement, as a linear combination of the Fock states with just zero, one, and two photons. The purpose of the displacement transformation is that it allows us to make this approximation.

When the dynamics of one system is damped much faster than the speed at which another evolves, we can use the approximation method called *adiabatic elimination*. This involves assuming that the state of the slow system is fixed, and determining the steady state of the fast system for each fixed state of the slow system. We can then replace the variables of the fast system with their steady-state values, being functions of the slow variables. This eliminates the fast system from the equations of motion, and we are left with an equation of motion for the slow system alone. One speaks of the variables of the fast system as being "slaved" to the variables of the slow system.

We now apply adiabatic elimination to the cavity, assuming that the damping rate κ of the cavity is much larger than the dynamical timescale of the target and the coupling between the system and the target. To do this we write the joint density matrix of the two systems as

$$\rho = \sum_{j=0}^{\infty} \sum_{k=0}^{\infty} \sigma_{jk} |j\rangle \langle k| \approx \sum_{j=0}^{2} \sum_{k=0}^{2} \sigma_{jk} |j\rangle \langle k|. \tag{7.98}$$

Here $|j\rangle$ is the Fock state of the cavity with j photons, and σ_{jk} is the block of the joint density matrix that corresponds to the cavity density matrix element with indices jk in the Fock basis. Note that $\sigma_{kj} = \sigma_{jk}^{\dagger}$. If we define the small parameter $\epsilon = g\alpha^2 \langle A \rangle / \kappa \sim \langle H_{\text{tg}} \rangle / \kappa$, then we will find that the scaling of the blocks is $\sigma_{jk} \propto \epsilon^{j+k}$. Thus the block σ_{22} is fourth-order in ϵ, and it will turn out that we will not need it in our adiabatic elimination. If we trace ρ over the cavity, we obtain the density matrix for the target, being

$$\sigma = \text{Tr}_{\text{cavity}}[\rho] \approx \sigma_{00} + \sigma_{11}. \tag{7.99}$$

We now substitute the expression for the joint density matrix, Eq. (7.98), into the master equation, and this gives the following set of differential equations for the blocks:

$$\dot{\sigma}_{00} = -\frac{i}{\hbar}[H_{\text{tg}}', \sigma_{00}] + ig\alpha(A\sigma_{10} - \sigma_{10}^{\dagger}A) + \kappa\sigma_{11},$$

$$\dot{\sigma}_{10} = -\frac{i}{\hbar}[H_{\text{tg}}', \sigma_{10}] - \frac{\kappa}{2}\sigma_{10} + ig\alpha\left[A\left(\sigma_{00} + \frac{\sigma_{10}}{\alpha} + \sqrt{2}\sigma_{20}\right) - \sigma_{11}A\right],$$

$$\dot{\sigma}_{11} = -\frac{i}{\hbar}[H_{\text{tg}}', \sigma_{11}] + ig\alpha(A\sigma_{10}^{\dagger} - \sigma_{10}A) - \kappa\sigma_{11},$$

$$\dot{\sigma}_{20} = -\frac{i}{\hbar}[H_{\text{tg}}', \sigma_{20}] - \kappa\sigma_{20} + igA(2\sigma_{20} + \sqrt{2}\alpha\sigma_{10}),$$

where we have defined $H_{\text{tg}}' = H_{\text{tg}} + \hbar g\alpha^2 A$. We see that the equations for the off-diagonal blocks contain fast damping, and so we wish to solve for their steady states. To do this we need to be careful about the orders of the various terms, so as to obtain the result to leading order. We illustrate how this is done using one of the above equations as an example.

Consider the equation of motion for σ_{20}. We first set the derivative to zero, divide the equation by κ, and move all terms involving σ_{20} to the left-hand side. The result is

$$i\left[\frac{H'_{\text{tg}}}{\hbar\kappa},\sigma_{20}\right] - i2\frac{g}{\kappa}A\sigma_{20} + \sigma_{20} = i\sqrt{2}\frac{g\alpha}{\kappa}A\sigma_{10}. \tag{7.100}$$

We now examine the order of the various terms. In this case, with the benefit of knowing how the blocks scale in advance, we can see that the first two terms on the left-hand side are higher order than all other terms. Dropping these we can immediately solve for σ_{20}. Nevertheless we will encounter situations in which more than one term on the left-hand side is the same order as the terms on the right-hand side. Fortunately, it is only the leading-order terms on the left-hand side that contribute to the solution to leading order, as the following analysis shows. With this knowledge we will not need to know anything about the orders of the blocks to solve for the steady states. To show that only the leading-order term in σ_{20} is important, we first write the above equation in the form

$$(1+\varepsilon\mathcal{L}_1)\sigma_{20} = \epsilon\mathcal{L}_0\sigma_{10}, \tag{7.101}$$

where we have defined the linear operators $\mathcal{L}_1$ and $\mathcal{L}_0$ by

$$\mathcal{L}_1\sigma_{20} = i\left[\frac{H'_{\text{tg}}}{\langle H'_{\text{tg}}\rangle},\sigma_{20}\right] - i2\frac{\hbar g}{\langle H'_{\text{tg}}\rangle}A\sigma_{20}, \tag{7.102}$$

$$\mathcal{L}_0\sigma_{10} = i\sqrt{2}\frac{A}{\langle A\rangle}\sigma_{10}, \tag{7.103}$$

and the small parameters

$$\epsilon = \frac{g\alpha\langle A\rangle}{\kappa} \sim \varepsilon = \frac{\langle H'_{\text{tg}}\rangle}{\hbar\kappa}. \tag{7.104}$$

The operators $\mathcal{L}_j$ have been scaled so that they are zeroth-order. Since all the operations are linear, we can think of $\mathcal{L}_1$ and $\mathcal{L}_0$ as matrices, and σ_{20} and σ_{10} as vectors. Solving the equation for σ_{20} gives

$$\sigma_{20} = (1+\varepsilon\mathcal{L}_1)^{-1}\epsilon\mathcal{L}_0\sigma_{10} \tag{7.105}$$

$$= (1+\varepsilon\mathcal{M}_1+\varepsilon^2\mathcal{M}_2+\cdots)\epsilon\mathcal{L}_0\sigma_{10}. \tag{7.106}$$

If the reader is unfamiliar with power-series expansions like the one we use here, Exercise 21 gives the details. The form of this solution makes it clear that only the leading-order term on the left-hand side contributes to the steady-state solution to leading order. With this knowledge we can readily obtain the steady states for σ_{20} and σ_{10} to leading order, being

$$\sigma_{20} = i\sqrt{2}\left(\frac{g\alpha}{\kappa}\right)A\sigma_{10}, \tag{7.107}$$

$$\sigma_{10} = 2i\left(\frac{g\alpha}{\kappa}\right)[A\sigma_{00} - \sigma_{11}A]. \tag{7.108}$$

Substituting these expressions into the equations of motion for σ_{00} and σ_{11}, we obtain

$$\dot{\sigma}_{00} = -\frac{i}{\hbar}[H'_{\text{tg}}, \sigma_{00}] - k(A^2\sigma_{00} - 2A\sigma_{11}A + \sigma_{00}A^2) + \kappa\sigma_{11}, \tag{7.109}$$

$$\dot{\sigma}_{11} = -\frac{i}{\hbar}[H'_{\text{tg}}, \sigma_{11}] - k(A^2\sigma_{11} - 2A\sigma_{00}A + \sigma_{11}A^2) - \kappa\sigma_{11}, \tag{7.110}$$

where we have defined

$$k \equiv 2\frac{(g\alpha)^2}{\kappa}. \tag{7.111}$$

Since the density matrix for the target is given by $\sigma \approx \sigma_{00} + \sigma_{11}$, the equation of motion for σ is given by adding the above two differential equations together. The result is

$$\dot{\sigma} = -\frac{i}{\hbar}[H'_{\text{tg}}, \sigma] - k[A, [A, \sigma]]. \tag{7.112}$$

This is the master equation for a continuous measurement of the target operator A, if the observer ignores – and thus averages over – the measurement results. We obtained this master equation because we started from the master equation for the cavity in which the homodyne measurement of the phase of the output light was averaged over. To derive the full stochastic master equation arising from the measurement, we need to explicitly include the photon-detections.

The effective jump operator for the target system

Recall from Eq. (3.174) in Section 3.4 that the operator that gives the effect of a single photon detection, for a homodyne measurement, is

$$\sqrt{\kappa}\tilde{a} = \sqrt{\kappa}(a + i\beta). \tag{7.113}$$

We can now determine the effect of this operator on the state of the target. To do this we first note that the equation of motion of the block σ_{11} contains fast damping, and its state is therefore slaved to the block σ_{00}. We obtain the steady-state solution for σ_{11} to leading order using the method discussed above. The result is

$$\sigma_{11} = \epsilon^2 \mathcal{L}_0 \sigma_{00} = -2\left(\frac{k}{\kappa}\right)A\sigma_{00}A. \tag{7.114}$$

This expression for σ_{11}, along with Eq. (7.99), allows us to write all the blocks σ_{jk} in terms of the target density matrix, σ. Note that to solve Eq. (7.99) to leading order we have to once again invert a matrix power series as we did in Eq. (7.106) (see Exercise 21). We can now write the state of the joint density matrix in terms of the target density matrix, and the result is

$$\rho = \left(\sigma + 2\left(\frac{k}{\kappa}\right)A\sigma A\right) \otimes |0\rangle\langle 0| + 2\left(\frac{k}{\kappa}\right)A\sigma A \otimes |1\rangle\langle 1|$$

$$+ i\sqrt{\frac{2k}{\kappa}}A\sigma \otimes |1\rangle\langle 0| + \frac{\sqrt{2k}}{\kappa}A^2\sigma \otimes |2\rangle\langle 0| + \text{H.c.}, \tag{7.115}$$

where the abbreviation H.c. denotes the Hermitian conjugates of the previous terms on the line. This expression for ρ allows us to determine the action of the jump operator on the density matrix of the target. Recall that upon detecting a photon, the state of the joint system is acted upon by the jump operator, given in Eq. (7.113), and normalized. Leaving out the normalization, the transformation is

$$\rho \to \kappa(a + i\beta)\rho(a^\dagger - i\beta^*). \tag{7.116}$$

Substituting into the right-hand side the expression above for ρ, and keeping only leading-order terms, we obtain

$$\sigma \to 2k(A + i\beta)\sigma(A - i\beta^*). \tag{7.117}$$

The effective jump operator for the target is therefore

$$J = \sqrt{2k}(A + i\beta). \tag{7.118}$$

Now that we have the master equation for the target, Eq. (7.112), and the jump operator, we can determine the full stochastic master equation for the target. Following the method shown in Section 3.4, we take the limit for homodyne detection in which the measurement is driven by Gaussian noise. The result is the stochastic master equation for a continuous measurement of the operator A, being

$$d\sigma = -\frac{i}{\hbar}[H'_{\text{tg}}, \sigma]\,dt - k[A, [A, \sigma]]\,dt + \sqrt{2k}(A\sigma + \sigma A - 2\langle A\rangle\sigma)\,dW. \tag{7.119}$$

The efficiency of an oscillator-mediated measurement

There are two primary sources of inefficiency in measurements mediated by a cavity or oscillator. The first is that the measured output may not be the only source of damping for the cavity or oscillator. These additional sources of damping are effectively output channels that cannot be measured, and are usually referred to as "internal loss mechanisms," or "internal damping." The total damping rate, κ_{tot}, is simply the sum of the output damping rate, κ_{out}, and the internal damping rate, κ_{int}. It is the total damping rate that gives us the measurement rate k via the formula $k = 2(g\alpha)^2/\kappa_{\text{tot}}$. But since it is only the contribution from κ_{out} that is measured, the efficiency is $\eta_{\text{out}} = \kappa_{\text{out}}/(\kappa_{\text{out}} + \kappa_{\text{int}})$.

The second source of inefficiency for an optical cavity is the inefficiency of the photo-detection. This can be modeled by saying that the photo-detector collects some fraction η_{det} of the output light, and so the total efficiency of the measurement is

$$\eta = \eta_{\text{det}}\eta_{\text{out}} = \eta_{\text{det}}(1 + \kappa_{\text{int}}/\kappa_{\text{out}})^{-1}. \tag{7.120}$$

The second source of inefficiency for a superconducting measurement is any additional classical noise added by the amplifier. In fact, superconducting measurements usually have more than one amplification stage, so what concerns us is the overall extra classical noise added by the amplifiers. Referring to Eq. (7.89) above, if the extra noise is P_{xtr} then the amplifier efficiency is $\eta_{\text{amp}} = (1 + 4kP_{\text{xtr}})^{-1}$. The overall efficiency is

$$\eta = \eta_{\text{amp}}\eta_{\text{out}} = (1 + 4kP_{\text{xtr}})^{-1}(1 + \kappa_{\text{int}}/\kappa_{\text{out}})^{-1}. \tag{7.121}$$

In both these situations the measurement strength is

$$k = 2(g\alpha)^2/(\kappa_{\text{out}} + \kappa_{\text{int}}). \tag{7.122}$$

An oscillator-mediated measurement using a linear coupling

When we use a linear coupling between the target system and the oscillator to measure the target operator A, the total Hamiltonian is

$$H = H_{\text{tg}} + \hbar gA(a + a^\dagger) + \hbar\omega a^\dagger a, \tag{7.123}$$

where H_{tg} is the Hamiltonian of the target and ω is the frequency of the oscillator. In the interaction picture with respect to the oscillator motion this is

$$H = H_{\text{tg}} + \hbar gA(ae^{-i\omega t} + a^\dagger e^{i\omega t}). \tag{7.124}$$

Recall now that for the measurement to be effective the motion of the target system must be imprinted on the cavity. Our discussion above regarding resonance is important here. The target will only have an appreciable effect on the probe if the frequencies of the two systems are sufficiently well matched. In fact, this is largely accounted for by setting the damping rate of the cavity to be much faster than the motion of the system. While an undamped cavity oscillates only at a single frequency, the damping broadens the range of frequencies at which the cavity will respond. The spectrum of a driven and damped cavity has a width on the order of κ (see, e.g., Section 3.1.4), so the range of frequencies that it will pick up from the target is approximately $[\omega - \kappa, \omega + \kappa]$. To faithfully copy the target motion, the transfer function of the cavity needs to be uniform over the range of frequencies contained in this motion. That is why it is ideal that κ be much larger than the timescale of the target motion, and this allows the cavity to be adiabatically eliminated.

Even though we choose κ to be large compared to the speed of the target dynamics, we can also use frequency conversion to convert the average speed of the target to the oscillation frequency of the cavity if this is necessary to ensure that the target motion is mapped to the response window of the cavity. When the coupling of the target to the cavity is linear, we may need to modulate the strength of the coupling to perform the frequency conversion. Including this modulation the Hamiltonian in the interaction picture becomes

$$H = H_{tg} + \hbar g \cos(\nu t) A(a e^{-i\omega t} + a^\dagger e^{i\omega t}), \tag{7.125}$$

where ν is the frequency of the modulation. The spectrum of the dynamics for the operator A will be centered on some frequency ω_{tg}. If we could analytically solve for the motion of A, then we could put this motion into the Hamiltonian explicitly by moving to the interaction picture with respect to H_{tg}. But there is no need to do this. We merely choose $\nu = |\omega - \omega_{tg}|$, where ω_{tg} is the central frequency of the spectrum of A. Expanding the expression for the Hamiltonian and dropping terms oscillating at ω (we assume that ω is much greater than all other frequencies) we obtain

$$H = H_{tg} + \hbar g \cos(\nu t) A(a e^{-i\omega_{tg} t} + a^\dagger e^{i\omega_{tg} t}). \tag{7.126}$$

In our analysis of the oscillator-mediated measurement above, in which the coupling to the cavity is proportional to $a^\dagger a$, we can achieve the same conversion, if necessary, by changing the driving frequency from ω to $\omega - \omega_{tg}$. This is referred to as *detuning* the drive from the cavity resonance. With this addition the analysis of the measurement remains essentially unchanged.

7.7.4 Quantum transducers and nonlinear measurements

In this section we describe a general procedure that can be used to implement a measurement, continuous or otherwise. This procedure was first elucidated by von Neumann [621]. It is often the case that we wish to measure an observable that is difficult to connect directly to an amplifier. In this case we can couple the system to another mesoscopic quantum system, for which a means of amplification is available. The conversion of a signal from one physical medium to another is called *transduction*. If a quantum system is used as an intermediary between the target system and an amplifier, then the system is being used as a *transducer*. In the measurement of position via an optical cavity discussed above, the cavity is a transducer from position to optical phase, and the photo-detector used to measure the light is a transducer from optical phase to an electrical signal.

To use a quantum system B as a transducer to measure an observable A of quantum system A, one needs an interaction between A and B that is proportional to A. Let us assume that the interaction is $H_I = \hbar g A B$, where B is a Hermitian operator of system B. Let us assume that A has a set of eigenvectors $\{|a_n\rangle\}$, $n = 1, \ldots, N$, such that $A|a_n\rangle = a_n|n\rangle$. If we write the interaction in the eigenbasis of A we have

$$H_I = \hbar g \sum_n a_n |a_n\rangle\langle a_n| \otimes B. \tag{7.127}$$

This means that we can write the evolution due to H_I as

$$U_I = e^{-igABt} = \sum_n |a_n\rangle\langle a_n| \otimes \exp\left[-i(ga_nt)B\right]. \qquad (7.128)$$

This form for the interaction shows us that for each eigenstate of the operator A, the action on system B is the evolution generated by B, but for each state the speed of this evolution is different. To make a measurement on B that reveals the eigenstate of A, we must measure an observable of B whose value is *changed* by the action of B. This will be true of any observable that does not commute with B, but observables that almost commute with B will be less sensitive to the eigenvalue of A.

The simplest example of two noncommuting observables that generate a change in each other is that of the canonical coordinates position x and momentum p. The commutation relation $[x,p] = i\hbar$ implies that evolution generated by x simply shifts the value of p, and vice versa. The precise relations are

$$e^{iap/\hbar} x e^{-iap/\hbar} = x + a, \qquad (7.129)$$

$$e^{ibx/\hbar} p e^{-ibx/\hbar} = p - b. \qquad (7.130)$$

Thus if the transducer system interacts with the target via its momentum operator, then position would be a good transducer observable to measure. The amplitude and phase quadratures of a single optical mode are equivalent to position and momentum, having essentially the same commutation relation (see Appendix D).

Another set of noncommuting observables is that of the Pauli spin operators for a spin-1/2 system. The spin operators satisfy the commutation relation $[\sigma_x,\sigma_y] = 2i\sigma_z$ for any cyclic permutation of the subscripts x, y, and z. As a result of these commutators any one Pauli operator generates a periodic change in the other two operators. For example, using the general expression given in Eq. (D.5), we have

$$e^{-i\omega t\sigma_z} = \cos(\omega t)\sigma_x + \sin(\omega t)\sigma_y. \qquad (7.131)$$

The action of σ_z is to periodically transform σ_x into σ_y and vice versa. This is circular motion on the equator of the Bloch sphere, at a frequency of 2ω.

Even though the change induced by a Pauli operator is periodic, a single qubit is still good for use as a probe, both for single-shot and continuous measurements. Since a qubit has only two states, if we use a qubit as a probe for a single-shot measurement, then the von Neumann measurement of the probe only has two outcomes, and can thus obtain only one bit of information. If we use a qubit to make a continuous measurement, this limitation is no longer apparent, since the rate at which we obtain information depends on the rate at which the target rotates the qubit, and the rate at which we observe the qubit. There is still a limitation, but if the rates of interaction and measurement are matched appropriately, the qubit provides a near-ideal continuous measurement at a moderately reduced rate [297].

Measuring nonlinear observables

From the analysis of a quantum transducer above it is clear that to make a measurement of an observable A, we need an interaction that contains A. Because of this not all observables are equally easy to measure. For example, the position of an object is simple, since it can interact with other objects by applying forces, and also interacts with light in a similar way. The energy of a mechanical resonator is not nearly so easy to measure directly. To do so requires an interaction containing $b^\dagger b$, where b is the annihilation operator. While we have already encountered an interaction of this form for an optical mode or superconducting oscillator (Eqs. 7.70 and 7.80), to realize this for a mechanical oscillator is not as easy. The membrane-in-the-middle cavity discussed in Section 7.6 does provide one way to do this, as the x^2 interaction is transformed into an effective energy interaction by the rotating wave-approximation. It is possible to construct interactions of even more complex functions of x and/or p, but in the author's experience there are no methods that have the power or flexibility to create strong interactions for arbitrary nonlinear functions.

Nonlinear interactions can be engineered in superconducting systems using the nonlinearity inherent in the Josephson junctions (see, e.g., [348]). Here we present another method for generating nonlinear interactions that involves coupling a qubit to a resonator. In this case the interaction is chosen to be weak compared to the frequency difference between the qubit and the resonator, so that the effect of the resonator on the qubit is merely perturbative. Effective nonlinear interactions can then be obtained by using time-independent perturbation theory (TIPT) to diagonalize the Hamiltonian of the qubit using a power-series expansion.

We now examine the process of diagonalizing a Hamiltonian perturbatively. We do this for two reasons. First, traditional TIPT considers only a single system rather than a coupling to a second system. Second, for some Hamiltonians it is possible to bypass the general method and write the required transformation in a closed form. To begin with, in traditional TIPT [86] the Hamiltonian of the system has the form

$$\tilde{H}_\varepsilon = H + \varepsilon V \tag{7.132}$$

where the difference between the maximum and minimum eigenvalues of λV is much smaller than the smallest energy gap between the levels of H. In this case we can determine the eigenvectors and eigenvalues of $\tilde{H}$ as a power series in ε. This means that we can write

$$\tilde{H}_\varepsilon = H + \varepsilon H_1 + \varepsilon^2 H_2 + \cdots = \sum_{n=0}^{\infty} \varepsilon^n H_n, \tag{7.133}$$

and determine explicit expressions for H_1, H_2, etc.

The TIPT procedure changes very little when the perturbation is an interaction with another system. This would be

$$\tilde{H} = H + V\Lambda + F \tag{7.134}$$

where Λ and F are operators of the second system. We will call the first system the *primary*, and the second the *auxiliary*. The connection with traditional TIPT becomes clear if we write Λ in its eigenbasis, which we denote by $\Lambda|\lambda\rangle = \lambda|\lambda\rangle$. The Hamiltonian becomes

$$\tilde{H} = H + \sum_{\lambda=0}^{N-1} \lambda V \otimes |\lambda\rangle\langle\lambda| + F. \tag{7.135}$$

Each eigenvector of Λ defines a subblock of the full Hamiltonian, in which the Hamiltonian of the primary is

$$\tilde{H}_\lambda = H + \lambda V. \tag{7.136}$$

To diagonalize the Hamiltonian of the primary, we therefore need to diagonalize H_λ for each value of λ, but this is what traditional TIPT does. Note that to diagonalize H_λ we need only perform a transformation of the primary, so the auxiliary is left untouched. The perturbative solution to Eq. (7.134) is given by substituting the perturbative solution to Eq. (7.132) into Eq. (7.135). This gives

$$\tilde{H}' = H + \Lambda H_1 + \Lambda^2 H_2 + \cdots + F, \tag{7.137}$$

which is just the usual perturbative solution with ϵ replaced with the operator Λ.

Standard TIPT gives explicit expressions for the elements of the perturbed eigenvectors and eigenvalues in terms of the matrix elements of the operators H and V, and we give a complete set of formulae for these elements in Appendix G. However, if H and V satisfy sufficiently simple commutation relations, then we can sometimes determine explicitly the unitary that transforms $\tilde{H}_\lambda$ to its diagonal power-series form. Since this unitary transformation is necessarily a function of λ, the unitary transformation in the joint space of both systems is a function of Λ, so we will write it as $U(\Lambda)$. Note that this unitary has no effect on the operator Λ since it commutes with it. Note also that to obtain the perturbation expansion, $U(\Lambda)$ is applied only to the Hamiltonian $K \equiv H + \Lambda V$, and not to any other auxiliary operators that might appear in the joint Hamiltonian $\tilde{H}$.

To determine the unitary $U(\Lambda)$, we note that if $U(\lambda) = e^{-\Lambda A}$, then

$$\tilde{H}' = e^{\Lambda A} K e^{-\Lambda A}$$
$$= K + \Lambda[A, K] + \frac{\Lambda^2}{2}[A, [A, K]] + \frac{\Lambda^3}{3}[A, [A, [A, K]] + \cdots, \tag{7.138}$$

where

$$K = H + V\Lambda. \tag{7.139}$$

This means that we will obtain the perturbative power series to at least second order in Λ if we choose $U(\lambda) = e^{-\Lambda A}$, and the operator A so that the second term in the above expansion, $\Lambda[A, \tilde{H}]$, cancels the interaction term ΛV. We therefore need to find an operator A such that

$$[\Lambda A, K] = -V\Lambda + \mathcal{O}(\Lambda^2). \tag{7.140}$$

which is equivalent to

$$[A, H] = -V. \tag{7.141}$$

We now show how the above procedure is applied to a resonator perturbatively coupled to a qubit. The joint Hamiltonian of the two systems is

$$\tilde{H} = \hbar\Omega\sigma_z + \hbar g\sigma_x x + \hbar\omega a^\dagger a, \tag{7.142}$$

which can be realized readily in superconducting circuits. Here we define $x = a + a^\dagger$. Now we assume that Ω is large enough, and the energy of the resonator is small enough, that

$$\left\langle gx + \omega a^\dagger a \right\rangle \ll \Omega, \tag{7.143}$$

so that the resonator is a perturbation of the qubit.

We now make the identification $\Lambda = gx$, and we need to find the operator A that satisfies

$$[A, \Omega\sigma_z] = -\sigma_x. \tag{7.144}$$

Noting that $[\sigma_y, \sigma_z] = 2i\sigma_x$, the above relation is true if we choose

$$A = \frac{1}{2\Omega} i\sigma_y. \tag{7.145}$$

Transforming the qubit part of the Hamiltonian using $U = e^{-ig\sigma_y x/(2\Omega)}$, we obtain the power series expansion to second order:

$$\tilde{H}' = \hbar\Omega\sigma_z + \hbar\left(\frac{g^2}{\Omega}\right)\sigma_z x^2 + \hbar\omega a^\dagger a \tag{7.146}$$

This new form for the Hamiltonian is called an *effective Hamiltonian*, and we see that it contains the nonlinear interaction $\sigma_z x^2$.

To obtain the perturbative expansion to third (or higher) order, we find the lowest-order term that (i) appears in $\tilde{H}'$, and (ii) does not commute with H. Call this term C. We now find a new operator A' that when applied to H cancels C. In the example above C is a third-order term that we have ignored. Applying a transformation A' that cancels this third-order term generates a fourth-order term that completes the fourth-order contribution to the perturbative power series. To obtain the expansion to higher orders we repeat this procedure.

History and further reading

The theory of superconductors is a deep subject, and has a rather long history; the theory we require is an effective theory that emerges from the many-body dynamics of the electrons in a solid. The idea that superconductivity could be considered as a macroscopic quantum effect, in which the electrons could be described by a single effective wavefunction, emerged in the 1950s around the same time that the BCS theory of superconductivity was proposed [31, 32]. The quantization of the flux around a superconducting loop could be derived by applying the Landau–Ginzburg theory to the collective electrons, in which the order parameter becomes the effective macroscopic wavefunction [193, 373, 501]). But it was Gorkov in 1959 who made the connection between the BCS theory and Landau–Ginzburg, showing that the phase of the BCS wavefunction was proportional to the Landau–Ginzburg order parameter [221]. Further details regarding the connection between BCS and Landau–Ginzburg can be found in [2], and regarding the derivation of the condensate wavefunction in [193, 373, 501]. We built our treatment of the rules for obtaining the Hamiltonians of superconducting circuits mainly from the work of Devoret [143], who in turn generalized the work of Yurke and Denker [709].

The key theoretical step toward observing superpositions of macroscopic quantum states, which is what powers all the superconducting devices we have described in this chapter, was taken by Widom in 1979 [659]. He introduced the idea that one could observe superpositions of macroscopically distinct quantum states of the collective electrons ("the electron condensate"), and introduced the quantum operator corresponding to macroscopic flux [659]. Widom's suggestion that macroscopic superpositions could exist, and be seen in the laboratory, was difficult for many to accept. Macroscopic superpositions of flux states were thought to be suppressed by irreducible environmental effects or other "macro-realistic" constraints [361, 362]. Now that it is commonplace to observe these macroscopic superpositions in the laboratory, it is easy to forget the dramatic change in thinking that occurred over a period of about a decade and a half (see, e.g., the sequence [361, 362, 164, 478, 594]). Despite the difficulties, Terry Clark's group at Sussex pioneered experiments to observe these superpositions [489], but the results were not unambiguous enough to reduce the controversy [571]. Widom *et al.* [660] and Prances *et al.* [490] also showed that charge states would be quantized for small capacitances, which is the origin of the Coulomb blockade [623, 391]. In 1985 Martinis, Devoret, and John Clarke [408, 145] and Washburn *et al.* [652] performed experiments that very clearly demonstrated macroscopic quantum effects in a superconducting device. It was not until 1999, however, that superpositions of macroscopic quantum states were observed for the first time, by Nakamura, Pashkin, and Tsai [439], twenty years after Widom's theory.

Over the next five years quite a number of groups realized macroscopic superpositions of single qubits in a range of superconducting circuits, and were able to manipulate and observe these systems with increasing fidelity [183, 625, 640, 702, 409, 121, 565, 531, 162, 24, 129]. Circuits in which more than one qubit are coupled together were realized in [468, 696, 390, 415, 576, 225, 254]. These advances were due to improved experimental techniques, as well as new theoretical designs for superconducting circuits, the latter of which we have referenced throughout the chapter.

Combining a qubit with a superconducting resonator opened up many new possibilities. Pioneered primarily by the groups of Schökopf, Girvin, and Devoret from 2004 onward [645, 184, 389, 270, 552, 148, 463, 518], this combination of systems is often referred to as "circuit quantum electrodynamics" (circuit QED) because its dynamics is equivalent to a two-level atom interacting with a single mode of an optical cavity [67]. Quantum state-engineering was demonstrated by the group of Martinis using circuit QED in 2009 [256].

In parallel with the development of quantum superconducting circuits, experimentalists were refining techniques to construct and measure the quantum mechanical behavior of tiny "nano"-mechanical oscillators. Measurements of these oscillators are made by coupling them to superconducting circuits, and in particular to superconducting resonators as described in Section 7.5. Experimental advances in this endeavor have been made by the groups of Schwab [349, 437], Lehnert [507, 595, 465], Martinis and Cleland [461], Simmonds [596, 597], and Silanpää [410, 411]. Theoretical analyses of various ways to couple superconducting and mechanical systems can be found in [666, 610, 69, 719].

Techniques were also developed to observe the quantum mechanical behavior of mechanical resonators interacting with optical cavities, a field now referred to as quantum optomechanics [23]. In this case the mechanical systems are measured via the optical cavities, as described in Section 7.7.3. A sampling of early work on optical cavities with oscillating mirrors, before it became possible to probe quantum mechanical effects, can be found in [175, 294, 355, 356, 397, 77, 211, 398, 608, 79, 405]. Advances in the development of optomechanics in the quantum regime have been made by the groups of Heidemann [18], Kippenberg [542, 14, 543, 654], Harris [309, 599], Aspelmeyer [229, 230], Painter [166, 165, 527, 112], and Stamper-Kurn [434, 495].

At first, measurements on superconducting quantum systems were swamped by noise, and the systems had to be re-prepared many times so that the results of many measurements could be averaged. The ability now to make single-shot measurements on these systems, and thus to monitor them in real time, was made possible by the development of extremely low-noise amplifiers. This was a result of rapid progress by a number of groups between 2007 and 2010 [103, 494, 422, 80, 569, 334, 104, 697, 51, 50, 1, 243], building on earlier work on superconducting parametric amplifiers [707, 708, 433]. References to experiments in which nano-electromechanical systems are monitored in real time are given at the end of Chapter 5.

In Section 7.7.3 we analyzed a technique for measuring a quantum system by coupling it to an electrical or optical resonator, which is then measured by homodyne detection. This technique, which has proven to be useful in superconducting systems, was first analyzed fully quantum mechanically by Wiseman and Milburn in the setting of quantum optics [679], and we used their method of analysis here. Examples of experiments which employ this measurement technique are [507, 595, 465, 596, 597, 51, 50, 1, 243, 637, 638, 517]. Schemes for measuring nonlinear observables can be found in [407, 296, 92, 297, 304, 302, 630].

Exercises

1. Show that if a single link consists of a current source I in parallel with an inductor L, then the effective energy of the inductor in terms of the total flux across the link, ϕ, is $E = (\phi + IL)^2/(2L)$. Hint: rather than considering the flux, begin by writing down the total current through the link in terms of I and the current through the inductor. From this current determine the contribution to the equation of motion, and from this the term required in the Lagrangian.

2. Consider a circuit consisting of a real Josephson junction and a current source generating current I. This circuit is referred to as a "current biased Josephson junction." Calculate the Hamiltonian for this circuit. Since the nonlinear term in the Hamiltonian is a function of the phase, $\hat{\theta}$, it is natural to think of θ as the equivalent of the position coordinate, and N_J the momentum coordinate, so that the nonlinear function is the equivalent of a potential. Using this terminology, what does the potential for the current-biased Josephson junction look like? (It is usually referred to as a "washboard" potential.)

3. Derive the Hamiltonian for the circuit shown below, in which two LC-resonators are coupled by a JJ. Write the Hamiltonian in terms of the annihilation and creation operators for the oscillators. What is the interaction Hamiltonian that couples the oscillators?

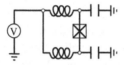

4. Use the formula in Eq. (7.12) to obtain the Hamiltonian in Eq. (7.59) from the Lagrangian in Eq. (7.58).

5. (i) Show that if a voltage is applied to a superconducting LC-resonator, in series with the capacitor, and the voltage oscillates at the frequency of the resonator (resonant driving), then the Hamiltonian for the resonator is given by Eq. (7.73). (ii) Determine the Hamiltonian when the LC-resonator is driven by a current source, also on resonance, where the current is in parallel with the inductor.

6. Three harmonic oscillators are coupled via the three-system (three-body) interaction

$$H_1 = \hbar g(a + a^\dagger)(b + b^\dagger)(c + c^\dagger), \tag{7.147}$$

where a, b, and c are the respective annihilation operators for the oscillators. By using energy conservation, determine the relationships between the frequencies of the oscillators that must be satisfied for them to readily exchange quanta. Assume that g is much smaller than all the frequencies of the oscillators.

7. Apply the rotating-wave approximation to the Hamiltonian in Eq. (G.10) to obtain the approximate Hamiltonian of Eq. (G.11). The rotating-wave approximation is described in Appendix G.

8. Show that the evolution given by $H = \chi(a^\dagger a)^2$ for a time $t = \pi/(2\chi)$ transforms the coherent state $|\alpha\rangle$ into the state

$$|\psi\rangle = [|\alpha\rangle - i|-\alpha\rangle]/\sqrt{2} \qquad (7.148)$$

up to an overall phase factor. The number-state representation for a coherent state is given in Appendix D.

9. (i) Determine the operator A that satisfies $[A, b^\dagger b] = -a^\dagger a(b + b^\dagger)$. (ii) Apply the unitary transformation $U = e^{-\eta A}$ to the optomechanical Hamiltonian given by

$$H = \hbar\Omega a^\dagger a + \hbar g a^\dagger a(b + b^\dagger) + \hbar\omega b^\dagger b, \qquad (7.149)$$

and choose η so that the interaction term between the two oscillators vanishes. Show that the transformed Hamiltonian has a term $H' = \hbar(g^2/\omega)(a^\dagger a)^2$. This term is called a Kerr, or $\chi^{(3)}$, nonlinearity. The necessary operator relations can be found in Section G.7.

10. By using the operator re-ordering relations in Section G.7, show that the evolution of the Hamiltonian $H = \hbar\omega b^\dagger b + \hbar g(b + b^\dagger)a^\dagger a + \hbar\Omega a^\dagger a$ can be written as

$$U(t) = e^{-i\Omega a^\dagger a t} e^{i(g^2/\omega)(a^\dagger a)^2(t - \sin\omega t)} e^{(g/\omega)a^\dagger a(\eta b^\dagger - \eta^* b)} e^{-i\omega b^\dagger b t}, \qquad (7.150)$$

where $\eta = 1 - e^{-i\omega t}$. Hint: begin by using the transformation in Exercise 9 above.

11. Read the paragraph just above Eq. (7.72), and refer to the parts of this chapter indicated there to help you determine the coupling constant g in Eq. (7.72).

12. The electromechanical coupling, given by the Hamiltonian in Eq. (7.149), can be viewed as being off-resonant. This is because the operator $a^\dagger a$ is constant under the Hamiltonian of the electrical oscillator, while the operator $b + b^\dagger$ oscillates at the mechanical frequency ω. If you make the rotating-wave approximation (RWA), the interaction is eliminated. (In fact, the Kerr nonlinearity derived in Exercise 9 is the first correction to the RWA. This can be understood from Sections G.4 and 7.7.4.) We can effectively bring the interaction into resonance by modulating the coupling rate g at the mechanical frequency ω. Your task is: modulate g at the frequency $\omega - \delta$, move into the interaction picture with respect to the Hamiltonian $H_0 = \hbar(\omega - \delta)b^\dagger b$, and make the rotating-wave approximation assuming g is much smaller than all the other rate constants. Once you have done this, what is the new effective frequency of the mechanical resonator? How will this modify the answer to Exercise 9? How will it modify the answer to Exercise 10?

13. Two mechanical resonators described by the annihilation operators a_1 and a_2, and with respective frequencies ω and $\omega + \delta$, are coupled to an LC-oscillator with annihilation operator b. The Hamiltonian for the three oscillators is

$$\frac{H}{\hbar} = \omega a_1^\dagger a_1 + (\omega + \delta)a_2^\dagger a_2 + \sum_{j=1,2} g b^\dagger b(a_j + a_j^\dagger) + \Omega b^\dagger b. \qquad (7.151)$$

Transform to two new mode operators c_1 and c_2, defined by $c_1 = (a_1 + a_2)/\sqrt{2}$ and $c_2 = (a_1 - a_2)/\sqrt{2}$. Now transform to the interaction picture, and make the rotating-wave approximation assuming that ω and Ω are both much greater than δ and g but that δ and g are similar in size. Finally, transform back to the Schrödinger picture. (i) How many oscillators is the LC-oscillator coupled to now? (ii) What is the frequency of the oscillator to which it is coupled?

14. The electromechanical coupling given by the Hamiltonian in Eq. (7.149) is a linear drive applied to the mechanical resonator. (i) Start the LC-resonator in a photon-number state $|n\rangle$. Since $|n\rangle$ is an eigenstate of the Hamiltonian, the joint system remains in the subspace in which the LC-resonator is in this state. Thus the operator $a^\dagger a$ is equivalent to the number n. What is the size of the force on the mechanical resonator? (ii) Since the force on the mechanical resonator is constant, it merely shifts the equilibrium point of the mechanical resonator. What is the new equilibrium point in the original phase space of the mechanical resonator? (iii) What coherent state of the original mechanical resonator is equal to the vacuum state of the shifted resonator? (iv) Now modulate the coupling constant g at the mechanical frequency, and transform to the interaction picture with respect to the Hamiltonian $H_0 = \hbar \omega b^\dagger b$. If the mechanical resonator starts in the coherent state $|\alpha\rangle$, what is its state as a function of time (in the interaction picture)? (v) Keep everything from question (iv), but change the initial state of the LC-oscillator to $|\psi\rangle = (|0\rangle + |1\rangle)/\sqrt{2}$. What is the state of the joint system at time t now?

15. Consider a mechanical resonator coupled to two optical or electrical modes, where the latter are coupled via a linear interaction. The Hamiltonian is H where

$$\frac{H}{\hbar} = \omega_m b^\dagger b + g(b + b^\dagger)(\tilde{c}^\dagger \tilde{c} + \tilde{d}^\dagger \tilde{d}) + \Omega \tilde{c}^\dagger \tilde{c} + \Omega \tilde{d}^\dagger \tilde{d} + \lambda(\tilde{c}\tilde{d}^\dagger + \tilde{d}\tilde{c}^\dagger). \quad (7.152)$$

Here the electromagnetic modes are $\tilde{c}$ and $\tilde{d}$, with λ their coupling rate. An example of this Hamiltonian may be found in [232]. (i) Make the transformation $c = (\tilde{c} + \tilde{d})/\sqrt{2}$, $d = (\tilde{c} - \tilde{d})/\sqrt{2}$, to new annihilation operators c and d. The result should be a three-way coupling between the modes b, c, and d. What are the frequencies of the modes c and d? (ii) Move into the interaction picture with respect to the Hamiltonians of the individual oscillators. What relationship is required between λ and ω_m so that the three-way interaction is on resonance? (iii) Choose λ to make the three-way interaction resonant, and apply the RWA. (iv) Can we effectively switch the interaction between modes c and d on and off by changing the mechanical frequency?

16. If there are N linear amplifiers connected in series (that is, the output of the first is connected to the input of the second, etc.), each with gain G_n and noise temperature T_n, then what is the noise temperature of the whole series considered as a single amplifier?

17. Determine the value of the measurement strength that gives the minimum value of Eq. (7.88), and this minimum value. This latter is the minimum noise power for an inefficient measurement of the position of a quantum oscillator.

18. Consider amplifying the state of a quantum oscillator to turn it into a classical number. This amplification process constitutes a measurement of the position and momentum, or equivalently the amplitude and phase, of the oscillator. This process can be captured in a simple way by imagining that it maps the oscillator, described by annihilation operator $a = (x_a + p_a)/\sqrt{2}$, to a second oscillator, described by annihilation operator $b = (x_b + p_b)/\sqrt{2}$, as $b = \sqrt{G}a$. Note that here $[x,p] = i$. This mapping is equivalent to mapping the state of the oscillator to a classical number, because when G is large the second oscillator has many quanta, and so the quantum uncertainty drops to zero in relation to the mean values of x_b and p_b. The initial oscillator is the input to the amplifier, and the second oscillator is the output. (i) show that the relationship $b = \sqrt{G}a$ is incompatible with the commutation relation $[b,b^\dagger] = 1$. (ii) Modify the amplification relationship to $b = \sqrt{G} + c$, for some operator c of a third system. Show that the commutation relation for b implies that $[c,c^\dagger] = G - 1$. (iii) Show that the variance of b, $V(b) \equiv V(x_b)/2 + V(p_b)/2$, must now be larger than $GV(a)$ by $(G-1)/2$. (Hint: this is not completely straightforward: you have to use the inequality $\Delta x \Delta p \geq (1/2)|\langle[x,p]\rangle|$.) This result is called the "quantum limit to amplification." Further details, and justification for the above steps, are given in [106, 244].

19. It is possible to amplify the oscillation of an oscillator in a phase-selective way by modulating the frequency of the oscillator at twice this frequency. This procedure is called parametric amplification. To analyze this scenario, first consider a harmonic oscillator with Hamiltonian $H = p^2/(2m) + m\omega^2 x^2/2$. Note that the frequency ω appears in the expression for the annihilation operator, so we will write it as a_ω. Now write the Hamiltonian $H' = p^2/(2m) + m\Omega^2 x^2/2$ in terms of a_ω and $a_\omega^\dagger$. Use what you learn from this exercise to write the Hamiltonian for an oscillator whose frequency is given by $\Omega(t) = \omega[1 + \lambda\sin(2\omega t)]$ in terms of a_ω and $a_\omega^\dagger$. Now move into the interaction picture with respect to the unmodulated Hamiltonian $H_0 = \hbar\omega a_\omega^\dagger a_\omega$, assume that λ is much less than unity, and make the rotating-wave approximation. The resulting Hamiltonian, in the interaction picture, should be time-independent. Now solve the equations of motion for a_ω and $a_\omega^\dagger$ in the interaction picture. The quadrature operators are $X = (a_\omega + a_\omega^\dagger)/2$ and $Y = -i(a_\omega - a_\omega^\dagger)/2$. Show that one of these quadratures is amplified by the evolution.

20. Consider the measurement of a target system mediated by an optical or superconducting resonator as described in Section 7.7.3. Take the interaction with the cavity to be $L^\dagger a e^{i\omega t} + La^\dagger e^{-i\omega t}$, where L is an operator of the target, a is the annihilation operator for the cavity, and ω is the frequency of the cavity mode. Derive the stochastic master equation for the target for this case by adiabatically eliminating the cavity.

21. Here you will calculate the power-series expansion of $(1 + \varepsilon\mathcal{L}_1)^{-1}$, where $\mathcal{L}_1$ may be a matrix. To do this we first consider the expression $E = 1 + \varepsilon\mathcal{L}_1$ as the power series $S = 1 + \sum_{n=1}^{\infty} \varepsilon^n \mathcal{L}_n$, where the matrices $\mathcal{L}_n$ are equal to zero for $n \geq 2$. Next we assume

that there is another power series, $P = 1 + \sum_{n=1}^{\infty} \varepsilon^n \mathcal{M}_n$, such that

$$SP = \left(1 + \sum_{n=1}^{\infty} \varepsilon^n \mathcal{M}_n\right)\left(1 + \sum_{n=1}^{\infty} \varepsilon^n \mathcal{L}_n\right) = 1 + \sum_{n=1}^{\infty} \varepsilon^n \mathcal{V}_n = 1, \qquad (7.153)$$

where all the matrices $\mathcal{V}_n = 0$. We now equate the coefficients of the powers of ε on each side of this equation to systematically determine the matrices $\mathcal{M}_n$. Show that

$$\mathcal{M}_1 = -\mathcal{L}_1, \quad \text{and} \quad \mathcal{M}_2 = (\mathcal{L}_1)^2. \qquad (7.154)$$

22. Consider a system B used as a transducer to measure observable A of a system A. Show that if the interaction between systems A and B is $H_1 = \hbar g A B$, then a measurement of B on system B will contain no information about the value of A.

23. Consider a single qubit coupled to a resonator via the Hamiltonian

$$\tilde{H} = \hbar \Omega \sigma_z + \hbar g \sigma_z (a + a^\dagger) + \hbar \omega a^\dagger a. \qquad (7.155)$$

If we want to measure the qubit to determine, with certainty, whether the resonator contains zero or one photons: (i) what state should we start the qubit in? (ii) How long should we wait before measuring the qubit? (iii) What measurement should we make on the qubit?

24. Determine the perturbative expansion for the Hamiltonian in Eq. (7.146) to fourth order in (g/Ω).

25. (i) Consider the Hamiltonian in Eq. (7.146). Assume that $g \ll \omega$ and make the rotating-wave approximation. (ii) If the qubit was used as a probe to measure the oscillator, what observable would it measure?

26. A qubit is to be used as a probe to make a measurement on a resonator. The Hamiltonian of the qubit and resonator is

$$H = \hbar g \sigma_z a^\dagger a + \hbar \omega a^\dagger a. \qquad (7.156)$$

The qubit is started in an eigenstate of σ_x, and then allowed to interact with the oscillator for a time $\tau = \pi/(2g)$. A von Neumann measurement is then made on the qubit that projects onto the σ_x basis. (i) What information does the measurement provide about the number of photons in the resonator? (ii) Add to the Hamiltonian the additional term $H_0 = \hbar \Omega \sigma_z$. In what basis do we now need to measure the qubit so that the measurement provides the same information as before? (iii) Can the measurement distinguish between the two states $|\psi_\pm\rangle = (|\alpha\rangle \pm |-\alpha\rangle)/\mathcal{N}_\pm$, where $|\alpha\rangle$ denotes a coherent state with amplitude α, and $\mathcal{N}_\pm$ are the normalizations for the respective states?

27. Consider the Hamiltonian for a single qubit given by $H_q = \hbar \omega \sigma_z$. (i) Determine the Heisenberg equation of motion for the lowering operator $\sigma = \sigma_x - i\sigma_y$. Solve this

equation to determine σ as a function of time. (ii) Now consider the Hamiltonian for a single qubit coupled to a resonator, given in Eq. (7.142). Assume that $g \ll \omega$ so as to make the rotating-wave approximation, converting the interaction to the form $\sigma a^\dagger + \sigma^\dagger a$. You will need the results in (i), and to modulate the interaction rate g at the right frequency. (iii) Now use the procedure in Section 7.7.4 to diagonalize this Hamiltonian perturbatively. The oscillator is a perturbation on the qubit Hamiltonian, and the small parameter is $\varepsilon = g/\Omega \ll 1$. The transformation you will need is $U = \exp[-\varepsilon(\sigma a^\dagger - \sigma^\dagger a)]$. Determine the transformed Hamiltonian (the effective Hamiltonian) to second order in ε.

28. Consider two Kerr resonators described by annihilation operators a and b, and which are linearly coupled. The Hamiltonian for the two resonators is

$$H = \hbar \omega a^\dagger a + \hbar \chi (a^\dagger a)^2 + \hbar \omega b^\dagger b + \hbar \chi (b^\dagger b)^2 + \lambda (ab^\dagger + ba^\dagger). \tag{7.157}$$

Now transform to two new mode operators c and d, defined by $c = (a+b)/\sqrt{2}$ and $d = (a-b)/\sqrt{2}$. Make the rotating-wave approximation, which means dropping all terms that do not contain an equal number of annihilation and creation operators. (i) What parameter must be small to justify the rotating-wave approximation? (ii) What are the frequencies of the new modes? (iii) How are the new modes coupled together? (This nonlinear coupling is called a "cross-Kerr" term.)

8

Quantum mesoscopic systems II: measurement and control

In this chapter we discuss ways to control the mesoscopic systems we introduced in Chapter 7. We will be concerned more with creating and controlling the *states* of these systems, rather than engineering entire unitary transformations, although we will consider the latter in Section 8.1. The difference between these two tasks, state-control and full control, is that in the latter one must realize an operation that will map *every* initial state to a given final state, without knowing what state the system is actually in. Such a mapping, realized by a unitary transformation, can be used to process information.

8.1 Open-loop control

The evolution of a sufficiently well isolated mesoscopic system, which may consist of a number of interacting subsystems, is determined by its Hamiltonian. Usually this Hamiltonian will contain one or more parameters that can be changed by the experimenter. For example, in a superconducting circuit one can change externally applied voltages or magnetic fields. In an optomechanical system one can change the power of the laser driving the optical cavity, and sometimes the frequency of the mechanical oscillator. These parameters are real numbers, and are often referred to as "classical fields" or classical controls. The word "classical" refers to the fact that in describing the voltages and other applied forces as numbers in the Hamiltonian we are treating them as determined by classical systems. Ultimately all the systems that apply these forces are quantum systems, and the forces are due to interactions between them and the primary system. The description of the system in terms of parameters that can be externally varied is a (very useful) *effective* theory.

The ability to vary the Hamiltonian with time is a powerful tool. The reason for this is that the range of dynamics that can be created in this way is much greater than that of all the individual Hamiltonians that we can select by choosing the parameters. To explain what we mean by this, consider a single qubit with the Hamiltonian

$$H = \hbar(f\sigma_z + g\sigma_x), \qquad (8.1)$$

in which we are free to vary f and g with time. Let us take two consecutive segments of time, each of length T, and choose $f = 0$ in the first segment, and $g = 0$ in the second.

The unitary operator that gives the total evolution resulting from both time-segments is $U = U_2 U_1$, where $U_1 = \exp(-ig\sigma_x T)$ and $U_2 = \exp(-if\sigma_z T)$. If the operators σ_x and σ_z commuted then we would have $U = \exp(-i[f\sigma_z + g\sigma_x]T)$, but they do not. Instead, to determine the effective Hamiltonian over the full time period of $2T$, we must use the relations for the Pauli matrices in Eqs. (D.4) and (D.5). The result is

$$U = e^{-i2\lambda(n_x\sigma_x + n_y\sigma_y + n_z\sigma_z)T}, \qquad (8.2)$$

where $\cos(2\lambda T) = \cos(fT)\cos(gT)$ and

$$n_x = \frac{\cos(fT)\sin(gT)}{\sin(2\lambda T)}, \quad n_y = \frac{\sin(fT)\sin(gT)}{\sin(2\lambda T)}, \quad n_x = \frac{\sin(fT)\cos(gT)}{\sin(2\lambda T)}. \qquad (8.3)$$

The effective Hamiltonian over the total time $2T$ is thus $H = \hbar\lambda(\mathbf{n}\cdot\boldsymbol{\sigma})$, where $\mathbf{n} = (\sigma_x, \sigma_y, \sigma_z)$ and $\boldsymbol{\sigma} = (\sigma_x, \sigma_y, \sigma_z)$. So by varying the parameters f and g over time we are able to create essentially any effective Hamiltonian. The caveat is that it may take a longer time to implement this Hamiltonian than it would take if we were able to implement the Hamiltonian directly. For large systems this could be very much longer. The ability to create an effective Hamiltonian that includes σ_y from one that includes only σ_x and σ_z is equivalent to the fact that in three dimensions one can generate rotations around the y axis by using only rotations about the x and z axes.

For a finite-dimensional system it is a somewhat remarkable fact that, given enough time, *any* unitary transformation can be created from a Hamiltonian with only two controllable parameters. In fact, for any two operators A and B picked at random, the Hamiltonian $H = \hbar(fA + gB)$ can generate every unitary operator in the space with probability one [377]. Given finite maximum values for $|fA|$ and $|gB|$, the time required to generate any unitary in an N-dimensional space scales with N^2, since the number of parameters that define an arbitrary unitary scales with N^2. However, for a fixed N, the time required to approximate an arbitrary unitary with error ε scales only as a polynomial in $\ln(1/\varepsilon)$, a result known as the Solovay–Kitaev theorem [137].

When we generate a unitary transformation by using time-dependent control, we do pay a price in speed. The size of the forces we can apply to a system determines the maximum speed of the motion in Hilbert space. One way to set a limit on the size of the forces is to set a limit on the eigenvalues of the Hamiltonian. Note that since one can always add a multiple of the identity operator to the Hamiltonian without affecting the dynamics, it is not the size of the eigenvalues that determines the maximal speed over evolution in Hilbert space but the difference between the maximum and minimum eigenvalues. If we set the minimum eigenvalue to be zero, then the speed is bounded by the maximum eigenvalue. The maximum eigenvalue is, in fact, the "H-infinity" (H_∞) norm of an operator. For finite-dimensional vector spaces, all "norms" that can be defined for operators have been shown to be equivalent, meaning that they all place the operators in the same order. Thus the speed of evolution can be bounded, for example, by bounding the "H-two" (H_2) norm, given by $\text{Tr}[H^2]$, or the H_∞ norm.

If we want to implement a unitary transformation U, then a fundamental question arrises. If the forces we can apply are bounded, quantified by a bound on the norm of the Hamiltonian, then what is the shortest time in which we can implement U, and what choice of Hamiltonian achieves this time? The following theorem answers this question.

Theorem 15 If the norm is the only restriction on the Hamiltonian, then any unitary U is implemented in the minimum time if and only if the Hamiltonian is time-independent. All Hamiltonians that achieve the minimum time are therefore proportional to $i\hbar \ln [U]$.

The proof of the above theorem employs differential geometry and so is beyond our scope, save to say it involves showing that the time taken is the length of a path between two points in a curved space, and the shortest path – a geodesic – is realized by a constant Hamiltonian [648, 647]. This theorem tells us a number of things. First, control protocols that use time-dependent Hamiltonians incur a penalty in speed, and we can expect that the more the Hamiltonian varies with time the longer the protocol will take. Second, the fastest way to perform any unitary is given by the simplest possible evolution. This is useful if we are only constrained by the norm of the Hamiltonian, or if we wish to obtain an upper bound on the speed of all protocols under a constraint on the norm. Third, we can use the above theorem to determine the maximum speed at which a state can evolve in Hilbert space for a given value of the norm. If we want to transform one state $|0\rangle$, to an orthogonal state $|1\rangle$, then we know that a Hamiltonian that does this in the least time is a constant Hamiltonian that generates the required unitary. Let us first consider performing the transformation just within the two-dimensional space spanned by $|0\rangle$ and $|1\rangle$. Since all unitary transformations must preserve the inner product, the unitary that transforms $|0\rangle$ to $|1\rangle$ must also transform $|1\rangle$ to $|0\rangle$ (up to an arbitrary phase), and therefore has the form

$$U_2 = |1\rangle\langle 0| + e^{i\phi}|0\rangle\langle 1|, \tag{8.4}$$

for some phase ϕ, up to a global phase factor. The Hamiltonians that generate this class of unitaries in a time $t = \pi/(2\lambda)$ are, up to the addition of a multiple of the identity,

$$H = \hbar\lambda \left(e^{i\theta}|1\rangle\langle 0| + e^{-i\theta}|0\rangle\langle 1| \right), \tag{8.5}$$

for some phase θ. The eigenvalues of these Hamiltonians are $\pm\hbar\lambda$, and thus their H_∞ norm is $2\hbar\lambda$. This Hamiltonian also tells us the fastest evolution from any state to any other (non-orthogonal) state, because in order to take the shortest path from $|0\rangle$ to $|1\rangle$, it must also take the shortest path between any states on this path. Under a constraint on the norm, the shortest path(s) between two states $|0\rangle$ and $|\tilde{0}\rangle = \sqrt{1-p}|0\rangle + \sqrt{p}|1\rangle$ are thus

$$|\psi(t)\rangle = \cos(t)|0\rangle + \sin(t)e^{i\phi}|1\rangle, \tag{8.6}$$

where t parametrizes the path and ϕ is an arbitrary phase. We refer to these paths as the *geodesics*, and they are great circles on the hypersphere of states (vectors with unit norm).

Note that to arrive at these paths we considered only two-dimensional unitaries. To really show that these paths are the shortest under a constraint on the norm of the Hamiltonian we would need to consider all N-dimensional unitaries that transform $|0\rangle$ to $|1\rangle$. We explore these unitaries in Exercise 2. That the fastest evolution between two states is given by the Hamiltonian in Eq. (8.5), under constraints that are equivalent to a constraint on the norm (all norms are equivalent for finite-dimensional vector spaces), is proved in [9, 401, 99].

Obtaining control protocols using numerical optimization

When we wish to control a quantum system we are usually faced with the problem of working out how to vary the available parameters in the Hamiltonian to achieve the state or evolution that we want. If we have N parameters f_j, $j = 1,\ldots,N$, then we will refer to the N functions of time $f_j(t)$ as the *control functions*. Together the full set of control functions is a *control protocol*. Sometimes there is an obvious and simple way to choose the control functions, if we assume that they can be switched instantaneously from one value to another. Often this problem is so complex that there is no analytic solution, and we must resort to numerical methods to search for one. To do this, we must write the control functions in some discrete basis, so that there is a finite number of parameters over which to search. There are many such choices of basis, of which piecewise-constant, piecewise-linear, or Fourier bases truncated to some maximum frequency are examples. Once we have selected in this way a finite set of real numbers to represent the control functions, we write computer code to solve for the evolution given these numbers. The code can then calculate the fidelity with which the resulting state of the system realizes the desired state, and we can then use a search method to find a set of numbers that maximizes this fidelity. Search methods cannot guarantee to find a global maximum, just local maxima. One can increase the likelihood of finding a global maximum by starting the search with a number of different random initial sets of values for the numbers defining the control functions, and looking at the spread of the resulting fidelities. The use of a numerical search method to find a control protocol is often referred to loosely as "optimal control," because of the maximization process used in the numerical search.

Before using a numerical search to find a good set of control functions, there is an analytic procedure one can use to check whether the available set of parameters will in theory allow any unitary transformation to be realized. If this is the case the control problem is referred to as "controllable." Let us say that we want to control an M-dimensional system, and the Hamiltonian for the system has the form

$$H = H_0 + \sum_{j=1}^{N} a_j H_j, \qquad (8.7)$$

where the parameters a_j can be varied in time by the controller. The procedure involves calculating the commutators of the $N+1$ Hermitian operators $\{H_0, H_j\}$, and each time a commutator results in an operator not already in the set $\{H_0, H_j\}$, one adds the new commutator to this set. One then continues to calculate all the mutual commutators of the operators in the enlarged set, continuing to add any new operators that result to the set. If at some point this set of operators spans the space of all Hermitian operators of dimension M, then the system is controllable. In saying that a set of operators "spans" the space of Hermitian operators, we mean that any M-dimensional Hermitian operator can be written as a linear combination of the operators in the set. Since there are $M(M+1)/2$ linearly independent M-dimensional Hermitian operators, for the system to be controllable the set of operators obtained by taking repeated commutators of $\{H_0, H_j\}$ must contain at least $M(M+1)/2$ distinct operators (including the identity, which we are always free to add to the set).

Preparing states or realizing unitary transformations by varying control parameters is often called *open-loop* control, and we use this terminology here. The purpose of this terminology is to distinguish it from feedback control: as discussed in Chapter 5, feedback control is called "closed-loop" control because the information flow from the system to controller and back to the system is thought of as a loop. Time-dependent control of the kind we consider here is called open-loop because there is no feedback, and thus no loop.

8.1.1 Fast state-swapping for oscillators

We now consider an example of using a numerical search to find an open-loop control protocol. The problem is that of transferring the information stored in one resonator into another, and doing so in the shortest time. We usually want to perform quantum control processes as fast as possible because of the ever-present decoherence and/or thermalization that results from interactions with the environment – the longer the process takes, the more the effects of decoherence.

The simplest way to couple two harmonic oscillators together is usually via a linear interaction, proportional to the positions of each oscillator. Even in the case of the nonlinear optomechanical interaction between an optical cavity mode and a mechanical resonator, it is the effective linear interaction that results from driving the cavity mode that provides the faster (stronger) interaction (see Chapter 7).

If we have two oscillators labeled A and B, linearly coupled, whose annihilation operators are a and b respectively, then their joint Hamiltonian is

$$H = \hbar\omega a^\dagger a + \hbar g(a + a^\dagger)(b + b^\dagger) + \hbar\Omega b^\dagger b, \tag{8.8}$$

where ω and Ω are the frequencies of the respective oscillators and g is the strength (or rate) of the coupling. It turns out that if we modulate the coupling rate g at the frequency $\nu = |\Omega - \omega|$, limit the maximum value of g to be small (slow) compared to the frequencies ω and Ω, and make the rotating wave approximation (see Section 7.2 or Appendix G), then the Hamiltonian that results, namely,

$$H = \hbar\omega a^\dagger a + \hbar g(ab^\dagger e^{-i(\Omega-\omega)t} + ba^\dagger e^{i(\Omega-\omega)t}) + \hbar\Omega b^\dagger b, \tag{8.9}$$

realizes a perfect state-swap between the two oscillators in a time $\tau = \pi/(2g)$. By this we mean that the joint unitary implemented at the time τ takes every initial state of the form $|\psi\rangle|\phi\rangle$, where $|\psi\rangle$ is a state of resonator A, and $|\phi\rangle$ of resonator B, and maps it to $|\phi\rangle|\psi\rangle$.

While we can realize a swap in the above way, the speed of the swap is limited by the requirement that $g \ll \min(\omega, \Omega)$. It turns out that the swap can be realized at a much faster rate by using the original Hamiltonian, Eq. (8.8), and varying g and the frequency of one of the oscillators with time. It is also useful to broaden our definition of a swap to a unitary that maps any initial state $|\psi\rangle|\phi\rangle$ to the state $U_a|\phi\rangle U_b|\psi\rangle$, where U_a and U_b are fixed unitaries that act only on the first and second resonators, respectively. If we define the swap in this way, then it still achieves a full transfer of information, up to a change of the basis in which this information is encoded. In this case it is sufficient to vary only g, because the terms in the Hamiltonian proportional to the oscillators' frequencies only change the states of each oscillator separately (that is, they adjust U_a and U_b).

Numerical optimization

To realize a fast swap, and determine the fastest swap time, we pick a time T for which we will evolve the system, and divide this time into M segments of duration $\Delta t = T/M$. We make g constant on each segment, but allow it to be different from segment to segment, so that g is piecewise-constant. We denote the value of g on the mth segment by g_m. To simulate the evolution of the oscillators, we must truncate the state space so that it has a finite dimension. We choose the finite basis for each oscillator to be given by the number states $|n\rangle$, for $n = 0, 1, \ldots, N-1$. We choose the initial state of oscillator A to be the state that is completely mixed in the space spanned by $\{|0\rangle, |1\rangle, \ldots, |J\rangle\}$, with $J < N-1$. This is the state $\rho = (1/J)\sum_{n=1}^{J} |n\rangle\langle n|$. We choose the initial state of oscillator B to be pure. This choice of initial states is a simple way to capture the extent to which a swap has been realized: the state of A has been swapped into B only if the final state of A is pure. If the final state of A is pure, then the unitary must successfully swap every state in the space spanned by $\{|0\rangle, |1\rangle, \ldots, |J\rangle\}$. We can therefore search for the set of values $\{g_m\}$ that achieves a swap in time T by searching for the set that maximizes the purity of oscillator A.

To find the fastest time in which the swap can be performed, we run a gradient search for each of a range of values of T so as to determine the smallest value of T for which one can obtain a purity arbitrarily close to unity. Gradient search methods involve calculating the gradient of the function to be maximized, and using it to decide on a direction (a line) along which to climb to reach a local maximum in that direction. Once a local maximum of the function along this line has been found, the gradient is calculated at this new point, and a new direction is chosen. Different gradient search methods vary in the way they choose the search direction. The most naive method is to choose this direction to be the one in which the function increases fastest, which is the direction of the gradient vector itself. This is often not the best choice however, and a number of more sophisticated methods have been developed. To obtain the results presented here we used the BFGS method, named after Broyden, Fletcher, Goldfarb, and Shanno [451].

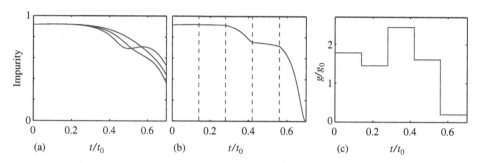

Figure 8.1 Here we show a control protocol in which the linear coupling rate g between two oscillators is varied with time in order to swap the states of the two in a time of 0.7 of the oscillation period, t_0. Oscillator A starts in a mixed state, while the state of oscillator B is pure. Plot (b) shows the evolution of the impurity of oscillator A under the control protocol, and (c) shows the way in which g is varied with time. For comparison, plot (a) displays the evolution of the impurity of oscillator A when g is held constant, for $g = 1.8g_0, 2g_0,$ and $2.2g_0$, where $g_0 = 1/t_0$.

In performing the numerical search it is important to remember that some minimum number of segments will be required to achieve the desired unitary. The more segments we use, the larger the search space and the longer the search takes, but if we use too few there will be no protocol to find. In Fig. 8.1 we show the results of a numerical search for a control protocol with five segments, and a total time of $t = 0.7t_0$, where $t_0 = 2\pi/\omega$ is the period of oscillator A. For simplicity in this example we have set $\Omega = \omega$. The resulting control protocol performs an essentially perfect swap, reducing the impurity of oscillator A to zero as shown in Fig. 8.1b. The time-dependent interaction strength that performs the swap is shown in Fig. 8.1c. For comparison we display in Fig. 8.1a the evolution of the impurity when g is constant, for three values of g.

8.1.2 Preparing non-classical states

Here we give a number of examples of the use of open-loop control, in which some parameters in a Hamiltonian are varied with time to prepare highly non-classical states. These states are useful in experiments to probe the predictions of quantum mechanics and explore the quantum-to-classical transition (see Section 4.4), and have potential applications in metrology (see Chapter 6). Since open-loop control applied to a single quantum system cannot usually reduce a system's entropy, we assume that some cooling procedure has been used first to prepare the system in a pure state, usually the ground state. The task is then to transform this state into one that has highly non-classical features. Two features of particular interest are (i) the ability to exist in a superposition of states that can be clearly distinguished by the experimenter, and (ii) the ability of two separated systems to exist in an "entangled state," in which the correlations between observables of the two systems cannot be explained by a local classical theory. In creating states with the first feature, one seeks to produce superpositions of states that are as distinct as possible. For example, if our goal were to place an object in a superposition of two different positions, then we would try to make the distance between them as large as possible, and the mass of the object as large as possible. In this way we push the boundaries between quantum mechanics and the classical macroscopic world.

Creating a superposition of mesoscopic states

In this example we will use a superconducting circuit to place a mechanical resonator in a superposition of two coherent states. Consider an LC-oscillator coupled to a mechanical resonator as in Fig. 7.5, but with one change: a Josephson junction (JJ) is placed in series with the inductor and capacitor, adding a nonlinear inductance. The Hamiltonian for the superconducting oscillator is then given by Eq. (7.65), and we give it again here for convenience:

$$H = \frac{q^2}{2C_J} - E_J \cos\left[\frac{2\pi}{\phi_0}(\phi + \Phi)\right] + \frac{\phi^2}{2L}. \tag{8.10}$$

If we write this Hamiltonian in terms of the annihilation and creation operators defined for the oscillator when there is no JJ (see Eq. 7.26) then we have

$$H = \hbar\Omega a^\dagger a + \hbar\lambda \cos(k(a + a^\dagger) + \Phi), \tag{8.11}$$

for two constants λ and k.

We can now choose the parameters of the JJ so that, for the energy range (photon numbers) that we will have in our system, only the first three terms in the Taylor series expansion for the cosine will be important. In expanding the cosine in a power series, we obtain terms that are products of powers of a and $a^\dagger$. If we move into the interaction picture with respect to the harmonic oscillator Hamiltonian $H_0 = \hbar\Omega a^\dagger a$, then the annihilation and creation operators oscillate at frequency Ω: $a(t) = ae^{-i\omega t}$. If λ is small compared to Ω (and assuming $k \lesssim 1$), then any term in the power series that does not contain an equal number of a and $a^\dagger$ operators will average to zero on the timescale at which these terms generate motion. As a result, the rotating-wave approximation allows us to drop all terms in the power series that cannot be written as powers of $a^\dagger a$.

Choosing $\Phi = \pi$ so that the bottom of the cosine is at the minimum of the parabolic potential, expanding to cosine in a Taylor series, dropping constant terms, and making the rotating-wave approximation ($\lambda \ll \Omega$), the Hamiltonian becomes

$$H = \hbar\Omega a^\dagger a + \hbar\lambda\left(\frac{k^2}{2}\right)(a+a^\dagger)^2 + \hbar\lambda\left(\frac{k^4}{24}\right)(a+a^\dagger)^4 + \cdots$$

$$\approx \hbar(\Omega + \lambda k^2)a^\dagger a + \hbar\lambda\frac{k^4}{4}(a^\dagger a)^2 \equiv \hbar\tilde\Omega a^\dagger a + \hbar\chi(a^\dagger a)^2. \tag{8.12}$$

Finally, including the capacitative interaction with the mechanical resonator, the Hamiltonian of the two interacting resonators is

$$H = \hbar\tilde{\Omega}a^\dagger a + \hbar\chi(a^\dagger a)^2 + \hbar g a^\dagger a(b + b^\dagger) + \hbar\omega b^\dagger b, \tag{8.13}$$

where ω is the mechanical frequency and b is the mechanical annihilation operator. The term containing $(a^\dagger a)^2$ is called a "$\chi^{(3)}$" (pronounced "kai-three") nonlinearity.

Our goal here is to prepare the resonator in a superposition of two spatially separated coherent states. The latter are quantum states of a single particle that have a well-defined position and momentum, at least to the extent that this is possible in quantum mechanics, and are defined in Appendix D. We will denote a coherent state by $|\alpha\rangle$ where α is a complex number. The coherent state is localized in phase space around the (dimensionless) mean position $\langle b + b^\dagger\rangle = 2\text{Re}[\alpha]$ and the dimensionless mean momentum $-i\langle b - b^\dagger\rangle = 2\text{Im}[\alpha]$. A superposition of coherent states is often referred to as a "Schrödinger cat" state, or even just as a "cat."

The technique we will use to prepare the mechanical resonator in the desired state is to use the nonlinearity in the superconducting resonator to first create this state in the latter. We will then use the interaction to "swap" the states of the two resonators, which places the cat state in the mechanical resonator.

We begin by preparing the superconducting resonator in a coherent state $|\alpha\rangle$. This is achieved by first allowing it to settle into its ground state, and then driving it with a classical resonant voltage signal. This is performed on a timescale fast compared to the rate χ, so that the nonlinearity does not have time to modify the coherent state while it is being created.

The second step is to allow the coherent state to sit in the superconducting resonator for the time needed for the nonlinearity to turn it into a superposition of two coherent states. This time is $\tau = \pi/(2\chi)$ because

$$\exp\left[-i\chi\tau(a^\dagger a)^2\right]|\alpha\rangle = \frac{1}{\sqrt{2}}\left[|\alpha\rangle - i|-\alpha\rangle\right]. \tag{8.14}$$

The third and final step is to drive the superconducting resonator at the right frequency to create an effective resonant linear interaction between the two resonators. This linear interaction will swap the states of the resonators when it acts for a given time. We showed how to create this effective linear interaction in Section 7.5. The superconducting resonator is driven at the frequency $\nu = \Omega - \omega$, and the resulting effective interaction Hamiltonian in the interaction picture is

$$H_I = \hbar\lambda(ab^\dagger + ba^\dagger). \tag{8.15}$$

This Hamiltonian is valid under the rotating-wave approximation assuming that $\lambda \ll \omega$, where $\lambda = g\sqrt{\langle a^\dagger a\rangle_{ss}}$ and $\langle a^\dagger a\rangle_{ss}$ denotes the steady-state average number of photons in the superconducting cavity. Allowing H_I to act for a time $\tau_{swap} = \pi/(2\lambda)$ swaps the states of the superconducting and mechanical resonators. The cat state is loaded into the mechanical resonator and at the same time the initial thermal state of the mechanical resonator is dumped into the superconducting resonator.

While we perform steps 1 and 2 the two resonators are coupled via the nonlinear interaction. We are able to ignore its effect during these steps by choosing g to be sufficiently small. The linear interaction is then made sufficiently strong (so that the swap is fast) by strongly driving the superconducting resonator so that $\sqrt{\langle a^\dagger a\rangle_{ss}} \gg 1$, and thus $\lambda \gg g$.

Creating an entangled state between a qubit and a mechanical resonator

We now show how to create a state of the form

$$|\psi\rangle = \frac{1}{\sqrt{2}}\left(|0\rangle_q|\alpha\rangle_r + e^{-i\phi}|1\rangle_q|-\alpha\rangle_r\right) \tag{8.16}$$

where $|0\rangle_q$ and $|1\rangle_q$ are two orthonormal states of a qubit, and $|\alpha\rangle_r$ denotes a coherent state of a resonator with amplitude α. Note that the coherent state with amplitude $\alpha = 0$ is also the ground state of the resonator.

When the two systems are in the joint state $|\psi\rangle$ various observables of the two systems are correlated. This correlation results purely from the superposition and not from any mixing (as it would in a classical system), and so the joint state is pure. This kind of state has correlations that cannot be replicated in classical systems, and is referred to as being *entangled*. We do not discuss the property of entanglement in any detail in this text, but the interested reader can find extensive treatments in [447, 631, 484]).

First we consider a superconducting qubit (a Cooper-pair box or transmon) coupled to a mechanical resonator as described in Section 7.5. The Hamiltonian is H, where

$$\frac{H}{\hbar} = \Delta\sigma_z + \lambda\sigma_x + g\sigma_z(b + b^\dagger) + \omega b^\dagger b. \tag{8.17}$$

Here g is the coupling rate between the qubit and the resonator, b is the resonator's annihilation operator, and ω is its frequency. To begin we prepare the qubit and the oscillator separately in their respective ground states. For the qubit this is simple, since it is in its ground state at the temperature produced by a dilution refrigerator (~ 20 mK). To prepare the resonator in the ground state we must use a more sophisticated cooling scheme, such as one of those described below in Section 8.3. Here we merely assume that we are able to prepare both systems in their ground states.

To best see how to prepare the state we want, it is useful to change the basis of Pauli operators that we use to describe the qubit. Since the Pauli operators correspond to orthogonal directions in space, they act like the components of a three-dimensional vector. If we rotate the x and z axes by $-\theta$ in the xz-plane, then the two new Pauli operators that correspond to the new axes are

$$\tilde{\sigma}_z = \cos\theta\sigma_z + \sin\theta\sigma_x, \tag{8.18}$$

$$\tilde{\sigma}_x = -\sin\theta\sigma_z + \cos\theta\sigma_x. \tag{8.19}$$

If we choose $\cos\theta = \Delta/\sqrt{\Delta^2 + \lambda^2}$, then $\Delta\sigma_z + \lambda\sigma_x = \sqrt{\Delta^2 + \lambda^2}\,\tilde{\sigma}_z$, and the Hamiltonian is

$$H = \hbar\Lambda\tilde{\sigma}_z + \hbar g(\cos\theta\,\tilde{\sigma}_z - \sin\theta\,\tilde{\sigma}_x)(b + b^\dagger) + \hbar\omega b^\dagger b, \qquad (8.20)$$

where we have defined $\Lambda^2 \equiv \Delta^2 + \lambda^2$. The two eigenstates of $\tilde{\sigma}_z$ have eigenvalues ± 1. We will denote the positive eigenstate of $\tilde{\sigma}_z$ by $|1\rangle_q$ and the negative one by $|0\rangle_q$, and we note that the latter is the ground state of the qubit.

We now modulate the interaction strength g at the frequency of the resonator, ω, and move into the interaction picture with respect to the Hamiltonian $H_0 = \hbar\Lambda\tilde{\sigma}_z + \hbar\omega b^\dagger b$. The Hamiltonian in the interaction picture is then

$$H_I = \hbar g\cos\theta\cos(\omega t)\tilde{\sigma}_z(be^{-i\omega t} + b^\dagger e^{-i\omega t})$$
$$- \hbar g\sin\theta\cos(\omega t)(\cos(\Lambda t)\tilde{\sigma}_x + \sin(\Lambda t)\sigma_y)(be^{-i\omega t} + b^\dagger e^{-i\omega t}). \qquad (8.21)$$

For this nano-electromechanical system we typically have the timescale separation $g \ll \omega \ll \Lambda$. We now exploit this fact to make the rotating-wave approximation described in Appendix G: since oscillations at the frequencies ω and Λ are fast on the timescale of the changes induced by the interaction at rate g, any terms in the interaction Hamiltonian that oscillate at frequency ω or higher will cancel themselves out over a time of $1/g$. Dropping all the oscillating terms from H_I we obtain the effective interaction Hamiltonian

$$H_I = \hbar f\tilde{\sigma}_z(b + b^\dagger), \qquad f = \frac{g\Delta}{2\sqrt{\Delta^2 + \lambda^2}}. \qquad (8.22)$$

To begin the protocol, we set $g = 0$ and increase λ so as to rotate the qubit to the state $|+\rangle_q = (|0\rangle_q + e^{-i\phi}|1\rangle_q)/\sqrt{2}$. To see how this works, upon changing $\lambda \to \lambda + \delta$, the Hamiltonian for the qubit (now back in the Schrödinger picture, since $g = 0$) is

$$H_{qb} = \hbar\Lambda\tilde{\sigma}_z + \hbar\delta\sigma_x = \hbar a\tilde{\sigma}_z + \hbar b\tilde{\sigma}_x, \qquad (8.23)$$

where $a = \Lambda + \delta\sin\theta$ and $b = \delta\cos\theta$. In the Bloch-sphere representation this Hamiltonian rotates the qubit state around an axis that lies in the $\tilde{x}\tilde{z}$-plane, at an angle ψ to the $\tilde{z}$-axis, where $\cos\psi = a/\sqrt{a^2 + b^2}$. If we let this Hamiltonian act for the right amount of time, it will rotate the initial state $|0\rangle_q$ down to the equator of the Bloch sphere (the $\tilde{x}\tilde{y}$-plane), and thus to a state of the form $|\psi\rangle_q = (|0\rangle_q + e^{-i\phi}|1\rangle_q)/\sqrt{2}$ for some value of ϕ.

Once we have prepared the state $|\psi\rangle_q$, we return λ to its original value so that the qubit Hamiltonian is $H_{qb} = \hbar\Lambda\tilde{\sigma}_z$. If g is still zero then the state $|\psi\rangle_q$ stays on the equator of the Bloch sphere, evolving as $|\psi(t)\rangle_q = (|0\rangle_q + e^{-i(\phi+\Lambda t)}|1\rangle_q)/\sqrt{2}$.

We now turn on the modulated interaction so as to obtain the Hamiltonian H_I in Eq. (8.22). To determine the state of the qubit and resonator generated by this Hamiltonian we can use the fact that the states $|0\rangle_q$ and $|1\rangle_q$ are eigenstates of the $\tilde{\sigma}_z$ operator. When applying a function of an operator to an eigenstate we can replace the operator with the corresponding eigenvalue. Denoting the ground state of the resonator by $|0\rangle_r$, we have

$$|\Psi(t)\rangle = \exp[-iH_1 t/\hbar]|\psi\rangle_q|0\rangle_r = \exp[-ift\tilde{\sigma}_z(b+b^\dagger)]\frac{(|0\rangle_q + e^{-i\phi}|1\rangle_q)}{\sqrt{2}}|0\rangle_r$$

$$= \frac{\exp[ift(b+b^\dagger)]|0\rangle_q|0\rangle_r}{\sqrt{2}} + \frac{e^{-i\phi}\exp[-ift(b+b^\dagger)]|1\rangle_q|0\rangle_r}{\sqrt{2}}. \tag{8.24}$$

Since the operator $e^{\alpha b - \alpha^* b^\dagger}$ transforms the resonator ground state to a coherent state with amplitude α (see Appendix D), we have

$$|\Psi(t)\rangle = \frac{1}{\sqrt{2}}[|0\rangle_q|{-}ft\rangle_r + e^{-i\phi}|1\rangle_q|ft\rangle_r]. \tag{8.25}$$

Engineering resonator states with a qubit

It is possible to create any pure state of an oscillator with a finite number of quanta by repeatedly interacting the oscillator with a qubit. This was shown by Law and Eberly [357] and we now describe the scheme they devised to do it. Consider a single qubit interacting with a quantum oscillator, examples of which are the Hamiltonians in Eqs. (7.62) and Eq. (7.72). If we take the former, a superconducting qubit interacting with an LC-resonator, write it in terms of the resonator annihilation operator (defined in Eq. (7.27)) and make the rotating-wave approximation, then the Hamiltonian is H, where

$$\frac{H}{\hbar} = \Delta\sigma_z + \lambda\sigma_x - i\mu(\sigma^\dagger a - \sigma a^\dagger) + \omega a^\dagger a, \tag{8.26}$$

where $\mu = g\sqrt{C/(8\hbar L)}$ and g is defined in Eq. (7.61). We now modulate μ at the frequency $\nu = \Delta - \delta - \omega$, move into the interaction picture, and continue to make the rotating-wave approximation assuming $\mu \ll \omega$. The Hamiltonian becomes

$$H_I = H_I^q + H_I^{int} \tag{8.27}$$

with

$$H_I^q = \hbar\delta\sigma_z + \hbar\lambda\sigma_x, \tag{8.28}$$

$$H_I^{int} = -i\hbar\mu(\sigma^\dagger a - \sigma a^\dagger). \tag{8.29}$$

To develop the method of Law and Eberly, we first consider an arbitrary state of the resonator, $|\psi_0\rangle$, with a maximum number of quanta equal to M. With the qubit in its ground state, which we denote by $|0\rangle_q$, the joint state of the two systems is

$$|\psi_0\rangle = \sum_{m=0}^{M} c_m|m\rangle_r|0\rangle_q, \tag{8.30}$$

where $|m\rangle_r$ denotes the Fock state with m photons. We now examine a two-step operation on the joint system that will transform $|\psi_0\rangle$ to another state that has at most $M-1$ photons. The reason this operation is useful will become clear below. In the first step of the operation we will set $\delta = \lambda = 0$, and thus apply the Hamiltonian H_I^{int} for a time τ_a. In the second step we will set $\mu = 0$ and apply H_I^q for a time τ_b.

In the first step the interaction Hamiltonian H_I^{int} couples the state $|M\rangle_r|0\rangle_q$ to $|M-1\rangle_r|1\rangle_q$, and does not couple these two states to any others. This is because the interaction can take a photon out of the resonator only by raising the qubit to its excited state. We can therefore calculate analytically the 2×2 evolution operator for these two states. If we write the coefficients of the states $|M\rangle_r|0\rangle_q$ and $|M-1\rangle_r|1\rangle_q$ in the initial state $|\psi_0\rangle$ as the column vector $(c_M, c_{M-1})^T$, then the evolution of this vector over the time τ_a is given by the unitary matrix

$$U_0^{(M)} = \begin{pmatrix} \cos\left(\mu\sqrt{M}\tau_a\right) & \sin\left(\mu\sqrt{M}\tau_a\right)/\sqrt{M} \\ -\sin\left(\mu\sqrt{M}\tau_a\right)/\sqrt{M} & \cos\left(\mu\sqrt{M}\tau_a\right) \end{pmatrix}. \tag{8.31}$$

Choosing μ such that

$$\cos\left(\mu\sqrt{M}\tau_a\right)c_M + \left[\sin\left(\mu\sqrt{M}\tau_a\right)/\sqrt{M}\right]c_{M-1} = 0, \tag{8.32}$$

then after the evolution the coefficient of $|M\rangle_r|0\rangle_q$ is zero.

The 2×2 unitary $U_0^{(M)}$ is a sub-block of the $2(M+1) \times 2(M+1)$ unitary, U_0, that acts on $|\psi\rangle$. The state $U_0|\psi_0\rangle$ no longer has M quanta, but we note that in general it contains the states $|m\rangle_r|1\rangle_q$, for $m = 0, 1, \ldots, M-1$, in which the qubit is in its excited state.

In the second part of our operation, we apply a Hamiltonian that acts only on the qubit. Consider the coefficients of the states $|M-1\rangle_r|0\rangle_q$ and $|M-1\rangle_r|1\rangle_q$, which we will denote by c_{M-1} and d_{M-1}. If we write these coefficients as the column vector $(c_{M-1}, d_{M-1})^T$, then the Hamiltonian transforms this vector exactly as it does the state vector of a single qubit. We can therefore use the Hamiltonian to transform $(c_{M-1}, d_{M-1})^T$ to $(1, 0)$, thus reducing to zero the coefficient of $|M-1\rangle_r|1\rangle_q$. This can be achieved by applying the Hamiltonian H_I^q twice, with two different choices of δ and λ. First we set $\lambda = 0$ and rotate the qubit state around the z-axis on the Bloch sphere, until it lies in the yz-plane. We then set $\delta = 0$ so as to rotate the vector around the x-axis, which allows use to rotate it to the σ_z eigenstate $|0\rangle_q$.

Let us now summarize what our two-step operation has achieved. The first part eliminates the coefficient c_M, and in doing so places some population in the state $|M-1\rangle_r|1\rangle_q$, in which the qubit is in its excited state. The second part eliminates the coefficient of this state. As a result, the final joint state of the resonator and qubit has the form

$$|\psi_1\rangle = a_{M-1}|M-1\rangle_r|0\rangle_q + \sum_{m=0}^{M-2} a_m|m\rangle_r|0\rangle_q + b_m|m\rangle_r|1\rangle_q, \tag{8.33}$$

for some coefficients $\{a_m\}$ and $\{b_m\}$. If we now examine the state $|\psi_1\rangle$, we see that the same two-step operation can be applied to this state, with different choices of μ, δ, and λ, so as to again reduce the value of M by 1, obtaining a state of the form

$$|\psi_2\rangle = e_{M-2}|M-2\rangle_{\rm r}|0\rangle_{\rm q} + \sum_{m=0}^{M-3} e_m|m\rangle_{\rm r}|0\rangle_{\rm q} + f_m|m\rangle_{\rm r}|1\rangle_{\rm q}, \tag{8.34}$$

for some coefficients $\{e_m\}$ and $\{f_m\}$. It is clear now that repeating the operation M times will reduce the resonator to its ground state. If we denote the joint state after j operations by $|\psi_j\rangle$, then we have $|\psi_M\rangle = |0\rangle_{\rm r}|0\rangle_{\rm q}$. If we denote the unitary that takes us from the state $|\psi_j\rangle$ to state $|\psi_{j+1}\rangle$ by V_{j+1}, then we have

$$|0\rangle_{\rm r}|0\rangle_{\rm q} = |\psi_M\rangle = V_M V_{M-1}\cdots V_1|\psi_0\rangle = \sum_{m=0}^{M} c_m|m\rangle_{\rm r}|0\rangle_{\rm q}. \tag{8.35}$$

Since we can calculate analytically the values of μ, δ, and λ required to generate each of the V_j, the above procedure provides an explicit protocol to transform an arbitrary state of the resonator to its ground state.

The protocol above also provides us with an explicit protocol to generate an arbitrary state of the resonator. By multiplying both sides of Eq. (8.35) by the product of the Hermitian conjugates of the V_j, we obtain

$$\sum_{m=0}^{M} c_m|m\rangle_{\rm r}|0\rangle_{\rm q} = V_1^{\dagger}\cdots V_{M-1}^{\dagger} V_M^{\dagger}|0\rangle_{\rm r}|0\rangle_{\rm q}. \tag{8.36}$$

That is, since any unitary can be performed in reverse, all we have to do is prepare the two systems in their ground states, and then apply the reverse protocol to generate an arbitrary state of the resonator. The above equation is the Law–Eberly protocol.

To perform the Law–Eberly protocol we therefore first work out how to reduce the state we want to create to the ground state. The Hermitian conjugates of the unitaries required to do this are the operations we perform to generate the state. These conjugate operations are obtained simply by negating the values of μ, δ, and λ required to perform the forward operations.

8.2 Measurement-based feedback control

In Chapter 3 we introduced continuous measurements, and in Chapter 5 we introduced feedback control and discussed continuous time measurement-based feedback in quite some depth. In this section we discuss two applications of this feedback technique to mesoscopic systems. To fully understand this section you will need to have understood Sections 3.1.2, 3.1.4, and 3.7.2. The continuous measurement part of Section 5.3 may also be useful. Both the examples we consider here involve linear feedback control of a resonator, and both are simple applications of the optimal linear control protocol in Section 5.5.2, although the second has an additional twist.

8.2.1 Cooling using linear feedback control

Here we consider cooling a mechanical resonator via feedback from a continuous measurement of the position of the resonator. We measure the position by coupling the mechanical resonator to a superconducting resonator as described in Section 7.7.3. The circuit for this set-up is the same as that depicted in Fig. 7.7, for a cavity-mediated measurement of the position of a mechanical oscillator, with the sole addition of a means to apply a force to the mechanics.[1] We can apply a force to the resonator using the same kind of capacitive coupling that we use to couple a resonator to an LC circuit: we place an electrode on the resonator, and another close by so that there is a force between them. We then vary the force by varying the voltage on the electrode.

The stochastic master equation (SME) that describes the mechanics is

$$dp = -i[\omega b^\dagger b - f(t)x, \rho]\,dt - \tilde{k}[x, [x, \rho]]\,dt + \sqrt{2\eta\tilde{k}}\,(x\rho + \rho x - 2\langle x\rangle \rho)\,dW$$

$$- \frac{\gamma}{2}(n_T + 1)\left(a^\dagger a\rho + \rho a^\dagger a - 2a\rho a^\dagger\right)dt$$

$$- \frac{\gamma}{2}n_T\left(aa^\dagger \rho + \rho aa^\dagger - 2a^\dagger \rho a\right)dt. \tag{8.37}$$

Here b and x are respectively the annihilation and position operator for the mechanics, so that $x = \sqrt{\hbar/(2m\omega)}(b + b^\dagger)$, where ω and m are as usual the mechanical frequency and mass. The damping or thermalization rate is γ, and n_T is the average number of thermal phonons in the resonator when it is at the temperature of the bath. Recall from Chapter 4 that this master equation is only valid when the damping is weak, so that $\gamma \ll \omega$. The measurement strength is $\tilde{k}$, the efficiency of the measurement is η, and $f(t)$ is the time-dependent force that we will apply to the resonator. In order to cool the resonator we will choose the value of this force at time t to be a function of the state ρ at that time t. Since ρ is determined by the measurement results, this choice of f is a feedback protocol. (Recall from Chapter 3 that the noise increment, dW, that appears in Eq. (8.37) is determined in real time from the measurement record.)

To make our example concrete, we recall how the measurement strength k and efficiency η are determined by the physical parameters of the circuit elements. In Section 7.7.3 we derived the measurement strength k by assuming that the interaction between the mechanical resonator and the LC-oscillator/superconducting cavity had the form given in Eq. (7.93). By comparing this with the optomechanical interaction that we use in our circuit here, the operator A defined in Eq. (7.93) is in our case $A = b + b^\dagger = x/\Delta x$, a dimensionless version of x. Here $\Delta x = \sqrt{\hbar/(2\omega m)}$ is the standard deviation of the position of the mechanical resonator when it is in its ground state. By comparing Eq. (8.37) above with Eq. (7.119), and using $A = x/\Delta x$, we see that the relationship between k and our measurement strength $\tilde{k}$ is

$$k = (\Delta x)^2 \tilde{k}. \tag{8.38}$$

---

[1] We will use the term "mechanics" as shorthand for "mechanical resonator," since it is now in common usage.

Thus k is a rate constant, and is the measurement strength when we express the position in units of Δx.

From Eq. (7.111) the measurement strength (in units of Δx) is $k = 2(g\alpha)/\kappa$, where g is the optomechanical coupling constant, α^2 is the steady-state average number of photons in the LC-oscillaor, and κ is its damping rate. To go one step further the optomechanical coupling rate is given by $g = \sqrt{\hbar/(2m\omega)}\Omega\xi/(2d_0)$. This is often as far as we need to go in modeling an experiment. Here Ω is the frequency of the LC-oscillator, d_0 is the mean distance between the electrodes that create the capacitative coupling between the LC-oscillator and the mechanics, and ξ is a dimensionless "fudge factor" that accounts for the geometry of the electrodes. We will take concrete values for these parameters from the experiment reported in [465]. The mechanical resonator in this experiment is a circular membrane similar to the head of a drum, a configuration first introduced in [596]. The membrane is 100 nm thick and has a diameter of 15 μm. We use the fundamental ("drumhead") mode of the membrane as our mechanical resonator. The frequency of this mode is $f = (\omega/2\pi) = 10.5$ MHz, the membrane's mass is $m = 48$ pg, and the damping rate is $\gamma = 220$ s^{-1} (for the relationship between Hz and s^{-1} see Appendix G). With these parameters the standard deviation of the position of the resonator when it is in its ground state, often referred to as the *zero point motion*, is $\Delta x = \sqrt{\hbar/(2\omega m)} = 4.1$ fm.

The drumhead forms one plate of a capacitor whose gap is $d_0 = 50$ nm, and the geometrical fudge factor is not required ($\xi \approx 1$). The frequency of the LC-oscillator is $\Omega = 2\pi \times 7.5$ GHz $\equiv 47.1 \times 10^9$ s^{-1}, and the resulting optomechanical coupling constant is $g = \Delta x \Omega \xi/(2d_0) = 2\pi \times 200$ Hz $\equiv 1.26 \times 10^3$ s^{-1}. The LC-oscillator is coupled to a transmission line, to which energy damps at the rate $\kappa_{\text{out}} = 2\pi \times 275$ kHz. The LC-oscillator also experiences "internal damping" that does not go to the transmission line, and is therefore a source of measurement inefficiency. The internal damping rate is $\kappa_{\text{int}} = 2\pi \times 45$ kHz, and the total damping rate of the cavity is therefore $\kappa = \kappa_{\text{int}} + \kappa_{\text{out}} = 2\pi \times 320$ kHz. The resulting efficiency of the measurement is $\eta_{\text{out}} = \kappa_{\text{out}}/(\kappa_{\text{int}} + \kappa_{\text{out}}) = 0.86$. By driving the LC-oscillator we can easily obtain a large range of values for the steady-state number of photons, $|\alpha|^2$. If we set $|\alpha|^2 = 10^9$ the measurement rate is $k = 2(g\alpha)/\kappa = 68$ s^{-1}.

The transmission line is connected to an amplifier, and the measurement efficiency of this amplification stage depends on the combination of the losses in the transmission line and the extra noise added by the amplifier. As an example, the amplification efficiency for the experiment reported in [638] is $\eta_{\text{amp}} = 0.46$. Referring to Eq. (7.121), the total efficiency of the oscillator-mediated measurement is then $\eta = \eta_{\text{amp}}\eta_{\text{out}} = 0.4$.

It is clear that the way to cool the resonator is to apply a force that opposes its velocity so as to slow it down. But we have freedom as to how we choose the size of this force as a function of the speed of the resonator. Further, we don't know whether or not it is useful to have the force depend on the position of the resonator as well as its velocity.

Using an LQG control protocol

The LQG (linear, quadratic, Gaussian) control method described in Section 5.5.2 will tell us how we should choose the force so as to minimize the average of a quadratic function of the coordinates and the applied force. Since it turns out that in this case the optimal force is a linear function of the coordinates, the function that is minimized can be chosen to be the sum of the energy of the resonator and a fictitious potential energy associated with the applied force. The relative weighting that we give to the resonator energy and that of the applied force is arbitrary; the less weighting we give to the energy of the force, the larger the force will be in the resulting feedback protocol. Note that if we do not include the force in the function to minimize, then the optimization problem is trivial in the sense that we can always control the resonator arbitrarily well by applying an arbitrarily strong force. While we might ideally wish to know how to minimize the resonator energy given that there is some maximum force we can apply, LQG control theory does not answer that question.

To obtain the LQG control protocol for our problem, we first write the equation of motion for the mean values of the dynamical variables, including the feedback force:

$$\frac{d}{dt}\begin{pmatrix} \langle x \rangle \\ \langle p \rangle \end{pmatrix} = A \begin{pmatrix} \langle x \rangle \\ \langle p \rangle \end{pmatrix} + F\mathbf{u}. \tag{8.39}$$

Here A gives the equations of motion for the oscillator ($A_{12} = p/m$, $A_{21} = -m\omega^2$), and $\mathbf{u}$ is the vector containing any parameters that will be changed with time as part of the feedback protocol. In the language of control theory, the elements of $\mathbf{u}$ are referred to as the *control inputs*. The control inputs are a linear function of the mean values of the dynamical variables, and so we can write these as

$$\mathbf{u} = -G \begin{pmatrix} \langle x \rangle \\ \langle p \rangle \end{pmatrix}, \tag{8.40}$$

in which G is a matrix. Since we will apply a force to the system, the effect of this force can only appear in the equation of motion for the momentum, $\langle p \rangle$. This restriction is enforced by the matrix F, by setting some of its elements to zero so that the inputs cannot appear in the equation of motion for $\langle x \rangle$. Since G is arbitrary, setting

$$F = \begin{pmatrix} 0 & 0 \\ 0 & 1 \end{pmatrix} \tag{8.41}$$

does not prevent the force from being an arbitrary linear combination of the estimates $\langle x \rangle$ and $\langle p \rangle$.

To obtain the feedback protocol, which is now entirely determined by G, we must define the function we wish to minimize, f. The general form of this function is given in Eq. (5.166). Here we choose f to be a sum of the energy of the resonator and an equivalent fictitious energy for the inputs:

$$f = (x, p) Q \begin{pmatrix} x \\ p \end{pmatrix} + \mathbf{u}^{\mathsf{T}} R \mathbf{u}, \tag{8.42}$$

where

$$Q = \frac{1}{2}\begin{pmatrix} m\omega^2 & 0 \\ 0 & 1/m \end{pmatrix}, \quad R = \alpha^2 Q. \tag{8.43}$$

Here the constant α sets the relative weight given to minimizing the oscillator energy as opposed to the "energy" of the control force. Note that since we define F to be dimensionless, the units of the elements of the input vector $\mathbf{u}$ are those of the elements of the state vector $\mathbf{x}$ multiplied by a rate. Because of this α has units of time.

Now that we have defined the matrices A, F, G, Q, and R, the optimal steady-state LQG feedback protocol is

$$G = R^{-1}F^{\mathrm{T}}S_{ss}, \tag{8.44}$$

where S_{ss} is the solution to the set of quadratic equations

$$0 = -A^{\mathrm{T}}S_{ss} - S_{ss}A + S_{ss}FR^{-1}F^{\mathrm{T}}S_{ss} - Q. \tag{8.45}$$

Solving the above equation and calculating G is the subject of Exercise 23. The first row of G is zero, and the second row is

$$G_{21} = m\omega^2\left(\sqrt{1 + \frac{1}{(\omega\alpha)^2}} - 1\right), \tag{8.46}$$

$$G_{22} = \omega\sqrt{\frac{1}{(\omega\alpha)^2} + 2\sqrt{1 + \frac{1}{(\omega\alpha)^2}} - 2}. \tag{8.47}$$

We can now solve the equations of motion for the oscillator, including the feedback, and determine the phase space distribution for the position and momentum in the steady state. To do this, we first obtain the equations of motion for the means and variances of the dynamical variables x and p. These can be derived from the stochastic master equation, Eq. (8.37), or by referring to the equations for a general linear system given in Section 5.5.2. In our case the equations of motion for the means are

$$d\langle x\rangle = \frac{\langle p\rangle}{m}dt - \frac{\gamma}{2}\langle x\rangle dt + \sqrt{8\eta\tilde{k}}V_x dW, \tag{8.48}$$

$$d\langle p\rangle = -(m\omega^2 + G_{21})\langle x\rangle dt - \left(\frac{\gamma}{2} + G_{22}\right)\langle p\rangle dt + \sqrt{8\eta\tilde{k}}C dW, \tag{8.49}$$

and those for the variances are

$$\dot{V}_x = \left(\frac{2}{m}\right)C - 8\eta\tilde{k}V_x^2 - \gamma(V_x - V_x^T), \tag{8.50}$$

$$\dot{C} = \left(\frac{1}{m}\right)V_p - m\omega^2 V_x - 8\eta\tilde{k}CV_x - \gamma C, \tag{8.51}$$

$$\dot{V}_p = -2m\omega^2 C - 8\eta\tilde{k}C^2 + 2\tilde{k}\hbar^2 - \gamma(V_p - V_p^T). \tag{8.52}$$

Here $C = \langle xp + px \rangle/2 - \langle x \rangle \langle p \rangle$ is the symmetrized "covariance" of x and p.

From the equations of motion for the variances it is clear that the effect of G_{21} is to modify the frequency of the oscillator. Because of this, in what follows we will absorb G_{21} into the definition of ω. The effect of G_{22} is to add a frictional damping term into the equation of motion for $\langle p \rangle$, and from now on we will denote G_{22} by Γ. We will allow Γ to be arbitrary, and in so doing we can determine the performance of all linear feedback protocols, not merely the optimal LQG strategy.

It is important to remember that the variances above are those of the state-of-knowledge of an observer who is continually receiving the information from the stream of measurement results. We refer to them as the *conditional* variances. Since the means of x and p will be randomly fluctuating, the total variances averaged over all possible trajectories that the system may take while it is being controlled are given by adding the variances of the means of x and p to the conditional variances. That is, if we denote the total variances by $\mathbb{V}_x$, $\mathbb{V}_p$, and $\mathbb{C}$, and the variances of the means by $V_{\bar{x}}$, $V_{\bar{p}}$, and C, then

$$\mathbb{V}_x = V_x + V_{\bar{x}}, \quad \mathbb{V}_p = V_p + V_{\bar{p}}, \quad \mathbb{C} = C + C. \tag{8.53}$$

We can derive the equations of motion of the variances of the means by first deriving the differential equations for $\langle x \rangle^2$, $\langle p \rangle^2$, and $\langle x \rangle \langle p \rangle$ from Eqs. (8.48) and (8.49). To do this we need to use Ito calculus (see Appendix C). Taking averages on both sides of the differential equations for the square means gives us the differential equations for the second moments of the means. From these we can obtain the equations of motion for the variances of the means, and these are

$$\frac{d}{dt}\begin{pmatrix} \tilde{V}_{\bar{x}} \\ \tilde{V}_{\bar{p}} \\ \tilde{C} \end{pmatrix} = -\begin{pmatrix} \gamma & 0 & -2\omega \\ 0 & \gamma + 2\Gamma & 2\omega \\ \omega & -\omega & \gamma + \Gamma \end{pmatrix}\begin{pmatrix} \tilde{V}_{\bar{x}} \\ \tilde{V}_{\bar{p}} \\ \tilde{C} \end{pmatrix} + 8k\begin{pmatrix} (\tilde{V}_x)^2 \\ (\tilde{C})^2 \\ \tilde{C}\tilde{V}_x \end{pmatrix}. \tag{8.54}$$

Here we have written the equations in terms of dimensionless (scaled) versions of the variances, defined by

$$\tilde{V}_x \equiv \frac{V_x}{(\Delta x)^2}, \quad \tilde{V}_p \equiv \frac{V_p}{(\Delta p)^2}, \quad \tilde{C} \equiv \frac{C}{\Delta x \Delta p}, \tag{8.55}$$

where Δx is defined above in Eq. (8.38), and $\Delta p = \sqrt{\hbar m\omega/2}$. These scaled variances are those of the scaled variables $\tilde{x} = x/\Delta x$ and $\tilde{p} = p/\Delta p$. Using the scaled variances simplifies the equations, and exposes the important rate constants in the dynamics. From now on, any variance with a tilde will indicate the dimensionless version of that variance (e.g., $\tilde{V}_x \equiv V_x/(\Delta x)^2$). The scaled versions of the thermal variances are

$$\tilde{V}_x^T = \tilde{V}_p^T = (1 + 2n_T) \equiv \tilde{V}^T. \tag{8.56}$$

The harmonic oscillator ground state has $\tilde{V}_x = \tilde{V}_p = 1$.

To calculate the total variances in the steady state we need to determine the steady states of both the conditional variances and the variances of the means. This can be done by setting the left-hand sides of the equations of motion to zero, and solving the resulting algebraic equations. There is a big difference between the differential equations for the conditional variances and those for the variances of the means: we have written the equations for the latter in matrix form because they are linear, whereas the equations for the former are not, and as they stand do not have analytic solutions.

If the harmonic oscillator has no damping, so that γ equals zero, then there is an analytic solution for the steady states of the conditional variances, and this is

$$\tilde{V}_x^0 = \frac{\sqrt{2}}{\sqrt{\eta(\xi+1)}}, \quad \tilde{V}_p^0 = \xi \tilde{V}_x^0, \quad \tilde{C}^0 = \frac{\sqrt{\xi-1}}{\sqrt{\eta(\xi+1)}}, \tag{8.57}$$

where

$$\xi = \sqrt{1+\eta r^2}, \quad r = \frac{8k}{\omega}. \tag{8.58}$$

We see from these solutions that if we want to keep the oscillator close to the ground state, for which $\tilde{V}_x = \tilde{V}_p = 1$, then ξ must be close to unity, and thus $(8k)^2 \ll \omega^2$ (assuming that $\eta \sim 1$).

While we cannot obtain an analytic solution for the steady states of the conditional variances for all values of γ, we note that a very useful regime is one in which the damping is weak compared to the measurement, so that $\gamma \ll k$ and $\gamma n_T \ll k$. In this case the feedback can keep the oscillator very close to the ground state. For simplicity we will also assume that $\gamma(n_T + 1) \ll \omega$. In this regime we can obtain an approximate solution for the steady states of the conditional variances, valid to first order in the small parameter $\varepsilon = \gamma/k \sim \gamma n_T/\omega$. To do this we write the steady-state variances as $\tilde{V}_x^0 + c_x\varepsilon$, $\tilde{V}_p^0 + c_p\varepsilon$, and $\tilde{C}^0 + c\varepsilon$, and write the small parameters as proportional to ε. We then expand the equations for the steady states to first order in ε to determine c_x, c_p, and c. The result is

$$\tilde{V}_x^{\text{ss}} = \tilde{V}_x^0 + \left(\frac{\gamma}{8\eta k}\right) \left(\frac{1+\eta r\tilde{V}_x^0/2}{1+\eta r\tilde{V}_x^0}\right) \frac{(\tilde{V}^T - \tilde{V}_x^0)}{\tilde{V}_x^0}, \tag{8.59}$$

$$\tilde{C}^{\text{ss}} = \tilde{C}^0 + \left(\frac{\gamma}{2\omega}\right) \frac{(\tilde{V}^T - \tilde{V}_x^0)}{1+\eta r\tilde{V}_x^0}, \tag{8.60}$$

$$\tilde{V}_p^{\text{ss}} = \tilde{V}_p^0 + \left(1 + \eta r\tilde{C}^0\right)(\tilde{V}_x^{\text{ss}} - \tilde{V}_x^0) + \eta r\tilde{V}_x^0(\tilde{C}^{\text{ss}} - \tilde{C}^0). \tag{8.61}$$

So far we have not assumed any particular relationship between k and ω. If we want to cool close to the ground state, then we need $(8k)^2 \ll \omega^2$ or $r^2 \ll 1$. Expanding the above expressions to second order in r, we have

$$\tilde{V}_x^{ss} = \tilde{V}_x^0 + \left(\frac{\gamma}{8\eta k}\right)\left(1 - \frac{1}{2}\sqrt{\eta}r\right)\left(\sqrt{\eta}[1 + \eta r^2/8]\tilde{V}^T - 1\right), \qquad (8.62)$$

$$\tilde{C}^{ss} = \tilde{C}^0 + \left(\frac{\gamma}{2\omega}\right)(1 - \sqrt{\eta}r)\left(\tilde{V}^T - \frac{[1 - \eta r^2/8]}{\sqrt{\eta}}\right), \qquad (8.63)$$

$$\tilde{V}_p^{ss} = \tilde{V}_p^0 + (\tilde{V}_x^{ss} - \tilde{V}_x^0) + \sqrt{\eta}r(\tilde{C}^{ss} - \tilde{C}^0). \qquad (8.64)$$

For efficient detection these equations simplify considerably, and we see more clearly the effects of the measurement and thermal noise:[2]

$$\tilde{V}_x^{ss} = \tilde{V}_x^0 + \left(\frac{\gamma}{4k}\right)\left[n_T\left(1 - \frac{4k}{\omega}\right) + (2n_T + 1)\left(\frac{2k}{\omega}\right)^2\right], \qquad (8.65)$$

$$\tilde{C}^{ss} = \tilde{C}^0 + \left(\frac{\gamma}{\omega}\right)\left[n_T\left(1 - \frac{8k}{\omega}\right) + \left(\frac{2k}{\omega}\right)^2\right], \qquad (8.66)$$

$$\tilde{V}_p^{ss} = \tilde{V}_p^0 + (\tilde{V}_x^{ss} - \tilde{V}_x^0) + r(\tilde{C}^{ss} - \tilde{C}^0). \qquad (8.67)$$

Calculating the steady-state variances of the means is straightforward because the equations of motion are linear, although the resulting expressions are rather cumbersome. Since we are focusing on the regime in which we can obtain good ground-state cooling, $k \gg \gamma$. In solving for the steady states of the variances of the means we find that these are only small when $G_{22} \equiv \Gamma \gg k$. This makes sense because the momentum noise from the back-action of the measurement is proportional to k, and it is the job of the feedback damping at rate Γ to counteract it. We therefore expand the solutions for the variances of the means in the small parameter $k/\Gamma \sim \varepsilon$, and keep all terms up to second order in ε. Note that we assume no particular relationship between Γ and ω. The resulting expressions for the variances of the means are

$$\tilde{V}_{\bar{x}}^{ss} = \frac{4k}{\Gamma}\left[\left(1 + \frac{\Gamma^2}{\omega^2}\right)(\tilde{V}_x^{ss})^2 + (\tilde{C}^{ss})^2 + 2\left(\frac{\Gamma}{\omega}\right)\tilde{C}^{ss}\tilde{V}_x^{ss}\right], \qquad (8.68)$$

$$\tilde{V}_{\bar{p}}^{ss} = \frac{4k}{\Gamma}\left[(\tilde{V}_x^{ss})^2 + (\tilde{C}^{ss})^2\right], \qquad (8.69)$$

$$\tilde{C}^{ss} = -\frac{4k}{\omega}(\tilde{V}_x^{ss})^2. \qquad (8.70)$$

We can see from these expressions that we cannot achieve good cooling if we make Γ too large. This is because our feedback force damps only the momentum, and so to confine

---

[2] Since we are expanding to first order in γ/k and second order in r, we should drop terms proportional to $(\gamma/k)r^2$, since they contribute no more than the other second-order terms that have already been dropped. We have left these in simply to show how the second-order terms in r^2 affect the solution.

the position as well as the momentum we need the oscillation of the oscillator to transform position into momentum (and vice versa) on a timescale at least as fast as the damping rate Γ.

Now we have the steady-state solutions for the conditional variances and the variances of the means, we can combine them to obtain the total variances as per Eq. (8.53). We now use the fact that $\langle b^\dagger b \rangle = \tilde{V}_x/4 + \tilde{V}_p/4 - 1/2$ to obtain the mean steady-state phonon number under the feedback control to second order in our small parameter:

$$\langle b^\dagger b \rangle = n_T \left(\frac{\gamma}{8k} - \frac{\gamma}{2\omega} \right) + \frac{4k^2}{\omega^2} + \frac{2k}{\Gamma} \left(1 + \frac{\Gamma^2}{2\omega^2} \right) \left(1 + n_T \frac{\gamma}{4k} \right). \tag{8.71}$$

It is now simple to find the value of Γ that minimizes $\langle b^\dagger b \rangle$. This optimal value is $\Gamma = \sqrt{2}\omega$, and the resulting steady-state phonon number is

$$\langle b^\dagger b \rangle = n_T \left(\frac{\gamma}{8k} - \frac{\gamma}{2\omega} \right) + \frac{4k^2}{\omega^2} + \frac{\sqrt{8}k}{\omega} \left(1 + n_T \frac{\gamma}{4k} \right). \tag{8.72}$$

Finally, we can ask what value of the measurement strength now gives us the minimal value of $\langle b^\dagger b \rangle$, and this will give us the maximum cooling that can be achieved with linear feedback control when the auxiliary interacts with the oscillator via position. Because we now have a term proportional to k/ω we can drop all second-order terms and obtain the optimal cooling to first-order in ε. The resulting optimal value of the measurement strength is

$$k = \sqrt{n_T \gamma \omega / (\sqrt{2}16)}, \tag{8.73}$$

and the optimal cooling is

$$\langle b^\dagger b \rangle = \sqrt{n_T} \sqrt{\frac{\sqrt{2}\gamma}{\omega}} = \sqrt{n_T} \sqrt{\frac{\sqrt{2}}{Q}}, \tag{8.74}$$

where $Q \equiv \omega/\gamma$ is the quality factor of the resonator. We will compare this optimal cooling below to the cooling that can be achieved using a simple coherent feedback protocol.

8.2.2 Squeezing using linear feedback control

We are now going to do something a little more clever, from a quantum mechanical point of view, than in the previous section. There we fixed the kind of measurement that we made on the oscillator, and asked what was the best control protocol, under the LQG quadratic constraint, given that measurement. Now we are going to choose the measurement more carefully to achieve a specific goal. To explain our goal we need the concepts of the amplitude and phase quadratures of an oscillator, described in Appendix D. To summarize, the amplitude and phase quadratures are scaled versions of the time-dependent position and momentum operators under the oscillator Hamiltonian $H_0 = \hbar \omega a^\dagger a$. In particular, the amplitude quadrature is $X \equiv \tilde{x}(t)/2$, and the phase quadrature is $Y \equiv \tilde{p}(t)/2$, where $\tilde{x}$ and $\tilde{p}$ are defined below Eq. (8.55). If the oscillator starts in a coherent state in which the average momentum is zero, then the "x" quadrature is a linear approximation to the amplitude of the oscillation, and the "p" quadrature is a linear approximation to the phase of the oscillation. The Heisenberg uncertainty relation for the quadratures is $V(X)V(Y) \geq 1/16$. Under the oscillator Hamiltonian, because of their definition, the quadrature operators are time-independent (all the time-dependence due to simple oscillatory motion has been absorbed into their definition). The quadrature operators are constant under simple oscillatory motion for precisely the same reason that the amplitude and phase are time-independent. Of course, if the oscillator is driven by an external force, then the amplitude and phase will change with time, and thus the quadratures also change with time.

If we measure one of the quadratures, then because they are unaffected by the free evolution, we can narrow the distribution of one quadrature at the expense of the other, and this narrowing is maintained as the oscillator evolves. This is a quantum non-demolition measurement, as described in Appendix D. A state in which one quadrature has variance less than 1/4 is referred to as a *squeezed* state. When we measure one quadrature we can reduce its variance below 1/4, but this is the *conditional* variance, the variance we have because we are continually processing information provided by the measurement. While the conditional variance is squeezed, the means still fluctuate and these fluctuations eliminate the squeezing for any observer who is not processing the measurement results. If we want to produce an unconditional squeezed state, then we must use feedback to reduce the fluctuations of the means. This is our goal here.

The basic interaction we have at our disposal is proportional to $\tilde{x}$. While we cannot easily obtain a coupling proportional to the quadrature $X(t)$, we can do something that is almost as good. If we modulate the strength of the coupling at the frequency of the resonator, ω, then the interaction Hamiltonian is

$$H_{\text{int}} = g \cos(\omega t) A \tilde{x}, \tag{8.75}$$

in which A is an operator of the probe system. We now move into the interaction picture with respect to the oscillator, and the interaction Hamiltonian becomes

$$H_{\text{int}} = g \cos(\omega t) A \tilde{x}_{\text{I}}(t) = \hbar g \cos(\omega t) A \left[\tilde{x} \cos(\omega t) + \frac{\tilde{p} \sin(\omega t)}{m \omega} \right]$$

$$= \left(\frac{g}{2} \right) A \left[\tilde{x} + \tilde{x} \cos(2\omega t) + \frac{\tilde{p} \sin(2\omega t)}{m \omega} \right]. \tag{8.76}$$

So long as any additional dynamics of the system is slow compared to 2ω, then we can make the rotating-wave approximation (RWA) and average over the oscillating terms to obtain

$$H_{\text{int}} = \left(\frac{g}{2} \right) A \tilde{x}. \tag{8.77}$$

But now the oscillatory motion due to H_0 has been removed, so that $\tilde{x}$ is constant, and is thus a quantum non-demolition observable. Another way to say this is that $\tilde{x}$ is now proportional to the quadrature operator X, and this is what want to measure. Yet a third way to describe this measurement is to say that the measurement involves a demodulation, so that we measure a signal that is contained on a carrier at frequency ω. We can modulate the coupling strength as in Eq. (8.75) if we use an optical cavity or electrical oscillator as the probe, and employ the technique discussed in the final subsection of Section 7.7.3.

It is now simple to describe the evolution of the oscillator under our modulated measurement, because it is equivalent to the measurement of position analyzed in the previous section, but with the oscillation frequency set to zero, since H_0 has been removed. The equations of motion for the means of position and momentum, assuming that the oscillator is damped at the rate γ, are therefore

$$d\langle\tilde{x}\rangle = -\frac{\gamma}{2}\langle\tilde{x}\rangle\,dt + \sqrt{8\eta k}\,\tilde{V}_x\,dW,$$

$$d\langle\tilde{p}\rangle = -\frac{\gamma}{2}\langle\tilde{p}\rangle\,dt + \sqrt{8\eta k}\,\tilde{C}\,dW, \tag{8.78}$$

and those for the conditional variances are

$$\dot{\tilde{V}}_x = -8\eta k\tilde{V}_x^2 - \gamma(\tilde{V}_x - \tilde{V}_x^T), \tag{8.79}$$

$$\dot{\tilde{C}} = -8\eta k\tilde{C}\tilde{V}_x - \gamma\tilde{C}, \tag{8.80}$$

$$\dot{\tilde{V}}_p = -8\eta k\tilde{C}^2 + 8k - \gamma(\tilde{V}_p - \tilde{V}_p^T). \tag{8.81}$$

Here we have written the equations in terms of the scaled variables and variances defined in, and just below, Eq. (8.55). The scaled measurement strength, k, is defined in Eq. (8.38), and the efficiency η is defined below Eq. (8.37).

The steady-state solutions for the conditional variances are

$$\tilde{V}_x^{ss} = \frac{\gamma}{16\eta k}\left[\sqrt{\frac{32\eta k}{\gamma}(1+2n_T)+1} - 1\right] \simeq \sqrt{\frac{\gamma(1+2n_T)}{8\eta k}}, \quad k \gg \gamma, \tag{8.82}$$

$$\tilde{C}^{ss} = 0, \tag{8.83}$$

$$\tilde{V}_p^{ss} = \frac{8k}{\gamma} + (1+2n_T). \tag{8.84}$$

We see from this that the conditional variance of the position decreases only as the square root of k, and so the standard deviation scales only as the fourth root of k. The measurement must be much stronger than the damping rate to achieve significant squeezing.

To determine the unconditional squeezing we need to solve the equations of motion for the means, because the total variance is the sum of the conditional variance and the variance of the means. From the equations of motion for the means we derive the equations of motion for the variances of the means as described in the previous section. Once the conditional variances have reached their steady states, these equations of motion are

$$\dot{\tilde{V}}_{\tilde{x}} = -\gamma \tilde{V}_{\tilde{x}} + 8\eta k \left(\tilde{V}_x^{ss}\right)^2, \qquad \dot{\tilde{V}}_{\tilde{p}} = -\gamma \tilde{V}_{\tilde{p}}, \tag{8.85}$$

to which the steady-state solutions are

$$\tilde{V}_{\tilde{p}}^{ss} = 0 \quad \text{and} \quad \tilde{V}_{\tilde{x}}^{ss} = \left(\frac{8\eta k}{\gamma}\right) \left(\tilde{V}_x^{ss}\right)^2 \simeq 1 + 2n_T, \quad \text{for } k \gg \gamma. \tag{8.86}$$

We see that there is no unconditional squeezing, because the total variance of the quadrature X is

$$\tilde{\mathbb{V}}_x/4 = \tilde{V}_x^{ss}/4 + \tilde{V}_{\tilde{x}}^{ss}/4 \geq 1/4. \tag{8.87}$$

As discussed above, we must apply a feedback force to reduce the fluctuations in the means if we want to realize unconditional squeezing. To do this we recall that our equations of motion are in the "rotating frame," in which phase space does not rotate. Just as we modulated the strength of the measurement interaction to measure an operator that is constant in this frame, we must modulate the feedback force. If we apply a force

$$F_{fb}(t) = -F \left[\frac{\langle p \rangle}{\Delta p} \cos(\phi) \cos(\omega t) + \frac{\langle x \rangle}{\Delta x} \sin(\phi) \sin(\omega t) \right], \tag{8.88}$$

where as usual $\Delta x = \sqrt{\hbar/(2m\omega)}$ and $\Delta p = \sqrt{\hbar m\omega/2}$, the resulting term in the Hamiltonian is $H_F = F_{fb}(t)x$. Moving into the interaction picture with respect to H_0 gives

$$H_F^I = F \left[\frac{\langle p \rangle}{\Delta p} \cos(\phi) \cos(\omega t) + \frac{\langle x \rangle}{\Delta x} \sin(\phi) \sin(\omega t) \right] \left[x \cos(\omega t) + \frac{p \sin(\omega t)}{m\omega} \right]. \tag{8.89}$$

Averaging over the terms oscillating at 2ω (performing the rotating-wave approximation) we have

$$H_F = \left(\frac{F \cos(\phi)}{2\Delta p} \right) \langle p \rangle x - \left(\frac{F \sin(\phi)}{2\Delta p} \right) \langle x \rangle p, \tag{8.90}$$

and the resulting equations of motion for the (scaled) means are

$$d\langle \tilde{x} \rangle = -\frac{1}{2} (\gamma + \sin(\phi)\Gamma) \langle \tilde{x} \rangle \, dt + \sqrt{8\eta k} \tilde{V}_x \, dW, \tag{8.91}$$

$$d\langle \tilde{p} \rangle = -\frac{1}{2} (\gamma + \cos(\phi)\Gamma) \langle \tilde{p} \rangle \, dt + \sqrt{8\eta k} \tilde{C} \, dW, \tag{8.92}$$

where the damping rate $\Gamma = F/(2\Delta)$. We can therefore realize an effective damping for both the position and the momentum by applying a feedback force.

Since the variance of the mean momentum is zero, we need only damp the mean position. We therefore choose $\cos(\phi) = 1$, and we can calculate the resulting variance of the mean position simply by replacing γ with $\gamma + \Gamma$ in Eq. (8.86). The total position variance, for $k \gg \gamma$ and $\Gamma \gg \gamma$, is then

$$\tilde{V}_x = \left(\frac{8\eta k}{\gamma + \Gamma}\right)(\tilde{V}_x^{ss})^2 \simeq \sqrt{\frac{\gamma(1 + 2n_T)}{8\eta k}} + \left(\frac{\gamma}{\Gamma}\right)(1 + 2n_T). \tag{8.93}$$

If the feedback is very strong, so that we can ignore the second term, then the measurement strength required to achieve squeezing, meaning that $\tilde{V}_x < 1$, is

$$\eta k > \frac{\gamma}{8}(1 + 2n_T). \tag{8.94}$$

8.3 Coherent feedback control

We are now going to explore ways of controlling quantum systems by coupling them to other quantum systems. Of course most of the examples of quantum control that we have considered so far already involve coupling the system to be controlled (the "primary") to another quantum system. Here we are specifically interested in control procedures in which the following two things are true: (i) the purpose is not only to control the unitary dynamics of the primary but also to reduce its entropy, and (ii) the procedure does not use explicit quantum measurements. We refer to such control procedures as exploiting "coherent feedback." The concept of coherent feedback, and the origin of this terminology, are discussed in Section 5.2.

8.3.1 The "resolved-sideband" cooling method

Resolved-sideband cooling, often abbreviated to "sideband cooling," is a method for reducing the entropy of a quantum system, and is possibly the simplest practical example of coherent feedback control. This technique was originally devised for cooling the motion of trapped ions, but is broadly applicable and has a simple underlying mechanism. This mechanism exploits the fact that the entropy of a system in thermal equilibrium depends on the energy differences between its energy levels. For example, for a two-level system in which the energy gap between the two levels is ΔE, the population of the excited state is $P = 1/(1 + e^{\Delta E/(kT)})$, in which k is Boltzmann's constant. When the energy gap is high enough, $P \simeq e^{-\Delta E/(kT)}$, and so the entropy drops exponentially with ΔE. Because of this, a system with a large energy gap can be used to extract the entropy from one with a lower energy gap.

To transfer entropy from one system to another, we can use a "swap" operation. Given two systems A and B, both of dimension N, whose respective bases we denote by $\{|n\rangle_A\}$ and $\{|n\rangle_B\}$, the joint unitary that applies the mapping

$$|n\rangle_A |m\rangle_B \rightarrow |m\rangle_A |n\rangle_B, \quad \text{for all } n, m, \tag{8.95}$$

will swap the states of the two systems. Up to a unitary that is the tensor product of unitaries acting separately on each system, the interactions

$$H = \frac{\hbar}{2}\lambda(\sigma_x \otimes \sigma_x + \sigma_y \otimes \sigma_y + \sigma_z \otimes \sigma_z) \qquad (8.96)$$

and

$$H = \hbar\lambda(ab^{\dagger} + ba^{\dagger}) \qquad (8.97)$$

will swap the states of two qubits and two oscillators, respectively, where as usual a and b are the annihilation operators of the respective oscillators. In both cases the time required for the swap is $\tau = \pi/(2\lambda)$. The interaction in Eq. (8.97) is sometimes referred to as the "rotating-wave" interaction, since it results from applying the rotating-wave approximation (RWA) to the linear interaction $H_{\text{lin}} = gx_a \otimes x_b$ between two oscillators, where x_a, x_b are the position operators for the respective oscillators. The RWA is valid so long as the interaction rate λ is much smaller than the frequencies of both oscillators.

Because sideband cooling requires the rotating-wave interaction, the interaction rate λ, and thus the rate at which it can extract energy from the system, is limited to $\lambda \ll \omega$. This translates to a limitation in the amount of cooling, because the faster the energy can be extracted, the lower the steady-state phonon number. In Section 8.1.1 we saw that if we varied the interaction rate g with time, then we could engineer a swap operation within a single period of the oscillator. This technique can be used to increase the cooling rate, and thus to achieve lower steady-state temperatures with sideband cooling. If we are limited to linear interactions with a linear auxiliary system, then the maximum rate at which a swap can be performed is limited by ω. To go beyond this we must use a nonlinear auxiliary system such as a qubit [386]. In Section 8.3.3 we will find that the minimum value of the phonon number, and in fact the maximum fidelity at which any state can be prepared, cannot be achieved in the steady state but only at a single instant.

If we couple two systems together via a time-independent interaction, both will remain at thermal equilibrium even if they have quite different energy gaps. The trick to sideband cooling is the following. If the energy gap for the system to be cooled (the "primary" system) is $\Delta\varepsilon$, and that of the auxiliary is ΔE, then the frequencies of their respective evolutions are $\omega = \Delta\varepsilon/\hbar$ and $\Omega = \Delta E/\hbar$. The difference in their frequencies, and the amount by which they are off-resonance, is $\nu = \Omega - \omega$. To cool the primary we modulate the coupling force between the two systems at ν. As discussed in Section 7.2, this involves pumping energy into the joint system, because the modulation provides the energy to make up the difference between the energy gaps of the two systems, allowing a downward transition in the primary to excite an upward transition in the auxiliary, and vice versa.

From a mechanical point of view, the modulation at the difference frequency fools the interaction into thinking that the systems are on resonance. The interaction can rapidly swap entropy from the primary to the auxiliary as if they were on resonance. So long as the damping of the auxiliary is fast compared to that of the primary, entropy is carried away from the primary faster than the thermal bath can pump it in, and the primary is cooled. Of course the systems are no longer at equilibrium, and in that sense their temperatures are no longer well defined. Referring to the process as "cooling" is a little strange. In actual fact the purpose of "cooling" mesoscopic quantum systems is to increase their purity, and thus the precision of the control that we can realize. For micro- and mesoscopic quantum systems, cooling is merely another name for preparing a system close to its ground state.

Resolved-sideband cooling of an oscillator

Mesoscopic mechanical oscillators tend to have frequencies in the range 1–500 MHz, and so contain a number of thermal phonons even at temperatures of 10–100 mK that can be reached with dilution refrigerators. We can use sideband cooling to cool a mechanical resonator by coupling it to a superconducting or optical auxiliary resonator, because the latter have much higher frequencies, and as a result have essentially zero entropy at 100 mK. To perform sideband cooling in this way, we use the method described in Section 7.5 to transform the nonlinear coupling between the mechanical oscillator and the superconducting or optical oscillator into an effectively linear coupling.

From Eq. (7.79) in Section 7.5, the Hamiltonian for the two oscillators has the form

$$H = \hbar\omega b^\dagger b + \hbar\Delta a^\dagger a - \hbar\frac{\lambda}{2}(b+b^\dagger)(a+a^\dagger), \tag{8.98}$$

where a and b are the annihilation operators of the mechanical and auxiliary resonators, respectively, and we have set the driving phase to zero. This Hamiltonian already contains a modulation of the linear interaction strength at a frequency $\nu = \Omega - \Delta$, where Ω is the frequency of the auxiliary oscillator. The effect of this modulation is to change the effective frequency of the auxiliary resonator to Δ. If we set $\Delta = \omega$ then the two resonators are effectively on resonance, giving the fastest entropy transfer and the best cooling. The mechanical resonator is subject to damping at rate γ and thermalization to a temperature T. The auxiliary resonator is subject only to damping at rate κ, because it is effectively coupled to a bath at zero temperature. We assume that both damping rates are much smaller than the frequencies of their respective oscillators, so that the joint system can be described by the master equation

$$\dot{\rho} = -i[H,\rho]\,dt - \frac{\kappa}{2}\left(a^\dagger a\rho + \rho a^\dagger a - 2a\rho a^\dagger\right)$$
$$-\frac{\gamma}{2}(n_T+1)\left(b^\dagger b\rho + \rho b^\dagger b - 2b\rho b^\dagger\right) - \frac{\gamma}{2}n_T\left(bb^\dagger\rho + \rho bb^\dagger - 2b^\dagger \rho b\right). \tag{8.99}$$

We can now write down the equivalent quantum Langevin equations for the operators a and b using the input–output formalism described in Section 3.6. These equations are

$$\dot{a} = -[i\omega + \kappa/2]a + i\lambda(b+b^\dagger)/2 + \sqrt{\kappa}\,a_{\text{in}}(t), \tag{8.100}$$

$$\dot{b} = -[i\omega + \gamma/2]b + i\lambda(a+a^\dagger)/2 + \sqrt{\gamma}\,b_{\text{in}}(t). \tag{8.101}$$

where the input fields have the correlation functions

$$
\begin{array}{rclrcl}
\langle b_{\mathrm{in}}(t)b_{\mathrm{in}}^{\dagger}(t+\tau)\rangle &=& [n_T+1]\delta(\tau), & \langle b_{\mathrm{in}}^{\dagger}(t)b_{\mathrm{in}}(t+\tau)\rangle &=& n_T\,\delta(\tau), \\
\langle a_{\mathrm{in}}(t)a_{\mathrm{in}}^{\dagger}(t+\tau)\rangle &=& \delta(\tau), & \langle a_{\mathrm{in}}^{\dagger}(t)a_{\mathrm{in}}(t+\tau)\rangle &=& 0.
\end{array}
\tag{8.102}
$$

Since the quantum Langevin equations for the coupled oscillators are linear, they can be solved exactly, and the steady-state solutions for the mean-square of the mechanical position and momentum obtained. In Section G.10 we describe the method used to calculate the steady state and give the full solution. We can then obtain the steady-state average phonon number using the fact that $\langle b^{\dagger}b\rangle = \langle x^2\rangle/4 + \langle p^2\rangle/4 - 1/2$, where as usual we define $x = b + b^{\dagger}$ and $p = -i(b - b^{\dagger})$. Since we are interested in the regime of good ground-state cooling, we consider here the parameter regime in which $\gamma \ll \kappa \sim \lambda \ll \omega$. To do so we define a small parameter ε and assume that $\gamma/\kappa \sim \kappa/\omega \sim \lambda/\omega \sim \varepsilon \ll 1$ and $\gamma/\omega \sim \varepsilon^2$. We then expand the steady-state solution for the average number of photons in the resonator to second order in ε. The result is

$$
\langle b^{\dagger}b\rangle_{\mathrm{ss}} = n_T\left(\frac{\gamma}{\kappa}\left[1+\frac{\kappa^2}{\lambda^2}\right] - \frac{1}{2}\left(\frac{\gamma}{\kappa}\right)^2\left[1+\frac{\kappa^2}{\lambda^2}+\frac{\kappa^4}{\lambda^4}\right]\right) + \frac{\kappa^2}{16\omega^2} + \frac{\lambda^2}{8\omega^2}.
\tag{8.103}
$$

Here the first term is the dominant term, since it is first-order in ε while the other three terms are second-order in ε. We have assumed that $\lambda \sim \kappa$, because both must be much smaller than ω to achieve $\langle b^{\dagger}b\rangle \ll 1$. As a result, all powers of κ/λ are on the order of unity. If we kept only the first-order term in the expression for $\langle b^{\dagger}b\rangle_{\mathrm{ss}}$, then we would think that the best cooling was obtained by setting $\lambda \gg \kappa$. The third term shows that this is not the case, since we must also satisfy $\lambda \ll \omega$.

The parameters γ and ω are properties of the oscillator we want to cool, while λ and κ are parameters that we may have the freedom to choose as part of designing our controller. It is therefore natural to ask what values of λ and κ will give us the best cooling. We can, for example, determine the optimal value of κ for a given interaction rate λ by using just the first-order terms and minimizing with respect to κ. The resulting minimum phonon number is

$$
\langle b^{\dagger}b\rangle_{\mathrm{ss}} = 2n_T\left(\frac{\gamma}{\lambda}\right),
\tag{8.104}
$$

which is achieved at $\kappa = \lambda$. Note that we have dropped the second-order terms in this expression, and thus the terms that limit the cooling as λ is increased.

Alternatively, we can determine the optimal value of λ for a given value of κ. It turns out that we can answer this question within our approximate treatment, because the optimal value of λ preserves the validity of our approximation. To see this we first discard the term proportional to κ^4/λ^4 in Eq. (8.103) above (we will justify this shortly), and then differentiate $\langle b^{\dagger}b\rangle$ with respect to λ to find the minimum. The value of λ that minimizes

$\langle b^\dagger b \rangle$ is given by

$$\lambda_{\text{opt}} = \sqrt{\omega\sqrt{8\gamma\kappa n_T}\left(1 - \frac{\gamma}{4\kappa}\right)}, \quad \text{and so} \quad \frac{\lambda_{\text{opt}}}{\omega} \sim \varepsilon^{3/4}. \tag{8.105}$$

The assumption that we used in our expansion in powers of ε/ω was that $\lambda \sim \varepsilon$, but $\lambda_{\text{opt}}/\omega$ is larger than this by a factor of $(1/\varepsilon)^{1/4}$. We must therefore check that terms in the expansion that we dropped, and that are proportional to positive powers of λ, are still significantly higher than second-order in ε. By inspecting the exact solution for $\langle b^\dagger b \rangle_{\text{ss}}$, we find that the only terms of this nature are powers of λ/ω, and further that the expansion only contains even powers of λ/ω. The lowest-order term we have dropped is therefore $(\lambda/\omega)^4 \sim \varepsilon^3$, so our approximation remains valid. In addition, $\kappa^4/\lambda^4 \sim \varepsilon$, and so the total contribution of this term to $\langle b^\dagger b \rangle$ is $\sim \varepsilon^3$, which is why we could drop it before determining λ_{opt}, above.

Substituting λ_{opt} into Eq. (8.103), and keeping terms only up to second order in ε, the minimum average phonon number is

$$\langle b^\dagger b \rangle_{\text{ss}} = n_T\frac{\gamma}{\kappa} + \sqrt{\frac{\gamma\kappa n_T}{2\omega^2}} - \frac{n_T}{2}\left(\frac{\gamma}{\kappa}\right)^2 + \frac{\kappa^2}{16\omega^2}. \tag{8.106}$$

The first term is the dominant term, proportional to ε, the second term is proportional to $\varepsilon^{1.5}$, and the last two terms are proportional to ε^2. The lowest-order term that we have discarded is $\mathcal{O}(\varepsilon^{2.5})$. One would now like to minimize with respect to κ, but this is not possible analytically. We consider a way to do this approximately in Section 8.3.2 below.

8.3.2 Resolved-sideband cooling via one-way fields

We now consider implementing sideband cooling by coupling a mechanical oscillator to a superconducting oscillator (or equivalently an optical resonator) via one-way (unidirectional) traveling modes of the electromagnetic field. We described this method of coupling two systems together in Section 5.3.3. The full configuration is shown on the right in Fig. 8.2. The mechanical oscillator is coupled to two superconducting "probe" oscillators that interact with the mechanics via its position, as described in Section 7.5. This interaction is denoted by H_I in Fig. 8.2. The two probe oscillators are each coupled to a transmission line via a circulator, also known as a unidirectional coupler. This input–output coupling is shown explicitly in terms of superconducting circuit elements in Fig. 5.2. The auxiliary oscillator is also coupled to a transmission line, and all the transmission lines are connected together to form a single line as shown in Fig. 8.2.

To realize sideband cooling we first adiabatically eliminate the probe oscillators as described in Section 7.7.3. The result of each of these adiabatic eliminations is to replace the probe oscillator with a direct coupling between the mechanics and the transmission line, where this coupling is mediated by the mechanical position operator, x. The resulting configuration is that depicted at the bottom right in Fig. 5.2, in which the operators M and F are proportional to the mechanical position, and the operator P is proportional to the annihilation operator of the auxiliary cavity.

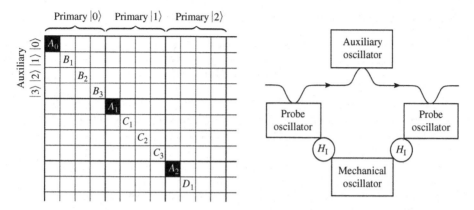

Figure 8.2 On the left is a diagram depicting the elements of the density matrix discussed in Section 8.3.3. On the right is a diagram of the field-mediated implementation of resolved-sideband cooling discussed in Section 8.3.2.

To derive the quantum Langevin equations for the mechanical and auxiliary oscillators, we first write down the equations of motion for each oscillator separately, given their respective input–output coupling to their own transmission lines. We will then connect the inputs and outputs together to obtain the configuration in Fig. 5.2. The Langevin equations for the mechanical resonator, given that it is coupled at damping rate γ to a thermal bath and has two input–output couplings via its position operator, are

$$\dot{\tilde{x}} = -(\gamma/2)\tilde{x} + \omega\tilde{p} + \sqrt{\gamma}\,x_{\text{in},\gamma}, \tag{8.107}$$

$$\dot{\tilde{p}} = -(\gamma/2)\tilde{p} - \omega\tilde{x} + \sqrt{\gamma}\,p_{\text{in},\gamma} + \sqrt{2k}\,p_{\text{in},k} + \sqrt{2f}\,p_{\text{in},f}. \tag{8.108}$$

Here, as in Eq. (3.241) in Chapter 3, the input fields for the damping at rate γ are defined by $x_{\text{in},\gamma} = b_{\text{in},\gamma} + b_{\text{in},\gamma}^{\dagger}$ and $p_{\text{in},\gamma} = -i(b_{\text{in},\gamma} - b_{\text{in},\gamma}^{\dagger})$, and the fields for the two couplings to the mechanical position are defined similarly in terms of input operators $b_{\text{in},k}$ and $b_{\text{in},f}$. The parameters k and f are the respective rates for these two couplings, which are equivalent to continuous measurements of the position at rates k and f. The correlation functions for the input field $b_{\text{in},\gamma}$ are

$$\langle b_{\text{in},\gamma}^{\dagger}(t)b_{\text{in},\gamma}(t+\tau)\rangle = \langle b_{\text{in},\gamma}(t)b_{\text{in},\gamma}^{\dagger}(t+\tau)\rangle + \delta(\tau) = (n_T + 1)\delta(\tau), \tag{8.109}$$

where n_T is defined in Eq. (4.116). The correlation functions for the inputs $b_{\text{in},k}$ and $b_{\text{in},f}$ are the same as those for $b_{\text{in},\gamma}$ but with $n_T = 0$. The output fields are

$$b_{\text{out},\gamma} = b_{\text{in},\gamma} - \sqrt{\gamma}\,b, \quad b_{\text{out},k} = -(b_{\text{in},k} + \sqrt{2k}\,\tilde{x}), \quad b_{\text{out},f} = b_{\text{in},f} - \sqrt{2f}\,\tilde{x}. \tag{8.110}$$

Note that we have imposed a π phase shift on the output light $b_{\text{out},k}$. This effectively flips the sign of the interaction with x. As usual $\tilde{x} = b + b^{\dagger}$ and $\tilde{p} = -i(b - b^{\dagger})$.

The auxiliary oscillator is described by the quantum Langevin equations

$$\dot{X} = -(\kappa/2)X + \Omega P + \sqrt{\kappa}\, X_{\text{in}}, \qquad (8.111)$$

$$\dot{P} = -(\kappa/2)P - \Omega X + \sqrt{\kappa}\, P_{\text{in}}, \qquad (8.112)$$

where we have defined the operators $X = a + a^{\dagger}$ and $P = -i(a - a^{\dagger})$, and input fields $X_{\text{in}} = a_{\text{in}} + a_{\text{in}}^{\dagger}$ and $P_{\text{in}} = -i(a_{\text{in}} - a_{\text{in}}^{\dagger})$. The input field correlation functions are the same as those in Eq. (8.109) with $n_T = 0$, and the output field is

$$a_{\text{out}} = a_{\text{in}} - \sqrt{\kappa}\, a. \qquad (8.113)$$

We note that in reality the auxiliary may also have some significant additional losses that would need to be modeled by a second vacuum output.

We now send the output of the measurement at rate k into the input of the auxiliary, and the output of the auxiliary into the input channel for the measurement at rate f. That is, we set

$$a_{\text{in}} = b_{\text{out},k} \qquad \text{and} \qquad b_{\text{in},f} = a_{\text{out}}. \qquad (8.114)$$

This achieves the coupling configuration shown at the bottom right in Fig. 5.2, with $M = -\sqrt{k}\tilde{x}$, $F = \sqrt{f}\tilde{x}$, and $P = \sqrt{\kappa/2}\,a$. Note that the measurement interaction with the operator x at rate f is not being used to obtain information about x but to apply a force to the resonator. Recall that a term in the Hamiltonian proportional to x corresponds to a simple (spatially independent) force.

The Langevin equations for the two coupled oscillators can be obtained by substituting the equalities in Eq. (8.114) into the Langevin equations for both oscillators, or alternatively by substituting the above choices for M, P, and F into Eqs. (5.188) and (5.189). Setting $f = k$, the Langevin equations for the coupled oscillators become

$$\dot{\tilde{x}} = -(\gamma/2)\tilde{x} + \omega\tilde{p} + \sqrt{\gamma}\, x_{\text{in},\gamma}, \qquad (8.115)$$

$$\dot{\tilde{p}} = -(\gamma/2)\tilde{p} - \omega\tilde{x} + \sqrt{\gamma}\, p_{\text{in},\gamma} + \sqrt{8k\kappa}\, P, \qquad (8.116)$$

$$\dot{X} = -(\kappa/2)X + \Omega P + \sqrt{\kappa}\, p_{\text{in},k} - \sqrt{8k\kappa}\, \tilde{x}, \qquad (8.117)$$

$$\dot{P} = -(\kappa/2)P - \Omega X + \sqrt{\kappa}\, x_{\text{in},k}. \qquad (8.118)$$

The mechanical oscillator no longer experiences any noise from the input–output channels that couple to the auxiliary, because they cancel each other out.

The above equations of motion are exactly those that would result from coupling the two oscillators together via the interaction Hamiltonian

$$H_{\text{I}} = -\hbar\frac{\lambda}{2}\tilde{x}P, \qquad (8.119)$$

where $\lambda = \sqrt{8k\kappa}$. Further, since the dynamics of the auxiliary is essentially symmetric in its two quadratures X and P, from the point of view of the primary this is equivalent to the coupling $H = \hbar\lambda\tilde{x}X/2$ that we used above in sideband cooling. If we modulate the coupling strength λ at the frequency difference between the two oscillators, then the field-mediated feedback loop implements sideband cooling as analyzed above, with $\lambda = \sqrt{8k\kappa}$. The modulation of the coupling strength can be obtained by modulating the strength of the effective linear interaction between the mechanics and the transduction oscillators. Alternatively, it can be achieved by imprinting a modulation on the fields that couple the auxiliary to the other components.

Comparing measurement-based and coherent feedback cooling for oscillators

The measurement-based feedback cooling method, discussed in Section 8.2.1 and sideband cooling implemented via one-way fields have something in common: both extract information from the resonator in the same way. In both techniques the resonator is coupled to a bath of electromagnetic modes via its position. In the measurement-based cooling method the electromagnetic modes are then measured to realize a continuous measurement of the position of the resonator. In sideband cooling the electromagnetic modes interact with an auxiliary system which then acts back on the oscillator via another bath of electromagnetic modes, being an example of coherent feedback. Since both methods extract information in the same way, we can compare the cooling that each can achieve for a given interaction rate (measurement strength) k, to see which is able to make the best use of the information.

The average mean phonon number achievable with field-mediated sideband cooling is obtained from Eq. (8.106) by substituting in for λ using $\lambda = \sqrt{8\kappa k}$. This gives

$$\langle b^\dagger b \rangle_{\text{sb}} = n_T \left(\frac{\gamma}{\kappa} \left[1 + \frac{\kappa}{8k} \right] - \frac{1}{2} \left(\frac{\gamma}{\kappa} \right)^2 \left[1 + \frac{\kappa}{8k} + \frac{\kappa^2}{(8k)^2} \right] \right) + \frac{\kappa^2}{16\omega^2} + \frac{k\kappa}{\omega^2}. \qquad (8.120)$$

To determine $\langle b^\dagger b \rangle_{\text{sb}}$ as a function of k we need to find the optimal value of κ for each value of k. It is not possible to do this analytically, but we can can do something nearly as good. We anticipate that the optimal value of κ will be at least as large as k, since it must keep the first term as small as it can without making the second-to-last term too large. Based on this assumption, some terms will contribute much more to the value of $\langle b^\dagger b \rangle_{\text{sb}}$ than others. We can perform an approximate optimization by optimizing κ with respect to the most important terms only. This gives us an answer that is not only close to optimal but is also an upper bound on the achievable minimum value of $\langle b^\dagger b \rangle_{\text{sb}}$. Inspecting the relative sizes of the terms in Eq. (8.120), we discard all but the first and last terms and find the value of κ that minimizes their sum. The approximately optimal value we obtain for κ is

$$\hat{\kappa} = \left(\frac{n_T \gamma \omega^2}{2} \right)^{1/3}, \qquad (8.121)$$

with the result that $\gamma/\kappa \propto \varepsilon^{4/3}$ and $\kappa/\omega \propto \varepsilon^{2/3}$. Substituting this into Eq. (8.120), and keeping only the leading-order terms we obtain

$$\langle b^\dagger b \rangle_{\text{sb}} = \frac{n_T}{8} \left(\frac{\gamma}{k} \right) + A \left(\frac{k}{\omega} \right) + B, \qquad (8.122)$$

with

$$A = \left(\frac{n_T}{2}\right)^{1/3} \left(\frac{\gamma}{\omega}\right)^{1/3} \propto \varepsilon^{2/3}, \quad B = \frac{33}{32} \left(\sqrt{2}n_T\right)^{2/3} \left(\frac{\gamma}{\omega}\right)^{2/3} \propto \varepsilon^{4/3}.$$

We have already calculated the performance of measurement-based feedback as a function of k, given in Eq. (8.72). If we keep only the terms that are first-order in the small parameter ε, then we have

$$\langle b^{\dagger}b\rangle_{\mathrm{mf}} = \frac{n_T}{8} \left(\frac{\gamma}{k}\right) + \sqrt{8} \left(\frac{k}{\omega}\right). \tag{8.123}$$

We see that the two expressions for $\langle b^{\dagger}b\rangle$ are very similar. Since A and B are small there is a range of k for which the coherent scheme, sideband cooling, performs better than the measurement-based feedback scheme. It is the second term in each expression that limits the cooling, but this term is not fundamental. It is due to the way in which the controlling device couples to the resonator. In the case of the measurement-based control protocol, the resulting continuous measurement of x squeezes the state of the resonator, forcing it away from the ground state, and this is the origin of the second term. In the coherent feedback protocol, it is also the linear coupling to x that generates the second term. This fact can be seen by replacing the x interaction by the interaction in Eq. (8.9), and solving again for the steady-state phonon number.

In the coherent cooling scheme the limitation due to the x interaction can be greatly reduced by making the scheme *time-dependent*. In Section 8.1 we showed that by varying the interaction strength with time, the entropy in the hot resonator can be swapped into the cold resonator within a single period of the hot oscillator. This means that we can set $k \approx g \approx \omega$ in the coherent scheme while effectively maintaining the interaction in Eq. (8.9), and thus achieve a minimum phonon number $\langle b^{\dagger}b\rangle \approx n_T\gamma/\omega$ [650, 388]. If we move to a nonlinear auxiliary system, then we can increase the rate at which we extract entropy beyond that of the oscillation frequency ω [386], and in that case the cooling is only limited by the rate of the interaction. In the next section we will explore the fundamental limit to cooling and state-preparation when we are able to employ any interaction with the system, and the only restriction is on the interaction rate.

8.3.3 Optimal cooling and state-preparation

In the previous sections we have discussed how to cool a resonator by making a continuous measurement of its position, or by coupling it, via its position, to another quantum system that has zero entropy at the ambient temperature. These methods are useful because the strongest interactions that are available for an oscillator are proportional to the oscillator's position. Nevertheless, we now ask a more fundamental question. If we have the ability to interact with a system in any way we want, and are limited only by the strength of the interaction, how well can we prepare a system in pure state?

The complexity of optimal state-preparation comes from the fact that the thermal environment acts on the system simultaneously with whatever control protocol we apply. It is the interplay between the two dynamical processes that determines how well we can do. At the simplest level, the less time our control protocol takes, the better the state preparation, because the thermal noise will have less time to mess it up.

To begin we must chose a precise definition for "the strength" of the interaction with the system. Since the maximum speed of evolution in Hilbert space generated by a Hamiltonian, H, is proportional to $|\lambda_{\max}(H) - \lambda_{\min}(H)|$, where $\lambda_{\max}(H)$ and $\lambda_{\min}(H)$ are the maximum and minimum eigenvalues of the Hamiltonian, we define the strength of an interaction H_{int} by

$$\mathfrak{s}(H_{\text{int}}) \equiv |\lambda_{\max}(H_{\text{int}}) - \lambda_{\min}(H_{\text{int}})|. \tag{8.124}$$

Note that $\mathfrak{s}(H_{\text{int}})$ does not change if a multiple of the identity is added to H_{int}. Because of this, we can always add such a term to the Hamiltonian so as to set $\lambda_{\min}(H_{\text{int}}) = 0$ so that $\mathfrak{s}(H_{\text{int}}) = |\lambda_{\max}(H_{\text{int}})|$.

We want to place a restriction only on the speed of the interaction, and not on that of the Hamiltonian of the system itself, so we need to clarify the difference between the two. Let us call the system we wish to control the *primary*, also to be known as system A, and denote its basis states by $\{|n\rangle\}$. We will call the system that interacts with it the *auxiliary*, also to be known as system B, and denote its basis states by $\{|k\rangle\}$. The joint states of A and B are $|nk\rangle \equiv |n\rangle|k\rangle \equiv |n\rangle \otimes |k\rangle$. If we denote the elements of the joint density matrix of A and B by $\rho_{nk,n'k'}$, and the diagonal elements as $\rho_{nk} \equiv \rho_{nk,nk}$, then ρ_{nk} is the population of state $|nk\rangle$. The populations of the states of A and B will be denoted, respectively, by $p_n = \sum_k \rho_{nk}$ and $q_k = \sum_n \rho_{nk}$. The difference between the interaction Hamiltonian and the Hamiltonians of the subsystems is that the interaction Hamiltonian simultaneously changes the populations of the states of both systems, whereas the latter change only the populations of one of the systems.

Consider the elements of the joint density matrix, depicted on the diagram on the left in Fig. 8.2 (in which the auxiliary has four states). The population of the state $|0\rangle$ of the primary is the sum of all the diagonal elements in the top left 4-by-4 box: $p_0 = A_0 + B_1 + B_2 + B_3$. The population of state $|0\rangle$ of the auxiliary is the sum of all the elements labeled by A: $q_0 = A_0 + A_1 + A_2 + \cdots$. The Hamiltonian of the auxiliary cannot change p_0, so it can only rearrange the populations within each box. And whenever it acts it rearranges the populations within every box in exactly the same way. Similarly, the Hamiltonian of the primary cannot change q_0, so it can only mix together the populations labeled by A in Fig. 8.2. And once again, the primary Hamiltonian simultaneously rearranges all the elements that contribute to q_1 (and q_2, q_3, etc.) in the same way as it does the elements that make up q_0. Neither the primary nor the auxiliary Hamiltonian can transfer population from A_1 to B_1, which can only be done by an interaction Hamiltonian.

Bounding the maximum difference of the eigenvalues of the interaction Hamiltonian means bounding the rate at which population can be transferred between two elements,

ρ_{nk} and ρ_{mj}, that are (i) in different 4-by-4 boxes (so that $m \neq n$), and (ii) have different locations within their respective 4-by-4 boxes (so that $j \neq k$). It is the matrix element $H_{nk,mj}$ of an interaction Hamiltonian H that generates the population transfer between ρ_{nk} and ρ_{mj}.

Note that because we bound the eigenvalues of the interaction Hamiltonian, our analysis here is appropriate for control in which the auxiliary interacts in a coherent way with the controller, but not for that in which the controller interacts via an irreversible Markovian coupling, such as a continuous measurement of the system.

The regimes of weak coupling and good control

We will not attempt to find the optimal preparation protocol in all generality, but restrict ourselves to a regime that allows two simplifications. First, we assume that the strength of the interaction is much larger than the rate at which noise is being fed into the system. This means that the protocol will be able to prepare states that are very close to pure, and is the regime of "good control." If the maximum value of the interaction strength is μ, and the rate at which the noise induces dynamics is Γ (to be defined precisely below), then the regime of good control is

$$\max\left[\mathfrak{s}(H_{\mathrm{int}})\right] = \mu \gg \Gamma. \tag{8.125}$$

In our analysis we will calculate the optimal state-preparation protocol to first-order in $\varepsilon \equiv \Gamma/\mu$.

Since we assume here that the master equation describing the noise is Markovian, it is a rate equation, giving transition rates between the populations of various states. We define the noise rate, Γ, as the largest of these transition rates. More precisely, it is the largest transition rate between any two states that are appreciably populated during the preparation protocol. As an example, consider a harmonic oscillator subjected to thermal noise, where the master equation is given by Eq. (4.82). In this case the transition rate into the state with n photons is $\gamma_n = \gamma(n_T + 1)n$, so the largest transition rate is determined by the largest number of photons that the resonator may contain, $n_{\max}$. If our purpose is to prepare the ground state, then $n_{\max}$ is determined by n_T. If $n_T \ll 1$ then $n_{\max} = 1$, and if $n_T \gg 1$ then $n_{\max} \approx n_T$.

The second simplification we employ is that the noise is independent of the control protocol. If the noise is thermal noise, then as we have seen in Section 4.3.2 the master equation that describes this noise, and thus the noise itself, depends in the energy-eigenstates of the system. If the control protocol changes these eigenstates, which it does necessarily because of the interaction with the system, then the protocol will modify the noise. Treating the noise as if it is time-independent will only be a good approximation if the magnitude of the interaction Hamiltonian is small compared to the energy gaps between adjacent energy levels of the system. If the energy gaps are on the order of $\Delta E = \hbar\Delta$, then the regime of weak coupling is

$$\max\left[\mathfrak{s}(H_{\text{int}})\right] = \mu \ll \Delta. \tag{8.126}$$

For a harmonic oscillator $\Delta = \omega$, where ω is the oscillator frequency. Since we are already solving the problem to first order in the noise, a change to the master equation to first order in μ/Δ will only change the answer to second order in $\varepsilon = \Gamma/\mu \sim \mu/\Delta$. In our analysis we will assume that there is no change to the master equation that describes the noise during the control protocol, and thus solve the problem to first order in μ/Δ.

An ideal controller

Since we want to find the best possible state-preparation that can be achieved for a given value of μ, we should place no additional restriction on the auxiliary system. We therefore allow the auxiliary to have a dimension M which is as large as we like. We also allow the auxiliary to undergo any linear quantum evolution during the control protocol. That is, it can undergo any evolution that does not involve multiple measurement outcomes. The reason that we exclude multiple outcomes is that we are interested in a single outcome for the primary, regardless of what happens to the auxiliary, so we must trace out the latter. The auxiliary can undergo damping or any other interaction with an environment. In fact, we will find below that because the size of the auxiliary Hilbert space is not restricted, no damping or other trace-preserving operation on the auxiliary can improve the state-preparation of the primary. The ideal controller will therefore have no noise or damping, and will be prepared initially in its ground state.

The fastest cooling protocol in the absence of noise

Cooling is the special case of state-preparation in which we want to prepare the ground state. If we examine the density matrix in Fig. 8.2, we see that because the auxiliary starts in its ground state, it is only the elements labeled by A that are initially populated. Since the ground state of the system is the top left 4-by-4 box, to prepare the system in the ground state we need to transfer the population in the elements A_1, A_2, etc., to this box. Since a unitary transformation cannot map two states to a single state, and since changing basis is unnecessary in maximizing the population in any box (see Horn's lemma in Chapter 2), we need only consider operations that rearrange the diagonal elements. We can transfer all the population to the top left box if we transfer $A_1 \to B_1$, $A_2 \to B_2$, etc. To do this we must have at least as many elements in the top left box as there are states of the system, and thus the auxiliary must be at least as large as the system. From our discussion in Section 8.1, we know that the fastest way to transform from state $|a\rangle$ to state $|b\rangle$, given that $\mathfrak{s}(H_{\text{int}}) = \mu$, is via the Hamiltonian

$$H_{ab} = \frac{\mu}{2}\left(|b\rangle\langle a| + |a\rangle\langle b|\right). \tag{8.127}$$

If we start in state $|a\rangle$ the evolution is

$$|\psi(t)\rangle = \cos[\theta(t)]|a\rangle + \sin[\theta(t)]|b\rangle, \tag{8.128}$$

with $\theta = \mu t/2$. The evolution of the population in state $|b\rangle$ is then

$$P_b(t) = 1 - P_a(t) = \frac{1}{2}[1 - \cos(\mu t)]. \tag{8.129}$$

So long as all the initial states are mutually orthogonal, we can transform any number of initial states to any number of final states simultaneously at this maximum speed, while still satisfying $\mathfrak{s}(H_{\text{int}}) = \mu$. We need to transform each joint state $|m0\rangle$ respectively to the final state $|0(m+1)\rangle$, and so the Hamiltonian that prepares the system in the ground state in the minimum time is

$$H_{\text{int}} = \frac{\mu}{2} \sum_{m=0}^{N} |0(m+1)\rangle\langle m0| + |m0\rangle\langle 0(m+1)|. \tag{8.130}$$

The time taken is $\tau = \pi/\mu$.

Local operations are unnecessary

We now show that no local Hamiltonian of the auxiliary can increase the rate of state-preparation, and neither can any other quantum evolution that affects the auxiliary alone. Consider the initial states $|m0\rangle$ as they evolve toward their respective final states, $|0(m+1)\rangle$, and denote the state that $|m0\rangle$ has evolved to at time t by $|\psi_m(t)\rangle$. We now recall that local Hamiltonians can only transform states within the respective spaces discussed above. Since for each initial state the destination states are orthogonal to these spaces, the geometry of vector spaces tells us that the action of local Hamiltonians does not change the angle $\theta(t)$ between $|\psi_m(t)\rangle$ and its destination state. The local Hamiltonians therefore cannot change the populations of the destination states, and thus cannot increase the population of the ground state of the primary.

Now we consider all quantum evolutions of the auxiliary. As we know from Chapter 1, all quantum evolutions of a system can be obtained by having that system interact with another system (e.g., an environment) which is then traced over. We therefore encompass all evolutions of the auxiliary by including a third system in our control scenario, where this system interacts with the auxiliary, but not the primary, via an arbitrary Hamiltonian H_3. We will call the third system the *tertiary*. Now all we have to do to show that the interaction H_3 cannot improve the state-preparation, is to note that since the auxiliary can be as large we like, we can subsume the tertiary within the auxiliary. The Hamiltonian H_3 is now merely part of the local auxiliary Hamiltonian, which cannot improve the state-preparation. It is worth noting, however, that the Hamiltonian H_3 can degrade the state preparation. The reason for this is that the interaction Hamiltonian H_{int} that implements the preparation is not allowed to act on the tertiary, by our definition of the tertiary. The

Hamiltonian H_3 can therefore take the partially rotated states $|\psi_m(t)\rangle$ and rotate them to states that H_{int} cannot access. It is because we have allowed the auxiliary to have unlimited size that any interaction of the auxiliary with an environment is unnecessary.

A local Hamiltonian of the primary cannot improve the protocol we have introduced above, but we note that this is because our goal is to prepare the ground state and we assumed that the initial state of the primary was thermal. As a result, the primary initially has some probability of being in the ground state, and so we do not need to touch that state. If instead we wanted to prepare the primary in the state $(|0\rangle + |1\rangle)/\sqrt{2}$, which is not an eigenstate of the initial density matrix, then the fastest protocol would require the action of a local primary Hamiltonian. What a local primary Hamiltonian cannot do is to change the entropy of the primary. Since the primary starts in a mixed state, the preparation protocol reduces this entropy to zero, and the rate at which this entropy is reduced cannot be increased by a local Hamiltonian.

The effect of weak noise

The reason that we have constructed the fastest preparation protocol in the absence of noise is the intuition that the faster the preparation the better it will be, because the noise will have the least time to increase the entropy of the primary. We must now examine the effect of the noise on the preparation process. The preparation time is on the order of $1/\mu$, and the master equation describing the noise is a rate equation giving transition rates between the populations of the states. Since these transition rates are on the order of Γ, the amount of noise added to the system (or more precisely, the change in the state of the system due to the noise) during the preparation is $\sim \Gamma/\mu = \varepsilon$ to first order in ε.

It is reasonable to choose as the initial state of the primary its steady state under the master equation. At least, this is one well-motivated scenario. In that case, at the initial time, the master equation does not induce any net rate into or out of any state. But this changes as soon as the protocol starts because the protocol takes the primary out of equilibrium. The effect of the master equation is therefore to produce transition rates that "leak" population out of the destination states while the protocol is increasing their populations. Where this population goes depends on the master equation.

Optimal state-preparation in the presence of weak noise

Given the analysis above, it seems likely that the optimal preparation protocol in the presence of noise, to first order in Γ/μ, is to apply the minimal-time protocol presented above, and to add to it a further set of minimal-time state-to-state transformations that retrieve the population from the states to which the master equation sends it during the protocol. In this way we optimally combat the noise process as it acts against our preparation protocol.

As a concrete example of applying the above procedure, consider the problem of preparing a harmonic oscillator in its ground state. As discussed above we begin by choosing the interaction Hamiltonian to transfer each of the populations labeled by A_j in Fig. 8.2 to the populations B_j. To determine what additional transformations we need to include in H_{int}, we examine where the thermal master equation places the population that it steals from the B_j.

For the harmonic oscillator the Redfield equation, in this case the quantum optics master equation, possesses transitions only between adjacent energy levels. The master equation therefore takes the population in B_j to C_j, since it does not change the state of the auxiliary. To apply "corrective" transformations to take the population from the elements C_j back to the top left box, we cannot use the B_j as destinations, because they are already being used. For any given temperature there is some value of j above which the initial populations A_j are zero for all practical purposes. If we call this value J, then we need only use J of the B elements as destinations for the A populations. We can therefore use the elements B_{J+1} to B_{2J} as destinations for the populations of the C_j. This will be possible so long as the top left box, and thus the auxiliary, has dimension $2J + 1$. Labeling the energy levels of the harmonic oscillator by the number of phonons they contain, the interaction Hamiltonian that performs the above protocol is

$$ H_{\text{int}} = G + G^{\dagger} \quad \text{with} \quad G = \frac{\mu}{2} \sum_{j=1}^{J} |0,j\rangle\langle j,0| + |0,j+J\rangle\langle 1,j|. \tag{8.131} $$

Strong evidence that this Hamiltonian is indeed optimal for state-preparation, in the regimes of weak coupling and good control, has been obtained by comparing its performance to that obtained by a numerical search [651]. In what follows we will assume that H_{int} is optimal.

The size of the auxiliary system

It is important to note that the size required for the auxiliary system is determined by the number of different Hamiltonians that we need to apply. For example, in the case of a three-level primary initially in a thermal state, and coupled to a thermal bath at $T > 0$, we must apply three different Hamiltonians to transform the initial thermal state to the ground state: the first of these does nothing so as to leave the ground state where it is, the second transforms the first excited state to the ground state, and the third transforms the second excited state to the ground state. We also need to transfer back to the ground state the population that leaks out during the preparation. This transfer requires the same number of unitary operations as the number of orthogonal states to which the population leaks. For a three-level system the maximum number is two, so the maximum size required for the auxiliary is five.

Calculating the performance: linear trajectories as a perturbative technique

In the section above we arrived at a procedure for constructing protocols for state preparation, using an ideal auxiliary, that analytical arguments and numerical optimization suggest is optimal. Now we need to know how to calculate the performance of the resulting protocols, to first order in Γ/μ. The performance is the probability that the system does end up in the desired state, and this is also the fidelity of the preparation. One way to calculate this probability is to use the linear stochastic Schrödinger equation derived in Section 3.3. A stochastic Schrödinger equation is a way of breaking a Markovian master equation into a number of possible "trajectories" for the evolution of pure states, such that the average over the trajectories correctly reproduces the evolution of the master equation. To recall Section 3.3, a master equation given by

$$\dot{\rho} = -\frac{i}{\hbar}[H, \rho] - \frac{\gamma}{2}\left(A^\dagger A \rho + \rho A^\dagger A - 2A\rho A^\dagger\right) \tag{8.132}$$

has the equivalent Poisson (or "quantum-jump") linear stochastic Schrödinger equation

$$d|\psi\rangle = \left[\frac{\gamma}{2}(1 - A^\dagger A) - \frac{i}{\hbar}H\right]dt|\psi\rangle + (A - 1)dN|\psi\rangle. \tag{8.133}$$

Here, in each time-step dt, the increment dN is either 0 or 1, the value of which is random, and governed by the Poisson probabilities

$$\text{Prob}(dN = 1) = \gamma\,dt, \quad \text{Prob}(dN = 0) = 1 - \gamma\,dt. \tag{8.134}$$

The value of dN is usually zero, but every now and then $dN = 1$, and when it does so the state of the system "jumps," undergoing the instantaneous evolution

$$|\psi\rangle \rightarrow A|\psi\rangle. \tag{8.135}$$

The norm of the state changes under the evolution. A given trajectory is described by a discrete set of times, $t_1, \ldots, t_n$, at which a jump has occurred, and both the total number of jumps and the jump times are random variables. The probability density that the final state is the one resulting from a given trajectory is the probability density for the set of jump times given by the Poisson process, multiplied by the square norm of the given final state. The Poisson process gives the number of jumps, n, in the interval τ as

$$P(n, \tau) = e^{-\gamma\tau}\sum_{n=0}^{\infty}\frac{(\gamma\tau)^n}{n}, \tag{8.136}$$

and the jump times t_k as mutually independent, and uniformly distributed over the evolution time.

In our control protocol the product of the evolution time τ and the noise rate γ is $\gamma\tau \sim \varepsilon \ll 1$, and because of this the probability that there are n jumps in time τ is proportional to ε^n. To determine the evolution to first order in ε we therefore need only consider trajectories that have at most a single jump. We thus calculate the final state for each of these trajectories and then average them to obtain the final state of the primary. If the initial state of the primary is mixed, then we must evolve each trajectory for each eigenstate of the initial density matrix, and average over all the resulting final states.

An example: optimal preparation of a weakly damped qubit

We now construct a control protocol for a simple example, and use the linear stochastic Schrödinger equation to determine its performance. We consider the problem of preparing a single qubit in the "target" state $|T\rangle = (|0\rangle + |1\rangle)/\sqrt{2}$, when the qubit is subjected to spontaneous decay at rate γ and dephasing at rate κ. The master equation describing this situation is

$$\dot{\rho} = -\frac{\gamma}{2}(\sigma^\dagger \sigma \rho + \rho \sigma^\dagger \sigma - 2\sigma \rho \sigma^\dagger) - \frac{\kappa}{2}[\sigma_z, [\sigma_z, \rho]], \qquad (8.137)$$

where $\sigma = |0\rangle\langle 1|$ is the decay operator. For simplicity we assume that the qubit has no Hamiltonian evolution except for the control protocol we will apply.

In this example we have two noise processes, and thus two kinds of jump that can occur. Just as the master equation is the sum of the master equations for each noise process, the linear stochastic equation is the sum of the stochastic equations for each noise process. This stochastic equation is

$$
\begin{aligned}
d|\psi\rangle &= \left[\frac{\gamma}{2}(1 - \sigma^\dagger \sigma) + \frac{\kappa}{2}(1 - \sigma_z^2)\right] dt |\psi\rangle + (\sigma - 1) dN_1 |\psi\rangle + (\sigma_z - 1) dN_2 |\psi\rangle \\
&= \frac{\gamma}{4}(1 - \sigma_z) dt |\psi\rangle + (\sigma - 1) dN_1 |\psi\rangle + (\sigma_z - 1) dN_2 |\psi\rangle, \qquad (8.138)
\end{aligned}
$$

where the jump rate for dN_1 is γ and that for dN_2 is κ, and to get to the second line we have used the fact that $\sigma_z^2 = I$ and $1 - \sigma^\dagger \sigma = (1 - \sigma_z)/2$.

Now we need to determine the size of the auxiliary, and for this we need to know how many different unitary operations we have to perform on the primary. Since the steady state of the master equation is the ground state, we will assume that the primary is in this state at the start of the protocol. This state is pure, so we need only a single unitary operation to transform it to the target state. This transformation is generated by the Hamiltonian $H = -\mu\sigma_y/2$. We now need to examine to what states the master equation transfers population from the target state. The damping process transfers this population to the ground state. It can be seen from the stochastic Schrödinger equation above that the dephasing process applies "phase flip" operations, being multiplication by σ_z, and these take the target state to the state $|0\rangle - |1\rangle)/\sqrt{2}$. We now observe that the Hamiltonian that rotates the ground state to the target state via the shortest path also rotates the state $|0\rangle - |1\rangle)/\sqrt{2}$ to the target by the shortest path. Thus we only need one Hamiltonian for the initial preparation, and the same Hamiltonian performs the correction for the leaked population. Since we only need to apply a single Hamiltonian for the entire optimal protocol, we do not need an auxiliary at all. The optimal protocol merely involves applying the Hamiltonian $H = -\mu\sigma_y/2$ for a time $\tau = \pi/(2\mu)$.

To determine the final state of the qubit at the end of our control protocol, we must solve the stochastic equation for each initial state, and for three different scenarios: (i) no jump, (ii) one jump of the decay process (one emission), and (iii) one jump of the dephasing process (one phase flip). Because the steady state of the master equation is pure, there is only one initial state.

Trajectory 1: no jumps

In this case we evolve only under the evolution equation given by

$$d|\psi\rangle = \left[i\frac{\mu}{2}\sigma_y + \frac{\gamma}{4}(1-\sigma_z)\right]dt|\psi\rangle. \tag{8.139}$$

Rewriting this as a differential equation, and exponentiating we have

$$|\psi(t)\rangle = e^{\gamma t/4}\exp\left[-(\gamma\sigma_z + i2\mu\sigma_y)t/4\right]|0\rangle. \tag{8.140}$$

Now using the formula in Eq. (D.4), and expanding to first order in ε we get

$$|\psi(t)\rangle = e^{\gamma t/4}\left(\cos(\mu t/2) - \frac{\gamma}{2\mu}\sin(\mu t/2)\right)|0\rangle + e^{\gamma t/4}\sin(\mu t/2)|1\rangle. \tag{8.141}$$

The probability that this is the final state at time τ is given by the Poisson probability that there are no jumps up until τ, multiplied by the square norm of $|\psi(\tau)\rangle$. But what we really want to know is the contribution from this trajectory to the probability that the system ends in the target state. This is the Poisson probability that there are no jumps multiplied by $|\langle\psi(\tau)|T\rangle|^2$, where $|T\rangle$ is the target state. From the Poisson distribution above, the probability that there is no emission is $e^{-\gamma\tau}$, and the probability that there is no phase flip is $e^{-\kappa\tau}$. Since these two Poisson processes are independent of each other, the probability that there is no jump at all is $e^{-\gamma\tau}e^{-\kappa\tau}$. Multiplying this by $|\langle\psi(\tau)|T\rangle|^2$, noting that $\tau = \pi/(2\mu)$ and expanding to first order in $\varepsilon \sim \gamma\tau \sim \kappa\tau$, the probability that the system takes this trajectory and ends in the target state is

$$P_0 = 1 - \frac{\pi\gamma}{4\mu} - \frac{\pi\kappa}{2\mu}. \tag{8.142}$$

Trajectory 2: one emission

It is useful to first calculate the contribution to the probability of this trajectory that comes from the Poisson process. The probability that there is one emission is $\gamma\tau e^{-\gamma\tau}$, and the probability that there is no phase flip is $e^{-\kappa\tau}$, to give a total joint probability $P_1 = \gamma\tau e^{-(\gamma+\kappa)\tau}$. Since the emission time t is uniformly distributed over the interval $[0,\tau]$, the probability density for the single jump is $P_1/\tau = \gamma e^{-(\gamma+\kappa)\tau}$. The Poisson contribution to the probability for the 1-emission trajectory, when integrated over the time t of the single jump, is therefore proportional to $\gamma\tau$. As a result we need only calculate the evolution of the state to zeroth order in ε, since all first-order terms will become second-order when multiplied by $\gamma\tau$. This in turn means that we can evolve the state between jumps using only the control Hamiltonian.

Since we do not know the time at which the emission occurs, we must calculate the evolution for an arbitrary emission time $t \in [0,\tau]$. Up until the emission at time t the evolution due to the control Hamiltonian is

$$|\psi(t)\rangle = \exp(i\mu\sigma_y/2)|0\rangle = \cos(\mu t/2)|0\rangle + \sin(\mu t/2)|1\rangle. \tag{8.143}$$

We now apply the jump, which means multiplying by σ, to give $|\psi(t_+)\rangle = \sin(\mu t/2)|0\rangle$. Finally we evolve again under the control Hamiltonian up until time τ, so the final state is

$$|\psi(\tau|t)\rangle = \sin(\mu t/2)\left[\cos(\mu[\tau - t]/2)|0\rangle + \sin(\mu[\tau - t]/2)|1\rangle\right]. \tag{8.144}$$

To determine the total probability that the system ends in the target state when it takes this trajectory, we must integrate the contribution over all the jump times t. Recalling from above that the Poisson probability density for the jump time t is P_1/τ, the total probability density for ending in the target state is $(P_1/\tau)\langle\psi(\tau|t)|T\rangle$. Integrating this expression over t, using $\tau = \pi/(2\mu)$, and expanding to first order in ε gives

$$P_1 = \frac{\gamma}{8\mu}\left(\pi - \frac{5}{4}\right). \tag{8.145}$$

Trajectory 3: one phase flip
The calculation for this trajectory is very similar to that above for the trajectory with one emission. We evolve the initial state using the control Hamiltonian up until time t, apply the phase-flip operator, and then evolve until the final time τ. The state at time t is given by Eq. (8.143), and we note that applying a phase flip is the same as negating the angle of the rotation:

$$\sigma_z|\psi(t)\rangle = \cos(-\mu t/2)|0\rangle + \sin(-\mu t/2)|1\rangle. \tag{8.146}$$

The rest of the evolution by the control Hamiltonian rotates the state by an additional angle $\theta = \mu[\tau - t]/2$, and so the final state is

$$|\psi(\tau|t)\rangle = \cos(\mu[\tau - 2t]/2)|0\rangle + \sin(\mu[\tau - 2t]/2)|1\rangle. \tag{8.147}$$

This time the Poisson probability density for the trajectory is $\kappa e^{-(\gamma+\kappa)\tau}$. Integrating the resulting probability density for the target state over the time t as for the previous trajectory, and expanding to first order in ε, we obtain

$$P_2 = \frac{\pi\kappa}{4\mu}. \tag{8.148}$$

The performance of the protocol
The total probability that the qubit is in the target state at the end of the protocol is the sum of the contributions from the three trajectories, and is thus

$$P(|T\rangle) = 1 - \frac{\pi\kappa}{4\mu} - \frac{\pi\gamma}{8\mu}\left[1 - \frac{5}{4\pi}\right]. \tag{8.149}$$

Since we used an ideal controller, and the optimal protocol, this is an absolute upper bound on the probability with which a physical qubit can be prepared, under Markovian dephasing and damping, given a control Hamiltonian with speed μ. Of course, we cannot know whether the preparation has succeeded or not, so $P(|T\rangle)$ is also an upper bound on the fidelity.

Limits to cooling of oscillators and qubits

We can use the method above to determine the probability for preparing an oscillator in the ground state, when it is coupled to a thermal bath at temperature T. This calculation is longer than that in the example above because we must evolve for each initial energy eigenstate. Nevertheless, one can sum all the contributions from each trajectory and each initial state, and obtain an analytic expression for the probability that the ground state is prepared. The result is

$$P(|0\rangle) = 1 - \frac{\pi\gamma}{2\mu}n_T\left(1 + n_T\frac{(3+n_T)}{4(1+n_T)^2} + n_T^2\frac{(3+n_T)}{2(1+n_T)^2}\right), \qquad (8.150)$$

where n_T is the average number of phonons in the resonator at the temperature of the bath, and γ is the damping rate of the resonator. When comparing the above expression with that obtained with sideband cooling, Eq. (8.104), one must remember that since the coupling in sideband cooling is proportional to the position x, the maximum eigenvalue of the interaction Hamiltonian depends effectively on the maximum number of phonons in the oscillator, and thus on the thermal occupation n_T. When $n_T \ll 1$, and within the RWA, the strength of the interaction is $\mathfrak{s}(H_{\text{int}}) = \lambda$. For $n_T \ll 1$ the effective strength is $\sim \lambda n_T$.

The upper bound on the fidelity for preparing a qubit in the ground state, when in contact with a thermal bath at temperature T, is

$$P(|0\rangle) = 1 - \frac{\pi\gamma}{2\mu}P_T\left(\frac{1 - P_T/4}{1 - 2P_T}\right), \qquad (8.151)$$

where P_T is the excited-state population at the temperature T.

Steady-state versus instantaneous state-preparation

We have considered above the preparation of a state at a single instant of time. State-preparation methods such as sideband cooling prepare instead a steady state. We now present an argument that says that steady-state preparation can never be as good as instantaneous preparation. The noise process that drives a system will have some steady state, and it is the goal of a control protocol to prepare the system in a different state. To do this the protocol must apply a dynamical process that takes the system out of the steady state and moves it toward the target state. This process takes some time, τ. During this time the noise process takes population out of the target state, and distributes it to other states as it tries to restore the steady state. It is this population that the control process cannot completely return to the target because doing so will again take a non-zero time, during which the noise process will continue to depopulate the target.

Now consider the difference between a control process that tries to maintain a system in the target state, and one that merely tries to place it in the target state at a single time. In the latter case, the initial state of the system is the steady state of the noise process. It is only as the protocol takes the system out of this steady state that the noise induces a net transition rate back to this state. The further the populations are from the steady state, the greater the net transition rates induced by the noise. So as the control protocol acts, the leakage out of the target increases, reaching a maximum at the preparation time τ. If we now wish to apply control operations to *keep* the system in the target state, then the leakage rates out of the target are at their maximum to start with. The effect of the noise is therefore larger when the system begins the control process out of equilibrium, which it must do when the control is acting to continuously maintain the target state.

We conclude from the above arguments that the highest possible fidelity can only be obtained at some fixed time τ after a system leaves its steady state. After this time the best achievable fidelity drops monotonically until it reaches the best achievable steady-state fidelity, whatever that is.

History and further reading

Further details regarding the mathematical structure of quantum control and the controllability of quantum systems can be found in the text by D'Alessandro [134]. Further details regarding numerical optimization methods can be found in Nocedal and Wright [451]. Further information relating specifically to the application of numerical optimization to quantum systems can be found in [557, 538, 387, 430, 182]. The fast state-swapping scheme we discussed in Section 8.1.1 was drawn from [650]. Many schemes have been devised for preparing single quantum systems in non-classical states. In Section 8.1.2 we drew from the work in [706, 357, 20, 605, 465]. The reader might also find the following papers interesting entry points into the subject [376, 432, 580].

Cooling of resonators by measurement-based quantum feedback has been considered in [398, 153, 264, 485, 204, 205, 712, 428, 240, 241, 28]. Our analysis followed that in [28]. The method for preparing a resonator in a squeezed state using feedback control was first introduced by Braginsky *et al.* [84, 83], and this and similar schemes have been analyzed further in [526, 125, 584, 585, 586]. Our analysis was drawn from [125].

The 2013 Nobel Prize in physics was awarded to David Wineland for his pioneering work in the precise control of quantum states of trapped ions [427, 364] (jointly with Serge Haroche who realized precise control of optical modes [91, 498]), and it was for preparing these ions in the ground state that resolved-sideband cooling was first devised [665, 149]. It was later realized that the same technique could be used for cooling resonators, and the first theoretical analyses of this cooling technique were given by Marquadt *et al.* [403] and Wilson-Rae *et al.* [664]. The first full and exact calculation of resolved sideband cooling for resonators was given by Genes *et al.* [204, 205]. Our exact calculation, detailed in Appendix G, follows that in [28] and is a little different to theirs. They used the Langevin equations for asymmetric damping (Eqs. 4.118 and 4.119), but in the high-temperature regime, whereas we used the symmetric Langevin equations valid for weak damping but for all temperatures. That sideband cooling could be understood as coupling the resonator to an auxiliary system at zero temperature was first elucidated by Tian [604]. Further cooling schemes that build on this view of sideband cooling can be found in [386, 650, 388].

Our comparison of measurement-based feedback cooling and sideband cooling was drawn from [28].

The method discussed in Section 8.3.3 for finding optimal preparation protocols for quantum systems was developed by Wang *et al.* [651].

Sideband cooling has been realized experimentally in nano-electromechanical and optomechanical systems [542, 543, 112, 597].

Exercises

1. Derive the expressions in Eq. (8.3) by using Eqs. (D.4) and (D.5).
2. Here we examine the norms of N-dimensional Hamiltonians that transform a state $|0\rangle$ to an orthogonal state $|1\rangle$. To do this we first determine the structure of all unitaries that perform this transformation. Since unitaries preserve the inner product, we know that the state $|1\rangle$ must be transformed to a state orthogonal to itself, $|\psi_2\rangle$ (but not necessarily orthogonal to $|0\rangle$). Similarly $|\psi_2\rangle$ must be transformed to a state orthogonal to itself, say $|\psi_3\rangle$, and so on. This chain of successive orthogonal states closes when the last state selected closes the space. For example, if we choose $|1\rangle \rightarrow |\psi_2\rangle = |0\rangle$, then this closes the space so that the 2D space $\{|0\rangle, |1\rangle\}$ is closed under the action of the unitary. If we choose the chain $|\psi_2\rangle = (|0\rangle + |2\rangle)/\sqrt{2}$ and $|\psi_3\rangle = (|0\rangle - |2\rangle)/\sqrt{2}$ then $|\psi_3\rangle$ closes the space so that the 3D space $\{|0\rangle, |1\rangle, |2\rangle\}$ is closed under the unitary. We also note that each mapping can involve multiplication by an arbitrary phase factor, so that there are also N independent phases.

   The above analysis shows that all unitaries that map a given state to an orthogonal state are composed of M unitaries that act on disjoint (mutually orthogonal) subspaces, where the mth unitrary has dimension N_m, and is completely characterized by a chain of N_m successively orthogonal states, along with N_m independent phase factors.

   (i) Consider the 2D unitary that transforms $|0\rangle \rightarrow |1\rangle \rightarrow |0\rangle \rightarrow |0\rangle$. Determine the constant Hamiltonians that generate this unitary, and determine their norms when they achieve the mapping $|0\rangle \rightarrow |1\rangle$ in the time $t = \pi/(2\lambda)$. You can do this easily with a computational tool such as Matlab.

   (ii) Consider the 3D unitary that transforms $|0\rangle \rightarrow |1\rangle \rightarrow |2\rangle \rightarrow |0\rangle$, and another that transforms $|0\rangle \rightarrow |1\rangle \rightarrow (|2\rangle + |0\rangle)/\sqrt{2} \rightarrow (|2\rangle - |0\rangle)/\sqrt{2}$. Do the same thing for these unitaries that you did in (i).

   (iii) What insight do the results from (i) and (ii) give you into the fastest way to transform $|0\rangle \rightarrow |1\rangle$ under a constraint on the norm of the Hamiltonian? The fastest path under a constraint on the norm has been determined in [9, 401, 99].
3. Show that the displacement operator, $D(\alpha) \equiv \exp(\alpha a - \alpha^* a^\dagger)$ transforms the ground state $|0\rangle$ to the coherent state $|\alpha\rangle$. (See Appendix D for the definition of a coherent state.) The neatest way to do this is to use Eq. (G.29) to calculate the action of a on $D(\alpha)|0\rangle$. determine the inner product of $D(\alpha)|0\rangle$ with the number state $|n\rangle$. You may also like to use the method described in Section G.7 to derive Eq. (G.29).

4. You want to apply the operator $\exp(-i\pi\sigma_y/4)$ to a single qubit, initially in the state $|+\rangle = (|0\rangle+|1\rangle)/\sqrt{2}$. But you only have the ability to apply the operations $\exp(-ia\sigma_x)$ and $\exp(-ia\sigma_z)$. You can apply each of these operations as many times as you like, and on each application you can choose any value for a. The total time required for a sequence of these operations is the sum of the values of a for all the applications. What is the shortest time in which you can realize the operation $\exp(-i\pi\sigma_y/4)$, and how many applications of each of the two operations does it require?

5. Just after Eqs. (8.18) and (8.19), we state that $\cos\theta = \Delta/\sqrt{\Delta^2+\lambda^2}$ implies that $\Delta\sigma_z + \lambda\sigma_x = \sqrt{\Delta^2+\lambda^2}\,\tilde{\sigma}_z$. Show that this is true.

6. Derive Eq. (8.22) from Eq. (8.21).

7. Derive Eqs. (8.57) and (8.58) from Eqs. (8.50)–(8.52) with $\gamma = 0$.

8. Derive Eqs. (8.54).

9. Derive Eqs. (8.59)–(8.61) using the perturbative procedure described.

10. Derive Eqs. (8.68) – Eqs. (8.69). This involves solving Eqs. (8.54) for their steady states, and then expanding the solutions to first-order in $\varepsilon \sim k/\Gamma \sim \gamma/k$.

11. Derive Eqs. (8.76) and (8.77).

12. Solve for the steady states of Eqs. (8.79)–(8.81) to obtain Eqs. (8.82)–(8.84).

13. Derive Eqs. (8.85) and (8.86).

14. Derive Eq. (8.90) from Eq. (8.89), and Eqs. (8.91) and (8.92) from Eq. (8.90).

15. Consider two oscillators described respectively by the annihilation operators a and b. Show that the interaction $H_{\text{int}} = \hbar g(ab^\dagger+ba^\dagger)$ will swap the states of the two oscillators in a time of $t = \pi/(2g)$.

16. Show that the interaction $H_{\text{int}} = \hbar g(\sigma_x\sigma_x + \sigma_y\sigma_y + \sigma_z\sigma_z)$ between two qubits will swap the states of the two qubits, for any initial states, and determine the time at which the swap is realized. Use Matlab if you wish. For an interesting extension, see [78].

17. Draw the circuit diagram for a superconducting implementation of the field-mediated resolved-sideband cooling scheme described in Section 8.3.2.

18. Find the value of k beyond which resolved-sideband cooling does better than measurement-based feedback cooling. The relevant equations are Eqs. (8.122) and (8.123).

19. Consider the performance of resolved-sideband cooling, given in Eq. (8.122), and that of measurement-based feedback cooling, given in Eq. (8.123). Find the values of k that give the best cooling, respectively, for both cooling schemes. Substitute these into the respective equations for the performance of each scheme. For what value of $Q \equiv \gamma/\omega$ does resolved-sideband cooling perform better than measurement-based feedback cooling? How does the value of $\langle b^\dagger b\rangle$ scale with Q for each scheme?

20. Fill in the steps in the calculation of the three trajectories in the example of optimal state-preparation that we worked out in Section 8.3.3.

21. For what values of μ is the expression in Eq. (8.150) a good approximation? (i) Answer this question for $n_T \lesssim 1$ and $n_T \gg 1$. Recall that the regime of validity is when $\mu \gg \gamma(n_T + 1)n_{\text{max}}$, as discussed below Eq. (8.125). (ii) From your answer in (i), is the expression in Eq. (8.150) only valid when $P(|0\rangle) \ll 1$?

22. Consider an optical cavity mode damped at the rate γ into a zero-temperature bath. The master equation for the mode is that given in Eq. (4.82) with $n_T = 0$. (i) Write down the linear stochastic Schrödinger equation for the mode. Given that the initial state of the mode is an equal mixture of 0 and 1 photons: (ii) for both initial states solve the evolution of the trajectory that has no jump; (iii) solve the evolution of the trajectory with one jump. (iv) Average over these trajectories to determine the average number of photons in the mode at time t. (v) Calculate the probability that there is a jump in the interval $[0, 1/\gamma]$.

23. Consider preparing a single qubit in the excited state $|1\rangle$, given that it is subject to depolarizing noise given by the master equation in Eq. (5.77). Assume that the control operations have a rate λ that is much greater than the rate γ of the depolarizing noise. (i) Elucidate the different Hamiltonians that must be applied to the system to prepare $|1\rangle$, and the Hamiltonians that must be subsequently applied to correct for the leakage to first order in γ/λ. (ii) From this list of Hamiltonians determine the size of auxiliary required to perform the optimal preparation.

24. Consider the optimal protocol for preparing a qubit in a pure state, as in the example we worked out in Section 8.3.3. But this time determine the optimal protocol for preparing an arbitrary pure state, $|\psi\rangle = \cos\theta|0\rangle + e^{-i\phi}\sin\theta|1\rangle$. (i) For what values of θ will an auxiliary system by required? (ii) Calculate the probability that the state preparation succeeds.

Appendix A

The tensor product and partial trace

Combining two systems: the tensor product

The state of a quantum system is a vector in a complex vector space. (Technically, if the dimension of the vector space is infinite, then it is a separable Hilbert space.) Here we will always assume that our systems are finite-dimensional. We do this because everything we will discuss transfers without change to infinite-dimensional systems. Further, when one actually simulates a system on a computer, one must always truncate an infinite-dimensional space so that it is finite.

Consider two separate quantum systems. Now consider the fact that if we take them together, then we should be able to describe them as one big quantum system. That is, instead of describing them by two separate state vectors, one for each system, we should be able to represent the states of both of them together using one single state vector. (This state vector will naturally need to have more elements in it than the separate state-vectors for each of the systems, and we will work out exactly how many below.) In fact, in order to describe a situation in which the two systems have affected each other – by interacting in some way – and have become correlated with each other as a result of this interaction, we will *need* to go beyond using a separate state vector for each. If each system is in a pure state, described by a state vector for it alone, then there are no correlations between the two systems. Using a single state vector to describe the joint state of the two systems is the natural way to describe all possible states that the two systems could be in (correlated or uncorrelated). We now show how this is done.

Let us say we have two quantum systems, A and B, both of which are two-dimensional. These could be, for example, the spins of two spin-1/2 particles. Let us denote the basis states of A by $|A_0\rangle$ and $|A_1\rangle$, and those of B as $|B_0\rangle$ and $|B_1\rangle$. We need to determine a vector-space that will describe both systems together. First, let us determine a set of basis vectors for this space. We know that each of the two systems can be in one of two mutually orthogonal (distinguishable) states. This means that the systems together can be in one of four possible distinguishable states: for each state of system A there are two possible states of system B, and $2 \times 2 = 4$. So these states must be a set of basis states for the combined system. Let us denote these four basis states as

$$|C_{00}\rangle = |A_0\rangle|B_0\rangle,$$
$$|C_{01}\rangle = |A_0\rangle|B_1\rangle,$$
$$|C_{10}\rangle = |A_1\rangle|B_0\rangle,$$
$$|C_{11}\rangle = |A_1\rangle|B_1\rangle. \tag{A.1}$$

A state of the combined system is thus a four-dimensional vector, being a complex linear combination of the four basis states. Let us denote the state of the combined system as

$$|\psi_C\rangle = \sum_{ij} c_{ij}|C_{ij}\rangle, \tag{A.2}$$

and write it as

$$|\psi_C\rangle = \begin{pmatrix} c_{00} \\ c_{01} \\ c_{10} \\ c_{11} \end{pmatrix}. \tag{A.3}$$

The next question we need to ask is, if A has been prepared in state $|\psi_A\rangle = a_0|A_0\rangle + a_1|A_1\rangle$, and B has been separately prepared in state $|\psi_B\rangle = b_0|B_0\rangle + b_1|B_1\rangle$, what is the state vector that describes the combined system? To work this out we first note that in this case, because A and B have been prepared separately, they are uncorrelated. That is, the state of A in no way depends on the state of B. In particular, the probabilities that A is in either of its basis states do not depend on which basis state B is in, and vice versa. We next note that the state of the combined system that we seek must give the same probabilities for the basis states of A and B as do the individual states of A and B. If we denote the probabilities for the A basis states as $P(A_i)$, and those of the B basis states as $P(B_j)$, then $P(A_i) = |a_i|^2$ and $P(B_j) = |b_j|^2$. Because the states of A and B are *uncorrelated*, the joint probability distribution for the basis states of A and B is just the product of the distributions for each. That is,

$$P(A_i, B_j) = P(A_i)P(B_j) = |a_i|^2|b_j|^2 = |a_i b_j|^2. \tag{A.4}$$

The probability distribution for the states of A and B as determined by $|\psi_C\rangle$ is

$$P(A_i, B_j) = |c_{ij}|^2. \tag{A.5}$$

Equating Eqs. (A.4) and (A.5) we see that we require $|c_{ij}|^2 = |a_i b_j|^2$, and we can satisfy this by choosing

$$c_{ij} = a_i b_j. \tag{A.6}$$

Using this relation, the state of the joint system is

$$|\psi_C\rangle = \begin{pmatrix} a_0 b_0 \\ a_0 b_1 \\ a_1 b_0 \\ a_1 b_1 \end{pmatrix} = \begin{pmatrix} a_0 \begin{pmatrix} b_0 \\ b_1 \end{pmatrix} \\ a_1 \begin{pmatrix} b_0 \\ b_1 \end{pmatrix} \end{pmatrix} = \begin{pmatrix} a_0 |\psi_B\rangle \\ a_1 |\psi_B\rangle \end{pmatrix}. \tag{A.7}$$

So to form the vector for the combined system, we have multiplied the vector for B by each of the elements of the vector for A, and stacked the resulting vectors end-to-end. This operation is called "taking the tensor product of the vectors of A and B," and is denoted by $\otimes$. We write

$$|\psi_C\rangle = |\psi_A\rangle \otimes |\psi_B\rangle. \tag{A.8}$$

Naturally the density matrix for the combined system is $\rho^C = |\psi_C\rangle\langle\psi_C|$. If we denote the density matrix of A by $\rho = |\psi_A\rangle\langle\psi_A|$ and that of B by $\sigma = |\psi_B\rangle\langle\psi_B|$, then a little calculation (this is a good exercise) shows that

$$\rho^C = \begin{pmatrix} \rho_{11}\begin{pmatrix} \sigma_{11} & \sigma_{12} \\ \sigma_{21} & \sigma_{22} \end{pmatrix} & \rho_{12}\begin{pmatrix} \sigma_{11} & \sigma_{12} \\ \sigma_{21} & \sigma_{22} \end{pmatrix} \\ \rho_{21}\begin{pmatrix} \sigma_{11} & \sigma_{12} \\ \sigma_{21} & \sigma_{22} \end{pmatrix} & \rho_{22}\begin{pmatrix} \sigma_{11} & \sigma_{12} \\ \sigma_{21} & \sigma_{22} \end{pmatrix} \end{pmatrix} = \begin{pmatrix} \rho_{11}\sigma & \rho_{12}\sigma \\ \rho_{21}\sigma & \rho_{22}\sigma \end{pmatrix}, \tag{A.9}$$

where the ρ_{ij} are the elements of the density matrix of A. We see that to form the density matrix of the combined system we multiply the density matrix of B with each of the elements of the density matrix of A. This is the same operation that we used to obtain the state vector for the combined system, but now generalized for matrices. We write it as

$$\rho^C = \rho \otimes \sigma. \tag{A.10}$$

We know now how to determine the state of the total system, when each of the subsystems is in a separate state, independent of the other. Next we need to know how to write the (two-dimensional) operators for each system as operators in the (four-dimensional) combined space. Let us consider first an operator U_B acting on B. We know that for a state such as $|\psi_C\rangle$ above, in which each system is in its "own" state, uncorrelated with the other system, the operator U_B must change the state of B, but leave the state of A alone. This means that U_B must change the coefficients b_0 and b_1 appearing in $|\psi^C\rangle$, while leaving the coefficients a_0 and a_1 unchanged. This is precisely the action of the operator

$$U_{full}^B = \begin{pmatrix} U^B & \begin{pmatrix} 0 & 0 \\ 0 & 0 \end{pmatrix} \\ \begin{pmatrix} 0 & 0 \\ 0 & 0 \end{pmatrix} & U^B \end{pmatrix} = I_2 \otimes U^B, \tag{A.11}$$

where I_2 is the two-dimensional identity operator. Using the same argument, an operator for system A must take the form

$$
U^{\mathrm{A}}_{\mathrm{full}} =
\begin{pmatrix}
U^{\mathrm{A}}_{11} \begin{pmatrix} 1 & 0 \\ 0 & 1 \end{pmatrix} & U^{\mathrm{A}}_{12} \begin{pmatrix} 1 & 0 \\ 0 & 1 \end{pmatrix} \\
U^{\mathrm{A}}_{21} \begin{pmatrix} 1 & 0 \\ 0 & 1 \end{pmatrix} & U^{\mathrm{A}}_{22} \begin{pmatrix} 1 & 0 \\ 0 & 1 \end{pmatrix}
\end{pmatrix}
= U^{\mathrm{A}} \otimes I_2,
\tag{A.12}
$$

where the U^{A}_{ij} are the matrix elements of U^{A}.

An interaction between systems A and B is described by a term in the Hamiltonian that depends jointly upon physical observables of both systems. As an example, the total energy of the combined system may contain a term that is the product of the values of the z-component of spin for the two systems. We can determine the four-dimensional operator that corresponds to this product by working in the basis in which the operators for the z-components of spin for the two systems are diagonal (the eigenbasis of both operators). Let us denote the σ_z operator for A by σ_z^{A}, and that for B by σ_z^{B}. Let us denote the respective eigenstates of these operators as $|\pm\rangle_{\mathrm{A}}$ and $|\pm\rangle_{\mathrm{B}}$, so that

$$
\sigma_z^{\mathrm{A}} |\alpha\rangle_{\mathrm{A}} = \alpha |\alpha\rangle_{\mathrm{A}}, \quad \text{where } \alpha = \pm 1,
\tag{A.13}
$$

$$
\sigma_z^{\mathrm{B}} |\beta\rangle_{\mathrm{B}} = \beta |\beta\rangle_{\mathrm{B}}, \quad \text{where } \beta = \pm 1.
\tag{A.14}
$$

If we denote the operator that corresponds to the product of σ_z^{A} and σ_z^{B} as $\sigma_z^{\mathrm{A}} \sigma_z^{\mathrm{B}}$, then it must have the property that

$$
\sigma_z^{\mathrm{A}} \sigma_z^{\mathrm{B}} |n\rangle_{\mathrm{A}} |m\rangle_{\mathrm{B}} = \alpha\beta |\alpha\rangle_{\mathrm{A}} |\beta\rangle_{\mathrm{B}}.
\tag{A.15}
$$

By writing the basis states of the joint system as products of the bases $\{|\pm\rangle_{\mathrm{A}}\}$ and $\{|\pm\rangle_{\mathrm{B}}\}$, it is clear that $\sigma_z^{\mathrm{A}} \sigma_z^{\mathrm{B}}$ is given by

$$
\sigma_z^{\mathrm{A}} \sigma_z^{\mathrm{B}} = \sigma_z^{\mathrm{A}} \otimes \sigma_z^{\mathrm{B}}.
\tag{A.16}
$$

It is important to note that when a joint system evolves under a Hamiltonian that includes interactions between the subsystems, the resulting state of the joint system can, in general, no longer be written as a tensor product of the states of each of the subsystems. When two systems interact their states becomes correlated. This means that the joint probability distribution giving the probabilities that each of them are in either of two basis states no longer factors into the respective probability distributions for each subsystem. Further, the state of the two systems will, in general, become *entangled*, such that the correlations between observables of both subsystems can no longer be described by classical correlations. In this book we do not deal with the properties of entanglement – this interesting subject is discussed in detail in, e.g., [484, 267, 631, 447].

Note that if we reorder the basis states of the combined system we can swap the roles of system A and B in our representation. That is, we can alternatively use the convention

$$\rho^C = \begin{pmatrix} \sigma_{11}\, \rho & \sigma_{12}\, \rho \\ \sigma_{21}\, \rho & \sigma_{22}\, \rho \end{pmatrix} \equiv \rho \otimes \sigma \qquad (A.17)$$

for the density matrix of the combined system, and similarly for other operators.

There are certain shorthand notations that are useful when dealing with combined systems. We have, in fact, already used one: now that we have defined the tensor product, we see that $|a_0\rangle |b_0\rangle$ is actually a shorthand notation for $|a_0\rangle \otimes |b_0\rangle$. Since operators in the combined space consist of a matrix of sub-blocks of operators that act in one of the spaces alone, it is useful to have a ket-style notation for these sub-blocks. Consider a general operator in the full space given by

$$V = \sum_{nn'jj'} v_{nj,n'j'} |c_{nj}\rangle \langle c_{n'j'}| = \sum_{nn'jj'} v_{nj,n'j'} |a_n\rangle |b_j\rangle \langle a_{n'}| \langle b_{j'}|. \qquad (A.18)$$

Here the numbers $v_{nj,n'j'}$ are the matrix elements of V. If we sandwich V between the basis states $|a_n\rangle |b_j\rangle$ and $|a_{n'}\rangle |b_{j'}\rangle$, then we pick out one of its elements:

$$\langle a_n| \langle b_j|V|a_{n'}\rangle |b_{j'}\rangle = v_{nj,n'j'}. \qquad (A.19)$$

It is convenient to define $\langle a_n|V|a_{n'}\rangle$ as the sub-block of V corresponding to the (n,n') element of system A. That is,

$$\langle a_n|V|a_{n'}\rangle \equiv \sum_{jj'} v_{nj,n'j'} |b_j\rangle \langle b_{j'}|, \qquad (A.20)$$

and this is an operator acting in the two-dimensional space of system B alone. This is handy notation because it means that we can write

$$
\begin{aligned}
v_{nj,n'j'} &= \langle b_j| \langle a_n|V|b_{j'}\rangle |a_{n'}\rangle \\
&= \langle b_j| \left(\langle a_n|V|a_{n'}\rangle \right) |b_{j'}\rangle \\
&= \langle b_j| \left(\sum_{kk'} v_{nk,n'k'} |b_k\rangle \langle b_{k'}| \right) |b_{j'}\rangle.
\end{aligned} \qquad (A.21)
$$

This notation is equivalent to the partial inner product defined in Chapter 1.

The partial inner product notation is also useful for describing von Neumann measurements on one of the subsystems. A von Neumann measurement projects a system onto one of a set of basis states (see Sections 1.3.1 and 1.3.3). Let us say that our system is in the state $|\psi\rangle = \sum_n c_n |n\rangle$ (where $\{|n\rangle\}$ is an orthonormal basis), so that the density matrix is $\rho = |\psi\rangle \langle \psi|$. A von Neumann measurement in the basis $\{|n\rangle\}$ will project the system onto the state

$$\tilde{\rho}_n = |n\rangle \langle n| = \frac{1}{\mathcal{N}} \left(|n\rangle \langle n| \right) \rho \left(|n\rangle \langle n| \right), \qquad (A.22)$$

(where $\mathcal{N}$ is chosen so that the resulting state is normalized) and this occurs with probability

$$|c_n|^2 = \langle n|\rho|n\rangle. \tag{A.23}$$

So how do we describe a von Neumann measurement on system A in the combined space of A and B? If the von Neumann measurement tells us that the state of system A is $|a_0\rangle$, then since the measurement must not have any direct action on system B, the state of system B must simply be the sub-block of the density matrix that corresponds to $|a_0\rangle$. So the state of system B after the measurement is, in our convenient notation,

$$\rho^B = \frac{1}{\mathcal{N}}\langle a_0|\rho^C|a_0\rangle. \tag{A.24}$$

The state of the joint system after the measurement is

$$\frac{1}{\mathcal{N}}(|a_0\rangle\langle a_0|)\rho^C(|a_0\rangle\langle a_0|) = |a_0\rangle\langle a_0| \otimes \rho^B. \tag{A.25}$$

The probability of getting this result is the sum over all the diagonal elements of ρ^C that correspond to system A being in state $|a_0\rangle$. This is

$$P(|a_0\rangle) = \mathrm{Tr}[\langle a_0|\rho^C|a_0\rangle] = \mathcal{N}. \tag{A.26}$$

Discarding a system: the partial trace

Now that we understand how to represent the states and operators of two systems in a joint space, another question arises: how do we retrieve the state of one of the subsystems from the state of the joint system? To answer this, first we must explain what we mean by "the state of one system." If the two systems are together in a pure state, but this state is *correlated*, meaning that measurements on one system will in general be correlated with measurements on the other, then each of the systems cannot be represented by a pure state.

We can nevertheless define the state of one of the systems alone as an object from which we can determine the probabilities of the outcomes of all measurements on that system alone, and that does not refer to the other system. It turns out that it is possible to obtain such an object, and it is simply a density matrix for the single system. That is, while the state of a single system cannot necessarily be written as a state vector, it can be written as a density matrix, and this is true for *any* joint state of the combined systems, pure or mixed. We now explain how to calculate the density matrix for a single system, when it is part of a larger system.

First let us consider the trivial case in which each system has been separately prepared in its own state. In this case the joint state is simply $|\psi_C\rangle = |\psi_A\rangle \otimes |\psi_B\rangle$. Writing the joint state as a density matrix we have

$$\rho^C = \rho \otimes \sigma = \begin{pmatrix} \rho_{11}\,\sigma & \rho_{12}\,\sigma \\ \rho_{21}\,\sigma & \rho_{22}\,\sigma \end{pmatrix}. \tag{A.27}$$

Because $\rho_{11} + \rho_{22} = 1$, we see that we retrieve the density matrix for system B by summing the diagonal sub-blocks corresponding to the states of system A:

$$\rho_{11}\sigma + \rho_{22}\sigma = (\rho_{11} + \rho_{22})\sigma = \text{Tr}[\rho]\sigma = \sigma. \tag{A.28}$$

Summing the diagonal elements of a matrix is called the "trace" operation. Summing the diagonal sub-blocks that correspond to a state of system A, so as to obtain the density matrix of system B, is call taking the "partial trace over system A." If we take the partial trace over system A, and then take the trace of the resulting density matrix, the result is the same as taking the full trace over the combined system. Similarly, we can take the partial trace over system B, to obtain the density matrix for system A. This is

$$\text{Tr}_B[\rho^C] = \begin{pmatrix} \rho_{11}\,\text{Tr}[\sigma] & \rho_{12}\,\text{Tr}[\sigma] \\ \rho_{21}\,\text{Tr}[\sigma] & \rho_{22}\,\text{Tr}[\sigma] \end{pmatrix} = \begin{pmatrix} \rho_{11} & \rho_{12} \\ \rho_{21} & \rho_{22} \end{pmatrix} = \rho. \tag{A.29}$$

While we have only shown that the partial trace gives the correct answer when the sub-systems are in separate (unentangled) states, it seems reasonable that this should remain true when they are not. We now present two arguments that show that it does. The first is obtained by noting that the density matrix for system B alone must give the same expectation value for all observables of system B as does the joint density matrix. Because the full trace over the joint system is the same as taking the partial trace over A followed by the trace over system B, this is guaranteed to be true if the density matrix of B is equal to the partial trace over A:

$$\langle X_B \rangle = \text{Tr}[X_B \rho^B] = \text{Tr}[X_B(\text{Tr}_A[\rho^C])] = \text{Tr}[(I \otimes X_B)\rho^C]. \tag{A.30}$$

That the last equality here is true can be easily seen by writing out ρ^C in terms of its sub-blocks. Since Eq. (A.30) is true for *every* observable of system B, ρ^B defined as the partial trace $\text{Tr}_A[\rho^C]$ must be the correct state of system B.

A second way to see that the partial trace $\text{Tr}_A[\rho^C]$ gives the correct density matrix for system B employs measurements and the resulting states-of-knowledge, and so is very much in the theme of this book. We note that if an observer, whom we will call Alice, only has access to system B, then she cannot know what has happened to system A. In particular, she cannot know what measurements have been performed on A. If another observer has made a von Neumann measurement on A, then Alice's state-of-knowledge is given by averaging over the possible outcomes of this measurement. If A is measured in the basis $\{|a_n\rangle\}$, then the probability of getting the final state

$$\rho_n^B = \frac{1}{\mathcal{N}_n} \langle a_n | \rho^C | a_n \rangle \tag{A.31}$$

is

$$P(n) = \text{Tr}[\langle a_n | \rho^C | a_n \rangle] = \mathcal{N}_n. \tag{A.32}$$

So Alice's state-of-knowledge is

$$\rho^{\mathrm{B}} = \sum_n P(n)\rho_n^{\mathrm{B}} = \sum_n \langle a_n|\rho^{\mathrm{C}}|a_n\rangle = \mathrm{Tr_A}[\rho^{\mathrm{C}}], \tag{A.33}$$

being precisely the partial trace over A. However, since Alice cannot know what measurement has been made on A, this argument only works (and quantum mechanics is only consistent!) if Alice's state-of-knowledge is the same for *every* measurement that can be made on A. So if Alice's state is indeed given by the partial trace, this trace must be derivable by choosing *any* measurement that can be made on A.

In Section 1.3.3 we show that all quantum measurements are described by a set of operators $\{A_n\}$ that satisfy $\sum_n A_n^\dagger A_n = I$. The probability that result n will occur is $P(n) = \mathrm{Tr}[A_n^\dagger A_n \rho]$, where ρ is the initial density matrix of the measured system. The final state given by result n is $\rho_n = A_n \rho A_n^\dagger / P(n)$. If a general measurement is made on system A, then the state of the combined system, from Alice's point of view, is

$$\rho_{\mathrm{Alice}}^{\mathrm{C}} = \sum_n P(n)\rho_n^{\mathrm{C}} = \sum_n (A_n \otimes I)\rho^{\mathrm{C}}(A_n^\dagger \otimes I). \tag{A.34}$$

Alice's state-of-knowledge of B is the partial trace over this, which is

$$\rho^{\mathrm{B}} = \mathrm{Tr_A}\left[\sum_n (A_n \otimes I)\rho^{\mathrm{C}}(A_n^\dagger \otimes I)\right] = \mathrm{Tr_A}\left[\sum_n (A_n^\dagger \otimes I)(A_n \otimes I)\rho^{\mathrm{C}}\right]$$

$$= \mathrm{Tr_A}\left[\left(\sum_n A_n^\dagger A_n \otimes I\right)\rho^{\mathrm{C}}\right] = \mathrm{Tr_A}\left[(I \otimes I)\rho^{\mathrm{C}}\right] = \mathrm{Tr_A}\left[\rho^{\mathrm{C}}\right], \tag{A.35}$$

and so it is independent of the measurement as required. Here we have used the (very useful!) cyclic property of the trace: $\mathrm{Tr}[ABC] = \mathrm{Tr}[CAB] = \mathrm{Tr}[BCA]$. Note, however, that because we are taking the *partial* trace over A, we can only cycle operators that act purely on A inside the trace – operators that act on the subspace for system B cannot be cycled in this way.

Two examples of correlated systems

There are two fundamentally different ways that two quantum systems can be correlated. The first is the same way in which classical systems can be correlated. An example for two qubits is the mixed state

$$\rho^{\mathrm{C}} = \frac{1}{2}|0\rangle\langle 0| \otimes |0\rangle\langle 0| + \frac{1}{2}|1\rangle\langle 1| \otimes |1\rangle\langle 1|. \tag{A.36}$$

In this case if we measure system B and get $|0\rangle$, then if we measure system A in the basis $\{|0\rangle, |1\rangle\}$, we will always get the result $|0\rangle$. Thus in the basis $\{|0\rangle, |1\rangle\}$, the two systems are perfectly correlated. Classical systems can also be correlated in precisely this way, since the origin of the correlation is merely the classical probability to be in the joint state $|0\rangle\langle 0| \otimes |0\rangle\langle 0|$, or the joint state $|1\rangle\langle 1| \otimes |1\rangle\langle 1|$.

The other way in which two quantum systems can be correlated is when the correlation is due to the joint system being in a *superposition* of various joint states of the two subsystems. For example, if the joint system is in the state $\rho_C = |\psi\rangle\langle\psi|$, where

$$|\psi\rangle = \frac{1}{\sqrt{2}}|0\rangle \otimes |0\rangle + \frac{1}{\sqrt{2}}|1\rangle \otimes |1\rangle, \tag{A.37}$$

then the two systems are again perfectly correlated in the basis $\{|0\rangle, |1\rangle\}$. This time there is an entirely new feature to the correlation, however: if we transform both systems in the same way to a new basis, then we find that they are perfectly correlated in the new basis! This means that they are perfectly correlated in *all bases*. This uniquely quantum mechanical type of correlation also has other remarkable properties that cannot be replicated by classical systems. It is the subject of Bell's inequalities [44, 45, 421, 350, 16, 17, 293] and entanglement [484, 267, 631, 447].

The origin of the term "tensor product"

As a final note, the reason that the product operation "⊗" is called the "tensor product" is due to its relationship with tensors. From a mathematical point of view, a tensor is defined as a linear functional on one or more vector spaces. A functional is something that maps elements in its domain (in this case a set of vectors in a set of vector spaces) to a scalar. Without going into the details, the representation of a vector that acts on two vector spaces is a matrix. When representing tensors as matrices, the natural product of two tensors is given by taking the "tensor product" of their respective matrices.

Appendix B

A fast-track introduction for experimentalists

It is not necessary to understand the full structure of quantum measurement theory to understand quantum measurements in a wide range of experiments. This pedagogical approach was pointed out to me by Alexander Korotkov, and is the one he uses in his work on continuous measurements [343, 524, 638]. When we measure a single observable, and when this observable is not being changed during the measurement by any dynamics other than the measurement process, then only a small addition to Bayesian inference is required to describe quantum measurements. In fact, once we have made this addition, for infinitesimal time-steps we can include a Hamiltonian under which the measured observable changes with time, and obtain a full description of the continuous measurement of any quantum observable.

There is also another situation in which quantum measurement theory simplifies: when the system is linear, and when the observable being measured is a linear combination of the canonical coordinates. In this case, even when the observable is undergoing linear dynamics, a continuous quantum measurement reduces to a classical continuous measurement of a classical linear system, with the addition of a specified amount of white (flat-spectrum) noise. This noise is the "quantum back-action" of the measurement.

Let us examine what happens when we measure a single observable of a quantum system where the observable is not undergoing any dynamics other than the measurement process. In this case the measurement changes the probability distribution for the observable in *precisely* the same way as does a classical measurement. Note that the probability distribution for the observable is the observer's state-of-knowledge of the observable. The change induced in the state-of-knowledge by the measurement (quantum or classical) is determined by the rules of Bayesian inference. (If you are not familiar with Bayesian inference, it is discussed in detail in Section 1.2, "Classical measurement theory.") For example, consider a classical measurement of a variable x, in which x can take any value on the real line. If the error in the measurement is Gaussian, then the probability distribution for the measurement result, y, given the value of x, is

$$P(y|x) = \frac{1}{\sqrt{2\pi\sigma^2}} e^{(y-x)^2/(2\sigma^2)},$$ (B.1)

where σ is the standard deviation of the measurement error. The function $P(y|x)$ is called the likelihood function. If the observer's initial state-of-knowledge of x is given by the probability distribution $P(x)$, then Bayes' theorem tells us that after obtaining the measurement result y the observer's new state-of-knowledge is

$$P(x|y) = \left[\frac{P(y|x)}{\mathcal{N}} \right] P(x).$$ (B.2)

Here $\mathcal{N}$ is the scaling required to normalize $P(x|y)$. We see that the multiplication of $P(x)$ by the likelihood function narrows the probability distribution for x about the value of the measurement result y.

A quantum measurement of a real observable x with a Gaussian error changes the probability distribution for x in the same way as Eq. (B.2). If we use the density matrix to describe the state of the system, then the probability density for x is simply the diagonal of the density matrix. If we denote the density by ρ, and its elements by $\rho_{x,x'}$, then $P(x) = \rho_{x,x}$. The diagonal of ρ therefore changes under a measurement using the usual rules of Bayesian inference. Of course in quantum mechanics the system can be in a superposition over different eigenstates of x, and this is reflected in the values of the off-diagonal elements of the density matrix. So we also need to know how these off-diagonal elements change when a measurement is made. Fortunately the rule is simple.

First we create a diagonal matrix whose diagonal is the likelihood function. More precisely, we define a matrix for each value of the measurement result y. If we denote this matrix by $D(y)$, and its elements by $D(y)_{x,x'}$, then the diagonal of $D(y)$ is chosen to be $D(y)_{x,x} = P(y|x)$. The density matrix after the measurement is given by the rule

$$\tilde{\rho} = \frac{\sqrt{D(y)}\rho\sqrt{D(y)}}{\mathcal{N}},$$ (B.3)

where $\mathcal{N}$ is the normalization. By multiplying out the matrices in the above expression we see immediately that the change in the diagonal is merely that given by Bayesian inference:

$$\tilde{\rho}_{x,x} = \frac{\sqrt{P(y|x)}\rho_{x,x}\sqrt{P(y|x)}}{\mathcal{N}} = \left[\frac{P(y|x)}{\mathcal{N}} \right] \rho_{x,x}.$$ (B.4)

The change induced in the off-diagonal elements is similar:

$$\tilde{\rho}_{x,x'} = \left[\frac{\sqrt{P(y|x)P(y|x')}}{\mathcal{N}} \right] \rho_{x,x'}.$$ (B.5)

Now we have a Bayesian update rule for both the diagonal and the off-diagonal elements of ρ.

Let us apply the above rule for updating the density matrix to a situation in which we make a weak measurement of a qubit. Consider a measurement of qubit that has two outcomes labeled as 0 and 1. If the measurement is weak, then when we obtain outcome 0 there is still some probability that the qubit is in the state 1. If the measurement gives the wrong outcome with probability p, then the classical likelihood function is $P(0|0) = 1 - p$, and $P(0|1) = p$, and similarly for outcome 1 we have $P(1|1) = 1 - p$ and $P(1|0) = p$. If the initial density matrix is

$$\rho = \begin{pmatrix} 1-q & z \\ z^* & q \end{pmatrix},$$
(B.6)

where the first diagonal element gives the probability that the qubit is in state 0, the density matrix after we have made the measurement and obtained the result 0 is then

$$\tilde{\rho} = \frac{1}{\mathcal{N}} \begin{pmatrix} \sqrt{1-p} & 0 \\ 0 & \sqrt{p} \end{pmatrix} \begin{pmatrix} 1-q & z \\ z^* & q \end{pmatrix} \begin{pmatrix} \sqrt{1-p} & 0 \\ 0 & \sqrt{p} \end{pmatrix}$$

$$= \frac{1}{(1-p)(1-q)+pq} \begin{pmatrix} (1-p)(1-q) & \sqrt{(1-p)p}\,z \\ \sqrt{(1-p)p}\,z^* & pq \end{pmatrix}.$$
(B.7)

If the error in the measurement is small ($p \ll 1$), then

$$\tilde{\rho} \approx \begin{pmatrix} 1-pq & \sqrt{p}z \\ \sqrt{p}z^* & pq. \end{pmatrix}.$$
(B.8)

A continuous measurement of a qubit

A continuous measurement is a sequence of weak measurements, each of which takes a small amount of time Δt. The measurements are chosen so that each one obtains an amount of information proportional to Δt. This allows us to take the limit as $\Delta t \to 0$ to obtain a continuous measurement process.

The weak measurement that we described above is exactly the kind of measurement that we need to make in each time-step Δt to obtain a continuous measurement. The amount of information obtained by the measurement is determined by the error-probability p. If $p = 1/2$ then the measurement obtains no information. So a measurement that obtains only a little information is one that has $p = (1/2)(1 + \epsilon)$, with $\epsilon \ll 1$. We now determine what happens when we make this weak measurement on an initial density matrix ρ. Since the density matrix is defined entirely by one of its diagonal elements and one of its off-diagonal elements, we only need keep track of two elements. If we write the elements of the initial density matrix as ρ_{ij}, then the density matrix after the measurement is

$$\tilde{\rho}_{11} = \frac{(1 \pm \epsilon)\rho_{11}}{1 \pm \epsilon(\rho_{11} - \rho_{00})} \approx \rho_{11} \left[1 + 2\rho_{00} \left\{ \pm\epsilon - \epsilon^2(\rho_{11} - \rho_{00}) \right\} \right],$$
(B.9)

$$\tilde{\rho}_{01} = \frac{\sqrt{1-\epsilon^2}\rho_{01}}{1 \pm \epsilon(\rho_{11} - \rho_{00})} \approx \rho_{01} \left[1 - \epsilon^2/2 + (\rho_{11} - \rho_{00}) \left\{ \pm\epsilon - \epsilon^2(\rho_{11} - \rho_{00}) \right\} \right], \text{(B.10)}$$

where the positive sign corresponds to outcome 1 and the negative sign to outcome 0. In each time-step we obtain either a positive or negative change to ρ_{11}. We can associate with each of the two measurement outcomes any numerical value and call it the measurement "result" for that outcome. We will call the measurement result that occurs r, and choose its values to be ϵ for outcome 1, and $-\epsilon$ for outcome 0, since it is these values that appear in the equations above for the elements of ρ. The average value of r (averaged over the two outcomes) is then

$$\langle r \rangle = \epsilon \times \text{Prob}(r = \epsilon) + (-\epsilon) \times \text{Prob}(r = -\epsilon)$$

$$= \frac{\epsilon}{2}[\rho_{00}(1-\epsilon) + \rho_{11}(1+\epsilon)] - \frac{\epsilon}{2}[\rho_{00}(1+\epsilon) + \rho_{11}(1-\epsilon)]$$

$$= \epsilon^2(\rho_{11} - \rho_{00}), \tag{B.11}$$

and its variance is

$$V[r] = \langle r^2 \rangle - \langle r \rangle^2 = \epsilon^2 - \epsilon^4(\rho_{11} - \rho_{00})^2. \tag{B.12}$$

We now consider a time ΔT in which there are many measurements, and thus many measurement results, but with ϵ small enough that the change induced in ρ in the time ΔT is small. Let us say that in the time ΔT there are N measurements, and denote the nth measurement result by r_n, where $r_n = \pm\epsilon$. Since there are N measurement results in the time interval ΔT, the value of ϵ must scale in the right way so that the amount of information we get stays fixed as $N \to \infty$. Since our goal is to obtain a continuous measurement, ϵ must also scale with ΔT so that the amount of information obtained is proportional to ΔT. Both these requirements are satisfied by setting $\epsilon = \sqrt{8k\Delta T/N}$ for some real constant k (the factor of 8 is purely for convenience later on). The square root in this expression comes directly from the fact that the error in a set of independent measurements is proportional to the inverse square root of the number of measurements. This is discussed further in Section 3.1.1.

We now apply the N measurements to ρ, one after the other, and the result is

$$\tilde{\rho}_{11} \approx \rho_{11}\left[1 + 2\rho_{00}\left\{\sum_{n=1}^{N} r_n - N\epsilon^2(\rho_{11} - \rho_{00})\right\}\right] + \mathcal{O}(\epsilon^2), \tag{B.13}$$

$$\tilde{\rho}_{01} \approx \rho_{01}\left[1 - N\frac{\epsilon^2}{2} + (\rho_{11} - \rho_{00})\left\{\sum_{n=1}^{N} r_n - N\epsilon^2(\rho_{11} - \rho_{00})\right\}\right] + \mathcal{O}(\epsilon^2). \tag{B.14}$$

Here the symbol $\mathcal{O}(\epsilon^2)$ represents terms containing products of two or more different measurement results r_n. We can discard these terms because they vanish in the limit of small ΔT. (It is not obvious why this is true, but it can be made clear using consistency arguments that we will not go into here.)

The fact that we can determine the change in ρ solely from the sum of the N measurement results means that this sum serves as the single measurement result obtained in the time ΔT. This measurement result, $\sum_n r_n$, is a random variable, and due to the central limit theorem it is Gaussian in the limit of large N. Its mean and variance are

$$\left\langle \sum_n r_n \right\rangle = \sum_n \langle r_n \rangle = N\langle r \rangle = N\epsilon^2(\rho_{11} - \rho_{00}) = 8k\Delta T\langle \sigma_z \rangle, \qquad (B.15)$$

$$V\left[\sum_n r_n\right] = \sum_n V[r_n] = NV[r] = N\epsilon^2 + \mathcal{O}(\epsilon^4) \approx 8k\Delta T. \qquad (B.16)$$

So if we define ΔW as a Gaussian random variable with mean zero and variance equal to ΔT, then we can write

$$\sum_n r_n = 8k\Delta T\langle \sigma_z \rangle + \sqrt{8k}\Delta W. \qquad (B.17)$$

Substituting this expression for the sum of the measurement results into Eqs. (B.13) and (B.14) we have

$$\rho_{11}(t + \Delta T) = \rho_{11}(t) + 4\sqrt{2k}\rho_{11}(t)\rho_{00}(t)\Delta W, \qquad (B.18)$$

$$\rho_{01}(t + \Delta T) = \rho_{01}(t) + \rho_{01}(t)[-4k\Delta T + \sqrt{8k}\{\rho_{11}(t) - \rho_{00}(t)\}\Delta W]. \qquad (B.19)$$

This is an evolution equation for the density matrix under a measurement in which we obtain the measurement result $\sum_n r_n$ in each time-step ΔT. To see the meaning of the constant k it is helpful to define the measurement result as $R(t) \equiv \sum_n r_n/(8k\Delta T)$, so that

$$R(t) = (\rho_{11} - \rho_{00}) + \frac{1}{\sqrt{8k}}\frac{\Delta W}{\Delta T} = \langle \sigma_z \rangle + \frac{1}{\sqrt{8k}}\frac{\Delta W}{\Delta T}. \qquad (B.20)$$

Here we see that the measurement result is $\langle \sigma_z \rangle$ plus an error proportional to $1/\sqrt{k}$. The larger the value of k, the smaller the error in the measurement result at each time-step, and thus the faster the rate at which the measurement extracts information about σ_z. The constant k is often referred to as the strength of the measurement. Since the standard deviation of ΔW is $\sqrt{\Delta T}$, we also see that the error in $R(t)$ is proportional to $1/\sqrt{\Delta T}$, and thus the information obtained tends to zero as ΔT tends to zero, which makes sense.

When we take the continuum limit so that $\Delta T \rightarrow dt$, we write the random variable as dW, which now has variance dt, and is given a special name: the Wiener process. When writing the evolution of ρ we usually define the quantity $d\rho_{11}$ by the relation $\rho_{11}(t + dt) = \rho_{11}(t) + d\rho_{11}$, and so write the continuum version of the equations as

$$d\rho_{11} = \rho_{11}\rho_{00}4\sqrt{2k}\,dW, \qquad (B.21)$$

$$d\rho_{01} = \rho_{01}[-4k\,dt + \sqrt{8k}(\rho_{11} - \rho_{00})\,dW]. \qquad (B.22)$$

We can also write them in the more compact form

$$d\rho = -k[\sigma_z,[\sigma_z,\rho]]\,dt + \sqrt{2k}(\sigma_z\rho + \rho\sigma_z - 2\text{Tr}[\sigma_z\rho]\rho)\,dW. \qquad (B.23)$$

The reason that we defined k with the extra factor of 8 above was merely to remove any numerical factors in front of the double commutator in this stochastic master equation.

Note that since $V[\Delta W/\Delta T] \to \infty$ as $\Delta T \to 0$, the measurement result $r(t)$ has infinite fluctuations in the continuum limit. Because of this we instead define the measurement result after time t as

$$y(t) = \int_0^t R(t')\,dt', \tag{B.24}$$

and $y(t)$ is not infinite. Its mean is proportional to t and its fluctuations are proportional to $\sqrt{t}$. With this definition the increment in the value of y at each time-step is finite, and is given by

$$dy = R(t)\,dt = \langle \sigma_z \rangle\,dt + \frac{dW}{\sqrt{8k}}. \tag{B.25}$$

For a given run of the experiment, the values of dW in Eq. (B.25) are the same as those in Eqs. (B.21), (B.22), and (B.23). In Appendix C we give a quick introduction to Ito calculus, the rules required to manipulate differential equations that contain dW.

In an experiment the measuring device spits out the stream of results $dy(t)$, often referred to as the *measurement record*. In order to track the evolution of the system from these results one must evolve ρ using Eq. (B.23). To do this all one has to do is to calculate dW from dy at each time-step using

$$dW = \sqrt{8k}(dy - \mathrm{Tr}[\sigma_z \rho]\,dt). \tag{B.26}$$

Thus in each time-step one calculates dW from the measurement result dy and the current density matrix ρ, and then puts dW into Eq. (B.23) to obtain the new value of ρ. it doesn't matter especially what initial value we choose for ρ at the start of the experiment, since ρ will converge to the correct value as time goes by. This is the very nature of the measurement process.

Continuous measurements on linear quantum systems

A linear quantum system is one that involves one or more single particles, each described by a position and momentum operator, where the Hamiltonian for the system is a polynomial that is no more than quadratic in the position and momentum operators. The nice thing about linear systems is that if one makes a continuous measurement of position (or in fact any linear combination of position and momentum operators) the dynamics of the system is that of the equivalent classical system, with only one addition. This addition is white noise (noise with a flat spectrum) that drives the system, and it is referred to as the "back-action" noise of the measurement.

If we continuously measure the position x of a particle, then the system is driven by a force whose size fluctuates as white noise. For a measurement with strength k, the stream of measurement results is

$$dy = x\,dt + \frac{dW}{\sqrt{8k}},$$
(B.27)

where x is simply a number, since our model is now classical, k is defined by this equation, and dW is defined above. (dW is a random Gaussian variable with mean zero and variance dt, for which a new value is chosen at each time-step.) The resulting contribution to the dynamics of the system is

$$dp = \sqrt{2k}\hbar\,dW,$$
(B.28)

which is the effect of a white-noise force. Note that for a given experiment, the stream of random values for dW in Eq. (B.28) is the same as the stream for dW in Eq. (B.27); the same noise drives both equations. Using this classical model, we can solve in a straightforward manner for the dynamics of the quantum system and the spectrum of the measurement results. We can also use classical measurement theory to evolve the probability distribution for the coordinates of the system (x and p), as is discussed in Section 3.1.1. An example of calculating the spectrum for a measurement of the position of a harmonic oscillator using this classical model is given in Section 3.1.4.

More generally, if we have a linear system with N particles labeled by $n = 1,\dots,N$, and we make a continuous measurement of the operator $X = \sum_n c_n x_n + d_n p_n$ with strength k, then the contribution of the back-action noise to the dynamics of the classical model is

$$dx_n = \sqrt{2k}\,d_n\hbar\,dW,$$
(B.29)

$$dp_n = \sqrt{2k}\,c_n\hbar\,dW.$$
(B.30)

For simplicity it is best to chose the units of c_n and d_n so that X is dimensionless, and thus k is a rate constant. The stream of measurement results is

$$dy = \left(\sum_n c_n x_n + d_n p_n\right) dt + \frac{dW}{\sqrt{8k}}.$$
(B.31)

With the information in this appendix, there are a number of sections of this book that are accessible to you, and these are listed in Section 1.1.

Appendix C

A quick introduction to Ito calculus

Ito calculus is the set of rules needed to manipulate differential equations driven by Gaussian noise, also called Wiener noise. In fact, there is only *one* additional rule required over the usual rules of calculus. But if someone is not familiar with it, it is easy to forget to apply it correctly because we are so used to applying standard calculus.

Differential equations containing noise are called *stochastic* equations. Stochastic equations that contain a single Gaussian noise stream have the form

$$d\mathbf{x} = \mathbf{f}(\mathbf{x},t)\,dt + \mathbf{g}(\mathbf{x},t)\,dW, \tag{C.1}$$

where dW is a "random increment." This means that in each time-step dt a new real number dW is chosen from the Gaussian probability distribution

$$P(dW) = \frac{1}{\sqrt{2\pi\,dt}}e^{-(dW)^2/(2\,dt)}. \tag{C.2}$$

The mean of dW is zero and the variance, $V(dW)$, is equal to dt. This dependence of the variance on the time-step is essential – any other scaling with time does not produce a valid noise stream. The reason for this stems from the fact that when independent random variables are added together, the variances add together, as do the means. Note that the value of dW in each time-step is independent of its values in the other time-steps, and so the variance for the total random increment over two time-steps is simply the sum of those for each time-step. So the total increment over two time-steps will only have a variance proportional to the total time-step, $2\,dt$, if each of the increments has a variance of dt. The reason why the variance of the total increment must scale linearly with the time increment is because the number of increments scales linearly with the total time: as we take the limit $dt \to 0$, the variance of the total increment for a non-infinitesimal time T will go to zero if the variance scales as a higher power of dt, and infinity if the variance scales with a fractional power of dt. In the continuum limit, the only scaling that produces a non-zero non-infinite variance is $V(dW) \propto dt$.

It is the fact that $V(dW) = dt$ that leads to the additional rule required to manipulate dW. Note that since there are an infinite number of increments in any non-infinitesimal time, the total effect of the increments in any non-infinitesimal time is obtained by summing an

448

infinite number of infinitely small random numbers. In view of this, consider the effect of the increment $(dW)^2$ in the equation

$$dy = h(\mathbf{x},t)(dW)^2. \tag{C.3}$$

The mean of $(dW)^2$ is dt, and its variance is proportional to $(dt)^2$. If we consider dt to be equal to $dt = N/T$, then we obtain the continuum limit by taking $N \to \infty$. Denoting the value of dW in the nth time-step by dW_n, the mean of the total increment for the time T is

$$\left\langle \sum_{n=1}^{N}(dW_n)^2 \right\rangle = \sum_{n=1}^{N}\left\langle (dW_n)^2 \right\rangle = \sum_{n=1}^{N}dt = N\,dt = T. \tag{C.4}$$

But the variance of the total increment is

$$V\left(\sum_{n=1}^{N}(dW_n)^2 \right) = \sum_{n=1}^{N}V\left[(dW_n)^2\right] \propto \sum_{n=1}^{N}(dt)^2 = N\left(\frac{T}{N}\right)^2 = T^2/N \to 0. \tag{C.5}$$

The sum of all the random increments $(dW)^2$ is not random but deterministic! And it is equal exactly to T. The reason why the randomness vanishes is the same reason why the more independent measurements we make of a quantity, the more accurate (the less random) is the average of the measurement results. The mean stays the same but the variance drops as the number of samples increases.

The above result means that we can replace all occurrences of $(dW)^2$ in a stochastic differential equation (SDE) by dt, since $\sum_n dt = \sum_n (dW_n)^2$. When does $(dW)^2$ appear in an SDE? It does so when we change variables. For example, let us say that we want to calculate the SDE for $y = x^2$ when the equation for x is $dx = f(x,t)\,dt + g(x,t)\,dW$. The increment of y in a single time-step dt is

$$dy = y(t+dt) - y(t) = [x(t+dt)]^2 - [x(t)]^2 = [x(t)+dx]^2 - [x(t)^2]$$
$$= 2x\,dx + (dx)^2. \tag{C.6}$$

So the square of the increment of x appears in the increment for y. Because of this the increment for y is

$$dy = 2x\,dx + (dx)^2 = 2x(f\,dt + g\,dW) + (f\,dt + g\,dW)^2$$
$$= 2xf\,dt + 2xg\,dW + f^2(dt)^2 + 2fg\,dt\,dW + g^2(dW)^2$$
$$= 2xf\,dt + 2xg\,dW + g^2\,dt, \tag{C.7}$$

where in the last line we have made the replacements $(dt)^2 = 0$, $dt\,dW = 0$, and $(dW)^2 = dt$. More generally, if $z = f(x)$, then

$$dz = \frac{df}{dx}dx + \frac{1}{2}\frac{d^2f}{dx^2}(dx)^2 + \cdots \tag{C.8}$$

This is how one changes variables in an SDE.

It is important to note that keeping the square of an infinitesimal increment contradicts the usual rules of calculus. To determine the derivative of one quantity, y, with respect to another, x, we take the limit as $dx \to 0$, and in doing so we assume $(dx)^2$ to vanish with respect to dx. For example the derivative of $y = x^2$ is equal to $2x$ only if one drops the term $(dx)^2$. If we do not, then

$$dy = 2x\,dx + (dx)^2 \Rightarrow \frac{dy}{dx} = 2x + \text{extra}. \tag{C.9}$$

How is it possible for dW to break the usual rules of calculus? For any smooth (differentiable) function, the square of an infinitesimal *does* vanish with respect to the infinitesimal. The noise stream can break the rules because dW is not differentiable. In each infinitesimal increment, dW takes a random value that is not related to the value in the previous increment. Because of this, dW is rough on infinitely small scales, and does not have a finite derivative. For linear systems the derivative of dW can in fact be usefully defined as being infinite, and the result is that dW/dt is white noise. This is discussed in Section 3.1.1. A much fuller explanation of Ito calculus and the origin of the Ito rule $(dW)^2 = dt$ can be found in, e.g., [288].

Appendix D

Operators for qubits and modes

The Bloch-sphere representation of a qubit

Arguably the simplest example of a nonlinear quantum system is one with only two states, often referred to as a *qubit*, a term coined by Benjamin Schumacher [549]. The density matrix for a qubit can be conveniently written in terms of the three Pauli matrices and the identity, which together form a basis for complex 2-by-2 matrices. The Pauli matrices are

$$\sigma_x = \begin{pmatrix} 0 & 1 \\ 1 & 0 \end{pmatrix}, \quad \sigma_y = \begin{pmatrix} 0 & -i \\ i & 0 \end{pmatrix}, \quad \sigma_z = \begin{pmatrix} 1 & 0 \\ 0 & -1 \end{pmatrix}, \tag{D.1}$$

and satisfy the equations

$$\sigma_x \sigma_y = i\sigma_z, \quad [\sigma_x, \sigma_y] = 2i\sigma_z, \tag{D.2}$$

which are true for any cyclic permutation of the subscripts x, y, and z. (To check if a permutation is cyclic, you read the subscripts by starting at the subscript x and moving to the right. When you reach the end of the equation you go to the start and continue moving to the right. If this procedure produces the sequence xyz then the permutation is cyclic.)

We denote the positive and negative eigenstates of the σ_z operator by $|1\rangle$ and $|0\rangle$, respectively. It is also useful to define the lowering operator for a qubit by

$$\sigma = |0\rangle\langle 1| = \begin{pmatrix} 0 & 0 \\ 1 & 0 \end{pmatrix} = \frac{1}{2}(\sigma_x - i\sigma_y). \tag{D.3}$$

One also has $\sigma_x = \sigma + \sigma^\dagger$.

Two very handy relations for the Pauli matrices are

$$e^{-i\lambda \mathbf{n}\cdot\sigma} = I\cos(\lambda) - i(\mathbf{n}\cdot\sigma)\sin(\lambda), \tag{D.4}$$

$$(\mathbf{b}\cdot\sigma)(\mathbf{c}\cdot\sigma) = (\mathbf{b}\cdot\mathbf{c})I + i\sigma\cdot(\mathbf{b}\times\mathbf{c}). \tag{D.5}$$

Here $\sigma = (\sigma_x, \sigma_y, \sigma_z)$ is the vector of Pauli matrices, $\mathbf{n}$ is a real, three-dimensional vector with unit length, $\mathbf{b}$ and $\mathbf{c}$ are real three-dimensional vectors, and λ is a real number.

451

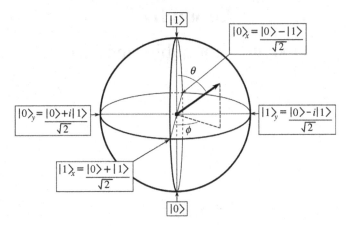

Figure D.1 The Bloch sphere, showing the spherical polar angles, θ and ϕ, along with the locations of the eigenstates of the Pauli spin operators: $|0\rangle$ and $|1\rangle$ are the eigenstates of σ_z, and those with the x and y subscripts are the eigenstates of σ_x and σ_y, respectively.

The density matrix can be expressed using the basis of Pauli matrices as

$$\rho = \frac{1}{2}\left(I + a_x\sigma_x + a_y\sigma_y + a_z\sigma_z\right) = \frac{1}{2}(I + \mathbf{a}\cdot\boldsymbol{\sigma}), \tag{D.6}$$

where $\mathbf{a} = (a_x, a_y, a_z)$ is a vector of real numbers satisfying $|\mathbf{a}|^2 = a_x^2 + a_y^2 + a_z^2 \leq 1$. The state is pure if and only if $|\mathbf{a}|^2 = 1$, and the purity is given by $P = (1 + |\mathbf{a}|^2)/2$.

The above expression for ρ constitutes the famous "Bloch-sphere" representation of a two-level system. For pure states the vector $\mathbf{a}$, called the *Bloch vector*, is a unit vector, and so all the possible values of $\mathbf{a}$ define a sphere. Mixed states have $|\mathbf{a}| < 1$, and so fill the interior of the sphere to give a "ball." Fig. D.1 shows the Bloch sphere, and gives the locations of the eigenstates of the Pauli operators.

In the Bloch-sphere representation the two states that make up any given basis always have their Bloch vectors lying on a single line. We can therefore characterize any basis by the two angles that denote the direction of this line. The line of the z-basis is vertical, and that for the x-basis is horizontal. When we speak of "the angle between" two bases, we will mean the angle between their respective lines. The angle between the x and z bases is $90°$.

It is also useful to use spherical polar coordinates for representing the Bloch sphere. These are denoted by (r, θ, ϕ), where $r = |\mathbf{a}|$ is the distance from the center, and θ and ϕ give the latitude and longitude, respectively. The relationship between the Cartesian coordinates a_x, a_y, a_z, and the polar coordinates is

$$a_z = r\cos\theta, \tag{D.7}$$

$$a_x = r\sin\theta\cos\phi, \tag{D.8}$$

$$a_y = r\sin\theta\sin\phi. \tag{D.9}$$

It is also useful to define s as the length of the projection of the Bloch vector on the xy-plane, so that $s = \sqrt{a_x^2 + a_y^2}$. Figure D.1 shows the spherical-polar angles θ and ϕ on the Bloch sphere. Note that if the state is pure, then it can be written in terms of these angles as $|\psi\rangle = \cos\theta|0\rangle + \sin\theta e^{-i\phi}|1\rangle$.

Modes: number, phase, and quadratures

A harmonic oscillator with (angular) frequency ω is described by the Hamiltonian $H = \hbar\omega(a^\dagger a + 1/2)$, where the operator a satisfies the commutation relation $[a, a^\dagger] = 1$. The term equal to $\hbar\omega/2$ is often dropped, since it does not affect the dynamics. The operator $a^\dagger a$ is called the number operator, and has the discrete set of eigenstates $\{|n\rangle : n = 0, 1, 2, \ldots\}$, where $a^\dagger a|n\rangle = n|n\rangle$. These states are called the "number states" or "Fock states." The position and momentum operators for the oscillator are

$$x = \sqrt{\frac{\hbar}{2m\omega}}(a + a^\dagger), \qquad p = \sqrt{\frac{\hbar m\omega}{2}}(-i)(a - a^\dagger), \qquad (\text{D}.10)$$

where m is the mass of the oscillator. The canonical commutation relation is $[x, p] = i\hbar$.

The reason the harmonic oscillator is so ubiquitous in quantum theory is that anything that oscillates harmonically is described by the same Hamiltonian. Such is true of the modes of the electromagnetic field, corresponding to either traveling or standing waves. If one considers standing waves trapped between two mirrors, the set of modes that can exist within this "optical cavity" is discrete. In this case each mode is described precisely by the harmonic oscillator Hamiltonian above. The (angular) frequency of a mode is still given by ω, but the energy quanta are now referred to as photons rather than phonons. Since the motion does not involve position and momentum, the operators x and p no longer have the same physical meaning. For field modes it is more natural to think in terms of the amplitude and phase of the oscillation. While the amplitude is described by the number operator, obtaining a Hermitian operator corresponding to phase is not so simple. The most complete solution to this problem was obtained by Pegg and Barnett [471, 472], and their phase operator is referred to as the canonical phase operator. To define the canonical phase operator, one truncates the state space of the oscillator, keeping number states up to $n = N - 1$. The N eigenstates of the phase operator are then defined as

$$|\phi_k\rangle = \frac{1}{N}\sum_{n=0}^{N-1} e^{i\phi_k n}|n\rangle, \quad \text{for } k = 0, \ldots, N - 1, \qquad (\text{D}.11)$$

and the corresponding phase eigenvalues as $\phi_k = 2\pi k/N$. The phase operator is then

$$\Phi = \sum_k \phi_k |\phi_k\rangle\langle\phi_k|. \qquad (\text{D}.12)$$

To perform calculations with the phase operator one works with the truncated state space until a final expression is obtained for an expectation value or the probability of a measurement outcome, and then takes the limit as $N \to \infty$.

The "quadrature" operators

The operators

$$X = \tfrac{1}{2}(a + a^\dagger), \qquad Y = \tfrac{-i}{2}(a - a^\dagger), \tag{D.13}$$

are dimensionless equivalents of position and momentum. As such their Heisenberg equations of motion are linear, and they are algebraically simple to deal with. Further, if we absorb the oscillatory motion of the oscillator into the definition of the quadrature, then they act as linearized versions of amplitude and phase. To explain why this is true we now introduce the *coherent state* $|\alpha\rangle$. This is an eigenstate of the annihilation operator a with a complex eigenvalue α. In terms of number states the coherent state is given by

$$|\alpha\rangle = \exp(-|\alpha|^2/2) \sum_{n=0}^{\infty} \frac{\alpha^n}{\sqrt{n}} |n\rangle. \tag{D.14}$$

The coherent state acts very much like a classical state with a given position and momentum (a given value of X and Y). Recall that if we plot the position on one axis, and the momentum on the other, then the motion of a classical harmonic oscillator draws out a circle. This depiction of the motion is called "phase space." The expectation values of X and Y for the coherent state are given respectively by the real and imaginary parts of α, and the evolution of the coherent state is

$$|\alpha(t)\rangle = |\alpha e^{-i\omega t}\rangle. \tag{D.15}$$

Thus the complex number $\alpha = \langle X \rangle + i\langle Y \rangle$ rotates in the complex plane, drawing out the same circle as the classical oscillator. Thus the expectation values of position and momentum oscillate just as if they were the position and momentum of a classical oscillator. The coherent state is like a little blob in phase space with the same dynamics as a classical point.

We now take a snapshot of the motion at the point when α is purely real, and this is depicted in Fig. D.2. From the figure we see that changes in the value of X are also changes in the amplitude of the oscillation, α, and changes in Y are closely related to changes in the phase when these changes are small. For this reason the operators X and Y can often be used as approximations for the amplitude and phase, and are referred to respectively as the *amplitude quadrature* and *phase quadrature*.

As the oscillator evolves, and the state rotates in phase space, then X is no longer an approximation to the amplitude. We therefore define the amplitude quadrature and phase quadrature as rotating versions of X and Y that do give linear approximations to the phase and amplitude as the system evolves, assuming that the initial value of α is real. All we have to do to obtain these rotating quadratures is to put the evolution of a and $a^\dagger$ into the definitions of X and Y. The evolution of the annihilation operator under the oscillator

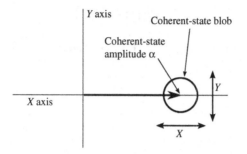

Figure D.2 The relationship between the variables X and Y (two "quadratures") and the amplitude and phase of an oscillator, when the phase is 0.

Hamiltonian is $a(t) = ae^{-i\omega t}$, and so the quadratures are given by

$$X(t) = \tfrac{1}{2}(ae^{-i\omega t} + a^\dagger e^{i\omega t}) = X\cos(\omega t) + Y\sin(\omega t), \tag{D.16}$$

$$Y(t) = \tfrac{-i}{2}(ae^{-i\omega t} - a^\dagger e^{i\omega t}) = Y\cos(\omega t) - X\sin(\omega t). \tag{D.17}$$

If we move into the interaction picture with respect to the oscillator Hamiltonian, then the time-dependence of the quadrature operators is eliminated.

Note that the dimensionless versions of x and p that we use in many places throughout this book do not have the same scaling as X and Y. The former, denoted by $\tilde{x}$ and $\tilde{p}$, are defined so that they each have unit variance when the oscillator is in its ground state. Specifically $\tilde{x} \equiv x/\Delta x$ and $\tilde{p} \equiv p/\Delta p$, where $\Delta x \equiv \sqrt{\hbar/(2m\omega)}$ and $\Delta p \equiv \sqrt{\hbar m\omega/2}$, and thus

$$\tilde{x} = a + a^\dagger = 2X(0), \qquad \tilde{p} = -i(a - a^\dagger) = 2Y(0). \tag{D.18}$$

The *displacement operator*, defined by

$$\mathcal{D}(\alpha) = \exp(\alpha a^\dagger - \alpha^* a), \tag{D.19}$$

is a unitary operator that shifts the state of a mode (or single particle) in phase space. That is

$$\mathcal{D}(\beta)|\alpha\rangle = |\alpha + \beta\rangle \quad \text{and} \quad [\mathcal{D}(\beta)]^\dagger a \mathcal{D}(\beta) = a + \beta. \tag{D.20}$$

Quantum non-demolition measurements

A quantum non-demolition (QND) measurement is a measurement of any observable that commutes with itself at all times. In the interaction picture with respect to the oscillator Hamiltonian, the quadratures as defined above are constant, and can therefore be treated as quantum non-demolition observables. If we measure one of these time-dependent observables then we effectively make a quantum non-demolition measurement of the (linearized) amplitude or phase. More generally, any time we measure an explicitly time-dependent observable that is constant in the Heisenberg picture, we effectively make a QND measurement.

Appendix E

Dictionary of measurements

The following is an alphabetical list of names used to refer to various classes of measurement. The Venn diagram in Fig. E.1 shows the relationships between the classes that the author has found the most useful.

Bare: Measurements for which all the measurement operators A_n are positive operators.

Complete: Measurements for which all the measurement operators A_n are rank-one matrices.

Continuous: A continuum of measurements, one for each time t, in which all the measurement operators tend to the identity (or vanish) as $t \to 0$.

Efficient: Measurements for which, on any single outcome, only one measurement operator acts on the system. That is, for every outcome n, the final state can be written as $\tilde{\rho} = A_n \rho A_n^\dagger / \mathcal{N}$, where ρ is the initial state and $\mathcal{N}$ is a positive number.

First kind: This terminology is due to Pauli, who divided measurements into "first kind" that do not disturb the observable they are measuring, and "second kind" that do. These terms have fallen out of use, partly because there is no single precise definition that fits the intuitive notion. In [83] measurements of the first kind are defined as bare measurements, and those of the second kind as everything else.

Incomplete: The complement of complete measurements. That is, measurements for which at least one of the measurement operators is a matrix that has rank 2 or greater.

Inefficient: Measurements for which there is at least one outcome for which two or more distinct measurement operators act on the system. That is, for at least one outcome, the final state is $\tilde{\rho} = (1/\mathcal{N}) \sum_{n=1}^{k} A_n \rho A_n^\dagger$, where $k \geq 2$ and $A_1 \neq A_2$. Here ρ is the initial state and $\mathcal{N}$ is a positive number.

Measurements of observables: Semi-classical measurements in which, for a given observable: (i) all the measurement operators commute with the observable; (ii) the resolving power for a pair of eigenstates of the observable increases monotonically with the separation between their eigenvalues. (See Section 1.4.2.)

456

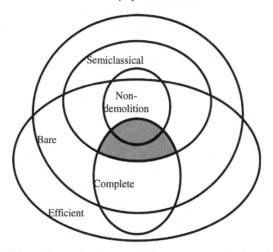

Figure E.1 A Venn diagram showing the relationship between various classes of measurement. The class represented by each ellipse is indicated by the name that touches it. The grey area corresponds to von Neumann measurements.

Minimally disturbing: An alternative name for measurements that are both bare and efficient.

Non-demolition: Measurements for which all the measurement operators A_n commute with each other in the Heisenberg picture at all times.

Projection: Another name for von Neumann measurements.

Second kind: A terminology introduced by Pauli (see the entry *First kind*, above).

Semiclassical: Measurements for which all the measurement operators A_n commute with one another.

Unitarily covariant: A measurement whose set of measurement operators, $\{A_n\}$, remains the same when every A_n is transformed by $A_n \to U A_n U^\dagger$, for any unitary operator U.

von Neumann: Complete measurements in which all the measurement operators A_n satisfy $A_n A_m = A_n \delta_{nm}$.

Weak: Measurements in which every measurement operator is a matrix that has rank 2 or higher.

Appendix F

Input–output theory

F.1 A mode of an optical or electrical cavity

Input–output theory, developed by Collett and Gardiner in 1984, is a way to treat the interaction of a system with a thermal bath, in which the bath is modeled as a quantum field [128, 199]. It applies to exactly the same situation as that of Lindblad Markovian master equations, such as the Redfield equation, and these master equations can easily be derived from it. But input–output theory goes much further than the equivalent master equations: (i) it allows one to calculate the output that flows from the system into the bath; (ii) it makes the physical connection between thermal baths and continuous measurements; (iii) its derivation avoids the obscure approximations used in the standard derivation of master equations; (iv) it can be used to connect systems together in networks by connecting the outputs of some systems to the inputs of others; (v) more generally the quantum Langevin approach of input–output theory can be applied outside the regime of validity of the Markovian master equations, to explicitly treat open systems in which non-Markovian effects are significant [211, 715].

Input–output theory was originally derived by considering the damping of a mode of an optical cavity. We use this example here in our derivation as a concrete reference, but the theory provides a model for damping and thermalization of any quantum system that is weakly coupled to a large environment. If you are not familiar with the concept of an optical cavity, a quick review is given in Section 3.3. The transmission of light through one of the end-mirrors of the optical cavity will be modeled as an interaction of the cavity mode with the electromagnetic field outside the cavity. In the original treatment the cavity just has two mirrors, and thus the output light comes through the same mirror as the input light. Because of this the electromagnetic field with which the cavity interacts fills only half of the real line (see Fig. F.1). There are other treatments in which the output field goes in the same direction as the input field [100, 669, 686]. While the original method requires a bit more work, we prefer it because it teaches one how to analyze this experimentally natural situation. Our derivation is a pedagogically expanded version of that given by Gardiner and Zoller [201].

Since we are only interested in light propagating along a line, we need consider only the modes of the electric field (actually the vector potential) whose wavenumber vectors $\mathbf{k}$ lie along a single line. This gives us a vector potential that changes only along this line, and so is effectively a one-dimensional scalar field. Denoting the position along the line by x, we write the vector potential as $A(x)$. One usually decomposes this potential into spatial modes by taking the Fourier transform of $A(x)$ [126], but we need to do this a little differently. As depicted in Fig. F.1, the field input to the cavity comes from the right-hand side ($x > 0$), and the output field is also emitted back into this side. We therefore need to consider the vector potential in only half of real space, $x \in [0, \infty)$. It will make our life much easier if we choose a mode representation that contains only the modes necessary to describe the degrees of freedom in this half-space. We can do this by choosing a representation on the full line that is symmetric (or anti-symmetric) about $x = 0$. Choosing the symmetric option means using spatial modes that are cosine functions, so that we write

$$A(x,t) = \frac{1}{\sqrt{\pi L^2/2}} \int_0^\infty \mathcal{A}(k,t)\cos(kx)\,dk, \tag{F.1}$$

where $\mathcal{A}(k,t)$ is the vector potential in the cosine mode representation. Here L is a length that comes from the fact that the full transform is over three dimensions rather than one. Assuming that the vector potential is constant along y and z, the transform in each of those directions gives us a length. We can interpret L^2 as the cross section of the laser that drives the cavity. The inverse transform is

$$\mathcal{A}(k,t) = \frac{L}{\sqrt{\pi/2}} \int_0^\infty A(x,t)\cos(kx)\,dx. \tag{F.2}$$

The equations of motion for each of the mode operators $\mathcal{A}(k,t)$ are the same as those for the usual Fourier modes, since the cosine modes are really just a subset of the Fourier modes. Given the equations of motion for the free field, we can as usual define mode operators

$$b(\omega,t) \equiv \sqrt{\frac{\varepsilon_0 \omega}{2\hbar}} \left(\mathcal{A}(k,t) + \frac{i\dot{\mathcal{A}}(k,t)}{\omega} \right), \quad k = \frac{\omega}{c} \tag{F.3}$$

that have the simple harmonic motion

$$b(\omega,t) = b(\omega,t_0)e^{-i\omega(t-t_0)} \tag{F.4}$$

and the commutation relations

$$[b(\omega,t),b^\dagger(\omega',t)] = \delta(\omega - \omega'), \quad [b(\omega,t),b(\omega',t)] = 0. \tag{F.5}$$

Note that the mode operators have units of the square root of time, and are annihilation operators. In terms of these operators the vector potential is

$$A(x,t) = \int_0^\infty \zeta(\omega) \left[b(\omega,t)\cos(kx) + b^\dagger(\omega,t)\cos(kx) \right] d\omega \tag{F.6}$$

$$= \frac{1}{2} \int_0^\infty \zeta(\omega) \left[b(\omega,t)(e^{ikx} + e^{-ikx}) + \text{h.c.} \right] d\omega, \tag{F.7}$$

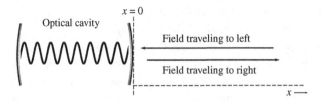

Figure F.1 Diagram of an optical cavity interacting with the electromagnetic field via transmission through its end-mirror on the right-hand side. The electromagnetic field with which the cavity mode interacts fills the half-line from $x = 0$ to $x = \infty$, and can be broken up into a field traveling to the left, and one traveling to the right.

where $\zeta(\omega) \equiv \sqrt{\hbar/(\pi \varepsilon_0 c^2 \omega L^2)}$. The vector potential may be written in terms of the mode operators at some arbitrary time t_0 as

$$A(x,t) = \frac{1}{2} \int_0^\infty \zeta(\omega) \left[b(\omega, t_0)(e^{i(kx - \omega(t - t_0))} + e^{-i(kx + \omega(t - t_0))}) + \text{h.c.} \right] d\omega. \qquad \text{(F.8)}$$

Since $e^{i(kx \pm \omega(t - t_0))}$ is a wave traveling in the positive (negative) x direction, we see that each mode operator $b(k)$ is associated with *two* waves of frequency ω traveling in opposite directions.

Choosing to represent the field by a cosine expansion, rather than a sine expansion, means that we have set the boundary condition at $x = 0$ so that an incoming wave is reflected without a phase change. This can be seen by noting that in this case the field for positive x is effectively half of a symmetric field extending over all x. Therefore, a wave traveling toward $x = 0$ from positive x has a twin wave traveling in from negative x. As the first wave passes the origin the twin wave also passes the origin, and appears to an observer in the positive x region as a reflection of the original wave with zero phase change. If we had used a sine expansion instead this would have resulted in a π phase change upon reflection. However, we may use any phase change we wish, as the correct phase change in any experimental realization can be correctly accounted for simply by multiplying the output field by the correct phase factor. Because of this, different conventions are used by different authors. Here we follow Gardiner [201] in setting the phase change to zero for the purposes of the derivation.

We now introduce a somewhat phenomenological interaction between the system (in our example this is an optical cavity, but it could be any system for which we want to model damping) and the electric field. We choose the interaction Hamiltonian to be the product of a Hermitian operator X for the system, which we define to be dimensionless, and the electric field $E(x,t) = \dot{A}(x,t)$, where the interaction is non-zero only in a relatively small region of space near $x = 0$. This interaction is inspired by the interaction between the dipole moment of an atom and the electric field [126]. The spatial profile of the interaction is given by a dimensionless positive function $f(x)$, that has a maximum of unity, and is non-zero only in the interval $[0, \Delta x]$. The interaction Hamiltonian is then

$$H_{\mathrm{I}} = -gX \int_0^{\Delta x} f(x) E(x,t)\, dx = i\hbar\kappa X \int_0^\infty \sqrt{\omega} \tilde{f}(\omega)[b(\omega) - b^\dagger(\omega)]\, d\omega, \qquad (\mathrm{F}.9)$$

where g takes the units required to make X dimensionless, and determines the overall strength of the coupling,

$$\kappa = \frac{g}{\sqrt{2\hbar^3 \varepsilon_0 c^2 L^2}}, \qquad (\mathrm{F}.10)$$

and

$$\tilde{f}(\omega) = \frac{1}{\sqrt{\pi/2}} \int_0^\infty f(x) \cos(\omega x / c)\, dx \qquad (\mathrm{F}.11)$$

is the inverse cosine transform of $f(x)$. Note that since the field mode operators have units of root time, κ is dimensionless. We are free to choose the overall sign of the interaction Hamiltonian, and we do so to give us the sign "convention" for input–output theory that we prefer (see below).

With this interaction Hamiltonian the equations of motion for an arbitrary system operator Y, and for the field mode operators, are

$$\dot{Y} = \frac{-i}{\hbar}[Y, H] + \kappa [Y, X] \int_0^\infty \sqrt{\omega}\tilde{f}(\omega)[b(\omega) - b^\dagger(\omega)]\, d\omega, \qquad (\mathrm{F}.12)$$

$$\dot{b}(\omega) = -i\omega b(\omega) - \kappa \sqrt{\omega}\tilde{f}(\omega)X, \qquad (\mathrm{F}.13)$$

where H is the Hamiltonian for the system. Because the equation of motion for the mode operator $b(\omega)$ is a linear first-order differential equation driven by X, we can write an explicit solution for this equation in terms of $x(t)$. This solution is

$$b(\omega, t) = b(\omega, t_0) e^{-i\omega(t - t_0)} - \kappa \sqrt{\omega}\tilde{f}(\omega) \int_{t_0}^t e^{-i\omega(t-s)} X(s)\, ds. \qquad (\mathrm{F}.14)$$

Note that once $b(\omega)$ is specified at a single time t_0, then it is determined for all past and future times, given $X(t)$.

There are two things we want to know. The first is the nature of the field that is output from the cavity, and how it relates to the input field. The second is the dynamics of the system, and how it is affected by the input field. We now examine the first question without making any approximations. When we answer the second question we will assume that the coupling to the field is relatively weak so that we can make the rotating-wave approximation. It is this that results in a simple dynamics for the system. As we will show, this dynamics is that given by the Redfield master equation, but written as a Heisenberg equation for the operators.

F.2 The traveling-wave fields at $x = 0$: the input and output signals

We first examine the field without the interaction with the system (the so-called "free field"). The free field is given by Eq. (F.8). At $x = 0$, the point at which the leftward-traveling (incoming) part of the field reflects off the boundary, this field is

$$A^{\text{free}}(0,t) = \frac{1}{2} \int_0^\infty \zeta \left[b(k,t_0)e^{-i\omega(t-t_0)} + \text{h.c.} \right] dk. \tag{F.15}$$

By a simple rearrangement of $A(x,t)$ it is seen that

$$A^{\text{free}}(x,t) = A\left(0, t - \frac{x}{c}\right) + A\left(0, t + \frac{x}{c}\right). \tag{F.16}$$

Here the field is written explicitly as the sum of an inward-traveling field and an outward-traveling field. It also shows that these inward and outward propagating fields may be written in terms of the field at $x = 0$: if we know the field at $x = 0$ for all time, then we may construct the entire field for all time. This may be understood as follows. The field at $x = 0$ is the result of a field that has propagated along the x-line from positive infinity to $x = 0$. Also, now that the field has this value at $x = 0$, this will propagate in the positive x direction back out to infinity. The field at a particular position x and time t is therefore the sum of a field which at a time x/c later will reach $x = 0$, and is therefore $A(0, t + \frac{x}{c})$, and another field which will have come from $x = 0$ and taken a time x/c to reach position x, and so is therefore $A(0, t - \frac{x}{c})$.

Let us denote the free field at $x = 0$ by $A_0^{\text{free}}(t)$. With no system interaction, this operator is the total field at $x = 0$ for all time. But when we include the system interaction the evolution of the field is no longer given merely by multiplying the mode operators by an exponential, but is instead given by Eq. (F.14). This means that $A_0^{\text{free}}(t)$ is only equal to the field at $x = 0$ when $t = t_0$, and will in general be different from it for all other times.

Writing $A(x,t)$ in terms of the mode operators at $t = t_0$, but using the full evolution of the field in the presence of the system, it is possible to write, after some careful manipulation [201],

$$A(x,t) = A_0^{\text{free}}\left(t + \frac{x}{c}\right) + A_0^{\text{free}}\left(t - \frac{x}{c}\right) - \frac{g}{2c} \int_{x-c(t-t_0)}^{x+c(t-t_0)} f(z)X\left(t - \left|\frac{z-x}{c}\right|\right) dz. \tag{F.17}$$

We will now examine closely the third term in this expression, which is the contribution of the system to the field. Recall first of all that we take the interaction of the system with the field to be non-zero only over the small range $x \in [0, \Delta x]$. (That is, $f(x) = 0$ for $x > \Delta x$.) Now let us examine the third term for values of x outside the interaction region. In this case z is always less than x in the integral, so that we may write the system contribution as

$$A_{\text{sys}}(x,t) = -\frac{g}{2c} \int_{x-c(t-t_0)}^{x+c(t-t_0)} f(z)X\left(t - \frac{x-z}{c}\right) dz. \tag{F.18}$$

Taking t_0 to be far enough in the past so that the interval $[x - c(t - t_0), x + c(t - t_0)]$ covers completely the interval $[0, \Delta x]$, we may simplify the expression for the system contribution further to obtain

$$A(x,t) = A_0^{\text{free}}\left(t + \frac{x}{c}\right) + A_0^{\text{free}}\left(t - \frac{x}{c}\right) - \frac{g}{2c}\int_0^{\Delta x} f(z)X\left(t - \frac{x-z}{c}\right)dz. \tag{F.19}$$

The system contribution to the field at position x and time t is clearly an integral over the range over which the system operator X interacts with the system. The contribution from each point of the interaction is the value that the system operator had at a time which is earlier than the current time t by the amount required for the field to propagate from that point to x. Outside the interaction region the field generated by the system is propagating in the positive x direction, which is clear since we may write A_{sys} as a function of $t - x/c$. Returning to Eq. (F.18), it is clear that the lower bound on the integration, namely $x - c(t - t_0)$, is there because the contribution from points which are further away from x than $c(t - t_0)$ should not be included, as the field will not have had time to propagate to x in the time interval $t - t_0$.

What Eq. (F.17) tells us is that if we choose the state of the field at $t = t_0$, then for times *later* than t_0, the field is what it would have been from free evolution (given by calculating $A_0 = A(0,t)$ using free evolution and writing $A(x,t) = A_0^{\text{free}}(t + x/c) + A_0^{\text{free}}(t - x/c)$), plus a term generated by the system, which is propagating outwards. This is a very natural result, given that Maxwell's equations generate propagating solutions.

To summarize, the input is the left-traveling field evaluated at $x = 0$, $A_{\text{in}}(t) = A_0^{\text{free}}(t)$, and the output is the right-traveling filed evaluated at $x > \Delta x$. But since the outward field is merely propagating, we may sit at any distance away from the system to observe how it changes with time. So just by shifting the time axis a little for the output we can write it as if it were evaluated at $x = 0$. This is cheating a little, but later we will take the limit as Δx tends to zero. Then we really can evaluate it at $x = 0$, and the result will be

$$A_{\text{out}}(t) = A_{\text{in}}(t) - rX(t) \tag{F.20}$$

for some constant r. This is the celebrated input–output relation, and allows us to calculate the properties of the output field in terms of the input field and the system evolution. Since the input and output to the system are functions of time, and not position, we can refer to them as *signals*. The minus sign that appears in the input–output relation is there because of the particular convention we have chosen, and this will be explained below. Switching conventions is very simple.

F.3 The Heisenberg equations of motion for the system

As mentioned above, to derive the Heisenberg equations we want to make the rotating-wave approximation. To do this we have to split the evolution of the system operator X into its distinct frequency components. This in turn can be done by writing X in the energy eigenbasis of the system. In this basis each off-diagonal element of X then oscillates at the frequency-difference between the two eigenstates it connects. (As it happens, any diagonal elements of X that are non-zero merely induce shifts of the system's energy levels, so we set them all to zero.)

If we denote the energy eigenstates by $|j\rangle$, with $H|j\rangle = \hbar\omega_j|j\rangle$, then we can write X as

$$X = \sum_{j>k} X_{jk}|k\rangle\langle j| + X^*_{jk}|j\rangle\langle k|, \tag{F.21}$$

where X_{jk} is the off-diagonal element $\langle j|X|k\rangle$. If we now define the "downward transition" operators $L_{jk} = X_{jk}|k\rangle\langle j|$ for $\omega_j > \omega_k$, then under the system Hamiltonian they have the simple evolution

$$L_{jk}(t) = L_{jk}e^{-i(\omega_j-\omega_k)t} \equiv L_{jk}e^{-i\Delta_{jk}t}, \tag{F.22}$$

and

$$X = \sum_{jk} L_{jk} + L^\dagger_{jk}. \tag{F.23}$$

We will now make one more modification to our notation. Some of the transition frequencies Δ_{jk} may be the same. In this case the system is referred to as having degenerate energy gaps. To handle this possibility in what follows, we will label each of the distinct values of the Δ_{jk} by a single index n, and denote them by Δ_n, $n = 1,\dots,N$. We then define N operators L_n, where each L_n is the sum of all the L_{jk} whose transition frequencies are equal to Δ_n. That is,

$$L_n = \sum_{\omega_j-\omega_k=\Delta_n>0} L_{jk}. \tag{F.24}$$

We have defined the L_n specifically so that each L_n oscillates at a single frequency Δ_n, and no two Δ_n are the same:

$$L_n(t) = L_n e^{-i\Delta_n t} \quad \text{and} \quad X = \sum_n L_n + L^\dagger_n. \tag{F.25}$$

If we now move into the interaction picture, then the interaction Hamiltonian becomes

$$H_{\mathrm{I}} = i\hbar\kappa \sum_n \int_0^\infty \sqrt{\omega}\tilde{f}(\omega)\left(L_n e^{-i\Delta_n t} + L^\dagger_n e^{i\Delta_n t}\right)\left[b(\omega)e^{-i\omega t} - b^\dagger(\omega)e^{-i\omega t}\right]d\omega. \tag{F.26}$$

If the coupling rate κ is small compared to all the transition frequencies Δ_n between the system's energy states, then we can make the rotating-wave approximation (RWA). This means dropping all terms in the interaction Hamiltonian that are oscillating at a frequency much larger than κ. The result is that for each operator L_n, only the field operators in the frequency band $[\Delta_n - \lambda_n \kappa, \Delta_n + \lambda_n \kappa]$ contribute to the dynamics, where λ is a number such that $\kappa \ll \lambda \kappa \ll \Delta_n$. This means that the integral splits up into N separate integrals over non-overlapping frequency bands. The interaction Hamiltonian becomes

$$H_{\mathrm{I}} = \sum_n i\hbar\kappa \int_{\Delta_n - \lambda_n \kappa}^{\Delta_n + \lambda_n \kappa} \sqrt{\omega} \tilde{f}(\omega) \left[L_n^\dagger b(\omega) e^{i(\Delta_n - \omega)t} - b^\dagger(\omega) L_n e^{-i(\Delta_n - \omega)t} \right] d\omega. \tag{F.27}$$

Here we have swapped the order of $b(\omega)^\dagger$ and L_n. Since the former is a bath operator, and the latter is a system operator, and they are both evaluated at the same time, they commute and their ordering does not matter. We choose to write them in the above order to follow the convention of Collett and Gardiner, so as to write the Hamiltonian in an explicitly Hermitian symmetric form, regardless of commutation relations.

Because of the weak coupling/RWA we can now make the crucial approximations that render simple equations of motion for the system. To do so we note that each operator L_n is coupled to a set of field modes to which no other operator L_m is coupled. We can therefore pretend that the L_n are coupled to N completely separate fields, whose mode operators we will denote by $b_n(\omega)$. For the nth field it is only the field modes in a finite frequency band about Δ_n that affect the motion of the system. Because of this we can extend the integrals over the frequency bands from $-\infty$ to ∞ (even though there are no modes with negative frequency, so we just have to make them up!). This is the key approximation. Gardiner and Collett refer to it as the first Markov approximation [199]. We now judiciously choose $\tilde{f}(\omega) = 1/\sqrt{\omega}$, which will give us a damping rate that is independent of ω.

Moving back into the Schrödinger picture, the new approximate form of the interaction Hamiltonian is

$$H_{\mathrm{I}} = i\hbar\kappa \sum_n \int_{-\infty}^{\infty} L_n^\dagger b_n(\omega) - b_n^\dagger(\omega) L_n \, d\omega. \tag{F.28}$$

The resulting equations of motion for the field modes, and for an arbitrary system operator Y are

$$\dot{Y} = \frac{[Y, H]}{i\hbar} + \sum_n \kappa [Y, L_n^\dagger] \int_{-\infty}^{\infty} b_n(\omega) d\omega - \kappa \int_{-\infty}^{\infty} b_n^\dagger(\omega) d\omega [Y, L_n], \tag{F.29}$$

$$\dot{b}_n(\omega) = -i\omega b_n(\omega) - \kappa L_n. \tag{F.30}$$

We now solve the equations of motion for the mode operators, as in Eq. (F.14), which gives

$$b_n(\omega, t) = b_n(\omega, t_0) e^{-i\omega(t - t_0)} - \kappa \int_{t_0}^{t} e^{-i\omega(t - s)} L_n(s) \, ds. \tag{F.31}$$

We now want to substitute this solution into the equation of motion for the system operator Y. In doing so we find the following integral:

$$\int_{-\infty}^{\infty} b_n(\omega)\, d\omega = \int_{-\infty}^{\infty} b_n(\omega,t_0) e^{-i\omega(t-t_0)}\, d\omega - \kappa \int_{-\infty}^{t} \int_{t_0}^{t} e^{-i\omega(t-s)} L_n(s)\, ds\, d\omega.$$

$$(F.32)$$

We now note from Eq. (F.20), and the analysis above it, that the first integral on the right-hand side above is the annihilation part of the input field $L_{in}(t)$, modified slightly due to the RWA. We therefore denote this integral by

$$b_{in}^{(n)}(t) \equiv \frac{1}{\sqrt{2\pi}} \int_{-\infty}^{\infty} b_n(\omega,t_0) e^{-i\omega(t-t_0)}\, d\omega,$$

$$(F.33)$$

where the scaling factor is chosen for later convenience. The second part of the integral is

$$\kappa \int_{t_0}^{t} \left[\int_{-\infty}^{\infty} e^{-i\omega(t-s)}\, d\omega \right] L_n(s)\, ds = \kappa \int_{t_0}^{t} 2\pi \delta(t-s) L_n(s)\, ds$$

$$= \pi\kappa L_n(t),$$

$$(F.34)$$

where we have used the fact that each end of the integral contains half the weight of the δ-function. Now defining the "damping rates"

$$\gamma = 2\pi\kappa^2,$$

$$(F.35)$$

and substituting the solution for the field mode operators into the equation of motion for Y, we obtain

$$\dot{Y} = \frac{1}{i\hbar}[Y,H] + \sum_n [Y,L_n^\dagger]\left[-\frac{\gamma}{2}L_n + \sqrt{\gamma}\, b_{in}^{(n)}(t) \right]$$

$$- \sum_n \left[-\frac{\gamma}{2}L_n^\dagger + \sqrt{\gamma}\, b_{in}^{(n)\dagger}(t) \right][Y,L_n].$$

$$(F.36)$$

We see that the equation of motion for a system operator Y is driven by the input field. The input–output relation, Eq. (F.20), written in terms of the annihilation part of the field, is

$$b_{out}^{(n)}(t) = b_{in}^{(n)}(t) - \sqrt{\gamma}\, L_n.$$

$$(F.37)$$

The commutation relations for the input and output fields are also very simple:

$$[b_{out}^{(n)}(t), b_{out}^{(n)\dagger}(t')] = [b_{in}^{(n)}(t), b_{in}^{(n)\dagger}(t')] = \delta(t-t').$$

$$(F.38)$$

The above equations encapsulate input–output theory, showing how the system is driven by the input field, and how the output field contains a contribution from the system. But we are not yet done. We must analyze the equation of motion for the system operators further, because it turns out that they must be treated as stochastic equations. But before we do that we explain the possible sign conventions for the above equations.

F.4 A weakly damped oscillator

If the system is an oscillator with annihilation operator a and frequency ω, and there is a single coupling operator $L_1 = a$, then the input–output equations for the oscillator are

$$\dot{a} = -i\omega a - \frac{\gamma}{2}a + \sqrt{\gamma}b_{\text{in}}(t) \tag{F.39}$$

and

$$b_{\text{out}}(t) = b_{\text{in}}(t) - \sqrt{\gamma}a(t). \tag{F.40}$$

In this case, because the quantum noise operator does not appear in the equation of motion multiplied by any system operator, the Ito and Stratonovich versions of the equation of motion are the same, and we do not need to write the equation in terms of differentials. The difference between Ito and Stratonovich equations is discussed below in Section F.6. Note that because input–output theory uses the rotating-wave approximation, the equations above are only appropriate for a weakly damped oscillator, by which we mean that $\gamma \ll \omega$.

F.5 Sign conventions for input–output theory

You will note that there is a minus sign that appears in front of the system operator L_n in Eq. (F.37). This sign can be changed, if one wishes, by making the replacement $\sqrt{\gamma} \rightarrow -\sqrt{\gamma}$ (or alternatively $L_n \rightarrow -L_n$), but one must do so in *both* Eqs. (F.36) and (F.37). The result is a minus sign in front of the input fields $b_{\text{in}}^{(n)}(t)$ in the quantum Langevin equation for the system operator Y.

The input–output relation as stated in Eq. (F.36) contains no phase change upon reflection at the cavity mirror. We can easily include a phase change θ simply by multiplying the output field by $e^{-i\theta}$. This gives us the more general input–output relation

$$b_{\text{out}}^{(n)}(t) = e^{-i\theta}\left(b_{\text{in}}^{(n)}(t) - \sqrt{\gamma}L_n\right). \tag{F.41}$$

If we include a π phase change, which is usual for a mirror, then the input–output relation becomes

$$b_{\text{out}}^{(n)}(t) = \sqrt{\gamma}L_n - b_{\text{in}}^{(n)}(t). \tag{F.42}$$

This is a rather intuitive way to write it, because it shows the output field as the output of the system plus the reflected (and thus inverted) input field. (Of course this intuition only goes so far, because it would suggest that the input field should be multiplied by the reflection coefficient of the mirror.) The above possibilities cover all the sign conventions for input–output theory.

F.6 The quantum noise equations for the system: Ito calculus

We now consider how to solve the Heisenberg equation of motion for the system, Eq. (F.36). It contains an external driving signal given by $b_{\text{in}}^{(n)}(t)$, defined in Eq. (F.33). We will now see that this driving has a very odd property. From the commutation relation Eq. (F.38), we see that $b_{\text{in}}^{(n)}(t)$ has units of the square root of inverse time. This means that if we write the differential equation in the form of an infinitesimal increment,

$$dY = \frac{1}{i\hbar}[Y,H]\,dt + \sum_n [Y,L_n^\dagger]\left[-\frac{\gamma}{2}L_n\,dt + \sqrt{\gamma}\,b_{\text{in}}^{(n)}(t)\,dt\right]$$
$$-\sum_n \left[-\frac{\gamma}{2}L_n^\dagger\,dt + \sqrt{\gamma}\,b_{\text{in}}^{(n)\dagger}(t)\,dt\right][Y,L_n], \qquad (\text{F.43})$$

then the term $b_{\text{in}}^{(n)}(t)\,dt$ has units of the square root of time. This increment is therefore proportional to $\sqrt{dt}$, and not to dt as is usually the case in differential equations. But we have encountered infinitesimal increments with this property before. In Chapter 3 we needed to use differential equations driven by Gaussian white noise, whose increment is proportional to $\sqrt{dt}$. As a result of this the noise increment, denoted by dW, obeys the rule of Ito calculus, $(dW)^2 = dt$. The scaling of $b_{\text{in}}^{(n)}(t)\,dt$ tells us that the input field has the properties of white noise. And it is precisely the *quantum* nature of the field from which this comes, since it is due to the non-vanishing commutator, the fact that the quantum field describes probabilistic outcomes. The white-noise nature of the input is also revealed by its autocorrelation function. If the input field is in the vacuum state, then $\langle b_{\text{in}}^{(n)\dagger}(t)b_{\text{in}}^{(n)}(t')\rangle = 0$, and using the commutator we have

$$\langle b_{\text{in}}^{(n)}(t)b_{\text{in}}^{(n)\dagger}(t')\rangle = \langle [b_{\text{out}}^{(n)}(t), b_{\text{out}}^{(n)\dagger}(t')]\rangle = \delta(t - t'). \qquad (\text{F.44})$$

This is the autocorrelation function of white noise (see Section 3.1.4).

To write the equation of motion for Y in a way that allows us to use the rules of Ito calculus, there is a subtlety that we must take into account. If the white noise in a differential equation has been obtained as the broad-band limit of a real noise process, as is the case here, then it obeys the normal rules of calculus. This is possible because the noise increments are, in fact, not independent of the state of the system, and therefore quite different from Ito noise. In this case the differential equation is referred to as being in the *Stratonovich* form, rather than the Ito equations that we usually use. Full details regarding the difference between Ito and Stratonovich equations, and how to convert between them, are given in, e.g., [288, 197]. To obtain an equation that can be more easily manipulated, we need to rewrite it in the form of an Ito stochastic equation, in which the noise increments are independent of the state of the system, and obey the Ito rule. To transform to the Ito form we need first to specify the correlation functions of the input noise operators. To be able to model a thermal field at a non-zero temperature, we must give the modes some average number of photons. If we set the average photon number of the nth input field to the average number of photons contained by an oscillator with frequency Δ_n at temperature T, then we obtain the correct thermal steady state for the system at temperature T. This means that

$$\langle b_{\text{in}}^{(n)}(t) b_{\text{in}}^{(n)\dagger}(t') \rangle = [1 + N(\Delta_n)] \delta(t' - t), \tag{F.45}$$

$$\langle b_{\text{in}}^{(n)\dagger}(t) b_{\text{in}}^{(n)}(t') \rangle = N(\Delta_n) \delta(t' - t), \tag{F.46}$$

with $N(\Delta_n) = [\exp(\hbar\Delta_n/kT) - 1]^{-1}$. The corresponding Ito noise increments, $db_{\text{in}}^{(n)}$, satisfy

$$db_{\text{in}}^{(n)} db_{\text{in}}^{(n)\dagger} = [1 + N(\Delta_n)] dt, \tag{F.47}$$

$$db_{\text{in}}^{(n)\dagger} db_{\text{in}}^{(n)} = N(\Delta_n) dt. \tag{F.48}$$

The resulting Ito stochastic differential equation for a system operator Y is

$$
\begin{aligned}
dY ={}& \frac{1}{i\hbar}[Y, H]\,dt + \sum_n \sqrt{\gamma}\left([Y, L_n^\dagger]\,db_{\text{in}}^{(n)} - db_{\text{in}}^{(n)\dagger}[Y, L_n]\right) \\
&+ \sum_n \frac{\gamma}{2}[N(\Delta_n) + 1]\left(2 L_n^\dagger Y L_n - L_n^\dagger L_n Y - Y L_n^\dagger L_n\right) dt \\
&+ \sum_n \frac{\gamma}{2} N(\Delta_n)\left(2 L_n Y L_n^\dagger - L_n L_n^\dagger Y - Y L_n L_n^\dagger\right) dt.
\end{aligned} \tag{F.49}
$$

We may choose the state of each input field so that $\langle db_{\text{in}}^{(n)} \rangle = 0$, or we may include a coherent part in any of these fields so that $\langle db_{\text{in}}^{(n)} \rangle = \alpha_n\,dt$. Each α_n describes "coherent driving." For example, if the system is a cavity mode with frequency ω, and L_n is the mode annihilation operator a, then $\alpha_n = E_n e^{-i\theta_n}$ describes a laser beam with power $P_n = \hbar\omega E_n^2/\gamma$, phase θ_n, and frequency ω incident on the cavity.

F.7 Obtaining the Redfield master equation

To obtain the master equation equivalent to the above quantum stochastic differential equation (QSDE), we calculate the equation of motion for the expectation value of Y by taking averages on both sides. To make the presentation a bit more compact, let us define the shorthand notation

$$dY = \mathcal{L}(Y)\,dt + \sum_n \sqrt{\gamma}\left([Y, L_n^\dagger]\,db_{\text{in}}^{(n)} - db_{\text{in}}^{(n)\dagger}[Y, L_n]\right), \tag{F.50}$$

so that $\mathcal{L}(Y)$ stands for all the terms in the equation of motion for Y that are proportional to dt. If we choose the input fields so that $\langle db_{\text{in}}^{(n)} \rangle = 0$, then

$$\frac{d\langle Y \rangle}{dt} = \langle \mathcal{L}(Y) \rangle = \text{tr}[\mathcal{L}(Y)\rho]. \tag{F.51}$$

But it is also true that $\langle \dot{Y} \rangle = \text{tr}[Y\dot{\rho}]$, and hence

$$\text{tr}[Y\dot{\rho}] = \text{tr}[\mathcal{L}(Y)\rho]. \tag{F.52}$$

To determine the equation of motion for ρ we use the cyclic property of the trace to rearrange the expression inside the trace on the right-hand side, to write it as

$$\text{tr}[\mathcal{L}(Y)\rho] = \text{tr}[Y\mathcal{B}(\rho)], \tag{F.53}$$

where $\mathcal{B}(\rho)$ is some operator expression containing ρ. Substituting this into Eq. (F.52) gives

$$\text{tr}[Y\dot{\rho}] = \text{tr}[Y\mathcal{B}(\rho)]. \tag{F.54}$$

Since this is true for any system operator Y, it must be true that

$$\dot{\rho} = \mathcal{B}(\rho). \tag{F.55}$$

The result is the Redfield master equation for ρ, given in Eq. (4.79).

F.8 Spectrum of the measurement signal

We show here that the two-time correlation function of the measurement signal for a Gaussian continuous measurement of an operator A is equal to the two-time correlation function of the relevant output field for a system coupled to an input field via the operator A. For generality we allow the operator A to be non-Hermitian, so we can describe homodyne detection (in which $A = a$) as well as a measurement of an observable x (in which $A = A^{\dagger} = x$). The stochastic master equation for a measurement of A is given in Eq. (3.184). The corresponding measurement record is given in Eq. (3.185), and is

$$r(t) = dr/dt = \text{Tr}[X\rho(t)]/2 + \xi(t)/\sqrt{8k}. \tag{F.56}$$

Here $\xi(t) = dW(t)/dt$ is white noise with correlation function $\langle \xi(t)\xi(t+\tau) \rangle = \delta(\tau)$.

The equivalent input–output equation of motion for a system operator Y is

$$dY = -\frac{i}{\hbar}[Y,H]\,dt - k\left[YA^{\dagger}A + A^{\dagger}AY - 2A^{\dagger}YA\right]dt$$
$$+ \sqrt{2k}\left(\left[Y,A^{\dagger}\right]db_{\text{in}} - db_{\text{in}}^{\dagger}[Y,A]\right). \tag{F.57}$$

The input–output relation for the field is $b_{\text{out}}(t) = b_{\text{in}}(t) - \sqrt{2k}A$. The output field that corresponds to a measurement of A is $R = (b_{\text{out}} + b_{\text{out}}^{\dagger})/\sqrt{8k}$. Using the method that we used to derive Eq. (3.220) in Section 3.6, the autocorrelation function of R can be written as

$$\langle R(t)R(t+\tau) \rangle = \frac{\langle X(t+\tau)A(t) \rangle}{4} + \frac{\langle A^{\dagger}(t)X(t+\tau) \rangle}{4} + \frac{\delta(\tau)}{8k}, \tag{F.58}$$

where we have defined $X(t) = A(t) + A^\dagger(t)$. Our task is to show that $\langle r(t)r(t+\tau)\rangle = \langle R(t)R(t+\tau)\rangle$.

Part 1

We first write the correlation functions in the above equation explicitly in terms of the system density matrix and the master equation. To do this we consider the fact that the master equation for ρ, given by averaging over the noise in the stochastic master equation given in Eq. (3.184), is derived from the unitary evolution of the system and the bath of electromagnetic modes. If we denote the operator that gives this unitary evolution over a time τ by U, and the density matrix for the joint state of the system and bath by Σ, then for any two operators B and C the two-time correlation function is

$$\langle B(t+\tau)C(t)\rangle = \mathrm{Tr}_{\mathrm{all}}[B(t+\tau)C\Sigma(t)]. \tag{F.59}$$

Here on the RHS the operator $B(t+\tau)$ is the Heisenberg operator at time $t+\tau$, given by $B(t+\tau) = U^\dagger BU$, and $\mathrm{Tr}_{\mathrm{all}}[\cdots]$ denotes the trace over the system and bath. So we now have

$$\langle B(t+\tau)C(t)\rangle = \mathrm{Tr}_{\mathrm{all}}\left[U^\dagger BUC\Sigma(t)\right]$$
$$= \mathrm{Tr}_{\mathrm{all}}\left[BUC\sigma(t)U^\dagger\right] = \mathrm{Tr}\left[B\mathrm{Tr}_{\mathrm{bath}}[UC\Sigma(t)U^\dagger]\right], \tag{F.60}$$

where $\mathrm{Tr}[\cdots]$ denotes the trace over the system. Now note that if we define $K(t) = C\Sigma(t)$, then $UC\Sigma(t)U^\dagger = K(t+\tau)$. When we trace over the bath, the unitary evolution U gives the evolution of the master equation for the system operators, where the noise has been averaged over. That is, it evolves the operator $\kappa(t) = \mathrm{Tr}_{\mathrm{bath}}[K(t)]$ under a linear equation of motion. Because of this the evolution it induces over a time τ can be obtained by operating on $\kappa(t)$ with a linear super-operator. Denoting this linear super-operator by $\mathcal{M}$ we can write

$$\mathrm{Tr}_{\mathrm{bath}}[UKU^\dagger] = \kappa(t+\tau) = \mathcal{M}\kappa(t) = \mathcal{M}C\sigma(t), \tag{F.61}$$

where σ is the density matrix of the system. Substituting this into Eq. (F.60) we obtain

$$\langle B(t+\tau)C(t)\rangle = \mathrm{Tr}[B\mathcal{M}C\sigma(t)], \tag{F.62}$$

and a similar derivation shows that

$$\langle C(t)B(t+\tau)\rangle = \mathrm{Tr}[B\mathcal{M}\sigma(t)C]. \tag{F.63}$$

Substituting these expressions into the expression for $\langle R(t)R(t+\tau)\rangle$ above we obtain

$$\langle R(t)R(t+\tau)\rangle = \frac{1}{4}\mathrm{Tr}\left[X\mathcal{M}(A\sigma(t) + \sigma(t)A^\dagger)\right] + \frac{\delta(\tau)}{8k}. \tag{F.64}$$

Part 2

We now consider the correlation function of the measurement signal $r(t)$. Using Eq. (F.56) this correlation function is

$$\langle r(t)r(t+\tau)\rangle = \frac{\langle \xi(t)\xi(t+\tau)\rangle}{8k} + \frac{\langle \bar{X}(t)\xi(t+\tau)\rangle}{2\sqrt{8k}} + \frac{\langle \xi(t)\bar{X}(t+\tau)\rangle}{2\sqrt{8k}} + \frac{\langle \bar{X}(t)\bar{X}(t+\tau)\rangle}{4}, \quad (\text{F.65})$$

where we have defined $\bar{X}(t) \equiv \text{Tr}[X\rho(t)]$. Note that the angle brackets $\langle\cdots\rangle$ denote an average over the noise process $\xi(t)$.

We now rewrite each term in Eq. (F.65) explicitly in terms of the density matrix and the master equation, as we did for the terms in the expression for $\langle R(t)R(t+\tau)\rangle$ above. For the following analysis we will denote the density matrix of the system by ρ.

The first term: This is merely the correlation function of the noise: $\delta(\tau)/(8k)$.

The second term: Since the measurement noise $\xi(t+\tau)$ is independent of $A(t)$ for $\tau > 0$, $\langle \bar{X}(t)\xi(t+\tau)\rangle = 0$ for $\tau > 0$. For $\tau = 0$ this term can also be discarded: it is swamped by the δ-function of the measurement noise because $X(t)$ does not contain white noise (its spectrum is determined by the transfer function of the system).

The third term: This term is not zero, as the value of $\bar{X}$ at $t+\tau$ depends on $\xi(t)$. To evaluate it we will use the following notation:

$$\rho_0 \equiv \rho(t), \quad \rho_1 \equiv \rho(t+dt), \quad \rho_2 \equiv \rho(t+2dt), \quad dW_0 \equiv dW(t), \quad dW_1 \equiv dW(t+dt).$$

Writing $\rho(t+\tau)$ given the value of dW_0 as $\rho(t+\tau|dW_0)$, the third term is given by

$$\langle \bar{X}(t+\tau)\xi(t)\rangle = \left\langle \text{Tr}\left[X\rho(t+\tau|dW_0)\frac{dW_0}{dt}\right]\right\rangle = \int_{-\infty}^{\infty} \text{Tr}[X\rho(t+\tau|z)]\frac{z}{dt}P(z)dz, \quad (\text{F.66})$$

where $P(\cdot)$ is the probability distribution for dW_0. To write this more explicitly in terms of the stochastic master equation, we need to see how the evolution depends on dW_0. We can break this evolution into two parts: that in the interval $[t, t+dt]$, and that in the interval $[t+dt, t+\tau]$. The evolution in the first interval depends on dW_0, and is given by

$$\rho_1 = \rho_0 + d\rho = \rho_0 + \mathcal{L}\rho_0\, dt + \mathcal{H}[\rho_0]\, dW_0, \quad (\text{F.67})$$

where we have defined "super-operators" that act on ρ by

$$\mathcal{L}\rho = -(i/\hbar)[H,\rho] - k(A^\dagger A\rho + \rho A^\dagger A - 2A\rho A^\dagger), \quad (\text{F.68})$$

$$\mathcal{H}[\rho] = \sqrt{2k}(A\rho + \rho A^\dagger - \text{Tr}[(A+A^\dagger)\rho]\rho). \quad (\text{F.69})$$

Since $\mathcal{L}$ is a linear operator, we can think of it as a matrix if we write the elements of ρ as a vector. Thus we can treat the action of $\mathcal{L}$ as multiplication, which is why we do not need to write its action as $\mathcal{L}[\rho]$.

The evolution over the second interval is independent of dW_0, but does depend on all the noise increments between $t + dt$ and $t + \tau$. To see clearly how $\rho(t + \tau)$ is related to ρ_1, we consider the evolution over the next interval, $[t + dt, t + 2dt]$. This is

$$\rho_2 = \rho_1 + d\rho_1 = \rho_1 + \mathcal{L}\rho_1 \, dt + \mathcal{H}[\rho_1] \, dW_1. \tag{F.70}$$

Now let us see what happens when we calculate the correlation $\langle \bar{X}(t+2dt)\xi(t) \rangle$:

$$\langle \xi(t)\bar{X}(t+2dt) \rangle = \left\langle \text{Tr}[X\rho_2]\frac{dW_0}{dt} \right\rangle = \left\langle \text{Tr}[X([1 + \mathcal{L} \, dt]\rho_1 + \mathcal{H}[\rho_1] \, dW_1)]\frac{dW_0}{dt} \right\rangle.$$

Since dW_1 is independent of dW_0, all expectation values of the form $\langle (dW_0)^n \, dW_1 \rangle$ factor into $\langle (dW_0)^n \rangle \langle dW_1 \rangle$, and consequently vanish because $\langle dW_1 \rangle = 0$. The correlation thus reduces to

$$\langle \xi(t)\bar{X}(t+2dt) \rangle = \left\langle \text{Tr}[X(1 + \mathcal{L} \, dt)\,\rho_1]\frac{dW_0}{dt} \right\rangle. \tag{F.71}$$

This expression tells us that for the purposes of calculating the correlation function, ρ_2 is given by evolving ρ_1 using only the deterministic part of the evolution, due to the fact that the stochastic evolution averages to zero. This is similarly true for every time-step between $t + dt$ and $t + \tau$. Since the deterministic evolution is linear, $\rho(t+\tau)$ can be written as $\rho(t+\tau) = \mathcal{M}\rho_0$, where $\mathcal{M}$ is the super-operator that gives the evolution of the master equation over a time τ. The third term in Eq. (F.65) can therefore be written as

$$\langle \xi(t)\bar{X}(t+\tau) \rangle = \left\langle \text{Tr}[X\mathcal{M}\rho_1]\frac{dW_0}{dt} \right\rangle = \left\langle \text{Tr}[X\mathcal{M}([1 + \mathcal{L} \, dt]\rho_0 + \mathcal{H}[\rho_0] \, dW_0)]\frac{dW_0}{dt} \right\rangle.$$

We now note that since ρ_0 is determined by all the Wiener increments prior to time t, it is independent of dW_0. So we have

$$\langle \xi(t)\bar{X}(t+\tau) \rangle = \langle \text{Tr}[X\mathcal{M}([1 + \mathcal{L} \, dt]\rho_0 + \mathcal{H}[\rho_0])] \rangle \left\langle dW_0 \frac{dW_0}{dt} \right\rangle \tag{F.72}$$

$$= \sqrt{2k}\,\text{Tr}\left[X\mathcal{M}(A\langle\rho_0\rangle + \langle\rho_0\rangle A^\dagger)\right] - \sqrt{2k}\,\langle \text{Tr}[X\mathcal{M}\rho_0] \rangle \text{Tr}[X\rho_0] \rangle.$$

The fourth term: Since $\bar{X}(t) = \text{Tr}[X\rho_0]$ we have

$$\langle \bar{X}(t)\bar{X}(t+\tau) \rangle = \langle \text{Tr}[X\rho_0]\text{Tr}[X\rho(t+\tau)] \rangle. \tag{F.73}$$

Using similar analysis to that above, all the stochastic terms in $\rho(t + \tau)$ average to zero, and so we can replace $\rho(t + \tau)$ with $\mathcal{M}\rho_0$. Thus

$$\langle \bar{X}(t)\bar{X}(t+\tau) \rangle = \langle \text{Tr}[X\rho_0]\text{Tr}[X\mathcal{M}\rho_0] \rangle, \tag{F.74}$$

and we see that this will cancel the second term in Eq. (F.72).

Putting all the terms together we have

$$\langle r(t)r(t+\tau)\rangle = \frac{1}{4}\mathrm{Tr}\left[X\mathcal{M}(A\langle\rho_0\rangle + \langle\rho_0\rangle A^\dagger)\right] + \frac{\delta(\tau)}{8k}. \tag{F.75}$$

Now since the density matrix $\sigma(t)$ in Eqs. (F.62) and (F.63) is that given by tracing over the bath, which is equivalent to averaging over the noise process ξ up until time t, $\sigma(t) = \langle\rho_0\rangle$, the above equation is indeed equal to Eq. (F.64).   QED.

Appendix G
Various formulae and techniques

G.1 The relationship between Hz and s^{-1}, and writing decay rates in Hz

It is common to assume at first glance that Hz and s^{-1} are the same units, but this is not quite true. If we consider a complex exponential that contains both a decay rate γ and an angular frequency ω, then this is $e^{-(\gamma+i\omega)t}$. If we write this instead in terms of the frequency (in cycles per second or *Hertz*) the expression is $e^{-(\gamma+i2\pi f)t}$. Thus the decay rate γ is on an equal footing with ω, but there is a factor of 2π between γ and f. Because of this the most natural convention is to equate s^{-1} with rad s^{-1}.

A second commonly used convention is to write the angular frequency ω associated with the frequency $f = 10$ Hz (for example) as $\omega = 2\pi \times 10$ Hz, meaning that $\omega = 20\pi$ rad s^{-1}. This is a horrible abuse of notation, but it is commonly used. (It is an abuse of notation because one would expect to be able to multiply the 2π by the 10 in the expression $\omega = 2\pi \times 10$ Hz and thus obtain $\omega = 20\pi$ Hz, which is *not* what is meant.)

If we combine the two conventions above, that s^{-1} equates to rad s^{-1} and that we write 20π rad s^{-1} as $2\pi \times 10$ Hz, then we can write a decay rate $\gamma = 100\pi$ s^{-1} (for example) as $\gamma = 2\pi \times 50$ Hz. This notation for decay rates is standard terminology used by experimentalists in journal articles.

G.2 Position representation of a pure Gaussian state

A general pure Gaussian state is given by

$$\psi(x) = \left(\frac{\exp\left[-\mu_x^2/V_x\right]}{2\pi V_x}\right)^{1/4} \exp\left\{-\frac{(1-ic)}{4V_x}\left(x^2 - 2\mu_x x\right) + \left(\frac{i\mu_p}{\hbar}\right)x\right\} \tag{G.1}$$

with

$$V_p = \frac{1}{V_x}\left(\frac{\hbar^2}{4} + C_{xp}^2\right),$$

$$C_{xp} = \frac{\hbar c}{2}, \qquad c \in [-1, 1]. \tag{G.2}$$

Here μ_x and V_x are the mean and variance of x, μ_p and V_p are mean and variance of p, and C_{xp} is the covariance of x and p, defined as $C_{xp} \equiv \langle xp + px \rangle / 2 - \langle x \rangle \langle p \rangle$.

G.3 The multivariate Gaussian distribution

If we define a column vector of N variables, $\mathbf{x} = (x_1, \ldots, x_N)^{\mathrm{T}}$, the general form of the multivariate Gaussian probability density is

$$P(\mathbf{x}) = \frac{1}{\sqrt{(2\pi)^N \det[C]}} \exp\left[-\frac{1}{2}(\mathbf{x} - \boldsymbol{\mu})^{\mathrm{T}} C^{-1} (\mathbf{x} - \boldsymbol{\mu}) \right]. \tag{G.3}$$

Here $\boldsymbol{\mu}$ is the vector of the means of the random variables, and C is the matrix of the covariances of the variables,

$$C = \langle \mathbf{x}\mathbf{x}^{\mathrm{T}} \rangle - \langle \mathbf{x} \rangle \langle \mathbf{x}^{\mathrm{T}} \rangle = \langle \mathbf{x}\mathbf{x}^{\mathrm{T}} \rangle - \boldsymbol{\mu}\boldsymbol{\mu}^{\mathrm{T}}. \tag{G.4}$$

Note that the diagonal elements of C are the variances of the individual variables.

G.4 The rotating-wave approximation (RWA)

The rotating-wave approximation, or RWA, applies when the interaction between two systems is much slower that the oscillation frequency(ies) (the energy gaps) of both. A typical example is that of two coupled harmonic oscillators, with the Hamiltonian

$$H = H_0 + H_{\mathrm{int}} = \hbar\omega b^{\dagger} b + \hbar g (a + a^{\dagger})(b + b^{\dagger}) + \hbar\omega a^{\dagger} a. \tag{G.5}$$

Here the interaction Hamiltonian is $H_{\mathrm{int}} = \hbar g (a + a^{\dagger})(b + b^{\dagger})$. Both oscillators have frequency ω, and the rate of the dynamics generated by the coupling is characterized by g. At the simplest level, the RWA is obtained by moving into the interaction picture. In this case we place the evolution due to the oscillator Hamiltonians into the operators. The Hamiltonian that evolves the states is now the interaction Hamiltonian H_{int}, but with the time-dependent operators:

$$H^{\mathrm{I}} = \hbar g (ae^{-i\omega t} + a^{\dagger} e^{i\omega t})(be^{-i\omega t} + b^{\dagger} e^{i\omega t})$$
$$= \hbar g (ab^{\dagger} + ba^{\dagger} + abe^{-2i\omega t} + (ab)^{\dagger} e^{2i\omega t}). \tag{G.6}$$

Here we have written the time-dependence of the operators explicitly, so the operators a and b that appear in H^{I} are constant. Since the evolution of the states is generated only by H^{I} the states evolve at rate g. This means that on the timescale at which the states change, the terms in H^{I} oscillating at frequency ω go equally positive and negative, and their effect on the motion of the states approximately cancels out. The RWA is obtained by setting these rapidly oscillating terms to zero. Going back into the Schrödinger picture, the Hamiltonian under the RWA is

$$H_{\text{RWA}} = \hbar\omega b^\dagger b + \hbar g(ab^\dagger + ba^\dagger) + \hbar\omega a^\dagger a. \tag{G.7}$$

Note that the RWA has removed the terms in the interaction that do not conserve the energy as given by the Hamiltonians of the two systems alone, leaving only the energy-conserving terms.

In fact, the RWA is precisely an application of time-independent perturbation theory, although it is not obvious from the method we used above. The point is that if $g \ll \omega$, then the interaction H_{int} is a perturbation on the Hamiltonian H_0. If we use time-independent perturbation theory, then we can obtain an explicit expression for the size of the error, and/or take the approximation to higher powers in the perturbation. The simplest way to apply perturbation theory to the Hamiltonian of the coupled oscillators is to use the method explained in Section 7.7.4. To do this we apply a transformation $U = e^{-A}$ to the Hamiltonian, where the operator A is chosen so that $[A, H_0]$ cancels the same part of the Hamiltonian that is eliminated in the RWA. Playing around with the harmonic oscillator commutation relations, we find that the operator we need is $A = g/(2\omega)(a^\dagger b^\dagger - ab)$. Applying $U = e^{-A}$ to the Hamiltonian in Eq. (G.5), we obtain

$$H' = UHU^\dagger = \hbar\omega b^\dagger b + \hbar g(ab^\dagger + ba^\dagger) + \hbar\omega a^\dagger a - \hbar\frac{g^2}{2\omega}(a^2 + a^{\dagger 2} + b^2 + b^{\dagger 2}). \tag{G.8}$$

to first order in g/ω. The first three terms are the RWA Hamiltonian, and the fourth term is the lowest-order perturbative correction. This shows that in making the RWA, the lowest-order term that is ignored is proportional to g^2/ω. We see further that in the interaction picture the fourth term will contain rapid oscillations, and thus its real effect will be even smaller. This indicates that we could perform another transformation to eliminate it, leaving us with higher-order corrections.

G.5 Suppression of off-resonant transitions

A force oscillating at a frequency $\Delta E/\hbar$ will couple together energy levels whose energy difference is ΔE. Such a force will also couple energy levels separated by a different energy, but the effective rate of this coupling is suppressed by the mismatch between the frequency of the force and the energy separation. One can calculate this suppression by examining the dynamics of a two-level system in the rotating-wave approximation (RWA). If we consider a two-level system in which the energy gap is ΔE, and we denote the two levels as $|0\rangle$ and $|1\rangle$, then we can write the Hamiltonian as

$$H = \Delta E|1\rangle\langle 1| \equiv \frac{\hbar\omega}{2}\sigma_z, \tag{G.9}$$

where the equivalence follows because the dynamics is unchanged by adding the identity to the Hamiltonian, σ_z is the Pauli operator, and $\omega = \Delta E/\hbar$.

A force that couples the two levels together appears in the Hamiltonian as off-diagonal matrix elementss. If this force oscillates at an angular frequency v, then the Hamiltonian becomes

$$H = \frac{\hbar\omega}{2}\sigma_z + \frac{\hbar\lambda}{2}\cos(vt)\sigma_x, \tag{G.10}$$

for some coupling rate λ. If we restrict ourselves to the regime in which $\lambda \ll \omega$, then we can make the rotating-wave approximation (see the section above), and the Hamiltonian can be written approximately as

$$H = \frac{\hbar\omega}{2}\sigma_z + \frac{\hbar\lambda}{2}(\sigma_- e^{-ivt} + \sigma_+ e^{ivt}), \tag{G.11}$$

where $\sigma_\pm = \sigma_x \pm i\sigma_y$. Because the evolution of the operators $\sigma_\pm$ under the Hamiltonian $\hbar\omega\sigma_z$ is $\sigma_\pm(t) = \sigma_\pm e^{\pm i\omega t}$, changing to the interaction picture with respect to the Hamiltonian $H_0 = \hbar\omega\sigma_z/2$ gives

$$H = \frac{\hbar}{2}(\omega - v)\sigma_z + \frac{\hbar\lambda}{2}\sigma_x. \tag{G.12}$$

The evolution for two-dimensional time-independent Hamiltonians can be solved exactly using the handy relation

$$e^{-i\alpha \mathbf{n}\cdot\boldsymbol{\sigma}} = I\cos(\alpha t) - i\mathbf{n}\cdot\boldsymbol{\sigma}\sin(\alpha t), \tag{G.13}$$

where α is a real number, $\mathbf{n}$ is a unit vector, and $\boldsymbol{\sigma} = (\sigma_x, \sigma_y, \sigma_z)$ is the vector of Pauli matrices. For our Hamiltonian the evolution for the initial state $|\psi(0)\rangle = |1\rangle$ is

$$|\psi(t)\rangle = i\delta\sin(\mu t)|0\rangle + \left[\cos(\mu t) + i\sqrt{1-\delta^2}\sin(\mu t)\right]|1\rangle, \tag{G.14}$$

where

$$\mu = \sqrt{(\omega - v)^2 + \lambda^2}/2, \tag{G.15}$$

$$\delta = \frac{\lambda}{\sqrt{(\omega - v)^2 + \lambda^2}}. \tag{G.16}$$

We see from this that when δ is very small the coupling hardly results in any population transfer from $|1\rangle$ to $|0\rangle$, so the states are effectively uncoupled. As the mismatch between the frequencies ω and v increases with respect to the coupling rate λ, δ tends to zero.

G.6 Recursion relations for time-independent perturbation theory

Time-independent perturbation theory allows one to compute the eigenvalues and eigenvectors of a Hamiltonian $H_0 + \lambda V$ as a power series in λ. If $|n\rangle$ and $E_n^{(0)}$ are the N (non-degenerate) eigenvectors and eigenvalues of H_0, then the eigenvectors and eigenvalues of

$(H_0 + \lambda V)$, which we denote by $|\psi_n\rangle$ and E_n, are given by the expansions

$$|\psi_n\rangle = \sum_{k=0}^{\infty} \lambda^k |n^{(k)}\rangle \quad \text{and} \quad E_n = \sum_{k=0}^{\infty} \lambda^k E_n^{(k)}. \tag{G.17}$$

We will denote the elements of the vectors $|n^{(k)}\rangle$ in this expansion as $C_{mn}^{(k)}$, so that

$$|n^{(k)}\rangle = \sum_{m} C_{mn}^{(k)} |m\rangle. \tag{G.18}$$

The Hamiltonians H_0 and V are two operators of a given system and λ may contain one (or more) operators of other quantum system(s). Either way, for the purposes of the perturbation expansion, λ is still effectively a real number with respect to H_0 and V. The terms in the expansions for E_n and $|\psi_n\rangle$ are determined by three recursion relations. These are

$$E_n^{(k)} = \sum_{m=1}^{N} V_{nm} C_{mn}^{(k-1)} - \sum_{j=1}^{k-1} E_n^{(j)} C_{nn}^{(k-j)}, \tag{G.19}$$

$$C_{ln}^{(k)} = \sum_{m=1}^{N} \frac{V_{lm}}{\Delta_{ln}} C_{mn}^{(k-1)} - \sum_{j=1}^{k-1} \frac{E_n^{(j)}}{\Delta_{ln}} C_{ln}^{(k-j)}, \quad l \neq n, \tag{G.20}$$

$$C_{nn}^{(k)} = -\frac{1}{2} \sum_{m=1}^{N} \sum_{j=1}^{k-1} C_{mn}^{(j)} C_{mn}^{(k-j)*}, \tag{G.21}$$

with

$$C_{mn}^{(0)} = \delta_{mn}, \tag{G.22}$$

$$C_{mn}^{(1)} = (1 - \delta_{mn}) \frac{V_{mn}}{\Delta_{mn}}, \tag{G.23}$$

and $\Delta_{ln} \equiv E_n^{(0)} - E_l^{(0)}$. One can choose all the $C_{nn}^{(k)}$ to be real.

G.7 Finding operator transformation, reordering, and splitting relations

If the Hamiltonian of a system, H, is time-independent, then the unitary operator that gives the time-evolution is $U = e^{-iH/\hbar}$. In fact, for any linear system whose dynamics is $\dot{x} = Ax$, for a constant matrix A, the evolution is given by $\mathbf{x}(t) = e^{At}\mathbf{x}(0)$. For this reason it is sometimes useful to split the exponential of a matrix into products, and/or reorder those products. By "splitting" we mean finding the operator C for which

$$e^{A+B} = e^A e^B e^C, \tag{G.24}$$

and by reordering we mean finding an operator D for which

$$e^A e^B = e^B e^D. \tag{G.25}$$

Some techniques can be used for both purposes. We consider first operator transformations as these can be used to find reordering relations.

Operator transformations

Consider the problem of finding an explicit expression for $\tilde{A}$ where

$$\tilde{A} = e^{-\lambda B} A e^{\lambda B} \tag{G.26}$$

for two operators A and B. To determine $\tilde{A}$ it is useful to remember that this transformation is the solution to the differential equation

$$\frac{dA}{d\lambda} = [A, B]. \tag{G.27}$$

To remember that Eq. (G.27) implies Eq. (G.26), we need only recall the Heisenberg picture, in which the parameter λ is time, and the operator B is proportional to the Hamiltonian. Thus if one can solve Eq. (G.27) for A, one can write an explicit expression for $\tilde{A}$. This solution can often be found when: (i) the algebra of A and B is closed, meaning that repeated commutators of A and B generate only a finite number of distinct operators; (ii) the number of distinct operators in the algebra is small. When these two conditions hold, we solve Eq. (G.27) by first calculating $C = [A, B]$, and then determining the equation of motion for C using $dC/d\lambda = [C, B]$. Then we need to calculate the equation of motion for $D = [C, B]$. We continue in this way until the procedure stops generating new operators. At that point we have a closed set of linear differential equations for the operators in the algebra of A and B, and we can proceed to solve them.

Useful examples of operator transformations that can be derived using the above technique are

$$e^{i\lambda a^\dagger a} a e^{-i\lambda a^\dagger a} = a e^{-i\lambda}, \tag{G.28}$$

$$e^{-\lambda a^\dagger + \lambda^* a} a e^{\lambda a^\dagger - \lambda^* a} = a + \lambda, \tag{G.29}$$

$$e^{i\lambda x} p e^{-i\lambda x} = p - \hbar\lambda, \tag{G.30}$$

along with

$$e^{\lambda B} f(A) e^{-\lambda B} = f\left(e^{\lambda B} A e^{-\lambda B}\right), \tag{G.31}$$

for any analytic function f, operators A and B, and complex number λ. Note that λ may be replaced with any operator that commutes with all the other operators appearing in the expressions.

From operator transformations to reordering

The solutions to operator transformation problems also provide operator reordering relations. First, taking the exponential of both sides of Eq. (G.26) and using Eq. (G.31) we have

$$e^{\tilde{A}} = e^{-\lambda B} e^A e^{\lambda B}. \tag{G.32}$$

Multiplying on the left by $e^{\lambda B}$, and setting $\lambda = 1$ we have

$$e^A e^B = e^B e^{\tilde{A}}, \tag{G.33}$$

which is an operator reordering relation.

The Zassenhaus and BCH formulas

We now consider two formulas that will allow us to split an exponential approximately, and sometimes exactly. The Zassenhaus formula is

$$e^{A+B} = e^A e^B e^{Z_2} e^{Z_3} e^{Z_4} \cdots, \tag{G.34}$$

where

$$Z_2 = -\frac{1}{2}[A,B], \quad Z_3 = \frac{1}{3}[B,[A,B]] + \frac{1}{6}[A,[A,B]], \tag{G.35}$$

$$Z_4 = 3[B,[A,[A,B]]] + 3[B,[B,[A,B]]] - \frac{1}{24}[A,[A,[A,B]]]. \tag{G.36}$$

Methods for calculating higher Z_n can be found in [687, 661].

The Baker–Campbell–Hausdorff (BCH) formula is

$$e^A e^B = e^{A+B+C_2+C_3+C_4+\cdots}, \tag{G.37}$$

where

$$C_2 = \frac{1}{2}[A,B], \quad C_3 = \frac{1}{12}[A,[A,B]] + \frac{1}{12}[[A,B],B]. \tag{G.38}$$

Explicit expressions for higher C_n can be found in [512, 510], and methods for calculating these terms are given in [510, 661].

We note that there are also similar (although more complex) formulas for the case when a product involves an infinite number of infinitesimal operators. Such an expression is the solution to a time-dependent differential equation of the form $\dot{\mathbf{x}} = A(t)\mathbf{x}$. These formulas are called the Magnus and Fer expansions, and can be found in [661].

The Zassenhaus and BCH formulas will allow us to split the exponential e^{A+B} when the algebra of A and B terminates. By this we mean that if one takes repeated commutators of A and B then after a few iterations one gets zero. The simplest example is the algebra for the annihilation and creation operators, in which $[a, a^\dagger] = 1$ and so $[a, [a, a^\dagger]] = 0$ and $[a^\dagger, [a, a^\dagger]] = 0$. These two formulas are also useful if A and B are much less than unity, so that the terms in the expansion become increasingly small. (By this we mean that in the relevant state space $\langle \sqrt{A^2} \rangle \sim \langle \sqrt{B^2} \rangle \ll 1$.)

Splitting and reordering for closed sets of operators

There is another much more powerful method that will allow us to split and reorder operators whose algebra is small and finite [210]. This technique uses the fact that a set of operators that form a closed algebra can usually be represented by finite-dimensional matrices. As an example, consider the four operators $N = a^\dagger a$, a, $a^\dagger$, and I. The commutation relations for this closed algebra are reproduced by the matrices

$$a^\dagger a \equiv \begin{pmatrix} 0 & 0 & 0 \\ 0 & 1 & 0 \\ 0 & 0 & 0 \end{pmatrix}, \qquad a \equiv \begin{pmatrix} 0 & 1 & 0 \\ 0 & 0 & 0 \\ 0 & 0 & 0 \end{pmatrix}, \tag{G.39}$$

$$a^\dagger \equiv \begin{pmatrix} 0 & 0 & 0 \\ 0 & 0 & 1 \\ 0 & 0 & 0 \end{pmatrix}, \qquad I \equiv \begin{pmatrix} 0 & 0 & 1 \\ 0 & 0 & 0 \\ 0 & 0 & 0 \end{pmatrix}. \tag{G.40}$$

We can write this set of equivalences in a much more compact form by defining $M_{ij}^{(n)}$ to be the $n \times n$ matrix whose elements are zero except the element (i,j), which is equal to unity. With this notation we have $a^\dagger a \equiv M_{22}^{(3)}$, $a \equiv M_{12}^{(3)}$, $a \equiv M_{23}^{(3)}$, and $I \equiv M_{13}^{(3)}$.

Since all the matrices are finite-dimensional, we can calculate exponentials and products of exponentials analytically, and thus determine splitting and reordering relations just by equating matrix elements. As an example, we can determine whether $\exp(\alpha a^\dagger a + \beta a + \beta^* a^\dagger)$ can be factored as $\exp(\alpha a^\dagger a) \exp(ya + za^\dagger) \exp(wI)$ by calculating the matrices for the two cases, and equating their elements. If the forms of the matrices are compatible, then this procedure gives us equations that determine x, y, and z as functions of α and β.

Note that we can determine the exponential e^A by solving the set of linear differential equations given by $\dot{\mathbf{x}} = A\mathbf{x}$. If necessary these equations can be solved using the Laplace transform.

An example for the set $a^\dagger a, a, a^\dagger, I$

We now determine the parameters x, y, z, and w in the following relation,

$$e^{\alpha a^\dagger a + \beta a - \beta^* a^\dagger} = e^{x a^\dagger a} e^{ya + za^\dagger} e^{wI}. \tag{G.41}$$

Note that if α is purely imaginary then the operator on the LHS is unitary. We first calculate the matrix representation of $e^{\alpha a^\dagger a + \beta a + \gamma a^\dagger}$, as this will give us the representations for all but one of the exponentials appearing in Eq. (G.41). The set of differential equations that we need to solve is

$$\dot{\mathbf{x}} = \begin{pmatrix} 0 & \beta & 0 \\ 0 & \alpha & \gamma \\ 0 & 0 & 0 \end{pmatrix} \mathbf{x}. \tag{G.42}$$

The solution to these gives us the following matrix representation:

$$e^{\alpha a^\dagger a + \beta a + \gamma a^\dagger} \equiv \begin{pmatrix} 1 & \beta(e^\alpha - 1)/\alpha & \beta\gamma(e^\alpha - 1 - \alpha)/\alpha^2 \\ 0 & e^\alpha & \gamma(e^\alpha - 1)/\alpha \\ 0 & 0 & 1 \end{pmatrix}. \tag{G.43}$$

The exponential factor in Eq. (G.41) that is not covered by the above matrix is e^{wI}. We can obtain the representation for this easily by substituting the matrix $M_{13}^{(3)}$ directly into the Taylor series for the exponential:

$$\exp(wI) \equiv \sum_{n=0}^{\infty} \frac{z^n [M_{13}^{(3)}]^n}{n} = I + wM_{13}^{(3)} = \begin{pmatrix} 1 & 0 & w \\ 0 & 1 & 0 \\ 0 & 0 & 1 \end{pmatrix}. \tag{G.44}$$

Multiplying out the matrix representations for the exponentials on the RHS of Eq. (G.41), and equating them with the matrix representation for the LHS, we find that

$$x = \alpha, \tag{G.45}$$

$$y = \beta(e^\alpha - 1), \tag{G.46}$$

$$z = -\beta^*(e^{-\alpha} - 1), \tag{G.47}$$

$$w = \frac{|\beta|^2}{2\alpha^2}[\alpha - \sinh\alpha]. \tag{G.48}$$

Note that when α is purely imaginary $z = -y^*$ and w is purely imaginary.

Three more closed algebras for mode operators

The sets $\{a^\dagger a + 1/2, a^2/2, a^{\dagger 2}/2\}$ and $\{a^\dagger a - b^\dagger b, ab^\dagger, a^\dagger b\}$ both have the algebra of the 2×2 spin matrices $\{\sigma_z, \sigma, \sigma^\dagger\}$. That is,

$$a^\dagger a + 1/2 \equiv a^\dagger a - b^\dagger b \equiv \sigma_z = \begin{pmatrix} 1 & 0 \\ 0 & -1 \end{pmatrix}, \tag{G.49}$$

$$\frac{1}{2}a^2 \equiv ab^\dagger \equiv \sigma = \begin{pmatrix} 0 & 0 \\ 1 & 0 \end{pmatrix}, \tag{G.50}$$

$$\frac{1}{2}a^{\dagger 2} \equiv ba^\dagger \equiv \sigma^\dagger = \begin{pmatrix} 0 & 1 \\ 0 & 0 \end{pmatrix}. \tag{G.51}$$

The set $\{a^\dagger a + b^\dagger b + 1, ab, a^\dagger b^\dagger\}$ has the slightly different mapping

$$a^\dagger a + b^\dagger b + 1 \equiv \sigma_z, \quad ab \equiv -\sigma, \quad a^\dagger b^\dagger \equiv \sigma^\dagger. \tag{G.52}$$

Further information on the topics in this section can be found in [210, 661, 510].

G.8 The Haar measure

The Haar measure is the measure that gives an equal weight to all quantum states in a space of a given dimension, N. The definition of "equal weight" is that the measure should be invariant under a group of transformations that can map any state to any other state. In the case of quantum states this group is the set of all unitary transformations, so the Haar measure is the measure that is invariant under all unitary transformations. If we want to specify a probability distribution over pure quantum states that captures the notion of "knowing nothing" about the state, then this distribution is equal to unity when integrating with respect to the Haar measure.

If we parametrize all quantum states $|\psi\rangle$ of dimension N by

$$|\psi\rangle = \sum_{n=1}^{N} (x_n + iy_n)|n\rangle, \tag{G.53}$$

then integration of a function $f(|\psi\rangle)$ with respect to the Haar measure is

$$\int f(\{x_n\}, \{y_n\}) \left(\frac{2\pi^n}{(n-1)} \right) dx_1 \cdots dx_N \, dy_1 \cdots dy_N, \tag{G.54}$$

where the surface of integration is the unit hypersphere defined by

$$\sum_{n=1}^{N} x_n^2 + y_n^2 = 1. \tag{G.55}$$

We can alternatively write the quantum state in terms of the probabilities $P_n = x_n^2 + y_n^2$ and phases $\theta_n = \arg(x_n + iy_n)$, so that

$$|\psi\rangle = \sum_{n=1}^{N} \sqrt{P_n} e^{i\theta_n} |n\rangle. \tag{G.56}$$

The integral over the Haar measure becomes

$$\int f(\{P_n\}, \{\theta_n\}) \left(\frac{(2\pi)^n}{(n-1)} \right) dP_{N-1} \cdots dP_1 d\theta_N \cdots d\theta_1, \tag{G.57}$$

where $P_N = 1 - \sum_{n=1}^{N-1} P_n$. The region of integration over the phases is

$$\theta_N \in [-\pi/2, \pi/2], \tag{G.58}$$

$$\theta_n \in [0, 2\pi], \quad 1 \le n \le N - 1. \tag{G.59}$$

The region of integration over the probabilities P_n is the simplex

$$\sum_{i=1}^{N-1} P_n \le 1, \tag{G.60}$$

or explicitly,

$$\int_0^1 \int_0^{1-P_1} \cdots \int_0^{1-(P_1+\ldots+P_{N-2})} f(\{P_n\}, \{\theta_n\}) dP_{N-1} \cdots dP_1. \tag{G.61}$$

Examples of integrating over the Haar measure can be found in [316, 321].

G.9 General form of the Kushner–Stratonovich equation

Consider a classical dynamical system with the following equation of motion.

$$\mathbf{dx} = \mathbf{F}(\mathbf{x}, t) dt + G(\mathbf{x}, t) \mathbf{dW}. \tag{G.62}$$

Here $\mathbf{x}$ is the state of the system and $\mathbf{dW}$ is a set of mutually independent Wiener increments. Since $\mathbf{x}$ and $\mathbf{dW}$ are vectors, G is a matrix.

A set of M continuous measurements is made on the system. We denote the M streams of measurement results by vector $\mathbf{dy}$, which is given by

$$\mathbf{dy} = \mathbf{H}(\mathbf{x}, t) dt + R(t) \mathbf{dV}. \tag{G.63}$$

Here $\mathbf{dV}$ is a set of mutually independent Wiener increments that are independent of the noise driving the system, and R is an arbitrary matrix that determines the correlations between the noise on each of the M measurements.

The Kushner–Stratonovich equation for the observer's state-of-knowledge of the system is

$$dP = -\sum_{i=1}^n \frac{\partial}{\partial x_i}(F_i P) dt + \frac{1}{2} \sum_{i=1}^n \sum_{j=1}^n \frac{\partial^2}{\partial x_i \partial x_j}([GG^\mathrm{T}]_{ij} P) dt \tag{G.64}$$

$$+ [\mathbf{H}(\mathbf{x}, t) - \langle \mathbf{H}(\mathbf{x}, t) \rangle]^T (RR^\mathrm{T}) [\mathbf{dy} - \langle \mathbf{H}(\mathbf{x}, t) \rangle dt] P. \tag{G.65}$$

Here x_i and F_i are the elements of $\mathbf{x}$ and $\mathbf{F}$, respectively, the $[GG^T]_{ij}$ are the elements of the matrix GG^T, and $\langle \cdots \rangle$ denotes the expectation value with respect to P at the current time.

G.10 Obtaining steady states for linear open systems

Steady states for linear open quantum systems can be obtained by solving the equations of motion in the frequency domain, and then integrating the spectrum over all frequencies. This integration can be done with an integral formula. We demonstrate the method by calculating the steady-state mean photon number for resolved-sideband cooling (RSB). This cooling method is discussed in detail in Section 8.3.1.

The equations of motion for the RSB two-oscillator system are given in Eqs. (8.100) and (8.101). If we change variables to $X = a + a^\dagger$, $P = -i(a - a^\dagger)$, $x = b + b^\dagger$, and $p = -i(b - b^\dagger)$, and then define

$$
\mathbf{x} = \begin{pmatrix} x \\ p \\ X \\ P \end{pmatrix}, \; \boldsymbol{\xi} = \begin{pmatrix} \sqrt{2\Gamma}\, x_{\mathrm{in}} \\ \sqrt{2\Gamma}\, p_{\mathrm{in}} \\ \sqrt{2K}\, X_{\mathrm{in}} \\ \sqrt{2K}\, P_{\mathrm{in}} \end{pmatrix}, \; M = \begin{pmatrix} -\Gamma & \omega & 0 & 0 \\ -\omega & -\Gamma & -\lambda & 0 \\ 0 & 0 & -K & \omega \\ -\lambda & 0 & -\omega & -K \end{pmatrix}, \tag{G.66}
$$

the Langevin equations of motion for the system are $\dot{\mathbf{x}} = M\mathbf{x} + \boldsymbol{\xi}(t)$. The noise sources in the vector $\boldsymbol{\xi}$ are Hermitian noise sources just like those defined in Eq. (3.241) in Section 3.7.2. The relationships between the damping rates defined here and those in Section 8.3.1 are $\Gamma = \gamma/2$ and $K = \kappa/2$.

To solve the equations of motion for $\mathbf{x}$ in the frequency domain we take the Fourier transform of both sides of the equation. Denoting the frequency space variables with a tilde, e.g.,

$$
\tilde{\mathbf{x}}(v) = \frac{1}{\sqrt{2\pi}} \int_{-\infty}^{\infty} \mathbf{x}(t) e^{-ivt}\, dt, \tag{G.67}
$$

the equations of motion become $-iv\tilde{\mathbf{x}} = M\tilde{\mathbf{x}} + \tilde{\boldsymbol{\xi}}(v)$. Rearranging gives

$$
\tilde{\mathbf{x}} = -(M + ivI)^{-1}\tilde{\boldsymbol{\xi}}(v) \equiv A(v)\tilde{\boldsymbol{\xi}}(v). \tag{G.68}
$$

The dynamical variables are therefore given by a linear combination of the noise sources, where the coefficients are functions of v and therefore filter the noise. Inverting the matrix $M + ivI$, for which a software package such as Scientific Workplace or Mathematica is invaluable, we obtain the matrix

$$
A(v) = \frac{\begin{pmatrix} f(\Gamma)g(K) & \omega g(K) & \omega\lambda f(K) & \omega^2\lambda \\ -\omega g(K) - \lambda^2\omega & f(\Gamma)g(K) & \lambda f(\Gamma)f(K) & \omega\lambda f(\Gamma) \\ \omega\lambda f(\Gamma) & \omega^2\lambda & f(K)g(\Gamma) & \omega g(\Gamma) \\ \lambda f(\Gamma)f(K) & \omega\lambda f(K) & -\omega g(\Gamma) - \lambda^2\omega & f(K)g(\Gamma) \end{pmatrix}}{D(v)}, \tag{G.69}
$$

with

$$
f(\alpha) = \alpha - iv, \tag{G.70}
$$

$$
g(\alpha) = (i\omega + iv - \alpha)(i\omega - iv + \alpha), \tag{G.71}
$$

and

$$D(v) = \left[f(K)^2 + \omega^2\right]\left[f(\Gamma)^2 + \omega^2\right] - \lambda^2\omega^2.$$ (G.72)

Two important properties of the matrix A are (i) that each element is a ratio of polynomials, and (ii) that the imaginary unit i, and the frequency v always appear together (this can be seen from Eq. (G.68)). The second property means that taking the complex conjugate of any element of A is the same as replacing v with $-v$.

The steady-state variance of a dynamical variable is given by integrating the spectrum for that variable over all v. As stated in Eq. (3.77), the spectrum for x (for example) is given by

$$S_x(v) = F(v, -v) \quad \text{where} \quad \langle \tilde{x}(v)\tilde{x}(v')\rangle = F(v, v')\delta(v + v').$$ (G.73)

We can obtain the correlation functions for the dynamical variables $\tilde{\mathbf{x}}(v)$ directly from those of the noise sources:

$$\langle \tilde{\mathbf{x}}(v)\tilde{\mathbf{x}}(v')\rangle = A(v)\langle \tilde{\boldsymbol{\xi}}(v)\tilde{\boldsymbol{\xi}}(v')^{\mathrm{T}}\rangle A(v')^{\mathrm{T}} = A(v)GA(v')^{\mathrm{T}}\delta(v + v'),$$ (G.74)

where G is the correlation matrix for the noise sources. In our case G is (see Eqs. 3.242 – 3.244 in Section 3.7.2)

$$G = 2\begin{pmatrix} \Gamma(2n_T + 1) & 0 & 0 & 0 \\ 0 & \Gamma(2n_T + 1) & 0 & 0 \\ 0 & 0 & K & 0 \\ 0 & 0 & 0 & K \end{pmatrix}.$$ (G.75)

The spectrum for x is

$$S_x(v) = \frac{2\Gamma(2n_T + 1)|g(K)|^2\left(|f(\Gamma)|^2 + \omega^2\right) + 2K(\lambda\omega)^2\left(|f(K)|^2 + \omega^2\right)}{D(v)D(-v)}$$ (G.76)

and that for p is

$$S_p(v) = \frac{\Gamma(2n_T + 1)\left\{|g(K)|^2\left(|f(\Gamma)|^2 + \omega^2\right) + 2\lambda^2\omega^2\mathrm{Re}[g(K)] + (\lambda^2\omega)^2\right\}}{D(v)D(-v)}$$
$$+ \frac{K(\lambda\omega)^2|f(\Gamma)|^2\left(|f(K)|^2 + \omega^2\right)}{D(v)D(-v)}.$$ (G.77)

The expressions for the spectra contain 8th-order polynomials in the denominator. If these polynomials had no special structure, they would likely be impossible to integrate analytically. The fact that this is possible is due to the following remarkable integral formula, which is a slightly simplified version of a formula in Gradshteyn and Ryzhik:[1]

---

[1] In the "Corrected and Enlarged Edition" of Gradshteyn and Ryzhik, the edition that follows the fourth edition, the integral formula is number 3.112, on page 218.

$$I_n \equiv \int_{-\infty}^{\infty} \frac{y_n(v)}{z_n(v)z_n(-v)}dv = \frac{\pi}{a_0}\left|\frac{M_n}{L_n}\right|, \tag{G.78}$$

where $z(v)$ must satisfy $z(-v) = z^*(v)$,

$$y(v) = b_0 v^{2n-2} + b_1 v^{2n-4} + \cdots + b_{n-1}, \tag{G.79}$$

$$z(v) = a_0 v^n + a_1 v^{n-1} + \cdots + a_n, \tag{G.80}$$

and

$$M_n = \begin{vmatrix} b_0 & b_1 & b_2 & \cdots & b_{n-1} \\ u_0 & a_2 & a_4 & \cdots & 0 \\ 0 & a_1 & a_3 & \cdots & 0 \\ \vdots & \vdots & \vdots & \vdots & \vdots \\ 0 & 0 & 0 & \cdots & a_n \end{vmatrix}, \quad L_n = \begin{vmatrix} a_1 & a_3 & a_5 & \cdots & 0 \\ a_0 & a_2 & a_4 & \cdots & 0 \\ 0 & a_1 & a_3 & \cdots & 0 \\ \vdots & \vdots & \vdots & \vdots & \vdots \\ 0 & 0 & 0 & \cdots & a_n \end{vmatrix}. \tag{G.81}$$

Note that L_n and M_n are determinants of matrices that differ only by their first row.

We need the case $n = 4$, for which the integral is

$$I_4 = \pi \left| \frac{(b_0/a_0)(a_2 a_3 - a_1 a_4) - b_1 a_3 + b_2 a_1 + (b_3/a_4)(a_0 a_3 - a_1 a_2)}{a_0 a_3^2 + a_1^2 a_4 - a_1 a_2 a_3} \right|. \tag{G.82}$$

Using this integral formula, and the fact that the steady-state mean squares of x and p are

$$\langle x^2 \rangle_{\text{ss}} = \frac{1}{2\pi}\int_{-\infty}^{\infty} S_x(v)dv, \quad \langle p^2 \rangle_{\text{ss}} = \frac{1}{2\pi}\int_{-\infty}^{\infty} S_p(v)dv, \tag{G.83}$$

we obtain

$$\langle x^2 \rangle_{\text{ss}} = \frac{(2n_T + 1)A_1 + A_2}{F} + \frac{(2n_T + 1)B_1 + B_2}{CF} \tag{G.84}$$

and

$$\langle p^2 \rangle_{\text{ss}} = \frac{(2n_T + 1)G_1 + G_2}{F} + \frac{(2n_T + 1)J_1 + J_2}{CF}. \tag{G.85}$$

Here we have defined

$$A_1 = r\Gamma\left(b^2 + a(b-4\omega^2) + \lambda^2\omega^2\right) + 4\Gamma\left(2r^2 + b\right)(aK + b\Gamma), \tag{G.86}$$

$$A_2 = rK\lambda^2\omega^2, \tag{G.87}$$

$$B_1 = r\Gamma ab^2(r^2 + \Gamma K + \lambda^2\omega^2), \tag{G.88}$$

$$B_2 = r\Gamma b\lambda^2\omega^2(r^2 + \Gamma K + \lambda^2\omega^2), \tag{G.89}$$

$$G_1 = r\Gamma\left(b^2 + a(2c-b)(r+2K+a/r)(aK+b\Gamma) + 3\omega^2\lambda^2\right), \tag{G.90}$$

$$G_2 = \lambda^2 K\left(aK + b\Gamma + r(a+c+\omega^2)\right), \tag{G.91}$$

$$J_1 = r\Gamma(r^2 + \Gamma K + \omega^2)(\lambda^4\omega^2 + ab^2 - 2b\lambda^2\omega^2), \tag{G.92}$$

$$J_2 = rb\lambda^2\Gamma^2 K(r^2 + \Gamma K + \omega^2), \tag{G.93}$$

$$F = 2r^2[\Gamma K(r^2 + 4\omega^2) - \lambda^2\omega^2], \tag{G.94}$$

$$C = ab - \lambda^2\omega^2, \tag{G.95}$$

with

$$a = \Gamma^2 + \omega^2, \quad b = K^2 + \omega^2, \quad c = K^2 - \omega^2, \quad r = \Gamma + K. \tag{G.96}$$

Appendix H
Some proofs and derivations

H.1 The Schumacher–Westmoreland–Wootters theorem

We restate the theorem first for convenience:

A system is prepared in the ensemble $\{\rho_j, p_j\}$, whose Holevo χ-quantity we denote by χ. A measurement is then performed having N_1 outcomes labeled by n, in which each outcome has probability p_n. If, on the nth outcome, the measurement leaves the system in an ensemble whose Holevo χ-quantity is χ_n, then the mutual information is bounded by

$$I_{\text{ens}} \leq \chi - \sum_n p_n \chi_n. \tag{H.1}$$

The equality is achieved if and only if $[\rho, \rho_n] = 0, \forall n$, and the measurement is semiclassical. This result is also true for inefficient measurements. In this case the measurement will in general have operators $\{A_{nk}\}$, such that the observer knows n but not k. The measurement outcome labeled by n is the result of summing over the outcomes for the index k [548, 286, 30].

Proof We will denote the initial ensemble by $\varepsilon = \{\rho_j, p_j\}$, and the initial state of the system from the point of view of the observer as $\rho_S = \sum_j p_j \rho_j$. The measurement performed by the observer has operators A_{nk}, where she knows n but not k. We also note that the ensemble that remains after the observer has obtained result n is $\varepsilon_n = \{\sigma_{n|j}, p_{j|n}\}$, where $p_{j|n}$ is the probability that the initial state ρ_j was prepared given outcome n, and

$$\sigma_{n|j} = \sum_k p_{k|n,j} \sigma_{kn|j} = \frac{\sum_k A_{kn} \rho_j A_{kn}^\dagger}{p_{n|j}}. \tag{H.2}$$

In this expression $p_{k|n,j}$ is the probability for outcome k given that the outcome obtained is n and the initial state was ρ_j. The density matrix $\sigma_{kn|j}$ is the final state that would result if the initial state were ρ_j and the observer knew that both outcomes n and j had been obtained.

We now collect a few facts. The first is that any efficient measurement on a system S, described by $K = N_1 N_2$ operators, A_{nk} ($n = 1, \ldots, N_1$ and $k = 1, \ldots, N_2$), can be realized by bringing up an auxiliary system A of dimension K, performing a joint unitary operation on S and A, and then making a von Neumann measurement on A. If the initial state of S is

490

ρ^S, then the final joint state of $\mathcal{A}$ and $\mathcal{S}$ after the von Neumann measurement is

$$\sigma^{AS} = |nk\rangle\langle nk| \otimes \frac{A_{nk}\rho^S A_{nk}^\dagger}{p_{n,k}}. \qquad \text{(H.3)}$$

Here $|nk\rangle$ is the state of $\mathcal{A}$ selected by the von Neumann measurement.

The second fact is that the state which results from discarding all information about the measurement outcomes k and j can be obtained by performing a unitary operation between A and another system E which perfectly correlates the states $|kj\rangle$ of A with a set of orthogonal states of E, and then tracing out E.

The final fact is a result proven by Schumacher, Westmoreland, and Wootters (SWW) [548], which is that the Holevo χ quantity is non-increasing under partial trace. That is, if we have two quantum systems $\mathcal{A}$ and $\mathcal{B}$, and an ensemble of joint states of $\mathcal{A}$ and $\mathcal{B}$, ρ_i^{AB} with associated probabilities q_i, then

$$\chi^A = S(\rho^A) - \sum_i q_i S(\rho_i^A) \leq S(\rho^{AB}) - \sum_i q_i S(\rho_i^{AB}) = \chi^{AB}, \qquad \text{(H.4)}$$

where $\rho_i^A = \text{Tr}_B[\rho_i^{AB}]$. To prove this result SWW used the strong subadditivity of the von Neumann entropy [523, 371, 372].

We now encode information in system $\mathcal{S}$ using the ensemble ε, and consider the joint system that consists of the three systems $\mathcal{S}$, $\mathcal{A}$, $\mathcal{E}$, and a fourth system $\mathcal{M}$, where the latter has dimension N_1. We prepare the systems A, E, and M initially in pure states, so that the Holevo quantity for the joint system is $\chi^{MASE} = \chi^S$. We then perform the required unitary operation between $\mathcal{S}$ and $\mathcal{A}$, and a unitary operation between A and E that perfectly correlates the states $|nk\rangle$ of $\mathcal{A}$ with a set of orthogonal states of E. Unitary operations do not change the Holevo quantity. Then we trace over E, so that we are left with the state

$$|\psi\rangle\langle\psi| \otimes \sum_{nk} p_{n,k} |nk\rangle\langle nk| \otimes \sigma_{nk}^S. \qquad \text{(H.5)}$$

Here $|\psi\rangle$ is the state of system $\mathcal{M}$, and $\sigma_{nk}^S = A_{nk}\rho^S A_{nk}^\dagger/p_{nk}$ is the final state that would result from knowing both outcomes n and k. After the two unitaries and the partial trace over E, the Holevo quantity for the remaining systems, which we will denote by χ'^{MAS}, satisfies $\chi'^{MAS} \leq \chi^{MASE} = \chi^S$.

We now perform one more unitary operation, this time between $\mathcal{M}$ and $\mathcal{A}$, so that we correlate a basis of states of $\mathcal{M}$, which we denote by $|n\rangle$, with the index n of the states of $\mathcal{A}$, giving

$$\sum_n |n\rangle\langle n| \otimes \sum_k p_{nk} |nk\rangle\langle nk| \otimes \sigma_{nk}^S. \qquad \text{(H.6)}$$

Finally we trace out $\mathcal{A}$, leaving us with the state

$$\sigma^{MS} = \sum_n |n\rangle\langle n| \otimes \sum_k p_{nk}\sigma_{nk}^S. \qquad \text{(H.7)}$$

After this final unitary, and the partial trace over $\mathcal{A}$, the Holevo quantity for the remaining systems $\mathcal{S}$ and $\mathcal{M}$, which we will denote by $\chi''^{\mathcal{MS}}$, satisfies $\chi''^{\mathcal{MS}} \leq \chi'^{\mathcal{MAS}} \leq \chi^{\mathcal{S}}$. We have gone through the above process using the initial state ρ, but we could just as easily have started with any of the initial states, ρ_j, in the ensemble. Let us denote the final states which we obtain using the initial state ρ_j as $\sigma_i^{\mathcal{MS}}$. Calculating $\chi''^{\mathcal{MS}}$ we have

$$\chi''^{\mathcal{MS}} = S(\sigma^{\mathcal{MS}}) - \sum_j p_j S(\sigma_i^{\mathcal{MS}})$$

$$= H[\{p_n\}] - \sum_j p_j H[\{p_{n|j}\}] + \sum_n p_n \left[S(\sigma_n) - \sum_j p_{j|n} \sigma_{n|j} \right] \qquad \text{(H.8)}$$

$$= M(N{:}J) + \sum_n p_n \chi_n^{\mathcal{S}} \leq \chi^{\mathcal{S}}. \qquad \text{(H.9)}$$

Here, as usual, $M(N{:}J)$ is the mutual information between the random variables n and j. Rearranging the above expression gives the desired result.

H.2 The operator-sum representation for quantum evolution

Here we give Benjamin Schumacher's elegant proof that every quantum evolution (every valid transformation of one density matrix ρ to another $\tilde{\rho}$) for an N-dimensional system can be written in the form $\tilde{\rho} = \sum_n A_n \rho A_n^\dagger$ with no more than N^2 operators A_n [550]. This proof is also an excellent example of thinking outside the box with regard to the operations and structure of quantum mechanics. To do so we will need two N-dimensional systems, A and B. We will denote a basis for system A by $\{|n\rangle_A\}$ and one for B by $\{|n\rangle_B\}$. We will also find that the following unnormalized vector in the joint space of the two systems is very useful: $|v\rangle = \sum_n |n\rangle_A \otimes |n\rangle_B$. We refer to it as a vector rather than a state because it is not normalized. Of course, we could always normalize $|v\rangle$ if we wished by dividing it by $\sqrt{N}$, and then it would be a perfectly valid quantum state. The reason that the joint vector $|v\rangle$ is useful is that we can use it to define a linear map from the states of B to those of A (and vice versa). Let us denote this map by the function f, so that if the input state of B is $|\psi\rangle_B$, and we denote the corresponding output state of system A by $|\psi^*\rangle_A$, then we write $f(|\psi\rangle_B) = |\psi\rangle_B$. The mapping f is defined by taking the (partial) inner product of the input state with the joint vector $|v\rangle$ (the partial inner product is defined in Section 1.3.3). That is, if $|\psi\rangle_B = \sum_k c_k |k\rangle_B$, then

$$f(|\psi\rangle_B) = \langle\psi|_B|v\rangle = \sum_k c_k^* \langle k|_B \sum_n |n\rangle_A \otimes |n\rangle_B = \sum_k c_k^* |k\rangle_A \equiv |\psi^*\rangle_A. \qquad \text{(H.10)}$$

Thus the mapping maps each basis state $|k\rangle_A$ to the basis state $|k\rangle_B$, and conjugates the coefficients of the input state to give the output state. The mapping is conjugate linear, in that

$$f(a_1|\psi_1\rangle_A + a_2|\psi_2\rangle_A) = a_1^* f(|\psi_1\rangle_A) + a_2^* f(|\psi_2\rangle_A). \qquad \text{(H.11)}$$

Now consider a quantum evolution for system A that takes an initial state ρ_A to $\tilde{\rho}_A$. We will denote this evolution by $\mathcal{L}$, so that we write $\mathcal{L}(\rho) = \tilde{\rho}$. Recall that $\mathcal{L}$ is completely positive, which means that the joint evolution in which nothing happens to system B is positive. We can write the (super-)operator that gives this evolution as $\mathcal{L} \otimes \mathcal{I}_B$, where $\mathcal{I}_B$ is the linear (super-)operator on B that does nothing.

We now imagine that system A is in the initial state $|\psi\rangle_A$, and that $|\psi^*\rangle_B$ is the state of system B for which $\langle\psi^*|_B|v\rangle = |\psi\rangle_A$. We will also define $\tilde{\rho} = \mathcal{L}(|\psi\rangle_A\langle\psi|_A)$. With these definitions we can generate the final state of system A, $\tilde{\rho}_A$, by starting the joint system in the unnormalized state $|v\rangle$, taking the inner product with $|\psi^*\rangle_B$, and applying the evolution $\mathcal{L}$. In symbols:

$$\tilde{\rho} = \mathcal{L}\left(\langle\psi^*|_B(|v\rangle\langle v|)|\psi^*\rangle_B\right). \tag{H.12}$$

We now observe that there is another way to generate the final state $\tilde{\rho}_A$: because the operator $\mathcal{L} \otimes \mathcal{I}_B$ commutes with the operation of taking the partial inner product with a state of system B, we can apply the evolution $\mathcal{L} \otimes \mathcal{I}_B$ first, and then take the partial inner product. Schumacher points out that this is clear on physical grounds: since one operation acts non-trivially only on A, and the other on B (note that as far as the effect on system A is concerned, the partial inner product is the same as a measurement on B), they cannot interfere with each other. But we can also show more formally that we can apply the evolution first. To do so we first note that if we apply the operator $P = I_A \otimes |\psi^*\rangle_B\langle\psi^*|_B$ on both sides of the matrix $|v\rangle\langle v|$, where I_A is the identity operator for system A, then this applies the partial inner product on both sides:

$$\begin{aligned} P|v\rangle\langle v|P &= (I_A \otimes |\psi^*\rangle_B\langle\psi^*|_B)|v\rangle\langle v|(I_A \otimes |\psi^*\rangle_B\langle\psi^*|_B) \\ &= (\langle\psi^*|_B|v\rangle)(\langle v|\psi^*\rangle_B) \otimes |\psi^*\rangle_B\langle\psi^*|_B \\ &= |\psi\rangle_A\langle\psi|_A \otimes |\psi^*\rangle_B\langle\psi^*|_B. \end{aligned} \tag{H.13}$$

But since P acts as the identity on B it commutes with $\mathcal{L} \otimes \mathcal{I}_B$, so

$$\begin{aligned} P[\mathcal{L} \otimes \mathcal{I}_B(|v\rangle\langle v|)]P &= \mathcal{L} \otimes \mathcal{I}_B(P|v\rangle\langle v|P) \\ &= \mathcal{L} \otimes \mathcal{I}_B(|\psi\rangle_A\langle\psi|_A \otimes |\psi^*\rangle_B\langle\psi^*|_B) \tag{H.14} \\ &= \tilde{\rho} \otimes |\psi^*\rangle_B\langle\psi^*|_B. \tag{H.15} \end{aligned}$$

We now use the fact that $\mathcal{L} \otimes \mathcal{I}_B(|v\rangle\langle v|)$ is a positive operator, so that $D = \mathcal{L} \otimes \mathcal{I}_B(|v\rangle\langle v|)$ is a positive matrix. We can therefore write it in terms of its eigenvectors. Denoting these eigenvectors as $|D_n\rangle, n = 1,\ldots,N$, the corresponding eigenvalues as λ_n, and defining the unnormalized vectors $|\tilde{D}_n\rangle \equiv \sqrt{\lambda_n}|D_n\rangle$, we can write D as

$$D = \sum_n |\tilde{D}_n\rangle\langle\tilde{D}_n|. \tag{H.16}$$

To me it seems that in the analysis so far there has been a lot of "thinking outside the box," but from where Schumacher was coming from it seems that he regarded this as all fairly straighforward, since it is the next step that he describes as the essential trick.

We now observe that the operation that takes the joint system from the initial matrix $|v\rangle\langle v|$ to the final state $|\tilde{D}_n\rangle\langle\tilde{D}_n|$ is linear. This is because the operation is simply $\mathcal{L}\otimes\mathcal{I}_B$ followed by a projection onto the state $|D_n\rangle$. This means that if we further project the state using the projector P, which applies the partial inner product with $|\psi^*\rangle_B$, the operation is still linear, and this time it produces a final pure state of system A. The sequence of operations therefore defines a linear map from the state $|\psi\rangle_A$ to a final pure state of A, and there is a different map for each vector $|\tilde{D}_n\rangle$. We now define a set of operators $\{A_n\}$, so that A_n acting on $|\psi\rangle_A$ implements the map generated by the vector $|\tilde{D}_n\rangle$.

We can now write the full operation in terms of the operators $\{A_n\}$ as follows:

$$\tilde{\rho}\otimes|\psi^*\rangle_B\langle\psi^*|_B = P[\mathcal{L}\otimes\mathcal{I}_B(|v\rangle\langle v|)]P = PDP = \sum_n P|\tilde{D}_n\rangle\langle\tilde{D}_n|P$$

$$= \left(\sum_n A_n|\psi\rangle_A\langle\psi|_A A_n^\dagger\right)\otimes|\psi^*\rangle_B\langle\psi^*|_B. \qquad (H.17)$$

We have now obtained a way to write the evolution $\tilde{\rho}=\mathcal{L}(|\psi\rangle_A\langle\psi|_A)$ in terms of a set of operators $\{A_n\}$. But more importantly, since the matrix D has dimension N^2, there are no more than N^2 vectors $|\tilde{D}_n\rangle$, and thus no more than N^2 operators in this set. This means that every evolution can be realized with a measurement that has at most N^2 outcomes.

In Chapter 1 we showed that any measurement on a system A that is described by measurement operators $\{A_n\}$ could be implemented by applying a joint unitary, U, to the system A and a second "probe" system B. In this construction the dimension of the probe system is equal to the number of measurement operators. After applying the unitary U, if we trace out the probe the evolution of A is given by $\tilde{\rho}=\sum_n A_n\rho A_n^\dagger$, where ρ is the initial state of A. But we now know that every quantum evolution can be written in this form, with no more than N^2 operators in the sum. We therefore have the additional result that every quantum evolution of system A can be obtained by applying a joint unitary U to A and a second system with dimension N^2, and tracing out the second system. Schumacher refers to an operator U that realizes an evolution $\mathcal{L}$ as a "unitary representation" of $\mathcal{L}$ [550].

H.3 Derivation of the Wiseman–Milburn Markovian feedback SME

Wiseman and Milburn considered an especially simple kind of feedback in which a term in the Hamiltonian is made proportional to the stream of measurement results from a continuous measurement. That is, if we make a continuous measurement of the observable x, so that the stream of measurement results is

$$\frac{dr}{dt} = \langle x\rangle + \frac{1}{8\eta k}\xi(t), \qquad (H.18)$$

where $\xi(t)$ is effectively white noise, then Wiseman and Milburn considered feeding back this measurement signal by choosing the Hamiltonian of the system to be of the form

$$H = H_0 + \left(\frac{dr}{dt}\right) F, \tag{H.19}$$

where F is a Hermitian operator. Because of its simple form, this kind of feedback has the special property that, when one averages over all possible realizations of the measurement noise, one obtains a master equation that is still Markovian. (That is, in which the evolution of the density matrix is determined only by its value at the present time and not at earlier times.) Because of this property this kind of feedback is often referred to as Markovian feedback, and it is often simpler to analyze than other more sophisticated feedback protocols. If we process the measurement signal in a more complex manner, for example by using the measurement record to calculate the new density matrix at each time, and making the Hamiltonian some function of this density matrix, then the evolution, after averaging over all trajectories, cannot be described by any simple equation of motion.

Deriving the stochastic equation of motion for the density matrix under Markovian feedback is simple, except for one important subtlety. In reality the measurement signal has a finite bandwidth, and as such is continuous and differentiable. Our description of it as white noise is an idealization, and usually a convenient one. But for this we have to pay a price. If we assume that the random increment dW is independent of the system at each time-step, then the noise does not obey the usual rules of calculus (and strictly speaking, dW/dt does not exist). But if we make the Hamiltonian proportional to dr/dt we know that since it is a real signal it is differentiable, and the resulting equation of motion for the density matrix does obey the usual rules of calculus.

The apparent contradiction is resolved by realizing that to obtain an equation of motion that contains white noise, we need to take the broad-band limit of an equation involving real noise, such that the white-noise equation correctly approximates the real evolution. This is possible, but the resulting equation is not one in which the increment dW is independent of the current state of the system. This stochastic differential equation is called a *Stratonovich* equation, whereas a stochastic equation in which dW is independent of the current state is called an *Ito* equation. Stratonovich equations, while providing the bridge from real noise to white noise, are analytically and numerically difficult to deal with. Fortunately there is an elegant solution to this problem. There is a simple way to transform any Stratonovich equation into a different Ito equation, such that the Ito equation describes the same dynamics (has the same solution) as the Stratonovich equation. So the procedure for describing any physical system whose derivative is some function of a real broad-band signal is (i) replace the real signal with its white-noise limit (dW/dt); (ii) rewrite the equation in terms of the increment dW, but using the usual rules of calculus $((dW)^2 = 0)$ instead of Ito calculus $((dW)^2 = dt)$; (iii) transform the resulting Stratonovich equation to an Ito equation. The transformation rule is given in, e.g., [288, 197].

To make our expressions more compact, we will use the two super-operators defined by Wiseman. These are

$$\mathcal{H}(c)\rho \equiv c\rho + \rho c^\dagger - \langle c + c^\dagger \rangle \rho, \tag{H.20}$$

$$\mathcal{D}(c)\rho \equiv -\frac{1}{2}(c^\dagger c\rho + \rho c^\dagger c - 2c\rho c^\dagger). \tag{H.21}$$

It is useful to note that if F is Hermitian, then

$$\mathcal{H}(iF)\rho = i[F, \rho], \tag{H.22}$$

$$\mathcal{D}(F)\rho = \mathcal{D}(iF)\rho = -\frac{1}{2}[F, [F, \rho]]. \tag{H.23}$$

We now apply the procedure for treating broadband noise, described above, to Markovian feedback. To do this we first write the evolution equation for ρ under the feedback Hamiltonian, which is

$$\frac{d\rho_{fb}}{dt} = -\frac{i}{\hbar}\left(\frac{dr}{dt}\right)[F, \rho] = \frac{1}{\hbar}\left(\langle x\rangle + \frac{1}{\sqrt{8\eta k}}\frac{dW}{dt}\right)\mathcal{H}(-iF)\rho. \tag{H.24}$$

Next we write this evolution equation in terms of infinitesimal increments, which in this case means multiplying by dt:

$$d\rho_{fb}(\rho) = \frac{1}{\hbar}\left(\langle x\rangle\, dt + \frac{1}{\sqrt{8\eta k}}dW\right)\mathcal{H}(-iF)\rho. \tag{H.25}$$

Here we have given $d\rho_{fb}$ the explicit argument ρ, as this will aid us later on. The above equation for $d\rho_{fb}(\rho)$ is a Stratonovich equation. We now convert to an Ito equation (see, e.g., [288, 197]), and the result in this case is an extra deterministic term equal to half the square of the stochastic term:

$$d\rho_{fb}(\rho) = \left\{\frac{1}{\hbar}\left(\langle x\rangle\, dt + \frac{1}{\sqrt{8\eta k}}dW\right)\mathcal{H}(-iF)\rho + \frac{1}{8\eta\hbar^2 k}\mathcal{D}(iF)\,dt\right\}\rho. \tag{H.26}$$

This is now an Ito equation.

Since the feedback acts after the measurement, the full evolution of ρ in the time-step dt is given by first applying the measurement evolution to ρ, and then applying the feedback. The measurement evolution is (see Eq. 3.39)

$$\rho_m = \rho + d\rho_m = \rho + \left[2k\mathcal{D}(x)\,dt + \sqrt{2k}\mathcal{H}(x)\,dW\right]\rho. \tag{H.27}$$

So the state following the feedback is

$$\begin{aligned}
\rho(t + dt) &= \rho_m + d\rho_{fb}(\rho_m) = \rho + d\rho_m + d\rho_{fb}(\rho + d\rho_m)\\
&= \rho + d\rho_m + d\rho_{fb}(\rho) + d\rho_{fb}(d\rho_m)\\
&\equiv \rho + d\rho,
\end{aligned} \tag{H.28}$$

where $d\rho_{\text{fb}}(d\rho_{\text{m}})$ is given by Eq. (H.26) with ρ replaced with $d\rho_{\text{m}}$.

After using the Ito rule $(dW)^2 = dt$, and some judicious rearranging, we obtain the Wiseman–Milburn (WM) feedback SME:

$$d\rho_{\text{full}} = -\frac{i}{\hbar}[H_0 + H_{\text{fb}}, \rho]dt + 2k\left\{\mathcal{D}\left(x - \frac{iF}{4\hbar k}\right) - \left(\frac{1-\eta}{\eta}\right)\mathcal{D}\left(\frac{iF}{4\hbar k}\right)\right\}\rho\,dt$$

$$+\sqrt{2\eta k}\,\mathcal{H}\left(x - \frac{iF}{4\eta\hbar k}\right)\rho\,dW. \tag{H.29}$$

Here

$$H_{\text{fb}} = \frac{1}{4}(xF + Fx) \tag{H.30}$$

and we have included the evolution due to H_0.

For perfectly efficient detection $\eta = 1$, and the WM feedback SME reduces to

$$d\rho_{\text{full}} = -\frac{i}{\hbar}[H_0 + H_{\text{fb}}, \rho]dt + 2k\mathcal{D}(A)\rho\,dt + \sqrt{2k}\,\mathcal{H}(A)\rho\,dW, \tag{H.31}$$

with

$$A = x - \frac{iF}{4\hbar k}. \tag{H.32}$$

The expressions for the WM feedback SME are also correct when x is non-Hermitian, such as in homodyne detection.

References

[1] Abdo, B., Schackert, F., Hatridge, M., Rigetti, C., and Devoret, M. 2011. Josephson amplifier for qubit readout. *Appl. Phys. Lett.*, **99**, 162506.

[2] Abrikosov, A. A., Gorkov, L. P., and Dzyaloshinski, I. E. 1975. *Methods of Quantum Field Theory in Statistical Physics*. New York: Dover.

[3] Agarwal, A., and Lang, J. 2005. *Foundations of Analog and Digital Electronic Circuits*. Burlington: Morgan Kaufmann.

[4] Ahn, C., Doherty, A. C., and Landahl, A. J. 2002. Continuous quantum error correction via quantum feedback control. *Phys. Rev. A*, **65**, 042301.

[5] Ahn, C., Wiseman, H., and Jacobs, K. 2004. Quantum error correction for continuously detected errors with any number of error channels per qubit. *Phys. Rev. A*, **70**, 024302.

[6] Alicki, R., Horodecki, M., Horodecki, P., and Horodecki, R. 2004. Thermodynamics of quantum information systems – Hamiltonian description. *Open Sys. & Information Dyn.*, **11**, 205–217.

[7] Altafini, C. 2006. Homogeneous polynomial forms for simultaneous stabilizability of families of linear control systems: A tensor product approach. *IEEE T. Automat. Contr.*, **51**, 1566–1571.

[8] Altafini, C. 2007. Feedback stabilization of isospectral control systems on complex flag manifolds: Application to quantum ensembles. *IEEE T. Automat. Contr.*, **52**, 2019–2028.

[9] Anandan, J., and Aharonov, Y. 1990. Geometry of quantum evolution. *Phys. Rev. Lett.*, **65**(Oct), 1697–1700.

[10] Anders, F. B., and Schiller, A. 2005. Real-time dynamics in quantum-impurity systems: a time-dependent numerical renormalization-group approach. *Phys. Rev. Lett.*, **95**, 196801.

[11] Anderson, P. W. 1958. Absence of diffusion in certain random lattices. *Phys. Rev.*, **109**(Mar), 1492–1505.

[12] Ando, T. 1989. Majorizations, doubly stochastic matrices, and comparison of eigenvalues. *Linear Algebra Appl.*, **118**, 163–248.

[13] Andrei, N., Furuya, K., and Lowenstein, J. H. 1983. Solution of the Kondo problem. *Rev. Mod. Phys.*, **55**(Apr), 331–402.

[14] Anetsberger, G., Arcizet, O., Unterreithmeier, Q. P., Rivière, R., Schliesser, A., Weig, E. M., Kotthaus, J. P., and Kippenberg, T. J. 2009. Near-field cavity optomechanics with nanomechanical oscillators. *Nature Phys.*, **5**, 909–914.

[15] Araki, H., and Lieb, E. H. 1970. Entropy inequalities. *Commun. Math. Phy.*, **18**(2), 160–170.

[16] Aravind, P. K. 2002. Bell's theorem without inequalities and only two distant observers. *Found. Phys. Lett.*, **15**, 397–405.

[17] Aravind, P. K. 2002. A simple demonstration of Bell's theorem involving two observers and no probabilities or inequalities. Eprint:ArXiv:quant-ph/0206070.

[18] Arcizet, O., Cohadon, P.-F., Briant, T., Pinard, M., and Heidmann, A. 2006. Radiation-pressure cooling and optomechanical instability of a micromirror. *Nature*, **444**, 71–74.

[19] Armen, M. A., Au, J. K., Stockton, J. K., Doherty, A. C., and Mabuchi, H. 2002. Adaptive homodyne measurement of optical phase. *Phys. Rev. Lett.*, **89**, 133602.

[20] Armour, A. D., Blencowe, M. P., and Schwab, K. C. 2002. Entanglement and deco-herence of a micromechanical resonator via coupling to a Cooper-pair box. *Phys. Rev. Lett.*, **88**, 148301.

[21] Arnold, V. I. 1989. *Mathematical Methods of Classical Mechanics* (Graduate Texts in Mathematics, Vol. 60). New York: Springer.

[22] Ashhab, S., and Nori, F. 2010. Control-free control: manipulating a quantum system using only a limited set of measurements. *Phys. Rev. A*, **82**, 062103.

[23] Aspelmeyer, M., Kippenberg, T. J., and Marquardt, F. 2014. *Cavity Optomechanics*. Eprint: arXiv:1303.0733.

[24] Astafiev, O., Pashkin, Y. A., Yamamoto, T., Nakamura, Y., and Tsai, J. S. 2004. Single-shot measurement of the Josephson charge qubit. *Phys. Rev. B*, **69**(May), 180507.

[25] Bagad, V. S. 1978. *Linear Systems, Fourier Transforms, and Optics*. New York: Wiley-Interscience.

[26] Bakr, W. S., Gillen, J. I., Peng, A., Fölling, S., and Greiner, M. 2009. A quantum gas microscope for detecting single atoms in a Hubbard-regime optical lattice. *Nature*, **462**, 74–77.

[27] Balouchi, A., and Jacobs, K. 2013. Near-optimal measurement-based feedback control for a single qubit. In submission.

[28] Balouchi, A., Nurdin, H., James, M., Strauch, F. W., and Jacobs, K. 2014. Sideband cooling versus measurement-based feedback cooling: a direct comparison in the regime of good control. In preparation.

[29] Barchielli, A. 1993. Stochastic differential-equations and a posteriori states in quantum-mechanics. *Int. J. Theor. Phys.*, **32**, 2221.

[30] Barchielli, A., and Lupieri, G. 2006. Quantum measurements and entropic bounds on information transmission. *Quantum Inform. Comput.*, **6**, 16–45.

[31] Bardeen, J., Cooper, L. N., and Schrieffer, J. R. 1957. Microscopic theory of superconductivity. *Phys. Rev.*, **106**(Apr), 162–164.

[32] Bardeen, J., Cooper, L. N., and Schrieffer, J. R. 1957. Theory of superconductivity. *Phys. Rev.*, **108**(Dec), 1175–1204.

[33] Barkeshli, M. M. 2005. Dissipationless information erasure and Landauer's Principle. Eprint: arXiv:cond-mat/0504323.

[34] Barnett, A. H. 2006. Asymptotic rate of quantum ergodicity in chaotic euclidean billiards. *Commun. Pure Appl. Math.*, **59**, 1457–1488.

[35] Barnum, H. August 2000. Information-disturbance tradeoff in quantum measurement on the uniform ensemble. CSTR-00-013, Department of Computer Science, University of Bristol. Available as Eprint: Arxiv:quant-ph/0205155.

[36] Barnum, H., Nielsen, M. A., and Schumacher, B. 1998. Information transmission through a noisy quantum channel. *Phys. Rev. A*, **57**(Jun), 4153–4175.

[37] Basché, T., Kummer, S., and Bräuchle, C. 1995. Direct spectroscopic observation of quantum jumps of a single molecule. *Nature*, **373**, 132–134.

[38] Batchelor, M. T. 2007. The Bethe ansatz after 75 years. *Physics Today*, **60**, 36.

[39] Bayes, T. 1763. An Essay Towards Solving a Problem in the Doctrine of Chances. *Phil. Trans. Roy. Soc.*, **53**, 370.

[40] Beckner, W. 1975. Inequalities in Fourier analysis. *Ann. Math.*, **102**, 159–182.

[41] Belavkin, V. P. 1987. Non-demolition measurement and control in quantum dynamical systems. Pages 331–336 of: Blaquiere, A., Diner, S., and Lochak, G. (eds), *Information, Complexity and Control in Quantum Physics. Proceedings of the 4th International Seminar on Mathematical Theory of Dynamical Systems and Microphysics*. New York: Springer.

[42] Belavkin, V. P. 1994. Quantum diffusion, measurement and filtering. I. *Theor. Prob. Appl.*, **38**(4), 573–585.

[43] Belavkin, V. P. 1995. Measurement and filtering of quantum diffusion. II. *Theor. Prob. Appl.*, **39**(3), 363–378.

[44] Bell, J. S. 1964. On the Einstein Podolsky Rosen paradox. *Physics (N.Y.)*, **1**, 195–200.

[45] Bell, J. S. 1988. *Speakable and Unspeakable in Quantum Mechanics*. Cambridge University Press.

[46] Bennett, C. H. 1982. The thermodynamics of computation — a review. *Int. J. Theor. Phys.*, **21**, 905.

[47] Bennett, C. H., DiVincenzo, D. P., Fuchs, C. A., Mor, T., Rains, E., Shor, P. W., Smolin, J. A., and Wootters, W. K. 1999. Quantum nonlocality without entanglement. *Phys. Rev. A*, **59**, 1070–1091.

[48] Bennett, C. H., Shor, P. W., Smolin, J. A., and Thapliyal, A. V. 2002. Entanglement-assisted capacity of a quantum channel and the reverse Shannon theorem. *IEEE Trans. Inf. Theor.*, **48**, 2637.

[49] Bensoussan, A. 1992. *Stochastic Control of Partially Observable Systems*. Cambridge University Press.

[50] Bergeal, N., Vijay, R., Manucharyan, V. E., Siddiqi, I., Schoelkopf, R. J., Girvin, S. M., and Devoret, M. H. 2010. Analog information processing at the quantum limit with a Josephson ring modulator. *Nature Phys.*, **6**, 296–302.

[51] Bergeal, N., Schackert, F., Metcalfe, M., Vijay, R., Manucharyan, V. E., Frunzio, L., Prober, D. E., Schoelkopf, R. J., Girvin, S. M., and Devoret, M. H. 2010. Phase-preserving amplification near the quantum limit with a Josephson ring modulator. *Nature*, **465**, 64–68.

[52] Berger, J. O., Bernardo, J. M., and Sun, D. 2009. The formal definition of reference priors. *Ann. Statist.*, **37**, 905–938.

[53] Bergquist, J. C., Hulet, R. G., Itano, W. M., and Wineland, D. J. 1986. Observation of quantum jumps in a single atom. *Phys. Rev. Lett.*, **57**, 1699–1702.

[54] Berry, D. W., and Wiseman, H. M. 2000. Optimal states and almost optimal adaptive measurements for quantum interferometry. *Phys. Rev. Lett.*, **85**(Dec), 5098–5101.

[55] Berry, D. W., Wiseman, H. M., and Breslin, J. K. 2001. Optimal input states and feedback for interferometric phase estimation. *Phys. Rev. A*, **63**, 053804.

[56] Berry, D. W., Higgins, B. L., Bartlett, S. D., Mitchell, M. W., Pryde, G. J., and Wiseman, H. M. 2009. How to perform the most accurate possible phase measurements. *Phys. Rev. A*, **80**, 052114.

[57] Berry, M. V. 1977. Regular and irregular semiclassical wavefunctions. *J. Phys. A*, **10**, 2083–2091.

[58] Berta, M., Renes, J. M., and Wilde, M. M. 2013. Identifying the information gain of a quantum measurement. arXiv:1301.1594v1.

[59] Bertet, P., Harmans, C. J. P. M., and Mooij, J. E. 2006. Parametric coupling for superconducting qubits. *Phys. Rev. B*, **73**, 064512.

[60] Beugeling, W., Moessner, R., and Haque, M. 2013. Finite-size scaling of eigenstate thermalization: Power-law dependence on Hilbert-space dimension. Eprint: arXiv:1308.2862.

[61] Bhatia, R. 1997. *Matrix Inequalities*. Berlin: Springer.

[62] Bhattacharya, T., Habib, S., and Jacobs, K. 2000. Continuous quantum measurement and the emergence of classical chaos. *Phys. Rev. Lett.*, **85**, 4852.

[63] Bhattacharya, T., Habib, S., and Jacobs, K. 2003. Continuous quantum measurement and the quantum-to-classical transition. *Phys. Rev. A*, **67**(4), 042103.

[64] Bialynicki-Birula, I., and Madajczyk, J. 1985. Entropic uncertainty relations for angular distributions. *Phys. Lett. A*, **108**(8), 384–386.

[65] Bialynicki-Birula, I., and Mycielski, J. 1975. Uncertainty relations for information entropy in wave mechanics. *Commun. Math. Phys.*, **44**, 129.

[66] Bialynicki-Birula, I. 1984. Entropic uncertainty relations. *Phys. Lett. A*, **103**(5), 253–254.

[67] Blais, A., Huang, R.-S., Wallraff, A., Girvin, S. M., and Schoelkopf, R. J. 2004. Cavity quantum electrodynamics for superconducting electrical circuits: an architecture for quantum computation. *Phys. Rev. A*, **69**(6), 062320.

[68] Blatter, G., Geshkenbein, V. B., and Ioffe, L. B. 2001. Design aspects of superconducting-phase quantum bits. *Phys. Rev. B*, **63**(Apr), 174511.

[69] Blencowe, M. P., and Buks, E. 2007. Quantum analysis of a linear dc SQUID mechanical displacement detector. *Phys. Rev. B*, **76**, 014511.

[70] Bohm, D. 1952. A suggested interpretation of the quantum theory in terms of "hidden" variables. I. *Phys. Rev.*, **85**(Jan), 166–179.

[71] Bohm, D. 1952. A suggested interpretation of the quantum theory in terms of "hidden" variables. II. *Phys. Rev.*, **85**(Jan), 180–193.

[72] Boixo, S., Flammia, S. T., Caves, C. M., and Geremia, J. 2007. Generalized limits for single-parameter quantum estimation. *Phys. Rev. Lett.*, **98**, 090401.

[73] Boixo, S., Datta, A., Flammia, S. T., Shaji, A., Bagan, E., and Caves, C. M. 2008. Quantum-limited metrology with product states. *Phys. Rev. A*, **77**(Jan), 012317.

[74] Boixo, S., Datta, A., Davis, M. J., Flammia, S. T., Shaji, A., and Caves, C. M. 2008. Quantum metrology: dynamics versus entanglement. *Phys. Rev. Lett.*, **101**(Jul), 040403.

[75] Bondurant, R. S., and Shapiro, J. H. 1984. Squeezed states in phase-sensing interferometers. *Phys. Rev. D*, **30**, 2548–2556.

[76] Born, D., Shnyrkov, V. I., Krech, W., Wagner, T., Il'ichev, E., Grajcar, M., Hübner, U., and Meyer, H.-G. 2004. Reading out the state inductively and microwave spectroscopy of an interferometer-type charge qubit. *Phys. Rev. B*, **70**, 180501.

[77] Bose, S., Jacobs, K., and Knight, P. L. 1997. Preparation of nonclassical states in cavities with a moving mirror. *Phys. Rev. A*, **56**, 4175.

[78] Bose, S., Vedral, V., and Knight, P. L. 1998. Multiparticle generalization of entanglement swapping. *Phys. Rev. A*, **57**(Feb), 822–829.

[79] Bose, S., Jacobs, K., and Knight, P. L. 1999. Scheme to probe the decoherence of a macroscopic object. *Phys. Rev. A*, **59**(May), 3204–3210.

[80] Boulant, N., Ithier, G., Meeson, P., Nguyen, F., Vion, D., Esteve, D., Siddiqi, I., Vijay, R., Rigetti, C., Pierre, F., and Devoret, M. 2007. Quantum nondemolition readout using a Josephson bifurcation amplifier. *Phys. Rev. B*, **76**, 014525.

[81] Box, R., and Tiao, G. C. 1973. *Bayesian Inference in Statistical Analysis*. Sydney: Addison-Wesley.

[82] Boykin, P. O., Mor, T., Roychowdhury, V., Vatan, F., and Vrijen, R. 2002. Algorithmic cooling and scalable NMR quantum computers. *Proc. Natl. Acad. Sci.*, **99**, 3388–3393.

[83] Braginsky, V. B., Khalili, F. Y., and Thorne, K. S. 1995. *Quantum Measurement*. Cambridge University Press.

[84] Braginsky, V. B., Vorontsov, Y. I., and Thorne, K. S. 1980. Quantum nondemolition measurements. *Science*, **209**(4456), 547–557.

[85] Brakhane, S., Alt, W., Kampschulte, T., Martinez-Dorantes, M., Reimann, R., Yoon, S., Widera, A., and Meschede, D. 2012. Bayesian feedback control of a two-atom spin-state in an atom-cavity system. *Phys. Rev. Lett.*, **109**(Oct), 173601.

[86] Bransden, B., and Joachain, C. 2000. *Quantum Mechanics*. San Francisco: Benjamin Cummings.

[87] Braunstein, S. L., and Caves, C. M. 1994. Statistical distance and the geometry of quantum states. *Phys. Rev. Lett.*, **72**, 3439–3443.

[88] Braunstein, S. L., Caves, C. M., and Milburn, G. 1996. Generalized uncertainty relations: theory, examples, and Lorentz invariance. *Ann. Phys.*, **247**(1), 135 – 173.

[89] Breuer, H.-P., and Petruccione, F. 2007. *The Theory of Open Quantum Systems*. Oxford University Press.

[90] Brun, T. A., Percival, I. C., and Schack, R. 1996. Quantum chaos in open systems: a quantum state diffusion analysis. *J. Phys. A*, **29**, 2077.

[91] Brune, M., Hagley, E., Dreyer, J., Maître, X., Maali, A., Wunderlich, C., Raimond, J. M., and Haroche, S. 1996. Observing the progressive decoherence of the "meter" in a quantum measurement. *Phys. Rev. Lett.*, **77**, 4887–4890.

[92] Buks, E., Arbel-Segev, E., Zaitsev, S., Abdo, B., and Blencowe, M. P. 2008. Quantum nondemolition measurement of discrete Fock states of a nanomechanical resonator. *Europhy. Lett.*, **81**, 10001.

[93] Bulla, R., Tong, N.-H., and Vojta, M. 2003. Numerical renormalization group for bosonic systems and application to the subohmic spin-boson model. *Phys. Rev. Lett.*, **91**, 170601.

[94] Buscemi, F., Hayashi, M., and Horodecki, M. 2008. Global information balance in quantum measurements. *Phys. Rev. Lett.*, **100**(May), 210504.

[95] Bushev, P., Rotter, D., Wilson, A., Dubin, F., Becher, C., Eschner, J., Blatt, R., Steixner, V., Rabl, P., and Zoller, P. 2006. Feedback cooling of a single trapped ion. *Phys. Rev. Lett.*, **96**, 043003.

[96] Büttiker, M. 1987. Zero-current persistent potential drop across small-capacitance Josephson junctions. *Phys. Rev. B*, **36**(Sep), 3548–3555.

[97] Caldeira, A. O., and Leggett, A. J. 1983. Path integral approach to quantum Brownian motion. *Physica A*, **121**, 587.

[98] Caldeira, A. O., and Leggett, A. J. 1983. Quantum tunneling in a dissipative system. *Ann. Phys.*, **149**, 374–456.

[99] Carlini, A., Hosoya, A., Koike, T., and Okudaira, Y. 2006. Time-optimal quantum evolution. *Phys. Rev. Lett.*, **96**, 060503.

[100] Carmichael, H. 1993. *An Open Systems Approach to Quantum Optics*. Berlin: Springer.

[101] Carmichael, H. J. 1993. Quantum trajectory theory for cascaded open systems. *Phys. Rev. Lett.*, **70**(Apr), 2273–2276.

[102] Carvalho, A. R. R., and Hope, J. J. 2007. Stabilizing entanglement by quantum-jump-based feedback. *Phys. Rev. A*, **76**, 010301.

[103] Castellanos-Beltran, M. A., Irwin, K. D, Vale, L. R., Hilton, G. C., and Lehnert, K. W. 2009. *IEEE Trans. Appl. Supercond.* **19**, 944–947.

[104] Castellanos-Beltran, M. A., Irwin, K. D., Hilton, G. C., Vale, L. R., and Lehnert, K. W. 2008. Amplification and squeezing of quantum noise with a tunable Josephson metamaterial. *Nature Phys.*, **4**, 929–931.

504 *References*

[105] Caves, C. M. 1981. Quantum-mechanical noise in an interferometer. *Phys. Rev. D*, **23**, 1693–1708.

[106] Caves, C. M. 1982. Quantum limits on noise in linear amplifiers. *Phys. Rev. D*, **26**(Oct), 1817–1839.

[107] Caves, C. M. 1985. Defense of the standard quantum limit for free-mass position. *Phys. Rev. Lett.*, **54**, 2465–2468.

[108] Caves, C. M. 1990. Quantitative limits on the ability of a Maxwell demon to extract work from heat. *Phys. Rev. Lett.*, **64**(Apr), 2111–2114.

[109] Caves, C. M. 1993. Information and entropy. *Phys. Rev. E*, **47**, 4010–4017.

[110] Caves, C. M., Thorne, K. S., Drever, R. W. P., Sandberg, V. D., and Zimmermann, M. 1980. On the measurement of a weak classical force coupled to a quantum-mechanical oscillator. I. Issues of principle. *Rev. Mod. Phys.*, **52**(Apr), 341–392.

[111] Caves, C. M., Fuchs, C. A., and Schack, R. 2002. Quantum probabilities as Bayesian probabilities. *Phys. Rev. A*, **65**(Jan), 022305.

[112] Chan, J., Alegre, T. P. M., Safavi-Naeini, A. H., Hill, J. T., Krause, A., Groeblacher, S., Aspelmeyer, M., and Painter, O. 2011. Laser cooling of a nanomechanical oscillator into its quantum ground state. *Nature*, **478**, 89–92.

[113] Changlani, H. J., Kinder, J. M., Umrigar, C. J., and Chan, G. K.-L. 2009. Approximating strongly correlated wave functions with correlator product states. *Phys. Rev. B*, **80**, 245116.

[114] Chase, B. A., Landahl, A. J., and Geremia, J. M. 2008. Efficient feedback controllers for continuous-time quantum error correction. *Phys. Rev. A*, **77**, 032304.

[115] Chaturvedi, S., and Shibata, F. 1979. Time-convolutionless projection operator formalism for elimination of fast variables—applications to Brownian-motion. *Z. Phys. B*, **35**, 297.

[116] Chaudhury, S., Merkel, S., Herr, T., Silberfarb, A., Deutsch, I. H., and Jessen, P. S. 2007. Quantum control of the hyperfine spin of a Cs atom ensemble. *Phys. Rev. Lett.*, **99**, 163002.

[117] Chen, C., Dong, D., Lam, J., Chu, J., and Tarn, T. 2012. Control design of uncertain quantum systems with fuzzy estimators. *IEEE Trans. Fuzzy Sys.*, **20**(5), 820–831.

[118] Chia, A., and Wiseman, H. M. 2011. Complete parametrizations of diffusive quantum monitorings. *Phys. Rev. A*, **84**(Jul), 012119.

[119] Chia, A., and Wiseman, H. M. 2011. Quantum theory of multiple-input–multiple-output Markovian feedback with diffusive measurements. *Phys. Rev. A*, **84**(Jul), 012120.

[120] Chin, A. W., Rivas, Á., Huelga, S. F., and Plenio, M. B. 2010. Exact mapping between system-reservoir quantum models and semi-infinite discrete chains using orthogonal polynomials. *J. Math. Phys.*, **51**, 092109.

[121] Chiorescu, I., Nakamura, Y., Harmans, C. J. P. M., and Mooij, J. E. 2003. Coherent quantum dynamics of a superconducting flux qubit. *Science*, **299**(5614), 1869–1871.

[122] Chirikov, B. V. 1991. A theory of quantum diffusion localization. *Chaos*, **1**(1), 95.

[123] Chiruvelli, A., and Jacobs, K. 2008. Rapid-purification protocols for optical homodyning. *Phys. Rev. A*, **77**, 012102.

[124] Choi, M.-D. 1975. Completely positive linear maps on complex matrices. *Linear Algebra Appl.*, **10**, 285.

[125] Clerk, A. A., Marquardt, F., and Jacobs, K. 2008. Back-action evasion and squeezing of a mechanical resonator using a cavity detector. *New J. Phys.*, **10**, 095010.

[126] Cohen-Tannoudji, C., Dupont-Roc, J., and Grynberg, G. 1997. *Photons and Atoms: Introduction to Quantum Electrodynamics*. New York: Wiley-VCH.

[127] Coles, W. J. 1965. Matrix Riccati differential equations. *J. Soc. Ind. Appl. Math.*, **13**(3), 627–634.

[128] Collett, M. J., and Gardiner, C. W. 1984. Squeezing of intracavity and traveling-wave light fields produced in parametric amplification. *Phys. Rev. A*, **30**(3), 1386–1391.

[129] Collin, E., Ithier, G., Aassime, A., Joyez, P., Vion, D., and Esteve, D. 2004. NMR-like control of a quantum bit superconducting circuit. *Phys. Rev. Lett.*, **93**(Oct), 157005.

[130] Combes, J., and Jacobs, K. 2006. Rapid state-reduction of quantum systems using feedback control. *Phys. Rev. Lett.*, **96**, 010504.

[131] Combes, J., Wiseman, H. M., and Jacobs, K. 2008. Rapid measurement of quantum systems using feedback control. *Phys. Rev. Lett.*, **100**, 160503.

[132] Combes, J., and Wiseman, H. M. 2011. Maximum information gain in weak or continuous measurements of qudits: complementarity is not enough. *Phys. Rev. X*, **1**, 011012.

[133] Cover, T. M., and Thomas, J. A. 1991. *Elements of Information Theory*. New York: Wiley.

[134] D'Alessandro, D. 2007. *Introduction to Quantum Control and Dynamics*. Boca Raton, FL: Chapman and Hall/CRC.

[135] Daley, A. J., Kollath, C., Schollwoeck, U., and Vidal, G. 2004. Time-dependent density-matrix renormalization-group using adaptive effective Hilbert spaces. *J. Stat. Mech. Theor. Exp.*, P04005.

[136] Dalibard, J., Castin, Y., and Mølmer, K. 1992. A wave-function approach to dissipative processes in quantum optics. *Phys. Rev. Lett.*, **68**, 580.

[137] Dawson, C. M., and Nielsen, M. A. 2006. The Solovay–Kitaev algorithm. *Quant. Inf. Comput.*, **6**, 81–95.

[138] de Verdière, Y. C. 1985. Ergodicitè fonctions propres du Laplacien. *Commun. Math. Phys.*, **102**, 497–502.

[139] Deffner, S., and Jarzynski, C. 2013. Information processing and the second law of thermodynamics: an inclusive, Hamiltonian approach. Eprint: arXiv:1308.5001.

[140] Deutsch, D. 1983. Uncertainty in quantum measurements. *Phys. Rev. Lett.*, **50**(Feb), 631–633.

[141] Deutsch, J. M. 1991. Quantum statistical mechanics in a closed system. *Phys. Rev. A*, **43**, 2046.

[142] Devetak, I. 2005. The private classical capacity and quantum capacity of a quantum channel. *IEEE Trans. Inf. Theory*, **51**, 44–55.

[143] Devoret, M. H. 1997. Quantum fluctuations in electrical circuits. Pages 351–386 of: Reynaud, S., Giacobino, E., and Zinn-Justin, J. (eds), *Quantum Fluctuations, Les Houches Session LVIII*. Amsterdam: Elsevier.

[144] Devoret, M. H., Wallraff, A., and Martinis, J. M. 2004. Superconducting qubits: a short review. Eplint: arXiv:cond-mat/0411174.

[145] Devoret, M. H., Martinis, J. M., and Clarke, J. 1985. Measurements of macroscopic quantum tunneling out of the zero-voltage state of a current-biased Josephson junction. *Phys. Rev. Lett.*, **55**(Oct), 1908–1911.

[146] DeWitt, B. 1973. The many-universes interpretation of quantum mechanics. Pages 167–218 of: DeWitt, B., and Graham, N. (eds), *The Many-Worlds Interpretation of Quantum Mechanics*. Princeton University Press.

[147] D'Helon, C., and James, M. R. 2006. Stability, gain, and robustness in quantum feedback networks. *Phys. Rev. A*, **73**(5), 053803.

[148] DiCarlo, L., Chow, J. M., Gambetta, J. M., Bishop, L. S., Johnson, B. R., Schuster, D. I., Majer, J., Blais, A., Frunzio, L., Girvin, S. M., and Schoelkopf, R. J. 2009. Demonstration of two-qubit algorithms with a superconducting quantum processor. *Nature*, **460**, 240–244.

[149] Diedrich, F., Bergquist, J. C., Itano, W. M., and Wineland, D. J. 1989. Laser cooling to the zero-point energy of motion. *Phys. Rev. Lett.*, **62**, 403–406.

[150] Dijkstra, A. G., and Tanimura, Y. 2010. Non-Markovian entanglement dynamics in the presence of system-bath coherence. *Phys. Rev. Lett.*, **104**, 250401.

[151] Diosi, L. 1986. Stochastic pure state representation for open quantum systems. *Phys. Lett. A*, **114**, 451.

[152] Dodonov, V. V. 2002. 'Nonclassical' states in quantum optics: a 'squeezed' review of the first 75 years. *J. Opt. B*, **4**(1), R1.

[153] Doherty, A. C., and Jacobs, K. 1999. Feedback control of quantum systems using continuous state estimation. *Phys. Rev. A*, **60**, 2700–2711.

[154] Doherty, A. C., and Wiseman, H. M. 2004. Quantum limits to feedback control of linear systems. Page 322 of: Heszler, P. (ed.), *Proceedings of the SPIE Symposium on Fluctuations and Noise in Photonics and Quantum Optics II.*, vol. 5468. Bellingham, WA: SPIE.

[155] Doherty, A. C., Habib, S., Jacobs, K., Mabuchi, H., and Tan, S. M. 2000. Quantum feedback control and classical control theory. *Phys. Rev. A*, **62**, 012105.

[156] Doherty, A. C., Jacobs, K., and Jungman, G. 2001. Information, disturbance, and Hamiltonian quantum feedback control. *Phys. Rev. A*, **63**, 062306.

[157] Dolinar, S. 1973. *Tech. Rep. 111, Research Laboratory of Electronics*. Cambridge: MIT.

[158] Dong, D. Y., Chen, C. L., Chen, Z. H., and Zhang, C. B. 2006. Estimation-based information acquisition in quantum feedback control. *Dynam. Cont. Dis. Ser. B*, **13**, 1204–1208.

[159] Dowling, J. P. 1998. Correlated input-port, matter-wave interferometer: Quantum-noise limits to the atom-laser gyroscope. *Phys. Rev. A*, **57**, 4736–4746.

[160] Dowling, J. P. 2008. Quantum optical metrology – the lowdown on high-N00N states. *Contemp. Phys.*, **49**, 125–143.

[161] Dubey, S., Silvestri, L., Finn, J., Vinjanampathy, S., and Jacobs, K. 2012. Approach to typicality in many-body quantum systems. *Phys. Rev. E*, **85**, 011141.

[162] Duty, T., Gunnarsson, D., Bladh, K., and Delsing, P. 2004. Coherent dynamics of a Josephson charge qubit. *Phys. Rev. B*, **69**(Apr), 140503.

[163] Dziarmaga, J., Dalvit, D. A. R., and Zurek, W. H. 2004. Conditional quantum dynamics with several observers. *Phys. Rev. A*, **69**(2), 022109.

[164] Eckern, U. 1986. Quantum mechanics: is the theory applicable to macroscopic objects? *Nature*, **319**, 726.

[165] Eichenfield, M., Chan, J., Camacho, R. M., Vahala, K. J., and Painter, O. 2009. Optomechanical crystals. *Nature*, **462**, 78–82.

[166] Eichenfield, M., Camacho, R., Chan, J., Vahala, K. J., and Painter, O. 2009. A picogram- and nanometre-scale photonic-crystal optomechanical cavity. *Nature*, **459**, 550.

[167] Eisert, J., Cramer, M., and Plenio, M. B. 2010. Colloquium: Area laws for the entanglement entropy. *Rev. Mod. Phys.*, **82**, 277–306.

[168] Esposito, M., and den Broeck, C. V. 2011. Second law and Landauer principle far from equilibrium. *EPL (Europhysics Letters)*, **95**(4), 40004.

[169] (LIGO Scientific Collaboration). J. Abadie *et al.* 2011. A gravitational-wave observatory operating beyond the quantum shot-noise limit. *Nature Phys.*, **7**, 962.

[170] Everett, H. 1957. "Relative state" formulation of quantum mechanics. *Rev. Mod. Phys.*, **29**(Jul), 454–462.

[171] Everitt, M. J. 2007. Recovery of classical chaotic-like behavior in a conservative quantum three-body problem. *Phys. Rev. E*, **75**, 036217.

[172] Everitt, M. J. 2009. On the correspondence principle: implications from a study of the chaotic dynamics of a macroscopic quantum device. *New J. Phys.*, **11**, 013014.

[173] Everitt, M. J., Clark, T. D., Stiffell, P. B., Ralph, J. F., Bulsara, A., and Harland, C. 2005. Persistent entanglement in the classical limit. *New J. Phys.*, **7**, 64.

[174] Everitt, M. J., Munro, W., and Spiller, T. 2009. The quantum-classical crossover of a field mode. *Phys. Rev. E*, **79**, 032328.

[175] Fabre, C., Pinard, M., Bourzeix, S., Heidmann, A., Giacobino, E., and Reynaud, S. 1994. Quantum-noise reduction using a cavity with a movable mirror. *Phys. Rev. A*, **49**, 1337–1343.

[176] Fedorov, A., Feofanov, A. K., Macha, P., Forn-Díaz, P., Harmans, C. J. P. M., and Mooij, J. E. 2010. Strong coupling of a quantum oscillator to a flux qubit at Its symmetry point. *Phys. Rev. Lett.*, **105**, 060503.

[177] Feingold, M., and Peres, A. 1986. Distribution of matrix elements of chaotic systems. *Phys. Rev. A*, **34**, 591–595.

[178] Finn, J., Jacobs, K., and Sundaram, B. 2009. Comment on "nonmonotonicity in the quantum-classical transition: chaos induced by quantum effects". *Phys. Rev. Lett.*, **102**(Mar), 119401.

[179] Fischer, T., Maunz, P., Pinkse, P. W. H., Puppe, T., and Rempe, G. 2002. Feedback on the motion of a single atom in an optical cavity. *Phys. Rev. Lett.*, **88**, 163002.

[180] Fleming, C. H., and Cummings, N. I. 2011. Accuracy of perturbative master equations. *Phys. Rev. E*, **83**, 031117.

[181] Fleming, W. H., and Rishel, R. W. 1982. *Deterministic and Stochastic Optimal Control*. New York: Springer.

[182] Fouquieres, P. D., and Schirmer, S. G. 2013. A closer look at quantum control landscapes and their implication for control optimization. *Infin. Dimens. Anal., Quantum Probab. Rela. Top.*, **16**(03), 1350021.

[183] Friedman, J. R., Patel, V., Chen, W., Tolpygo, S. K., and Lukens, J. E. 2000. Quantum superposition of distinct macroscopic states. *Nature*, **406**, 43.

[184] Frunzio, L., Wallraff, A., Schuster, D., Majer, J., and Schoelkopf, R. 2005. Fabrication and characterization of superconducting circuit QED devices for quantum computation. *IEEE Trans. Appl. Supercond.*, **15**, 860–863.

[185] Fuchs, C. A. 1995. *Distinguishability and Accessible Information in Quantum Theory*. Ph.D. diss. Albuquerque: UNM.

[186] Fuchs, C. A. 1998. Information gain vs. state disturbance in quantum theory. *Fortschr. Phys.*, **46**, 535–565.

[187] Fuchs, C. A. 203. Quantum mechanics as quantum information, mostly. *J. Mod. Opt.*, **50**, 987.

[188] Fuchs, C. A., and Jacobs, K. 2001. Information-tradeoff relations for finite-strength quantum measurements. *Phys. Rev. A*, **63**, 062305.

[189] Fuchs, C. A., and van de Graaf, J. 1999. Cryptographic distinguishability measures for quantum-mechanical states. *IEEE Trans. Inform. Theory*, **45**, 1216–1227.

[190] Fuchs, C. A., and Caves, C. M. 1994. Ensemble-dependent bounds for accessible information in quantum mechanics. *Phys. Rev. Lett.*, **73**, 3047–3050.

[191] Fuchs, C. A., Mermin, N. D., and Schack, R. 2013. *An introduction to QBism with an application to the locality of quantum mechanics*. Eprint: arXiv:1311.5253.

[192] Fujii, M., Furuta, T., Nakamoto, R., and Takahasi, S.-E. 1997. Operator inequalities and covariance in noncommutative probability. *Math. Japon.*, **46**, 317–320.

[193] Gallop, J. C. 1991. *SQUIDs, the Josephson Effects and Superconducting Electronics*. New York: Taylor & Francis.

[194] Gambetta, J. M., and Wiseman, H. M. 2005. Stochastic simulations of conditional states of partially observed systems, quantum and classical. *J. Opt. B: Quantum Semiclass. Opt.*, **7**, S250.

[195] Gambetta, J., and Wiseman, H. M. 2001. State and dynamical parameter estimation for open quantum systems. *Phys. Rev. A*, **64**, 042105.

[196] Gambetta, J., Blais, A., Boissonneault, M., Houck, A. A., Schuster, D. I., and Girvin, S. M. 2008. Quantum trajectory approach to circuit QED: quantum jumps and the Zeno effect. *Phys. Rev. A*, **77**, 012112.

[197] Gardiner, C. W. 1985. *Handbook of Stochastic Methods*. Berlin: Springer.

[198] Gardiner, C. W. 1993. Driving a quantum system with the output field from another driven quantum system. *Phys. Rev. Lett.*, **70**(Apr), 2269–2272.

[199] Gardiner, C. W., and Collett, M. J. 1985. Input and output in damped quantum systems: quantum stochastic differential equations and the master equation. *Phys. Rev. A*, **31**(6), 3761–3774.

[200] Gardiner, C. W., Parkins, A. S., and Zoller, P. 1992. Wave-function quantum stochastic differential equations and quantum-jump simulation methods. *Phys. Rev. A*, **46**(Oct), 4363–4381.

[201] Gardiner, C., and Zoller, P. 2010. *Quantum Noise*. New York: Springer.

[202] Garraway, B. M., and Knight, P. L. 1994. Comparison of quantum-state diffusion and quantum-jump simulations of 2-photon processes in a dissipative environment. *Phys. Rev. A*, **50**, 2548–2563.

[203] Gaspard, P., and Nagaoka, M. 1999. Slippage of initial conditions for the Redfield master equation. *J. Chem. Phys.*, **111**, 5668.

[204] Genes, C., Vitali, D., Tombesi, P., Gigan, S., and Aspelmeyer, M. 2008. Ground-state cooling of a micromechanical oscillator: Comparing cold damping and cavity-assisted cooling schemes. *Phys. Rev. A*, **77**, 033804.

[205] Genes, C., Vitali, D., Tombesi, P., Gigan, S., and Aspelmeyer, M. 2009. Erratum: Ground-state cooling of a micromechanical oscillator: Comparing cold damping and cavity-assisted cooling schemes [Phys. Rev. A 77, 033804 (2008)]. *Phys. Rev. A*, **79**, 039903.

[206] Ghose, S., Alsing, P., Deutsch, I., Bhattacharya, T., Habib, S., and Jacobs, K. 2003. Recovering classical dynamics from coupled quantum systems through continuous measurement. *Phys. Rev. A*, **67**, 052102.

[207] Ghose, S., Alsing, P., Deutsch, I., Bhattacharya, T., and Habib, S. 2004. Transition to classical chaos in a coupled quantum system through continuous measurement. *Phys. Rev. A*, **69**(5), 052116.

[208] Ghose, S., Alsing, P. M., Sanders, B. C., and Deutsch, I. H. 2005. Entanglement and the quantum-to-classical transition. *Phys. Rev. A*, **72**(1), 014102.

[209] Gillett, G. G., Dalton, R. B., Lanyon, B. P., Almeida, M. P., Barbieri, M., Pryde, G. J., O'Brien, J. L., Resch, K. J., Bartlett, S. D., and White, A. G. 2010. Experimental feedback control of quantum systems using weak measurements. *Phys. Rev. Lett.*, **104**(Feb), 080503.

[210] Gilmore, R. 1974. Baker–Campbell–Hausdorff formulas. *J. Math Phys.*, **15**, 2090.

[211] Giovannetti, V., and Vitali, D. 2001. Phase-noise measurement in a cavity with a movable mirror undergoing quantum Brownian motion. *Phys. Rev. A*, **63**, 023812.

[212] Gisin, N. 1984. A model for the macroscopic description and continual observations in quantum mechanics. *Phys. Rev. Lett.*, **52**, 1657.

[213] Gleyzes, S., Kuhr, S., Guerlin, C., Bernu, J., Deléglise, S., Hoff, U. B., Brune, M., Raimond, J.-M., and Haroche, S. 2007. Quantum jumps of light recording the birth and death of a photon in a cavity. *Nature*, **446**, 297.

[214] Gnutzmann, S., and Haake, F. 1996. Positivity violation and initial slips in open systems. *Z. Phys. B*, **101**, 263.

[215] Goan, H.-S., and Milburn, G. J. 2001. Dynamics of a mesoscopic qubit under continuous quantum measurement. *Phys. Rev. B*, **64**, 235307.

[216] Goetsch, P., and Graham, R. 1994. Linear stochastic wave-equations for continuously measured quantum systems. *Phys. Rev. A*, **50**, 5242.

[217] Gogolin, C., Müller, M. P., and Eisert, J. 2011. Absence of thermalization in nonintegrable systems. *Phys. Rev. Lett.*, **106**(Jan), 040401.

[218] Goldstein, S., Lebowitz, J. L., Tumulka, R., and Zanghì, N. 2006. Canonical typicality. *Phys. Rev. Lett.*, **96**, 050403.

[219] Goldstein, S., Lebowitz, J. L., Mastrodonato, C., Tumulka, R., and Zanghì, N. 2010. Approach to thermal equilibrium of macroscopic quantum systems. *Phys. Rev. E*, **81**(1), 011109.

[220] Goldstein, S., Lebowitz, J. L., Mastrodonato, C., Tumulka, R., and Zanghì, N. 2010. Normal typicality and von Neumann's quantum ergodic theorem. *Proc. Roy. Soc. A*, **466**, 3203–3224.

[221] Gorkov, L. P. 1959. Microscopic derivation of the Ginzburg–Landau equations in the theory of superconductivity. *J. Exp. Theor. Phys. (U.S.S.R.)*, **36**, 1918–1923.

[222] Gough, J., and James, M. 2009. Quantum feedback networks: Hamiltonian formulation. *Commun. Mathe. Phys.*, **287**(3), 1109–1132.

[223] Gough, J., and James, M. 2009. The series product and its application to quantum feedforward and feedback networks. *IEEE Trans. Automat. Contr.*, **54**(11), 2530–2544.

[224] Gozzi, F., Swiech, A., and Zhou, X. Y. 2005. A corrected proof of the stochastic verification theorem within the framework of viscosity solutions. *SIAM J. Control. Optim.*, **43**, 2009–2019.

[225] Grajcar, M., Izmalkov, A., van der Ploeg, S. H. W., Linzen, S., Plecenik, T., Wagner, T., Hübner, U., Il'ichev, E., Meyer, H.-G., Smirnov, A. Y., Love, P. J., Maassen van den Brink, A., Amin, M. H. S., Uchaikin, S., and Zagoskin, A. M. 2006. Four-qubit device with mixed couplings. *Phys. Rev. Lett.*, **96**, 047006.

[226] Grajcar, M., Liu, Y.-x., Nori, F., and Zagoskin, A. M. 2006. Switchable resonant coupling of flux qubits. *Phys. Rev. B*, **74**(Nov), 172505.

[227] Greenbaum, B. D., Jacobs, K., and Sundaram, B. 2007. Conditions for the quantum-to-classical transition: trajectories versus phase-space distributions. *Phys. Rev. E*, **76**, 036213.

[228] Griffith, E. J., Hill, C. D., Ralph, J. F., Wiseman, H. M., and Jacobs, K. 2007. Rapid state purification in a superconducting charge qubit. *Phys. Rev. B*, **75**, 014511.

[229] Gröblacher, S., Hertzberg, J. B., Vanner, M. R., Cole, G. D., Gigan, S., Schwab, K. C., and Aspelmeyer, M. 2009. Demonstration of an ultra-cold micro-optomechanical oscillator in a cryogenic cavity. *Nature Phys.*, **5**, 485–488.

[230] Gröblacher, S., Hammerer, K., Vanner, M. R., and Aspelmeyer, M. 2009. Observation of strong coupling between a micromechanical resonator and an optical cavity field. *Nature*, **460**, 724–727.

[231] Groenewold, H. J. 1971. A problem of information gain by quantum measurements. *Int. J. Theor. Phys.*, **4**, 327.

[232] Grudinin, I. S., Lee, H., Painter, O., and Vahala, K. J. 2010. Phonon laser action in a tunable two-level system. *Phys. Rev. Lett.*, **104**(Feb), 083901.

[233] Guerlin, C., Bernu, J., Deléglise, S., Sayrin, C., Gleyzes, S., Kuhr, S., Brune, M., Raimond, J.-M., and Haroche, S. 2007. Progressive field-state collapse and quantum non-demolition photon counting. *Nature*, **448**, 889–893.

[234] Guhr, T., Müller-Groeling, A., and Weidenmüller, H. A. 1998. Random matrix theories in quantum physics: common concepts. *Phys. Rep.*, **299**, 189–425.

[235] Habib, S. 2004. Gaussian dynamics is classical dynamics. Eprint: avXiv: quant-ph/0406011.

[236] Habib, S., and Ryne, R. D. 1995. Symplectic calculation of Lyapunov exponents. *Phys. Rev. Lett.*, **74**(1), 70–73.

[237] Habib, S., Jacobs, K., and Shizume, K. 2006. Emergence of chaos in quantum systems far from the classical limit. *Phys. Rev. Lett.*, **96**, 010403.

[238] Hall, M. J. W. 1995. Information exclusion principle for complementary observables. *Phys. Rev. Lett.*, **74**, 3307.

[239] Hall, M. J. W. 1997. Quantum information and correlation bounds. *Phys. Rev. A*, **55**, 100–113.

[240] Hamerly, R., and Mabuchi, H. 2012. Advantages of coherent feedback for cooling quantum oscillators. *Phys. Rev. Lett.*, **109**(Oct), 173602.

[241] Hamerly, R., and Mabuchi, H. 2013. Coherent controllers for optical-feedback cooling of quantum oscillators. *Phys. Rev. A*, **87**(Jan), 013815.

[242] Hasegawa, H.-H., Ishikawa, J., Takara, K., and Driebe, D. 2010. Generalization of the second law for a nonequilibrium initial state. *Phys. Lett. A*, **374**, 1001–1004.

[243] Hatridge, M., Vijay, R., Slichter, D. H., Clarke, J., and Siddiqi, I. 2011. Dispersive magnetometry with a quantum limited SQUID parametric amplifier. *Phys. Rev. B*, **83**(Apr), 134501.

[244] Haus, H. A., and Mullen, J. A. 1962. Quantum noise in linear amplifiers. *Phys. Rev.*, **128**, 2407–2413.

[245] Hegerfeldt, G. C., and Wilser, T. S. 1992. Ensemble or individual system, collapse or no collapse: a description of a single radiating atom. Pages 104–115 of: Doebner, H. D., Scherer, W., and Schroeck, F. (eds), *Classical and Quantum Systems, Proceedings of the Second International Wigner Symposium*. Singapore: World Scientific.

[246] Heller, E. J., and Landry, B. R. 2007. Statistical properties of many-particle eigenfunctions. *J. Phys. A*, **40**, 9259–9274.

[247] Hellwig, K., and Kraus, K. 1970. Operations and Measurements. II. *Commun. Math. Phys.*, **16**, 142.

[248] Helstrom, C. W. 1976. *Quantum Detection and Estimation Theory*. Mathematics in Science and Egineering, vol. 123. New York: Academic Press.

[249] Hepp, K. 1972. Quantum theory of measurement and macroscopic observables. *Helv. Phys. Acta*, **45**, 237.

[250] Higgins, B. L., Berry, D. W., Bartlett, S. D., Wiseman, H. M., and Pryde, G. J. 2007. Entanglement-free Heisenberg-limited phase estimation. *Nature*, **450**, 393–396.

[251] Hill, C., and Ralph, J. F. 2007. Weak measurement and rapid state reduction in entangled bipartite quantum systems. *New J. Phys*, **9**, 151.

[252] Hillery, M., O'Connel, R. F., Scully, M. O., and Wigner, E. P. 1984. Distribution-functions in physics—fundamentals. *Phys. Rep.*, **106**, 121–167.

[253] Hills, R. L. 1996. *Power From Wind: A History of Windmill Technology*. Cambridge University Press.

[254] Hime, T., Reichardt, P. A., Plourde, B. L. T., Robertson, T. L., Wu, C.-E., Ustinov, A. V., and Clarke, J. 2006. Solid-state qubits with current-controlled coupling. *Science*, **314**(5804), 1427–1429.

[255] Hirschman, I. I. 1957. A note on entropy. *Am. J. Math.*, **79**, 152–156.

[256] Hofheinz, M., Wang, H., Ansmann, M., Bialczak, R. C., Lucero, E., Neeley, M., O'Connell, A. D., Sank, D., Wenner, J., Martinis, J. M., and Cleland, A. N. 2009. Synthesizing arbitrary quantum states in a superconducting resonator. *Nature*, **459**, 546–549.

[257] Holevo, A. S. 1973. Statistical decision theory for quantum systems. *J. Multivariate Anal.*, **3**, 337.

[258] Holevo, A. S. 1973. Statistical problems in quantum physics. Pages 104–119 of: Maruyama, G., and Prokhorov, J. V. (eds), *Proceedings of the Second Japan–USSR Symposium on Probability Theory, Lecture Notes in Mathematics*, vol. 330. Berlin: Springer.

[259] Holevo, A. S. 1982. *Probabitistic and Statistical Aspects of Quantum Theory*. Amsterdam: North-Holland.

[260] Holevo, A. S. 2002. On entanglement-assisted classical capacity. *J. Math. Phys.*, **43**, 4326.

[261] Holevo, A. S. 2012. Information capacity of quantum observable. *Prob. Inf. Transmiss*, **48**, 1.

[262] Holland, M. J., and Burnett, K. 1993. Interferometric detection of optical phase shifts at the Heisenberg limit. *Phys. Rev. Lett.*, **71**(Aug), 1355–1358.

[263] Hong, C. K., Ou, Z. Y., and Mandel, L. 1987. Measurement of subpicosecond time intervals between two photons by interference. *Phys. Rev. Lett.*, **59**(Nov), 2044–2046.

[264] Hopkins, A., Jacobs, K., Habib, S., and Schwab, K. 2003. Feedback cooling of a nanomechanical resonator. *Phys. Rev. B*, **68**, 235328.

[265] Horn, A. 1954. Doubly stochastic matrices and the diagonal of a rotation matrix. *Am. J. Math.*, **76**, 620–630.

[266] Horn, R. A., and Johnson, C. R. 1985. *Matrix Analysis*. Cambridge University Press.

[267] Horodecki, R., Horodecki, P., Horodecki, M., and Horodecki, K. 2009. Quantum entanglement. *Rev. Mod. Phys.*, **81**, 865–942.

[268] Horowitz, J. M., and Parrondo, J. M. R. 2011. Thermodynamic reversibility in feedback processes. *EPL (Europhysics Letters)*, **95**(1), 10005.

[269] Horowitz, J. M., Sagawa, T., and Parrondo, J. M. R. 2013. Imitating chemical motors with optimal information motors. *Phys. Rev. Lett.*, **111**(Jul), 010602.

[270] Houck, A. A., Schuster, D. I., Gambetta, J. M., Schreier, J. A., Johnson, B. R., Chow, J. M., Frunzio, L., Majer, J., Devoret, M. H., Girvin, S. M., and Schoelkopf, R. J. 2007. Generating single microwave photons in a circuit. *Nature*, **449**, 328–331.

[271] Howard, R. M. 1872. *Theory of Heat*. New York: D. Appleton & Co.

[272] Howard, R. M. 2002. *Principles of Random Signal Analysis and Low Noise Design: The Power Spectral Density and its Applications*. New York: Wiley-IEEE Press.

[273] Hu, B. L., Paz, J. P., and Zhang, Y. 1992. Quantum Brownian motion in a dissipative environment: exact master equation with nonlocal dissipation and colored noise. *Phys. Rev. D*, **45**, 2843.

[274] Hudson, R. L. 1974. When is the Wigner quasi-probability density non-negative? *Rep. Math. Phys.*, **6**, 249.

[275] Hudson, R. L., and Parthasarathy, K. R. 1984. Quantum Ito's formula and stochastic evolutions. *Commun. Math. Phys.*, **93**, 301–323.

[276] Hughston, L. P., Jozsa, R., and Wootters, W. K. 1993. A complete characterization of quantum ensembles having a given density-matrix. *Phys. Lett. A*, **183**, 14.

[277] Hush, M. R., Carvalho, A. R. R., and Hope, J. J. 2009. Scalable quantum field simulations of conditioned systems. Eprint: arXiv:0901.4391.

[278] Ishizaki, A., and Tanimura, Y. 2005. Quantum dynamics of system strongly coupled to low-temperature colored noise bath: reduced hierarchy equations approach. *J. Phys. Soc. Jpn.*, **74**, 3131.

[279] Ishizaki, A., and Fleming, G. R. 2009. Unified treatment of quantum coherent and incoherent hopping dynamics in electronic energy transfer: reduced hierarchy equation approach. *J. Chem. Phys.*, **130**, 234111.

[280] Ivonovic, I. D. 1981. Geometrical description of quantal state determination. *J. Phys. A: Mathe. Gen.*, **14**(12), 3241.

[281] Iwasawa, K., Makino, K., Yonezawa, H., Tsang, M., Davidovic, A., Huntington, E., and Furusawa, A. 2013. Quantum-limited mirror-motion estimation. *Phys. Rev. Lett.*, **111**, 163602.

[282] Jacobs, K. 1998. *Topics in Quantum Measurement and Quantum Noise*. Ph.D. diss. London: Imperial College.

[283] Jacobs, K. 2003. Efficient measurements, purification, and bounds on the mutual information. *Phys. Rev. A*, **68**, 054302.

[284] Jacobs, K. 2003. How to project qubits faster using quantum feedback. *Phys. Rev. A*, **67**, 030301(R).

[285] Jacobs, K. 2005. Deriving Landauer's erasure principle from statistical mechanics. Eprint: arXiv:quant-ph/0512105.

[286] Jacobs, K. 2006. A bound on the mutual information, and properties of entropy reduction, for quantum channels with inefficient measurements. *J. Math. Phys.*, **47**, 012102.

[287] Jacobs, K. 2007. Feedback control for communication with non-orthogonal states. *Quantum Inform. Comput.*, **7**, 127–138.

[288] Jacobs, K. 2010. *Stochastic Processes for Physicists: Understanding Noisy Systems*. Cambridge University Press.

[289] Jacobs, K. 2010. Wave-function Monte Carlo method for simulating conditional master equations. *Phys. Rev. A*, **81**, 042106.

[290] Jacobs, K., and Knight, P. L. 1998. Linear quantum trajectories: applications to continuous projection measurements. *Phys. Rev. A*, **57**, 2301.

[291] Jacobs, K., and Lund, A. P. 2007. Feedback control of nonlinear quantum systems: a rule of thumb. *Phys. Rev. Lett.*, **99**, 020501.

[292] Jacobs, K., and Steck, D. A. 2011. Engineering quantum states, nonlinear measurements and anomalous diffusion by imaging. *New J. Phys.*, **13**, 013016.

[293] Jacobs, K., and Wiseman, H. M. 2005. An entangled web of crime: Bell's theorem as a short story. *Am. J. Phys.*, **73**, 932.

[294] Jacobs, K., Tombesi, P., Collett, M. J., and Walls, D. F. 1994. Quantum-nondemolition measurement of photon number using radiation pressure. *Phys. Rev. A*, **49**, 1961–1966.

[295] Jacobs, K., Tittonen, I., Wiseman, H. M., and Schiller, S. 1999. Quantum noise in the position measurement of a cavity mirror undergoing Brownian motion. *Phys. Rev. A*, **60**, 538.

[296] Jacobs, K., Lougovski, P., and Blencowe, M. P. 2007. Continuous measurement of the energy eigenstates of a nanomechanical resonator without a nondemolition probe. *Phys. Rev. Lett.*, **98**, 147201.

[297] Jacobs, K., Jordan, A. N., and Irish, E. K. 2008. Energy measurements and preparation of canonical phase states of a nanomechanical resonator. *Europhys. Lett.*, **82**, 18003.

[298] Jacobs, K., Silvestri, L., Dunjko, V., and Olshanii, M. 2011. Typical, finite baths as a means of exact simulation of open quantum systems. *Phys. Rev. E*. In press.

[299] Jacobs, K. 2009. Second law of thermodynamics and quantum feedback control: Maxwell's demon with weak measurements. *Phys. Rev. A*, **80**(Jul), 012322.

[300] Jacobs, K. 2010. Feedback control using only quantum back-action. *New J. Phys.*, **12**(4), 043005.

[301] Jacobs, K. 2012. Quantum measurement and the first law of thermodynamics: The energy cost of measurement is the work value of the acquired information. *Phys. Rev. E*, **86**, 040106.

[302] Jacobs, K., and Landahl, A. J. 2009. Engineering giant nonlinearities in quantum nanosystems. *Phys. Rev. Lett.*, **103**(Aug), 067201.

[303] Jacobs, K., and Lougovski, P. 2007. Emergent quantum jumps in a nano-electro-mechanical system. *J. Phys. A: Math. Theor.*, **40**, F987–F993.

[304] Jacobs, K., Tian, L., and Finn, J. 2009. Engineering superposition states and tailored probes for nanoresonators via open-loop control. *Phys. Rev. Lett.*, **102**, 057208.

[305] Jacobs, K., Finn, J., and Vinjanampathy, S. 2011. Real-time feedback control of a mesoscopic superposition. *Phys. Rev. A*, **83**, 041801.

[306] Jaekel, M. T., and Reynaud, S. 1990. Quantum limits in interferometric measurements. *Europhys. Lett.*, **13**(4), 301–306.

[307] James, M. R. 2004. Risk-sensitive optimal control of quantum systems. *Phys. Rev. A*, **69**, 032108.

[308] James, M., and Gough, J. 2010. Quantum dissipative systems and feedback control design by interconnection. *IEEE Trans. Automat. Contr.*, **55**(8), 1806–1821.

[309] Jayich, A. M., Sankey, J. C., Zwickl, B. M., Yang, C., Thompson, J. D., Girvin, S. M., Clerk, A. A., Marquardt, F., and Harris, J. G. E. 2008. Dispersive optomechanics: a membrane inside a cavity. *New J. Phys.*, **10**(9), 095008.

[310] Jaynes, E. T. 1957. Information theory and statistical mechanics II. *Phys. Rev.*, **108**, 171.

[311] Jaynes, E. T. 2003. *Probability Theory: The Logic of Science*. Cambridge University Press.

[312] Jeffreys, H. 1931. *Scientific Inference*. Cambridge University Press.

[313] Jeffreys, H. 1931. *Theory of Probability*. Oxford: Clarendon Press.

[314] Jeffreys, H. 1932. On the theory of errors and least squares. *Proc. Roy. Soc.*, **138**, 48–55.

[315] Johnson, J. 1928. Thermal agitation of electricity in conductors. *Phys. Rev.*, **32**, 97.

[316] Jones, K. R. W. 1994. Fundamental limits upon the measurement of state vectors. *Phys. Rev. A*, **50**(Nov), 3682–3699.

[317] Joos, E., and Zeh, H. D. 1985. The emergence of classical properties through interaction with the environment. *Z. Phys. B*, **59**, 223–243.

[318] Jordan, A. N., and Korotkov, A. N. 2006. Qubit feedback and control with kicked quantum nondemolition measurements: a quantum Bayesian analysis. *Phys. Rev. B*, **74**(Aug), 085307.

[319] Jordan, A. N., and Korotkov, A. N. 2010. Uncollapsing the wavefunction by undoing quantum measurements. *Contemp. Phys.*, **51**(2), 125–147.

[320] Jozsa, R. 1995. Fidelity for mixed quantum states. *J. Mod. Opt.*, **41**, 2315.

[321] Jozsa, R., Robb, D., and Wootters, W. K. 1994. Lower bound for accessible information in quantum mechanics. *Phys. Rev. A*, **49**, 668–677.

[322] Jungmam, G. 2011. Private communication.

[323] Kac, M. 1959. *Statistical Independence in Probability, Analysis, and Number Theory*. New York: Wiley.

[324] Kalman, R., and Bucy, R. 1961. New results in linear filtering and prediction theory. *J. Basic Eng.*, **82**, 34–45.

[325] Kapulkin, A., and Pattanayak, A. K. 2008. Nonmonotonicity in the quantum-classical transition: chaos induced by quantum effects. *Phys. Rev. Lett.*, **101**(Aug), 074101.

[326] Karbach, M., and Müller, G. 1997. Introduction to the Bethe ansatz I. *Comput. Phys.*, **11**, 36–43.

[327] Kenyon, I. 2011. *The Light Fantastic: A Modern Introduction to Classical and Quantum Optics*. Oxford University Press.

[328] Kerckhoff, J., and Lehnert, K. W. 2012. Superconducting microwave multivibrator produced by coherent feedback. *Phys. Rev. Lett.*, **109**(Oct), 153602.

[329] Kerckhoff, J., Nurdin, H. I., Pavlichin, D. S., and Mabuchi, H. 2010. Designing quantum memories with embedded control: photonic circuits for autonomous quantum error correction. *Phys. Rev. Lett.*, **105**(Jul), 040502.

[330] Kerckhoff, J., Andrews, R. W., Ku, H. S., Kindel, W. F., Cicak, K., Simmonds, R. W., and Lehnert, K. W. 2013. Tunable coupling to a mechanical oscillator circuit using a coherent feedback network. *Phys. Rev. X*, **3**, 021013.

[331] Khalili, F. Y. 2010. Optimal configurations of filter cavity in future gravitational-wave detectors. *Phys. Rev. D*, **81**, 122002.

[332] Kieu, T. D. 2004. The second law, Maxwell's demon, and work derivable from quantum heat engines. *Phys. Rev. Lett.*, **93**, 140403.

[333] Kimble, H. J., Levin, Y., Matsko, A. B., Thorne, K. S., and Vyatchanin, S. P. 2001. Conversion of conventional gravitational-wave interferometers into quantum non-demolition interferometers by modifying their input and/or output optics. *Phys. Rev. D*, **65**, 022002.

[334] Kinion, D., and Clarke, J. 2008. Microstrip superconducting quantum interference device radio-frequency amplifier: Scattering parameters and input coupling. *Appl. Phys. Lett.*, **92**, 172503.

[335] Kirk, D. E. 2004. *Optimal Control Theory: An Introduction*. Mineola: Dover.

[336] Kleckner, D., and Bouwmeester, D. 2006. Sub-kelvin optical cooling of a micromechanical resonator. *Nature*, **444**, 75.

[337] Kleiner, R., Koelle, D., Ludwig, F., and Clarke, J. 2004. Superconducting quantum interference devices: state of the art and applications. *Proc. IEEE*, **92**(10), 1534–1548.

[338] Kloeden, P. E., and Platen, E. 1992. *Numerical Solution of Stochastic Differential Equations*. Berlin: Springer.

[339] Koch, J., Yu, T. M., Gambetta, J., Houck, A. A., Schuster, D. I., Majer, J., Blais, A., Devoret, M. H., Girvin, S. M., and Schoelkopf, R. J. 2007. Charge-insensitive qubit design derived from the Cooper-pair box. *Phys. Rev. A*, **76**, 042319.

[340] Koch, M., Sames, C., Kubanek, A., Apel, M., Balbach, M., Ourjoumtsev, A., Pinkse, P. W., and Rempe, G. 2010. Feedback cooling of a single neutral atom. *Phys. Rev. Lett.*, **105**, 173003.

[341] Kok, P., and Lovett, B. W. 2010. *Introduction to Optical Quantum Information Processing*. Cambridge University Press.

[342] Kok, P., Munro, W. J., Nemoto, K., Ralph, T. C., Dowling, J. P., and Milburn, G. J. 2007. Linear optical quantum computing. *Rev. Mod. Phys.*, **79**, 135–174.

[343] Korotkov, A. N. 2001. Selective quantum evolution of a qubit state due to continuous measurement. *Phys. Rev. B*, **63**, 115403.

[344] Korotkov, A. N., and Jordan, A. N. 2006. Undoing a weak quantum measurement of a solid-state qubit. *Phys. Rev. Lett.*, **97**(Oct), 166805.

[345] Kraus, K. 1983. *States, Effects and Operations: Fundamental Notions of Quantum Theory*. Lecture Notes in Physics, Vol. 190. Berlin: Springer.

[346] Kraus, K. 1987. Complementary observables and uncertainty relations. *Phys. Rev. D*, **35**(May), 3070–3075.

[347] Kubanek, A., Koch, M., Sames, C., Ourjoumtsev, A., Pinkse, P. W. H., Murr, K., and Rempe, G. 2009. Photon-by-photon feedback control of a single-atom trajectory. *Nature*, **462**, 898.

[348] Kumar, S., and DiVincenzo, D. P. 2010. Exploiting Kerr cross nonlinearity in circuit quantum electrodynamics for nondemolition measurements. *Phys. Rev. B*, **82**(Jul), 014512.

[349] LaHaye, M. D., Buu, O., Camarota, B., and Schwab, K. C. 2004. Approaching the quantum limit of a nanomechanical resonator. *Science*, **304**, 74.

[350] Laloe, F. 2001. Do we really understand quantum mechanics? Strange correlations, paradoxes, and theorems. *Am. J. Phy.*, **69**(6), 655–701.

[351] Landau, L. 1995. Das Dämpfungsproblem in der Wallenmechanik. *Z. Phys.*, **45**, 430–441.

[352] Landauer, R. 1961. Irreversibility and heat generation in the computing process. *IBM J. Res. Dev.*, **5**, 183.

[353] Lanford, O. E., and Robinson, D. 1968. Mean entropy of states in quantum-statistical mechanics. *J. Math. Phys.*, **9**, 1120–1125.

[354] Langevin, P. 1908. On the theory of Brownian motion. *C. R. Acad. Sci. (Paris)*, **146**, 530–533.

[355] Law, C. K. 1994. Effective Hamiltonian for the radiation in a cavity with a moving mirror and a time-varying dielectric medium. *Phys. Rev. A*, **49**, 433–437.

[356] Law, C. K. 1995. Interaction between a moving mirror and radiation pressure: a Hamiltonian formulation. *Phys. Rev. A*, **51**, 2537–2541.

[357] Law, C. K., and Eberly, J. H. 1996. Arbitrary control of a quantum electromagnetic field. *Phys. Rev. Lett.*, **76**, 1055.

[358] Leff, H. S., and Rex, A. F. 1994. Entropy of measurement and erasure—Szilard's membrane model revisited. *Am. J. Phys.*, **62**, 994–1000.

[359] Leff, H. S., and Rex, A. F. (eds). 2002. *Maxwell's Demon 2: Entropy, Classical and Quantum Information, Computing.* New York: CRC Press.

[360] Leff, H., and Rex, A. (eds). 1990. *Maxwell's Demon: Entropy, Information, Computing.* New Jersey: Princeton University Press.

[361] Leggett, A. J. 1980. Macroscopic quantum systems and the quantum theory of measurement. 1980. *Prog. Theor. Phys. (Suppl.)* **69**, 80; made freely available by Oxford Journals at http://ptps.oxfordjournals.org/content/69.toc. Accessed 11 April 2014.

[362] Leggett, A. J., and Garg, A. 1985. Quantum mechanics versus macroscopic realism: is the flux there when nobody looks? *Phys. Rev. Lett.*, **54**(Mar), 857–860.

[363] Lehmann, E. L., and Casella, G. 1998. *Theory of Point Estimation.* New York: Springer.

[364] Leibfried, D., Blatt, R., Monroe, C., and Wineland, D. 2003. Quantum dynamics of single trapped ions. *Rev. Mod. Phys.*, **75**, 281–324.

[365] Leifer, M. S., and Spekkens, R. W. 2013. Towards a formulation of quantum theory as a causally neutral theory of Bayesian inference. *Phys. Rev. A*, **88**, 052130.

[366] Leroux, I. D., Schleier-Smith, M. H., and Vuletić, V. 2010. Implementation of cavity squeezing of a collective atomic spin. Eprint: arXiv:0911.4065.

[367] Levitin, L. B. 1995. Optimal quantum measurements for two pure and mixed states. Pages 439–448 of: Belavkin, V. P., Hirota, O., and Hudson, R. L. (eds), *Quantum Communications and Measurement.* New York: Plenum.

[368] Levy, B. C. 2008. *Principles of Signal Detection and Parameter Estimation.* New York: Springer.

[369] Li, H., Shabani, A., Sarovar, M., and Whaley, K. B. 2013. Optimality of qubit purification protocols in the presence of imperfections. *Phys. Rev. A*, **87**(Mar), 032334.

[370] Lieb, E. H. 1973. Convex trace functions and the Wigner-Yanase-Dyson conjective. *Ad. Math.*, **11**, 267.

[371] Lieb, E. H., and Ruskai, M. B. 1973. A fundamental property of quantum-mechanical entropy. *Phys. Rev. Lett.*, **30**, 434.

[372] Lieb, E. H., and Ruskai, M. B. 1973. Proof of the strong subadditivity of quantum-mechanical entropy. *J. Math. Phys.*, **14**, 1938.

[373] Likharev, K. K. 1986. *Dynamics of Josephson Junctions and Circuits.* New York: CRC Press.

[374] Lindblad, G. 1976. On the generators of quantum dynamical semigroups. *Commun. Math. Phys.*, **48**, 119.

[375] Linden, N., Popescu, S., Short, A. J., and Winter, A. 2009. Quantum mechanical evolution towards thermal equilibrium. *Phys. Rev. E*, **79**, 061103.

[376] Lloyd, S., Landahl, A. J., and Slotine, J.-J. E. 2004. Universal quantum interfaces. *Phys. Rev. A*, **69**, 012305.

[377] Lloyd, S. 1995. Almost any quantum logic gate is universal. *Phys. Rev. Lett.*, **75**, 346–349.

[378] Lloyd, S. 2000. Coherent quantum feedback. *Phys. Rev. A*, **62**(Jul), 022108.

[379] Ludwig, G. 1968. Attempt of an axiomatic foundation of quantum mechanics and more general theories III. *Commun. Math. Phys.*, **9**, 1.

[380] Luis, A., and Sánchez-Soto, L. L. 1992. Multimode quantum analysis of an interferometer with moving mirrors. *Phys. Rev. A*, **45**, 8228–8234.

[381] Maalouf, A., and Petersen, I. 2011. Coherent H^∞ control for a class of annihilation operator linear quantum systems. *IEEE Trans. Automat. Contr.*, **56**(2), 309–319.

[382] Maassen, H., and Uffink, J. B. M. 1988. Generalized entropic uncertainty relations. *Phys. Rev. Lett.*, **60**, 1103.

[383] Mabuchi, H., and Wiseman, H. M. 1998. Retroactive quantum jumps in a strongly coupled atom-field system. *Phys. Rev. Lett.*, **81**, 4620.

[384] Mabuchi, H. 2008. Coherent-feedback quantum control with a dynamic compensator. *Phys. Rev. A*, **78**(Sep), 032323.

[385] Mabuchi, H. 2011. Coherent-feedback control strategy to suppress spontaneous switching in ultralow-power optical bistability. *Appl. Phys. Lett.*, **98**, 193109.

[386] Machnes, S., Plenio, M. B., Reznik, B., Steane, A. M., and Retzker, A. 2010. Superfast laser cooling. *Phys. Rev. Lett.*, **104**(May), 183001.

[387] Machnes, S., Sander, U., Glaser, S. J., de Fouquières, P., Gruslys, A., Schirmer, S., and Schulte-Herbrüggen, T. 2011. Comparing, optimizing, and benchmarking quantum-control algorithms in a unifying programming framework. *Phys. Rev. A*, **84**, 022305.

[388] Machnes, S., Cerrillo, J., Aspelmeyer, M., Wieczorek, W., Plenio, M. B., and Retzker, A. 2012. Pulsed laser cooling for cavity optomechanical resonators. *Phys. Rev. Lett.*, **108**(Apr), 153601.

[389] Majer, J., Chow, J. M., Gambetta, J. M., Koch, J., Johnson, B. R., Schreier, J. A., Frunzio, L., Schuster, D. I., Houck, A. A., Wallraff, A., Blais, A., Devoret, M. H., Girvin, S. M., and Schoelkopf, R. J. 2007. Coupling superconducting qubits via a cavity bus. *Nature*, **449**, 443.

[390] Majer, J. B., Paauw, F. G., ter Haar, A. C. J., Harmans, C. J. P. M., and Mooij, J. E. 2005. Spectroscopy on two coupled superconducting flux qubits. *Phys. Rev. Lett.*, **94**, 090501.

[391] Makhlin, Y., Schon, G., and Shnirman, A. 1999. Josephson-junction qubits with controlled couplings. *Nature*, **398**, 305–307.

[392] Makhlin, Y., Schön, G., and Shnirman, A. 2001. Quantum-state engineering with Josephson-junction devices. *Rev. Mod. Phys.*, **73**, 357.

[393] Makri, N. 1995. Numerical path integral techniques for long time dynamics of quantum dissipative systems. *J. Math. Phys.*, **36**, 2430.

[394] Makri, N., and Makarov, D. E. 1995. Tensor propagator for iterative quantum time evolution of reduced density matrices. I. Theory. *J. Chem. Phys.*, **102**, 4600.

[395] Makri, N., and Makarov, D. E. 1995. Tensor propagator for iterative quantum time evolution of reduced density matrices. II. Numerical methodology. *J. Chem. Phys.*, **102**, 4611.

[396] Mancini, S. 2006. Markovian feedback to control continuous variable entanglement. *Phys. Rev. A*, **73**, 010304(R).

[397] Mancini, S., Man'ko, V. I., and Tombesi, P. 1997. Ponderomotive control of quantum macroscopic coherence. *Phys. Rev. A*, **55**, 3042–3050.

[398] Mancini, S., Vitali, D., and Tombesi, P. 1998. Optomechanical cooling of a macroscopic oscillator by homodyne feedback. *Phys. Rev. Lett.*, **80**, 688.

[399] Mandal, D., and Jarzynski, C. 2012. Work and information processing in a solvable model of Maxwell's demon. *PNAS*, **109**, 11641.

[400] Mandal, D., Quan, H. T., and Jarzynski, C. 2013. Maxwell's refrigerator: an exactly solvable model. *Phys. Rev. Lett.*, **111**(Jul), 030602.

[401] Margolus, N., and Levitin, L. B. 1998. The maximum speed of dynamical evolution. *Physica D: Nonlinear Phenomena*, **120**, 188–195.

[402] Maroney, O. J. E. 2009. Generalizing Landauer's principle. *Phys. Rev. E*, **79**, 031105.

[403] Marquardt, F., Chen, J. P., Clerk, A. A., and Girvin, S. M. 2007. Quantum theory of cavity-assisted sideband cooling of mechanical motion. *Phys. Rev. Lett.*, **99**, 093902.

[404] Marshall, A. W., and Olkin, I. 1979. *Inequalities: Theory of Majorization and Its Applications*. New York: Academic Press.

[405] Marshall, W., Simon, C., Penrose, R., and Bouwmeester, D. 2003. Towards quantum superpositions of a mirror. *Phys. Rev. Lett.*, **91**(Sep), 130401.

[406] Marti, K. H., Bauer, B., Reiher, M., Troyer, M., and Verstraete, F. 2010. Complete-graph tensor network states: a new fermionic wave function ansatz for molecules. *New J. Phys.*, **12**(10), 103008.

[407] Martin, I., and Zurek, W. H. 2007. Measurement of energy eigenstates by a slow detector. *Phys. Rev. Lett.*, **98**, 120401.

[408] Martinis, J. M., Devoret, M. H., and Clarke, J. 1985. Energy-level quantization in the zero-voltage state of a current-biased Josephson junction. *Phys. Rev. Lett.*, **55**(Oct), 1543–1546.

[409] Martinis, J. M., Nam, S., Aumentado, J., and Urbina, C. 2002. Rabi oscillations in a large Josephson-junction qubit. *Phys. Rev. Lett.*, **89**(Aug), 117901.

[410] Massel, F., Heikkilä, T. T., Pirkkalainen, J.-M., Cho, S. U., Saloniemi, H., Hakonen, P. J., and Sillanpää, M. A. 2011. Microwave amplification with nanomechanical resonators. *Nature*, **480**, 351–354.

[411] Massel, F., Cho, S. U., Pirkkalainen, J.-M., Hakonen, P. J., Heikkilä, T. T., and Sillanpää, M. A. 2012. Multimode circuit optomechanics near the quantum limit. *Nature Commun.*, **3**, 987.

[412] Maurer, P. C., Kucsko, G., Latta, C., Jiang, L., Yao, N. Y., Bennett, S. D., Pastawski, F., Hunger, D., Chisholm, N., Markham, M., Twitchen, D. J., Cirac, J. I., and Lukin, M. D. 2012. Room-temperature quantum bit memory exceeding one second. *Science*, **336**, 1283–1286.

[413] Maxwell, J. C. 1868. On governors. *Proc. Roy. Soc. London*, **16**, 270–283.

[414] Maybeck, P. S. 1982. *Stochastic Models, Estimation and Control.* Vols. I and II. New York: Academic Press.

[415] McDermott, R., Simmonds, R. W., Steffen, M., Cooper, K. B., Cicak, K., Osborn, K. D., Oh, S., Pappas, D. P., and Martinis, J. M. 2005. Simultaneous state measurement of coupled Josephson phase qubits. *Science*, **307**, 1299.

[416] McDonough, R. N., and Whalen, A. D. 1995. *Detection of Signals in Noise.* New York: Academic Press.

[417] McKenzie, K., Shaddock, D. A., McClelland, D. E., Buchler, B. C., and Lam, P. K. 2002. Experimental demonstration of a squeezing-enhanced power-recycled Michelson interferometer for gravitational wave detection. *Phys. Rev. Lett.*, **88**, 231102.

[418] Mehmet, M., Vahlbruch, H., Lastzka, N., Danzmann, K., and Schnabel, R. 2010. Observation of squeezed states with strong photon-number oscillations. *Phys. Rev. A*, **81**(Jan), 013814.

[419] Mehta, M. L. 1967. *Random Matrices and the Statistical Theory of Energy Levels.* New York: Academic Press.

[420] Mehta, M. L. 2004. *Random Matrices.* Boston: Academic Press.

[421] Mermin, N. D. 1993. Hidden variables and the two theorems of John Bell. *Rev. Mod. Phys.*, **65**(Jul), 803–815.

[422] Metcalfe, M., Boaknin, E., Manucharyan, V., Vijay, R., Siddiqi, I., Rigetti, C., Frunzio, L., Schoelkopf, R. J., and Devoret, M. H. 2007. Measuring the decoherence of a quantronium qubit with the cavity bifurcation amplifier. *Phys. Rev. B*, **76**(Nov), 174516.

[423] Metz, J., Trupke, M., and Beige, A. 2006. Robust entanglement through macroscopic quantum jumps. *Phys. Rev. Lett.*, **97**, 040503.

[424] Mirrahimi, M., Rouchon, P., and Turinici, G. 2005. Lyapunov control of bilinear Schrödinger equations. *Automatica*, **41**, 1987–1994.

[425] Mirrahimi, M., Turinici, G., and Rouchon, P. 2005. Reference trajectory tracking for locally designed coherent quantum controls. *J. Phys. Chem. A*, **109**, 2631–2637.

[426] Mirrahimi, M., and vanHandel, R. 2007. Stabilizing feedback controls for quantum systems. *SIAM J. Control Optim.*, **46**, 445–467.

[427] Monroe, C., Meekhof, D. M., King, B. E., and Wineland, D. J. 1996. A "Schrödinger cat" superposition state of an atom. *Science*, **272**(5265), 1131–1136.

[428] Montinaro, M., Mehlin, A., Solanki, H. S., Peddibhotla, P., Mack, S., Awschalom, D. D., and Poggio, M. 2012. Feedback cooling of cantilever motion using a quantum point contact transducer. *Appl. Phys. Lett.*, **101**, 133104.

[429] Mooij, J. E., Orlando, T. P., Levitov, L. S., Tian, L., van der Wal, C. H., and Lloyd, S. 1999. Josephson persistent-current qubit. *Science*, **285**, 1036.

[430] Moore, K. W., and Rabitz, H. 2012. Exploring constrained quantum control landscapes. *J. Chem. Phys.*, **137**, 134113.

[431] Morrow, N. V., Dutta, S. K., and Raithel, G. 2002. Feedback control of atomic motion in an optical lattice. *Phys. Rev. Lett.*, **88**, 093003.

[432] Motzoi, F., Gambetta, J. M., Rebentrost, P., and Wilhelm, F. K. 2009. Simple pulses for elimination of leakage in weakly nonlinear qubits. *Phys. Rev. Lett.*, **103**, 110501.

[433] Movshovich, R., Yurke, B., Kaminsky, P. G., Smith, A. D., Silver, A. H., Simon, R. W., and Schneider, M. V. 1990. Observation of zero-point noise squeezing via a Josephson-parametric amplifier. *Phys. Rev. Lett.*, **65**, 1419–1422.

[434] Murch, K. W., Moore, K. L., Gupta, S., and Stamper-Kurn, D. M. 2008. Observation of quantum-measurement back-action with an ultracold atomic gas. *Nature Phys.*, **4**, 561–564.

[435] Murg, V., Verstraete, F., and Cirac, J. I. 2007. Variational study of hard-core bosons in a two-dimensional optical lattice using projected entangled pair states. *Phys. Rev. A*, **75**, 033605.

[436] Nagourney, W., Sandberg, J., and Dehmelt, H. 1986. Shelved optical electron amplifier: observation of quantum jumps. *Phys. Rev. Lett.*, **56**, 2797–2799.

[437] Naik, A., Buu, O., LaHaye, M. D., Armour, A. D., Clerk, A. A., Blencowe, M. P., and Schwab, K. C. 2006. Cooling a nanomechanical resonator with quantum back-action. *Nature*, **443**, 193–196.

[438] Nakajima, S. 1958. On quantum theory of transport phenomena. *Prog. Theor. Phys.*, **20**, 948.

[439] Nakamura, Y., Pashkin, Y. A., and Tsai, J. S. 1999. Coherent control of macroscopic quantum states in a single-Cooper-pair box. *Nature*, **398**, 786.

[440] Narnhofer, H., and Thirring, W. 1985. From relative entropy to entropy. *Fizika*, **17**, 257–265.

[441] Natarajan, C. M., Tanner, M. G., and Hadfield, R. H. 2012. Superconducting nanowire single-photon detectors: physics and applications. *Supercondr. Sci. and Technolo.*, **25**, 063001.

[442] Nelson, R. J., Weinstein, Y., Cory, D., and Lloyd, S. 2000. Experimental demonstration of fully coherent quantum feedback. *Phys. Rev. Lett.*, **85**, 3045–3048.

[443] Neuenhahn, C., and Marquardt, F. 2012. Thermalization of interacting fermions and delocalization in Fock space. *Phys. Rev. E*, **85**, 060101.

[444] Neumann, P., Beck, J., Steiner, M., Rempp, F., Fedder, H., Hemmer, P. R., Wrachtrup, J., and Jelezko, F. 2010. Single-shot readout of a single nuclear spin. *Science*, **329**, 542–544.

[445] Nielsen, M. A. 2000. Probability distributions consistent with a mixed state. *Phys. Rev. A*, **62**, 052308.

[446] Nielsen, M. A. 2001. Characterizing mixing and measurement in quantum mechanics. *Phys. Rev. A*, **63**, 022114.

[447] Nielsen, M. A., and Chuang, I. L. 2000. *Quantum Computation and Quantum Information*. Cambridge University Press.

[448] Nielsen, M. A., and Petz, D. 2005. A simple proof of the strong subadditivity inequality. *Quantum Inform. Comput.*, **5**, 507.

[449] Nielsen, M. A. 2002. A simple formula for the average gate fidelity of a quantum dynamical operation. *Phys. Lett. A*, **303**(4), 249–252.

[450] Niskanen, A. O., Nakamura, Y., and Tsai, J.-S. 2006. Tunable coupling scheme for flux qubits at the optimal point. *Phys. Rev. B*, **73**(Mar), 094506.

[451] Nocedal, J., and Wright, S. J. 2006. *Numerical Optimization*. New York: Springer.

[452] Nurdin, H. I., James, M. R., and Petersen, I. R. 2009. Coherent quantum LQG control. *Automatica*, **45**(8), 1837–1846.

[453] Nurdin, H. I., James, M. R., and Doherty, A. C. 2009. Network synthesis of linear dynamical quantum stochastic systems. *SIAM J. Control Optim.*, **48**(4), 2686–2718.

[454] Nyquist, H. 1928. Thermal agitation of electric charge in conductors. *Phys. Rev.*, **32**, 110.

[455] Ocone, D., and Pardoux, E. 1996. Asymptotic stability of the optimal filter with respect to its initial condition. *SIAM J. Control Optim.*, **34**, 226–243.

[456] Oreshkov, O., and Brun, T. A. 2005. Weak measurements are universal. *Phys. Rev. Lett.*, **95**(Sep), 110409.

[457] Orlando, T. P., Mooij, J. E., Tian, L., van der Wal, C. H., Levitov, L. S., Lloyd, S., and Mazo, J. J. 1999. Superconducting persistent-current qubit. *Phys. Rev. B*, **60**, 15398–15413.

[458] Ozawa, M. 1986. On information gain by quantum measurements of continuous observables. *J. Math. Phys.*, **27**, 759–763.

[459] Ozawa, M. 2003. Universally valid reformulation of the Heisenberg uncertainty principle on noise and disturbance in measurement. *Phys. Rev. A*, **67**(Apr), 042105.

[460] Ozawa, M. 2004. Uncertainty relations for noise and disturbance in generalized quantum measurements. *Ann. Phys.*, **311**(2), 350–416.

[461] O'Connell, A. D., Hofheinz, M., Ansmann, M., Bialczak, R. C., Lenander, M., Lucero, E., Neeley, M., Sank, D., Wang, H., Weides, M., Wenner, J., Martinis, J. M., and Cleland, A. N. 2010. Quantum ground state and single-phonon control of a mechanical resonator. *Nature*, **464**, 697–703.

[462] Pace, A. F., Collett, M. J., and Walls, D. F. 1993. Quantum limits in interferometric detection of gravitational radiation. *Phys. Rev. A*, **47**, 3173–3189.

[463] Paik, H., Schuster, D. I., Bishop, L. S., Kirchmair, G., Catelani, G., Sears, A. P., Johnson, B. R., Reagor, M. J., Frunzio, L., Glazman, L. I., Girvin, S. M., Devoret, M. H., and Schoelkopf, R. J. 2011. Observation of high coherence in Josephson junction qubits measured in a three-dimensional circuit QED architecture. *Phys. Rev. Lett.*, **107**(Dec), 240501.

[464] Pal, A., and Huse, D. A. 2010. Many-body localization phase transition. *Phys. Rev. B*, **82**(Nov), 174411.

[465] Palomaki, T. A., Harlow, J. W., Teufel, J. D., Simmonds, R. W., and Lehnert, K. W. 2013. Coherent state transfer between itinerant microwave fields and a mechanical oscillator. *Nature*, **495**, 210–214.

[466] Paris, M. G. A. 2009. Quantum estimation for quantum technology. *Int. J. Quantum. Inform.*, **7**, 125.

[467] Partovi, M. H. 1983. Entropic formulation of uncertainty for quantum measurements. *Phys. Rev. Lett.*, **50**(Jun), 1883–1885.

[468] Pashkin, Y. A., Yamamoto, T., Astafiev, O., Nakamura, Y., Averin, D. V., and Tsai, J. S. 2003. Quantum oscillations in two coupled charge qubits. *Nature*, **421**, 823.

[469] Pavon, M., and D'Alessandro, D. 1997. Families of solutions of matrix Riccati differential equations. *SIAM J. Control Optim.*, **35**(1), 194.

[470] Pechen, A., Il'in, N., Shuang, F., and Rabitz, H. 2006. Quantum control by von Neumann measurements. *Phys. Rev. A*, **74**(Nov), 052102.

[471] Pegg, D. T., and Barnett, S. M. 1988. Unitary phase operator in quantum mechanics. *Europhy. Lett.*, **6**, 483.

[472] Pegg, D. T., and Barnett, S. M. 1989. Phase properties of the quantized single-mode electromagnetic-field. *Phys. Rev. A*, **39**, 1665.

[473] Peil, S., and Gabrielse, G. 1999. Observing the quantum limit of an electron cyclotron: QND measurements of quantum jumps between Fock states. *Phys. Rev. Lett.*, **83**, 1287–1290.

[474] Perales, A., and Vidal, G. 2008. Entanglement growth and simulation efficiency in one-dimensional quantum lattice systems. *Phys. Rev. A*, **78**, 042337.

[475] Percival, I. C. 1999. *Quantum State Diffusion*. Cambridge University Press.

[476] Percival, I. C., and Strunz, W. T. 1998. Classical dynamics of quantum localization. *J. Phys. A*, **31**, 1815.

[477] Percival, I. C., and Strunz, W. T. 1998. Classical mechanics from quantum state diffusion – a phase-space approach. *J. Phys. A*, **31**, 1801.

[478] Peres, A. 1988. Quantum limitations on measurement of magnetic flux. *Phys. Rev. Lett.*, **61**(Oct), 2019–2021.

[479] Perron, F., and Marchand, E. 2002. On the minimax estimator of a bounded normal mean. *Stat. Probabil. Lett.*, **58**, 327–333.

[480] Petz, D. 1986. Complementarity in quantum systems. *Rep. Math. Phys.*, **23**, 57.

[481] Piechocinska, B. 2000. Information erasure. *Phys. Rev. A*, **61**, 062314.

[482] Piilo, J., Maniscalco, S., Härkönen, K., and Suominen, K.-A. 2008. Non-Markovian quantum jumps. *Phys. Rev. Lett.*, **100**, 180402.

[483] Plenio, M. B., and Vitelli, V. 2001. The physics of forgetting: Landauer's erasure principle and information theory. *Contemp. Phys.*, **42**, 25–60.

[484] Plenio, M. B., and Vedral, V. 1998. Entanglement in quantum information theory. *Contemp. Phys.*, **39**, 431–466.

[485] Poggio, M., Degen, C. L., Mamin, H. J., and Rugar, D. 2007. Feedback cooling of a cantilever's fundamental mode below 5 mK. *Phys. Rev. Lett.*, **99**, 017201.

[486] Poot, M., and van der Zant, H. S. J. 2012. Mechanical systems in the quantum regime. *Phys. Rep.*, **511**, 273–335.

[487] Popescu, S., Short, A. J., and Winter, A. 2006. Entanglement and foundations of statistical mechanics. *Nature Phys.*, **2**, 754.

[488] Pozar, D. M. 2011. *Microwave Engineering*. New York: Wiley.

[489] Prance, R. J., Long, A. P., Clark, T. D., Widom, A., Mutton, J. E., Sacco, J., Potts, M. W., Megaloudis, G., and Goodall, F. 1981. Coherent control of macroscopic quantum states in a single-Cooper-pair box. *Nature*, **289**, 543.

[490] Prance, R. J., Clark, T. D., Mutton, J. E., Prance, H., Spiller, T. P., and Nest, R. 1985. Localization of pair charge states in a superconducting weak link constriction ring. *Phys. Lett. A*, **107**, 133–138.

[491] Prather, D. W., Sharkawy, A., Shi, S., Murakowski, J., and Schneider, G. 2009. *Photonic Crystals: Theory, Applications and Fabrication*. New York: Wiley.

[492] Pregnell, K. L., and Pegg, D. T. 2002. Single-shot measurement of quantum optical phase. *Phys. Rev. Lett.*, **89**, 173601.

[493] Prior, J., Chin, A. W., Huelga, S. F., and Plenio, M. B. 2010. Efficient simulation of strong system–environment interactions. *Phys. Rev. Lett.*, **105**, 050404.

[494] Prober, D., Teufel, J., Wilson, C., Frunzio, L., Shen, M., Schoelkopf, R., Stevenson, T., and Wollack, E. 2007. Ultrasensitive quantum-limited far-infrared STJ detectors. *IEEE Trans. Appl. Supercond.*, **17**(2), 241–245.

[495] Purdy, T. P., Brooks, D. W. C., Botter, T., Brahms, N., Ma, Z.-Y., and Stamper-Kurn, D. M. 2010. Tunable cavity optomechanics with ultracold atoms. *Phys. Rev. Lett.*, **105**, 133602.

[496] Qi, B., Pan, H., and Guo, L. 2013. Further results on stabilizing control of quantum systems. *IEEE Trans. Automat. Contr.*, **58**(5), 1349–1354.

[497] Quan, H. T., Wang, Y. D., Xi Liu, Y., Sun, C. P., and Nori, F. 2006. Maxwell's demon-assisted thermodynamic cycle in superconducting quantum circuits. *Phys. Rev. Lett.*, **97**, 180402.

[498] Raimond, J. M., Brune, M., and Haroche, S. 2001. Manipulating quantum entanglement with atoms and photons in a cavity. *Rev. Mod. Phys.*, **73**, 565–582.

[499] Ralph, J. F., Griffith, E. J., Clark, T. D., and Everitt, M. J. 2004. Guidance and control in a Josephson charge qubit. *Phys. Rev. B*, **70**, 214521.

[500] Ralph, J. F., Jacobs, K., and Hill, C. D. 2009. Frequency tracking and parameter estimation for robust quantum state-estimation. Eprint: arXiv:0907.5034.

[501] Ralph, J., Clark, T., Prance, R., and Prance, H. 1996. Self-capacitance of the superconducting condensate. *Physica B: Condensed Matter*, **226**(4), 355–362.

[502] Rangarajan, G., Habib, S., and Ryne, R. D. 1998. Lyapunov exponents without rescaling and reorthogonalization. *Phys. Rev. Lett.*, **80**(17), 3747–3750.

[503] Raynal, P., Lü, X., and Englert, B.-G. 2011. Mutually unbiased bases in six dimensions: the four most distant bases. *Phys. Rev. A*, **83**, 062303.

[504] Redfield, A. G. 1957. On the theory of relaxation processes. *IBM J. Res. Dev.*, **1**, 19–31.

[505] Redfield, A. G. 1965. On the theory of relaxation processes. *Reprinted in: Adv. Magn. Reson.*, **1**, 1–32.

[506] Reed, M., and Simon, B. 1972. *Methods of Modern Mathematical Physics—Part I: Functional Analysis*. New York: Academic Press.

[507] Regal, C. A., Teufel, J. D., and Lehnert, K. W. 2008. Measuring nanomechanical motion with a microwave cavity interferometer. *Nature Phys.*, **4**, 555.

[508] Reid, W. T. 1972. *Riccati Differential Equations*. Waltham: Academic Press.

[509] Reimann, P. 2008. Foundation of statistical mechanics under experimentally realistic conditions. *Phys. Rev. Lett.*, **101**(19), 190403.

[510] Reinsch, M. W. 2000. A simple expression for the terms in the Baker-Campbell-Hausdorff series. *J. Math. Phys.*, **41**, 2434.

[511] Renes, J. M., Blume-Kohout, R., Scott, A. J., and Caves, C. M. 2004. Symmetric informationally complete quantum measurements. *J. Math. Phys.*, **45**(6), 2171–2180.

[512] Richtmyer, R. D., and Greenspan, S. 1965. Expansion of the Barker–Campbell–Hausdorff formula by computer. *Commun. Pure Appl. Math.*, **18**, 107–108.

[513] Riera, A., Gogolin, C., and Eisert, J. 2011. Thermalization in nature and on a quantum computer. Eprint: arXiv:1102.2389.

[514] Rigol, M., and Santos, L. F. 2010. Quantum chaos and thermalization in gapped systems. *Phys. Rev. A*, **82**(Jul), 011604.

[515] Rigol, M., Dunjko, V., Yurovsky, V., and Olshanii, M. 2007. Relaxation in a completely integrable many-body quantum system: an *ab initio* study of the dynamics of the highly excited states of 1D lattice hard-core bosons. *Phys. Rev. Lett.*, **98**, 050405.

[516] Rigol, M., Dunjko, V., and Olshanii, M. 2008. Thermalization and its mechanism for generic isolated quantum systems. *Nature*, **452**, 854.

[517] Ristè, D., Bultink, C. C., Lehnert, K. W., and DiCarlo, L. 2012. Feedback control of a solid-state qubit using high-fidelity projective measurement. *Phys. Rev. Lett.*, **109**(Dec), 240502.

[518] Ristè, D., van Leeuwen, J. G., Ku, H.-S., Lehnert, K. W., and DiCarlo, L. 2012. Initialization by measurement of a superconducting quantum bit circuit. *Phys. Rev. Lett.*, **109**(Aug), 050507.

[519] Rivas, A., Douglas, A., Plato, K., Huelga, S. F., and Plenio, M. B. 2010. Markovian master equations: a critical study. *New J. Phys.*, **12**(11), 113032.

[520] Robertson, H. P. 1929. The uncertainty principle. *Phys. Rev.*, **34**, 163–164.

[521] Robinson, D. W., and Ruelle, D. 1967. Mean entropy of states in classical statistical mechanics. *Commun. Math. Phys.*, **5**, 288.

[522] Rosenkrantz, R. D. 1989. *E.T. Jaynes: Papers on Probability, Statistics, and Statistical Physics*. New York: Springer.

[523] Ruskai, M. B. 2007. Another short and elementary proof of strong subadditivity of quantum entropy. *Rep. Math. Phys.*, **60**, 1–12.

[524] Ruskov, R., and Korotkov, A. N. 2002. Quantum feedback control of a solid-state qubit. *Phys. Rev. B*, **66**, 041401.

[525] Ruskov, R., Combes, J., Mølmer, K., and Wiseman, H. M. 2012. Qubit purification speed-up for three complementary continuous measurements. *Phil. Trans. Roy. Soc. A*, **370**, 5291–5307.

[526] Ruskov, R., Schwab, K., and Korotkov, A. N. 2005. Squeezing of a nanomechanical resonator by quantum nondemolition measurement and feedback. *Phys. Rev. B*, **71**, 235407.

[527] Safavi-Naeini, A. H., Alegre, T. P. M., Chan, J., Eichenfield, M., Winger, M., Lin, Q., Hill, J. T., Chang, D., and Painter, O. 2011. Electromagnetically induced transparency and slow light with optomechanics. *Nature*, **472**, 69–73.

[528] Sagawa, T., and Ueda, M. 2008. Second law of thermodynamics with discrete quantum feedback control. *Phys. Rev. Lett.*, **100**, 080403.

[529] Sagawa, T., and Ueda, M. 2009. Minimal energy cost for thermodynamic information processing: measurement and information erasure. *Phys. Rev. Lett.*, **102**, 250602.

[530] Sagawa, T., and Ueda, M. 2010. Generalized Jarzynski equality under nonequilibrium feedback control. *Phys. Rev. Lett.*, **104**(Mar), 090602.

[531] Saito, S., Thorwart, M., Tanaka, H., Ueda, M., Nakano, H., Semba, K., and Takayanagi, H. 2004. Multiphoton transitions in a macroscopic quantum two-state system. *Phys. Rev. Lett.*, **93**(Jul), 037001.

[532] Santos, L. F., and Rigol, M. 2010. Onset of quantum chaos in one-dimensional bosonic and fermionic systems and its relation to thermalization. *Phys. Rev. E*, **81**, 036206.

[533] Santos, L. F., Polkovnikov, A., and Rigol, M. 2012. Weak and strong typicality in quantum systems. *Phys. Rev. E*, **86**(Jul), 010102.

[534] Sarovar, M., Ahn, C., Jacobs, K., and Milburn, G. J. 2004. Practical scheme for error control using feedback. *Phys. Rev. A*, **69**(May), 052324.

[535] Sauter, T., Neuhauser, W., Blatt, R., and Toschek, P. E. 1986. Observation of quantum jumps. *Phys. Rev. Lett.*, **57**(Oct), 1696–1698.

[536] Sayrin, C., Dotsenko, I., Zhou, X., Peaudecerf, B., Théo Rybarczyk, S. G., Rouchon, P., Mirrahimi, M., Amini, H., Brune, M., Raimond, J.-M., and Haroche, S. 2011. Real-time quantum feedback prepares and stabilizes photon number states. *Nature*, **477**, 73–77.

[537] Schack, R., Brun, T. A., and Percival, I. C. 1995. Quantum state diffusion, localization and computation. *J. Phys. A*, **28**, 5401.

[538] Schirmer, S. G., and de Fouquieres, P. 2011. Efficient algorithms for optimal control of quantum dynamics: the Krotov method unencumbered. *New J. Phy.*, **13**(7), 073029.

[539] Schleier-Smith, M., Leroux, I. D., and Vuletić, V. 2009. Implementation of cavity squeezing of a collective atomic spin. Eprint: arXiv:0810.2582v2.

[540] Schleier-Smith, M. H., Leroux, I. D., and Vuletić, V. 2010. Squeezing the collective spin of a dilute atomic ensemble by cavity feedback. Eprint: arXiv:0911.3936.

[541] Schleier-Smith, M. H., Leroux, I. D., and Vuletić, V. 2010. Squeezing the collective spin of a dilute atomic ensemble by cavity feedback. *Phys. Rev. A*, **81**, 021804.

[542] Schliesser, A., Rivière, R., Anetsberger, G., Arcizet, O., and Kippenberg, T. J. 2008. Resolved-sideband cooling of a micromechanical oscillator. *Nature Phys.*, **5**, 415–419.

[543] Schliesser, A., Arcizet, O., Rivière, R., Anetsberger, G., and Kippenberg, T. J. 2009. Resolved-sideband cooling and position measurement of a micromechanical oscillator close to the Heisenberg uncertainty limit. *Nature Phys.*, **5**, 509–514.

[544] Schollwöck, U. 2005. The density-matrix renormalization group. *Rev. Mod. Phys.*, **77**, 259–315.

[545] Schrader, D., Dotsenko, I., Khudaverdyan, M., Miroshnychenko, Y., Rauschenbeutel, A., and Meschede, D. 2004. Neutral atom quantum register. *Phys. Rev. Lett.*, **93**(Oct), 150501.

[546] Schulman, L. J., and Vazirani, U. 1999. Molecular scale heat engines and scalable quantum computation. Pages 322–329 of: *Proceedings of the Thirty-First Annual ACM Symposium on the Theory of Computing*. New York: ACM.

[547] Schulman, L. J., Mor, T., and Weinstein, Y. 2005. Physical limits of heat-bath algorithmic cooling. *Phys. Rev. Lett.*, **94**(Apr), 120501.

[548] Schumacher, B., Westmoreland, M., and Wootters, W. K. 1996. Limitation on the amount of accessible information in a quantum channel. *Phys. Rev. Lett.*, **76**, 3452.

[549] Schumacher, B. 1995. Quantum coding. *Phys. Rev. A*, **51**, 2738–2747.

[550] Schumacher, B. 1996. Sending entanglement through noisy quantum channels. *Phys. Rev. A*, **54**(Oct), 2614–2628.

[551] Schumacher, B., and Nielsen, M. A. 1996. Quantum data processing and error correction. *Phys. Rev. A*, **54**(Oct), 2629–2635.

[552] Schuster, D. I., Houck, A. A., Schreier, J. A., Wallraff, A., Gambetta, J. M., Blais, A., Frunzio, L., Majer, J., Johnson, B., Devoret, M. H., Girvin, S. M., and Schoelkopf, R. J. 2007. Resolving photon number states in a superconducting circuit. *Nature*, **445**, 515–518.

[553] Scott, A. J., and Milburn, G. J. 2001. Quantum nonlinear dynamics of continuously measured systems. *Phys. Rev. A*, **63**, 042101.

[554] Shabani, A., and Jacobs, K. 2008. Locally optimal control of quantum systems with strong feedback. *Phys. Rev. Lett.*, **101**, 230403.

[555] Shaiju, A. J., and Petersen, I. 2012. A frequency domain condition for the physical realizability of linear quantum systems. *IEEE Trans. Automat. Contr.*, **57**(8), 2033–2044.

[556] Shannon, C. E., and Weaver, W. 1963. *The Mathematical Theory of Communication*. Chicago: University of Illinois Press.

[557] Shen, Z., Hsieh, M., and Rabitz, H. 2006. Quantum optimal control: Hessian analysis of the control landscape. *J. Chem. Phys.*, **124**, 204106.

[558] Shizume, K. 1995. Heat-generation required by information erasure. *Phys. Rev. E*, **52**, 3495.

[559] Shnirelman, A. I. 1974. Ergodic properties of eigenfunctions. *Usp. Mat. Nauk*, **29**, 181–182.

[560] Shnirman, A., Schön, G., and Hermon, Z. 1997. Quantum manipulations of small Josephson junctions. *Phys. Rev. Lett.*, **79**(Sep), 2371–2374.

[561] Shor, P. 2004. unpublished presentation.

[562] Short, A. J., and Farrelly, T. C. 2012. Quantum equilibration in finite time. *New J. Phys.*, **14**, 013063.

[563] Shuang, F., Zhou, M., Pechen, A., Wu, R., Shir, O. M., and Rabitz, H. 2008. Control of quantum dynamics by optimized measurements. *Phys. Rev. A*, **78**(Dec), 063422.

[564] Silberfarb, A., Jessen, P. S., and Deutsch, I. H. 2005. Quantum state reconstruction via continuous measurement. *Phys. Rev. Lett.*, **95**, 030402.

[565] Simmonds, R. W., Lang, K. M., Hite, D. A., Nam, S., Pappas, D. P., and Martinis, J. M. 2004. Decoherence in Josephson phase qubits from junction resonators. *Phys. Rev. Lett.*, **93**(Aug), 077003.

[566] Smith, G. A., Chaudhury, S., Silberfarb, A., Deutsch, I. H., and Jessen, P. S. 2004. Continuous weak measurement and nonlinear dynamics in a cold spin ensemble. *Phys. Rev. Lett.*, **93**, 163602.

[567] Smith, W. P., Reiner, J. E., Orozco, L. A., Kuhr, S., and Wiseman, H. M. 2002. Capture and release of a conditional state of a cavity QED system by quantum feedback. *Phys. Rev. Lett.*, **89**, 133601.

[568] Spekkens, R. W. 2007. Evidence for the epistemic view of quantum states: a toy theory. *Phys. Rev. A*, **75**(Mar), 032110.

[569] Spietz, L., Irwin, K., and Aumentado, J. 2008. Input impedance and gain of a gigahertz amplifier using a dc superconducting quantum interference device in a quarter wave resonator. *Appl. Phy. Lett.*, **93**, 082506.

[570] Spiller, T. P., and Ralph, J. F. 1994. The emergence of chaos in an open quantum system. *Phys. Lett. A*, **194**, 235.

[571] Spiller, T. P., Clark, T. D., Prance, R. J., and Widom, A. 1992. Chapter 4: Quantum phenomena in circuits at low temperatures. In: Brewer, D. F. (ed), *Progress in Low Temperature Physics*, vol. 13. Amsterdam: Elsevier.

[572] Srednicki, M. 1994. Chaos and quantum thermalization. *Phys. Rev. E*, **50**, 888.

[573] Srinivas, M. D., and Davies, E. B. 1981. Photon-counting probabilities in quantum optics. *Optica Acta*, **28**, 981.

[574] Steck, D., Jacobs, K., Mabuchi, H., Bhattacharya, T., and Habib, S. 2004. Quantum feedback control of atomic motion in an optical cavity. *Phys. Rev. Lett.*, **92**, 223004.

[575] Steck, D., Jacobs, K., Mabuchi, H., Habib, S., and Bhattacharya, T. 2006. Feedback cooling of atomic motion in cavity QED. *Phys. Rev. A*, **74**, 012322.

[576] Steffen, M., Ansmann, M., Bialczak, R. C., Katz, N., Lucero, E., McDermott, R., Neeley, M., Weig, E. M., Cleland, A. N., and Martinis, J. M. 2006. Measurement of

the entanglement of two superconducting qubits via state tomography. *Science*, **313**, 1423.

[577] Steixner, V., Rabl, P., and Zoller, P. 2005. Quantum feedback cooling of a single trapped ion in front of a mirror. *Phys. Rev. A*, **72**, 043826.

[578] Stockton, J. K., van Handel, R., and Mabuchi, H. 2004. Deterministic Dicke-state preparation with continuous measurement and control. *Phys. Rev. A*, **70**, 022106.

[579] Stockton, J. K., Geremia, J. M., Doherty, A. C., and Mabuchi, H. 2004. Robust quantum parameter estimation: coherent magnetometry with feedback. *Phys. Rev. A*, **69**, 032109.

[580] Strauch, F. W. 2012. All-resonant control of superconducting resonators. *Phys. Rev. Lett.*, **109**, 210501.

[581] Strunz, W. T., and Yu, T. 2004. Convolutionless non-Markovian master equations and quantum trajectories: Brownian motion. *Phys. Rev. A*, **69**, 052115.

[582] Suzuki, H., and Fujitani, Y. 2012. Numerical study on the generalized second law for a Brownian particle under the linear feedback control. *J. Phys. Soc. Jpn.*, **81**(8), 084003.

[583] Szilárd, L. 1929. Über die Entropieverminderung in einem thermodynamischen System bei Eingriffen intelligenter Wesen. *Z. Phys.*, **53**, 840–856.

[584] Szorkovszky, A., Doherty, A. C., Harris, G. I., and Bowen, W. P. 2011. Mechanical squeezing via parametric amplification and weak measurement. *Phys. Rev. Lett.*, **107**, 213603.

[585] Szorkovszky, A., Doherty, A. C., Harris, G. I., and Bowen, W. P. 2012. Position estimation of a parametrically driven optomechanical system. *New J. Phys.*, **14**(9), 095026.

[586] Szorkovszky, A., Brawley, G. A., Doherty, A. C., and Bowen, W. P. 2013. Strong thermomechanical squeezing via weak measurement. *Phys. Rev. Lett.*, **110**, 184301.

[587] Tabor, M. 1989. *Chaos and Integrability in Nonlinear Dynamics: An Introduction.* Hoboken: Wiley-Interscience.

[588] Takano, T., Fuyama, M., Namiki, R., and Takahashi, Y. 2009. Spin squeezing of a cold atomic ensemble with the nuclear spin of one-half. Eprint: arXiv:0808. 2353.

[589] Takano, T., Tanaka, S., Namiki, R., and Takahashi, Y. 2010. Manipulation of non-classical atomic spin states. Eprint: arXiv:0909.2423.

[590] Takara, K., Hasegawa, H.-H., and Driebe, D. 2010. Generalization of the second law for a transition between nonequilibrium states. *Phys. Lett. A*, **375**(2), 88–92.

[591] Tan, S. M. 1996. *Linear Systems.* This text is made freely available by the author at: http://home.comcast.net/~szemengtan/. Accessed March 10, 2014.

[592] Tan, S. M., Fox, C., and Nicholls, G. 1996. *Inverse Problems.* This text is freely available at: http://home.comcast.net/~szemengtan/. Accessed March 10, 2014.

[593] Tanaka, S., and Yamamoto, N. 2012. Robust adaptive measurement scheme for qubit-state preparation. *Phys. Rev. A*, **86**(Dec), 062331.

[594] Tesche, C. D. 1990. Can a noninvasive measurement of magnetic flux be performed with superconducting circuits? *Phys. Rev. Lett.*, **64**, 2358–2361.

[595] Teufel, J. D., Donner, T., Castellanos-Beltran, M. A., Harlow, J. W., and Lehnert, K. W. 2009. Nanomechanical motion measured with precision beyond the standard quantum limit. *Nature Nanotechnol.*, **4**, 820–823.

[596] Teufel, J. D., Li, D., Allman, M. S., Cicak, K., Sirois, A. J., Whittaker, J. D., and Simmonds, R. W. 2011. Circuit cavity electromechanics in the strong-coupling regime. *Nature*, **471**, 204–208.

[597] Teufel, J. D., Donner, T., Li, D., Harlow, J. W., Allman, M. S., Cicak, K., Sirois, A. J., Whittaker, J. D., Lehnert, K. W., and Simmonds, R. W. 2011. Sideband cooling of micromechanical motion to the quantum ground state. *Nature*, **475**, 359–363.

[598] Thacker, H. B. 1981. Exact integrability in quantum field theory and statistical systems. *Rev. Mod. Phys.*, **53**(Apr), 253–285.

[599] Thompson, J. D., Zwickl, B. M., Jayich, A. M., Marquardt, F., Girvin, S. M., and Harris, J. G. E. 2008. Strong dispersive coupling of a high-finesse cavity to a micromechanical membrane. *Nature*, **452**, 72.

[600] Thomsen, L. K., Mancini, S., and Wiseman, H. M. 2002. Spin squeezing via quantum feedback. *Phys. Rev. A*, **65**, 061801.

[601] Thorwart, M., Reimann, P., Jung, P., and Fox, R. F. 1998. Quantum hysteresis and resonant tunneling in bistable systems. *Chem. Phys.*, **235**(1-3), 61–80.

[602] Thorwart, M., Eckel, J., Reina, J., Nalbach, P., and Weiss, S. 2009. Enhanced quantum entanglement in the non-Markovian dynamics of biomolecular excitons. *Chem. Phys. Lett.*, **478**(4-6), 234–237.

[603] Tian, L. 2007. Correcting low-frequency noise with continuous measurement. *Phys. Rev. Lett.*, **98**, 153602.

[604] Tian, L. 2009. Ground state cooling of a nanomechanical resonator via parametric linear coupling. *Phys. Rev. B*, **79**, 193407.

[605] Tian, L., Allman, M. S., and Simmonds, R. W. 2008. Parametric coupling between macroscopic quantum resonators. *New J. Phys.*, **10**, 115001.

[606] Ticozzi, F., Nishio, K., and Altafini, C. 2013. Stabilization of stochastic quantum dynamics via open- and closed-loop control. *IEEE Trans. Automat. Contr.*, **58**(1), 74–85.

[607] Ticozzi, F., and Viola, L. 2008. Quantum Markovian subsystems: invariance, attractivity, and control. *IEEE Trans. Automat. Contr.*, **53**, 2048–2063.

[608] Tittonen, I., Breitenbach, G., Kalkbrenner, T., Müller, T., Conradt, R., Schiller, S., Steinsland, E., Blanc, N., and de Rooij, N. F. 1999. Interferometric measurements of the position of a macroscopic body: towards observation of quantum limits. *Phys. Rev. A*, **59**, 1038–1044.

[609] Tolman, R. C. 2010. *The Principles of Statistical Mechanics*. New York: Dover.

[610] Truitt, P. A., Hertzberg, J. B., Huang, C. C., Ekinci, K. L., and Schwab, K. C. 2007. Efficient and sensitive capacitive readout of nanomechanical resonator arrays. *Nano Lett*, **7**(1), 120–126.

[611] Tsang, M. 2013. Testing quantum mechanics. Eprint: arXiv:1306.2699.

[612] Tsang, M. 2009. Time-symmetric quantum theory of smoothing. *Phys. Rev. Lett.*, **102**, 250403.

[613] Tsang, M. 2012. Continuous quantum hypothesis testing. *Phys. Rev. Lett.*, **108**(Apr), 170502.

[614] Tsang, M., and Nair, R. 2012. Fundamental quantum limits to waveform detection. *Phys. Rev. A*, **86**(Oct), 042115.

[615] Tsang, M., Wiseman, H. M., and Caves, C. M. 2011. Fundamental quantum limit to waveform estimation. *Phys. Rev. Lett.*, **106**(Mar), 090401.

[616] Turgut, S. 2009. Relations between entropies produced in nondeterministic thermo-dynamic processes. *Phys. Rev. E*, **79**(Apr), 041102.

[617] Tyson, J. 2010. Estimates of non-optimality of quantum measurements and a simple iterative method for computing optimal measurements. Eprint: arXiv:0902.0395.

[618] Tyson, J. 2010. Two-sided bounds on minimum-error quantum measurement, on the reversibility of quantum dynamics, and on maximum overlap using directional iterates. Eprint: arXiv:0907.3386.

[619] Uhlmann, A. 2006. The 'transmission probability' in the state space of a *-algebra. *Rep. Math. Phys.*, **9**, 273–279.

[620] Unruh, W. G. 1982. Page 647 of: Meystre, P., and Scully, M. O. (eds), *Quantum Optics, Experimental Gravitation, and Measurement Theory*. New York: Plenum.

[621] von Neumann, J. 1955. *Mathematical Foundations of Quantum Mechanics*. Princeton University Press.

[622] Vamivakas, A. N., Lu, C.-Y., Matthiesen, C., Zhao, Y., Fält, S., Badolato, A., and Atatüre, M. 2010. Observation of spin-dependent quantum jumps via quantum dot resonance fluorescence. *Nature*, **467**, 297–300.

[623] van den Brink, A. M., Schön, G., and Geerligs, L. J. 1991. Combined single-electron and coherent-Cooper-pair tunneling in voltage-biased Josephson junctions. *Phys. Rev. Lett.*, **67**(Nov), 3030–3033.

[624] van der Ploeg, S., Izmalkov, A., van den Brink, A. M., Huebner, U., Grajcar, M., Il'ichev, E., Meyer, H.-G., and Zagoskin, A. 2007. Controllable coupling of superconducting flux qubits. *Phys. Rev. Lett.*, **98**, 057004.

[625] van der Wal, C. H., ter Haar, A. C. J., Wilhelm, F. K., Schouten, R. N., Harmans, C. J. P. M., Orlando, T. P., Lloyd, S., and Mooij, J. E. 2000. Quantum superposition of macroscopic persistent-current states. *Science*, **290**(5492), 773–777.

[626] van Handel, R. 2009. The stability of quantum Markov filters. *Infin. Dimens. Anal. Quantum Probab. Relat. Top.*, **12**, 153–172.

[627] Van Trees, H. L. 2001. *Detection, Estimation, and Modulation Theory, Parts I – III*. New York: Wiley.

[628] vanHandel, R., and Mabuchi, H. 2005. Quantum projection filter for a highly nonlinear model in cavity QED. *J. Opt. B: Quantum Semiclass. Opt.*, **7**, S226–S236.

[629] vanHandel, R., Stockton, J. K., and Mabuchi, H. 2005. Feedback control of quantum state reduction. *IEEE Trans. Automat. Contr.*, **50**, 768–780.

[630] Vanner, M. R. 2011. Selective linear or quadratic optomechanical coupling via measurement. *Phys. Rev. X*, **1**(Nov), 021011.

[631] Vedral, V. 2007. *Introduction to Quantum Information Science.* Oxford University Press.

[632] Verstraete, F., Doherty, A. C., and Mabuchi, H. 2001. Sensitivity optimization in quantum parameter estimation. *Phys. Rev. A*, **64**, 032111.

[633] Verstraete, F., Cirac, J. I., and Murg, V. 2008. Matrix product states, projected entangled pair states, and variational renormalization group methods for quantum spin systems. *Adv. Phys.*, **57**, 143.

[634] Vidal, G. 2008. Class of quantum many-body states that can be efficiently simulated. *Phys. Rev. Lett.*, **101**(11), 110501.

[635] Vidal, G. 2003. Efficient classical simulation of slightly entangled quantum computations. *Phys. Rev. Lett.*, **91**, 147902.

[636] Vidal, G. 2004. Efficient simulation of one-dimensional quantum many-body systems. *Phys. Rev. Lett.*, **93**, 040502.

[637] Vijay, R., Slichter, D. H., and Siddiqi, I. 2011. Observation of quantum jumps in a superconducting artificial atom. *Phys. Rev. Lett.*, **106**, 110502.

[638] Vijay, R., Macklin, C., Slichter, D. H., Weber, S. J., Murch, K. W., Naik, R., Korotkov, A. N., and Siddiqi, I. 2012. Stabilizing Rabi oscillations in a superconducting qubit using quantum feedback. *Nature*, **490**, 77–80.

[639] Vion, D. 2004. Josephson quantum bits based on a Cooper-pair box. Pages 521–560 of: Esteve, D., Raimond, J.-M., and Dalibard, J. (eds), *Quantum Entanglement and Information Processing, Lecture Notes of the Les Houches Summer School 2003.* Amsterdam: Elsevier.

[640] Vion, D., Aassime, A., Cottet, A., Joyez, P., Pothier, H., Urbina, C., Esteve, D., and Devoret, M. H. 2002. Manipulating the quantum state of an electrical circuit. *Science*, **296**, 886.

[641] Vitali, D., Zippilli, S., Tombesi, P., and Raimond, J.-M. 2004. Decoherence control with fully quantum feedback schemes. *J. Mod. Opt.*, **51**(6-7), 799–809.

[642] von Neumann, J. 1927. Wahrscheinlichkeitstheoretischer Aufbau der Quantenmechanik. *Göttinger Nachrichten*, 245–272.

[643] Voros, A. 1979. *Stochastic Behavior in Classical and Quantum Hamiltonian Systems.* Berlin: Springer.

[644] Žnidarič, M., Prosen, T., and Prelovšek, P. 2008. Many-body localization in the Heisenberg *XXZ* magnet in a random field. *Phys. Rev. B*, **77**(Feb), 064426.

[645] Wallraff, A., Schuster, D. I., Blais, A., Frunzio, L., Majer, J., Devoret, M. H., Girvin, S. M., and Schoelkopf, R. J. 2005. Approaching unit visibility for control of a superconducting qubit with dispersive readout. *Phys. Rev. Lett.*, **95**(Aug), 060501.

[646] Wang, S. K., Jin, J. S., and Li, X. Q. 2007. Continuous weak measurement and feedback control of a solid-state charge qubit: a physical unravelling of non-Lindblad master equation. *Phys. Rev. B*, **75**, 155304.

[647] Wang, X. 2014 in preparation.

[648] Wang, X., Allegra, M., Mohseni, M., Lloyd, S., Jacobs, K., and Lupo, C. 2014. A numerical method to solve the quantum brachistochrone equation: obtaining exact control protocols for maximum-speed quantum gates, in preparation.

[649] Wang, X., and Jacobs, K. 2013. Coherent feedback that beats all measurement-based feedback protocols. Eprint: arXiv: 1211.1724.

[650] Wang, X., Vinjanampathy, S., Strauch, F. W., and Jacobs, K. 2011. Ultraefficient cooling of resonators: beating sideband cooling with quantum control. *Phys. Rev. Lett.*, **107**(Oct), 177204.

[651] Wang, X., Vinjanampathy, S., Strauch, F. W., and Jacobs, K. 2013. Absolute dynamical limit to cooling weakly coupled quantum systems. *Phys. Rev. Lett.*, **110**(Apr), 157207.

[652] Washburn, S., Webb, R. A., Voss, R. F., and Faris, S. M. 1985. Effects of dissipation and temperature on macroscopic quantum tunneling. *Phys. Rev. Lett.*, **54**(Jun), 2712–2715.

[653] Wehrl, A. 1978. General properties of entropy. *Rev. Mod. Phys.*, **50**, 221–260.

[654] Weis, S., Rivière, R., Deléglise, S., Gavartin, E., Arcizet, O., Schliesser, A., and Kippenberg, T. J. 2010. Optomechanically induced transparency. *Science*, **330**, 1520–1523.

[655] Wheatley, T. A., Berry, D. W., Yonezawa, H., Nakane, D., Arao, H., Pope, D. T., Ralph, T. C., Wiseman, H. M., Furusawa, A., and Huntington, E. H. 2010. Adaptive optical phase estimation using time-symmetric quantum smoothing. *Phys. Rev. Lett.*, **104**, 093601.

[656] Whittle, P. 1981. Risk-sensitive linear/quadratic/Gaussian control. *Adv. Appl. Prob.*, **13**, 764–777.

[657] Whittle, P. 1991. A risk-sensitive maximum principle: the case of imperfect state observation. *IEEE Trans. Automat. Contr.*, **36**(7), 793–801.

[658] Whittle, P. 1996. *Optimal Control*. Chichester: Wiley.

[659] Widom, A. 1979. Quantum electrodynamic circuits at ultralow temperature. *J. Low Temp. Phys.*, **37**, 449–460.

[660] Widom, A., Megaloudis, G., Clark, T. D., Prance, H., and Prance, R. J. 1982. Quantum electrodynamic charge space energy bands in singly connected superconducting weak links. *J. Phys. A: Math. Gen.*, **15**(12), 3877.

[661] Wilcox, R. M. 1967. Exponential operators and parameter differentiation in quantum physics. *J. Math. Phys.*, **8**, 962.

[662] Wilson, K. G. 1975. The renormalization group: critical phenomena and the Kondo problem. *Rev. Mod. Phys.*, **47**, 773–840.

[663] Wilson, S. D., Carvalho, A. R. P., Hope, J. J., and James, M. R. 2007. Effects of measurement back-action in the stabilization of a Bose-Einstein condensate through feedback. *Phys. Rev. A*, **76**, 013610.

[664] Wilson-Rae, I., Nooshi, N., Zwerger, W., and Kippenberg, T. J. 2007. Theory of ground state cooling of a mechanical oscillator using dynamical back-action. *Phys. Rev. Lett.*, **99**, 093901.

[665] Wineland, D., and Dehmelt, H. 1975. Proposed $10^{14}\Delta\nu < \nu$ laser fluorescence spectroscopy on Tl^{+} mono-ion oscillator. *Bull. Am. Phys. Soc.*, **20**, 637.

[666] Wineland, D. J., Britton, J., Epstein, R. J., Leibfried, D., Blakestad, R. B., Brown, K., Jost, J. D., Langer, C., Ozeri, R., Seidelin, S., and Wesenberg, J. 2006. Cantilever cooling with radio frequency circuits. Eprint: arXiv:quant-ph/0606180.

[667] Winter, A. 2004. Extrinsic and intrinsic data in quantum measurements: asymptotic convex decomposition of positive operator valued measures. *Commun. Math. Phys.*, **244**, 157–185.

[668] Wiseman, H. M. 1994. Quantum theory of continuous feedback. *Phys. Rev. A*, **49**(Mar), 2133–2150.

[669] Wiseman, H. M. 1994. *Quantum Trajectories and Feedback*. Ph.D. diss. Brisbane: The University of Queensland.

[670] Wiseman, H. M. 1995. Adaptive phase measurements of optical modes: going beyond the marginal Q distribution. *Phys. Rev. Lett.*, **75**, 4587–4590.

[671] Wiseman, H. M. 1995. Using feedback to eliminate back-action in quantum measurements. *Phys. Rev. A*, **51**, 2459–2468.

[672] Wiseman, H. M. 1995. Feedback in open quantum-systems. *Mod. Phys. Lett.*, **B9**, 629–654.

[673] Wiseman, H. M. 1996. Quantum trajectories and quantum measurement theory. *Quantum Semiclass. Opt.*, **8**, 205.

[674] Wiseman, H. M. 2007. Grounding Bohmian mechanics in weak values and Bayesianism. *New J. Phys.*, **9**(6), 165.

[675] Wiseman, H. M., and Bouten, L. 2008. Optimality of feedback control strategies for qubit purification. *Quantum. Inform. Process*, **7**, 71–83.

[676] Wiseman, H. M., and Diosi, L. 2001. Complete parameterization, and invariance, of diffusive quantum trajectories for Markovian open systems. *Chem. Phys.*, **91**, 268.

[677] Wiseman, H. M., and Doherty, A. C. 2005. Optimal unravellings for feedback control in linear quantum systems. *Phys. Rev. Lett.*, **94**(7), 070405.

[678] Wiseman, H. M., and Killip, R. B. 1998. Adaptive single-shot phase measurements: the full quantum theory. *Phys. Rev. A*, **57**, 2169.

[679] Wiseman, H. M., and Milburn, G. J. 1993. Quantum theory of field-quadrature measurements. *Phys. Rev. A*, **47**, 642.

[680] Wiseman, H. M., and Milburn, G. J. 1993. Interpretation of quantum jump and diffusion processes illustrated on the Bloch sphere. *Phys. Rev. A*, **47**, 1652.

[681] Wiseman, H. M., and Milburn, G. J. 1993. Quantum theory of optical feedback via homodyne detection. *Phys. Rev. Lett.*, **70**(Feb), 548–551.

[682] Wiseman, H. M., and Milburn, G. J. 1994. All-optical versus electro-optical quantum-limited feedback. *Phys. Rev. A*, **49**(May), 4110–4125.

[683] Wiseman, H. M., and Milburn, G. J. 1994. Squeezing via feedback. *Phys. Rev. A*, **49**(Feb), 1350–1366.

[684] Wiseman, H. M., and Ralph, J. F. 2006. Reconsidering rapid qubit purification by feedback. *New. J. Phys*, **8**, 90.

[685] Wiseman, H. M., Harrison, F. E., Collett, M. J., Tan, S. M., Walls, D. F., and Killip, R. B. 1997. Nonlocal momentum transfer in Welcher Weg measurements. *Phys. Rev. A*, **56**, 55–75.

[686] Wiseman, H. M., and Milburn, G. J. 2010. *Quantum Measurement and Control*. Cambridge University Press.

[687] Witschel, W. 1975. Ordered operator expansions by comparison. *J. Phys. A: Math. Gen.*, **8**, 143.

[688] Wolf, A., Swift, J. B., Swinney, H. L., and Vastano, J. A. 1985. Determining Lyapunov exponents from a time-series. *Physica D*, **16**, 285.

[689] Wootters, W. K., and Fields, B. D. 1989. Optimal state determination by mutually unbiased measurements. *Ann. Phys.*, **191**(2), 363–381.

[690] Xiang, G. Y., Higgins, B. L., Berry, D. W., Wiseman, H. M., and Pryde, G. J. 2011. Entanglement-enhanced measurement of a completely unknown phase. *Nature Photonics*, **5**, 43.

[691] Xue, S.-B., Wu, R.-B., Zhang, W.-M., Zhang, J., Li, C.-W., and Tarn, T.-J. 2012. Decoherence suppression via non-Markovian coherent feedback control. *Phys. Rev. A*, **86**(Nov), 052304.

[692] Yamamoto, N. 2006. Robust observer for uncertain linear quantum systems. Pages 3138–3143 of: *2006 45th IEEE Conference on Decision and Control*. Washington DC: IEEE.

[693] Yamamoto, N., and Bouten, L. 2009. Quantum risk-sensitive estimation and robustness. *IEEE Trans. Automat. Contr.*, **54**(1), 92–107.

[694] Yamamoto, N., Tsumura, K., and Hara, S. 2007. Feedback control of quantum entanglement in a two-spin system. *Automatica*, **43**, 981–992.

[695] Yamamoto, N. 2005. Parametrization of the feedback Hamiltonian realizing a pure steady state. *Phys. Rev. A*, **72**, 024104.

[696] Yamamoto, T., Pashkin, Y. A., Astafiev, O., Nakamura, Y., and Tsai, J. S. 2003. Demonstration of conditional gate operation using superconducting charge qubits. *Nature*, **425**, 941.

[697] Yamamoto, T., Inomata, K., Watanabe, M., Matsuba, K., Miyazaki, T., Oliver, W. D., Nakamura, Y., and Tsai, J. S. 2008. Flux-driven Josephson parametric amplifier. *Appl. Phys. Lett.*, **93**, 042510.

[698] Yanagisawa, M., and Kimura, H. 1998. A control problem for Gaussian states. Pages 249–313 of: Yamamoto, Y., and Hara, S. (eds), *Learning, Control and Hybrid Systems, Lecture Notes in Control and Information Sciences*, vol. 241. New York: Springer.

[699] Yanagisawa, M., and Kimura, H. 2003. Transfer function approach to quantum control – Part I: Dynamics of quantum feedback systems. *IEEE Trans. Automat. Contr.*, **48**, 2107–2120.

[700] Yanagisawa, M., and Kimura, H. 2003. Transfer function approach to quantum control – Part II: control concepts and applications. *IEEE Trans. Automat. Contr.*, **48**, 2121–2132.

[701] You, J. Q., Hu, X., Ashhab, S., and Nori, F. 2007. Low-decoherence flux qubit. *Phys. Rev. B*, **75**(Apr), 140515.

[702] Yu, Y., Han, S., Chu, X., Chu, S.-I., and Wang, Z. 2002. Coherent temporal oscillations of macroscopic quantum states in a Josephson junction. *Science*, **296**(5569), 889–892.

[703] Yu, Y., Zhu, S.-L., Sun, G., Wen, X., Dong, N., Chen, J., Wu, P., and Han, S. 2008. Quantum jumps between macroscopic quantum states of a superconducting qubit coupled to a microscopic two-level system. *Phys. Rev. Lett.*, **101**, 157001.

[704] Yuen, H. P., Kennedy, R. S., and Lax, M. 1975. Optimum testing of multiple hypotheses in quantum detection theory. *IEEE Trans. Inf. Theory*, **IT-21**, 125.

[705] Yurke, B. 1986. Input states for enhancement of fermion interferometer sensitivity. *Phys. Rev. Lett.*, **56**, 1515–1517.

[706] Yurke, B., and Stoler, D. 1986. Generating quantum mechanical superpositions of macroscopically distinguishable states via amplitude dispersion. *Phys. Rev. Lett.*, **57**, 13–16.

[707] Yurke, B., Kaminsky, P. G., Miller, R. E., Whittaker, E. A., Smith, A. D., Silver, A. H., and Simon, R. W. 1988. Observation of 4.2-K equilibrium-noise squeezing via a Josephson-parametric amplifier. *Phys. Rev. Lett.*, **60**, 764–767.

[708] Yurke, B., Corruccini, L. R., Kaminsky, P. G., Rupp, L. W., Smith, A. D., Silver, A. H., Simon, R. W., and Whittaker, E. A. 1989. Observation of parametric amplification and deamplification in a Josephson parametric amplifier. *Phys. Rev. A*, **39**, 2519–2533.

[709] Yurke, B., and Denker, J. S. 1984. Quantum network theory. *Phys. Rev. A*, **29**, 1419–1437.

[710] Zelditch, S. 1987. Uniform distribution of eigenfunctions on compact hyperbolic surfaces. *Duke Math. J.*, **55**, 919–941.

[711] Zhang, G., and James, M. 2011. Direct and indirect couplings in coherent feedback control of linear quantum systems. *IEEE Trans. Automat. Contr.*, **56**(7), 1535–1550.

[712] Zhang, J., Liu, Y.-x., and Nori, F. 2009. Cooling and squeezing the fluctuations of a nanomechanical beam by indirect quantum feedback control. *Phys. Rev. A*, **79**, 052102.

[713] Zhang, J., Wu, R.-B., Li, C.-W., and Tarn, T. 2010. Protecting coherence and entanglement by quantum feedback controls. *IEEE Trans. Automat. Contr.*, **55**(3), 619–633.

[714] Zhang, J., Wu, R.-B., xi Liu, Y., Li, C.-W., and Tarn, T. 2012. Quantum coherent nonlinear feedback with applications to quantum optics on chip. *IEEE Trans. Automat. Contr.*, **57**(8), 1997–2008.

[715] Zhang, J., Liu, Y.-x., Wu, R.-B., Jacobs, K., and Nori, F. 2013. Non-Markovian quantum input-output networks. *Phys. Rev. A*, **87**(Mar), 032117.

[716] Zhou, K., and Doyle, J. C. 1997. *Essentials of Robust Control*. Englewood Clifts: Prentice Hall.

[717] Zhou, K., Doyle, J. C., and Glover, K. 1995. *Robust and Optimal Control*. Englewood Cliffts: Prentice Hall.

[718] Zhou, X. Y., Yong, J., and Li, X. 1997. Stochastic verification theorems within the framework of viscosity solutions. *SIAM J. Control. Optim.*, **35**, 243–253.

[719] Zhou, X., and Mizel, A. 2006. Nonlinear coupling of nanomechanical resonators to Josephson quantum circuits. *Phys. Rev. Lett.*, **97**, 267201.

[720] Zhu, X., Kemp, A., Saito, S., and Semba, K. 2010. Coherent operation of a gap-tunable flux qubit. *Appl. Phys. Lett.*, **97**, 102503.

[721] Zurek, W. H. 1981. Pointer basis of quantum apparatus: into what mixture does the wave packet collapse? *Phys. Rev. D*, **24**(Sep), 1516–1525.

[722] Zurek, W. H. 1982. Environment-induced superselection rules. *Phys. Rev. D*, **26**(Oct), 1862–1880.

[723] Zurek, W. H. 1984. Szilard's engine, Maxwell's demon and quantum measurements. In: More, G. T., and Scully, M. O. (eds), *Frontiers of Non-Equilibrium Statistical Physics*. New York: Plenum.

[724] Zurek, W. H., and Paz, J. P. 1994. Decoherence, chaos, and the second law. *Phys. Rev. Lett.*, **72**(Apr), 2508–2511.

[725] Zwanzig, R. 1960. Ensemble method in the theory of irreversibility. *J. Chem. Phys.*, **33**, 1338.

[726] Zwanzig, R. 1961. Memory effects in irreversible thermodynamics. *Phys. Rev.*, **124**, 983–992.

[727] Zwolak, M., and Vidal, G. 2004. Mixed-state dynamics in one-dimensional quantum lattice systems: a time-dependent superoperator renormalization algorithm. *Phys. Rev. Lett.*, **93**, 207205.

Index

Printed in the United States
by Baker & Taylor Publisher Services